# World-Systems Evolution and Global Futures

This series seeks to promote understanding of large-scale and long-term processes of social change, in particular the many facets and implications of globalization. It critically explores the factors that affect the historical formation and current evolution of social systems, on both the regional and global level. Processes and factors that are examined include economies, technologies, geopolitics, institutions, conflicts, demographic trends, climate change, global culture, social movements, global inequalities, etc. Building on world-systems analysis, the series addresses topics such as globalization from historical and comparative perspectives, trends in global inequalities, core-periphery relations and the rise and fall of hegemonic core states, transnational institutions, and the long-term energy transition. This ambitious interdisciplinary and international series presents cutting-edge research by social scientists who study whole human systems and is relevant for all readers interested in systems approaches to the emerging world society, especially historians, political scientists, economists, sociologists, geographers and anthropologists.

This book series is indexed in Scopus.

All titles in this series are peer-reviewed.

Leonid Grinin · Anton Grinin ·
Andrey Korotayev

# Cybernetic Revolution and Global Aging

## Humankind on the Way to Cybernetic Society, or the Next Hundred Years

Springer

Leonid Grinin
HSE University
Moscow, Russia

Institute of Oriental Studies of the Russian
Academy of Sciences
Moscow, Russia

Anton Grinin
HSE University
Moscow, Russia

Moscow State University
Moscow, Russia

Andrey Korotayev
HSE University
Moscow, Russia

Moscow State University
Moscow, Russia

Institute of Oriental Studies of the Russian
Academy of Sciences
Moscow, Russia

ISSN 2522-0985 ISSN 2522-0993 (electronic)
World-Systems Evolution and Global Futures
ISBN 978-3-031-56763-6 ISBN 978-3-031-56764-3 (eBook)
https://doi.org/10.1007/978-3-031-56764-3



This Springer imprint is published by the registered company Springer Nature Switzerland AG
The registered company address is: Gewerbestrasse 11, 6330 Cham, Switzerland

Paper in this product is recyclable.

# Foreword by Jack A. Goldstone

The motto of this book by Leonid Grinin, Anton Grinin, and Andrey Korotayev should be: "The future will be here before we know it—better plan now!" They address two of the most pervasive global trends likely to continue through the end of the century: the global aging of the human population, and the rapidly accelerating development of technological innovations, including a vast increase in the capabilities of non-human information processing, analysis and communication, better known as "Artificial Intelligence" or AI.

How humanity responds to these two trends—biological aging and the rise of highly capable autonomous systems of production and services—will shape the future of our species. Not surprisingly, both utopian and dystopian futures have been imagined based on these trends. But until now, these have remained in the world of science fiction and entertainment. No longer: the technologies that are capable of human-like interaction have become part of everyday apps, and the leading aging societies, such as Japan, already have more people over age 65 than under 15.

We thus have little time to decide how to shape a future guided by these two global trends. How best to marry them? As populations become more elderly, can we use cybernetic and technological innovations to enhance humans, adding mental capacity and physical strength and skill, to older people to enable them to work longer and live more vigorously? Or do we use cybernetic workers to replace the elderly, shunting them aside?

What older humans will need more than anything else is human companionship and attentive support and care. If cybernetic systems can do much of the work now being done by humans, will that release human workers to engage in caring and compassionate work, like elder care? Or will societies seek to turn over aging citizens to automated, robotic care—perhaps cheaper, but ultimately dehumanizing to those who receive it?

Potentially more exciting, but also more frightening, are the prospects for greatly extending lifespans by medical advances and cyber-engineering of new heart, lung and other vital systems. If we can radically extend the human lifespan, people will no doubt embrace it. But what if only the rich can afford it? Will democracy survive a future where the rich live for centuries and the poor only for decades? And if lifespans are extended, will people choose to have lengthier retirements—if so, who will pay for them? Or if people remain at work, will that block opportunities for younger workers? More radically—will we need workers at all? Or will we move to a society where production and services function under automated systems, serving people who cultivate their own interests?

Let me posit two over-arching questions that will guide the use of the products of the Cybernetic Revolution and their interaction with an aging human population: (1) Is the purpose of cybernetic innovations to make the lives of human beings more interesting, healthier, productive and enjoyable while remaining essentially human, or is it to replace humanity as we know it with something post-human, enhanced or partially or fully replaced by robotic and cybernetic/cyborg entities? And (2) Is the purpose of cybernetic innovations to make businesses more profitable to innovators and owners, or to create such widespread prosperity that ALL people can increasingly be released from necessary income-producing work to instead take up work that is socially valuable and quality-of-life enhancing?

It is interesting to me that in most futuristic science fiction, such as Star Trek and Star Wars, the human protagonists hundreds of years in the future do not differ noticeably from present-day humans. Aliens aside, the main actors are no stronger, have no better vision or hearing, no better or differently wired brains than humans today. This is extremely odd for societies that have mastered interstellar travel, anti-matter-based propulsion and energy weapons. It seems that at some point in their development, these societies decided that it would be a grave error to change what it means to be human, and instead to preserve humanity in its current genetic form, although equipping humans with more and more advanced external computing and mechanical power. Indeed, there are often references to the explicit prohibition of genetic modification after past failed and dangerous periods of genetic enhancement, while cyborgs and robots are often presented as evil and dangerous villains. Given the opportunities described in this book, it seems that our societies will have to make that choice at some point, and likely soon: do we use technological advances to enhance the lives we lead as humans, or do we choose to leave humanity behind and become something else? (And who do we trust to make that choice?)

In a world where technological advance is carried out mainly by private corporations (though usually with government funding), providing billions of dollars in profits to a relatively small cadre of business executives and investors, we also have to ask: should the financial and material gains from advances in new technologies accrue mainly to the few, or should they be widespread? The market, as we have seen, tends to produce the former outcome. In the last few decades, in which societies acted as though new technologies need various kinds of protection from taxation, regulation and redistribution, we have seen the creation of a new class of tech titans, with fortunes in the hundreds of billions of dollars, whose collective wealth approaches that of smaller countries, and eclipses the combined wealth of a third of humanity. Is it tolerable for this concentration of wealth to continue even further? How do we ensure that our societies avoid the dystopia of a small number of wealthy investors and executives profiting from the widespread deployment of labor-reducing technologies while former workers are left with ever smaller economic resources?

Grinin et al. have the courage to peer a century into the future and try to determine how aging and the Cybernetic Revolution will shape that future. They expect a society that is older, but more stable, less prone to conflict, but more regulated. Will human beings adapt to this new "cybernetic society?" Or will they insist on a degree of novelty, unpredictability, and freedom from regulation? Perhaps the price to be paid for longer and healthier lives in an aging society is adaptation to more autonomous systems and dependence on medical/nanotechnology/robotics systems to manage our lives. This book makes clear both immense opportunities and challenges that lie ahead. Readers may agree or disagree with these projections, but making the effort to envisage in detail how global society may be recast by ongoing technological and biological trends is essential to gaining control over our future.

It is clear that new technologies and new conditions for humanity will likely require a new set of social contracts, perhaps entirely new constitutions to protect or enable new conceptions of humanity and human societies. To prepare, we need to understand the prospects ahead of us, especially the extent to which population aging and Cybernetic Revolutions will set new conditions, previously unknown, for human development. This book will help readers understand the prospects, and what is at stake.

January 2024

Prof. Jack A. Goldstone
Schar School of Policy
and Government
George Mason University
Fairfax, USA

# Foreword: What Could Happen in the Next One Hundred Years?

"Today, many of us are convinced that the world is in a state of a major change, and the awareness that we are living in a period of transition to something new and unknown, is constantly growing". It is difficult to disagree with this idea of the authors of the present monograph. However, it is much more difficult to find an answer to the question, which is increasingly worrying many of us: why has humanity reached such a dangerous bifurcation point when technological progress begins to frighten us? And it is even more difficult to answer the questions that interest everyone: What kind of future is awaiting us and will it be better or worse than the present? Which forces are shaping our future? What changes are likely to take place in the near future and by the end of this century, and what dangers threaten us? The authors of this monograph—Leonid Grinin, Anton Grinin and Andrey Korotayev—have made a profound, conscientious and scrupulous attempt to answer these and many other questions. They characterize the past and the present state of the world in many aspects and details, but the monograph is mainly devoted to the future, presenting the analysis and descriptions of how and why our world will change in the next hundred years.

Before starting to read this substantial and serious book, a reader may reasonably ask: Why is this book worth devoting time and effort? What is new in it? What distinguishes it? What does it offer that cannot be found in other books? I am pleased to give readers a summary of my reading experiences. First of all, such books are rare, because it is challenging to comprehensively cover vast periods of time in a single approach, combining past, present and future organically. The description covers almost the entire period of the historical process starting from the emergence of Homo sapiens up to the beginning of the twenty-second century.

The authors base their global research on two leading trends of our time and the future—technological and demographic development. These trends are reflected in the title: 'Cybernetic Revolution and Global Aging'. Moreover, it is very important and at the same time unusual that these trends are interrelated, because the authors convincingly demonstrate that global aging is likely to be one of the most important factors affecting technological growth in the near future, as well as why aging will become a driver of the technological process in the coming decades, and then will begin to slow it down.

However, although these two trends are certainly among the most important ones, both today and in the future, there is a surprisingly small number of works that present a consistent forecast of technological development in a systematic and coherent way. The situation is even worse for global aging research. There is a serious lack of systematic research and analysis on this topic. This monograph is one of the few that present the issue of global aging, the impact of social and societal aging on different aspects and spheres of life, society, the World System and our future, in a systematic way and with different scenarios. Thus, the volume fills both gaps, offering a unique combination of intersecting global trends.

The authors provide a broad panorama of the current state and future development of many innovative fields, including breakthroughs in medicine, additive (3D printing), nano-, and biotechnologies, robotics (including drones and driverless vehicles), AI and cognitive technologies, which they group under the term MANBRIC technologies (see below).

Moreover, the monograph touches on other urgent important issues for our present and future, such as: Which occupations may change radically or even disappear in the future, and why; what advantages and dangers the development of technology will bring in the future; why the modern state will be radically transformed in the next five decades and in what forms; how global aging may threaten democracy; whether the threat of men and women living under the complete control of AI regulatory systems will be realized; is a post-human revolution possible, is the threat of becoming cyborgs real, etc.?

The monograph is mainly devoted to forecasts for the next few decades, up to the beginning of the twenty-second century. Of course, there can be several forecasts, but the future will be realized in the singular. Only time will tell which predictions will come true and to what extent. One thing is clear—our tomorrow will be different in many ways from today's lives, but how and in what ways it will be different is, of course, unknown. But is it a complete mystery? A discerning eye may offer a good opportunity to identify trends that are not yet obvious or clear today, but which will reveal themselves and become decisive tomorrow. The authors of this monograph have attempted to present a fairly comprehensive picture of the future. You may or may not agree with it, some may like such a future, and others may be outraged to imagine it. But it is extremely important that this monograph is a rare case of describing the future in such a comprehensive way, in a detailed and systematic manner, touching on aspects that most of us do not think about today. And it is extremely important that the book allows you to think deeply about where mankind is heading. Is it possible to change the course to the future? While the outcome is still unknown today, the authors strive for maximum objectivity, both in showing what concepts, tendencies and trends they base their predictions on, using in-depth historical research to do so, and in objectively assessing these changes.

So, Leonid Grinin, Anton Grinin and Andrey Korotayev share their vision of a cautious optimism regarding what the world will look like in hundred years. As is mentioned above, their view is based on two related phenomena: first the Cybernetic Revolution that will have reached its final phase and penetrate all aspects of everyday and social life in the course of the twenty-first century; second the advancement of the global aging process.

The Cybernetic Revolution is the third crucial technological revolution in the history of humankind. It follows the Agrarian Revolution which lasted from the tenth to the third millennia BC and the Industrial Revolution of the sixteenth to nineteenth centuries.

The Cybernetic Revolution started in the middle of the twentieth century and is still ongoing. The authors expect the final phase of the Cybernetic Revolution to start very soon, in the 2030s, and to end in the 2060s–2070s. During this period, there will appear a large number of technologies that will allow various systems to function without direct human involvement. The authors call such systems self-regulating, since they will become autonomous and will be able to function almost without human control. The most advanced of them can be called self-managing systems. Self-regulating/self-managing systems will become a major part of the technological process and of our entire lifestyle.

The completion of the Cybernetic Revolution will put social and economic life on new grounds. Relying on omnipresent data collection from web service users, smart solutions will replace industrial standards of production and democratic standards of politics. Growing amount of data, permanent feedback and algorithmic procedures to carve out the best possible solutions will pave the way for a new type of techno-social regulation. Random sampling, trial and error will give way to a purely rational way of planning. Today, the transition toward the Scientific-Cybernetic production principle is already manifested in a shift in leading sectors, with medicine, additive, nano, biotech, robotic, information-communication and cognitive technologies as core drivers of change, summarized in the acronym MANBRIC by the authors. They point out the crucial role of medical and health care in implementing self-regulating systems as leading, converging principles of the cybernetic production principle. These trends have been analyzed by the trio in previous works (Grinin & Grinin 2016; Grinin, Grinin and Korotayev 2017), both in historical perspective, tracing the patterns of development from the Agrarian and the Industrial to the Cybernetic Revolution, and with regard to future outcomes. They are further elaborated on the last topical developments in Parts 1 and 3. In this book, the authors underline their prognosis by combining the cybernetic transition with transitions in the field of global demography, in particular global aging.

As has been mentioned above, global aging, an extremely important trend, is equally portrayed in front of a *longue durée* historical analysis of demographic change, which reflects impacts of technological and scientific progress. "By 2050, the proportion of the population aged 65 and over will be more than double that of under-five children", they write (Chaps. 1, 6 and 7). According to the UN Population Division more than a quarter of the world's population will be over 65, exceeding the number under 20. The relevance of such a "demographic crossing" has been widely neglected in prognostic scholarly discourse so far. It will further attribute importance to the Med-Biotech-Cognitive industries complex in order to ensure a decent and healthy life for the growing share of elderly persons. The elderly will promote unprecedented demand for the health complex and stimulate its leading role not only for economic growth, but for the cybernetic way of life in general with its general vector, defined as a "special medical-biotechnological environment" with "rapidly expanding opportunities for correction or even modification of human biological nature" (Chaps. 8, 11, and 15).

The new demographic pattern will trigger a shift from social to generational conflicts, however, affecting the regional as well as the global political order. The authors are convinced of the great possibilities the cybernetic mode of life might offer in life expectancy, quality of life and well-being. In some parts of the text they can hardly hide their enthusiasm with, as the reviewer would call it according to Morozov (2013), "technological solutionism". They believe that technological progress would improve human life, as it did in the past. While depicting the chances, they also warn about the dangers, the uncontrolled technological development might lead to, in all fields of social life, including the utopian trend to leave behind the age of Homo sapiens and enter a post-human revolution and age. Although much of the post- or transhuman ideas are still dreams, the authors clearly point out the possibilities to realize them along with the cybernetic transformation, with global aging as an important push-factor to intervene in the biological nature of mankind.

In their Introduction they maintain: "And although we have no choice but to go forward, the maximum of caution, wisdom, prudence and even some humility before the greatness of the Universe and the world, and deep respect for the heritage left to us by billions of years of biological evolution, are absolutely necessary in order to successfully travel along this path". The ongoing tension between socio-technological progress and its uncontrollable turn into technocratic, authoritarian societies, where dataism, complete surveillance and cyborgization become dominant, overcoming human dignity, human rights and democratic order, represent one of the leitmotifs of the book.

Grinin, Grinin and Korotayev do not only call for prudence, but they clearly speak out, what the cybernetic accomplishments would look like in case of successful implementation: We would permanently deliver our data and live under the control of algorithms, artificial intelligence and all sorts of programs that monitor and supervise human performance, including health, body and brain control. Optimized surveillance would affect governance, which is likely to follow the direction and guidance of complexity sciences, instead of gauging interests in democratic decision-making, for the sake of the best compromise. It would affect political freedom, which is likely to be out-ruled by surveillance capitalism. Last but not least, we—the human beings—might face a major transformation of our existence and undergo a fusion with machines that govern our daily routine. Some techno-freaks cannot await transforming men and women into sexless cyborgs, experimenting with virtual realities or even dreaming of the eternal transplantation of human brain into the collective cyber-space (Harari 2017; Kurzweil 2006). Grinin, Grinin and Korotayev reject such kind of transhumanist phantasies, which they classify as myths. However, the question has to be raised, whether or not their own visions of prolonged, protected and monitored life also risk transcending the entitlements of human rights and individual freedom, represented in the political anthropology of the Enlightenment, and replace it with a technocratic paradigm.

This is why I take note of the authors' cautious optimism with certain reservations. Neither am I convinced of the inevitability of the Cybernetic Revolution to progress in the predicted way nor am I willing to share the entire admiration of smart solutionism and its promise to make life better, more comfortable and happier. I miss more attention to the limits of allegedly self-improving systems and to the contradictions, unintended consequences as well as to individual rejection and public resistance these innovations will generate. These hesitations do not prevent me from highly appreciating the scenarios and forecasts the three authors are delivering in an unprecedented way. The book is structured into 15 chapters, focusing on four big themes: technological transformations, demographic transformations, the prospects of MANBRIC technologies in the forthcoming epoch of self-regulating systems, to culminate in characterizing the Cybernetic Revolution in the light of overall technological progress and aging. A conclusion is taking up the questions of the introduction.

The authors are well aware of the dangers and fallacies of the surveillance regime, resulting from the growth imperative of data extraction on the one hand and the tendency to move from paternalistic betterment to authoritarian control practices. Their main argument is the following: We still live in a period of transition; the Cybernetic Revolution is approaching its final phase, revealing the main features of the future production regime, but it is not yet accomplished. This is why political, legal and ethical regulations must be discussed and implemented now, i.e., before the new principle will have acquired dominance. This may include R&D moratoria or restrictions on AI research and provisions, how much competence shall be delegated from political steering to the technological self-regulating imperative. The authors do not whatsoever doubt, that the dynamic to finalize the Cybernetic Revolution is under way and will affect all societal realms until it is fully implemented. The more it is crucial for the years to come to anticipate the social, economic, political and ethical implications and introduce as much regulation as necessary and as much social and democratic control as possible. The authors see that we (human beings) face a difficult choice—to gain something, but in return to lose something. Therefore, the book aims at contributing to find the optimal solution to this dilemma.

Frankly speaking, one has to be aware of the restricted maneuvering space for regulations. Regulations often rather serve as motors of acceleration. According to the authors it is not too late, and their clear descriptions of what technological change and self-regulating systems will presumably look like is an appeal for social and political actors to act now, instead of being overmanned by the pace of progress and its unwanted results.

To make it clear in advance: There is hardly a more precise and comprehensive forecast of how the transformation from industrial to cybernetic capitalism would happen. The authors restrain from using the term 'capitalism', however. Although they avoid any categorization of the social order or the property relations for the forthcoming cybernetic age, they seem to perpetuate the operation of the capitalist world-system. The strong influence of machine intelligence, which is going to determine social life in an unprecedented way, will further reduce the autonomy of the individual as well as the political domain. If politics and administration have to adapt to the practical constraints, derived from data analysis, self-regulation and self-management systems, (neo)liberal conceptions of the lean state will become obsolete. This is why the shift to authoritarian forms, or at least the subordination to a technocratic paradigm, is highly probable. Against the reviewer's rather pessimistic fear, the authors introduce a different trajectory for the future of the World System in the twenty-second century. Without referring to the political label of such a new system, they argue, that the "new demographic and technological system will promote the formation of a new societal type, which can be denoted as the cybernetic society. I will be extremely high-tech society with a stable population size, a very large proportion of the older population and total technological support for societal health and the quality of biological life, achieved by creating of a continuous medical-biotechnological environment" (Chaps. 7 and 15). In Chap. 6 they plea for a significant change in financial distribution in the direction of increasing the share of benefits and funds allocated to the elderly and disabled people. And in Chaps. 13 and 15 they also talk about the fact that a new consumption system will emerge and consumerism will decrease.

Theoretically, the argument of a flattening curve of innovation, economic growth and demographic reproduction, is based on the authors' model of the sequence of production revolutions as well as mathematical operationalizations, elaborated in Chap. 12. More easily comprehensible for those, who do not rely on mathematics, is the age-related change in mentality, the authors deduce from the growing share of old people. Accordingly, an aging population would contribute to lowering the pressure on permanent innovation and consumption by a more conservative, relaxed, less speedy and greedy elder generation, who—at the same time—exercise demand for those existing and novel health and longevity products that represent the chance for growth and investment return. Overall, an optimistic prognosis for a new societal model, that could replace today's global capitalism or reduce the imperative of accumulation.

Today, the dynamics and efforts to introduce a new production principle cannot ignore the fact that the international order is facing major restructuring. The hegemonic role of the West, led by the USA, does not hold out against the changing geography of global commodity chains, that have been transforming developing countries into new sites of manufacture, moving away from low end production to more profitable operations, including R&D in new lead industries. We face successful emerging nations, forging new initiatives and alliances, allowing the Global South to turn into a major force of transformation. In an earlier book, Grinin and Korotayev have been identifying a paradigmatic reversal from the "Great Divergence", a code for the growing North–South divide since European expansion, toward closing the gap, denominated "Great Convergence", since the second half of the twentieth century (Grinin & Korotayev 2015). The fact that the political architecture of international relations does not reflect global economic restructuring including its demographic impacts, has been giving way to serious conflicts; currently (January 2024) they are escalating militarily in Ukraine and in the Near East, with other eventual theaters of war lurking behind the curtain.

Renewal of leading sectors according to the Scientific-Cybernetic Production Principle is intrinsically tied to the changing role of the Global South, which gains momentum in the course of Western warfare, or weapon and logistical supply to war-faring parties. The consolidation of South–South relations supports the endeavors to set up international relations beyond the US-Dollar as the global currency and financial instrument of Western dominance. While at some points the authors' visions seem to resemble Klaus Schwab's and the World Economic Forum's visions of a "Great Reset" (Schwab & Malleret 2020), the strong commitment to a politically and socially inclusive international order, representing a new societal type, opens new dimensions of a multipolar order beyond Western hegemony and their institutions.

Focusing the prospects for a new role of the Global South in international relations leads back to global demographic issues. On the one hand, the growing share of emerging and developing countries designates these populations to be the consumers of the MANBRIC industries' products. The high-income societies of former industrial countries will not suffice allowing those branches a global breakthrough. On the other hand, Grinin, Grinin and Korotayev can show that economic upgrading will change demographic patterns and turn societies with an overwhelming percentage of young people into aging societies—still young compared to the Global North, but approaching a transition, that will require more and more care, repair, in order to qualify people to stay longer a part of the active workforce, and allow retired people to participate in social life until a much older age.

The prognostic horizon of this book spans hundred years. The type of Big History, that the author-trio represents, does not rely on empirical historical methods only, but on identified patterns of technological, demographic and social development and on their own theories of production revolutions and production principles that cover the entire historical process. This allows them to carry out extrapolations of time-series analyses into the future. They are outstanding experts in this field, including the price, one has to pay when using statistical macro-perspectives, that is losing sight of individual actors or single historical events.

The reviewer was able to make use of the three author's expertise in her own socio-economic analysis of the Corona lockdown period 2020–2022, a relatively short, but more influential historical moment (Komlosy 2022). The lockdowns, contact and movement restrictions came as a shock wave, disrupting social and economic relations by far-reaching closures of the economy, trade, education, travel, cultural and sport events, etc. Against all statistical evidence, governments, international health organizations and media constructed a health crisis, turning a fairly curable mass disease with a low mortality risk into a high mortality pandemic. The Corona moment served as a catalyst. It accelerated the implementation of the MANBRIC sectors: digitalization and remote services opened a way for people to take part in social interaction without personal contacts; health products like tests and vaccines were decreed under compulsion as well as subsidized by governments; control and tracking apps gained grounds as did the use of body monitoring to improve lifestyle and health status (Chap. 14).

The Corona moment as a single historical moment served as a prism, in which elements of the ongoing cybernetic transformation showed up in a condensed form. To interpret them beyond the daily grievances, Grinin, Grinin and Korotayev provide the long-range, wide-angle explanatory framework of cyclical sectoral renewal and shift in production principles. It is possible that the predicted transformations toward a cybernetic society would have proceeded, without the accelerating impact of the Corona regimes. History takes its course, like rivers shape their beds. Maybe. Such a perspective does not take into account that Big History is mirrored in Small History. Only if we consider long-term and short-term, event and structure, we can approach the driving forces of history. Therefore, we can assume, as Grinin, Grinin and Korotayev acknowledge in Chap. 4 as well as in previous works (2021), that the disruptive effect of the Corona management accelerated the speeding-up of the Cybernetic Revolution.

Once events are taken seriously, we have to consider unintended consequences, however. In the case of Corona, not only did the measurements pave the way for the acceptance of new health and control regimes, which are at the heart of MANBRIC convergence and the introduction of self-regulating systems of surveillance, health monitoring and control. They also provoked a strong popular sentiment against data expropriation and digital surveillance, giving rise to a mass movement for human self-determination and sovereign conduct of life. This sentiment might change the course, or pace of the cybernetic transformation; it might as well contribute to being more attentive vis-à-vis the authors' appeal to be vigilant against the dangers of authoritarian takeovers of the cybernetic society.

January 2024

Prof. Andrea Komlosy
University of Vienna
Vienna, Austria

## References

Grinin, L. E., & Grinin, A. L. (2016). *The cybernetic revolution and the forth-coming epoch of self-regulating systems*. Moscow: Uchitel Publishing House.

Grinin, L.E., Grinin, A.L., & Korotayev, A. (2017). Forthcoming Kondratieff wave, cybernetic revolution, and global ageing. *Technological Forecasting & Social Change, 115*, 52–68.

Grinin, L. E., & Korotayev A. V. (2015). *Great divergence and great convergence: a global perspective*. Springer.

Grinin, L., Grinin, A., & Korotayev, A. (2022). COVID-19 pandemic as a trigger for the acceleration of the Cybernetic Revolution, transition from e-government to e-state, and change in social relations. *Technological Forecasting & Social Change, 175*, 121348. https://doi.org/10.1016/j.techfore.2021.121348.

Harari, Y. (2017). *Homo deus: a brief history of tomorrow*. Harper.

Komlosy, A. (2022). *Zeitenwende. corona, big data und die kybernetische Zukunft*. Promedia.

Kurzweil, R. (2006). *The singularity is near: when humans transcend biology*. Penguin Books.
Morozov, E. (2013). *To save everything, click here: the folly of technological solutionism*. Public Affairs.
Schwab, K., & Malleret, T. (2020). *COVID-19: the great reset*. Forum Publishing.

# Acknowledgments

First and foremost, we would like to express our gratitude to Peter Herrmann (the Human Rights Centre at the Law School, Central South University, Changsha, PRC), Victor de Munck (Vilnius University, Lithuania and State University of New York—New Paltz, USA), Anthony Harper (Benedictine University, Lisle, IL, USA), and Elena Emanova (Uchitel Publishing House, Volgograd, Russia) for their brilliant editing of the manuscript of this monograph. We also express our gratitude to the team of Publishing House "Uchitel" (Volgograd, Russia) for its invaluable technical help.

Our deepest gratitude also goes to Jack A. Goldstone (George Mason University, Fairfax, USA) and Andrea Komlosy (University of Vienna, Vienna, Austria) for their wonderful forewords to this volume.

We would also like to thank our colleagues from HSE University—Moscow, Lomonosov Moscow State University, Russian Presidential Academy of National Economy and Public Administration (RANEPA), Russian State University for the Humanities, International Center for Education and Social and Humanitarian Studies—Moscow, Volgograd Center for Social Studies, as well as the Eurasian Center for Big History and System Forecasting at the Institute of Oriental Studies and Institute for African Studies of the Russian Academy of Sciences for their continuous support. Their invaluable comments helped us very much in our work on the present report.

# About This Book

This monograph is primarily about the future that awaits our world and all of us in this century and at the beginning of the next century. The desire to predict the future is inherent in people. But no less acute is the desire to understand our contemporary problems. Moreover, today, many of us are convinced that the world is in a state of major change, and the awareness that we are living in a period of transition to something new and unknown is constantly growing. In this book, you will find in-depth explanations of today's pressing issues and their historical roots.

This monograph, while characterizing in many aspects and details the present state of the world, is mostly devoted to the future, presenting analysis and description of how our world will change over the next hundred years. Yet, the authors do not fantasize, but show how and why humanity can reach a new state, what trends lead us to the described future. They also show the dangers along the way and talk about how these dangers can be minimized.

In this monograph the authors consider in detail the two most powerful trends which characterize the twenty-first century and have a major and constantly increasing impact on all walks of life. These trends make up the title: *Cybernetic Revolution and Global Aging* and, as it will be shown, they are tightly interconnected. These trends are (1) the continuous technological wave, named the Cybernetic Revolution, which, according to Grinin et al., will experience a new rise over the next two decades. It will transform all walks of life and widely introduce self-regulating and self-managing systems which will be able to function without human participation; (2) the increasing process of global aging. Over the next several decades, extending perhaps into the twenty-second century, global aging will become one of the strongest and most powerful factors, influencing the life and future transformations of society and the World System as a whole, although in different ways and at different times.

The authors believe that the Cybernetic Revolution will continue to evolve, enabling current and future generations to live longer and healthier lives. This is an extremely challenging process and far from a smooth path; and it will bring us to many crucially important transformations including our fundamental rights and freedoms as well as our biological nature. According to the authors, as a result of these two trends, a new type of society will emerge. This prediction can be found in the subtitle of the monograph—"Humankind on the Way to Cybernetic Society, or the Next Hundred Years".

Since the monograph touches on the burning and very important problems for our present and future, it will be relevant to a wide readership—including not only experts in technological forecasting and its social consequences, but also to many other experts. Futurologists will find a brain provoking comprehensive forecast of future development and its negative and positive aspects, while for economists it can be useful for the analyses of economic projections in light of technological change; as well as for demographers it might suggest a new approach for the analyses of global aging and its multiple consequences. Political scientists and politicians can find answers about political changes that may be triggered by the Cybernetic Revolution, while sociologists and anthropologists may find some useful ideas about possible social changes that may take place under the influence of technological development. Philosophers can also find a wide range of relevant topics discussed in the monograph, including issues of privacy and identity.

**Keywords**: Cybernetic Revolution, future technologies, societal aging, medicine, biomedicine, gerontology, biotechnologies, life expectancy, global aging, artificial intelligence, electronic state, future society, pandemics, cybernetic society, robotics, driverless vehicles, forecasts, the twenty-second century, demographic trends, demographic revolution, singularity, global trends, post-human revolution, future trends, young South and aging North, ageism, future scenarios.

# Contents

# About the Authors

**Leonid Grinin** is Senior Research Professor at the Center for Stability and Risk Analysis at the HSE University in Moscow, as well as Senior Research Professor of the Institute of Oriental Studies, Russian Academy of Sciences. He is co-editor of the international journals *Social Evolution & History* and the *Journal of Globalization Studies*, as well as co-editor of the international almanacs *Evolution, History & Mathematics*, and *Kondratieff Waves*. He is author of more than 600 publications, including 30 monographs in Russian, English, and Chinese. His current research focuses on comparative political studies, theory of revolution, political anthropology, global economy, global history, historical sociology, and futurology. Among his monographs are *The Cybernetic Revolution and the Forthcoming Epoch of Self-Regulating Systems* (Uchitel, 2016, with Anton Grinin), *Big History of Globalization. The Emergence of a Global World System* (Springer 2019, in co-authorship with Julia Zinkina, David Christian and others), and *Islamism, Arab Spring, and the Future of Democracy. World System and World Values Perspectives* (Springer, 2019; with Andrey Korotayev and Arno Tausch). In 2004 he was awarded as a winner of the best publication of *Socis* journal during the year for the article "People of Publicity—a New Social Layer?". In 2012 he was awarded the Gold Kondratieff Medal by the International N. D. Kondratieff Foundation. In 2016 he received the Belyaev award for the book *From Bifaces to Nanorobots* (the book published in Russian in co-authorship with Anton Grinin). 12, Rozhdestvenka, Moscow, 107031, Russian Federation. e-mail: leonid.grinin@gmail.com

**Anton Grinin** is a Russian scientist specializing in the study of modern technological trends and future studies, as well as an evolutionist and philosopher. He has a Ph.D. in Biology and works as a Research Fellow for Lomonosov Moscow State University and HSE University, Moscow, Russia. He is the author of more than 80 publications in Russian and English, including 2 monographs. Awards: Gold Kondratieff Young Scholars Medal and Alexander Belyaev Literature Award. Anton Grinin focuses on the issues of the technological revolution (the Cybernetic Revolution) evolving in the current decades of the twenty-first century and the resultant profound transformations in economy and society as well as related economic, political, ethic-legal and other risks, including political revolutions. He also investigates the correlation between the Cybernetic Revolution and Kondratieff waves as well as between the Cybernetic Revolution and global aging. 1, Leninskie Gory, Moscow, 119991, Russian Federation. e-mail: algrinin@gmail.com

**Andrey Korotayev** heads the Center for Stability and Risk Analysis at the HSE University in Moscow, Russia. He is also Senior Research Professor at the Institute of Oriental Studies and Institute for African Studies, Russian Academy of Sciences, as well as at the Faculty of Global Studies of the Lomonosov Moscow State University and the Russian Presidential Academy of National Economy and Public Administration. He has authored or co-authored over 650 scholarly publications, including such monographs as *Ancient Yemen* (Oxford University Press, 1995), *World Religions and Social Evolution of the Old World Oikumene Civilizations: A Cross-Cultural Perspective* (The Edwin Mellen Press, 2004), *Introduction to Social Macrodynamics: Compact Macromodels of the World System Growth* (URSS, 2006; with Daria Khaltorina and Artemy Malkov), *Introduction to Social Macrodynamics: Secular Cycles and Millennial Trends* (URSS, 2006; with Daria Khaltorina and Artemy Malkov), *Great Divergence and Great Convergence. A Global Perspective* (Springer, 2015; with Leonid Grinin), *Economic Cycles, Crises, and the Global Periphery* (Springer, 2016; with Leonid Grinin), *Islamism, Arab Spring, and the Future of Democracy. World System and World Values Perspectives* (Springer, 2019; with Leonid Grinin and Arno Tausch). He is a laureate of a Russian Science Support Foundation in 'The Best Economists of the Russian Academy of Sciences' Nomination (2006); in 2012 he was awarded the Gold Kondratieff Medal by the International N. D. Kondratieff Foundation. 20, Myasnitskaya, Moscow 101000, Russian Federation. e-mail: akorotayev@gmail.com

# 1 Introduction: Between Human and Post-human Revolutions, or What Future Awaits Us?

**Abstract**

In this chapter, the authors introduce the monograph titled 'Cybernetic Revolution and Global Aging', present its main ideas and findings and describe its structure which consists of four parts and fifteen chapters. Grinin et al. point out that the current volume, while characterizing in many aspects and details the present state of the world, is mostly devoted to the future, presenting analysis and description of how our world will change over the next hundred years. They deeply analyze two powerful trends, which make up the title: 'Cybernetic Revolution and Global Aging' and, as it will be shown, they are tightly interconnected. The authors hope that this research will be interesting for those who study the issues of technological development, global aging, and society, which in their forecasts, by the end of the twenty-first century, will transform into cybernetic society. In this chapter the authors describe the main problems and issues of the monograph. Then they discuss about the increasing possibilities to transform the biological nature of the human organism (all this is defined as the Post-Human Revolution) and point out great and unpredictable dangers of this process. Finally, authors present extended annotations to every part and chapter.

## 1.1 What This Chapter is About

In this chapter we introduce the monograph titled 'Cybernetic Revolution and Global Aging', present its main ideas and findings and describe its structure which consists of four parts and fifteen chapters. We point out that the current volume, while characterizing in many aspects and details the present state of the world, is mostly devoted to the future, presenting analysis and description of how our world will change over the next hundred years.

In this monograph we consider in detail the two most powerful trends which characterize the twenty-first century and have a major and constantly increasing



L. Grinin et al., *Cybernetic Revolution and Global Aging*, World-Systems Evolution and Global Futures, https://doi.org/10.1007/978-3-031-56764-3_1

impact on all walks of life. These trends make up the title: 'Cybernetic Revolution and Global Aging' and, as will be shown, they are tightly interconnected. These trends are (1) the continuous technological wave, named the Cybernetic Revolution, which, according to our point of view, will experience a new rise over the next two decades. It will transform all walks of life and widely introduce self-regulating and self-managing systems which will be able to function without human participation; (2) the increasing process of global aging. Over the next several decades, extending perhaps into the twenty-second century, global aging will become one of the strongest and most powerful factors, influencing the life and future transformations of society and the World System as a whole, although in different ways and at different times.

We believe that the Cybernetic Revolution will continue to evolve, enabling current and future generations to live longer and healthier lives. This is an extremely challenging process and far from a smooth path; and it will bring us to many crucially important transformations including our fundamental rights and freedoms as well as our biological nature.

This introductory chapter is titled 'Between Human and Post-Human Revolutions, or What Future Awaits Us?' The 'Human revolution' happened when Homo sapiens culture appeared (50–40,000 years ago). In the coming decades, technological progress will develop to such a level that it will change the biological nature of humans, and we can therefore speak of a post-human revolution. However, there is considerable anxiety regarding what the significant problems are that humanity will face. In the chapter (as well as in the volume) we discuss the following important issues: Are such post-human transformations possible? What will our future be like then? How can we prevent negative consequences? So, rapid progress on its way will cause not only great triumphs but also great perils.

That is why we assume that we have to balance technological optimism with reasonable caution. On the one hand, we should not be afraid of the future. However, on the other hand, the stakes are high and because of possible strong negative consequences we have to think about preventive measures to minimize these consequences. In this connection the authors also investigate whether it is possible that the rate of technological progress will accelerate to singularity? Because the Cybernetic Revolution will provide opportunities for the intervention and transformation of the biological nature of the human body and also enhance the impact of those interventions and transformations on it, including on our brain, in a special section, the authors discuss an important dilemma which they formulate as "to Be or not to Be for the Human Beings".

We first touch upon problems and risks associated with future scientific and technological progress. Then we are concerned about whether the time will come when a human will mainly or completely consist not of biological, but of artificial, material. In other words, they discuss whether we will become cyborgs, or will be in the process of cyborgization, i.e., changing biological tissues and organs for artificial ones and will go too far in this process. There are also other important questions and problems touched on in this chapter and monograph, and with them associated numerous forecasts.

We hope that this research will be interesting for those who study the issues of technological development, global aging and society which in our forecasts, by the end of the twenty-first century, will transform into cybernetic society, i.e., elderly society with the institutionalization of age differences, relying almost entirely on smart (cybernetic) technologies and self-regulating systems. That is why the subtitle of the monograph is 'Humankind on the way to cybernetic society, or how our world will change over the next hundred years'.

This chapter consists of eight sections; however, it can be presented as three main blocks. First, we describe the main problems and issues of the monograph. Then we discuss very important questions about the increasing possibilities to transform the biological nature of the human organism including the brain as well as a formation of the trend toward cyborgization (all this is defined as the Post-Human Revolution) and point out great and unpredictable dangers of this process. Finally, we present extended annotations to every part and chapter.

## 1.2 Preliminary Notes on the Book and Its Main Ideas

The Introduction may contain surprises for the reader as a consequence of some of the unique but nonetheless serious points and topics it contains; items not expected in otherwise scholarly discourse. This introductory chapter is titled 'Between Human and Post-Human Revolutions or What Future Awaits Us?' The 'Human (also known as 'Upper Paleolithic') Revolution' happened when *Homo sapiens sapiens* culture appeared (50–40,000 YA). Now technological progress is occurring on the eve of a possible transformation of human biological nature, and this transformation may be called the 'Post-Human Revolution'; although there is much apprehension that if this revolution happens, human beings will face countless difficulties and serious losses. In this chapter (as well as in the volume as a whole) the following important issues will be discussed: are such post-human transformations possible? If so, what then will be our future? How can we prevent any negative consequences associated with this Post-Human Revolution from occurring?

So, there will be not only great triumphs during this revolutionary transformation but also great perils associated with such expected rapid development. That is why the authors suppose that we have to find a place between technological development associated with negative consequences; we have to think about preventively minimizing them. In this connection we will also investigate whether it is possible that the rate of technological progress will approach singularity? Because the Cybernetic Revolution will provide opportunities for the intervention in and changing of the biological nature of the human organism and will also enhance the impact on both the human body and the brain, the authors discuss in a special section the very important dilemma which they formulate as 'to Be or not to Be for the Human Beings'. We also touch upon problems and risks associated with future scientific and technological progress. In particular, we will then be concerned about the question of whether or not the time will come when humans will mainly or completely consist not of biological, but of artificial material. In other words, they discuss whether we will become cyborgs. Will the process of cyborgization, i.e., the replacement of biological tissues and organs with artificial ones, ever go too far?

Thus, we are presumably standing at the unprecedented point where it will become possible to change the human body and even human brain and consciousness in some new ways. However, the expansion of such opportunities, with obvious positive aspects, cannot but be a concern. Consequently, future development, while solving many difficult problems, will simultaneously create new ones, often completely unfamiliar to the humanity. This is due to the fact that development is always a struggle between opposites, between progress and degeneration, between benefits and detriments, successes and failures. That is why it is so important to think in advance not only about the positive side of development, but also about its negative side. This is so in order to be ready for complex and unexpected events, problems and difficulties and also to mitigate them to the maximum possible extent, so that the price of development is made as low as possible. That is why we tried in our monograph to objectively show the positive aspects of changes as well as those future problems, difficulties and dangers that can be perceived already today.

As we have mentioned above, there are some points which are incredibly serious, nonetheless they are currently absent from active scholarly discourse. One of them is the discussion of the burning question of a possibility that human evolution will develop in the future in a somewhat unpleasant and quite dangerous way. The problem is that the development of humankind is now at a very important stage, beyond which its previous trends can change radically and rapidly.

Today, many of us are convinced that the world is in a state of a major change, and the awareness that we are living in a period of transition to something new and unknown is constantly growing. Recently, especially in the last several years, technological changes in the world have been occurring with extreme speed. Remote work, online education, e-passports or ubiquitous and more and more 'clever and cool' artificial intelligence (AI), which begins to stifle the freedom and privacy of people and now even deceives them, are quite evident, just to name a few examples (see Chap. 10)[1]; not to mention automation and technological substantiation of many professions and occupations. Technological progress changes not only our life, but also government and even society itself. The development of remote technologies and AI causes transformations in functions and apparatus of the state and leads to the development of what we call socio-technical self-regulating systems (SSSs). Their development pushes society toward the formation of an electronic state (e-state) with a significantly reduced number of the state officials as well as those of the state supervisory bodies. Governments will gradually delegate to these

[1] On the suppression of people's freedom and privacy see, e.g., Ashman et al., (2014); Alharbi (2020); Brammer et al., (2020); Cecere et al., (2015); Moustaka et al., (2019); Schwartz (1999); Schwab and Malleret (2020); Solove (2008).

technologies the right to automatically manage a variety of regulatory and administrative functions and social relations (about this see Chap. 14, as well as Grinin et al., 2021a). As a salient example of recently experienced rapid change in which human control has been reduced, we would mention the coronavirus pandemic that has dramatically changed our everyday life and our perceptions of the future, having caused anxiety in many hearts. In short, the world is changing rapidly. One should also mention that rapid and sharp changes are simultaneously happening in the political, ideological, economic and other spheres. That is why many of us are scared of the future.

## 1.3 Technology and Aging

So, our volume is devoted to the future. Of course, we can only consider some of the changes in the world that could happen in the next hundred years, especially the two most powerful trends that will accelerate in the twenty-first century.[2] The title of our volume mentions both of these trends—'Cybernetic Revolution and Global Aging'. The Cybernetic Revolution (mostly, its final phase) will bring a new technological tide, which, according to our forecast, will start during the next two decades. And it will unfold against a backdrop of an intensifying process of global aging, which will change dramatically many social, political and economic relations and the overall shape of society itself.

For deep insight into both subjects, we have to describe precisely the technological aspect of the historical process, because the modern world is a product of technological and scientific progress (its history will be presented in Chap. 2). In general, the whole of human history, especially the last few centuries, has been the story of the triumph of science and technology. Today, this technological advancement requires people spending considerable time simply to keep up with progress. As a result, humanity is becoming increasingly dependent on it. Modern technologies permeate almost every aspect of our lives as well as our mentality. So, the fact is that technology has become one of the most powerful forces of development. It is just this development of technologies that has made our life much safer and more prosperous than it used to be. It was technology (together with the development of science and education) that gave humankind the power of medicine, increasing life expectancy up to 80 years in many countries. And we believe that the Cybernetic Revolution will make even more progress in this direction, enabling the current and future generations to live longer and have a healthier life. Thus, global aging as a process is the result of technological and scientific progress. Hence, technological progress and global aging are closely related, and we will show in detail how they influence each other (see Chaps. 7, 11 and 13). This fact of their interrelationship, however unexpected it may sound, is very important, and

[2] That is why the subtitle of the monograph is *Humankind on the way to cybernetic society or how our world will change over the next hundred years*. About the cybernetic society see below and Chap. 15.

it explains the main reason behind our decision to consider technologies and aging as two leading themes for our volume.

However let us repeat, the path to the future is far from smooth, it is full of new kinds of dangers because complex medical devices, AI and other technologies will significantly affect the human body, consciousness, lifestyle and human rights (see below for possible negative consequences). The excess mortality of the elderly from the COVID-19[3] as well as increasing number of people choosing euthanasia (see Chap. 7 for details) also demonstrate that humanity will face quite a difficult struggle along this path.

Over the next few decades, perhaps, even over part of the next century, global aging will become one of the strongest and powerful trends, influencing life and future transformations of society and the World System as a whole, although in different ways and in different periods. In general, all population forecasts unanimously claim that in the coming decades we will face a rapid aging of the world population (Alkema et al., 2011; Kaneda et al., 2016; Korotayev et al., 2023a, 2023b; Lutz et al., 2018; Raftery et al., 2012; UN Population Division, 2019, 2022a, 2022b; Vollset et al., 2020; Wittgenstein Center, 2022).

The population aging projections, even the inertial ones, are quite impressive. In particular, people aged 65 and over will outnumber children under the age of five. By 2050, the proportion of the population aged 65 and older (15.6%) will be more than double that of under-five children (7.2%). This unique demographic phenomenon of the 'crossing' is unprecedented (He et al., 2016: 3, Table 2–3, P.5). The growth rate of the proportion of people 60+ is several times higher than the overall growth rate of the Earth's population (UN Population Division, 2013, 2022a, 2022b). In the early 1980s, there were no 'aging countries' in the world where older people would consume more than young people. In 2010, there were already 23 'aging countries' and by 2040 there will be 89 (UN Population Division, 2022b; UNFPA, 2012). According to the most recent UN medium projection, by 2100, 2.47 billion people, more than a quarter of the world's population, will be over 65 and only 2.30 billion will be under 20 (UN Population Division, 2013, 2022a, 2022b).[4] The number of those over 80 will increase sixfold, from about 140 million in 2020 to over 860 million by the end of the twenty-first century (Gallagher, 2020); thus, compared to 1950, the number of the 80 year olds should increase by almost two orders of magnitude by 2100! (for more details see Chaps. 6 and 7).

The aging process in such proportions had never been attested in human history before. We know societies that could be defined as gerontocratic, that is, those in which the ruling layer is constituted by older people (for each era, of course, old age itself could be defined differently). These were, for example, some hunter-gatherer societies, primarily the aborigines of Australia (Artemova, 1988;

---

[3] See, e.g., Grinin (2020); Grinin et al., (2021a); Irshad (2020); Widdowson (2021).

[4] The proportion of the population over the age of 60 has already reached 36% in Japan (UN Population Division, 2022a, 2022b; cp. WHO, 2021).

Chudinova, 1981; Woodburn, 1980, 1988; see also Grinin & Korotayev, 2009, 2013; Grinin, 2000). In Australian aboriginal communities, senior hunters had many privileges, including the right to have several wives, while they often married young girls who helped older wives and were in preparation for marital duties in the future. Consequently, young men were left without women or were content with wives much older than them. In addition, young hunters had to share prey with older men. In many hunting and archaic agricultural societies, the more wives a man had, the richer he was, since women worked more than men, collected food, engaged in horticulture, carried weights and arranged life. Archaic agriculturalists in some regions, such as, for example, New Guinea or Melanesia, had so-called "big men", that is, highly influential individuals who had great authority in the communities, in organizing common affairs, in inter-communal exchange and so on (e.g., Sahlins, 1963, 1972; Service, 1971, 1975). There are cases where a big man had up to six to ten wives and, accordingly, many children; this provided the family with wealth, working hands, many connections with various communities, prestige and power.

There were historical societies in which there were special governing bodies consisting only of older people (councils of the oldest). In particular, we can mention Sparta with its *Gerousia*, consisting of 30 members—28 *gerontes* of at least 60 years of age and two kings. The age of the latter did not matter. Thus, there were many societies where the elderly were the guiding force of society, and sometimes quite dynamic, but the strength of such societies still relied on youth. There was always a very small percentage of elderly people.

However, humanity has no experience of dealing with this kind of process on a modern scale.

There has never been a time in the history of humankind when elderly people would make such a high proportion of the population as they do today, but this proportion will be even higher in the future, which is likely to have various and unpredictable consequences.[5] The course of further social evolution will largely depend on humanity's response to this challenge. It is also important to take into account that the impact of global aging will become increasingly tangible, but at different stages global aging will have different (positive or negative) effects on scientific-technological progress (see Chaps. 12 and 13 for details).

The above-mentioned forecasts should make politicians think much more about aging and start changing their course, especially in view of the fact that aging is one of the few processes about which it can be said that it is likely to occur with a probability close to one hundred percent. However, politicians, the media, economists and many people are taking this trend lightly and ignore the urgent need to take serious steps that could prepare society for the growing negative

[5] See Chaps. 5–7; see also Coleman and Rowthorn (2015); Goldstone (2015); Goldstone et al., (2015); Grinin and Korotayev (2016a, 2016b, 2016c); Grinin et al., (2017, 2023a, 2023b); Haas (2015); Harper (2006); Korotayev et al., (2023a, 2023b); Powell and Khan (2013); Park and Shin (2015); Romanov et al., (2021); Sawyer et al., (2022); Ustyuzhanin et al., (2022, 2023); Zimmer (2016).

impact of global aging on the economy, the financial burden, the labor market, the public funds, the healthcare system, the education system, the military system, etc., or at least to mitigate these trends and tensions in the future (about this see, e.g., Gallagher, 2020). However, unfortunately, actions are not taken in this respect.

Therefore, we believe that *the level of knowledge of this problem—namely, of the problem of the growing influence of aging on all aspects of societal life every year and in the future—is clearly disproportionately small to its significance.* Such a situation of strange silence on this point has remained for several decades. Thus, in 1999 Peter Peterson reasonably argued that the aging of the developed world would change our collective future more than anything else ('the graying of the developed world's population—that may actually do more to reshape our collective future than any of the above') and at the same time he wondered why so little is said about the future manifestations of the global aging process compared to the climate problem (Peterson, 1999: 42). And ten years before he made this statement, Dieter Bös and Robert K. von Weizsäcker wrote about ignoring the problem of aging (Bös & Weizsäcker, 1989: 345). Meanwhile, let us repeat that the global aging is a completely new phenomenon in human history, and we have no experience in coping with this and related problems and risks.

This strange but dangerous ignorance of an important and growing problem underlies our decision to make aging one of the two leading themes of our volume.

An integral part of this unity between aging and technology is medicine (see Chaps. 8 and 11). Precisely, it is a combination of medicine and other technologies that we named the MANBRIC complex (see below and Chap. 4 for details). This complex will be the most important area of technological innovation, and it is clear that without its powerful impact, humanity will in all probability not succeed in drastically extending human life and improving its biological quality (which is the central point of the global aging process).

## 1.4 Singularity

Many researchers suppose that we have been moving toward considerable transformations and that human civilization will experience dramatic changes in the next decades. Some even consider it possible that the very nature of the development of humankind as a biological species will change qualitatively (Moiseev, 2000). They often use the term singularity to point out a certain unprecedented rate of technological growth that will result in an unprecedented phase transition in human evolution.[6] Indeed, the mathematicians denote the state which we are

[6] For more details about views on singularity see Chap. 12. The issue of a global history singularity has been discussed quite actively for more than two decades (see, e.g., Ayres 2006; Callaghan et al., 2017; Diakonoff 1999: 348; Eden et al., 2012; Grinin & Grinin, 2015a; Korotayev 2005, 2018, 2020a, 2020b; Kurzweil 2005; LePoire 2020; Modis 2006, 2020; Muehlhauser & Salamon 2012; Magee and Devezas 2011; Nazaretyan 2015, 2017, 2018, 2020; Panov 2005, 2017, 2020; Shanahan 2015; Tsirel 2014).

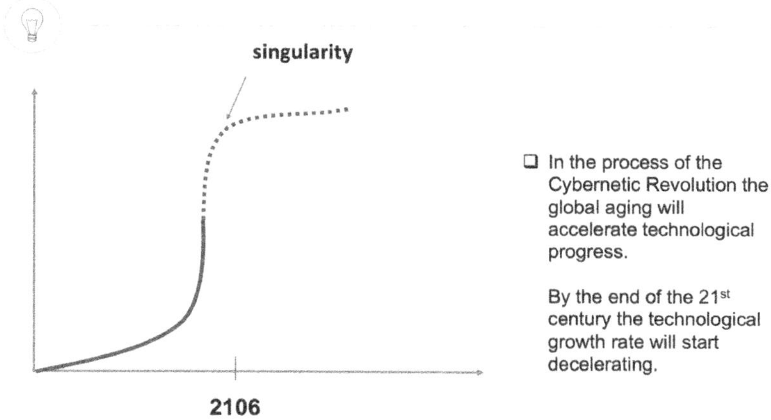


**Fig. 1.1** Global aging and global technological growth rate. See also Figs. 6, 7 and pages further in Chap. 12 for more details

approaching as a result of accelerating changes as a singularity: the rate of technological progress will increase so that the denominator in the formula will contain zero, and, when divided by zero, the rate of progress will be pushed—theoretically—to infinity. No one knows how this will work in practice. However, this gives rise to many speculative theories about possible super transformations, not least of which would be the transition of humankind to immortality. On the one hand, it is worth separating mysticism from reality, in the sense that one should not expect anything supernatural after reaching the conditional singularity point. But, on the other hand, one should understand that the change rate cannot increase endlessly, and therefore, the trend curve will change its direction after a while. This means that the processes will proceed in a significantly different way, that is, it will proceed along some new (mostly unknown and possibly even opposite) trajectories (see Fig. 1.1, for details see Chap. 12).[7]

[7] The subject of singularity became especially popular in 2005 after the publication of Raymond Kurzweil's book *The Singularity is Near* (Kurzweil, 2005). Kurzweil (Google's director of engineering) is probably the most famous supporter of the idea of the singularity and, moreover, heads the well-known Singularity Institute he has created to study the possibility and direction of such a development. However, he gives this concept a semi-mystical meaning, speaking of the matrix and other things which we do not agree on (see more on this below). However, we think that the "Singularity Hypothesis" may be quite useful for the theory of the historical process. The singularity does not indicate the point where the value of a given variable actually becomes infinite; rather, it indicates the point before which the hyperbolic shape of the curve should change to a different trajectory implying a certain slowdown of the processes that have been observed in recent decades (2021b; Grinin et al., 2013; Huebner, 2005; Korotayev & Malkov, 2016; Korotayev 2005, 2008, 2018, 2020a, 2020b, 2024; Korotayev et al., 2006; LePoire, 2005, 2020; LePoire & Korotayev, 2020; Panov et al., 2020; Phillips, 2011).

Indeed, the singularity in technological development is an intriguing topic. However, one should realize that in the development of society and all the more generally of the World System the achievement of the maximum change rate and the change of trajectory occur in different spheres of life and at different periods. Moreover, some processes have already passed this peak point of singularity. Take, for example, the population growth rate. Earlier, a conservative forecast of the 1950–1970s (which used to extrapolate the increasingly accelerating growth rate into the future) seemed to anticipate that population growth would lead to an absolute catastrophe (for more details see Chap. 5). For example, Heinz Von Foerster with co-authors (Von Foerster et al., 1960) called their famous article 'Doomsday. Friday 13 November A.D. 2026'. Today the population growth is slowing down, and in developed countries it has already stopped or even becomes negative. At the same time, we witness the process of global aging which is likely to continue for another century. In 1972, in the famous report to the Club of Rome 'The Limits of the Growth' (Meadows et al., 1972) its authors also extrapolated the process of resource consumption into the future and argued that humanity would very soon exhaust the available resources (in a certain sense, reach the point of singularity) so at that point the demographic catastrophe would start. However, they did not take into account technological progress in their model (see Grinin & Grinin, 2021c); thus, fortunately, resources have not yet been depleted, even though the issue still remains acute (Sadovnichy et al., 2023).

The speed of other processes is also changing. Thus, until recently globalization has been involving sphere after sphere, region after region of the World System, but now it has sharply slowed down and has even begun to roll back. This becomes especially evident in connection with Trump's policies and the COVID-19 pandemic (see Grinin & Korotayev, 2020; Grinin et al., 2021b; Grinin, 2023). Nevertheless, we think that in relation to globalization, this is not yet a matter of reaching a singularity but only a peculiar behavior of many processes which evolve in waves, with inflows and outflows, all of this suggesting another powerful tide of globalization in the coming two decades.

Already in the 1950s, the most farsighted academics spoke about a certain point of inevitable trend change—an assessment based on the speed of technological breakthroughs within the initial stages of the technological revolution which we call the Cybernetic Revolution (see Chap. 2 and below). The rapid change leading to 'some essential singularity' was articulated earlier by John von Neumann in the 1950s. He maintained that the ever-accelerating progress of technology and changes in the mode of human life conveyed 'the impression of approaching some essential singularity in the history of the race beyond which human affairs, as we know them, could not continue' (presented in Ulam, 1958: 5). However, later (after the 1980s) the pace of technological change slightly slowed down. Today, many scientists believe that the growth rate of technological progress has already passed its culmination (for recent research about the decrease of disruptive publications and patents see Park et al., 2023; see also Cowen & Southwood, 2019, as well as Chap. 12 below). However, in our interpretation, the current slowdown is just 'a calm before the storm' and we all should expect a new powerful technological wave and acceleration in the pace of technological progress. In fact, this wave is already starting (for details see Chaps. 3 and 4). That is why we have a strong feeling that the world is in a state of dramatic change, which we discussed above. We assume that only by the late twenty-first to early twenty-second century will technological development pass its singularity point and experience an essential slowing down of the growth rate (see Fig. 1.1 above; for more details, see Chap. 12, where we will return to the aspects of the singularity, and Chapter 13; see also Grinin et al., 2020). Moreover, since technological development is in many respects a key point for social advancement, we believe that it will be followed by a slowdown of other processes as well.[8]

## 1.5 Cybernetic and Other Production Revolutions

There are a lot of publications about different technological breakthroughs and at the same time a surprisingly small number of studies which in a systemic and consistent way describe technological development and can explain in scientific terms why and how technological revolutions occur. Meanwhile, it is extremely important to reveal certain regularities in the history of technological development, and make an effort to anticipate at least the coming transformations in technologies and society. However, there are several studies about the forthcoming technological breakthrough,[9] there is a small number of works that present a consistent forecast of technological development based on apparent developmental trends.

In our volume we propose to narrow these gaps. Our study is devoted to the development of technology, to the analysis of its past and current state and reflections on its future. We will discuss extensively the revolutionary changes which

[8] The above said, however, does not mean that a new acceleration in the twenty-second or twenty-third centuries is impossible. In any case, those will be processes of a different nature in many respects.

[9] Here one should also mention "Third Industrial Revolution" (Rifkin, 2011) and "The Fourth Industrial Revolution" (Schwab, 2016). The books present interesting predictions, but their authors simply rely on empirical observations of current trends or reproduce someone else's forecasts and do not rely on a scientifically derived paradigm. Meanwhile, forecasts based mainly on contemporary data or on profound and long-term trends and regularities are completely different things. From the titles it can be deduced that these authors actually believe that all three or four revolutions have similar nature, which, as the reader will see (in Chaps. 2–4), is not the case at all since the nature of the last technological revolution is different from the previous ones. Moreover, these theories talk only about Industrial Revolutions, but not technological ones (in particular they did not take into account the ancient Agrarian Revolution, which resulted in tremendous changes). All this is strongly narrowed them. In general, all these useful books are more journalistic than scientific. The same refers to Ralf Fücks' *Green revolution* (Fücks, 2013). Francis Fukuyama's *Our Post-Human Future: Consequences of the Biotechnology Revolution* (Fukuyama, 2002) is quite scientific but it does not rely on long-term evolutionary trends as well. There are also forecasts about the Fifth Industrial Revolution (Bröring, 2021; Peccoud, 2016).

the world will face in the coming decades with a special focus on the Cybernetic Revolution. What kind of revolution is this and why is it important? As a matter of fact, this revolution will largely determine our lives within the next 50–70 years, and perhaps even over a hundred years. The Cybernetic Revolution is the third crucial technological (or production) revolution in the history of humankind. It follows the Agrarian Revolution which lasted from the tenth to the third millennia BC and the Industrial Revolution of the sixteenth to nineteenth centuries. In order to understand the meaning and peculiarities of this revolution, it is necessary to trace technological trends throughout the entire historical process.

The evolution of the historical process in terms of three production revolutions could be represented as follows:

- The Agrarian Revolution comprised the transition of humankind from foraging (hunting and gathering) to food production (agriculture and animal husbandry), which led to the formation of an agrarian society with its estate structure, the important role of religion and cyclical dynamics. We consider the Neolithic breakthrough only as an initial stage of this revolution.
- The Industrial Revolution was based on the replacement of manual labor with machine labor, which led to the formation of an industrial society with its internal dynamism, instability and a focus on constant change;
- The Cybernetic Revolution; currently, the invention of high-speed computing devices and AI, their introduction into all spheres of life and, on this basis, the gradual replacement of intellectual labor by machine labor, and in the final phase, the widespread use of self-regulating systems, which can act without human control. All this will lead (and is already leading) to the formation of a society whose functioning decisively depends on smart and super smart systems, their algorithms and capabilities. In this respect, it is logical to call this emerging society 'cybernetic' (see Chap. 15).[10]

In Chap. 2 'Global Technological Transformations: Theory and History', we will elaborate on the history and success of previous production revolutions, but above all, we will show important regularities in production revolutions (see also Grinin, 2007a, 2007b, 2012: ch. 1; 2013; Grinin & Grinin, 2013, 2015a, 2015b; Grinin et al., 2017, 2020, 2021b; Grinin & Korotayev, 2009, 2015). This provides a starting point for predictions for the coming decades.

The Cybernetic Revolution began in the mid-twentieth century and it has not finished yet. In the 1950s and 1960s, the world passed through the initial phase of this great technological revolution. At the end of the twentieth century, it manifested achievements in engineering, automation, energy production, synthetic materials production, exploration of space and sea, agriculture and especially in the field of information technologies which became widely spread in most countries

[10] More detail about the Cybernetic Revolution see below in annotations of Chaps. 2–4 and others.

of the world. This revolution is variously termed (scientific-technological, information, Technetronic, the Third or Fourth Industrial Revolution, etc., see above, note 6 and Chap. 2). However, we denote this revolution as the Cybernetic Revolution, since cybernetics is the science about information and its transformations in the process of regulating various complex systems. The current revolution has dramatically changed information processing and is likely to create a huge number of self-regulating systems as well as promote a breakthrough in the regulation and control of complex processes in a wide range of natural and artificial systems that have become a part of the production process.

We expect the final phase of the Cybernetic Revolution to start between 2030 and 2040. This period will be given a thorough consideration in our volume. During the coming phase of the Cybernetic Revolution in 2030–2070, there will appear not only a large number of smart things but also such technologies that will allow different systems to function in a necessary mode without direct human involvement. We call such systems self-regulating, since, as we mentioned before, they will become autonomous, self-regulating and can function without human control. The most advanced among them may be called self-managing systems.

As to the directions that will become the most groundbreaking, we present a number of ideas, in particular, we introduce the concept of MANBRIC convergence. This acronym is made up of the first letters of the names of the most promising areas of technology, which will eventually make up a powerful system, actually a convergence of different systems. These are respectively: Medicine, Additive technologies, Nanotechnologies, Biotechnologies, Robotics, Information technologies and Cognitive technologies. This complex will become the core of future technological development (for more details see Chaps. 3 and 4 below) and eventually will change the quality of our lives in different ways, including by modification of the human organism through the use of biological and non-biological materials and systems. We will discuss those very contradictory issues in the next section.

## 1.6 Between Technological Optimism and Reasonable Caution

The main topic of our volume is the present and especially future development of technologies, global aging and their impact on our future. But in the present introduction, we also want to touch upon the subject which is extremely important, closely related to the main one, and which, in our opinion, is not much discussed. This is the issue of a possible future change in the fundamental biological basis of humans, along with technological development and other domains of progress. Unfortunately, at present, this process is uncontrollable and this can cause a huge number of tragedies. In our opinion, great care and prudence are required when moving along the path of changing the biological nature of humans. Previously, it was said that development would lead to a change of the humanity as a species either as a result of evolution (e.g., Teilhard de Chardin, 1987; Vernadsky, 1997;

see also Fuchs-Kittowski & Krüger, 1997) or as a result of the development of human activity on the Earth and in space (e.g., Tsiolkovsky, 1964) and accelerated development of historical processes or growing crisis (e.g., Moiseev, 2000; see also Diakonoff, 1999; see above the references related to the idea of a singularity). For the moment we have reached a stage when technological and scientific progress allows people themselves to change their biological nature.

Since the emergence of Homo sapiens, people have been connected with technology. Earlier thinkers even directly connected the event of human origin to new technologies, arguing that labor transformed apes into humans ('labor' implied first of all the 'production' of stone tools). Now it is clear that this process was the result of a number of factors. The influence of material factors, such as the interaction between and effect of life and the environment on the genotype has also enormously contributed to this process. In any case, technology separated ancient humans from a number of other hominids and became a hallmark of this type of selection. The transition to Homo sapiens is often termed the Human Revolution (e.g., Mellars & Stringer, 1989).[11]

Having passed through the three major technological revolutions mentioned above and many other technological breakthroughs, we are now confronted with the threshold of the Post-Human Revolution. The matter is that, as we have already said above, the Cybernetic Revolution will provide opportunities for intervening in and changing of the biological nature of the human organism and enhancing the impact on the human body and even on the brain. Here humanity enters an absolutely unknown and very dangerous path. Perhaps, it will be less dramatic than the transhumanists and other adherents of practical immortality imagine it to be in their aspiration to part with the human biological body.[12] But anyway we are speaking about a considerable extension of life, increasingly common replacement of organs and cells of the human body with non-biological materials, introduction of electronic and other (nano-, etc.) elements into the human body for rehabilitation or improvement of individual functionality, and systematic influence on genome (see Chaps. 8–11). On the whole, quite dramatic changes are coming.

Thus, figuratively speaking, our research lies at the intersection of the Human Revolution and the new 'Post-Human' Revolution, whose consequences are unclear in many respects but which will obviously start the era of intensive impact on the human body.

When trying to describe the future, which is already knocking at our door, we would not like the reader to think that the authors assess every change described in the book only in a positive and optimistic way. Not at all.

Futurologists can be divided into several categories: impetuous optimists, alarmists and careful optimists.

---

[11] Or the Upper Paleolithic revolution.

[12] The fundamental ideas of this theory were first put forward in 1923 by the British geneticist John B. S. Haldane and then developed in a comprehensive theory (Fukuyama, 2002; Haldane, 1924; Huxley, 2015; Kurzweil, 2010; More, 2013; Schneider, 2008).

The first one may be also called technological hyperoptimists (Eden et al., 2012: 127; see also Linstone, 2014). These thinkers and researchers believe that the development of AI will soon result in a complete and radical transformation of humans as a biological species and of humanity in general (e.g., Kurzweil, 2004, 2005; Yampolskiy & Fox, 2012). For example, Bill Gates although trying to take a more objective position, is close to this category in regard to AI (e.g., Gates, 2023) or especially vaccination.

The second category is formed by technological pessimists—see, for example, Maddison (2007), or Teulings and Baldwin (2014)—who believe that at present we deal with a slowdown of scientific progress (see also LePoire, 2005; Phillips, 2011; the Huebner–Modis debate on the possible declining trend in worldwide innovation [Huebner, 2005; Modis, 2005]; for details see Chaps. 12 and 13).

The third category is the pessimist-alarmists who claim that technological progress proceeds too fast and this threatens the humanity with disasters. Among them are many authors of reports to the Club of Rome (e.g., Bardi, 2014; Meadows et al., 1972, 2004; Wijkman & Rockström, 2012). The emergence of ChatGPT and others like it has raised a wave of alarm, particularly with the open letter signed by many famous people calling on all AI labs to immediately pause for at least six months the training of AI systems more powerful than GPT-4 (Musk et al., 2023; see also Vincent, 2023) and the statement by Geoffrey Hinton, sometimes called 'The Godfather of A.I.' (for details see Elias, 2023 and Chap. 10).

The fourth category consists of cautious optimists who realize that technological progress brings not only benefits, but also serious problems which humanity has never faced before; they ask us to think about the negative consequences in advance, without disregarding progress that these changes will bring about (e.g., Berg, 2019; Fücks, 2013; Gabor et al., 1978; Lovins et al., 2018; Pestel, 1989; Randers, 2012; Randers et al., 2018; Rifkin, 2011; Schwab, 2016; Scliaff & Friedrichs, 1982; von Weizsäcker & Wijkman, 2018, possibly, Elon Musk in some respect has begun to move to this position [see Vincent, 2023]).

We belong to the fourth category and think that we should not be afraid of the future. However, one can hardly expect that the future will be definitely better than the present, and moreover, in every way. The stakes are always high.

Indeed, the achievements presented as solutions to problems can also cause them to exist on an even larger scale, thus, contributing to the increasing dependence on technologies and ignoring their negative consequences. Besides, in the future we will observe the emergence of many problems resulting in changes to the habitual lifestyles.

It is difficult and actually pointless to try to hinder progress. However, there is always the question of what one should consider as progress in any particular epoch and at what cost. Anyway, it is better not to rush into changes with vague consequences. When treading new ground, it is better to be careful than to rush. Science, innovations and changes rapidly drive a lot of new legal, moral and economic problems and cause sharp disputes, conflicts, trade wars and phobias. Public consciousness definitely lags. Uncontrollable technological progress can be compared with the Roc, the legendary bird from the Arabian Nights that quickly carries humankind but also demands human sacrifice. Are we ready for this? And if we are, what are we ready to sacrifice for the sake of progress? The answer to this question should become one of the most important issues for all of us.

One may ask: why discuss these dangers today if they are still so far in the future? The problem is that the future may turn out quite unexpected and may have extremely negative consequences, especially if one is not ready for it. We all expected, for example, miracles from biotechnology, but got COVID-19, which can only be a harbinger of even more serious pandemics—and still nobody can say for sure what the nature of this virus is. We have been waiting for miracles from AI, but we face a real possibility of a digital concentration camp. It is necessary to speak about it in anticipation, since there are numerous threats, and the number of these threats is increasing in a geometric progression due to the accelerating changes; just as a higher vehicle speed increases the risk of an accident.

Development usually starts with euphoria from implementing new tools or knowledge and only much later one comes to realize the pressing problems. Finally, restrictive measures are taken to reduce the revealed negative consequences. However, it is much better to change this sequence and even to find legal and appropriate restrictions before implementation.

## 1.7 Anticipating Dramatic Changes in Evolution and Human Nature: To Be or not to Be for the Human Beings?

Continuing the discussion of technological optimism, it is worth mentioning again technological hyperoptimists or 'machine superoptimizers' (Eden et al., 2012: 127). Recall that this approach is the belief that the development of AI will soon result in a complete and radical change in humans as biological beings and in humanity in general (e.g., Yampolskiy & Fox, 2012; and especially Raymond Kurzweil's works [e.g., Kurzweil, 2005, see also Kurzweil, 2001, 2004]). The possibility of this scenario is growing very rapidly (see about new ChatGPT in Chap. 10). However, such a boundless technological optimism (or rather a blind belief) lacks sufficient scientific grounding.

On the other hand, one can hardly deny the positive side of progress. Development of neural interfaces, which we will discuss in Chap. 10, can allow the disabled and paralyzed people to perform many movements and improve their quality of life. This is truly amazing. But immediately the apologists of unrestrained technological progress have the idea that with the help of neural interfaces it is possible to create a new communication system ("brain – computer – network") and connect our brain directly to the network. However, will not it turn out that tomorrow our thoughts will be under surveillance of 'Big Brother'? And if one can directly influence parts of the brain, causing relevant emotions and desires, then what range of possible exploitation will open up for politicians and business people...? Other apologists believe that with the help of neural interfaces it is possible to speed up the work of the brain by a million times. Already today, 20–25% of people cannot cope with the current pace of change and suffer from various mental disorders and neuroses, while depression globally will soon become the second most common disease after cardiovascular disease (see, e.g., Kekelidze, 2013). According to a great scholar in the field of the brain research, Dick Swaab (2013), 27% of Europeans suffer from one or more brain diseases. Will it not turn out in the end that neural interfaces will be able to drive humanity crazy in the truest sense of the word? As a result, achievements intended to be a means of solving problems, simultaneously create them on an expanding scale.

Though this perception of dramatic changes is quite natural, the driving forces of future changes are by no means mysterious or unique. And still there are many disturbing aspects... One should realize that these changes in human development not only provide new opportunities, but also pose serious risks that need to be anticipated. As we said above, in the twentieth and early twenty-first centuries there was a widespread idea of a possible or even inevitable transition of humankind to a new biological form in the near or distant future. What does this mean, and what consequences will this transition have? Is it possible to change this trajectory or are there any exceptionally challenging evolutionary trends? It is extremely difficult to answer these questions, especially since they are inherently ideologized and become a kind of religion whose main postulate is the future immortality of every individual.

Even sex affirmation surgery that has already become commonplace still creates serious difficulties for modern society (e.g., in women's sports). But these are transformations within human nature, not beyond it. However, today the focus is on the increasing opportunities of changing human nature, i.e., going beyond it. Consequently, the question is what is human nature itself, and to what extent can it be modified? We think that Francis Fukuyama has thoroughly analyzed this issue in his *Our Post-human Future* (Fukuyama, 2002) and thus, there is no need to consider it in our research in detail. However, in his opinion, human nature is a set of behaviors and properties typical for a human as a species and arising from genetic backgrounds, but not from the environment (Fukuyama, 2002). Besides, he often expressed the idea that human rights can be derived from the biological nature trying to prove the advantages of democratic political regimes over others. We find this approach rather disputable. Whether human rights are intrinsic to the biological nature or not, it is extremely important that these rights would be preserved in the conditions of technological development and transformation of human nature. We would also like to emphasize that twenty years have passed since Fukuyama wrote his book, and today the issue is much more pressing. There exist different philosophic approaches concerning humans as a social species, but today the topical matter is human biological nature, not in philosophical terms but in a concrete medical and biological sense. The emergence of artificial organs and cells has already brought about the issue of human material and biological nature, namely: what matter will a future human be made of—natural biological or artificially made biological substance or will it be a non-biological being at all? How will humans reproduce? How will the brain and consciousness operate? Any change will dramatically alter human fundamental institutions including morals and interpersonal relations. Indeed, what will become of morals and what will the result be if the matter concerns a change in biological nature? Morals and human relations do not exist separately from human psychology which in a great measure depends on human physiology, and in a broader sense, on the biological basis of humans. Our biological core is a result of complex sociobiological evolution, and subsequently morals can disappear after losing their material biopsychic shell.

## 1.8 From Human to Cyborg?

The process of changing the lost limbs for artificial ones began a long time ago. The earliest evidence of prosthetics is recorded in ancient Egypt and Iran about 3000 BC. Also, there is an early recorded mention found in South Asia around 1200 BC (Vanderwerker, 1976). There is especially much evidence of various prosthetic limbs, that have become quite common since the Middle Ages. A famous example is the prosthetic arm of the German Imperial Knight, a mercenary, and poet Götz von Berlichingen, made at the beginning of the sixteenth century, which had a complex mechanism.

In the twentieth century, humanity progressed farther in the development of prosthetic technology. In the 1950s, the DNA code was deciphered. Already by the end of the last century it became clear that the opportunities to alter human genotype can generate a lot of complicated and dramatic social, political, ethical and legal problems in the future. One should note that bioethics became a response to these future and already emerging dangers (Engelhardt, 1996; Veatch & Guidry-Grimes, 2019). Of course, this society's response has no serious implications, but after all it was meaningful. Fukuyama (2002) gives a rather detailed list of the threats of genetic and other alterations to humanity. In particular, he mentions the increasing opportunities to regulate human behavior (here he emphasizes that neuromediators made it possible by the end of the last century), the emergence of genetic castes, classes or social strata, since changing of genetic makeup can heavily depend on parents' wealth; otherwise, a more egalitarian society in terms of genetics might appear (in case of obligatory correction).[13] In fact, the attempt to improve human moral nature by means of genetic modifications seems rather dangerous. In this context this reduced diversity can significantly weaken society and its ability to respond to challenges.

[13] These fears have not decreased since then; on the contrary, the number of opponents of genetic modifications has grown, and their arguments became more convincing. At the same time it seems that the practical modification of human embryos has already started. Thus, in April 2015 China declared about the conducted work on modification of the human embryo (Field, 2015). In particular, in January 2019, the Chinese authorities confirmed the birth of the world's first genetically modified people in Shenzhen. However, biologist He Jiankui, who performed it and announced his research at genome-editing summit in Hong Kong in late 2018, was arrested and jailed (Browne, 2022; Harney, 2018).

However, so far we have hardly touched on new and to some extent more serious threats that have recently appeared. Let us consider some of them.

It has already been mentioned that the pace of scientific and technological progress generally continues to speed up (see for details Chaps. 2 and 12). The historical process is also gearing up, while neither individual nor public consciousness is able to keep up with it. This creates serious conflicts and frustrations, and moreover, it arouses a quite reasonable alarm about our physiological and biological future, since the number of people increases who wish to put their brains and consciousness into abiotic immortal bodies (made of metal, plastic or other material). One of the apologists for this approach to the future formulated the idea in the following way: 'Becoming a race of immortals is the main evolutionary and historical task of the humanity in the third millennium' (Ariyanov n.d.). So, one can see that the problem of preserving the biological nature of humans is extremely important.

Thus, we come up with the idea of cyborgization, which is a hot topic these days. A popular word 'cyborg' (short for 'cybernetic organism') derives from the word 'cybernetic'.[14] At present the term cyborg is often applied to an organism that has restored function or enhanced abilities due to the integration of some artificial component or technology that relies on some sort of feedback.

So the term cyborgization means that after some time a human being will consist mainly or entirely not of biological, but of artificial material. This is an intriguing process which is the subject of many futuristic novels and which at the same time takes place right before our eyes. Will humans turn into cyborgs, as a result of the rapidly developing directions of bionanotechnology, AI and cognitive sciences? It is not an idle or innocent matter at all, especially with the account of the increasing number of prophets of 'avatarization'[15] who call for discarding of the perishable biological body (Schneider, 2008).[16]

[14] 'Cyborg' (short for 'cybernetic organism') is defined as a theoretical or fictional being with both organic and biomechatronic parts (for restoration or enhancement).The term was coined in 1960 by Manfred Clynes and Nathan S. Kline (Halacy, 1965) in connection with their theory of the expansion of human capabilities to survive in space. They wrote in their article introducing the notion of 'cyborgs': 'Altering man's bodily functions to meet the requirements of extraterrestrial environments would be more logical than providing an earthly environment for him in space … Artifact-organism systems which would extend man's unconscious, self-regulatory controls are one possibility' (Clynes & Kline 1960: 26).

[15] The word 'Avatar' is mostly known today as a graphical representation of the user, representing him/her in blogs, chats, etc. Here we mean its different, original meaning. Avatar is the term used in Hinduism philosophy to denote the terrestrial embodiments of God (especially Vishnu). Respectively this term is used by some supporters of coming human immortality to determine the transformation of the human spirit (brain and consciousness) into the new (non-biological) body (e.g., Schneider 2008).

[16] There is a popular idea of transhumanism that cyborgization will develop by placing the brain and consciousness in an abiotic immortal body. Immortality in general is one of the main issues of transhumanism (Fukuyama, 2002; Haldane, 1924; Hansell, 2011; Huxley, 2015; More, 2013).

How much is this possible? On the one hand, this direction seems logical, as medicine has been moving this way for many decades. Currently, bioprinters can create different tissues and organs (Murphy & Atala, 2014), and neural interfaces allow controlling some devices and equipment "by power of thought" (Schalk et al., 2004). Besides, there is some kind of oncoming traffic of technologies in terms of the rapprochement of people with artificial systems, in particular in the construction of humanoids (see, e.g., Hirose & Ogawa, 2007). Since these robots will be used not only for work and entertainment, but also for very close or even intimate contact with people (Yeoman & Mars, 2012), the borders between the human and artificial anthropomorphous systems may start dissolving.[17] Not to mention technologies such as virtual reality where even now it is becoming difficult to distinguish reality from illusion (Burdea & Coiffet, 2003).[18] One also may mention a modern digital 'spiritualism', i.e., communication with digital copies of dead relatives (for details see Chap. 10).

All this may contribute significantly to the possible transformation of the human body into a kind of cyborg.

Let us give a typical example that shows how such medical cyborgization can progress. In Italy, scientists have succeeded to create an artificial eye that will perform far from only a cosmetic function. The prosthetic device will be able to transmit an image, like a real eyeball. For its production, scientists used a 3D bioprinter, while the technology itself will be put into active use only in 2027. According to the authors, this study will produce a real revolution in the field of ophthalmology, and people suffering from eye ailments will be able to see our world. This bionic eye can be easily installed in place of the real one. The invention will allow one to connect to a Wi-Fi network and apply various filters. The specialists from Italy believe that their breakthrough invention will easily overtake glasses and contact lenses in popularity. In addition, it is planned to equip the eyes with a special function that will allow taking pictures and printing them using a 3D printer. To change the settings of the eye, it will be necessary to take an appropriate pill. Each of them will be responsible for a separate pre-set. In addition, in case of an error, the eyes can be easily replaced even at home (Pevtsov, 2015). It is difficult to say whether such plans are feasible in the near future. But this is of minor importance since the direction of progress is more important here, and one can hardly resist it.

[17] Let us note that the production of sexual robots (generally 'female', so far) has already started. There were also statements that by the middle of the century such contacts between humans and robots will become commonplace. Not without reason a campaign for their prohibition has been initiated (see Griffin, 2015; see also Chap. 10). However, so far these requirements come from the feminists who are concerned with such a detraction of the female role, but we agree with them. It is better to forbid or to take control of this situation in advance, because if businessmen have a chance, sex robots for any sexual orientation might appear.

[18] It is already difficult or impossible to distinguish texts, poetry, pictures, etc., created by humans from those created by AI. So, millions of people will plunge deeper and deeper into the virtual-illusory world, and live there in parallel with the real world.

But, on the other hand, over millions of years, biological evolution has balanced all the elements of biological organisms and their functions in an optimal and efficient (but sensitive to change) way. It is quite doubtful that the human brain is able to work without the body, because the main purpose and function of the brain is to control the body. It also seems irrational to change all organs and parts of the body, most of which usually work fine. It might be much more efficient and cheaper to change only broken or less durable parts. It is more probable that the process of cyborgization will never go too far, it will always remain "supplementary" for the biological component of organism, capable of improving the quality of life and also to prolong it (see Grinin & Grinin, 2020, 2021a, 2021b, as well as Chap. 15 for details).

Today scientists are learning to create artificial biological tissues and bodies by means of stem cells or other biotechnologies (about this see Chaps. 8, 9 and 10). We suppose that this way of 'repairing' our body will be more promising. For example, at present, we know the case when a person's heart was replaced six times (and once a kidney) during his life. This is the billionaire David Rockefeller Sr., who underwent the last heart transplant at the age of 99 (he died in 2017 at the age of 101). But at present only a multi-millionaire can afford it (and nevertheless, this man was really lucky to survive with such a number of transplantations). However, in the future it will be possible 'to repair' many people by means of laboratory-grown organs. But, undoubtedly, this interference in the functioning of the human body has both physiological and social restrictions. Interference in human genome can also cause serious problems, especially if used for the creation of individuals with superhuman potential, for example, in the pursuit of athletic records.

Thus, hopefully it will become possible to prevent reckless interference in the human body regardless of its motives: whether it is motivated by an aspiration for scientific fame or profit or the realization of superman ideology. While welcoming scientific and technological progress, we must consider that we should preserve our biological heritage, a heritage that has been created over millions of years.

The process of cyborgization undoubtedly elicits a lot of fear and opposition. The fight against scientific-technological progress has a long history. Each manifestation of this fight was caused not only by obscurantism, but also by a real necessity and also by grounded fears, since scientific and technological progress would often exacerbate the situation, lead to many bankruptcies and throw overboard many professions; sometimes it would even desolate whole cities and territories, and would also often reduce the quality of production. Sometimes this reduction opened unexpected opportunities for abuses or was a source of desperate social fighting and oppression. Nevertheless, no one has been able to seriously slow down this process. A few years ago, at the forefront of the clash between technological progress and the cessation of that progress, one could find stricter requirements for new drugs, for the banning of GMOs or of cloning. Quite recently, an even fiercer fight was being carried out in connection with the hasty introduction of total vaccination and insufficient control over the process. Meanwhile, before the COVID-19 pandemic, it took many years for any vaccine to be approved (for details see Chaps. 8, 9, and 14). Recent investigations show that the damage caused by the use of vaccines that have not been fully tested is very high.

Thus, of course, many of these restrictions and bans are necessary. On the one hand, it is difficult to expect that the development of scientific and technological progress can be fully controlled.[19] On the other hand, progress in the fight for the environmentally friendly production and the production of safe drugs shows that it is quite possible to achieve a certain level of control. In general, the mechanism of minimizing the damage from innovations consists in establishing certain institutions and rules optimizing the control over technologies; but it is especially important to make these institutions and rules in advance.

## 1.9 The Structure of the Book

The monograph consists of four parts subdivided into 13 chapters, together with an introduction and conclusion there are fifteen chapters.

Initially, we will characterize every part of the book and then describe the content of each chapter.

## 1.10 Book's Parts

### 1.10.1 Technological Transformations in the Past, Present and Future

In this part we describe the main historical technological changes. The description covers almost the entire period of the historical process starting from the emergence of *Homo sapiens* up to the end of the twenty-first century. We did not point out every single invention or all of the separate innovations in detail; instead we made an attempt to show the general process of changes and to explain how and why technological epochs succeeded each other. We pay the greatest attention to the main technological breakthroughs, both to those already completed and to the future ones, as well as giving attention to the transformations in society and to the social consciousness these transformations have already brought about and will also cause in the future. We will consider some aspects of technological development in the historical process in Part 4.

According to our theory, the historical process can be subdivided more effectively into four major stages or four formations on the basis of changes in the major qualitative developmental stages of the world's productive forces. Because the transition from any of these stages to another means a change of all basic

[19] A recent (late March 2023) example of a such reasonable suggestion: A number of well-known AI researchers have signed an open letter calling on AI labs around the world to pause development of large-scale AI systems citing fears over the "profound risks to society and humanity" (Vincent, 2023).

characteristics of the respective system, we named these stages production principles. Four production principles are: 1. Hunter-Gatherer; 2. Craft-Agrarian; 3. Trade-Industrial; 4. Scientific-Cybernetic. The change in production principles is connected with production revolutions. Among all major technological breakthroughs in history the most important are the three production revolutions: (1) the Agrarian Revolution (10–3 millennia BCE); (2) the Industrial Revolution (sixteenth to nineteenth centuries) and (3) the Cybernetic Revolution (1950–2070).

In this part, based on the studies of the whole previous technological development, we try to outline the direction and logic for further technological progress. So, the main goal of the authors is to analyze in detail the final phase of Cybernetic Revolution (2030–2070) and the period after it (sometimes including the early twenty-second century) and formulate a whole system of forecasts for the next 30–80 years. In the chapters devoted to current and future changes we will pay considerable attention to the analysis of the features that will manifest in technology and to what we can expect from these manifestations in terms of the development of self-regulating systems as well as of other characteristics of the Cybernetic Revolution (resources and energy saving, miniaturization, individualization, etc.).

The authors' goals are to discuss and deeply analyze the following questions: What is the Cybernetic Revolution and why is it Cybernetic? What are self-regulating systems? How do they correlate with AI? How can they influence our future life? What are the characteristics of the Cybernetic Revolution? What may we expect in technological development in the next half of this century? And so on.

We will also study the development of those directions which in our opinion are most likely to trigger the future technological breakthrough, namely, medicine, additive (3D printers), nano-, bio-, robotics, information (including AI), and cognitive technologies which we unite under the designation of MANBRIC technologies. We will also discuss the correlation between the final phase of the Cybernetic Revolution and the forthcoming Sixth Kondratieff waves which we expect in the 2020s–2070s.

Of utmost importance is the fact that in the Cybernetic Revolution, self-regulating systems will become the major part of the technological process.

Self-regulating systems are systems which can operate independently in a wide range of situations, having opportunities for choosing optimum regimes in the context of certain goals and tasks. These are systems that operate with minimal to zero human intervention. We discuss self-regulating and self-managing systems in detail, along with examples of these systems, current and future implementation, their main characteristics, etc. A wide implementation of self-regulation in various systems will be the most important characteristic of the final phase of the Cybernetic Revolution.

Then we also discuss the logic of the Cybernetic Revolution to explain why the new technological wave will happen in the 2030s–2040s, as well as explaining why medicine is to become the breakthrough sphere in the beginning of the final phase of the Cybernetic Revolution. We also estimate the chances of the following branches of the Cybernetic Revolution being the initiator of that revolution: robotics, driverless vehicles and green energy.

So, we foresee great changes in the future both in technologies and in the socio-demographic structure of society as well as in whole mode of our life.

### 1.10.2 Demographic Transformations and Global Aging

In this part, we explore the close relationship between technological (production) revolutions, which we consider in Part 1, and demographic transformations during the historical process as well as we consider in detail the complex relationships between the development of the Cybernetic Revolution and the cybernetic production principle, on the one hand, and the process of global aging on the other. Throughout history, technological change has been among the most fundamental factors influencing demographic growth and sociocultural development; in fact, it is this connection that distinguishes the human reproduction from the reproduction of any other living beings. Analysis of the demographic component and the demographic dimension of the historical process, unfortunately, has not been sufficiently studied. Meanwhile, this is an extremely important aspect that can not only explain a lot in the development of the historical process, but also provide a basis for explaining current processes and forecasting our futures.

We consider the dynamics of the correlation between production revolutions—the Agrarian, Industrial and Cybernetic (which are discussed in Chap. 2)—and major demographic transformations in the historical process. We rely on the conclusions drawn from the analysis of long-term demographic trends. But our main focus is on the demographic transformations associated with the Cybernetic Revolution. And among these transformations, global aging appears as an increasingly important process that affects all spheres of life. In Part 1, we have argued that the Cybernetic Revolution will produce a huge impact on the health-related quality of life and provide opportunities to modify the biological nature of humanity. Of no less importance is the fact that the increasing life expectancy and global aging over the next several decades will increasingly affect individual societies and the World System in general as well as further changing of the type of population reproduction.

We analyze the process of global aging within the World System and the interconnections of the aging process with techno-economic development, social changes and other processes; we also make long-term forecasts about the development and impact on changes in society due to the aging process, its connection with technological transformations, and finally possible scenarios for the development of an elderly society in the future. This gives us the opportunity in the following parts of this book to show the close and deep relationship between both specific areas of technology (medicine, biotechnology, robotics, etc.) with the aging process and the mutual influence of aging and technological progress in general, as well as influencing the transformation of the economic model, political and social structures and institutions.

Technological (production) revolutions opened up gigantic ecological niches that made it possible to make qualitative breakthroughs in population growth. In other words, production revolutions involve major demographic transformations. Due to them, the population was increasing hyperbolically in the long term. During the Paleolithic, the population grew on average by 3% per millennium (see more below), whereas in the 1960s there were years when the population of the Earth grew by almost 3% per year. However, afterwards the trend started being reversed and now the rate of population growth is only a bit more than 1% per year. What has happened to cause this change?

The Cybernetic Revolution will have a huge impact on the quality of human biological life and the ability to modify the very biological nature of humans. No less important is the fact that in the process of increasing life expectancy and global aging the very type of population reproduction is changing very significantly.

Finally, we consider in more detail the complex relationship between the development of the Cybernetic Revolution and the cybernetic production principle, on the one hand, and the process of global aging, which with each decade will increasingly affect individual societies and the World System as a whole, further changing the type of population reproduction (TPR). After this we (a) show the impact of the Cybernetic Revolution on changes in the type of population reproduction (TPR), the diverse (both positive and negative) consequences of global aging, and the contradictions that the process of global aging and the technological transformations that contribute to it will inevitably cause in future societies; (b) we will analyze various forms of adaptation to aging that must be developed so that frictions and tensions in society do not impede its development; (c) we will show the main features of a new type of population reproduction, which, in our opinion, should take shape by the end of this century. This new TPR (type of population reproduction) will be characteristic of the future society, which we have called cybernetic (it will be discussed in the Conclusion).

We will also analyze some important consequences of the aging process in Part 4.

### 1.10.3 MANBRIC Technologies in the Forthcoming Epoch of Self-regulating Systems (2030s–2090s)

**Part 3** of our monograph is devoted to forecasts about the final phase of the Cybernetic Revolution and for some fields the forecasts are provided until the end of the twenty-first century. We present in Part 3 a detailed analysis of the future development of a whole range of technologies. These are breakthroughs in medicine, additive (3D printers), nano-, biotechnologies, robotics (including drones and driverless vehicles), information (AI) and cognitive technologies which we unite under the designation of MANBRIC technologies (MANBRIC is an acronym). Altogether these areas will support development at the final phase of the

Cybernetic Revolution, which can be also named the epoch of self-regulating systems. They are systems that can operate without human control. This will provide groundbreaking opportunities and also may pose new unprecedented problems.

As we show (in Chap. 4), due to some significant reasons medicine will be the first sphere to launch the final phase of the Cybernetic Revolution, but, of course, later on medicine will hardly remain the only breakthrough industry; so the breakthrough will be provided by the whole MANBRIC convergence. Nevertheless, medicine will remain some considerable time the key and integral element of this convergence. That is why, while trying to characterize every MANBRIC branch, we pay as much attention as we do to medicine (Chap. 8 is devoted to it). Further, the authors show how population aging can reorganize medicine and make care for the elderly its primary focus (Chap. 11 is dedicated to this theme). Another subject of the authors' attention is biotechnology, since it is particularly closely related to medicine (Chap. 9 is devoted to biotechnology). The other MANBRIC technologies are united in Chap. 10. While discussing them, the authors try to show not only the characteristics and prospects of every field, but also their strong interrelationship. Thus, such seemingly unrelated areas as the development of AI and cognitive technologies appear unexpectedly close. Nanotechnologies can work in almost all areas, and together with medicine and biotechnology they can literally perform miracles in the future in the sphere of the creation of artificial organs and tissues, as well as in pharmaceuticals, etc. Additive technologies also appear unexpectedly universal and capable of facilitating the development of all other areas of the Cybernetic Revolution and even reshaping future production in many ways. Robotics, in turn, promises to develop its universal solutions from caring for the disabled to replacing humans in dangerous jobs. The perspective of AI seems now almost unlimited and will influence every branch. Thus, we aspire to show that convergence into the MANBRIC complex is not a fantasy, but rather a real and dramatically important trend. We also show how the above-described characteristics of the Cybernetic Revolution (microtization, saving of energy and resources, transition to the control over deeper and more fundamental processes and levels, etc.) will manifest themselves in the future (in fact, they already exist); and also discuss the future self-regulating systems in general and in each of the analyzed areas.

### 1.10.4 Trends of the Cybernetic Revolution in the Light of Technological Progress and Aging

In this part, we continue to develop some of the topics that are raised in the previous parts. In particular, we study the development of technological process from the Upper Paleolithic, but now relying not only on our theories of production revolutions and production principles, but also involving mathematical methods. However, the main themes of our monograph: the themes of the Cybernetic Revolution and global aging remain our primary focus.

Our first two tasks in this part are: (1) to make forecasts about how technological progress will develop in the twenty-first century, with a slowdown or an acceleration; (2) to show how the change in the pace of technological progress is related to global aging and the ongoing development of this process.

To solve the first problem, we study changes in the speed of technological process from ancient times to the present and beyond. The idea of a slowdown in technological progress is now very popular, as is the idea that this trend that emerged in the 1970s will remain unchanged, that is, the slowdown will continue. This approach, however, looks far too linear, and therefore raises serious doubts regarding its reality. However, our research shows that from the 2030s to 2040s we will see a new acceleration in technological progress. And it will definitely be associated with the beginning and development of the final phase of the Cybernetic Revolution (the phase we call the epoch of self-regulating systems). This acceleration will last for several decades, until, toward the end of the twenty-first century, it will start to slow down again. It is worth adding that we originally wanted to call this part 'Rise and demise of explosive technological progress', which fully reflects the essence of our conclusions. Thus, the picture of changes in the rate of technological progress is by no means linear, and it looks much more plausible.

To solve the second problem, we investigate the effect of aging on the rate of technological progress and we come to the conclusion that this influence is nonlinear. Aging will initially contribute to an increase in the rate of technological progress, but then it will begin to influence in the opposite way, slowing it down.

The next two tasks in this part are: (3) to show what changes in the consumption pattern of future society will take place as a result of the aging of society; (4) to make forecasts about how much and how radically technological development along with the aging process will change the political structure of the future society. At the same time, the experience of anti-COVID measures allows us to better see the prospects for future transformations and make important forecasts until the end of the twenty-first century.

In both cases, we expect very radical changes, both in the economic model of society and in the political one, where what we have called the electronic state is being formed.

## 1.11 Concluding Chapter

The conclusion to the monograph sums up the main results. We believe that as a result of the largely symbiotic development of the global aging process and society's adaptation to it, on the one hand, and the completion of the Cybernetic Revolution, on the other hand, a new type of society will emerge—a cybernetic one.

Cybernetic society is a society, which will be (a) over-technological; (b) sociotechnologically regulated at all levels; and (c) a society where, as a result of the aging process, the division into age categories will become socially much more significant than today (that is, age will become a much more important social marker than it is today).

## 1.12 Book's Chapters

The monograph consists of 15 chapters, including this Introductory one.

**The first part** "Technological Transformations in the Past, Present and Future' consists of three chapters (Chaps. 2–4).

In **Chap.** 2 (**"Global Technological Transformations since the Stone Age: Theory and History"**) of the monograph we describe the main technological changes in history.

The first part of this chapter is dedicated to the theoretical basis of the technological aspect of historical process. The second and the greater part of this chapter is dedicated to the factual description of the technological history of humankind in connection with the theory. The description covers almost the entire period of social evolution[20] starting from the emergence of *Homo sapiens* up to the mid-twentieth century when the Cybernetic Revolution began. Of course, we do not describe every single invention or separate innovation in detail; instead we make an attempt to show the general process of changes and to explain how and why technological epochs succeeded each other. The chapter introduces two interconnected theories: theory of Production Revolutions and theory of Production Principles. These are new explanatory paradigms which are of value when analyzing causes and trends of global shifts in historical process. The chapter describes the course of technological transformations in history and demonstrates a possible application of the theory to explain the present and forthcoming technological changes. We single out four production principles: (1) Hunter-Gatherer; (2) Craft-Agrarian; (3) Trade-Industrial; (4) Scientific-Cybernetic. The change in production principles is connected with production revolutions. Among all major technological breakthroughs in history the most important are the following three production revolutions: (1) the Agrarian Revolution (10–3 millennia BCE); (2) the Industrial Revolution (sixteenth to nineteenth centuries) and (3) the Cybernetic Revolution (1950–2070). See above for more details about the first two of them.

We show important recurrences in technological revolutions, which give us a starting point for projections for the coming decades, i.e., the ability to forecast the course of the Cybernetic Revolution. There are three phases in it. The first phase of the Cybernetic Revolution took place between the 1950s and mid-1990s, when a vigorous development of information technologies and the start of real economic globalization were observed. The second phase began in the mid-1990s in conjunction with the development and wide diffusion of user-friendly computers, communication technologies, cell phones and so on. This phase has been going on up to the present. The final phase of the Cybernetic Revolution may begin

[20] For more details on social evolution see Grinin and Korotayev (2023).

approximately between the 2030s and 2040s. It will usher in the beginning of the epoch of 'self-regulating/managing systems', which work without human control. We also offer some forecasts about the end of the Cybernetic Revolution, that is the mature phases of the scientific-cybernetic production principle up to the late twenty-first century.

**Chapter** 3 ("**Cybernetic Revolution, its Characteristics and Self-Managing Systems**") is dedicated to the consideration of the Cybernetic Revolution, its phases and main features as well as characteristics of self-regulating systems and their correlation with AI. According to the authors' concept *The Cybernetic Revolution* is the last of the largest production (technological) revolutions in all history following the Agrarian and Industrial ones. It is a great transition from the industrial production to both the production and service sectors based on the implementation of self-regulating systems. The first phase of this revolution started in the 1950s and 1960s and brought about the development of powerful information technologies, new materials and sources of energy, as well as the distribution of automation. Between the 2030s and 2070s, the final phase of this revolution will lead to a new level of self-operated control in technical systems, namely, to the level of self-regulating systems. The chapter will explain why the Cybernetic Revolution is called Cybernetic. How can it influence our future life? What may we expect in technology development in the next half of this century? And so on. Self-regulating systems are systems that by means of the embedded programs and a number of smart components can regulate themselves, responding in an intelligent way to the feedback from the environment as well as to operate and make decisions or suggest the best solutions independently, i.e., without human control, in a wide range of variations or cases of evolving situations with minimal or no human intervention. Of utmost importance is the fact that these self-regulating systems will become the major part of the technological process. That is the reason why we denote the final (forthcoming) phase of the Cybernetic Revolution as the epoch of self-regulating systems. This chapter gives the classification of self-regulating systems and examples of different levels of smart systems with elements or limited programs of self-regulation to systems with artificial intelligence that allow making decisions within a fairly wide framework (self-driving cars of the future; housekeeping robots; a system that independently selects plant varieties suitable for particular regions, etc.). The most advanced among self-regulating systems may be called "self-managing". By self-regulating systems we mean not only machines and other technical systems but a wider range of systems and controlling processes, including ones of biological, physiological, techno-biological, social and other nature, with a high level of self-regulation. These systems will be implemented in different areas (including medicine, genetic engineering, robotics, social relations). Special sections discuss differences between self-regulating systems and artificial intelligence. We enumerate the most important characteristics of the Cybernetic Revolution. These features are closely interconnected and complement each other. The dominant directions of the final phase of the Cybernetic Revolution are considered. One of the central concepts of our monograph is the abbreviation MANBRIC, which is an acronym for the names of the seven most active new areas of technology development, which will eventually constitute a powerful system, even a convergence of systems. These are respectively: Medicine, Additive technologies, Nanotechnologies, Biotechnologies, Robotics, Information technologies (artificial intelligence) and Cognitive technologies. It is this complex that will become the core of future technological development.

In **Chap.** 4 (**"Cybernetic Revolution and MANBRIC Technologies. When, How and Where Will Forthcoming Breakthrough Start?"**), we study the main directions of the final phase of the Cybernetic Revolution. They will form a peculiar and closely related cluster of innovative directions for the development of technologies. There are different points of view concerning the technologies that will become of utmost importance in the coming decades and make up this specific cluster as well as the sixth technological paradigm. We propose our own vision. Among numerous technological directions of the future, we identify seven leading ones. These directions are forming and in the future will finally combine into a single complex, which we call the MANBRIC convergence (see above). Further we analyze every branch from MANBRIC convergence with some forecasts about their development in the 2030–2070s.

We investigate in detail the answers to important questions, namely: (1) Which of the directions will be integral in this complex? (2) In what sphere will the final phase of the Cybernetic Revolution start? Which one will be the first? We explore such different directions as green energy, robotics and pilotless (self-driving) cars. We show that there are a number of reasons that make us provide the following answer to the last of the two questions: medicine will be the first sphere to start the final phase of the Cybernetic Revolution, because among other areas, medicine meets the criteria specified in this chapter to the greatest extent. In particular the medical sphere has unique opportunities to combine the abovementioned technologies into a single complex. Even today, all these areas of the MANBRIC are very actively used in modern medicine, but in the future, their relationship with medicine will noticeably and somewhere fundamentally increase. Besides, the process of society's aging, which is currently developing rapidly, creates an acute demand for labor-saving technologies, as well as gives a powerful impulse to medicine. Progress in the latter will help to extend working age and improve health-related (biological) quality of life. It will make the issues of health a subject of interest for the whole society, putting medicine and biotechnology at the forefront of technological development. The leading role of medicine in the final phase of the Cybernetic Revolution (at least in the 2030–2040/50s; then other areas may come to the fore) is also explained through the resulting economic and social consequences, as well as a number of other factors, to which the pandemic has been added since 2020.

Finally, we show why the breakthrough will happen in the 2030s/ 2040s and how the development of self-regulating systems will involve the most diverse areas of production, services and life. The chapter discusses many other subjects, in particular, the one of a very tightly correlation between the final phase of the Cybernetic Revolution and the so-called six technological paradigms.

**The second part**. "**Demographic Transformations and Global Aging**" consists of three chapters (Chaps. 5–7).

**Chapter** 5 ("**Demographic Transformations in the Historical Process**") provides an analysis of the demographic development of humankind in its close connection with technological development, including the trends of the twenty-first century. We consider the dynamics of correlation between production revolutions—the Agrarian, Industrial and Cybernetic (which have been discussed in Chap. 2)—and the major demographic transformations in the historical process. This allows us to show the features of modern demographic processes, the role of the aging process in these demographic processes and to make forecasts through the end of the current century. This also will lay the basis for a better understanding of Chaps. 6 and 7.

The demographic structure of modern society is very different from that of fifty or seventy years ago. However, deep demographic trends are very long-term. Therefore, to understand modern demographic processes, it is necessary to make their deep analysis, in fact, a demographic cut through the entire historical process. Modern society, completing the so-called demographic transition has the following model of population reproduction: low birth rate, low mortality, low, almost zero (or even negative) natural increase. At the same time, which is especially important for the topic of this monograph, we live in a rapidly aging society. And the process of global aging, as it develops, will increasingly influence the life of society and its technologies.

Analysis of the demographic component, the demographic dimension of the historical process, unfortunately, has not been sufficiently researched. Meanwhile, this is an extremely important aspect that can not only explain a lot in the development of the historical process, but also provide a basis for explaining current processes and allow the forecasting of our futures. It is especially important to note that throughout history, technological change has been among the most essential factors influencing demographic growth and sociocultural development. The close correlation between technological revolutions and radical demographic changes emerged quite long ago; in fact, it is this connection that distinguishes human reproduction from the reproduction of any other living beings.

However, the relationship between demography and production is far from simple or linear, and besides, the correlation between the demographic and production components of the life of societies during the historical process, while remaining, have changed significantly. This chapter aims to create a theoretical framework for the relationship between the development of production and technology, on the one hand, and demographic transformations, on the other, during the historical process, and also to describe all the major demographic transformations during human history. The chapter also reflects important points that affect population growth and its limitations. The authors describe the historical types of population reproduction and the causes for their change. All this allows us to make a forecast about the vectors and main features of the coming demographic transformation in the twenty-first century.

It is shown that as a result of the production revolution and the development of the cycle of the principle of production as a whole, there is a change in the type of population reproduction; together they give a powerful impetus to the qualitative reorganization of the entire social structure and social relations of the world system in its entirety. We conclude that every production revolution is followed by a fundamental change in demography. Conversely, population growth causes such changes in society and intersocial relations, communications and diffusion of innovations that it significantly accelerates technological development. In this chapter, the relationship between each production revolution and each principle of production, on the one hand, and demographic transformations and the type of population reproduction, on the other hand, is analyzed not just within a theoretical model, but in connection with the specific course of the historical process and those quantitative data which science currently has at its disposal.

All these ideas about demographic trends from the Upper Paleolithic to the end of the twenty-first century have also been summarized by us in a convenient tabular form.

A special place is given to the analysis of the demographic transition in connection with the demographic transformations taking place in the last seven to eight decades; in addition, the connection between the ongoing Cybernetic Revolution and global aging is revealed. Predictions are made about future demographic transformations associated with the development of the global aging process and the formation of a new type of population reproduction.

**Chapter** 6 ("**Global Aging and Other Demographic Trends Within Cybernetic Production Principle**") analyzes global aging and other important demographic processes, current and future demographic transformations in the twenty-first century, and their correlation with the technological changes that will occur as result of the Cybernetic Revolution. Dramatic changes are to be expected in connection with future demographic transformations. They create a powerful impulse that will lead to a qualitative reformatting of the entire social structure and social relations of the World System.

Based on this analysis, we make forecasts about the type of population reproduction that will characterize the World System by the end of the twenty-first century. The first sections of the chapter analyze two opposing demographic processes that together will greatly affect our demographic future. The first one is a very rapid demographic growth in Sub-Saharan Africa, which may continue for a very long time, even while the world-system population by the end of the twenty-first century is likely to start declining. In any case, it is Africa that will contribute much to the population growth on the world scale in the twenty-first century. However, the rapid population growth is associated with considerable growth of the number of young people in immature states, strongly influenced by national-ethnic tensions, rapid urbanization and other factors. Such factors can result in serious upheavals and destabilization processes in Africa.

The second trend is an opposite trend, that of possible depopulation in the second half of the twenty-first century, which is very likely in most countries of

Europe, East Asia and many others. Worldwide, more than 100 countries are likely losing their population between 2022 and 2100.

Among the main demographic processes, we put particular emphasis on the process of global aging. This is all the more important since in general, studies of this issue are scarce in both the public sphere and also in scientific discourses. However, even now global aging is an extremely important problem and, we suppose, it will become the most crucial problem in the future. Already today, the share of the population aged 65 + amounts to 20–28% in 22 countries and regions. By 2050, the population aged 65 and older is projected to make 16 percent worldwide, i.e., more than twice the number of children under 5 and about the same as the number of children under 12. For the first time in human history, people aged 65 and over will outnumber children.

We consider in detail the complex relation between development of the Cybernetic Revolution and the process of global aging, which—over several of the coming decades—will increasingly influence both individual societies and the World System in general, as well as the further change in the type of population reproduction. It is important to foresee that the struggle for very long life expectancy will lead to the creation of very complex systems for supporting a continuous medical-biotechnological environment for elderly and disabled people. However, this will not only have positive effects, but will also create problems connected with the limitation of human will and freedom.

The chapter analyzes various socio-political risks and problems associated with population aging. Among them, one finds economic slowdowns, pension crises, problems of care for the growing numbers of disabled and frail people, maintaining the financial savings of the elderly, the issue of ageism, and the vulnerability of the elderly during periods of crisis and pandemics and consequently maintaining life expectancy as well as stable democracy even including world order.

These risks and problems are of particular concern given that by the mid-twenty-first century two-thirds of people over 65 will live in middle- and low-income countries. However, though the influence of the aging process on various areas of social activity will be dramatic in most countries, the ability of modern society to cope with increasing risks associated with this process is problematic.

**Chapter** 7 (“**Global Aging Process, Adaption to It, and Future Demographic Transformations**”) provides a thorough system analysis of how the process of population and society aging will develop during the twenty-first century, and what achievements, challenges and conflicts await us and the world as a whole in connection with the formation of an elderly society, with the intensification of the global aging process and with population stabilization. We hope that the study we propose will be all the more relevant and important since it is not only a systematic research of these topics and subjects, but also, in general, because the studies of the problem of population aging are limited in scientific discourse. However, even now the global aging is an extremely important problem, and it will become the most crucial issue in the future while insufficient attention to it threatens to escalate societal tensions, crisis and intergenerational conflicts as well as political and social instability.

The chapter examines the mutual influence between global aging and technological change, as well as their combined impact on the transformation of society during the twenty-first century. We also analyze the challenges that already face society and will become even worse in the future: ageism, a sharp increase in the number of people suffering from dementia and other incurable age-related diseases in less developed countries, the struggle for resources due to the increasing number of the elderly and sick, increasing conservatism and possible slowdown of technological progress, an aggravation of intergenerational contradictions and conflicts, a possible decline in democracy due to population aging, and many others.

We reveal the necessity to adapt to societal aging through the transformation of a number of its institutions; we also analyze various ways and forms of adaptation to aging that must be elaborated in the future in order to mitigate frictions and tensions that could impede societal development. A large section is devoted to various scenarios for the development of future society associated with its aging.

In this chapter we also show the impact of the Cybernetic Revolution on the development of technologies, which can be stimulated precisely by the process of global aging and the need to increase life expectancy. But on the other hand, this enlarged life duration will create new serious problems, tensions and contradictions in a more graying and technological society. Comprehensive forecasts are made about the future society. In particular, we show the main features of a new type of population reproduction, which, in our opinion, should take shape by the end of the present century. We argue that a new demographic and technological system will promote the formation of a new societal type, which can be denoted as the cybernetic society. It will be an extremely high-tech society with a stable population size, a very large proportion of the older population and total technological support for societal health and the quality of biological life (which will be achieved by creating a continuous medical-biotechnological environment).

**The third part**. "**MANBRIC Technologies in the Forthcoming Epoch of Self-Regulating Systems (the 2030s–2090s)" consists of four chapters** (Chaps. 8–11).

In **Chap.** 8 ("**Medicine and Cybernetic Revolution: On the Way to Control the Human Body**") we are going to focus on the current and especially future technological transformations in medicine during the final phase of the Cybernetic Revolution. These transformations will greatly increase the ability to affect our biological nature as well as the life expectancy and health-related quality of life. In the present chapter (and in the book in general) we often use the terms 'medicine' and 'health care' as synonyms. We show the spheres which represent a new, in a broad sense, medical system or realm of medicine, creating a complex of technologies and their application with other progressive directions. Thus, for example, at present one already observes a great interconnectedness between different branches of medicine and related areas. There is an especially close relation with the seven leading technological directions of the future, which we identify in Chap. 4 as the MANBRIC convergence. This acronym includes medicine, additive (3D printers), nano-and bio technologies, robotics, IT and cognitive sciences. These are emerging areas of technological expansion, and in the future they may finally combine into a single complex, and it is the medical sphere that will unite them due to its unique features and opportunities to combine the above-mentioned technologies. We have already witnessed the development of this process: the use of robots in surgery and the care of patients, information technologies in remote medical treatment, neural interfaces for the treatment of mental illness and brain research; gene therapy and engineering, nanotechnologies for the creation of artificial immunity and biochips which monitor organisms; new materials for growing artificial organs and many other things which are to become a powerful sector of the economy. In the present chapter we provide more examples of such interrelations. In Chap. 4 we also pointed out that the final phase of the Cybernetic Revolution will begin somewhere at the intersection of medicine with numerous other technologies.

So today medicine is a crucial sector of economy, and in the future it will become even more powerful. Along with the achievements in medicine, the authors also consider a number of serious problems and challenges (like the growing medicalization, stagnation in the pharmaceutical industry, and a number of other challenges). So, we generally argue that there is a wide range of prospects for the development of medicine in the twenty-first century. We can hardly predict all possible innovations. However, the general vector of the breakthrough can be defined as rapidly expanding opportunities for correction or even modification of human biological nature. In other words, it may become possible to extend our opportunities to alter human body, and perhaps, even its genome; to widen sharply our opportunities of minimally invasive influence and operations instead of the contemporary surgical ones; to widely use the means of cultivating separate biological materials, to alter human bodies or their parts and elements for the regeneration and rehabilitation of an organism, and also to have access to artificial analogues of biological material (tissues, sense organs, receptors, brain electrodes and chips, etc.), and so on.

Undoubtedly, the struggle with incurable (or fatal) diseases and the fight against aging and its consequences will become the most important trends (we consider this in detail also in Chap. 11). At the same time, both these trends are tightly integrated and will naturally merge, since the older the society, the more patients with incurable diseases will be diagnosed in the group of elderly and old people. These two trends will also be very closely related to the direction of rehabilitation of the disabled. Already today, the percentage of disabled people among the elderly is especially large, but as society ages, their number will increase even more; in addition, incurable diseases with a high risk of death, unfortunately, very often lead to disability prior to death, further increasing the number of aging disabled. We also believe that emerging trends like implantation, reproductive medicine, gene therapy and aesthetic medicine will remain among the most promising dimensions.

In this chapter we devote a significant amount of attention to our forecasts. According to these forecasts, by the end of the twenty-first century, the development of self-regulating systems in medicine alongside an increasing life expectancy and general aging of society will lead to the creation of a super-complex system, which we may define as a special techno-bio-medical environment. Moreover, this environment will become permanent. This means that life may become almost impossible beyond the range of this new medical environment, in a similar fashion to today we can hardly imagine our life without the Internet and gadgets. Just as many hundreds of millions of people cannot live without lifelong drugs, so in the future a huge part of population will be unable to stay outside this medical-biotechnological environment for a long time. This system implies not only a breakthrough in medical technologies but also a powerful breakthrough in care technologies, continuous monitoring of the condition of people (not only of the elderly and disabled), control, access to remote medicine, psychological assistance, prediction, prescriptions and urgent care. But along with the increasing duration and comfort of biological life, this, of course, will bring a lot of dangers and problems, which we discuss in detail in the present chapter.

In **Chap.** 9 ("**Biotechnologies in Perspective: Major Breakthroughs, Development of Self-Regulating Systems and Possible Social Confrontations**") we concentrate on describing the current and future transformations of biotechnology in different spheres and its fusion with medicine and other MANBRIC technologies in terms of their forecasted huge impact on our future. The role of biotechnology has been increasing since the 1940s. And it proves rather essential at present. We suggest that the most breakthrough innovations will probably result from the synthesis of biotechnologies with other MANBRIC technologies (in particular, with medicine, nanotechnologies and AI), as well as within the direction of synthetic innovation—bionics. The chapter places particular emphasis on the probable breakthroughs in biotechnology and genetic engineering related to medicine (including the opportunities for creating artificial organs and tissues). Thus, biotechnologies will play a crucial role in the future.

There are emphasized the four most important fields of biotechnology, which are often classified with color codes. Red biotechnology is a health-related branch, white biotechnology is linked to application of biotechnology in the industrial sector, for example, biocatalysis or fermentation; green biotechnology is related to food and environment-related areas; and blue is biotechnology which uses marine organisms and resources. The chapter describes each of these application areas, as well as the bioeconomy associated with these areas, and makes extended projections about the possible future developments of every branch. Much attention is also given to other fields, in particular to improving the environment, development of aquaculture and agriculture and solutions for the problem of food security. The chapter pays much attention to modern achievements, problems and challenges facing biotechnology, and also shows possible ways to solve these problems and the dangers that may arise as a result of the active implementation of these achievements in the future (based on the experience of the anti-COVID campaign and the fight against GMOs).

Our analysis shows that the role of biotechnology will be fundamental during the final phase of the Cybernetic Revolution (in the 2030s–2070s), especially with respect to the synthesis between the medical field and the development of pharmaceuticals; this synthesis will become an essential part of medical technologies. However, the role of biotechnology will also be vital in a number of other industries and fields. Already today biotechnologies provide numerous and vivid examples of self-regulating systems, but the development of such systems in the future will become especially notable (e.g., in genetic engineering, self-regulation of artificial physiological systems, growth of artificial tissues, etc.). Besides, in the future, the creation of huge databases of plant varieties and variations will allow the whole process of genetic modification of a plant to proceed without human participation, thus, to become self-regulated. So, one may assume that by the end of the final phase of the Cybernetic Revolution, agricultural biotechnologies will already be developed to the degree when modified products become able to respond even to the smallest fluctuations of local conditions.

Thus, we anticipate that the role of biotechnology will not only increase in the future, but its branches will become extremely important for human life and the environment. In particular, they have great hopes that biotechnologies (and bioeconomics in general): (a) will make a huge contribution to the victory over diseases that are currently incurable as well as more efficiently curing those diseases that are already curable, including viral ones; (b) will allow a large number of people suffering from genetic abnormalities, dysfunctions of internal organs or tissue damage to return to a more active life with the help of artificial tissues and organs; (c) will greatly contribute to the fight against aging; (d) will considerably mitigate environmental problems, especially in cities, as well as facilitate the solution of the problem of biodiversity loss; (e) will become the leading sector in solving the problem of food supply (similar as it solved it to a large extent during the 'Green Revolution' in the initial phase of the Cybernetic Revolution); (f) will significantly change technological and production processes, forming waste-free production; and (g) expand our ability to manage natural processes.

In a word, we believe that in three or four decades our life will be absolutely impossible without biotechnologies.

In **Chap.** 10 ("**Nanotechnologies, Robotics, Artificial Intelligence and Other MANBRIC Technologies in the Long-Term Development**") we present an analysis of the development and forecasts for the future of a number of technologies: nanotechnologies, robotics, additive and cognitive technologies, ICT and AI.

Each area of the Cybernetic Revolution has its own future, its own problems and its own coordinate system. For each of the technologies, we try to forecast major breakthroughs. In general, they have a bright future. So, additive technologies promise to change the economy, moving from a mass production economy to a small-scale or custom-made one. A versatile and intelligent robot will be able to alleviate the problems of a lack of labor by replacing a nanny, a nurse, a cook and even a miner, as well as enhancing many other professions. Cognitive technologies will be able to rehabilitate many millions of currently paralyzed people and other disabled people. Nanotechnologies have great prospects, although some common predictions about them raise doubts, in particular, that with the help of nanotechnologies, by combining atoms, we can get almost any substance. The chapter analyzes these and other likely advances in the field of various technologies, but also clearly shows the problems and difficulties along the way, as well as the arguments and reasons why certain forecasts may be untenable. At the same time, the possible negative consequences of the development of technologies are also analyzed, including the actively discussed threat to jobs from robots.

The described technologies are different, but there is a very important common thing that unites these different technological directions. What is it? First, all these different technologies show us excellent examples of self-regulating systems, that is, systems that can function and even make decisions with minimal human influence or even autonomously. Self-managing systems are the basis of the future technological paradigm, the main characteristic of the Cybernetic Revolution. Secondly, all the technologies described in the chapter are included in the complex of innovative directions MANBRIC, to which we forecast a great future. Thirdly, using the example of the technologies described below, one can see the convergence of MANBRIC technologies, which consists in a number of manifestations. Thus, certain innovations can function only on the basis of combining many advanced technologies into a single breakthrough system. Here we can talk about MANBRIC clustering; in particular, robotics is an excellent example of this, since robotics seeks to combine the maximum number of technological lines and directions. It is important to pay attention to the mutual influence of technological areas in terms of ideas that influence the choice of direction of development. In particular, the chapter shows how AI is learning from cognitive science, how inventors try to design robotics for the home Internet of robots (give opportunities to connect robots to home computers or an operating center). We trace the close connection between MANBRIC directions in ensuring the functioning of institutions and industries. Thus, the military sphere will be an important driver for the development of these MANBRIC technologies, but robotics, including drones, AI and other technologies, are also of particular interest.

Fourth, we try to show how various features of the Cybernetic Revolution and the complexity of the functions of self-regulating systems in each technology are manifested. In particular, the functionality of robots can be distinguished: active information processing (its receiving, analysis, distribution, transformation, etc.), management of the whole system (and of other objects) by means of information, flexible interaction with environment, existing contours of positive and negative feedback loops that allow fulfilling various functions.

We pay special attention to the relationship of robotics with medicine and biotechnology. It is shown that robotics and ICT, medicine, bio, nano, robotics (including unmanned vehicles), additive technologies and cognitive technologies are already quite visibly combined in medical institutions and in work with disabled people. In addition to medicine, AI plays a special role in this integration of MANBRIC. One can speak about rapid development and even predominance of artificial intelligence. We agree that the future is closely associated with the development of artificial intelligence, which will participate in a very large number of self-regulating systems. It becomes important to define similarities and differences between self-regulating systems and artificial intelligence.

MANBRIC technologies have different degrees of development and different eras in the future for their heyday. AI has already reached a very high development, but in relation to nano, additive technologies and robotics, we expect that their explosive development will happen only in the 2050–2060s. One can expect a mass development of new means of transport in the middle or end of the final phase of the Cybernetic Revolution (between the late 2040s and 2060s). However, the self-driving electric vehicles can become a powerful source of technological development during the mature phase of the scientific-cybernetic production principle. Like self-driving cars, drones will achieve real autonomy sooner or later. Due to the great economic effect and convenience of such robots, they will see a very promising future.

A comprehensive analysis of these technologies within one chapter gives us the opportunity to show that the final phase of the Cybernetic Revolution will not be a wave of diverse innovations; it will be a complexly integrated and interconnected set of new generation technologies that will create an era of self-managing systems.

Thus, in this chapter, we solve a two-pronged problem—on the one hand, to show the unique logic of these technologies and development prospects for each of the described technologies. On the other hand, we try to show how the development of each of the technologies fits into the MANBRIC convergence as a whole, how this convergence will strengthen their development in the future, how it will prepare medical technologies for the initial breakthrough that will launch the final phase of the Cybernetic Revolution and ensure the powerful development of self-managing systems.

The chapter reveals to the reader many new perspectives, ideas, facts and problems that will face humanity in the process of the final phase of the Cybernetic Revolution.

The title of **Chap.** 11 ("**Anti-Aging as a Key Challenge for the Medicine of the Future**") well expresses not only its main idea but also formulates one of the leading ideas of the book. We will give many arguments in favor of these ideas, and in the last paragraph of this chapter we will summarize them. Despite the fact that the problems associated with aging are receiving more and more attention in medicine, for the most part they are not yet considered as a single very important area but instead are scattered over many areas, each of which is associated with certain types of diseases. Therefore, today's idea of the medicine of the future, in principle, is that it will be approximately the same institute, only much more technical and robotic. Meanwhile, we believe that there will be in many respects new medicine, because anti-aging technologies, that is, technologies for slowing down and, if we may say so, ennobling aging, as well as stopping and alleviating the problems associated with it, will become an integral part of the future medicine and, moreover, form its system-organizing direction. This will also mean that more and more attention will be paid not only to the treatment of diseases of old age, but to the elimination or alleviation of the so-called geriatric syndromes (or age-associated conditions). A very noticeable part of medicine will revolve around rejuvenation, plastic surgery and other anti-aging corrective procedures (which, perhaps, will become commonplace and regular not unlike a vacation in a sanatorium). The perception of aging in society will also change (see Chap. 7). Thus, the altered focus of medicine on the problems of aging in line with innovations in anti-aging technologies will mean a process of medical adaptation to aging.

The latter will become an essential part of the overall process that we have called adaptation to aging.

But, of course, such a transformation will not happen immediately. However, we assume that in the last third of the twenty-first century, a special medical-biotechnological environment will form that will reshape medicine entirely (see Chaps. 6 and 8).[21]

**The fourth part**. "**Trends of Cybernetic Revolution in the Light of Technological Progress and Aging**", **consists** of three chapters (Chaps. 12–14).

**Chapter** 12 ("**Long-Term Dynamics of the Technological Growth. Would it Lead Us to a Singularity?**") covers a very wide time span between the Upper Paleolithic and the early twenty-second century, when a new era will probably start. In this chapter we consider the long-term dynamics of technological progress, presenting an option for measuring the speed of this progress throughout the entire historical process. Measuring the speed of technological progress is only one and not the main goal of this chapter. The main themes of the monograph: the Cybernetic Revolution and Aging remain the focus of our attention. However, measuring process speed allows us to make some forecasts. In this chapter, we set the following tasks (1) to make forecasts about how technological progress will develop in the twenty-first century, with a slowdown or an acceleration; (2) to show how the change in the pace of technological progress is related to global aging and the development of this process. These problems are solved based on the theory of production (or technological) revolutions and the theory of production principles, as well as involving mathematical methods.

We find that the general dynamics of accelerating technological growth over the past 40 thousand years can be described with amazing accuracy ($R^2 = 0.99$) by a simple hyperbolic equation: $y_t = C/t_0 - t$, where $y_t$ is the technological growth rate, measured as a number of technological phase transitions per unit of time, while $t_0$ and C are constants, whereas $t_0$ can be interpreted as a 'technological singularity' point.

We show that so far on a macroscale there has been a tendency for an increase in technological growth, albeit with fluctuations. However, the upcoming technological growth will not be infinite. In this chapter, we describe a scenario of how, when and why the technological growth rate will change. First, the slowdown in technological growth in recent decades, beginning in the 1970s, can be explained as a general fluctuation of technological growth. We view this slowdown as a necessary precondition for further rapid growth. Secondly, our analysis shows that there are a number of reasons to expect that in the forthcoming decades global technological growth rate will return for some time to a hyperbolic trajectory when the final phase of the Cybernetic Revolution begins. Thirdly, this acceleration will continue up to the late twenty-first century. Fourthly, technological growth at the end of the twenty-first century will gradually slow down to the singularity point in the year

[21] If by this time aging is recognized as a disease, then, accordingly, medicine will become more and more noticeable to focus on the treatment of aging and related diseases and deformities.

2106. As we have seen, the global aging factor will play a leading role here. After the singularity point the rate of technological progress will slow down compared to the previous epoch. Fifth, the rate of technological growth will experience a slowdown, and it is difficult to forecast any subsequent acceleration in the twenty-second century (at least in its first half). And, probably, the most important point is that the pattern of scientific and technological development itself will change dramatically. Thus, such a slowdown will mean that the old type of technological progress is beginning to transform into a new one, and, most likely, a transition to new forms of social relations will also begin.

**Chapter** 13 ("**Will Global Aging Change the Pace of Technological Progress and Create a New Consumption Model?**"). In this chapter, we discuss how the pace of technological progress and the economic model of society will change under the influence of the development of the aging process. Thus, the chapter continues the topics raised in the previous chapter.

We believe that global aging is one of the most important factors in the coming decades. This chapter is continuing the discussion on the ways and role of the global aging and will be connected with the research questions which we have tried to answer in our book, these are: What could be the driver of an acceleration of technological growth in the forthcoming decades? Is a slowdown of technological growth likely to be observed afterwards? What could cause such a slowdown? How could such a slowdown affect socio-economic and socio-political relations? What is the relationship between global aging and technological progress? Why is global aging likely to be one of the most important factors affecting technological growth in the near future?

We continue to develop the idea raised in the previous chapters that aging has a dual and contradictory effect on technological progress: positive and negative. On the one hand, the need to improve the biological quality of life of the elderly, the struggle to increase life expectancy, the fight against incurable diseases, the need to extend the working age of the elderly, etc., spur science and technology to search for new breakthrough solutions. On the other hand, older people tend to be more conservative than the younger and middle-aged categories, leading to a slowdown in the rate of technological progress. In this chapter, we analyze both of these trends and show how these trends and their relationship will affect the rate of technological progress in the twenty-first century. So, the authors obtain a nontrivial result, according to which the process of global aging can accelerate and change the direction of technological progress in the coming decades, and then the aging of society can slow down scientific-technological progress in the late twenty-first and early twenty-second centuries. However, the term, slow down, does not mean a stop; we argue that technological or economic development will hardly stop. It would be even unreasonable to let technological progress slow down too much. Otherwise, economic growth will critically slow down or stop, and then society could plunge into a crisis and slide into a rough distribution system.[22] This thought allows us to move on to the next topic.

The other goal of this chapter is to show that one can assume that the current consumption pattern may also change under the influence of the global aging process. And this change, in turn, will have a serious impact on the entire production structure and on scientific-technological progress (as we will discuss further). The conservatism of the elderly may lead not only to a slowdown in development, but also to a transition to a different economic system. The current model is associated with an increasing rate of consumption. Consume more today than yesterday, and tomorrow more than today, is largely an absurd model. Also, sometimes the desire for sustainable GDP growth seems absurd. However, it works and will work for decades, especially for poor countries whose populations are not satisfied with their level of consumption. Thus, aging can change people's needs, especially in conditions of the stabilization of the population or its reduction. As a result, under the influence of all above-mentioned future changes, the model that measures economic growth in GDP should be replaced. The modern consumption model will also change. Overall, the Cybernetic Revolution and aging should ultimately lead to the transition to a new economic model without an endless increase in consumption. In such a case, the growth model in the economy should differ from that of today, it is likely to include some parameters of quality and longevity. Accordingly, business models may change, although it is not very clear how this will happen.

**Chapter** 14 ("**The Cybernetic Revolution, Covid-19 and the E-State**"). In this chapter we discuss the impact of the pandemic on the course and the rate of the Cybernetic Revolution and the ability to make forecasts about very large changes not only in technology, but also in the socio-political sphere has been significant.

The 2020–2022 pandemic has become one of the very large and unexpected events of the twenty-first century. The COVID-19 pandemic has significantly changed the public's focus; these changes are likely to persist for a long time. Ultimately, this may lead to significant technological transformations, which, in our opinion, are rather underestimated. Besides, the pandemic of coronavirus will accelerate the start of a new technological wave—the final phase of the Cybernetic Revolution. So one of the main ideas of this chapter is that the COVID-19 pandemic has become a trigger for the acceleration of the Cybernetic Revolution.

The COVID-19 pandemic has become a phenomenon that for the first time in many decades or even centuries placed healthcare problems at the center of both intrasocial and global relations. The socio-political and economic role of medicine has sharply increased due to the large number of COVID infections. On the whole, the pandemic has not only confirmed our forecasts, but also allowed us to significantly refine them and make them more accurate and enrich them. In particular, the experience of coping with the COVID-19 pandemic has confirmed that the

[22] Only economic growth in conjunction with the right social policy is the way to raise living standards. And the growth of the economy with the stabilization of the population can only be due to the growth of technology (Grinin & Grinin 2021a, 2021b; Grinin et al., 2022).

final phase of the Cybernetic Revolution will very likely begin in the 2030s at the intersection of a number of medical, bio, digital and several other technologies. That the growing rate of technological research due to COVID has changed the role of medicine considerably confirms our ideas of the integrating role of the medical technologies in the MANBRIC convergence and implies practical recommendations regarding the most promising directions of technological development in the nearest future. As a consequence, the fight with the pandemic has made significantly clearer the mechanisms which can provide such a breakthrough in medicine and related technologies. This impact is determined by the growing need for the development of a number of areas in medicine, bio- and nanotechnology, artificial intelligence and others, which we denote as MANBRIC convergence.

It is very important, that COVID-19 becomes a powerful impetus not only in terms of accelerating technological development and approaching the final phase of the Cybernetic Revolution, but also in changing socio-political (and socio-administrative) relations in the forthcoming decades. Combining technological and social forecasts in a single study appears all the more important, since systematic research of the relationships and interactions remains insufficient. The COVID-19 pandemic has shown some mechanisms of such system connections, thus helping improve forecasts.

What do we mean? As already mentioned in many chapters of this monograph, the most important outcome of the Cybernetic Revolution will be the development of self-regulating systems in the economy and life, which will flourish during the Cybernetic Revolution. In this chapter, we pay great attention to the role of a special type of self-regulating systems, the importance and influence of which on society has grown dramatically in recent years and will continue to grow rapidly. We are talking about socio-technical self-regulating systems (SSSs).

SSSs with the help of artificial intelligence will regulate various social and administrative relations. SSSs perform social and administrative functions (i.e., control, verification, distribution, security, checking, identification, issue of documents, rating and other functions) using a set of technologies with little or no involvement of human officials and specialists. Thus, they can be used by authorities at different levels and by the state as a whole (as well as by the administration of service centers where the authorities consider such regulation necessary: airports, crowded places, etc.). That is why the development of SSSs in one way or another pushes society toward the formation of electronic state.

What is an electronic state/e-state? And how radical will the changes associated with it be? We understand the concept of e-state as a state with a significantly reduced number of state oversight bodies, mainly based on SSS technology. In politics and administration, the development of the e-state will entail very significant, one might even say revolutionary, changes in the field of governance, because many social and administrative relations will be regulated by these systems. This will take place both at the level of individual administrative units and cities (so-called smart cities), and at the level of the state as a whole. However, development of the SSSs and formation of e-state can lead to total unification, thereby suppressing the individuality and special abilities of people as well as to serious changes in the political and social spheres. This can accelerate the change in social relations, as well as increase confrontation in society, because the use of self-regulating systems in governance will lead to increased technological and political control. This is all the more dangerous because, although such systems have not yet become widespread in private life, they are very likely to become so. There are also many other new ideas and forecasts.

## 1.13 Conclusion

**Chapter** 15 (“**Conclusion. Toward Cybernetic Society**”) sums up the main forecasts offered in this monograph. We have seen that very serious and at the same time encouraging and risky changes await humanity ahead. Trying to foresee the forthcoming decades up to the late twenty-first century, we especially focus on two important processes of development. The first of these are technological changes connected with a new technological wave, i.e., the final phase of the Cybernetic Revolution, and possible future breakthroughs which we will discuss in detail. This technological revolution will result in the widespread introduction of self-regulating systems. The second one is the process of global aging, which will have an expanded influence on all spheres and relations in future society. Thus, this conclusion is dedicated to some summation of the numerous forecasts made in this monograph, which we would like to conceptualize in order to help the reader see the shape of future society, which we call “cybernetic society” in different areas: technological, demographic, economic, political and administrative, international relations, social and environmental.

As a result of these transformations, a new unprecedented type of society will be formed—the cybernetic society. The authors call it “cybernetic society” because its formation will be inextricably linked with the completion of the Cybernetic Revolution and the finished process of aging populations and the complex consequences of both processes.

Thus, it will be an elderly society, with the institutionalization of age differences, relying almost entirely on smart (cybernetic) technologies and AI-based self-regulating systems.

The authors forecast that it will be society, which will have fundamentally new characteristics and consequently it will be very different from all previous societies. On the one hand, it will be closely connected with very deep technological changes in the management and regulation of our life at all levels. On the other hand, it will be inextricably linked with the process of global aging, because aging becomes an integral part of society. As this new society develops and the institutions adapt to it, no sphere of society will remain unchanged, including technology, economy, consumption, social sphere, ideology, etc. In short, this cybernetic society is an elderly society with the institutionalization of age differences, relying almost entirely on smart (cybernetic) technologies and self-regulating systems. Accordingly, this society will have its strengths and weaknesses, successes and even triumphs, but it will not be free from some serious problems. Cybernetic society must go through a difficult path of social frictions, conflicts, optimizations before many things become institutionalized. Techno-social regulation has important advantages, but it also includes violence against the will and freedom of people. Therefore, a considerable time is required to optimize regulations in order to take into account a variety of needs, desires and problems. Even short-term regulation with the help of technology during the COVID-19 pandemic has shown how painful it could be for societies.

In several chapters of this volume, we discuss the possible social changes in a future society. Sometimes the forecast horizon is a long way off; our forecast even considers some possible changes that might fully occur in the first third of the twenty-second century.

Two main themes have been explored in our monograph: (1) the cybernetic revolution, the development of new technologies and innovative areas, the impact of this technological progress on the future society; (2) the process of an aging of society and, in general, global aging in the World System and the impact of these processes on future society. Both of these topics, as the reader saw, turned out to be closely intertwined, because both of these powerful processes are closely and sometimes unexpectedly interconnected and integrated.

In this conclusion, we would like to show our vision of the future society, which we called cybernetic, and which is based on both foundations: high technology and the aging process that has reached maturity, the synthesis of which gives a new type of society with its advantages and probable shortcomings and problems that we also tried to anticipate.

**To the future with cautious optimism and prudence**. In conclusion of this introductory chapter let us summarize some ideas discussed above. Human power increases with the growth of technologies, but with them many changes in human lifestyle and numerous previously unknown problems arise. That is why, if we do not want to give up new opportunities (and why do we need to give them up?), it is highly desirable to be able to forecast problems and minimize them. We must go to the system which will not allow uncontrolled technological progress to take us down a dangerous road and at the same time will not stop progress radically.

For the more serious the technological breakthrough, the more radical the changes and the "future shock" (if we use Toffler's term [1970]) will be. Unfortunately, society does not learn enough from its mistakes and pays little attention to future problems. At best, it acts like generals from aphorisms, who always prepare not for future, but for past wars. In this volume we discuss possible problems and risks arising from the logic of changes during the final phase of the Cybernetic Revolution. Forecasting such problems can help to work with the creation of optimal social, legal and other tools so that such changes do not take us by surprise, so that negative consequences can be minimized. Of course, much of what biomedical ethics has been doing for the last three decades can be helpful. However, although during this time it has turned into an established area of knowledge with a wide specialization, it has its own international centers, conferences and periodicals; this is clearly not enough. This is not so much about ethics and ethical and legal conflicts as about the future of the humans as biological organisms as well as about

AI's rapidly widening influence on human mentality. We suppose that it is correct to speak of a categorical biohumanitarian imperative, the development of fundamental principles and forms that should be followed on the path to a new state of being (and which it is desirable to consolidate in some kind of international legal form). We move forward faster and faster, but at the same time always along an uncharted path, very poorly comprehending the consequences of our innovations. And this cannot but cause concern. And although we have no choice but to go forward, the maximum of caution, wisdom, prudence and even some humility before the greatness of the Universe and the world, and deep respect for the heritage left to us by billions of years of biological evolution, are absolutely necessary in order to successfully travel along this path. And then our perseverance, knowledge and (albeit still weak) ability to foresee will allow us to safely reach new heights of human power and leave future offspring capable of preserving it.

We hope that our research will be interesting for those who study the issues of technological development, its role in historical process and societal advance and also for those who would like to learn more about the future development of technologies and society.

**Acknowledgements** This chapter is an output of a research project implemented as part of the Basic Research Program at HSE University in 2024 with support from the Russian Science Foundation (Project No. 23-18-00535).

## References

Alharbi, F. S. (2020). Dealing with data breaches amidst changes in technology. *International Journal of Computer Science and Security (IJCSS), 14*(3), 108–114.

Alkema, L., Raftery, A. E., Gerland, P., Clark, S. J., Pelletier, F., Buettner, T., & Heilig, G. K. (2011). Probabilistic projections of the total fertility rate for all countries. *Demography, 48*(3), 815–839. https://doi.org/10.1007/s13524-011-0040-5

Ariyanov, A. (N.d.). *Stat' rasoy bessmertnykh – glavnaya evolyutsionno-istoricheskaya zadacha chelovechestva v III tysyacheletii. Rossiya 2045*. http://2045.ru/news/32900.html

Artemova, O. Y. (1987). *Lichnost' i sotsial'nye normy v rannepervobytnoy obschine (Po avstraliyskim etnograficheskim dannym)*. Nauka.

Ashman, H., Brailsford, T., Cristea, A. I., Sheng, Q. Z., Stewart, C., Toms, E. G., & Wade, V. (2014). The ethical and social implications of personalization technologies for e-learning. *Information & Management, 51*(6), 819–832. https://doi.org/10.1016/j.im.2014.04.003

Ayres, R. U. (2006). The singularity is near: When humans transcend biology by Ray Kurzweil (Book review). *Technological Forecasting and Social Change, 73*(2), 95–127.

Bardi, U. (2014). *Extracted: How the quest for mineral wealth is plundering the planet. A report to the club of Rome*. Chelsea Green Publishing.

Berg, C. (2019). *Sustainable action: Overcoming the barriers*. Routledge.

Bös, D., & Von Weizsäcker, R. K. (1989). Economic consequences of an aging population. *European Economic Review, 33*(2–3), 345–354.

Brammer, S., Branicki, L., & Linnenluecke, M. (2020). COVID-19, societalization and the future of business in society. *Science of the Total Environment, 34*(4), 2–7. https://doi.org/10.5465/amp.2019.0053

Bröring, S. (2021). *Moving the bioeconomy from mind to market: What drives technology transfer and innovation adoption?* Presentation at IAMO Forum June 7th 2021.

Browne, E. (2022). The story of He Jiankui, now out of jail after editing DNA of unborn babies. *Newsweek.* April 11, 2022. https://www.newsweek.com/he-jiankui-out-jail-editing-genome-dna-unborn-babies-crispr-1696846

Burdea, G. C., & Coiffet, P. (2003). *Virtual reality technology*. John Wiley & Sons.

Callaghan, V., Miller, J., Yampolskiy, R., & Armstrong, S. (2017). *Technological singularity*. Springer, Dordrecht. https://doi.org/10.1007/978-3-662-54033-6

Cecere, G., Le Guel, F., & Soulié, N. (2015). Perceived internet privacy concerns on social networks in Europe. *Technological Forecasting and Social Change, 96*, 277–287. https://doi.org/10.1016/j.techfore.2015.01.021

Chudinova, O. Y. (1981). Muzhchiny i zhenschiny v obschestve aborigenov Avstralii. In K. V. Malahovsky (Ed.), *Puti razvitiya Avstralii i Okeanii. Istoria, ekonomika, etnografia* (Vol. 1, pp. 220–237). Nauka.

Clynes, M. E., & Kline, N. S. (1960). Cyborgs and space. *Astronautics, 14*(9), 26–27.

Coleman, D., & Rowthorn, R. (2015). Population decline—making the best of inevitable destiny? In L. E. Grinin & A. V. Korotayev (Eds.), *History & mathematics: Political demography and global ageing* (pp. 26–41). Uchitel.

Cowen, T., & Southwood, B. (2019). *Is the rate of scientific progress slowing down?* George Mason University.

Diakonoff, I. M. (1999). *The paths of history*. Cambridge University Press.

Eden, A. H., Moor, J. H., Soraker, J. H., & Steinhart, E. (Eds.) (2012). *Singularity hypotheses: A scientific and philosophical assessment.* Springer-Verlag. https://doi.org/10.1007/978-3-642-32560-1

Elias, J. (2023). Google CEO Sundar Pichai warns society to brace for impact of A.I. acceleration, says 'it's not for a company to decide'. *CNBS*, April 17, 2023. https://www.cnbc.com/2023/04/17/google-ceo-sundar-pichai-warns-society-to-brace-for-impact-of-ai-acceleration.html

Engelhardt, H. T., Jr. (1996). *The foundations of bioethics*. Oxford University Press.

Field, D. (2015). Perfect genetic knowledge. *AeonMagazine*, September 11, 2015. https://aeon.co/essays/what-might-we-do-with-the-genomics-of-the-entire-planet

Fuchs-Kittowski, K., & Krüger, P. (1997). The Noosphere vision of Pierre Teilhard de Chardin and Vladimir I. Vernadsky in the perspective of information and of world wide communication. *World Futures, 50*(1–4), 768.

Fücks, R. (2013). *Intelligent Wachsen*. Carl Hanser Verlag, GmbH & Co.

Fukuyama, F. (2002). *Our post-human future: Consequences of the bio-technology revolution*. Farrar, Straus, and Giroux, New York.

Gabor, D., Colombo, U., King, A., & Galli, R. (1978). *Beyond the age of waste*. Pergamon Press.

Gallagher, J. (2020). Fertility rate: 'Jaw-dropping' global crash in children being born. *BBC*. https://www.bbc.com/news/health-53409521

Gates, B. (2023). The age of AI has begun. Artificial intelligence is as revolutionary as mobile phones and the Inter-net. *Gatesnotes*, March 21, 2023. https://www.gatesnotes.com/The-Age-of-AI-Has-Begun

Goldstone, J. A. (2015). Population aging and global economic growth. In J. A. Goldstone, L. Grinin, & A. Korotayev (Eds.), *History & mathematics: Demography & aging* (pp. 147–155). Uchitel.

Goldstone, J. A., Grinin, L., & Korotayev, A. (2015). Research into global aging and its consequences. In J. A. Goldstone, L. Grinin & A. Korotayev (Eds.), *History & mathematics: Demography & aging* (pp. 5–9). Uchitel.

Griffin, A. (2015). Sex robots should be banned, say campaigners, as engineers look to add AI to sex toys. *Independent*, September 15. 2015. http://www.independent.co.uk/life-style/gadgets-and-tech/news/sex-robots-should-be-banned-say-campaigners-as-engineers-look-to-add-ai-to-sex-toys-10501622.html

Grinin, L. E. (2007a). Production revolutions and periodization of history: A comparative and theoretic-mathematical approach. *Social Evolution & History, 6*(2), 11–55.

Grinin, L. E. (2007b). Production revolutions and the periodization of history. *Herald of the Russian Academy of Sciences, 77*(2), 150–156.

Grinin, L. E. (2013). Dinamika kondrat'yevskikh voln v svete teorii proizvodstvennykh revolyutsiy. In L. E. Grinin, A. V. Korotayev, & S. Y. Malkov (Eds.), *Kondrat'yevskiye volny: Palitra vzglyadov* (pp. 31–83). Uchitel.

Grinin, L. (2020). How can COVID-19 change geopolitics and economy? *Journal of Globalization Studies, 11*(2), 121–134.

Grinin, L. E. (2023). Uskoreniye rekonfiguratsii mir-sistemy v svyazi s SVO i vozmozhnyye stsenarii budushchego. Stat'ya 2. Vozmozhnyye stsenarii rekonfiguratsii Mir-Sistemy i mirovogo poryadka. *Vek Globalizatsii, 3*, 87–109. https://doi.org/10.30884/vglob/2023.03.07

Grinin, L. E., & Grinin, A. L. (2013). Global technological transformations. In L. E. Grinin, I. V. Ilyin, & A. V. Korotayev (Eds.), *Globalistics and globalization studies: Theories, research & teaching* (pp. 98–128). Uchitel.

Grinin, A. L., & Grinin, L. E. (2015a). The cybernetic revolution and historical process. *Social Evolution and History, 14*(1), 125–184.

Grinin, A. L., & Grinin, L. E. (2020). Crossing the threshold of cyborgization. *Journal of Big History, 4*(3), 54–65.

Grinin, A. L., & Grinin, L. E. (2021b). Cyborgization: To be or not to be? In L. E. Grinin & A. V. Korotayev (Eds.), *Evolution: Trajectories of social evolution* (pp. 187–202). Uchitel Publishing House.

Grinin, L. E., & Korotayev, A. V. (2009). The epoch of the initial politogenesis. *Social Evolution & History, 8*(1), 52–91.

Grinin, L. E., & Korotayev, A. V. (2013). *Sotsial'naya makroevolyutsiya: Genezis i transformatsii Mir-Sistemy*. LIBROKOM.

Grinin, L., & Korotayev, A. (2015). Global'noye stareniye naseleniya, shestoy tekhnologicheskiy uklad i mirovaya finansovaya sistema. In L. Grinin, A. Korotayev, & V. Bondarenko (Eds.), *Kondrat'yevskiye volny: Naslediye i sovremennost* (pp. 107–132). Uchitel.

Grinin, L., & Korotayev, A. (2016a). MENA region and the possible beginning of world system reconfiguration. In M. Erdogudu & B. Christiansen (Eds.), *Comparative political and economic perspectives on the MENA region* (pp. 28–58). Information Science Reference.

Grinin, L., & Korotayev, A. (2016c). Revolyutsii kak osobaya stadiya razvitiya obshchestva i Arabskaya vesna. In L. Grinin, A. Korotayev, L. Issayev, & K. Meshcherina (Eds.), *Sistemniy monitoring global'nykh i regional'nykh riskov: Arabskaya vesna v global'nom kontekste* (pp. 157–190). Uchitel'.

Grinin, L. E., & Korotayev, A. V. (2023). Social evolution: Theoretical aspects and general outlines. *Social Evolution & History, 22*(2), 111–158. https://doi.org/10.30884/seh/2023.02.07

Grinin, L. E., Grinin, A. L., & Korotayev, A. (2017). Forthcoming Kondratieff Wave, cybernetic revolution, and global ageing. *Technological Forecasting & Social Change, 115*, 52–68.

Grinin, L., Grinin, A., & Korotayev, A. (2020). A quantitative analysis of worldwide long-term technology growth: From 40,000 BCE to the early 22nd century. *Technological Forecasting and Social Change, 155*, 119955. https://doi.org/10.1016/j.techfore.2020.119955

Grinin, L., Grinin, A., & Korotayev, A. (2021a). COVID-19 pandemic as a trigger for the acceleration of the cybernetic revolution, transition from E-government to E-state, and change in social relations. *Technological Forecasting & Social Change*. https://doi.org/10.1016/j.techfore.2021.121348

Grinin, L., Grinin, A., & Korotayev, A. (2021b). Global trends and forecasts of the 21st century. *World Futures, 77*(5), 335–370. https://doi.org/10.1080/02604027.2021.1949939

Grinin, L. E., Malkov, S. Y., Grinin, A. L., & Korotayev, A. V. (2022). Umret li kapitalizm? Razmyshleniya o kapitalizme proshlogo, nastoyashchego i budushchego. *Sotsiologicheskiy Zhurnal, 28*(2), 100–130.

Grinin, L., Grinin, A., & Korotayev, A. (2023b). Global aging and our futures. *World Futures, 79*(5), 536–556. https://doi.org/10.1080/02604027.2023.2204791

Grinin, L., & Grinin, A. L. (2015). *Ot rubil do nanorobotov. Mir na puti k epokhe samoupravlyaemykh sistem (istoriya tekhnologiy i opisaniye ikh budushchego)*. Moskovskaya redaktsiya izdatelstva 'Uchitel'.

Grinin, A.L., & Grinin, L.E. (2021a). Are we on the threshold of cyborgization? In L. E. Grinin, I. V. Ilyin, & A. V. Korotayev (Eds.), *Globalistics and globalization studies: Current and future trends in the big history perspective* (pp. 355–368). 'Uchitel' Publishing House.

Grinin, A., & Grinin, L. (2021c). Analyzing the global problems of the twenty-first century. A review and forecast based on the report to the club of Rome 'Come On!'. *Journal of Globalization Studies, 12*(2), 181–195. https://doi.org/10.30884/jogs/2021.02.10

Grinin, L., & Korotayev, A. (2016b). Revolution and democracy: Socio-political systems in the context of modernization. *Central European Journal of International and Security Studies, 10*(3), 118–139. http://www.cejiss.org/issue-detail/revolution-and-democracy-socio-political-systems-in-the-context-of-modernisation

Grinin, L., & Korotayev, A. (2020). Seven Weaknesses of the U.S., Donald Trump, and the future of American Hegemony. *World Futures*. https://doi.org/10.1080/02604027.2020.1801309

Grinin, L., Markov, A., & Korotayev, A. (2013). On similarities between biological and social evolutionary mechanisms: Mathematical modeling. *Cliodynamics, 4*(2), 185–228. https://doi.org/10.21237/C7clio4221334

Grinin, L. E., Grinin, A. L., & Korotayev, A. V. (2023a). Aging of the global population as an integral problem of the future. *Sotsiologicheskiy Zhurnal = Sociological Journal, 29*(2), 110–131. https://doi.org/10.19181/socjour.2023.29.2.6

Grinin, L. E. (2000). *Filosofiya, sotsiologiya i teoriya istorii. (Opyt filosofsko sotsiologicheskogo analiza nekotorykh obshchestvennykh zakonov i postroyeniya teorii vsemirno-istoricheskogo protsessa)*. Uchitel'.

Grinin, L. E. (2012). *Macrohistory and Globalization*. Uchitel.

Haas, M. L. (2015). Population aging and the future of the great powers. In J. A. Goldstone, L. Grinin, & A. Korotayev (Eds.), *History & mathematics: Demography & aging* (pp. 133–146). Uchitel.

Halacy, D. S., Jr. (1965). *Cyborg: Evolution of the superman*. Harper & Row.

Haldane, J. B. S. (1924). *Daedalus or science and the future*. EP Dutton.

Hansell, G. R. (2011). *H+/-: Transhumanism and its critics*. Xlibris Corporation.

Harney, A. (2018). China orders investigation after scientist claims first gene-edited babies. *Reuters*, sec. Healthcare & Pharma. https://www.reuters.com/article/us-health-china-babies-genes-idUSKCN1NV19T

Harper, S. (2006). Addressing the implications of global aging. *Journal of Population Research, 23*(2), 205–223.

He, W., Goodkind, D., & Kowal, P. (2016). *An aging world: 2015, International population reports. U.S.* Government Printing Office.

Hirose, M., & Ogawa, K. (2007). Honda humanoid robots development. *Philosophical Transactions of the Royal Society a: Mathematical, Physical and Engineering Sciences, 365*(1850), 11–19. https://doi.org/10.1098/rsta.2006.1917

Huebner, J. (2005). A possible declining trend for worldwide innovation. *Technological Forecasting and Social Change, 72*(8), 980–986. https://doi.org/10.1016/j.techfore.2005.01.003

Huxley, J. (2015). Transhumanism. *Ethics in Progress, 6*(1), 12–16.

Irshad, S. M. (2020). COVID-19 and global 'At-Risk Community': From benefit-sharing to risk-sharing of economic crisis. *Journal of Globalization Studies, 11*(2), 135–145.

Kaneda, T., Greenbaum, C., Patierno, K. (2016). *World population data sheet*. Population Reference Bureau. https://www.prb.org/2016-world-population-data-sheet/

Kekelidze, Z. (2013). V Rossii v pomoshchi psikhiatrov nuzhdayetsya 21 million chelovek. *Media Novosti*. December 11, 2013. http://medportal.ru/mednovosti/news/2013/12/11/178psiho/

Korotayev, A. (2005). A compact macromodel of world system evolution. *Journal of World-Systems Research, 11*(1), 79–93. https://doi.org/10.5195/jwsr.2005.401

Korotayev, A., & Malkov, A. (2016). A compact mathematical model of the World System economic and demographic growth, 1 CE—1973 CE. *International Journal of Mathematical Models and Methods in Applied Sciences, 10*, 200–209.

Korotayev, A., & Zinkina, J. (2017). Systemic boundary issues in the light of mathematical modeling of world-system evolution. *Journal of Globalization Studies, 8*(1), 78–96.

Korotayev, A. (2024). Complexity growth patterns in the Big History. Preliminary results of a quantitative analysis. *Journal of Big History, 7*(2), 9–27. https://doi.org/10.22339/jbh.v7i2.7203
Korotayev, A., Romanov, D., Zinkina, J., & Slav, M. (2023b). Urban youth and terrorism: A quantitative analysis (are youth bulges relevant anymore?). *Political Studies Review, 21*(3), 550–574. https://doi.org/10.1177/14789299221075908
Korotayev, A., Malkov, A., & Khaltourina, D. (2006). *Introduction to social macrodynamics. Compact macromodels of the World System growth.* KomKniga/URSS.
Korotayev, A., Sawyer, P., Gladyshev, M., Romanov, D., & Shishkina, A. (2021). Some sociodemographic factors of the intensity of anti-government demonstrations: Youth bulges, urbanization, and protests. *Sociologiceskoe Obozrenie = Russian Sociological Review, 20*(3), 98–128. https://doi.org/10.17323/1728-192x-2021-3-98-128
Korotayev, A., Butovskaya, M., Shulgin, S., & Zinkina, J. (2023). How can the global ageing affect the global value system? An evolutionary perspective. *Social Evolution & History, 22*(1), 57–76. https://doi.org/10.30884/seh/2023.01.03
Korotayev, A. (2008). Compact mathematical models of world system development, and how they can help us to clarify our understanding of globalization processes. In G. Modelski, T. Devezas, W. R. Thompson (Eds.), *Globalization as evolutionary process: Modeling global change* (pp. 133–160). Routledge. https://doi.org/10.4324/9780203937297
Korotayev, A. (2018). The 21st century singularity and its big history implications: A reanalysis. *Journal of Big History, 2*(3), 73–119. https://doi.org/10.22339/jbh.v2i3.2329
Korotayev, A. (2020a). How singular is the 21st century singularity? In A. Korotayev, & D. LePoire (Eds.), *The 21st century singularity and global futures. A big history perspective* (pp. 571–595). Springer. https://doi.org/10.1007/978-3-030-33730-8_26
Korotayev, A. (2020b). The 21st century singularity in the big history perspective. a reanalysis. In A. Korotayev, & D. LePoire (Eds.), *The 21st century singularity and global futures. A big history perspective* (pp. 19–75). Springer. https://doi.org/10.1007/978-3-030-33730-8_2
Kurzweil, R. (2005). *The singularity is near: When humans transcend biology.* Viking Penguin.
Kurzweil, R. (2001). The law of accelerating returns. *KurzweilAI.net,* July 3. http://www.kurzweilai.net/articles/art0134.html?printable=1
Kurzweil, R. (2004). The law of accelerating returns. In C. Teuscher (Ed.), *Alan turing. Life and legacy of a great thinker* (pp. 381–416). Springer.
Kurzweil, R. (2010). *The singularity is near.* Gerald Duckworth & Co.
LePoire, D. J. (2005). Application of logistic analysis to the history of physics. *Technological Forecasting and Social Change, 72*(4), 471–479. https://doi.org/10.1016/S0040-1625(03)00044-1
LePoire, D., & Korotayev, A. (2020). Conclusion. In A. Korotayev & LePoire, D. (Eds.), *The 21st century singularity and global futures. A big history perspective* (pp. 599–620). Springer. https://doi.org/10.1007/978-3-030-33730-8_27
LePoire, D. J. (2020). Exploring the singularity concept within big history. In A. Korotayev, & D. LePoire (Eds.), *The 21st century singularity and global futures. A big history perspective* (pp. 77–97). Springer. https://doi.org/10.1007/978-3-030-33730-8_3
Linstone, H. (2014). Review of: Singularity hypotheses: A scientific and philosophical assessment. In A. H. Eden, J. H. Moor, J. H. Søraker, & E. Steinhart (Eds.), Springer Verlag (2012). *Technological Forecasting and Social Change, 82*, 226–227. https://doi.org/10.1016/j.techfore.2013.06.011
Lovins, L. H., Wallis, S., Wijkman, A., & Fullerton, J. (2018). *A finer future. Creating an economy in service to life.* New Society Publishers.
Lutz, W., Goujon, A., Kc, S., Stonawski, M., & Stilianakis, N. (2018). *Demographic and human capital scenarios for the 21st century: 2018 assessment for 201 countries.* Publications Office of the European Union.
Maddison, A. (2007). *Contours of the world economy, 1–2030.* Oxford University Press.
Magee, C. L., & Devezas, T. C. (2011). How many singularities are near and how will they disrupt human history? *Technological Forecasting and Social Change, 78*(8), 1365–1378. https://doi.org/10.1016/j.techfore.2011.07.013

Meadows, D., Meadows, D., Randers, J., & Behrens, W.W. (1972). *The limits to growth*. Potomac Associates.
Meadows, D., Randers, J., & Meadows, D. (2004). *Limits to growth: The 30-year update*. Earthscan.
Mellars, P., & Stringer, C. (Eds.). (1989). *The human revolution: Behavioural and biological perspectives on the origins of modern humans*. Princeton University Press.
Modis, T. (2005). Discussion of Huebner article. *Technological Forecasting and Social Change, 72*, 987–1000.
Modis, T. (2006). The singularity myth. *Technological Forecasting & Social Change, 73*(2).
Modis, T. (2020). Forecasting the growth of complexity and change—an update. In A. V. Korotayev & D. LePoire (Eds.), *The 21st century singularity and global futures. A big history perspective* (pp. 101–104). Springer. https://doi.org/10.1007/978-3-030-33730-8_4
Moiseev, N. N. (2000). *Sudba tsivilizatsii. Put' razuma*. Yazyki russkoi literatury. Moscow.
More, M. (2013). The philosophy of transhumanism. *The Transhumanist Reader, 8*.
Moustaka, V., Zenonas, T., Vakali, A., Kounoudes, A., & Anthopoulos, L. G. (2019). Enhancing social networking in smart cities: Privacy and security borderlines. *Technological Forecasting and Social Change, 142*, 285–300. https://doi.org/10.1016/j.techfore.2018.10.026
Muehlhauser, L., & Salamon, A. (2012). *Singularity hypotheses, singularity hypotheses: A scientific and philosophical assessment*. https://doi.org/10.1007/978-3-642-32560-1
Murphy, S. V., & Atala, A. (2014). 3D bioprinting of tissues and organs. *Nature Biotechnology, 32*(8), 773–785. https://doi.org/10.1038/nbt.2958
Musk, E. et al. (2023). *Pause giant AI experiments: An open letter. We call on all AI labs to immediately pause for at least 6 months the training of AI systems more powerful than GPT-4*. https://futureoflife.org/open-letter/pause-giant-ai-experiments/
Nazaretyan, A. (2015). Megahistory and its mysterious singularity. *Herald of the Russian Academy of Sciences, 85*(4), 352–361. https://doi.org/10.1134/S1019331615040061
Nazaretyan, A. (2017). Mega-history and the twenty-first century singularity puzzle. *Social Evolution & History, 16*(1), 31–52.
Nazaretyan, A. (2018). The polyfurcation century: does the evolution on earth have a cosmological relevance? *Journal of Big History, 2*(1), 27–41. https://doi.org/10.22339/jbh.v2i1.2253
Nazaretyan, A. (2020). The 21st century's "mysterious singularity" in the light of the big history. In A. V. Korotayev & D. LePoire (Eds.), *The 21st century singularity and global futures. A big history perspective* (pp. 345–362). Springer. https://doi.org/10.1007/978-3-030-33730-8_15
Panov, A. (2005). Scaling law of the biological evolution and the hypothesis of the selfconsistent galaxy origin of life. *Advances in Space Research, 36*(2), 220–225. https://doi.org/10.1016/j.asr.200503.0.01
Panov, A., LePoire, D., & Korotayev, A. (2020). The 21st century singularity in the big history perspective. An overview. In A. V. Korotayev & D. LePoire (Eds.), *The 21st century singularity and global futures. A big history perspective* (pp. 1–18). Springer. https://doi.org/10.1007/978-3-030-33730-8_1
Panov, A. (2017). Singularity of evolution and post-singular development. In B. Rodrigue, L. Grinin & A. Korotayev (Eds.), *From big bang to galactic civilizations. A big history anthology. The ways that big history works: Cosmos, life, society and our future* (Vol. 3, pp. 370–402). Primus Books.
Panov, A.D. (2020). Singularity of evolution and post-singular development in the big history perspective. In A. V. Korotayev & D. LePoire (Eds.), *The 21st century singularity and global futures. A big history perspective* (pp. 439–465). Springer. https://doi.org/10.1007/978-3-030-33730-8_20
Park, D., & Shin, K. (2015). Impact of population aging on Asia's future growth. In J. A. Goldstone, L. Grinin, & A. Korotayev (Eds.), *History & mathematics: Political demography & global aging* (pp. 107–132). Uchitel.
Park, M., Leahey, E., & Funk, R. J. (2023). Papers and patents are becoming less disruptive over time. *Nature, 613*(7942), 138–144.
Peccoud, J. (2016). Synthetic biology: Fostering the cyber-biological revolution. *Synthetic Biology, 1*(1), ysw001.

Pestel, E. (1989). *Beyond the limit to growth*. The Club of Rome.
Peterson, P. G. (1999). Gray down: The global aging crisis. *Foreign Affairs, 78*(1), 42–55.
Pevtsov, A. (2015). Uchenyye: Bionicheskiye glaza budushchego smogut podklyuchat'sya k Wi-Fi i primenyat' fil'try. *RIA 'Svopi.ru'*. http://svopi.ru/nauka/39641-uchenye-bionicheskie-glaza-buduschego-smogut-podklyuchatsya-k-wi-fi-i-primenyat-filtry.html
Phillips, F. (2011). The state of technological and social change: Impressions. *Technological Forecasting and Social Change, 78*(6), 1072–1078. https://doi.org/10.1016/j.techfore.2011.03.020
Powell, J. L., & Khan, H. T. (2013). Aging and globalization: A global analysis. *Journal of Globalization Studies, 4*(1), 137–146.
Raftery, A. E., Li, N., Ševčíková, H., Gerland, P., & Heilig, G. K. (2012). Bayesian probabilistic population projections for all countries. *Proceedings of the National Academy of Sciences, 109*(35), 13915–13921. https://doi.org/10.1073/pnas.1211452109
Randers, J., Rockström, J., Stoknes, P.E., Golüke, U., Collste, D., & Cornell, S. (2018). *Transformation is feasible. A report to the club of Rome*. Stockholm Resilience Centre.
Randers, J. (2012). *2052. A report to the club of Rome*. Chelsea Green Publishing.
Rifkin, J. (2011). *The third industrial revolution. How lateral power is transforming energy, the economy, and the world*. Palgrave Macmillan.
Romanov, D. M., Meshcherina, K. V., & Korotayev, A. V. (2021). The share of youth in the total population as a factor of intensity of non-violent protests: A quantitative analysis. *Polis-Politicheskie Issledovaniya, 3*, 166–181. https://doi.org/10.17976/jpps/2021.03.11
Sadovnichy, V., Akaev, A., Ilyin, I., Malkov, S., Grinin, L., Sayamov, Y., & Korotayev, A. (2023), Introduction. Hoping for the future. In V. Sadovnichy et al. (eds.), *Re-considering the limits to growth. A report to the Russian association of the club of Rome*. Springer. https://doi.org/10.1007/978-3-031-34999-7_1
Sahlins, M. (1963). Poor man, rich man, big-man, chief: Political types in Melanesia and Polynesia. *Comparative Studies in Societies and History, 5*, 285–303.
Sahlins, M. (1972). *Social stratification in Polynesia*. University of Washington Press.
Sawyer, P., Romanov, D., Slav, M., & Korotayev, A. (2022). Urbanization, the youth, and protest: A cross-national analysis. *Cross-Cultural Research, 56*(2–3), 125–149. https://doi.org/10.1177/10693971211059762
Schalk, G., McFarland, D. J., Hinterberger, T., Birbaumer, N., & Wolpaw, J. R. (2004). BCI2000: A general-purpose brain-computer interface (BCI) system. *IEEE Transactions on Biomedical Engineering, 51*(6), 1034–1043. https://doi.org/10.1109/TBME.2004.827072.
Schneider, S. (2008). *Future minds: Transhumanism, cognitive enhancement and the nature of persons*. http://repository.upenn.edu/neuroethics_pubs/37
Schwab, K. (2016). *The fourth industrial revolution*. World Economic Forum.
Schwab, K., & Malleret, T. (2020). *COVID-19: The great reset*. Forum Publishing. http://reparti.free.fr/schwab2020.pdf
Schwartz, P. M. (1999). Internet privacy and the state. *Connecticut Law Review, 32*(2000), 815.
Scliaff, A., & Friedrichs, G. (1982). *Microelectronics and society: for better or for worse. A report to the club of Rome*. Pergamon Press.
Service, E. R. (1971). *Primitive social organization*, 2nd edn. Random House.
Service, E. R. (1975). *Origins of the state and civilization. The process of cultural evolution*. Norton.
Shanahan, M. (2015). *The technological singularity*. MIT Press.
Solove, D. J. (2008). *Understanding privacy*. Harvard University Press.
Swaab, D. (2013). *Wit Zijn Ons Brein Van baarmoeder tot Alzheimer*. Uitgeverij Atlas Contact.
Teilhard de Chardin, P. (1987). *The phenomenon of man*. Harper Perennial Modern Classics.
Teulings, C., & Baldwin, R. (Eds.) (2014). *Secular stagnation: Facts, causes, and cures*. CEPR.
Toffler, A. (1970). *Future shock*. Bodley Head.
Tsiolkovsky, K. E. (1964). *Zhizn' v mezhzvozdnoy srede*. Nauka.
Tsirel, S. V. (2014). O tom, chto budet posle tochki singulyarnosti. In L. E. Grinin, A. V. Korotayev, & A. V. Markov (Eds.), *Evolyutsiya: Ot protozvezd k singulyarnosti?* (pp. 340–381). Uchitel.

Ulam, S. (1958). Tribute to John von Neumann. *Bulletin of the American Mathematical Society, 64*(3), 1–49.
UN—United Nations. (2013). *World population aging*. United Nations. https://www.un.org/en/development/desa/population/publications/pdf/aging/WorldPopulationAging2013.pdf
UN Population Division. (2019). *World population prospects 2019*. United Nations: UN Population Division.
UN Population Division. (2022a). *United Nations population division database*. United Nations. http://www.un.org/esa/population
UN Population Division. (2022b). *World population prospects 2022*. United Nations.
UNFPA. (2012). *Aging in the twenty-first century: A celebration and a challenge*. UNFPA.
Ustyuzhanin, V, Zhodzishskaya, P., & Korotayev, A. (2022). Demographic factors as predictors of revolutionary situations: Experience in quantitative analysis. *Sotsiologicheskiy Zhurnal = Sociological Journal, 28*(4), 34–59. https://doi.org/10.19181/socjour.2022.28.4.9314
Ustyzhanin, V. V., Zinkina, J. V., & Korotayev, A. V. (2023). Dangerous youth: Why do political uprisings take an unarmed form? *Sotsiologicheskie issledovaniya [Sociological Studies], 5*, 82–96. https://doi.org/10.31857/S013216250025805-8
Vanderwerker, E. (1976). A brief review of the history of amputations and prostheses. *Inter Clinic Information Bulletin, 15*(1976), 25.
Veatch, R. M., & Guidry-Grimes, L. K. (2019). *The basics of bioethics*. Routledge.
Vernadsky, V. (1997). *Scientific thought as a planetary phenomenon*. Nongovernmental Ecological V.I'.Vernadsky Foundation.
Vincent, J. (2023). *Elon musk and top AI researchers call for pause on 'giant AI experiments'*. https://www.theverge.com/2023/3/29/23661374/elon-musk-ai-researchers-pause-research-open-letter
Vollset, S. E., Goren, E., Yuan, C.-W., Cao, J., Smith, A. E., Hsiao, T., Murray, C. J. L., et al. (2020). Fertility, mortality, migration, and population scenarios for 195 countries and territories from 2017 to 2100: A forecasting analysis for the Global Burden of Disease Study. *The Lancet, 396*(10258), 1285–1306. https://doi.org/10.1016/S0140-6736(20)30677-2
von Foerster, H., Mora, P. M., & Amiot, L. W. (1960). Doomsday: Friday, 13 November, AD 2026. *Science, 132*(3436), 1291–1295. https://doi.org/10.1126/science.132.3436.1291
von Weizsäcker, E. U. & Wijkman, A. (2018). *Come on! capitalism, short-termism, population and the destruction of the planet. A report to the club of Rome*. Springer. https://doi.org/10.1007/978-1-4939-7419-1
WHO—World Health Organization. (2021). *Ageism is a global challenge*. https://www.who.int/news/item/18-03-2021-ageism-is-a-global-challenge-un
Widdowson, M. (2021). From Covid-19 to zero-gravity: Complex crises and production revolutions. *Journal of Globalization Studies, 12*(1), 117–144.
Wijkman, A., & Rockström, J. (2012). *Bankrupting nature*. Routledge.
Wittgenstein Centre. (2022). *Wittgenstein centre human capital data explorer*. http://dataexplorer.wittgensteincentre.org/wcde-v2/
Woodburn, J. C. (1980). Hunters and gatherers today and reconstruction of the past. In E. Gellner (Ed.), *Soviet and western anthropology* (pp. 95–117). Duckworth.
Woodburn, J. C. (1988). Some connections between property, power, and ideology. In T. Ingold, D. Riches, & J. Woodburn (Eds.), *Hunters and gatherers. Property, power and ideology* (Vol. 2, pp. 10–31). Berg Publishers.
Yampolskiy, R. V., & Fox, J. (2012). Artificial general intelligence and the human mental model. In A. H. Eden, J. H. Moor, J. H. Søraker, & E. Steinhart (Eds.), *Singularity hypotheses: A scientific and philosophical assessment* (pp. 129–145). Springer. https://doi.org/10.1007/978-3-642-32560-1_7
Yeoman, I., & Mars, M. (2012). Robots, men and sex tourism. *Futures, 44*(4), 365–371. https://doi.org/10.1016/j.futures.2011.11.004
Zimmer, Z. (2016). *Global ageing in the twenty-first century: Challenges*. Routledge.

# Part I
# Technological Transformations in the Past, Present and Future

Summary of Part I. In this part the authors describe the main technological changes in history. The description covers almost the entire period of the historical process starting from the emergence of Homo sapiens up to the end of the twenty-first century. We do not point out every considerable invention or separate innovation in detail, instead we make an attempt to show the general process of changes and to explain how and why technological epochs succeeded each other. We pay the greatest attention to the main technological breakthroughs, both to those already completed and to the future ones, as well as to the transformations in society and social consciousness that these transformations have already brought about and will also cause in the future.

According to our theory, the historical process can be subdivided more effectively into four major stages on the basis of changes in the major qualitative developmental stages of the world's technological levels. Because the transition from any of these stages to another means a change of all basic characteristics of the respective system, we named these stages production principles. Four production principles are: 1. Hunter-Gatherer; 2. Craft-Agrarian; 3. Trade-Industrial; 4. Scientific-Cybernetic. The change in production principles is connected with production (or technological) revolutions. Among all major technological breakthroughs in history the most important are the three production revolutions: (1) the Agrarian Revolution (10–3 millennia BCE); (2) the Industrial Revolution (sixteenth to nineteenth centuries) and 3) the Cybernetic Revolution (1950–2070).

In this part based on the studies of the whole previous technological development we try to outline the direction and logic for further technological progress. So, the main goal of the authors is to analyze in detail the final phase of the Cybernetic Revolution (2030–2070) and the period after it (up to the early twenty-second century) and formulate forecasts for the next 30–80 years. In the chapters devoted to current and future changes we will pay considerable attention to the analysis of what will manifest in technology and to what we can expect from it.

We also study the development of those directions which in our opinion are most likely to trigger the future technological breakthrough, namely, medicine, additive (3D printing), nano-, bio-, robotics, information (including AI) and cognitive technologies which we unite under the designation of MANBRIC technologies. We also discuss the forthcoming Sixth Kondratieff wave.

The authors' goals are to discuss and deeply analyze the following questions: What is the Cybernetic Revolution and why is it Cybernetic? What are self-regulating and self-managing systems? How do they correspond with AI? How can they influence our future life? What are the characteristics of the Cybernetic Revolution? What can we expect from technological development in the next half of this century? And so on.

Of utmost importance is the fact that in the Cybernetic Revolution self-regulating systems will become the major part of the technological process. Self-regulating and self-managing systems are systems which can operate independently in a wide range of situations, choosing optimum regimes in the context of certain goals and tasks. These are systems that operate with minimal to no human intervention. In terms of the development of self-regulating systems with their growing ability to select optimum modes of functioning as well as of other characteristics of the Cybernetic Revolution (resources and energy saving, miniaturization, individualization, etc.). We discuss self-regulating systems in detail, along with examples of these systems, current and future implementation, as well as their main characteristics, etc. A wide implementation of self-regulation in various systems will be the most important characteristic of the final phase of the Cybernetic Revolution.

Then we also discuss the logic of the Cybernetic Revolution to explain why a new technological wave will happen in the 2030s and 2040s, as well as explain why medicine is to become the breakthrough sphere in the beginning of the final phase of the Cybernetic Revolution. We also estimate the chances of the initial breakthroughs in robotics, driverless vehicles and green energy.

So, we foresee great changes in the future both in technology and in the socio-demographic structure of society as well as in the whole mode of our life.

# 2 Global Technological Transformations Since the Stone Age: Theory and History

**Abstract**

In this chapter Grinin et al. describe the main technological changes within the historical process, show the general process of changes and to explain how and why technological epochs succeeded each other. The chapter introduces two interconnected theories: the ones of Production (or technological) Revolutions and of Production Principles. The authors single out four production principles: 1. Hunter-Gatherer; 2. Craft-Agrarian; 3. Trade-Industrial; 4. Scientific-Cybernetic. The change in production principles is connected with production revolutions. Among all major technological breakthroughs in history the most important are the three production revolutions: 1) the Agrarian Revolution (10–3 millennia BCE); 2) the Industrial Revolution (the 16th–19th centuries) and 3) the Cybernetic Revolution (1950 – up to 2070). The first phase of the Cybernetic Revolution took place between the 1950s and mid-1990s with a vigorous development of information technologies. The final phase of the Cybernetic Revolution may begin approximately between the 2030s and 2040s and will finish in 2060–2070. It will usher the beginning of the epoch of 'self-regulating systems', which can be working without human control. The authors also offer some forecasts about the periods after the end of the Cybernetic Revolution up to the end of the 21st century.

## 2.1 Introduction

**What this Chapter is about**

In this chapter we describe the main technological changes.

The first part of the Chapter is dedicated to the theoretical frame of the technological aspect of the historical process.

 
L. Grinin et al., *Cybernetic Revolution and Global Aging*, World-Systems Evolution and Global Futures, https://doi.org/10.1007/978-3-031-56764-3_2

The second and the greater part of this chapter is dedicated to the factual description of the technological history of humankind in connection with the theory. The description covers almost the entire period of social evolution starting from the emergence of Homo sapiens up to the mid-twentieth century when the Cybernetic Revolution began. We do not describe every considerable invention or separate innovations in detail; instead we make an attempt to show the general process of changes and to explain how and why technological epochs succeeded each other.

The chapter introduces two interconnected theories: the ones of Production (or Technological) Revolutions and of Production Principles. This is new explanatory paradigms which are of value when analyzing causes and trends of global shifts in the historical process. The chapter describes the course of technological transformations in history and demonstrates a possible application of the theory to explain the present and forthcoming technological changes.

We single out four production principles: 1. Hunter-Gatherer; 2. Craft-Agrarian; 3. Trade-Industrial; 4. Scientific-Cybernetic. The change in production principles is connected with production revolutions. Among all major technological breakthroughs in history the most important are the three production revolutions: 1) the Agrarian Revolution (10–3 millennia BCE); 2) the Industrial Revolution (the 16th–19th centuries) and 3) the Cybernetic Revolution (1950 – up to 2070).

The Agrarian Revolution comprised the transition of humankind from foraging (hunting and gathering) to food production (agriculture and animal husbandry), which led to the formation of an agrarian society with its emerging of states and civilizations, wealth accumulation and estate structure, large population etc.

The Industrial Revolution was based on the replacement of manual labor with machine labor, which led to the formation of an industrial society with its internal dynamism, instability, and focus on constant change.

The Cybernetic Revolution – currently, the invention of high-speed computing devices and AI, their introduction into all spheres of life and, on their basis, the gradual replacement of intellectual labor by machine labor, and in the final phase, the widespread use of self-regulating and self-managing systems, which can act without human control.

We show important recurrences in technological revolutions, which give us a starting point for projections for the coming decades, i.e. the ability to forecast the course of the Cybernetic Revolution. There are three phases in it.

The first phase of the Cybernetic Revolution took place between the 1950s and mid-1990s, when a vigorous development of information technologies and the start of real economic globalization were observed. In this period this great technological breakthrough sometimes was called the 'scientific-technical' revolution. We regard that it would be more appropriate to call it the Cybernetic Revolution after the science Cybernetics which can be defined as a study of general laws of receiving, storing, and processing information in complex systems.

The second phase began in the mid-1990s in conjunction with the development and wide diffusion of user-friendly computers, communication technologies, cell phones and so on. This phase has been going on up to the present.

The final phase of the Cybernetic Revolution may begin approximately between the 2030s and 2040s. It will usher the beginning of the epoch of 'self-regulating systems', which can be working without human control.

We also offer some forecasts about the periods after the end of the Cybernetic Revolution, that is about the mature phases of the Scientific-Cybernetic Production Principle up to the end of the 21st century.

Over millennia, humans have improved their tools as well as developed technologies, storage techniques, exchange practice and transportation means. In the history of technology, especially during the last two centuries, there have occurred many significant breakthroughs while production has been modernized over and over again. Nowadays, we constantly learn about new achievements in engineering, communications, banking, and technologies. But for the most part of human history, the matter was different. For centuries and even millennia, the transformations would pass slowly and that is why were almost undistinguished.[1] Many technologies would appear rather conservative. However, even for the ancient epochs technological changes were among the most fundamental drivers of development and increasing complexity of societies, demographic growth and cultural progress.

On this large scale it becomes especially evident that, using Fernand Braudel's words (1985), in reality, everything rested upon the very broad back of material life; when material life expanded, everything moved ahead. That is why the identification of the greatest technological revolutions also allows a general periodization of the historical process. In the course of time, such transformations became powerful, multidimensional, and sometimes even revolutionary. However, a relevant dominance of technology in historical processes can be recognized only within very large time spans and strict limits. Meanwhile, in human history, one can distinguish the three most powerful production (or technological) revolutions, the most dramatic and transformative breakthroughs, which are the thresholds of the respective four production principles or technological epochs. These revolutions are: (1) the Agrarian Revolution[2]; (2) the Industrial Revolution; and (3) the Cybernetic Revolution. In what follows we will discuss them in detail.

[1] On this see, e.g., Anuchin (1923), Lurie et al., (1939), Semenov (1968), Chernousov et al., (2005), Belkind et al., (1956); see also: Boas (1911), Kosven (1953), Kremkova (1936), Osipov (1959), Virginsky and Khoteenko (1993), Bunch and Hellemans (2004), Sheypak (2009), Grinin and Grinin (2015a, 2015b), Grinin and Korotayev (2015), Bernal (1965), Blockmans (1989), Braudel (1973), Boserup (1965), Carus-Wilson (1987), Childe (1952), Clark (1966), Clarkson (1971), Cohen (1977), Deng (1999), Fourastie (1958), Frank and Gills (1993), Gamble (1986), Wymer (1982), Cipolla (1975).

[2] The Neolithic breakthrough (revolution) can be regarded only as its initial phase.

## 2.2 The Production Principles and Production Revolutions

### 2.2.1 Periodization of the Historical Process

According to our theory, the historical process can be subdivided more effectively into four major stages. The transition from any of these stages to another means a change of all basic characteristics of the respective societal system. However, in addition to this principal basis of periodization (that determines the number of periods and their characteristics), we need an additional basis that will help us to elaborate a concise chronology.

As such additional foundation, we have proposed the notion of *production principle*,[3] describing the major qualitative developmental stages of the world's productive forces. Each production principle means a transition within the framework of the historical process to completely new, more productive systems of economy, which eventually restructure the whole range of economic activities and respective relations. In other words, we speak not only about new technologies and economic management but also about a (fundamentally) complete and worldwide change of modes of activity.

Though the qualitative transformations in some spheres of life are closely connected with changes in others (and, thus, no factors can be considered absolutely dominant), some spheres can be considered more significant with respect to their influence; therefore, changes within them are more likely to affect other spheres than the other way round.[4] Production principles apply to such spheres due to a number of reasons (see Grinin & Grinin, for details), one of them is that production technologies are implemented by society at large (and what is especially important, by the lower social strata), whereas culture, politics, law and even religion are systems developed by some of their participants (usually the elites).

Thus, we single out four production principles:

1. Hunter-Gatherer;
2. Craft-Agrarian;
3. Trade-Industrial;
4. Scientific-Cybernetic.

Every production principle can be presented as a peculiar developmental cycle consisting of six phases (for more details see below and in Chap. 12). The change in production principles is connected with production revolutions (= technological

[3] E.g., Grinin (2006b, 2007a, 2007b, 2012: Chap. 1, 2013a), Grinin and Grinin (2013a; 2015a; 2015b), Grinin et al. (2017b, 2020b, 2021), Grinin and Korotayev (2015), Grinin and Grinin (2023).

[4] Of course, we do not mean continuous and regular influence but rather a qualitative breakthrough. If after a breakthrough within a more fundamental sphere other spheres do not catch up with it, the development within the former slows down.

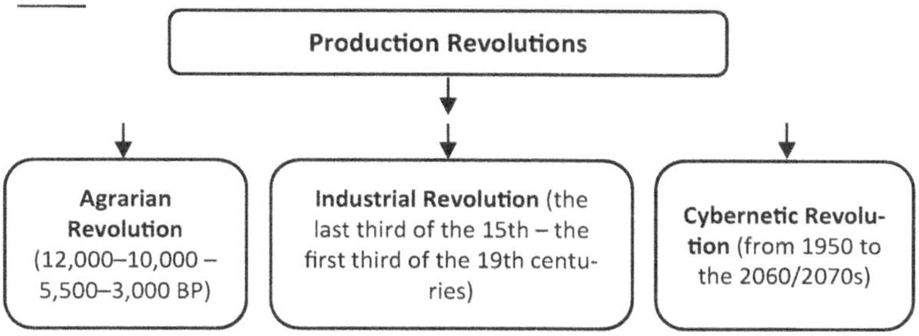


**Fig. 2.1** Production revolutions in the historical process

revolutions, so we will use these terms as synonyms). The starting point of such revolutions can be regarded as a convenient and natural point to establish the chronology of changing patterns.

Below we suggest a model of *periodization* of the historical process based on our theory of production principles. Its objective is to define a scale for measuring human development (or at least for the evolution of the World System[5]), to mark opportunities for intersocietal comparison as well as investigate a connection between technological and other types of development. In particular, we will dwell in detail on how the production revolutions and production principles, on the one hand, and major demographic transformations and types of population reproduction, on the other hand, are connected (see Chap. 5).

### 2.2.2 Production Revolutions

**Three production revolutions**. Let us emphasize again: among large technological breakthroughs in history, the most important are the three production (or technological) revolutions: (1) the Agrarian Revolution; (2) the Industrial Revolution, and (3) the Cybernetic Revolution. From our point of view, each revolution initiates a new stage of development of the world productive forces as well as in the final analysis a transition to a new stage of the historical process (see Fig. 2.1).

Each production revolution is a result of a long accumulation of quantitative and qualitative changes that eventually bring a great evolutionary breakthrough. Each one provokes an increasing complexity of the social division of labor and societal structure as well as the integration of humanity.

[5] Interpreted in the manner of Frank (1990, 1993), Frank and Gills (1993), Korotayev (2005, 2007), Korotayev et al., (2006a), Korotayev and Grinin (2006, 2013), Grinin and Korotayev (2006, 2009, 2012, 2014, 2015, 2017, 2018), Korotayev and Malkov (2016), Grinin et al. (2016, 2019), Grinin (2017).

1. *The Agrarian Revolution* (12,000/10,000–5,500 BP) was *a great transition from foraging subsistence pattern, that is from getting food by collecting what is available in nature (through hunting, gathering, and fishing) to food production/ farming.*
2. *The Industrial Revolution* (the last third of the fifteenth–the first third of the nineteenth centuries) was *a stupendous transition from the Craft-Agrarian Production Principle to a new pattern* which implied that the main production was concentrated in industry and performed by machines and mechanisms.
3. *The Cybernetic Revolution* is *a dramatic transition from the Trade-Industrial Production Principle to the production and service sector based on the implementation of self-regulating and self-managing systems.* This revolution started in the 1950s and 1960s and brought the development of powerful information technologies. Between the 2030s and 2070s, the final phase of this revolution is likely to unfold which will dramatically increase the opportunities for control over not only technical, but also some useful biological, ecological and even social systems which will be transformed into smart and independent working self-regulating and self-managing systems. Due to dramatic breakthroughs in medicine, there will appear opportunities to significantly increase life expectancy and expand the range of possible modifications of human biological nature (see discussion on this in Chap. 1; see also Chaps. 8 and 10).

**Production Revolutions: A Summary of Previous Research**. The above-mentioned technological thresholds in the history of societies have long attracted the attention of the academic community. The Industrial Revolution became the object of extensive research in the nineteenth and early twentieth centuries, both within the Marxist framework and within the non-Marxist paradigms (see, e.g., Engels, 1955 [1845]; Marx, 1960 [1867]; Plekhanov, 1956 [1895]; Labriola, 1986 [1896]; Toynbee, 1927 [1884], 1956 [1884]; Mantoux, 1929). The first ideas on the Agrarian revolution were introduced by Gordon Childe in the 1930s and between the 1940s and 1950s he developed the theory of the Neolithic revolution (Childe, 1934, 1944, 1948). From the 1940s, there was observed an increasing interest in the analysis of the impact of production on the historical development and historical process in general; meanwhile, rapid technological development received both optimistic and pessimistic assessments from philosophers and social thinkers. The interest became even more acute after it was perceived that the world had entered the Cybernetic Revolution.[6] It is not surprising that in the 1960s and 1980s the increasing interest in production revolutions found its expression in numerous works, including the publications of postindustrial theorists such as Bell (1973, 1978, 1990), Toffler (1980, 1985, 1990; Toffler & Toffler, 1995), Stonier (1983), Touraine (1974, 1983), Kahn (1982), and to a lesser extent other scholars (Brzezinski, 1970; Drucker, 1995, 1996; Thurow, 1996; see also Dizard, 1982; Martin,

[6] In the 1950s and 1980s it was denoted by different terms (see further).

1981; Castells, 1996), not to mention the philosophers of technology (Ellul, 1964, 1975, 1982, 1984; Mumford, 1966; etc.; see also Inozemtsev, 1999).

Much has been written about each of the three production revolutions (see, e.g., Allen, 2009, 2011; Bellwood, 2004; Benson & Lloyd, 1983; Bernal, 1965; Cauvin, 2000; Cipolla, 1976; Clark, 2007; Cohen, 1977; Cowan & Watson, 1992; Dietz, 1927; Goldstone, 2009a; Harris & Hillman, 1989; Henderson, 1961; Huang, 2002; Ingold, 1980; Knowles, 1937; Lieberman, 1972; Mokyr, 1985, 1990, 1993, 1999, 2010; Mokyr & Voth, 2010; More, 2000; North, 1981; Philipson, 1962; Phyllys, 1965; Pomeranz, 2000; Reed, 1977; Rindos, 1984; Sabo, 1979; Shnirelman, 1989, 2012a, 2012b; Smith, 1976; Stearns, 1993, 1998; Sylvester & Klotz, 1983).

Recently, there have appeared a number of works and books on the coming new revolution which different scholars define in different terms while we denote it as the Cybernetic Revolution. For example, much has been written on biotechnological revolution (Sylvester & Klotz, 1983; Fukuyama, 2002; Nefiodow, 1996; Avise, 2004; Das, 2007; Harris, 2010; Holdrege, 2010; McCabe & McCabe, 2008; Phillips & Su, 2009; Peccoud, 2016; see also Nefiodow & Nefiodow, 2014a, 2014b; see also Chap. 9). It is also named as "the green revolution" (e.g., Fücks, 2013). If numbers are used to denote revolutions then it is called the third industrial revolution (Rifkin, 2011),[7] but currently, following Klaus Schwab, it is more often numbered the fourth one (Schwab, 2016). Then an idea of the fifth industrial revolution appeared (Bröring, 2021; Peccoud, 2016).[8] To be honest, we consider these enumerations of industrial revolutions to be insufficiently scientific, since the number of revolutions does not reveal the basic theory explaining at least what an industrial revolution is.[9] We believe that creates a reason for the discord.[10] There

---

[7] The first, who discussed the third industrial revolution was Daniel Bell (1989 [see also Duff, 1998; Spassov, 2003]). However, he regarded it as an information revolution.

[8] On the basis of the so-called cyber-biological systems (see about them in Chap. 9).

[9] It is another matter to use the concept of the First Industrial Revolution to show that the first such revolution occurred in the UK, as Phyllys Deane does, also through naming his book *The First Industrial Revolution* (Deane, 1965).

[10] Thus, Rifkin considers the first industrial revolution the period of the eighteenth and nineteenth centuries that actually completed it, while the second one is the scientific and technological revolution or, in our terminology, the initial phase of the Cybernetic Revolution and the third, respectively, is the future revolution (the final phase of the Cybernetic Revolution) that starts today. Between the first and second revolutions of Rifkin's concept, Schwab adds another second industrial revolution of the late nineteenth and early twentieth centuries (associated with the transition to electric motors and internal combustion engines). Such a list of industrial revolutions comes from researchers of the Industrial Revolution itself, for example, Mokyr (2002). Mokyr defines three industrial revolutions prior to the coming revolution (which he completely ignores), only he dates the second to 1856 linking it to the invention of the Bessemer process of steelmaking. The Third Revolution has rather vague dating (probably, from 1914), and Mokyr himself poses the question in the headline of the respective chapter "Third Industrial Revolution?" (of course, in no way can we agree that the Third Industrial Revolution started in 1814). Mokyr actually defines revolutions as a new technological mode, which we will speak about in Chap. 4. But the mode is only one of the phases of the Industrial Revolution, which comprises three modes (Grinin, 2020; Grinin & Grinin, 2014, 2016, 2022).

are also differences in defining the leading directions of development of the coming revolution. Some researchers (e.g., Fücks, 2013; Rifkin, 2011) assume it is a new energy; Schwab (2016) believes there will be a complex of leading spheres, but primarily the AI (it is not original, see, e.g., Kurzweil, 2001, 2005). In the following chapters, we will discuss other ideas.

However, there is another issue that worries us even more than the superficial character of some concepts. There are a surprisingly small number of studies concerning these revolutions as recurrent phenomena, each representing an extremely important landmark in the history of humankind. We have developed a theory of production (technological) revolutions within the framework of the general theory of the world historical process (Grinin, 2007a, 2007b, 2012, 2019; Grinin & Grinin, 2013b, 2015a, 2015b, 2021; Grinin & Korotayev, 2015; Grinin et al., 2017b, 2020a, 2021).

**What Is a Production/Technological Revolution? What Are Its Characteristics and Phases?** *The production revolution can be defined as a radical breakthrough in the world's productive forces, connected with the transition to a new principle of economic management not only of technologies but also in the relation between society and nature. The difference between a production revolution and various technical breakthroughs is that it goes beyond particular essential branches, and involves the economy in general. And finally, the new trends of economic management become dominant. Such an overturn introduces some fundamentally new renewable or for a long time inexhaustible resources in the economical circulation, and the use of those resources diffuses widely enough within most territories. The labor (or/and land) productivity increases by orders of magnitude, which is also manifested in the creation of several orders greater volume of production and in the demographic revolution (or in the change in the type of population reproduction, TPR, see Chap. 5).*[11]

As a result, the most powerful impetus for the qualitative reorganization of the whole societal structure is generated. Although the production revolution originates geographically in one or a few places, in the long run, its beginning means a global turn of the world's productive forces and represents a long-term process that gradually involves more and more societies and territories. As a result a) the involved societies become substantially more developed in technological, economic, demographic, cultural and often military terms; b) the involvement of societies into a new production system becomes standard.

The production revolutions also bring:

1. Development of fundamentally new resources; 2. A vigorous growth of production output and population; 3. Substantial and comprehensive complications of society. Each production revolution implies a transition to a fundamentally new

[11] For the Agrarian Revolution, it was not so much by the growth of labor productivity that played a very important role, but the growth of the land productivity/carrying capacity (or the yield of useful products per unit of land) (see, e.g., Korotayev et al., 2006b).

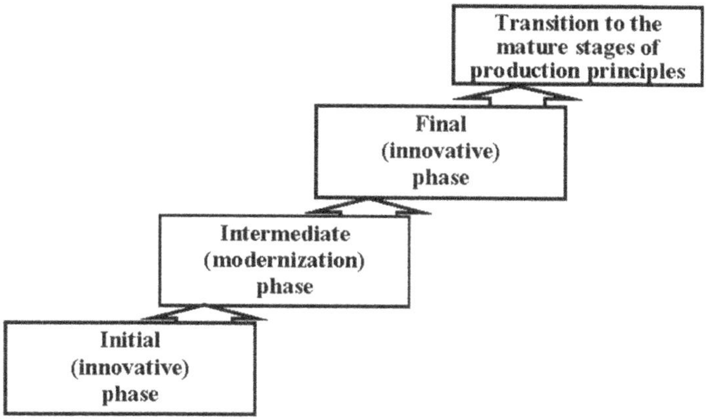


**Fig. 2.2** A cycle of a production revolution (phases and characteristics)

production system; the beginning of each production revolution marks thresholds between corresponding production principles.

It is obvious that each production revolution is unique and has absolutely peculiar characteristics. But at the same time, there are similarities in their development that allow the creation a model of production revolution as a global and recurrent phenomenon. Within the proposed theory we suggest a fundamentally new idea that each production revolution also has a common internal cycle consisting of three phases: two innovative and between them one modernization phase. Thus, the cycle of each production revolution looks as follows: *the initial innovative phase* (the emergence of a new revolutionary productive sector)—*the modernization phase* (diffusion, synthesis, and improvement of new technologies)—*the final innovative phase* (improving the potentials of new technologies up to the mature characteristics). See also Fig. 2.2.

So, at *the initial innovative phase* a new revolutionary productive sector emerges. The primary system for a new production principle is developed and for a long time, it co-exists with former technologies. *The modernization phase* is a long period of diffusion and development of innovations. It is a period of progressive innovations when conditions gradually emerge for the final innovative breakthrough. At *the final innovative phase,* a new wave of innovations dramatically expands and improves opportunities for a new production principle, which thus, gains full strength. The scheme and chronology of phases of each production revolution are presented in Tables 2.1 and 2.2 and Fig. 2.3.

*As the final phase of the production revolution unfolds, the 'essence' of the production principle, its opportunities and limitations are revealed*, as well as the geographical borders of its expansion.

Both innovative phases of a production revolution represent a breakthrough in production. During the initial innovative phase the seedbeds of the new production principle are formed; sectors concentrating the principally new production

**Table 2.1** Phases of the agrarian revolution

| Phase | Type | Name | Dates | Changes |
|---|---|---|---|---|
| Initial | Innovative | Manual farming | 12,000–9,000 BP | Transition to primitive manual (hoe) agriculture and animal husbandry |
| Intermediate | Modernization | Diffusion of agriculture | 9,000–5,500 BP | Emergence of new domesticated plants and animals, development of complex agriculture, emergence of a complete set of agricultural instruments |
| Final | Innovative | Irrigated and plough agriculture | 5,500–3,500 BP | Transition to irrigated or non-irrigated plow agriculture |

**Table 2.2** The phases of the industrial revolution

| Phase | Type | Name of the phase | Dates | Changes |
|---|---|---|---|---|
| Initial | Innovative | Manufacturing | 15th–16th centuries | Development of shipping, technology and mechanization on the basis of water engine, development of manufacture based on the division of labor and mechanization |
| Intermediate | Modernization | Primary industry | The 17th–early 18th centuries | Formation of a complex industrial sector and capitalist economy, increasing mechanization and division of labor |
| Final | Innovative | Machinery | 1730–1830s | Formation of sectors with the machine production cycle with steam energy |

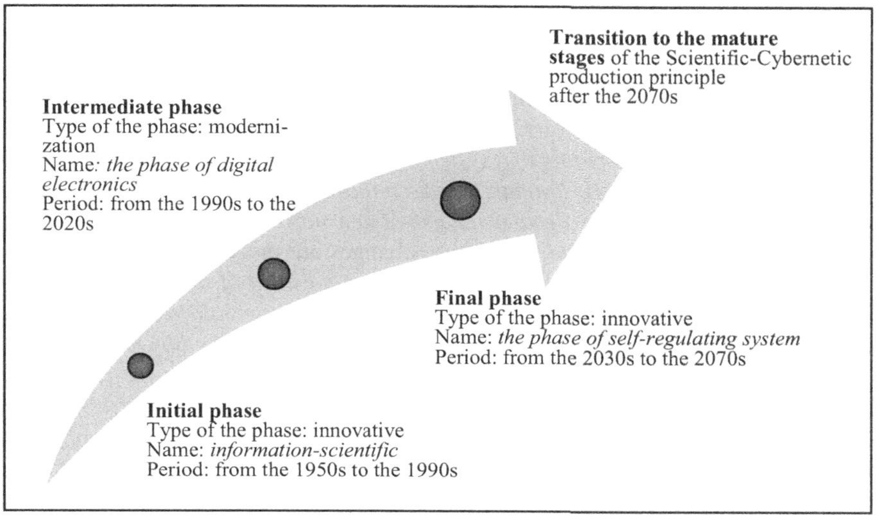


**Fig. 2.3** Phases of the cybernetic revolution

elements gain strength. Then, during the modernization phase, qualitatively new elements diffuse to other societies and territories. In countries with the most promising production and adequate social conditions, the transition to the last innovative phase of the production revolution occurs, marking the flourishing of the new production principle. So, the underdeveloped societies begin to catch up with the production revolution, actively integrating into it. We identify certain regularities in the phases of production revolutions. These regularities imply that within every production revolution, each of the three phases plays functionally the same role. Besides, we revealed an important ratio between the duration of phases which we found to *remain approximately the same within the framework of each cycle* (see Chaps. 4 and 12 for more details). This ratio allows for defining some regularities which can be employed to make predictions. In Chap. 4 we will add some important details about all three phases of the production revolution and the correlation between them (see also Chap. 12). Also, we will show how on the basis of the regularities identified within the Agrarian and Industrial Revolutions, we will make assumptions about the following points:

the duration of the intermediate (modernization) phase of the Cybernetic Revolution;
the beginning and approximate duration of the final phase of the revolution;
the sectors and directions that will be affected by the new technological breakthrough.

Further, we briefly describe the major characteristics of every production revolution and the general schemes of their three phases in Tables and Figures.

*The Agrarian Revolution (lasted from 12,000/10,000 to 5,500 BP) marked a dramatic breakthrough from the Hunter-Gatherer Production Principle to farming. Its initial innovative phase* was the transition to primitive hoe agriculture and animal husbandry (12,000–9,000 BP) while the final phase brought the transition to intensive agriculture (especially to irrigation [5300–3700 BP] or non-irrigation plough one). *It resulted in the transition to a systematic production of food, using a new type of energy (power of domestic animals), and, on this basis, in the transition to a complex social division of labor. Besides the use of new power sources (animal power), we witness the use of new materials.* These changes are presented in Table 2.1.

*The Industrial Revolution (lasted from the second third of the fifteenth century to the first third of the nineteenth centuries) constituted a great breakthrough from farming and manual craft production to machine industry, marked by an intentional search for and use of scientific and technological innovations in the production process. The Industrial Revolution was a process of active development of technology, especially designed to save labor in different areas, the most important of which was the mechanization of manual labor in manufacturing, which led to a shift from manual labor to machine labor and replacement of biological energy with abiotic energy (water first, and then steam).*

Overall, its *initial phase* started in the fifteenth and sixteenth centuries with rapid development of seafaring, trade, mechanization on the basis of water engines, the growing division of labor (Durkheim, 1997 [1893]; see also Xie & Cheng, 2016) and other processes. The *final phase* was the industrial breakthrough in the eighteenth century and the first third of the nineteenth century, associated with the introduction of various machines and steam energy. These changes are presented in Table 2.2.

*The Cybernetic Revolution is a great breakthrough from industrial production to production and services based on the implementation of self-regulating and self-managing systems.*

Its *initial* phase, which we call the *scientific-information epoch*, dates to the period between the 1950s and 1990s. The breakthroughs occurred in the spheres of automation, energy production, synthetic materials production, exploration of space and sea, agriculture and especially in the development of electronic control facilities, communication and information.

We assume that the *final* phase of the Cybernetic Revolution will start in the nearest decades, that is in the 2030s or a bit later, and will last until the 2070s. It can be considered the *epoch of self-regulating systems* i.e., these systems will become autonomous, self-regulating, and will be able to function without human control. On the other hand, the spread of self-regulating systems will increase our opportunities for direct interference and control over various natural, social, and production processes that are currently impossible or limited to control.

It is interesting to note that technologies of the so-called Fourth Industrial Revolution are defined by some researchers as cyber-physical systems (i.e., ICT and AI) while technologies of the so-called Fifth Industrial Revolution (which is predicted to start in the 2060s [Peccoud, 2016]) are defined as cyber-biological systems. We suppose that it is more reasonable to regard both as phases of the Cybernetic Revolution because both have a common technological base and correspond to their time intervals.

The drivers of the final phase of the Cybernetic Revolution will be medical technologies, additive manufacturing (3D printers), nano- and bio-technologies, robotics, IT, and cognitive sciences, which will together form a sophisticated system of self-regulated production. We can denote this complex as the MANBRIC[12]-convergence.[13]

Currently, we are in the modernization phase which will probably last until the 2030s. This intermediate phase is a period of rapid diffusion and improvement of the innovations of the previous phase (e.g., computers, Internet, cell phones, etc.). The technological and social conditions are also prepared for future breakthroughs.

The scheme of the Cybernetic Revolution is presented in Fig. 2.3.

### 2.2.3 Production Principles

**Phases of Production Principles** The production revolution is a fundamental technological breakthrough which is realized with the discovery of absolutely new principles for creating valuable mass products. These new ways (methods) of production mark the emergence and development of a new production principle that gradually changes social and economic relations. Thus, the production revolution is a long-term process that is an integral part of a production principle and can therefore be present as the first 'half' of a production principle after which the development of mature relations takes place.[14] So we can define the following large-scale (two-part) division of a production principle: the first part is the production revolution, and the second part is concerned with maximizing the structural, systemic and spatial potentials of new forms of production.

---

[12] The order of the letters in the acronym does not reflect our understanding of the relative importance of the areas included in the complex. For example, biotechnologies will be more important than nanotechnologies, let alone additive manufacturing. The order is determined simply by the convenience of pronunciation.

[13] With respect to the theory of the Sixth Kondratieff wave and the sixth technological paradigm (closely connected with it) there is a widely used notion of NBIC-convergence (Nano-Bio-Info-Cognitive; see Lynch, 2004; Bainbridge & Roco, 2005; Dator, 2006; Kovalchuk, 2011; Akayev, 2012). Other researchers (e.g., Jotterand, 2008) define another set of technological directions in this role—GRAIN (Genomics, Robotics, Artificial Intelligence, and Nano-technology). However, we believe that among other technologies, medical technologies will become the main integrating part (see Chap. 4 for detail; see also Grinin & Grinin, 2015a, 2015b, 2016; Grinin & Korotayev, 2016a, 2016b; Grinin et al., 2016a, 2016b; Grinin et al., 2017b, 2020b).

[14] However, these halves are equaled only in structural sense, not in duration. In fact, the first half (production revolution) is longer than the second half (mature phases of a production principle) (see Table 2.3).

**Table 2.3** Chronology of phases of production principle

| No. | Pro-duction principle | 1st phase | 2nd phase | 3rd phase | 4th phase | 5th phase | 6th phase | Total production principle |
|---|---|---|---|---|---|---|---|---|
| 1 | Hunter-Gatherer | 40,000–30,000 (38,000–28,000 BC) **10** | 30,000–22,000 (28,000–20,000 BC) **8** | 22,000–17,000 (20,000–15,000 BC) **5** | 17,000–14,000 (15,000–12,000 BC) **3** | 14,000–11,500 (12,000–9,500 BC) **2.5** | 11,500–10,000 (9,500–8,000 BC) **1.5** | 40,000–10,000 (38,000–8000 BC) **30** |
| 2 | Craft-Agrarian | 10,000–7,300 (8,000–5,300 BC) **2.7** | 7,300–5,000 (5,300–3,000 BC) **2.3** | 5,000–3,500 (3,000–1500 BC) **1.5** | 3500–2200 (1500–200 BC) **1.3** | 2200–1200 (200 BC –800 AD) **1.0** | 800–1430 AD **0.6** | 10,000–570 (8,000 BC–1430 AD) **9.4** |
| 3 | Indust-rial | 1430–1600 **0.17** | 1600–1730 **0.13** | 1730–1830 **0.1** | 1830–1890 **0.06** | 1890–1929 **0.04** | 1929–1955 **0.025** | 1430–1955 **0.525** |
| 4 | Scien-tific Cybernetic[15] | 1955–1995 **0.04** | 1995–2030/40 **0.035–0.04** | 2030/40–2055/70 **0.025–0.03** | 2055/70–2070/90 **0.015–0.02** | 2070/90–2080/105 **0.01–0.015** | 2080/2105–2090/2115 **0.01** | 1955–2090/2115 **0.135–0.160** |

[15] Due to the fact that starting from the second phase of the scientific-cybernetic production principal, the dates are tentative, we give two options for dating, according to which the total duration of this production principle may differ significantly. But given the many factors that will influence the unfolding of the scientific-cybernetic production principal, all dates should be considered as approximate.

This approach shows in a rather explicit way the main 'intrigue' of the cyclical pattern of the stages of the historical process. During the first period, we mostly observe dramatic changes in production, while in the second period, we deal with particularly profound changes in political and social relations, public consciousness and other spheres. Within these periods political-legal and sociocultural relations catch up with the more advanced productive forces while creating a new level that gives impetus to the formation of a new production principle.

However, the cycle of production principle can be also represented in terms of three conventional phases: *formation, maturity*, and *decline*.

However, for the purposes of this monograph it seems more appropriate to describe it in six phases, demonstrating the additional rhythm of change of qualitative and quantitative characteristics. This cycle looks as follows:

1. The first phase—starting—is the beginning of the production revolution. A new production principle emerges in one or a few places, albeit in rather undeveloped incomplete forms. That is why this phase can be called *transitional* from the old production principle to the new one.
2. The second phase is the stage of primary modernization (or *adolescence*). It is associated with a wider diffusion of new forms of production as well as with the strengthening and vigorous expansion of a new production principle.
3. The third phase is the period of *florescence* of a production principle or the completion stage of the production revolution. The production principle acquires advanced characteristics.
4. The fourth phase is the stage of *maturity* and expansion of a production principle. It is related to the diffusion of new technologies to most regions and spheres of production. The production principle acquires its mature forms and this leads to important changes in the socio-economic sphere.
5. The fifth phase is the stage of absolute dominance of a production principle in the world or *high maturity*. It leads to the intensification of production and the full realization of the potential of the principle.
6. The sixth phase is the stage of non-system phenomena or a *preparatory* phase for the transition to a new production principle. New inventions and improvements in technologies lead to the emergence of non-system elements that trigger and prepare the formation of a new production principle. Under favorable conditions, these elements form a new system. So, in some societies, the ongoing cycle ends and the transition to a new production principle begins so the cycle repeats, yet, at a new level.

The following paragraphs of the section are devoted to the description of the history of technological changes in the considered macroperiod. The correlation between the phases of production principles and the phases of production revolutions is shown in Fig. 2.4.

**Explication**:

/ / / / – temporal volume of production revolution,

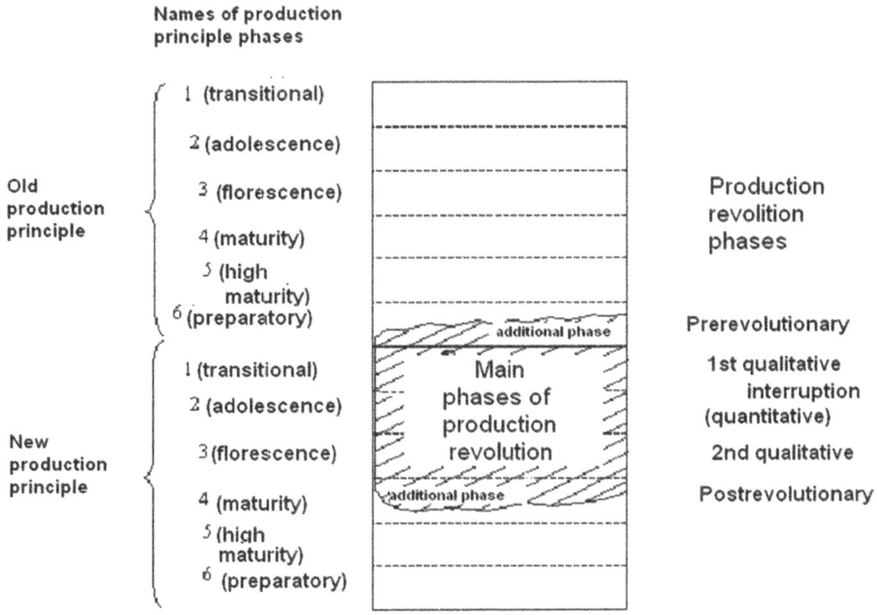


**Fig. 2.4** The correlation between phases of production principles and phases of production revolutions

—— –borders between production principles.

Similarly to the production revolution, within every production principle each of its six phases plays the same functional role and also the ratio between the duration of the phase within the framework of each cycle remains approximately the same. In Chap. 12 we will show how it is possible to predict the duration of every future phase of the Scientific-Cybernetic Production Principle.

## 2.3 The Development of the Historical Process in the Light of the Theory of Production Revolutions

### 2.3.1 When Did the Historical Process Start?

Now let us describe our chronology of production principles, production revolutions, and their phases. We start from the period about 40,000–50,000 years ago, that is, after the emergence of the first indisputable indications of genuine human culture and society.

Note that this date is not identical with the modern dating of the emergence of *Homo sapiens sapiens* (ca. 200,000 years ago). Although the discoveries of the recent decades have shifted the date of the origin of Homo sapiens back in time to 100–200 or even 300,000 years ago (see, e.g., Bar-Yosef, 2002; Bar-Yosef &

Vandermeersch, 1993; Culotta, 1999; Gibbons, 1997; Holden, 1998; Kaufman, 1999; Klima, 2003: 206; Lambert, 1991; Marks, 1993; Pääbo, 1995; Shea, 2007; Stringer, 1990; White et al., 2003; Zhdanko, 1999; see also Atkinson et al., 2009; Derevyanko, 2011; Grinin, 2009; Kazankov, 2012; Markov, 2011a, 2011b, 2012; Mellars, 2006; Wilkins et al., 2021), the landmark of 40,000–50,000 years ago still retains its major significance. This is the point when we can definitely speak about humans of modern cultural type, in particular, about the presence of developed languages and 'distinctly human' culture (Bar-Yosef & Vandermeersch, 1993: 94; Sachs, 2020; about language see further).[16] Although there are suggestions that developed languages appeared well before 40–50 thousand years ago, still these suggestions remain hypothetical. Most researchers suppose that the dependence on language appeared not earlier than 40,000 years ago (see Holden, 1998: 1455; Kellogg & Evans, 2019); meanwhile, as Richard Klein maintains, 'everybody would accept that 40,000 years ago language was everywhere' (*Ibid.*).[17] Klein, a paleoanthropologist from Stanford University, has offered a theory explaining this gap between the origin of anatomically modern Homo sapiens and delayed emergence of language and cultural artifacts: the modern mind is the result of a dramatic genetic change (see also an idea of a 'Great Leap Forward' [Diamond, 1999; Longrich, 2020]. He dates this change to around 50,000 years ago, pointing out that the rise of cultural artifacts comes after that date, as does the spread of modern humans from Africa (see Zimmer, 2003: 41 ff.). So, the period from 50,000 to 40,000 years ago is the origin of social evolution in the narrow sense (when different lines of Homo were united).[18]

---

[16] In addition, starting from this point we may speak about the final decline of the Neanderthals which however, could enrich the Cro-Magnons culture by crossbreeding (about the Neanderthal culture see e.g. Claessen, 2020). However new finds in 2022 from a cave in southeast France are complicating the story of human-Neanderthal co-occupation of Europe. A team from the University of Toulouse unearthed evidence of hominin occupation at a site called Grotte Mandrin in France: First Neanderthals were there, then modern humans, then Neanderthals again before modern humans became the only hominin in Europe. From both lithic and fossil evidence, this modern human occupation dates to older than 50,000 years ago, almost 10,000 years older than the previous record for modern humans in this region. This evidence tells us that not only did Neanderthals and modern humans live in the same area for a long span of time (potentially implying that our presence in Europe did not drive Neanderthals to extinction), but also that these two species occupied the same site alternately. This extended timespan of interaction could have implications for genetics as well, potentially adding another data point to the where and when of modern human-Neanderthal interbreeding (McRae & Pobiner, 2022).

[17] However, it must be borne in mind that there is no consensus on what is meant by speech for hominins and sapiences, as well as on when speech and language appeared, and these issues are very debatable (see, e.g., Chomsky, 1986; Turner & Machalek, 2018; Fitch, 2017; for an analysis of various approaches see Rozov, 2023).

[18] In the last ten to fifteen years, artifacts and remains of other Homo species (e.g., Denisovans) have been found, and their achievements are impressive, for example, Homo naledi buried their dead and carved symbols 100,000 years before humans (Strickland, 2023). Also, our understanding (towards recognizing a higher level of development than previously thought) about Neanderthals has been expanded …

Thus, the anthropogenesis had actively unfolded until the defined period. We agree with some scholars' idea to consider the period of anthropogenesis as a pre-history that can hardly be included in the history in its proper sense (Roginsky, 1977; see also: Boriskovsky, 1980: 171–173; Rumyantsev, 1987: 19; Yusoff, 2015; Bejan, 2020). Nevertheless, some new evidence can change this view since some traces of symbolic thinking and proto-art existed in Africa long before the Upper Paleolithic Age (see, e.g., Henshilwood et al., 2011; Lindauer, 2020). From time to time there appears information about sensational discoveries. Thus, in November 2006, Associate Professor Sheila Coulson, from the University of Oslo, announced the discovery of an artifact that is 70,000 years old and which points to the cult of a giant snake. This artifact is evidence of the human oldest known ritual. They used to think that the human mind had developed to the level of group rituals only by the period 40,000 years ago. However, in the mountain cave in the Kalahari Desert in Botswana the archeologists found a human-size stone figure of a python (Steiger, 2007; see also Helm et al., 2020). On the other hand, perhaps the existence of symbolic thinking and proto-art is not a criterion of only sapiens culture, because, as we show above (Note 17), proto-art existed in Africa long before Homo sapiens.

Meanwhile, to understand the reason for choosing this landmark, one should take into account that it is possible to speak about social evolution in its proper sense only after the social forces became the basic driving forces for the development of human communities. We suppose that the era of anthropogenesis should include not only the long period of time when our apelike ancestors (Ingold, 2002: 8) gradually acquires an anatomical resemblance to modern humans (i.e., up to about 100–200 thousand years ago), but also the subsequent rather long period (lasting for many thousands of years) when those creatures anatomically similar to us transformed into *Homo sapiens sapiens*, that is becoming humans in intellectual, social, mental, and linguistic terms (e.g., Stringer, 2011). Of course, during this second phase of anthropogenesis the role of social forces in the general balance of driving forces was much larger than it used to be during the first phase. However, we believe that in general, during the whole process of anthropogenesis the driving forces were primarily biological, and only to a rather small degree were they social (including influence of technologies and labor). Of course, it was a long-lasting process and one cannot point out a definite moment when a dramatic change occurred (and it is quite likely that there was no radical change in the literal sense). Nevertheless, we believe that after the above-mentioned landmark of 40,000–50,000 years ago, the social component of the evolutionary driving forces became dominant.[19] We also believe that for the same reasons, one can hardly speak about humankind as a set of societies prior to this time. Thus, the concepts serving as the basis for our periodization—the *formation of the historical process* and *production principles*—cannot be applied to the periods before

[19] Yet in some important respects the biological adaptation and anthropological transformation continued well beyond this threshold (see, e.g., Alexeev, 1984: 345–346, 1986: 137–145; Yarygin et al. 1999, vol. 2: 165; Diamond, 2002; Cochran & Harpending, 2009; Stringer, 2011).

40,000–50,000 years ago. Thus, our periodization starts from the most significant production revolution in human history; what is more, humans themselves are, undoubtedly, part of the productive forces.[20]

### 2.3.2 The Hunter-Gatherer Production Principle

Due to the paucity of information on this pattern, it appears reasonable to connect the phases of the Hunter-Gatherer Production Principle with the qualitative landmarks in human adaptation to and acquisition of nature. Indeed, during this period the community size, tools, economic modes, lifestyles—that is, virtually everything—depended almost exclusively on the natural environment. So, we define the level of productive forces of society during this production principle as natural, and many elements of the natural environment should be included in the theoretical model of its productive forces. And the more natural factors are included in them, the weaker is their technological component. For a deeper insight into social evolution and its relationship with the development of technology, it is important to take into account the rule that we formulated for the historical process (Grinin, 2006b, 2009, 2011a, 2013b; Grinin & Korotayev, 2009): the role of the natural-geographical environment for social evolution is the greater, the greater its place in the social system, especially in the system of productive forces. In other words, the earlier the period, the greater the role of the natural-geographical environment (about the notion of social evolution see Silverstein, 2021; Grinin & Korotayev, 2023).

If we correlate the phases of the Hunter-Gatherer Production Principle with the major changes in the environment, it appears possible to connect them with an absolute chronology on the panhuman scale.

As we will show in Chap. 5, the population estimates for the Upper Paleolithic vary greatly from 1 to 10 million people (Thomlinson, 1975; Kapitsa, 2009: 38; Livi-Bacci, 2017; Korotayev, 2020; for details see Korotayev et al., 2007: 14–15). But we believe that the real figures were much closer to the larger figure than to the smaller one. However, according to Livi-Bacchi, 12 thousand years ago it was 6 million (Livi-Bacci, 2017: 25, Table 1.2). Kapitsa (2006: 19, 2009: 38) estimates the population at about 10–15 million referring to the later date of 9–10 thousand years ago. Thus, the population of the Earth in the era of the Upper, or Late, Paleolithic (that is, after 40 thousand years ago) amounted to several million, gradually, but very slowly increasing. The world population generally grew so little that it could only be distinguished on a secular or even millennial scale. According to some data, the population during the Paleolithic increased only by 3–8% (per millennium!), that is, by 0.003–0.008% per year (Valentey, 1974: 236; Vishnevsky, 2005: 30; Livi-Bacci, 2017: 25). Of course, in some, particularly favorable, places

[20] Or, using the title of Paul Mellars and Chris Stringer's book, such a radical turn can be called 'the Human Revolution' (see Mellars & Stringer, 1989).

the population growth could be considerably faster, but on the other hand, many groups often experienced depopulation, and the climate during most of the Upper Paleolithic (40–15 thousand years ago) was quite severe over a large area, characterized by significant fluctuations (e.g., Bruxelles & Jarry, 2011; Clark et al., 2019). Nevertheless, by the beginning of the Agrarian revolution, this population had reached 6–10 million, and perhaps even more.

Its first phase [40,000–30,000 BP] may be related to the 'Upper Paleolithic' Revolution (for more details see Mellars & Stringer, 1989; Marks, 1993; Bar-Yosef, 2002; Shea, 2007, 2013; Markov, 2012; Mellars et al., 2007; Powell et al., 2009) and the formation of social productive forces (no matter how primitive they were at that time [see Grinin et al., 2012; Grinin & Grinin, 2015a, 2015b]). More than a hundred types of tools are already known for this period (Boriskovsky, 1980: 180; see also Tattersall, 2008: 150–158, 2012: 166–173; Jochim, 2011b; about the technological and instrumental 'legacy' of the anthropogenesis see d'Errico & Backwell, 2005; Anati, 2008; Markov, 2011a, 2011b; Jochim, 2011a).

The second phase (approximately and rather conditionally, from 30,000 to 22,000 BP) led to the final elimination of what could be called residue contradiction of anthropogenesis: between biological and social regulators of human activities (for details see Grinin & Korotayev, 2009; Grinin et al., 2012; Grinin & Grinin, 2015a, 2015b).[21] This phase is associated with a wide diffusion of humans, and their settlement in new places, including the people of Siberia (Doluhanov, 1979: 108) and, possibly, the first wave of the people of the New World (Zubov, 1963: 50; Sergeeva, 1983; Korotayev et al., 2017). Yet, the dates are very scattered and contradictory (Mochanov, 1977: 254; Sergeeva, 1983; Berezkin, 2007a, 2007b, 2013).[22]

The third phase lasted from 22,000 to 17,000 BP. This was the period of the maximum glacier expansion (known as the Last Glacial Maximum).[23] And although this was not the first glaciation, this time humans had sufficient levels of productive forces and sociality so that some groups were able to survive and even flourish under those severe conditions. Significant changes occurred regarding the variety and number of tools used (Chubarov, 1991: 94; Jochim, 2011b; Shea, 2013). It was at this time that the types of stone tools were rapidly changing; for example, in France and other European regions (Grigoriev, 1969: 213;

[21] Neanderthals were extinct between 34,000 and 25,000 years ago (Claessen, 2020; Mellars & Stringer, 1989) that is in this phase. Their extinction was one of demonstrations of the final elimination of this residue contradiction between biological and social regulators of human activities. Perhaps, their isolated groups are not the most probable surroundings for the development of new ideas or new techniques (Claessen, 2020: 15). However, we still know little about Neanderthals, so new findings often surprise us and can surprise us even more.

[22] The genetic data dates this period to 25–15 thousand years ago (Goebel et al., 2008). Still the settlement of the Americas was a complicated and long-lasting process (Berezkin, 2017; Korotayev, Berezkin et al., 2017).

[23] During the last glacial epoch, Würm III. The glacial maximum was observed about 20,000–17,000 BP when temperatures dropped by 5 or even 6° (Velichko, 1989: 13–15; Seltzer et al., 2021).

Jochim, 2011b), in the Levant (18,000 BP) microliths appeared (Doluhanov, 1979: 93; Shea, 2013). In many places, during these and forthcoming stages, the major evolutionary changes associated with the Epipaleolithic were increasing economic intensification and population growth (Shea, 2013: 162).[24]

During the subsequent fourth phase—c. 17,000–14,000 BP—the level of adaptation to the changing environment significantly increased (Jochim, 2011b, 2011c). In some places that avoided glaciation, an intensive gathering developed (Hall, 1986: 201; Harlan, 1986: 200; Fainberg, 1986: 185; Goring-Morris et al., 2009; Shea, 2013). This period also saw the development of proto-crafts such as sewing and weaving, making clothing and basketry (see Dyatchin, 2001: 37).

The fifth phase—from 14,000 to 11,500 BP, that is the threshold of the late Paleolithic and the early Mesolithic in Europe—may be related to the end of glaciation and climate warming (Yasamanov, 1985: 202–204; Koronovskij & Yakushova, 1991: 404–406; Shea, 2013; Goring-Morris & Belfer-Cohen, 2017). This warming together with the resulting changes in the landscape reduced the number of large mammals. For this reason, a transition to individual hunting was observed in respective areas, and also new tools appeared (Markov, 1979: 51; Childe, 1948: 40; Fainberg, 1986; Jochim, 2011c; Shea, 2013; for some later periods see Simmons, 2013). The technical means (bows, spear-throwers, traps, nets, harpoons, new types of axes, etc.) were developed to support the autonomous existence of smaller groups and even individual families (Markov, 1979: 51; Prido, 1979: 69; Avdusin, 1989: 47). Fishing in rivers, lakes and sea was developed and acquired major importance (Bergsvik & Ritchie, 2018; Lozovskaya et al., 2018; Matyushin, 1972; Ritchie et al., 2016). The following types of stone arrowheads developed: leaf-shaped, fluted, hollow-base and winged arrowheads. The bone and wood arrowheads took on a notched and later barbed and harpoon shape (Semenov, 1968: 323, 324; see also Semenov & Korobkova, 1983; Power & Williams, 2018).

The sixth phase (c. 11,500–10,000 BP) was also associated with continued climatic warming and environmental changes, culminating in the transition to the Holocene (see, e.g., Hotinskij, 1989: 39, 43; Wymer, 1982; Fernández-López de Pablo et al., 2019). In archaeological terms, this period is the transition to the Neolithic in connection with considerable progress in stone industries (Semenov, 1968; Monghite, 1973; Avdusin, 1989; Yanin, 2006; Milisauskas, 2011b). This period evidenced a large number of important innovations that, in general, opened the way to the new, craft-agrarian, production principle (see, e.g., Mellaart, 1975; Diamond, 2002; Shea, 2013; Ammerman & Cavalli-Sforza, 2014). Of particular interest are the methods of harvesting, which could potentially prove to be a more progressive development of the craft-agrarian mode as such harvesting can

[24] The term 'Epipaleolithic' is used by archaeologists to refer to cultures that were not fully or partially affected by the end of the Ice Age, as, for example in the Levant, North Africa and south-west Asia during the period after the Upper Paleolithic and before the Neolithic, between about 20,000 and 10,000 years BP archaeologists use the term 'Epipaleolithic', i.e., 'Final Paleolithic'. So, it overlaps with the late Upper Paleolithic and Mesolithic in Europe (Shea, 2013). In our periodization the Epipaleolithic is connected with the third–sixth phases.

be very productive (see, e.g., Antonov, 1982: 129; Shnirelman, 1989: 295–296, 2012a; Lips, 1956; Lamberg-Karlovsky & Sabloff, 1979; see also Tanno et al., 2013; March, 2013; Conte et al., 2018; Lozovskaya et al., 2018).

### 2.3.3 The Craft-Agrarian Production Principle

The first phase of the Craft-Agrarian Production Principle was at the same time the beginning of the Agrarian (or, to use the traditional term, Neolithic) revolution. The beginning of the Agricultural Revolution is often dated between 12,000 and 9,000 BP (although in some cases the traces of the first cultivated plants or bones of domesticated animals are even older, 14,000–15,000 years ago). As you can see, we take some intermediate dates for the initial innovative phase of the Agrarian Revolution/*first* phase of the Craft-Agrarian Production Principle, i.e., 10,000–7,300 BP. It is worth noting that the term Neolithic revolution can only be associated with this phase and the beginning of the next Agrarian Revolution. Whatever plants were cultivated, the independent invention of agriculture always took place in specific natural environments (with respect to Southeast Asia see, e.g., Deopik, 1977: 15). Correspondingly, the development of cereal production could only occur in certain natural and climate conditions (Gulyaev, 1972: 50–51; Shnirelman, 1989: 273, 2012a; Mellaart, 1982: 128; Harris & Hillman, 1989; Masson, 1967: 12; Lamberg-Karlovsky & Sabloff, 1979; Ammerman & Cavalli-Sforza, 2014; Milisauskas, 2011a, 2011b). The cultivation of cereal crops is supposed to have started somewhere in the Middle East: in the hills of Palestine (Mellaart, 1975, 1982; Özdoğan, 2022), in the Upper Euphrates area (Alexeev, 1984: 418; Hall, 1986: 202), or Egypt (Harlan, 1986: 200). A fairly large set of plants was domesticated and cultivated. Thus, according to some data in southern and eastern China, 97 different plants were cultivated (Londo et al., 2006). This period ends with the formation of the West Asian agricultural region, and in general, we can speak of the formation of the World System during this period (Grinin & Korotayev, 2009, 2012, 2013a, 2013b, 2014, 2018; Korotayev, 2005, 2007, 2012, 2013; Korotayev et al., 2006a), also including its first proto-cities (about proto-cities and first cities see Lamberg-Karlovsky & Sabloff, 1979; Masson, 1989; Schultz & Lavenda, 1998: 214–215; Balter, 2006; Korotayev, 2006; Korotayev & Grinin, 2006, 2012, 2013).[25]

As for population size, our knowledge of the agriculturalists is almost as imprecise as of the Hunter-Gatherer Production Principle. However, we can assume that the population grew from at least 6 to 10 million during this phase (Livi-Bacchi [2017: 25, Table 1.2]; Kapitsa (2006: 19, 2009: 38), but we assume that 12 thousand years ago the population was more than 6 million (8–10, as we indicate in

[25] It is worth noting that domestication has been by far the most important cause of changes in human gene frequencies over the past 10,000 years (Diamond, 2002: 704; Cochran & Harpending, 2009).

Chap. 5, Table 2.1); respectively, by the end of the initial phase of the Agrarian Revolution, the population could have reached 15 million or even more. But again, these are all very speculative estimates.

The *second* phase can be conventionally dated to 7,300–5,000 BP (from the sixth to the mid-to-late fourth millennia BCE), which is up to the formation of a unified state in Egypt and the development of a relatively sophisticated irrigation economy in Egypt and Mesopotamia. It includes the formation of new agricultural centers (Milisauskas, 2011b; Milisauskas & Kruk, 2011), the diffusion of domesticated animals from West Asia to other regions. The breeding of sheep, goats and the first draught animals developed (Shnirelman, 2012b; Meadows et al., 2007; see also Roberts, 1998; Gupta, 2004; Zeder & Hesse, 2000; Bryner, 2008). The active exchange of achievements (domesticates and their varieties, technologies, etc.) is observed (Korotayev & Zinkina, 2017; Zinkina et al., 2017, 2019). The first copper artifacts and tools in Egypt and Mesopotamia (and in Syria) date to this period (starting from the fifth millennium BCE) (Tylecote, 1976: 9). According to Childe, the so-called urban revolution took place at that time (Childe, 1952: Chap. 7; see also Lamberg-Karlovsky & Sabloff, 1979; Masson, 1980, 1989: 33–41; Oppenheim, 1968; Adams, 1981; Pollock, 2001: 45; Bernbeck & Pollock, 2005: 17; Zablotska, 1989: 34–38; Bondarenko, 2006: 50; Mellaart, 1975; Wenke, 1990: 326–330; Turnbaugh et al., 1993: 464–465; Harris, 1997: 146; Schultz & Lavenda, 1998: 214–215; Balter, 2006; Pradel-Miquel, 2019; Korotayev, 2020).[26]

By the end of this phase, the population had grown very significantly. According to some data, population growth accelerated by 8–40% over the millennium, i.e., five times, and in absolute terms the Earth's population reached 50 million (Valentey, 1974: 236; Vishnevsky, 2005: 31; Korotayev et al., 2006a; Korotayev, 2020). Of course, this figure may be overestimated, but one way or another, the population of the Earth already amounted to tens of millions of people.

During the *third* phase (which was the final phase of the Agrarian Revolution) from 5,000 to 3,500 BP, agriculture developed along with animal husbandry, crafts and trade which differentiated into separate branches of the economy (about craft specialization see Costin, 2005, 2015; Hruby & Flad, 2007). According to some data, at the end of the second phase and beginning of the third, a very wide diffusion of major innovations (wheel, plough, pottery wheel, harness [yoke], metallurgy, etc.) was observed.[27] This was the period when the first states, and later

[26] The formation of productive economies in Central Andes and Mesoamerica started in the seventh and sixth millennia BCE (see Berezkin, 2007b, 2013: 17; Dillehay et al., 2010; Quilter et al., 1991; Vega-Centeno, 2010).

[27] See Chubarov (1991), Grinin (2006b), Grinin and Grinin (2015a; 2015b), Zinkina et al., (2017); about plough see also in McNeill (1963: 24–25), Kramer (1965), Renfrew (2002), Bunch and Hellemans (2004), Milisauskas and Kruk (2011), Vandam et al., (2019) about bronze metallurgy see Tylecote (1976: 9), Chernykh (1992), Harding (2011); see also Duistermaat (2017), Li (2018), Roux (2017).

empires, rose in the Middle East.[28] Urbanization also expanded, reaching new regions (He, 2018; Chun & Xin, 2018). This period ended with a major economic, agrotechnical, and craft upsurge in Egypt at the beginning of the New Kingdom (Vinogradov, 2000).

The *fourth* phase (from 3,500 to 2,200 BP) is the period when systems of intensive agriculture (including those of non-irrigated ploughing) were developed in many parts of the world. We observe an unprecedented flourishing of crafts, cities, and trade, and the introduction and wide diffusion of iron metallurgy (Tylecote, 1976; Chubarov, 1991; Kolosovskaya & Shkunayev, 1988: 211–212; Davies, 2005: 61; Wells, 2011), as well as the formation of new civilizations and other processes indicating that the new production principle was approaching its maturity. This phase lasted till the formation of new vast world empires from Rome in the West to China in the East (Chase-Dunn & Hall, 1997; Chase-Dunn et al., 2010; Grinin et al., 2016a, 2016b; Grinin, 2010, 2011b; Korotayev & Grinin, 2021; Korotayev et al., 2022; Zinkina et al., 2019), which later led to major changes in productive forces and other social spheres. By the end of this phase, the Earth's population was already hundreds of millions of people.[29]

The *fifth* phase (from 2,200 to 1,200 BP, i.e., 200 BCE to 800 CE) was the period of the most complete development of the productive forces of the craft-agrarian economy, the period of flourishing and disintegration of the ancient civilizations and formation of civilizations of a new type (Arab, European, etc. [Chase-Dunn & Hall, 1997, 2011; Chase-Dunn & Manning, 2002; Grinin, 2011b; Korotayev et al., 1999; Wilkinson, 2022; Young, 2022; Zinkina et al., 2019]).

At the beginning of the sixth phase (800–1430 CE, i.e., from the ninth century to the first third of the fifteenth century), one could observe important changes in the production and other spheres in the Arab-Islamic world and in China; in particular, in the second half of the first millennium CE an extensive international trade network was developed in the Indian Ocean basin from the East African coast to Southeast Asia and China (Bentley, 1996; Chew, 2014, 2016; Boussac et al., 2016; about trans-Eurasian trade see Abu-Lughod, 1989; about diffusion of innovations see Grinin & Korotayev, 2015; Grinin & Grinin, 2015a, 2015b). Later the urban and economic growth started in Europe, eventually creating the first industrial centers of the Industrial Revolution, as well as preconditions for it (see also Grinin & Korotayev, 2013a, 2013b, 2015).[30]

From the twelfth century onwards, Western Europe witnessed important technological changes that prepared the ground for the subsequent breakthrough of

---

[28] Though, in most other parts of the world, states emerged significantly later (e.g., Jennings & Earle, 2023; Kradin, 2023; Romanchuk, 2023; Young, 2022).

[29] Toward the beginning of our era (i.e. after 200 years) it ranges from 170 (McEvedy & Jones, 1978) to 250 (Arab-Ogly, 1978: 88; Livi-Bacci, 2017: 25, Table 1.3; Korotayev, 2020: 590), 300–330 million people (Durand, 1977; Shcherbakova, 2011; Akimov, 1999), and sometimes even 400 mln (see Korotayev et al., 2007: 114; Korotayev et al., 2006a; Korotayev, 2020).

[30] Towards the end of this phase, the population fluctuated between 350 and 390 million for the earlier period of 1400 and between 425 and 540 million for the later period of 1500 (Durand,

the Industrial Revolution. In order to assess this unique breakthrough, it is important to keep in mind that until the early second millennium CE Europe lagged far behind the major Eastern countries in terms of the development of productive forces, statehood, urbanization, consumer culture, scientific achievements and other relevant parameters (e.g., Abu-Lughod, 1991; Christian, 2004; Crone, 1989; Goldstone, 2009a; Grinin & Korotayev, 2015; Lucas, 2005; Maddison, 2001, 2010; Pomeranz, 2000; Reinert, 2007; Saliba, 2007; Vries, 2013), whereas, according to some estimates, GDP per capita in the advanced economies of the East was at least twice as high as in Western Europe (e.g., Melyantsev, 1996: 74).

However, from the twelfth to early fifteenth century, Europe significantly caught up in its development (in various spheres), so that it succeeded in becoming a place of an evolutionary breakthrough. Thus, the period between 1100 and the 1430 s may be regarded as a period of preparation for the Industrial Revolution with a rather vivid manifestation of early capitalist relations and forms of production in some regions of Europe (northern Italy, southern Germany, the Netherlands, Southern France [see, e.g., Pirenne, 1920–1932; Wallerstein, 1974; Postan, 1987; Milskaya & Rutenburg, 1993; Lucas, 2005; Grinin & Grinin, 2015a, 2015b, 2016; Grinin & Korotayev, 2015]).

### 2.3.4 The Trade-Industrial Production Principle

The first phase of the Trade-Industrial Production Principle can be dated to the period from the second third of the fifteenth century to the late sixteenth century (1430–1600). It marks respectively the beginning of the initial phase of the Industrial Revolution. When the Industrial Revolution is associated only with its final phase (the Industrial Revolution in Great Britain in the second half of the eighteenth centuries), this seems unjustified from the point of view of the theory of production/technological revolutions.[31]

---

1974; Clark, 1967; Biraben, 1980; Maddison, 2003; McEvedy and Jones, 1978; Tanton, 1994; Livi-Bacci, 2017: 25, Table 1.3; Korotayev, 2020a: 590; see also Arab-Ogly, 1978: 88; Chesnais, 1992; Kapitsa, 2009: 38, Table 2.1; see also Kapitsa et al., 2003: 218 [with reference to Chesnais, 1992]).

[31] It is worth mentioning the viewpoint according to which there had also been an earlier (than the eighteenth century) industrial revolution (revolutions). This technological upswing that took place in Europe between 1100 and 1600 was noticed long ago—back in the 1930s—starting with the work of Mumford (1966 [1934]), Bloch (1935), Carus-Wilson (1941) and was actively studied by economic historians in around 1950–1980 (Lilley, 1976; Forbes, 1956; Armytage, 1961; Gille, 1969; White, 1978; Gimpel, 1992; see also Hill, 1955; Johnson, 1955; Bernal, 1965; Braudel, 1973; Islamov & Freidzon, 1986: 84; Gurevich, 1969: 68; Dmitriev, 1992: 140–141; Hoot, 2010; see Lucas, 2005 for more details). This period is also quite rightly considered as the time of scientific breakthrough, or rather a number of revolutionary breakthroughs in such areas as mathematics, astronomy, geography, cartography, etc. (see, e.g., Singer, 1941; Goldstone, 2009b). Though the idea of marking out the Early Modern Period (the late fifteenth—eighteenth centuries) has attracted a number of supporters, however, all these scholars do not associate the Early Modern Period with

This phase includes those types of activities that were both more open to innovations and capable of accumulating more surplus (trade [Mantoux, 1929; Bernal, 1965; Cameron, 1989; see also Acemoglu et al., 2005; Goldstone, 2009b; Grinin & Korotayev, 2015] and colonial activities [Baks, 1986], which had become more and more intertwined after the start of the sixteenth century). Besides, at that time, primitive industries (but still industries) developed in certain fields. It was during this period that, according to Wallerstein (1974, 1987), the capitalist world economy emerged.

So, the period from the second third of the fifteenth century to the end of the sixteenth century is the initial phase of the Industrial Revolution. It is associated with the development of navigation, engineering and mechanization based on watermills, the spread and improvement of various machines, and the development of the division of labor. At this time, significant breakthroughs in different directions could be observed in different parts of Europe, which by the end of the period were synthesized into the general Western European system (Johnson, 1955; Braudel, 1973; Wallerstein, 1974; Barg, 1991; Yastrebitskaya, 1993; Davies, 1996; Grinin & Grinin, 2015a, 2015b, 2016; Grinin & Korotayev, 2015; Grinin et al., 2020a, 2020b; see also Chap. 5). Changes in one country tended to produce substantial impact on the economy and the lives of others—through the spread of innovations, the publication of special technical books, the movement of technical experts to different countries, the introduction of various advances and innovations by kings and emperors in their realms, etc. Thus, we find impressive achievements in the field of mechanization in mining operations in Southern Germany and Bohemia; major contributions to the development of navigation, geographical discoveries and world trade accomplished by the Spanish and Portuguese, but also by the British; significant developments in manufacturing technologies in Italian and Flemish cities; significant changes in agriculture in northern France and the Netherlands; important scientific and mathematical discoveries made by scientists in Italy, France, Poland, England; new financial technologies developed in Italy and so on (Collins & Taylor, 2006; Davies, 1996, 2001; Ferguson, 2011; Goldstone, 2009b, 2012; Hale, 1993; Porter, 2012). But all of this, anyway, quickly became the common heritage of Europe.

The period from the early seventeenth century to the first third of the eighteenth century (1600–1730) is the second phase of the Trade-Industrial Production Principle (it can be also regarded as the middle phase of the Industrial Revolution). In this period one could observe the emergence of a complex industrial sector and capitalist economy, increasing mechanization and a deepening division of labor. This is the age of Dutch commercial leadership, the successor to the hegemony of Spain and Portugal. The Netherlands created an unprecedented industry of shipbuilding, mechanized port facilities and fishing (Allen, 2009; Boxer, 1965; de Vries & van der Woude, 1997; Goldstone, 2009a; Grinin & Grinin, 2015a, 2015b, 2016; Grinin & Korotayev, 2015; Israel, 1995; Jones, 1996; Rietbergen, 2002). But

---

an earlier industrial revolution. Meanwhile, it could give a great opportunity to deeper understand the logic of technological evolution as a whole.

the seventeenth century was a century of very large changes in military technology and science, engineering; while, as a result of wars and other processes, the Netherlands lost its leadership, which gradually passed to Britain (Rayner, 1964; Boxer, 1965; Snooks, 1997; Jones, 1996; de Vries & van der Woude, 1997; Rietbergen, 2002; see also Grinin, 2022). So, during this phase of the Industrial Revolution (and the new production principle) new sectors of industry became dominant in some countries (primarily in the Netherlands and Britain; about the development of innovations in different European countries in this period as well as in earlier and later ones (see Grinin & Korotayev, 2015, 2018).

Finally, the period between 1730 and 1830 can be identified as the third phase of the Trade-Industrial Production Principle (and at the same time as the final phase of the Industrial Revolution). This breakthrough was accompanied by the creation of sectors with a machine cycle of production and the use of steam power. The replacement of manual labor with machinery took place in the cotton textile production that developed in Britain (Berlanstein, 1992; Deane, 1965; Griffin, 2010; Mantoux, 1929; Mokyr, 1993, 1999). Watt's steam engine came into use in the 1760s and 1770s. A new powerful industry—machine production—developed. The industrial breakthrough was more or less complete in Britain by the 1830s. Although Britain was clearly the leader here, we also observe a number of important processes in this period that can be identified as pan-European (including the development of military technology, trade, science, the pan-European commercial and industrial crises of the second half of the eighteenth century, the beginning of the demographic revolution—see below). In this concept, we clearly see the Industrial Revolution as the result of the collective achievements of different European societies, a kind of relay race of achievements (see Grinin & Korotayev, 2015; Korotayev & Grinin, 2017). The successes of industrialization were evident in a number of countries by that time and this was also accompanied by significant demographic transformations (Deane, 1965; Armengaud, 1976; Minghinton, 1976: 85–89; Chesnais, 1992; Caldwell, 2006; Dyson, 2010; Livi-Bacci, 2017; Mokyr & Voth, 2010).[32] The decreasing mortality rate was noticeable in some countries of Western Europe from the late eighteenth century and it was associated with the development of science, medicine, education, the quick spread of knowledge and changes in public health and medical policy (e.g., Chesnais, 1992; Livi-Bacci, 2017; Urlanis, 1941; see also Deane, 1965: 259–261; Korotayev et al., 2015).

Many researchers, especially those of the California School, mark the end of this phase as the starting point of the so-called Great Divergence when the East obviously started to lag behind the West not only in terms of naval power and technology, but also in many other respects, such as fertility and mortality rates (see Chap. 5 for more details), the GDP per capita or calorie intake per capita, and

[32] In 1650, the population was 500 million, and in 1750–795 million (with variations are from 629 to 961 million people in 1750), and in 1800 it was 813–1,125 million people. In 1850, there lived 1,241–1,265 mln people (see Akimov, 1999; Scherbakova, 2011; Livi-Bacci [2017: 25, Table 1.3]; see also Korotayev, 2020a: 590; for details see Chap. 5).

yield and productivity.[33] And this gap would keep widening for a century and a half.

The fourth phase (from the 1830s to the late nineteenth century [1830–1890]) is the period of the victory of machine production and its powerful diffusion (see Grinin & Grinin, 2015a, 2015b, 2016 for more details). This is the period of development of mechanical engineering and the origin of heavy engineering. An exceptional breakthrough was made in the technologies of steelmaking and rolling, as well as in transportation. Steamboats developed at an accelerated pace and they could sail faster and faster; railways covered Europe and partially other continents with a rather dense network, and the power of steam locomotives increased dramatically.[34]

Let us give just a few figures to show the powerful development during this period. By 1850, after a nearly 20-fold increase in coal production, 18 million residents of Britain consumed half as much energy as 400 million Chinese (Smil, 1994: 186–187; Goldstone, 2009a: 164). Speaking about that period, Richard Jones points out that, in Britain, steam engines performed the work of 600 million people, while the actual number of employees was 4 million (Jones, 1937: 351).

The fifth phase (1890–1929) took place in the late nineteenth century—the early twentieth century up to the world economic crisis of the late 1920s and 1930s. During that period significant changes took place (see Grinin & Grinin, 2015a, 2015b, 2016 for more details). The chemical industries experienced rapid development including the production of man-made materials, and there was a breakthrough in

[33] In the 1970s and 1980s, many interesting works on the economic history of Asian countries were published, and many of them made valuable comparisons between Asia and Europe, including the economic development of the East in the context of global development. Some of these works have become classics (e.g., Braudel, 1973; Bairoch, 1975; Issawi, 1980; Maddison, 1983; Cameron, 1989). In these and many other works, one could observe the transition to truly scientific (and at the same time historical) methods because these problems (i.e., the question of the causes of the East lagging behind the West) were placed at the center of research in economic and demographic history with the appropriate use of quantitative methods. In fact, it is very difficult to compare different cultures and cultural codes, especially if one insists on their uniqueness and incomparability. However, it is possible to compare indicators such as fertility and mortality rates, GDP per capita or calorie intake per capita, and yield and productivity. Here we unquestionably see a clear common denominator which may help to perform a cross-national study of the economic and social history of all societies. Thus, there was a return to the idea of scientific approaches and common laws of historical development of different societies on a new basis. A particularly important contribution to the scientific study of the Great Divergence was made by the California School that formed in the late 1990s and the early 2000s (Blaut, 1993, 2000; Goody, 1996, 2004; Wong, 1997; Frank, 1998; Lee & Wang, 1999; Lieberman, 1999, 2003; Pomeranz, 2000, 2002; Goldstone, 1991, 2000, 2002, 2009a, 2013; Hobson, 2004; Rosenthal & Wong, 2011; Vries, 2013).

[34] In terms of the techno-economic paradigm, this period corresponded to the so-called second technological system or techno-economic paradigm (railway lines, coal, steel) and the beginning of formation of the third one (electricity, chemical industry and heavy engineering). On the nexus between Kondratieff waves and technological systems or techno-economic paradigms see Grinin et al. (2017a), Grinin (2019), Grinin and Grinin (2021). This is a period of unbelievable number of innovations (see Bunch & Hellemans, 2004; Korotayev & Grinin, 2017).

steel production (in particular, the invention of high-speed cutting steel and stainless steel). The invention of electric welding (1886) and thermite welding (1908) also played a huge role in the development of mechanical engineering, eliminating the need for riveting or bolting. Oxy-fuel cutting (oxy-acetylene torches and other similar devices) also came into use at the turn of the nineteenth and twentieth centuries. The extensive use of electricity (along with oil) gradually began to replace coal. Electric engines replaced the steam ones, therefore transforming factories and everyday life. The development of the internal combustion engine led to the widespread diffusion of automobiles. The introduction of assembly lines led to a huge increase in car production. It was a period of the first innovations in electronics. Thus, as a result of the fifth phase, a huge sector of heavy industry was developed in the economy, which eventually became its leading industry.

The sixth phase lasted till the mid-twentieth century (1929–1955). The period of the 1930s gave a large number of fundamental innovations, many of which were implemented from 1940 to the 1970s. There were especially many achievements in the military sphere, in aviation, in missiles and later in space rockets, in nuclear energy. It was a period of great growth in automobile, chemical production, and the beginning of electronics, including the first computers. During this period, production was greatly intensified and scientific methods of the organization were introduced. There was an unprecedented development of standardization and enlargement of production units. The signs of the forthcoming Cybernetic revolution became more and more evident (see Chap. 3 for details; see also Grinin, 2022).

### 2.3.5 Scientific-Cybernetic Production Principle and the Cybernetic Revolution

The Scientific-Cybernetic production principle is only at its beginning (see Fig. 2.3); only its first phase has been completed and the second phase is still developing. Hence, all the calculations of the length of forthcoming phases are highly hypothetical (these calculations are presented in Chap. 12).

The *first* phase of the Scientific-Cybernetic Production Principle took place between the 1950s and mid-1990s when a vigorous development of information technologies and the start of real economic globalization were observed. It is also associated with the transition to scientific methods of production and distribution management. The first phase of the Scientific-Cybernetic production principle is also the initial phase of the Cybernetic revolution. Particularly important changes took place in information technologies. In addition, the Cybernetic Revolution had a few other directions: in energy technologies, in the production of synthetic materials, in automation, in space exploration, and in agriculture. However, its main results are still to come.

The Cybernetic Revolution of the 1950s–1970s has sometimes been called the 'scientific-technical' revolution (e.g., Bernal, 1965; Benson & Lloyd, 1983). However, in any case, it would be more appropriate to call it the Cybernetic Revolution since its main changes will imply rapidly increasing opportunities to control various processes by means of creating self-regulating autonomous systems or by influencing the key parameters and elements capable of launching a necessary process, etc.

The *second* phase of the Scientific-Cybernetic production principle (= the intermediate phase of the Cybernetic Revolution, see Fig. 2.3) began in the mid-1990s in conjunction with the development and wide diffusion of user-friendly computers, communication technologies, cell phones and so on. Medicine and biotechnologies have also made great advances (see Chaps. 8 and 9) as well as some other areas of innovation (see Chap. 10). Much progress has been made and continues to be made in the field of AI (see Chap. 10). This phase continues to the present day. It is a period of a kind of slowdown in the rate of technological progress (for details see Chap. 12). We believe that this slowdown will not change until the mid-2030s and, for a number of reasons, may even persist until the early 2040s (see Chaps. 3, 4, and 12). However, later technological growth will experience a new acceleration. The major changes in the energy sector have been triggered by the threat of climate change.[35] However, despite the rapid development of green energy, it has not yet been possible to make a powerful innovative breakthrough in this technological direction, and there are doubts that it will be possible in the near future (see Chap. 4 for more details).

The *third* phase may begin approximately between the 2030s and 2040s. It will be the beginning of the final phase of the Cybernetic Revolution which in our view may become the epoch of '*self-regulating and self-managing systems*', that is, the vast expansion of opportunities to purposefully influence and direct various natural processes and production processes. There is a great number of different assumptions concerning changes in relevance, analyzed and discussed by intellectuals in different fields, starting from scientists and philosophers to fantasists (see e.g., Fukuyama, 2002; Sterling, 2005; de Grey & Rae, 2008). But as we will show in the following chapters, the final phase of this revolution may begin in the sphere of medicine and will be connected with its innovative branches; thus, this will lead to a serious modification of the human organism and, perhaps, its biological nature (see Grinin & Grinin, 2015a, 2015b, 2016; Grinin et al., 2016a, 2016b; Grinin et al., 2017a, 2020a, 2020b, 2021 for more details).

**Scientific-Cybernetic Production Principle in Its Maturity**. After the completion of the Cybernetic Revolution, its innovative technologies based on self-regulating/managing systems will be both widespread and improved. The mature phases of the Scientific-Cybernetic production principle will begin. Its fourth phase[36] will begin in the 2060s–2080s (depending on different scenarios of the completion

[35] See Rifkin (2011), Schwab (2016), Fücks (2013), Plakitkin (2006), Sidorovich (2015), see also Randers (2012), Wijkman and Skånberg (2017), Bardi (2017), Randers et al., (2018), von Weizsäcker and Wijkman (2018).

[36] The first three phases are the period of the Cybernetic Revolution.

of the Cybernetic Revolution [see above, Table 2.3]).[37] The fourth phase is that of maturity and expansion. It refers to the diffusion of new technologies into most regions and production sectors. The production principle will then acquire its mature form, and this leads to important changes in the socio-economic sphere. So this phase implies that, in its time, the sector of self-regulating systems will rapidly improve and spread to various regions at an enormous speed. MANBRIC technologies will finally be formed and will occupy a central place in the new production principle. We believe that during this period, some technologies, especially biotechnologies, will achieve significant success, thanks to which many environmental problems can be solved (see Chap. 9). The process of reducing manual and unskilled labor in the service sector (and, accordingly, reducing jobs) will continue due to the development of robotics and other MANBRIC technologies. However, we do not expect excessive unemployment, as by this time, due to the declining share of the working population (see Chap. 6), the shortage of labor will compensate for the loss of jobs. Such processes will be even more active in the next phase, where we expect a very active introduction of robots and other self-managed systems, particularly in the fields of medicine, and care for the sick and the elderly.[38] We also expect (under the influence of various factors, including the new system of technologies) that during this phase the formation of a system of a new world order based on greater stability and a broader consensus of the leading world forces will be completed (see Grinin et al., 2023, 2021). Cybernetic society will actively form during this period, although many of its features will not yet be sufficiently expressed, especially since such regions as Africa will still remain relatively young.

The fifth phase will be that of absolute dominance of a production principle. It leads to an intensification of production and the full realization of the potential of the principle. It will probably take place in the 2070s and the 2080s. Accordingly, it can be expected that during this period those areas that may lag behind during the Cybernetic Revolution period, in particular nanotechnology, possibly nanomechanics and nanomachine building (which were discussed in Chap. 10), and robotics, will achieve significant development. Self-driving cars will become mainstream. The fifth phase is that of non-system phenomena. Here, the major subsystems and institutions of the Cybernetic Society will begin to actively improve. In particular, medicine in symbiosis with other technologies will increasingly focus on anti-aging. During this period, the influence of conservatism, which will increase in connection with the growing proportion of the elderly population, will become noticeably stronger than before. As a result, consumerism will begin to decline, and

[37] The expected lengths of the fourth, fifth and sixth phases of the Scientific-Cybernetic Production Principle are summarized in Table 2.3. The dating of every particular phase depends on different scenarios of the unfolding of this production principle. In general, the Scientific-Cybernetic Production Principle may last until the end of this century, or by the beginning of the next (for more details see Chap. 12; see also Grinin, 2006a; Grinin et al., 2020b).

[38] Of course, these processes, as well as those described below, will not start suddenly, but they will increase starting from 2040 to 2050. But they will be particularly strong at the time described.

a new model of consumption will be actively formed (with less emphasis on market relations and a larger share of public funds, including medical ones).[39] Accordingly, the pace of technological progress will begin to slow down (see Chap. 13). As a result of the development of socio-technical self-managing systems, the electronic state will finally emerge (see Chap. 14; see also below).

The sixth phase will be even more completely characterized by the emergence of non-systemic elements; accordingly, it is during this period that cybernetic society will acquire its most mature features. There will be new inventions and technological improvements leading to the emergence of non-systemic elements (which pave the way for the formation of a new production and social system). Perhaps, this will be the time of major breakthroughs in the energy sector, and it will finally become possible to master new types of energy. It is expected to happen in the 2080s and 2090s.[40] Together with the continued improvement of self-regulating systems, which will eventually become self-managing, all this will greatly increase the level of complexity and autonomy in terms of technology, the amount of energy and the level of resource saving, which—together with the rise in the level of development of currently underdeveloped societies—will significantly reduce the demands on the pace of economic growth; accordingly, a further slowdown in the pace of technological progress can be compensated for by the rate of savings, non-wastefulness and depth of processing, as well as cheaper (but not free) energy. The latter will make it possible to implement a number of grandiose projects, it will also expand the possibilities of space exploration.

The fifth and sixth phases imply the beginning of the transition to a new economic system (see Chap. 13 and below), due to the increasing level of complexity of self-managing systems and serious advances in medicine. By this time, the process of global aging will reach every country in the world. There will be different scenarios of society's changes under the influence of aging. In any case, it will be a period of strong adaptation to aging, which will greatly contribute to the creation of a cybernetic society. There will also be a growing number of socio-technical self-regulating systems that will operate mostly autonomously and regulate the behavior of a large number of people in certain situations (for details see Chaps. 3 and 12).

During the fourth, and especially the fifth and sixth phases of the Scientific-Cybernetic Production Principle, we expect a significant increase in life expectancy and rapid global aging that will also involve 'young' regions, including sub-Saharan Africa and South Asia (see Chaps. 6 and 7; Grinin & Grinin, 2015a, 2015b, 2016; Grinin et al., 2016a, 2016b; Grinin et al., 2017a, 2017b, 2020a, 2020b, 2021).

As a result, due to the decline in the birth rate, the trend toward stabilization in the world's population (and in many countries toward depopulation), as well

[39] See Chap. 13 for details; see also below.

[40] The dates of the fourth–sixth phases will depend on when the Cybernetic Revolution will begin and finish, and consequently the dates the previous phases.

as the deepening of the process of global aging, a new TPR (type of population reproduction) will finally develop (see Chaps. 5 and 7; see also below). This TPR will include total technological support for health and quality of biological life with the stabilization (or reduction) of the population, with a very large proportion of elderly people.

All this suggests that the late twenty-first and early twenty-second centuries will be decisive for human civilization. Fundamentally new socio-economic relationships will begin to emerge, the outlines of which are not yet clear.

The next chapter is devoted to the analysis of the main features and characteristics of the Cybernetic Revolution while the chapters in Part 2 discuss its main innovative branches and directions.

**Acknowledgements** This chapter has been prepared with support from the Russian Science Foundation (Project No. **23-11-00160**).

## References

Abu-Lughod, J. (1989). *Before European hegemony: The world system AD 1250–1350*. Oxford University Press.

Abu-Lughod, J. (1991). *Before European hegemony*. Oxford University Press.

Acemoglu, D., Johnson, S., & Robinson, J. A. (2005). The rise of Europe: Atlantic trade, institutional change and economic growth. *American Economic Review, 95*, 546–579.

Adams, R. M. (1981). *Heartland of cities*. University of Chicago Press.

Akayev, A. A., et al. (2012). Matematicheskiye osnovy innovatsionno-tsiklicheskoy teorii ekonomicheskogo razvitiya Shumpetera – Kondrat′yeva. In A. A. Akayev (Ed.), *Kondratyevskiye volny: Aspekty i perspektivy* (pp. 314–341). Uchitel.

Akimov, A. (1999). Na poroge sed'mogo milliarda. Naseleniye i obshchestvo. *Informatsionnyy byulleten' tsentra demografii i ekologii cheloveka instituta narodokhozyaystvennogo prognozirovaniya RAN* № 39, September.

Alexeev, V. P. (1984). *Stanovleniye chelovechestva*. Politizdat.

Alexeev, V. P. (1986). *Etnogenez*. Vysshaya shkola.

Allen, R. C. (2009). *The British industrial revolution in global perspective*. Cambridge University Press.

Allen, R. C. (2011). *Global economic history: A very short introduction*. Oxford University Press.

Ammerman, A. J., & Cavalli-Sforza, L. L. (2014). *The Neolithic transition and the genetics of populations in Europe*. Princeton University Press.

Anati, E. (2008). *Palestina do drevnikh yevreyev*. Tsentrpoligraf.

Antonov, Y. V. (1982). *Primechaniya k kn.: Mellart Dzh. Drevneyshiye tsivili-zatsii Blizhnego Vostoka*. Nauka.

Anuchin, D. N. (1923). *Otkrytiye ognya i sposoby yego dobyvaniya*. Gosudarstvennoye izdatelstvo.

Arab-Ogly, E. A. (1978). *Demograficheskiye i ekologicheskiye prognozy. Kritika sovremennykh burzhuaznykh kontseptsiy*. Statistika.

Armengaud, A. (1976). Population in Europe 1700–1914. In C. M. Cipolla (Ed.), *The industrial revolution, 1700–1914*. Harvester Press, Barnes & Noble.

Armytage, W. H. G. (1961). *A social history of engineering*. Faber and Faber.

Atkinson, Q., Gray, R., & Drummond, A. (2009). Bayesian coalescent inference of major human mitochondrial DNA haplogroup expansions in Africa. *Proceedings of the Royal Society, B, 276*, 367–373.

Avdusin, D. A. (1989). *Osnovy arkheologii*. Vysshaya shkola.

Avise, J. C. (2004). *The hope, hype, and reality of genetic engineering: Remarkable stories from agriculture, industry, medicine, and the environment*. Oxford University Press.
Bainbridge, M. S., & Roco, M. C. (2005). *Managing nano-bio-info-cogno innovations: Converging technologies in society*. Springer.
Bairoch, P. (1975). *The economic development of the Third World countries since 1900*. University of California Press.
Baks, K. (1986). *Bogat·stva zemnykh nedr*. Progress.
Balter, M. (2006). *The goddess and the Bull. Catalhoyuk: An archaeological journey to the dawn of civilization*. Left Coast Press.
Bar-Yosef, O. (2002). The upper paleolithic revolution. *Annual Review of Anthropology, 31*, 363–393.
Bar-Yosef, O., & Vandermeersch, B. (1993). Modern humans in the levant. *Scientific American, 268*(4), 94–100.
Bardi, U. (2017). *The Seneca effect. Why growth is slow but collapse is rapid. A report to the club of Rome*. Springer.
Barg, M. A. (1991). Tsivilizatsionnyy podkhod k istorii. *Kommunist, 3*, 27–35.
Bejan, A. (2020). Human evolution is biological & technological evolution. *Bio Systems, 195*, 104156. https://doi.org/10.1016/j.biosystems.2020.104156
Belkind, L. D., Konfederatov, I. Y., & Shneiberg, Y. A. (1956). *Istoriya tekh-niki: uchebnik dlya energeticheskikh i elektrotekhnicheskikh vysshikh uchebnykh zavedeniy i fakul'tetov*. Gosenergoizdat.
Bell, D. (1973). *The coming of post-industrial society*. Basic Books.
Bell, D. (1978). *The cultural contradictions of capitalism*. Basic Books.
Bell, D. (1989). The third technological revolution and its possible socio-economic consequences. *Dissent, 36*(2), 164–176.
Bell, D. (1990). *The third technological revolution and its possible socio-economic consequences*. Tokyo.
Bellwood, P. (2004). *First farmers: The origins of agricultural societies*. Wiley-Blackwell.
Benson, I., & Lloyd, J. (1983). *New technology and industrial change: The impact of the scientific-technical revolution on labour and industry*. Kogan Pages.
Bentley, J. H. (1996). *Shapes of world history in twentieth century scholarship. Essays on global and comparative history*. American Historical Association.
Berezkin, Y. E. (2007a). *Mify zaselyayut Ameriku. Arealnoye rasprostraneniye fol'klornykh motivov i ranniye migratsii v Novyy Svet*. OGI.
Berezkin, Y. E. (2007b). O strukture istorii: vremennyye i prostranstvennyye sostavlyayushchiye. In P. V. Turchin, et al. (Eds.), *Istoriya i matematika: kontseptual'noye prostranstvo i napravleniya poiska* (pp. 88–98). LKI/URSS.
Berezkin, Y. E. (2013). *Mezhdu obshchinoy i gosudarstvom*. Nauka.
Berezkin, Y. E. (2017). Peopling of the new world from data on distributions of folklore motifs. In *Maths meets myths: Quantitative approaches to ancient narratives* (pp. 71–89). Springer.
Bergsvik, K. A., & Ritchie, K. (2018). Mesolithic fishing in western Norway. In L. O. Lozovskaya, A. Vybornov, & E. Dolbunova (Eds.), *Subsistence strategies in the stone age, direct evidence of fishing and gathering. Materials of the international conference dedicated to the 50th anniversary of Vladimir Mikhailovich Lozovski 15–18 May, 2018* (pp. 35–36). Institute for the History of Material Culture and the State Hermitage Museum.
Berlanstein, L. R. (Ed.). (1992). *The industrial revolution and work in nineteenth-century Europe*. Routledge.
Bernal, J. D. (1965). *Science in history* (3rd ed.). Hawthorn Books.
Bernbeck, R., & Pollock, S. (2005). A cultural-historical framework. In S. Pollock & R. Bernbeck (Eds.), *Archaeologies of the Middle East: Critical Perspectives* (pp. 11–40). Blackwell.
Biraben, J.-N. (1980). An Essay concerning Mankind's evolution. *Population, 4*, 1–13.
Blaut, J. (1993). *The colonizer's model of the world*. Guilford.
Blaut, J. (2000). *Eight Eurocentric historians*. Guilford.

Bloch, M. (1935). Avènement et conquêtes du moulin à eau. *Annales d'histoire économique et sociale, 7*, 538–563.

Blockmans, W. T. (1989). Preindustrial Europe. *Theory and Society, 18*(5), 733–755.

Boas, F. (1911). *The mind of primitive man.* A course of lectures delivered before the Lowell institute, Boston, Mass., and the National university of Mexico, 1910–1911. The Macmillan Company.

Bondarenko, Y. S. (2006). Informatsionnoye pole neolita Blizhnego Vostoka. *Istoriya i sovremennost, 2*, 47–66.

Boriskovsky, P. I. (1980). *Drevneysheye proshloye chelovechestva.* Nauka.

Boserup, E. (1965). *The conditions of agricultural growth: The economics of agrarian change under population pressure.* Aldine.

Boussac, M.-F., Salles, J.-F., & Yon, J.-B. (Eds.). (2016). *Ports of the ancient Indian ocean.* Primus Books.

Boxer, C. R. (1965). *The Dutch seaborne empire 1600–1800.* Knopf.

Braudel, F. (1973). *Capitalism and material life, 1400–1800.* Harper and Row.

Braudel, F. (1985). *La dynamique du capitalism.* Flammarion.

Bröring, S. (2021). Moving the BioEconomy from mind to market: What drives technology transfer and innovation adoption? Presetation at IAMO Forum June 7th 2021.

Bruxelles, L., & Jarry, M. (2011). Climatic conditions, settlement patterns and cultures in the Paleolithic: The example of the Garonne Valley (southwest France). *Journal of Human Evolution, 61*(5), 538–548.

Bryner, J. (2008). Egyptian tomb holds first known domesticated donkeys. *Live Science*, March 11. http://www.foxnews.com/story/2008/03/11/egyptian-tomb-holds-first-known-domesticated-donkeys/

Brzezinski, Z. (1970). *Between two ages. America's role in the technetronic era.* The Viking Press.

Bunch, B.H., & Hellemans, A. (2004). *The history of science and technology: A browser's guide to the great discoveries, inventions, and the people who made them, from the dawn of time to today.* Houghton Mifflin Harcourt.

Caldwell, J. C. (2006). *Demographic transition theory.* Springer.

Cameron, R. (1989). *A concise economic history of the world: From paleolithic times to present.* Hill Wang.

Carus-Wilson, E. (1941). An industrial revolution of the thirteenth century. *Economic History Review, 11*, 39–60.

Carus-Wilson, E. (1987). The woolen industry. In M. M. Postan & E. Miller (Eds.), *The Cambridge economic history of Europe. Trade and industry in the middle ages* (Vol. 2, pp. 614–692). Cambridge University Press.

Castells, M. (1996). *The rise of the network society, the information age: Economy, society and culture* (Vol. I). Blackwell Publishers.

Cauvin, J. (2000). *The birth of the gods and the origins of agriculture.* Cambridge University Press.

Chase-Dunn, C., & Hall, T. D. (2011). East and west in world-systems evolution. In P. Manning & B. Gills (Eds.), *Andre Gunder Frank and global development: Visions, remembrances, and explorations* (pp. 97–119). Routledge.

Chase-Dunn, C., & Hall, T. D. (1997). *Rise and demise: Comparing world-systems.* Westview Press.

Chase-Dunn, C., & Manning, S. (2002). City systems and world-systems: Four millennia of city growth and decline. *Cross-Cultural Research, 36*(4), 379–398.

Chase-Dunn, C., Niemeyer, R., Alvarez, A., Inoue, H., & Love, J. (2010). Cycles of rise and fall, upsweeps and collapses: Changes in the scale of settlements and polities since the Bronze Age. In L. E. Grinin, P. Herrmann, A. V. Korotayev, & A. Tausch (Eds.), *History & mathematics: Processes and models of global dynamics* (pp. 64–91). Uchitel.

Chernousov, P. I., Mapelman, V. M., & Golubev, O. V. (2005). *Metallurgiya zheleza v istorii tsivilizatsiy.* MISiS.

Chernykh, E. N. (1992). *Ancient metallurgy in the USSR: The early metal age.* Cambridge University Press.

Chun, C., & Xin, G. (2018). Erlitou and Xia: A dispute between Chinese and overseas scholars. *Social Evolution & History, 17*(2), 205–234.

Chesnais, J. C. (1992). *The demographic transition: Stages, patterns, and economic implications*. Clarendon Press.

Chew, S. C. (2014). The Southeast Asian connection in the first Eurasian world economy 200 BC–AD 500. *Journal of Globalization Studies, 5*(1), 82–109.

Chew Sing, C. (2016). From the Nanhai to the Indian Ocean and Beyond: South-east Asia in the Maritime 'Silk' Roads of the Eurasian World Economy 200 BC–AD 500. Paper presented at IROWS-ISA Workshop: Systemic Boundaries, March 5, Riverside. https://www.google.ru/url?sa=t&rct=j&q=&esrc=s&source=web&cd=1&ved=0ahUKEwjZ3qLgo_XMAhVCiSwKHSVfBysQFggcMAA&url=http%3A%2F%2Fwsarch.ucr.edu%2Farchive%2Fconferences%2Fconfname%2Fpapsing.docx&usg=AFQjCNGkevowK6M2Sh0SroQBh7oFzCJPFg&bvm=bv.122852650.d.bGg

Childe, V. G. (1934). *New light on the most ancient east*. Routledge & Kegan.

Childe, V. G. (1944). *Progress and archaeology*. Watts.

Childe, V. G. (1948). *What happened in history*. Penguin Books.

Childe, G. (1952). *New light on the most ancient east* (4th ed.). Routledge & Paul.

Chomsky, N. (1986). *Knowledge of language: Its nature, origin, and use*. Praeger.

Christian, D. (2004). *Maps of time: An introduction to big history*. University of California Press.

Chubarov, V. V. (1991). Blizhnevostochnyy lokomotiv: tempy razvitiya tekhniki i tekhnologii v drevnem mire. In A. V. Korotayev & V. V. Chubarov (Eds.), *Arkhaicheskoye obshchestvo: uzlovyye problemy sotsiologii* (pp. 92–135). Institute of USSR History, USSR Academy of Science.

Cipolla, C. M. (Ed.). (1975). *The Fontana economic history of Europe*. Collins/Fontana Books.

Cipolla, C. M. (Ed.). (1976). *The industrial revolution, 1700–1914*. Harvester Press, Barnes & Noble.

Claessen, H. J. M. (2020). Neandertals—life, work, and vanishing. *Social Evolution & History, 19*(1), 3–20. https://doi.org/10.30884/seh/2020.01.01

Clark, G. (2007). *A farewell to alms: A brief economic history of the world*. Princeton University Press.

Clark, J. G. D. (1966). *Prehistoric Europe: The economic basis*. Stanford University Press.

Clark, C. (1967). *Population growth and land use*. Martin's Press.

Clark, G. A., Barton, C. M., & Straus, L. G. (2019). Landscapes, climate change & forager mobility in the Upper Paleolithic of northern Spain. *Quaternary International, 515*, 176–187.

Clarkson, L. A. (1971). *The pre-industrial economy, 1500–1750*. Batsford.

Cochran, G., & Harpending, H. (2009). *The 10,000 year explosion: How civilization accelerated human evolution*. Basic Books.

Cohen, M. N. (1977). *The food crisis in prehistory*. Yale University Press.

Collins, J. B., & Taylor, K. L. (Eds.). (2006). *Early modern Europe*. Blackwell.

Conte, C., Ibáñez Estévez, J. J., Gibaja Bao, J. F., Mazzucco, N., Terradas, X., Mozota Holgueras, M., & Borrell, F. (2018). Cereal use-wear traces and harvesting methods. In O. V. Lozovskaya, A. A. Vybornov, & E. V. Dolbunova (Eds.), *Subsistence strategies in the stone age, direct evidence of fishing and gathering*. Materials of the international conference dedicated to the 50th anniversary of Vladimir Mikhailovich Lozovski 15–18 May 2018 (pp. 192–193). Institute for the History of Material Culture and the State Hermitage Museum.

Costin, C. L. (2005). The study of craft production. In H. D. Maschner & C. Chippindale (Eds.), *Handbook of archaelogical methods* (pp. 1032–1105). AltaMira Press.

Costin, C. L. (2015). Craft Specialization. In P. Whelehan & A. Bolin (Eds.), *The international encyclopedia of human sexuality* (pp. 1–5). Wiley.

Cowan, S. W., & Watson, P. J. (Eds.). (1992). *The origins of agriculture*. Smithsonian Institution Press.

Crone, P. (1989). *Pre-industrial societies*. Blackwell.

Culotta, E. (1999). A new human ancestor? *Science, 284*, 572–573.

d'Errico, F., & Backwell, L. (2005). *From tools to symbols: From early hominids to modern humans*. Wits University Press.

Das, G. G. (2007). Information age to genetic revolution: Embodied technology transfer and assimilation—a tale of two technologies. *Technological Forecasting and Social Change, 74*(6), 819–842. https://doi.org/10.1016/j.techfore.2006.07.001

Dator, J. (2006). Alternative futures for K-waves. In T. C. Devezas (Ed.), *Kondratieff waves, warfare and world security* (pp. 311–317). IOS Press.

Davies, N. (1996). *Europe: A history*. Oxford University Press.

Davies, N. (2001). *Heart of Europe: The past in Poland's present*. Oxford University Press.

Davies, N. (2005). *Istoria Evropy*. AST, Transitkniga.

De Grey, A., & Rae, M. (2008). *Ending aging: The rejuvenation breakthroughs that could reverse human aging in our lifetime*. Martin's Press.

Deane, P. (1965). *The first industrial revolution*. University Press.

Deng, G. (1999). *The premodern Chinese economy: Structural equilibrium and capitalist sterility*. Routledge.

Deopik, D. V. (1977). *Yugo-Vostochnaya Aziya v mirovoy istorii*. Nauka.

Derevyanko, A. P. (2011). *Verkhniy paleolit v Afrike i Yevrazii i formirovaniye cheloveka sovremennogo anatomicheskogo tipa*. Izd-vo IAET SO RAN.

Diamond, J. (1999). *Guns, germs, and steel: The fates of human societies*. W.W. Norton & Company.

Diamond, J. (2002). Evolution, consequences and future of plant and animal domestication. *Nature, 418*(8), 700–707.

Dietz, F. (1927). *The industrial revolution*. Holt.

Dillehay, T. D., Rossen, J., Ugent, D., Karathanasis, A., Vásquez, V., & Netherly, P. J. (2010). Early Holocene coca chewing in northern Peru. *Antiquity, 84*(326), 939–953.

Dizard, W. (1982). *The coming information age. An overview of technology, economics and politics*. Longman, Dordick, H. S.

Dmitriev, M. V. (1992). The genesis of capitalism as an alternative to historical development. In R. V. Manekin (Ed.), *Alternative history* (pp. 132–216). SAMI Donetsk Branch.

Doluhanov, P. M. (1979). *Geografiya kamennogo veka*. Nauka.

Drucker, P. F. (1995). *Post-capitalist society*. Harper Business.

Drucker, P. F. (1996). *Landmarks of tomorrow*. Transaction Publishers.

Duff, A. S. (1998). Daniel Bell's theory of the information society. *Journal of Information Science, 24*(6), 373–393.

Duistermaat, K. (2017). The organization of pottery production, toward a relational approach. In A. Hunt (Ed.), *The Oxford handbook of archeological ceramic analysis* (pp. 114–147). Oxford University Press.

Durand, J. D. (1974). *Historical estimates of world population: An evaluation* (p. 10). University of Pennsylvania.

Durand, J. D. (1977). Historical estimates of world population. An evaluation. *Population and Development Review, 3*(3), 253–296.

Durkheim, E. (1997) [1893]. *The division of labour in society*. Free Press.

Dyatchin, N. I. (2001). *Istoriya razvitiya tekhniki*. Feniks.

Dyson, T. (2010). *Population and development*. Zed Books.

Ellul, J. (1964). *The technological society*. Vintage Books.

Ellul, J. (1975). *Trahison de l'Occident*. Calmann-Levy.

Ellul, J. (1982). *Changer de revolution: L'ineluctable proletariat*. Seuil.

Ellul, J. (1984). *La subversion du Christianisme*. Seuil.

Engels, F. (1955) [1845]. Polozheniye rabochego klassa v Anglii. In Marx K., Engels F., *Sochineniya* (Vol. 2, pp. 231–517). Politoizdat.

Fainberg, L. A. (1986). Rannepervobytnaya obshchina okhotnikov, sobirateley, rybolovov. In Y. V. Bromley (Ed.), *Istoriya pervobytnogo obshchestva. Epokha pervobytnoy rodovoy obshchiny* (pp. 130–235). Nauka.

Ferguson, N. (2011). *Civilization: The west and the rest*. The Penguin Press.

Fernández-López de Pablo, J., Gutiérrez-Roig, M., Gómez-Puche, M., McLaughlin, R., Silva, F., & Lozano, S. (2019). Palaeodemographic modelling supports a population bottleneck during the Pleistocene-Holocene transition in Iberia. *Nature Communications, 10*(1), 1–13.

Fitch, W.T. (2017). Empirical approaches to the study of language evolution. *Psychonomic Bulletin & Review,* 1–31.

Forbes, R. J. (1956). Power. In C. Singer, et al. (Eds.), *A history of technology. The Mediterranean civilizations and the middle ages, c. 700 B.C. to c. A.D. 1500* (Vol. 2, pp. 601–606). Oxford University Press.

Fourastie, J. (1958). *La progres technique et revolution economique.* A. Colin.

Frank, A. G. (1990). A theoretical introduction to 5,000 years of world system history. *Review, 13*(2), 155–248.

Frank, A. G. (1993). The bronze age world system and its cycles. *Current Anthropology, 34*, 383–413.

Frank, A. G., & Gills, B. K. (Eds.). (1993). *The world system: Five hundred years of five thousand.* Routledge.

Frank, A. G. (1998). *ReOrient: Global economy in the Asian age.* University of California Press.

Fücks, R. (2013). *Intelligent wachsen. Die grune revolution.* Carl Hanser Verlag.

Fukuyama, F. (2002). *Our post-human future: Consequences of the bio-technology revolution.* Farrar, Straus, and Giroux.

Gamble, C. (1986). *The Paleolithic settlement of Europe.* Cambridge University Press.

Gibbons, A. Y. (1997). Chromosome shows that Adam was an African. *Science, 278*, 804–805.

Gille, B. (1969). The problems of power and mechanization. In M. Daumas (Ed.), *A history of technology and invention: Progress through the ages. The origins of technological civilization* (Vol. 1). Crown.

Gimpel, J. (1992). *The medieval machine: The industrial revolution of the middle ages* (2nd ed.). Pimlico.

Goebel, T., Waters, M. R., & O'Rourke, D. H. (2008). The late Pleistocene dispersal of modern humans in the Americas. *Science, 319*, 1497–1502.

Goldstone, J. A. (1991). *Revolution and rebellion in the early modern world.* University of California Press.

Goldstone, J. A. (2000). The rise of the West—or not? A revision to socio-economic history. *Sociological Theory, 18*(2), 173–194.

Goldstone, J. A. (2002). Efflorescences and economic growth in world history: Rethinking the 'rise of the West' and the British industrial revolution. *Journal of World History, 13*, 323–389.

Goldstone, J. A. (2009a). *Why Europe? The rise of the west in world history. 1500–1850.* McGraw-Hill.

Goldstone, J. A. (2009b). The origins of capitalism and the rise of the west. *The British Journal of Sociology, 60*(3), 651–653.

Goldstone, J. A. (2012). Is Islam bad for business? *Perspectives on Politics, 10*(1), 97–102.

Goldstone, J. A. (2013). The origins of western superiority—a comment on modes of meta-history and Duchesne's Indo-Europeans. *Cliodynamics: The Journal of Theoretical & Mathematical History, 4*(1), 54–66.

Goody, J. (1996). *The east in the west.* Cambridge University Press.

Goody, J. (2004). *Capitalism and modernity: The great debate.* Polity Press.

Goring-Morris, A. N., & Belfer-Cohen, A. (2017). The early and middle epipalaeolithic of cisjordan. In Y. Enzel & O. Bar-Yosef (Eds.), *Quaternary environments, climate change, and humans in the levant* (pp. 639–649). Cambridge University Press.

Goring-Morris, A. N., Hovers, E., & Belfer-Cohen, A. (2009). The dynamics of Pleistocene and early Holocene settlement patterns and human adaptations in the levant—an overview. In J. J. Shea & D. E. Lieberman (Eds.), *Transitions in prehistory: Essays in Honor of Ofer bar-yosef* (pp. 185–252). Oxbow Books for the American School of Prehistoric Research.

Griffin, E. (2010). *Short history of the British industrial revolution.* Palgrave Macmillan.

Grigoriev, G. P. (1969). Pervobytnoye obshchestvo i yego kul'tura v must'ye i nachale pozdnego paleolita. In I. P. Gerasimov (Ed.), *Priroda i razvitiye pervobytnogo obshchestva* (pp. 196–215). Nauka.
Grinin, L. E. (2006a). Periodization of history: A theoretic-mathematical analysis. In L. E. Grinin, V. de Munck, & A. Korotayev (Eds.), *History & mathematics: Analyzing and modeling global development* (pp. 10–38). KomKniga.
Grinin, L. E. (2006b). *Proizvoditel'nye sily i istoricheskiy protsess* (3rd ed.). KomKniga.
Grinin, L. E. (2007a). Production revolutions and periodization of history: A comparative and theoretic-mathematical approach. *Social Evolution & History, 6*(2), 11–55.
Grinin, L. E. (2007b). Production revolutions and the periodization of history. *Herald of the Russian Academy of Sciences, 77*(2), 150–156.
Grinin, L. E. (2009). *Gosudarstvo i istoricheskiy protsess: Politicheskiy srez istoricheskogo protsessa* (2nd ed.). LIBROKOM.
Grinin, L. E. (2010). *Gosudarstvo i istoricheskiy protsess: Evolyutsiya gosudarstvennosti: Ot rannego gosudarstva k zrelomu* (2nd ed.). LIBROKOM.
Grinin, L. E. (2011a). Prirodnyy faktor v aspekte teorii istorii. *Filosofia i Obschestvo, 2*, 168–199.
Grinin, L. E. (2011b). *Gosudarstvo i istoricheskiy protsess: Epokha formirovaniya gosudarstva: Obshchiy kontekst sotsial'noy evolyutsii pri obrazovanii gosudarstva.* LIBROKOM.
Grinin, L. E. (2012). *Macrohistory and Globalization.* Uchitel.
Grinin, L. E. (2013a). Dinamika kondrat'yevskikh voln v svete teorii proizvodstvennykh revolyutsiy. In L. E. Grinin, A. V. Korotayev, S. Y. Malkov (Eds.), *Kondrat'yevskiye volny: Palitra vzglyadov* (pp. 31–83). Uchitel.
Grinin, L. E. (2013b). Prirodno-geograficheskiy faktor v aspekte teorii istorii. *Vestnik KIGIT, 6*(36), 30–57.
Grinin, L. E. (2017). The processes of systemic integration in the world system. *Journal of Globalization Studies, 8*(1), 97–118.
Grinin, L. E. (2019). Kondratieff waves, technological modes, and the theory of production revolutions. In L. Grinin & A. Korotayev (Eds.), *Kondratieff waves: Spectrum of opinions* (pp. 95–144). Uchitel.
Grinin, L. (2020). Kondratieff waves, technological modes, and the theory of production revolutions. In L. Grinin & A. Korotayev (Eds.), *History & mathematics: Investigating past and future* (pp. 30–79). Uchitel.
Grinin, L. E. (2022). The third technological paradigm: Electricity, chemical industry and heavy engineering. In L. E. Grinin & A. V. Korotayev (Eds.), *Kondratieff waves: Processes, cycles, triggers, and technological paradigms* (pp. 197–203). Uchitel.
Grinin, L. E., & Grinin, A. L. (2013a). Macroevolution of technology. In L. E. Grinin & A. V. Korotayev (Eds.), *Evolution: Development within big history, evolutionary and world-system paradigms* (pp. 143–178). Uchitel.
Grinin, L. E., & Grinin, A. L. (2013b). Global technological transformations. In L. E. Grinin, I. V. Ilyin, & A. V. Korotayev (Eds.), *Globalistics and globalization studies: Theories, research & teaching* (pp. 98–128). Uchitel.
Grinin, L. E., & Grinin, A. L. (2014). The sixth kondratieff wave and the cybernetic revolution. In L. E. Grinin, T. Devezas, & A. V. Korotayev (Eds.), *Kondratieff Waves. Juglar – Kuznets – Kondratieff* (pp. 354–378). Uchitel.
Grinin, A. L., & Grinin, L. E. (2015a). The cybernetic revolution and historical process. *Social Evolution and History, 14*(1), 125–184.
Grinin, L. E., & Grinin, A. L. (2015b). *Ot rubil do nanorobotov. Mir na puti k epokhe samoupravlyaemykh sistem (istoriya tekhnologiy i opisanie ikh budushchego).* Uchitel.
Grinin, L. E., & Grinin, A. L. (2016). *The cybernetic revolution and the forth-coming epoch of self-regulating systems.* Moscow Branch of the Uchitel Publishing House.
Grinin, L., & Grinin, A. (2021). The dynamics of Kondratieff waves in the light of the theory of production revolutions. In L. Grinin & A. Korotayev (Eds.), *Kondratieff waves. Historical and theoretical aspects* (pp. 59–113). Uchitel.

Grinin, L., & Grinin, A. (2022). The dynamics of kondratieff waves in the light of the theory of production revolutions. In L. Grinin & A. Korotayev (Eds.), *History & mathematics: Historical and technological dynamics: Factors, cycles, trends* (pp. 201–255). Uchitel.

Grinin, L., & Grinin, A. (2023). Historical materialism: Does the concept have a future? *Social Evolution & History, 22*(1), 143–178.

Grinin, L. E., Grinin, A. L., & Korotayev, A. (2017a). Forthcoming Kondratieff wave, cybernetic revolution, and global aging. *Technological Forecasting & Social Change, 115*, 52–68.

Grinin, L., Grinin, A., & Korotayev, A., et al. (2017b). The MANBRIC-technologies in the forthcoming technological revolution. In T. Devezas (Ed.), *Industry 4,0—Entrepreneur-ship and structural change in the new digital landscape: What is coming on along with the fourth industrial revolution* (pp. 243–261). Springer.

Grinin, L., Grinin, A., & Korotayev, A. V. (2020a) Dynamics of technological growth rate and the forthcoming singularity. In A. V. Korotayev & D. LePoire (Eds.), *The 21st century Singularity and global futures. A big history perspective* (pp. 287–344). Springer. https://doi.org/10.1007/978-3-030-33730-8_14

Grinin, L., Grinin, A., & Korotayev, A. (2020b). A quantitative analysis of worldwide long-term technology growth: From 40,000 BCE to the early 22nd century. *Technological Forecasting and Social Change, 155*, 1–15. https://doi.org/10.1016/j.techfore.2020.119955

Grinin, L., Grinin, A., & Korotayev, A. (2021). Global trends and forecasts of the 21st century. *World Futures, 77*(5), 335–370. https://doi.org/10.1080/02604027.2021.1949939

Grinin, L., Grinin, A., & Korotayev, A. (2023). Future political change. Toward a more efficient world order. In V. Sadovnichy, et al. (Eds.), *Reconsidering the limits to growth. A report to the Russian Association of the Club of Rome* (pp. 191–206). Springer. https://doi.org/10.1007/978-3-031-34999-7_11

Grinin, L. E., Ilyin, I. V., & Andreev, A. I. (2016). World order in the past, present, and future. *Social Evolution& History, 15*(1), 58–84.

Grinin, L. E., & Korotayev, A. V. (2006). Political development of the world system: A formal quantitative analysis. In S. Y. Malkov, L. E. Grinin, & A. V. Korotayev (Eds.), *History & mathematics: Historical dynamics and development of complex societies* (pp. 49–101). KomKniga.

Grinin, L. E., & Korotayev, A. V. (2009). *Sotsial'naya makroevolyutsiya. Genezis i transformatsii Mir-Sistemy.* LIBROKOM.

Grinin, L., & Korotayev, A. (2012). The afroeurasian world-system: Genesis, transformations, characteristics. In S. J. Babones & C. Chase-Dunn (Eds.), *Routledge handbook ofworld-systems analysis* (pp. 30–41). Routledge.

Grinin, L. E., & Korotayev, A. V. (2013a). Globalization and the world system evolution. In L. E. Grinin & A. V. Korotayev (Eds.), *Evolution: Development within big history, evolutionary and world-system paradigms* (pp. 30–68). Uchitel.

Grinin, L. E., & Korotayev, A. V. (2013b). The origins of globalization. In J. Sheffield, A. Korotayev, & L. Grinin (Eds.), *Globalization, yesterday, today, and tomorrow* (pp. 1–29). Emergent Publications.

Grinin, L., & Korotayev, A. (2014). Origins of globalization in the framework of the Afroeurasian world-system history. *Journal of Globalization Studies, 5*(1), 32–64.

Grinin, L. E., & Korotayev, A. V. (2015). *Great divergence and great convergence: A global perspective*. Springer.

Grinin, L., & Korotayev, A. (2016a). Global population aging, the sixth Kondratieff wave, and the global financial system. *Journal of Globalization Studies, 7*(2), 11–31.

Grinin, L. E., & Korotayev, A. V. (2016b). Introduction, global evolution and global aging. In L. E. Grinin & A. V. Korotayev (Eds.), *Evolution and big history: Dimensions, trends, and forecasts* (pp. 5–17). Uchitel.

Grinin, L. E., & Korotayev, A. V. (2018). Origins of globalization in the framework of the Afroeurasian world-system history. In: T. D. Hall (Ed.), *Comparing globalizations, world-systems evolution and global futures* (pp. 37–70). Springer. https://doi.org/10.1007/978-3-319-68219-8_3

Grinin, L. E., & Korotayev, A. V. (2023). Social evolution: Theoretical aspects and general outlines. *Social Evolution & History, 22*(2), 111–158. https://doi.org/10.30884/seh/2023.02.07

Grinin, L. E., Korotayev, A. V., & Markov, A. V. (2012). Biologicheskaya i sotsial'naya fazy makroevolyutsii: skhodstva i razlichiya evolyutsionnykh printsipov i mekhanizmov. In: L. E. Grinin, A. V. Korotayev, & A. V. Markov (eds.), *Evolyutsiya: aspekty sovremennogo evolyutsionizma* (pp. 130–176). LKI.

Grinin, L. E., Korotayev, A. V., & Tausch, A. (2016b). *Economic cycles, crises, and the global periphery*. Springer.

Grinin, L. E., Korotayev, A. V., & Tausch, A. (2019). *Islamism, Arab Spring, and the future of democracy*. Springer.

Gulyaev, V. I. (1972). *Drevneyshiye tsivilizatsii Mezoameriki*. Nauka.

Gupta, A. K. (2004). Origin of agriculture and domestication of plants and animals linked to early Holocene climate amelioration. *Current Science, 87*(1), 54–59.

Gurevich, A. Y. (1969). Ob istoricheskoy zakonomernosti. In A. V. Gulyga & Yu. A. Levada (Eds.), *Filosofskiye problemy istoricheskoy nauki* (pp. 51–79). Nauka.

Hale, J. (1993). *The civilization of Europe in the renaissance*. HarperCollins.

Hall, F. (1986). Proiskhozhdeniye i razvitiye zemledeliya. In V. M. Masson (Ed.), *Drevniye tsivilizatsii Vostoka* (pp. 201–204). Tashkent FAN.

Harding, A. F. (2011). The bronze age, Ch. 10. In: S. Milisauskas (Ed.), *European history* (pp. 405–460). Springer.

Harlan, D. R. (1986). Resursnaya baza osnovnykh rastitel'nykh kul'tur Iranskogo plato i sosednikh regionov. In V. M. Masson (Ed.), *Drevniye tsivilizatsii Vostoka* (pp. 199–201). Tashkent FAN.

Harris, D., & Hillman, G. (1989). An evolutionary continuum of people-plant interaction. In: D. R. Harris & G. C. Hillman (Eds.), *Foraging and farming. The evolution of plant exploitation* (pp. 11–27). Unwin Hyman.

Harris, M. (1997). *Culture, people, nature: An introduction to general anthropology* (7th ed.). Longman.

Harris, J. (2010). *Enhancing evolution: The ethical case for making better people*. Princeton University Press.

He, N. (2018). Archaeological indicators for Chinese early states: A case study of Taosi in Shanxi. *Social Evolution & History, 17*(2), 205–234.

Helm, C. W., Cawthra, H. C., De Vynck, J. C., Helm, C. J., Rust, R., & Stear, W. (2020). Palaeoart on South Africa's Cape south coast. *South African Journal of Art History, 35*(2), 53–66.

Henderson, W. O. (1961). *The industrial revolution on the continent: Germany, France, Russia, 1800–1914*. F. Cass.

Henshilwood, C. S., et al. (2011). A 100,000-year-old ochre processing workshop at Blombos cave, south Africa. *Science, 334*(6053), 219–222.

Hill, C. (1955). *The English revolution, 1640* (3rd ed.). Lawrence & Wishart.

Hobson, J. M. (2004). *The eastern origins of western civilization*. Cambridge University Press.

Holden, C. (1998). No last word on language origins. *Science, 282*, 1455–1458.

Holdrege, C. (2010). *Beyond biotechnology: The barren promise of genetic engineering*. University Press of Kentucky.

Hoot, L. R. (2010). *Teoretiko-metodologicheskiye problemy izucheniya istorii Novogo vremeni v otechestvennoy istoriografii rubezha XX–XXI vv*. MGPU.

Hotinskij, K. A. (1989). Landshaftno-klimaticheskiye izmeneniya v pozd-lednikovoye vremya na territorii SSSR. In: A. A. Velichko, Y. Y. Gurtovaya, & M. A. Faustova (Eds.), *Paleoklimaty i oledeneniya v pleystotsene* (pp. 39–46). Nauka.

Hruby, Z., & Flad, R. (Eds.). (2007). Rethinking craft specialization in complex societies: Archaeological analyses of social meaning of production. In *Archaelogical papers of the American anthropological association* (Vol. 17, NO. 1). Wiley-Blackwell.

Huang, P. C. C. (2002). Development or involution in eighteenth-century Britain and China? A review of Kenneth Pomeranz's 'The Great Divergence: China, Europe, and the Making of the Modern World Economy.' *The Journal of Asian Studies, 61*(2), 501–538.

Ingold, T. (1980). *Hunters, pastoralists, and ranchers: Reindeer economies and their transformations*. Cambridge University Press.
Ingold, T. (2002). On the distinction between evolution and history. *Social Evolution and History, 1*(1), 5–24.
Inozemtsev, V. L. (Ed.). (1999). *Novaya post-industrial'naya volna na Zapade*. Academia.
Islamov, T. M., & Freidzon, V. I. (1986). Perekhod ot feodalizma k kapitalizmu v Zapadnoy, Tsentral'noy i Yugo-Vostochnoy Yevrope. *Novaya i Noveyshaya Istoriya, 1*, 83–96.
Israel, J. I. (1995). *The Dutch republic, its rise, greatness, and fall, 1477–1806*. Clarendon Press.
Issawi, C. (1980). Europe, the middle east and the shift in power: Reflections on a theme by Marshall Hodgson. *Comparative Studies in Society and History, 22*, 487–504.
Jennings, J., & Earle, T. (2023). Social stratification, genealogical distance, and state formation in early Hawaii and the central Andes: Steps, slopes, and off-ramps in social evolution. *Social Evolution & History, 22*(2), 44–66.
Jochim, M. A. (2011a). The lower and middle paleolithic (ch. 4). In: S. Milisauskas (Ed.), *European history* (pp. 31–65). Springer.
Jochim, M. A. (2011b). The misolithic (ch. 6). In S. Milisauskas (Ed.), *European history* (pp. 125–151). Springer.
Jochim, M. A. (2011c). The upper paleolithic. In S. Milisauskas (Ed.), *European history* (pp. 67–124). Springer.
Johnson, A. H. (1955). *Europe in the sixteenth century: 1494–1598*. Rivingtons.
Jones, J. R. (1996). *The Anglo-Dutch wars of the seventeenth century*. Longman.
Jones, R. (1937). *Ekonomicheskiye sochineniya*. Sotsekgiz.
Jotterand, F. (2008). *Emerging conceptual, ethical and policy issues in bionanotechnology* (Vol. 101). Springer Science & Business Media.
Kahn, H. (1982). *The coming boom: Economic*. Simon & Schuster.
Kapitsa, S. P. (2006). Ob uskorenii istoricheskogo vremeni. In L. E. Grinin, A. V. Korotayev, & S. Y. Malkov (Eds.), *Istoriya i Matematika: Problemy periodizatsii istoricheskikh makroprotsessov* (pp. 12–30). Yeditorial URSS.
Kapitsa, S. P. (2009). *Obshchaya teoriya rosta chelovechestva. Kak ros i kuda idet mir cheloveka*. Al'pina non-fikshn.
Kapitsa, S. P., Kurdyumov, S. P., & Malinetskiy, G. G. (2003). *Sinergetika i prognozy budushchego* (2 ed.). Yeditorial URSS.
Kaufman, D. (1999). *Archeological perspectives on the origins of modern humans. A view from levant*. Bergin & Garvey.
Kazankov, A. A. (2012). Kommentarii po povodu stat'i A. M. Burovskogo '"Zapad" epokhi pleystotsena'. In: L. Grinin, A. V. Korotayev, & A. V. Markov (Eds.), *Evolyutsiya: Aspekty sovremennogo evolyutsionizma* (pp. 271–279). LIBROKOM.
Kellogg, R. T., & Evans, L. (2019). The ensemble hypothesis of human cognitive evolution. *Evolutionary Psychological Science, 5*(1), 1–12.
Klima, B. (2003). Period cheloveka razumnogo sovremennogo vida do nachala proizvodstva pishchi (proizvodyashchego khozyaystva): obshchiy obzor (za isklyucheniyem iskusstva). *Istoriya Chelovechestva. T. 1. Doistoricheskiye vremena i nachala tsivilizatsii* (pp. 198–207). UNESCO.
Knowles, L. C. A. (1937). *The industrial and commercial revolutions in great britain during the nineteenth century*. Routledge–Dutton.
Kolosovskaya, Y. K., & Shkunayev, S. V. (1988). Kel'ty v Yevrope v pervoy polovine I tys. do n. e. In Y. S. Golubtsova (Ed.), *Istoriya Yevropy. Drevnyaya Yevropa* (Vol. 1, pp. 203–212). Nauka.
Koronovskij, N. V., & Yakushova, A. F. (1991). *Osnovy geologii*. Vysshaya shkola.
Korotayev, A. (2005). A compact macromodel of world system evolution. *Journal of World-Systems Research, 11*(1), 79–93. https://doi.org/10.5195/jwsr.2005.401
Korotayev, A. (2006). The world system urbanization dynamics: A quantitative analysis. In P. Turchin, L. Grinin, A. Korotayev, & V. C. de Munck (Eds.), *History & mathematics: Historical dynamics and development of complex societies* (pp. 44–62). KomKniga/URSS.

Korotayev, A. (2007). Compact mathematical models of world system development, and how they can help us to clarify our understanding of globalization processes. In: Modelski, T. Devezas, & W. R. Thompson (Eds.), *Globalization as evolutionary process: Modeling global change* (pp. 133–160). Routledge.

Korotayev, A. (2012). Globalization and mathematical modeling of global development. In L. E. Grinin, I. V. Ilyin, & A. V. Korotayev (Eds.), *Globalistics and globalization studies* (pp. 148–158). Uchitel.

Korotayev, A. (2013). Globalization and mathematical modeling of global evolution. In L. E. Grinin & A. V. Korotayev (Eds.), *Evolution: Development within big history, evolutionary and world-system paradigms* (pp. 69–83). Uchitel.

Korotayev, A. V. (2020). How singular is the 21st century Singularity? In A. V. Korotayev & D. LePoire (Eds.), *The 21st century Singularity and global futures. A big history perspective* (pp. 571–595). Springer. https://doi.org/10.1007/978-3-030-33730-8_26

Korotayev, A., Berezkin, Y., Borinskaya, S., Davletshin, A., & Khaltourina, D. (2017). Genes and myths: Which genes and myths did the different waves of the peopling of Americas bring to the new world? In L. E. Grinin & A. V. Korotayev (Eds.), *History & mathematics: Economy, demography, culture, and cosmic civilizations* (pp. 9–77). Uchitel.

Korotayev, A., Goldstone, J. A., & Zinkina, J. (2015). Phases of global demographic transition correlate with phases of the Great Divergence and Great Convergence. *Technological Forecasting and Social Change, 95*, 163–169. https://doi.org/10.1016/j.techfore.2015.01.017

Korotayev, A. V., & Grinin, L. E. (2006). Urbanization and political development of the world system: A comparative quantitative analysis. In P. Turchin, L. Grinin, V. C. de Munck, & A. Korotayev (Eds.), *History and mathematics. Historical dynamics and development of complex societies* (pp. 115–153). URSS.

Korotayev, A. V., & Grinin, L. E. (2012). Kondratieff waves in the world system perspective. In L. E. Grinin, T. C. Devezas, & A. V. Korotayev (Eds.), *Kondratieff waves, dimensions and prospects at the dawn of the 21st century* (pp. 23–64). Uchitel.

Korotayev, A., & Grinin, L. (2013). Urbanization and political development of the world system. *ENTELEQUIA Revista Interdisciplinary, 15*, 197–254.

Korotayev, A. V., & Grinin, L. E. (2017). The technological activity and competition in the Middle Ages and Modern history: A quantitative analysis. In L. E. Grinin & A. V. Korotayev (Eds.), *History & mathematics: Economy, demography, culture, and cosmic civilizations* (pp. 78–102). Uchitel.

Korotayev, A. V., & Grinin, L. E. (2021). Mathematical model of interaction between civilization center and tribal periphery: A description. *Social Evolution & History, 20*(2), 50–78. https://doi.org/10.30884/seh/2021.02.03

Korotayev, A. V., Grinin, L. E., & Grinin A. L. (2022). Mathematical model of interaction between civilization center and tribal periphery: An analysis. *Social Evolution & History, 21*(1), 65–97. https://doi.org/10.30884/seh/2022.01.03

Korotayev, A., Klimenko, V., & Proussakov, D. (1999). Origins of Islam: Political-anthropological and environmental context. *Acta Orientalia Academiae Scientiarum Hungaricae, 53*(3–4), 243–276.

Korotayev, A. V., Komarova, N. L., & Khalturina, D. A. (2007). *Zakony istorii: Vekovyye tsikly i tysyacheletniye trendy. Demografiya, ekonomika, voyny*. KomKniga.

Korotayev, A., Malkov, A., & Khaltourina, D. (2006a). *Introduction to Social Macrodynamics: Compact Macromodels of the World System Growth*. KomKniga/URSS.

Korotayev, A., Malkov, A., & Khaltourina, D. (2006b). *Introduction to social Macrodynamics: Secular cycles and millennial trends*. KomKniga/URSS.

Korotayev, A., & Malkov, A. (2016). A compact mathematical model of the World System economic and demographic growth, 1 CE–1973 CE. *International Journal of Mathematical Models and Methods in Applied Sciences, 10*, 200–209.

Korotayev, A., & Zinkina, J. (2017). Systemic boundary issues in the light of mathematical modeling of world-system evolution. *Journal of Globalization Studies, 8*(1), 78–96.

Kosven, M. O. (1953). *Ocherki istorii pervobytnoy kul′tury*. Izdatel′stvo Akademii Nauk SSSR.

Kovalchuk, M. V. (2011). Konvergentsiya nauk i tekhnologiy—proryv v budushcheye. *Rossiyskiye nanotekhnologii, 6*(1–2). http://www.nrcki.ru/files/nbik01.pdf).
Kradin, N. (2023). Origins of the early Russian state: Anthropological perspectives. *Social Evolution & History, 22*(2), 67–93.
Kramer, S. N. (1965). *Istoriya nachinayet·sya v Shumere*. Nauka.
Kremkova, V. M. (1936). Veka X–XII. Iz istorii sel′skokhozyaystvennoy tekhniki Anglii i Frantsii. In N. Y. Vavilov & V. P. Williams (Eds.), *Istoriya agrikul′tury* (pp. 125–163). Izd-vo Akademii nauk.
Kurzweil, R. (2001). The law of accelerating returns. *KurzweilAI.net* 3-7-2001. http://www.kurzweilai.net/articles/art0134.html?printable=1
Kurzweil, R. (2005). *The singularity is near.* Viking Penguin.
Labriola, A. (1986) [1896]. *Essays on the materialistic conception of history*. Charles H. Kerr & Company.
Lamberg-Karlovsky, C. C., & Sabloff, J. A. (1979). *Ancient civilizations*. Benjamin/Cummings Publishing.
Lambert, D. (1991). *Doistoricheskiy chelovek*. Nedra.
Lee, J., & Wang, F. (1999). *One quarter of humanity*. Harvard University Press.
Li, S. (2018). The mace-head: A significant evidence of the early cultural interaction between west and east. *Social Evolution & History, 17*(2), 258–272.
Lieberman, S. (Ed.). (1972). *Europe and the industrial revolution*. Schenkman.
Lieberman, V. (Ed.). (1999). *Beyond binary histories: Re-imagining Eurasia to c. 1830*. University of Michigan Press.
Lieberman, V. (2003). *Strange parallels: Southeast Asia in global context, c. 800–1830*. Cambridge University Press.
Lilley, S. (1976). Technological progress and the industrial revolution, 1700–1914. In C. M. Cipolla (Ed.), *The industrial revolution, 1700–1914. The fontana economic history of Europe* (Vol. 3, pp. 187–254). The Harvest Press Limited; Barnes & Noble.
Lindauer, M. S. (2020). Evolutionary roots. In *Mass-produced original paintings, the psychology of art, and an everyday aesthetics*. Palgrave Macmillan. https://doi.org/10.1007/978-3-030-51641-3_8
Lips, J. E. (1956). *The origin of things*. Fawcett.
Livi-Bacci, M. (2017). *A concise history of world population*. Wiley-Blackwell.
Londo, J. P., Chiang, Y.-C., Hung, K.-H., Chiang, T.-Y., & Schaal, B. A. (2006). Phylogeography of Asian wild rice, *Oryza rufipogon*, reveals multiple independent domestications of cultivated rice, *Oryza sativa*. In *Proceedings of the National Academy of Sciences of the United States of America (PNAS)* (Vol. 103/25, pp. 9578–9583). http://www.pnas.org/content/103/25/9578.long
Longrich, N. (2020). Evolution's 'great leap forward': When did humans cross the intelligence Rubicon? https://geneticliteracyproject.org/2020/11/04/evolutions-great-leap-forward-when-did-humans-cross-the-intelligence-rubicon/
Lozovskaya, O. V., Vybornov, A. A., & Dolbunova, E. V. (Eds.). (2018). *Materialy mezhdunarodnoy konferentsii, posvyashchennoy 50-letiyu V. M. Lozovskogo, 15–18 maya*. Institut Istorii Material′noy Kul′tury RAN; Gosudarstvennyy Ermitazh.
Lucas, A. R. (2005). Industrial milling in the ancient and medieval worlds. A survey of the evidence for an industrial revolution in medieval Europe. *Technology and Culture, 46*(1), 1–30.
Lurie, I., Lyapunova, K., Mathieu, M., Piotrovsky, B., & Flitner, M. (1939). *Ocherki po istorii tekhniki drevnego Vostoka*. Izd-vo AN SSSR.
Lynch, Z. (2004). Neurotechnology and society 2010–2060. *Annals of the New York Academy of Sciences, 1031*, 229–233.
Maddison, A. (1983). A comparison of levels of GDP per capita in developed and developing countries, 1700–1980. *Journal of Economic History, 43*(1), 27–41.
Maddison, A. (2001). *Monitoring the world economy: A millennial perspective*. OECD.
Maddison, A. (2003). *The world economy: Historical statistics*. OECD.
Maddison, A. (2010). *World population, GDP and per capita GDP*, A.D. 1–2008. http://www.ggdc.net/maddison

Mantoux, P. (1929). *The industrial revolution in the eighteenth century: An outline of the beginnings of the modern factory system in England*. Jonathan Cape.
March, R. J. (2013). Searching for the functions of fire structures in Eynan (Malahha) and their formation processes; A geochemical approach. In O. Bar-Yosef & F. R. Valla (Eds.), *Natufian foragers in levant: Terminal Pleistocene social changes in western Asia. International monographs in prehistory* (pp. 227–283). Archaelogical Series 19.
Markov, A. V. (2011a). *Evolyutsiya cheloveka:* v 2 t. T. 1. *Obez'yany, kosti i geny*. Astrel', Corpus.
Markov, A. V. (2011b). *Evolyutsiya cheloveka:* v 2 t. T. 2. *Obez'yany, neirony i dusha*. Astrel', Corpus.
Markov, A. V. (2012). Antropogenez—osobaya glava global'noy istorii. In L. Ye. Grinin, I. V. Ilyin, & A. V. Korotayev (Eds.), *Universal'naya i global'naya istoriya. Evolyutsiya Vselennoy, Zemli, zhizni i obshchestva* (pp. 295–314). Uchitel.
Markov, G. E. (1979). *Istoriya khozyaystva i pervobytnoy kul'tury v pervobytnoobshchinnom i ranneklassovom obshchestve*. MGU.
Marks, A. E. (1993). The Early upper paleolithic: The view from the levant. In H. Knecht, A. Pike-Tay, & R. White (Eds.), *Before Lascaux: The complete record of the early upper paleolithic* (pp. 5–22). CRC Press.
Martin, J. (1981). *Telematic society*. Prentice Hall.
Marx, K. (1960) [1867]. Kapital. In K. Marks & F. Engels, *Sochineniya* (Vols. 1 and 23). Politizdat.
Masson, V. M. (1967). Pervobytnoye zemledeliye. In V. D. Blavatskiy & A. V. Nikitin (Eds.), *Vozniknoveniye i razvitiye zemledeliya* (pp. 5–32). Nauka.
Masson, V. M. (1980). Rannezemledel'cheskiye obshchestva i formirovaniye poseleniy gorodskogo tipa. In N. A. Butinov & A. M. Reshetov (Eds.), *Ranniye zemledel'tsy. Etnograficheskiye ocherki* (pp. 178–185). Nauka.
Masson, V. M. (1989). *Pervyye tsivilizatsii*. Nauka.
Matyushin, G. N. (1972). *U kolybeli istorii*. Prosveshcheniye.
McCabe, L. L., & McCabe, E. R. B. (2008). *DNA: Promise and peril*. University of California Press.
McEvedy, C., & Jones, R. (1978). *Atlas of world population history*. Facts on File.
McNeill, W. H. (1963). *The rise of the west*. University of Chicago Press.
McRae, R., & Pobiner, B. (2022). *Fourteen discoveries made about human evolution in 2022*. https://www.smithsonianmag.com/smithsonian-institution/fourteen-discoveries-made-about-human-evolution-in-2022-180981344/
Meadows, J. R. S., Cemal, I., Karaca, O., et al. (2007). Five ovine mitochondrial lineages identified from sheep breeds of the near east. *Genetics, 175*, 1371–1379. http://www.genetics.org/content/175/3/1371.full
Mellaart, J. (1975). *The Neolithic of the near east*. Thames and Hudson.
Mellaart, J. (1982). *Drevnejshie tsivilizatsii Blizhnego Vostoka*. Nauka.
Mellars, P. (2006). Why did human populations disperse from Africa ca. 60,000 years ago? A new model. *Proceedings of the National Academy of Sciences, 103*, 9381–9386.
Mellars, P., & Stringer, C. (Eds.). (1989). *The human revolution: Behavioural and biological perspectives on the origins of modern humans*. Princeton University Press.
Mellars, P., Boyle, K., Bar-Yosef, O., & Stringer, C. (Eds.). (2007). *Rethinking the human revolution*. McDonald Institute for Archaeological Research; University of Cambridge.
Melyantsev, V. A. (1996). *Vostok i Zapad vo vtorom tysyacheletii: ekonomika, istoriya i sovremennost'*. Moskovskiy gosudarstvennyy universitet.
Milisauskas, S. (Ed.). (2011a). *European prehistory a survey*. Springer.
Milisauskas, S. (Ed.). (2011b). *Early Neolithic, the first farmers in Europe, 7000–7500/5000 BC. European prehistory*. Springer.
Milisauskas, S., & Kruk, J. (2011). Late Neolithic, crisis, collapse, new ideologies, and economics, 3500/3000–2200/2000 BC (ch. 9). In S. Milisauskas (Ed.), *European history* (pp. 327–403). Springer.
Milskaya, L. T., & Rutenburg, V. I. (Eds.). (1993). *Istoriya Yevropy*. In 8 vols. Vol. 3. *Ot srednevekov'ya k Novomu vremeni*. Nauka.

Minghinton, W. (1976). Patterns of demand 1750–1914. In C. M. Cipolla (Ed.), *The industrial revolution, 1700–1914* (pp. 77–186). Harvester Press, Barnes & Noble.
Mochanov, Y. L (1977). *Drevneyshiye etapy zaseleniya chelovekom Severo-Vostochnoy Azii*. Nauka.
Mokyr, J. (1985). *The economics of the industrial revolution*. Allen & Unwin.
Mokyr, J. (1990). *The lever of riches: Technological creativity and economic progress*. Oxford University Press.
Mokyr, J. (Ed.). (1993). *The British industrial revolution: An economic perspective*. Westview Press.
Mokyr, J. (1999). *The British industrial revolution: An economic perspective* (2nd ed.). Westview Press.
Mokyr, J. (2002). *The gift of Athena*. Princeton University Press.
Mokyr, J. (2010). *The enlightened economy: An economic history of Britain, 1700–1850*. Yale University Press.
Mokyr, J., & Voth, H.-J. (2010). Understanding growth in Europe, 1700–1870: Theory and evidence. In S. Broadberry & K. H. O'Rourke (Eds.), *The Cambridge economic history of modern Europe 1700–1870* (Vol. 1, pp. 7–42). Cambridge University Press.
Monghite, A. L. (1973). *Arkheologiya Zapadnoy Yevropy*. Nauka.
More, C. (2000). *Understanding the industrial revolution*. Routledge.
Mumford, L. (1966) [1934]. *Technics and civilization*. Harcourt, Brace & Co.
Nefiodow, L. (1996). *Der sechste Kondratieff. Wege zur Produktivität und Vollbeschäftigung im Zeitalter der Information*. 1 Auflage/Edition. Sankt Augustin.
Nefiodow, L., & Nefiodow, S. (2014a). *The sixth Kondratieff. The new long wave of the world economy*. Sankt Augustin.
Nefiodow, L., & Nefiodow, S. (2014b). The sixth Kondratieff. The growth engine of the 21st century. In L. E. Grinin, T. C. Devezas, & A. V. Korotayev (Eds.), *Kondratieff waves. Juglar – Kuznets – Kondratieff* (pp. 326–354). 'Uchitel' Publishing House.
North, D. C. (1981). *Structure and change in economic history*. Norton.
Oppenheim, L. A. (1968). *Ancient Mesopotamia, portrait of a dead civilization*. The University of Chicago Press.
Osipov, G. V. (1959). *Tekhnika i obshchestvennyy progress*. Izdatel′stvo AN SSSR.
Özdoğan, M. (2022). Reconsidering the early Neolithic of Anatolia. Recent recoveries, some excerpts and generalities. *L'Anthropologie, 126*(3), 103033.
Pääbo, S. (1995). The Y-chromosome and the origin of all of us (Men). *Science, 268*, 1141–1142.
Peccoud, J. (2016). Synthetic biology: Fostering the cyber-biological revolution. *Synthetic Biology, 1*(1), 1–7. https://doi.org/10.1093/synbio/ysw001
Philipson, M. (Ed.) (1962). *Automation: Implications for the future*. Vintage.
Phillips, F., Su, Y. S. (2009). Advances in evolution and genetics: Implications for technology strategy. *Technological Forecasting and Social Change, 76*(5): 597–607. https://doi.org/10.1016/j.techfore.2008.08.006
Phyllys, D. (1965). *The first industrial revolution*. Cambridge University Press.
Pirenne, H. (1920–1932). *Histoire de Belgique* (Vol. I–VII). H. Lamertin.
Plakitkin, Y. A. (2006). *Zakonomernosti razvitiya mirovoy energetiki i ikh vliyaniye na energetiku Rossii*. IAI 'Energiya'.
Plekhanov, G. V. (1956) [1895]. K voprosu o razvitii monisticheskogo vzglyada na istoriyu. In G. V. Plekhanov, *Izbrannyye filosofskiye proizvedeniya* (pp. 507–730). Politizdat.
Pollock, S. (2001). *Ancient Mesopotamia*. Cambridge University Press.
Pomeranz, K. (2000). *The great divergence: China, Europe, and the making of the modern world economy*. Princeton University Press.
Pomeranz, K. (2002). Beyond the East-West binary: Resituating development paths in the eighteenth-century world. *Journal of Asian Studies, 61*(2), 539–590.
Porter, D. (Ed.). (2012). *Comparative Early Modernities 1100–1800*. Palgrave-Macmillan.
Postan, M. (1987). The trade of medieval Europe: The north. In M. M. Postan & E. Miller (Eds.), *The Cambridge economic history of Europe. Trade and industry in the middle ages* (Vol. II, pp. 168–305). Cambridge University Press.

Powell, A., Shennan, S., & Thomas, M. G. (2009). Late Pleistocene demography and the appearance of modern human behavior. *Science, 324*(5932), 1298–1301.

Power, R. C., & Williams, F. L. E. (2018). Evidence of increasing intensity of food processing during the Upper Paleolithic of Western Eurasia. *Journal of Paleolithic Archaeology, 1*(4), 281–301.

Pradel-Miquel, M. (2019). Urban revolution. *The Wiley Blackwell Encyclopedia of Urban and Regional Studies*, 1–6.

Prido, T. (1979). *Kroman'onskiy chelovek*. Mir.

Quilter, K., Ojeda, B. E., Pearsall, D. M., Sandweiss, D. H., Jones, J. G., & Wing, E. S. (1991). Subsistence economy in El Paraiso, an early Peruvian site. *Science, 251*, 277–283.

Randers, J. (2012). 2052. *A global forecast for the next forty years. A report to the Club of Rome*. Chelsea Green Publishing.

Randers, J., Rockström, J., Stoknes, P. E., Golüke, U., Collste, D., & Cornell, S. (2018). *Transformation is feasible. A report to the Club of Rome*. Stockholm Resilience Centre.

Rayner, R. M. (1964). *European history 1648–1789*. David McKay Company, Inc.

Reed, C. A. (Ed.). (1977). *The origins of agriculture*. Mouton.

Reinert, E. S. (2007). *How rich countries got rich and why poor countries stay poor*. Constable.

Renfrew, C. (2002). Indoyevropeyskaya problema i osvoyeniye Yevraziyskikh stepey: voprosy khronologii. *Vestnik Drevney istorii*, 2, 20–32.

Rietbergen, P. J. A. N. (2002). *A short history of The Netherlands. From prehistory to the present day* (5th ed.). Bekking.

Rifkin, J. (2011). *The third industrial revolution. How lateral power is transforming. Energy, the economy, and the world*. Palgrave Macmillan.

Rindos, D. (1984). *The origins of agriculture: An evolutionary perspective*. Academic Press.

Ritchie, K., Hufthammer, A. K., & Bergsvik, K. A. (2016). Fjord fishing in Mesolithic western Norway. *Environmental Archaeology, 21*(4), 309–316.

Roberts, N. (1998). *The Holocene: An environmental history*. Blackwell.

Roginsky, Y. Y. (1977). *Problemy antropogeneza*. Vysshaya shkola.

Romanchuk, A. (2023). 'Blind Spots' of the 'Scandinavian-Centric' hypothesis on the origin of the old Russian state. *Social Evolution & History, 22*(2), 94–110.

Rosenthal, J.-L., & Wong, R. B. (2011). *Before and beyond divergence: The politics of economic change in China and Europe*. Harvard University Press.

Rozov, N. S. (2023). *The origin of language and consciousness. How social order and communicative concern gave rise to speech and cognitive abilities*. Springer. In print.

Roux, V. (2017). Ceramic manufacture. The Chaîn Operátory Approach. In A. M. W. Hunt (Ed.), *The Oxford handbook of archeological ceramic analysis*. Chapter 8 (pp. 103–113). Oxford University Press.

Rumyantsev, A. M. (1987). *Pervobytnyy sposob proizvodstva (politiko-ekonomicheskiye ocherki)*. Nauka.

Sabo, Y. P. (1979). *Revolyutsiya mashin. Istoriya promyshlennogo perevorota*. Korvina.

Sachs, J. D. (2020). 2. The paleolithic age (70.000–10.000 BCE). In *The Ages of Globalization* (pp. 33–40). Columbia University Press.

Saliba, G. (2007). *Islamic science and the making of the European renaissance*. MIT Press.

Shcherbakova, Y. M. (2011). V oktyabre 2011 goda chislennost' naseleniye mira prevysit 7 milliardov chelovek. *Demoskop Weekly*, 481–482, October 10–23. http://www.demoscope.ru/weekly/2011/0481/barom01.php

Schultz, E. A., & Lavenda, R. H. (1998). *Anthropology. A perspective on the human condition* (2nd ed.). Mountain View.

Schwab, K. (2016). *The forth industrial revolution*. World Economic Forum.

Seltzer, A. M., Ng, J., Aeschbach, W., Kipfer, R., Kulongoski, J. T., Severinghaus, J. P., & Stute, M. (2021). Widespread six degrees Celsius cooling on land during the Last Glacial Maximum. *Nature, 593*(7858), 228–232.

Semenov, S. A. (1968). *Razvitiye tekhniki v kamennom veke*. Nauka.

Semenov, S. A., & Korobkova, G. F. (1983). *Tekhnologiya drevneyshikh proizvodstv*. Nauka.

Sergeeva, V. G. (1983). Voprosy zaseleniya Ameriki i transokeanskikh kontaktov v trudakh Khuana Komasa. In Yu. V. Bromley (Ed.), *Puti razvitiya zarubezhnoy etnologii* (pp. 138–151). Nauka.
Shea, J. J. (2007). Behavioral differences between middle and upper Paleolithic homo sapiens in the east Mediterranean levant: The role of intraspecific competition and dispersal from Africa. *Journal of Anthropological Research, 63*, 449–488.
Shea, J. (2013). *Stone tools in the paleolithic and Neolithic near east: A guide.* Stony Brook; Stony Brook University.
Sheypak, A. A. (2009). *Istoriya nauki i tekhniki. Materialy i tekhnologii.* Part 1. MGIU.
Shnirelman, V. A. (1989). *Vozniknoveniye proizvodyashchego khozyaystva.* Nauka.
Shnirelman, V. A. (2012a). *Vozniknoveniye proizvodyashchego khozyaystva: Rychagi drevneyshego zemledeliya* (2nd ed.). LIBROKOM.
Shnirelman, V. A. (2012b). *Proiskhozhdeniye skotovodstva: Kul'turno-istoricheskaya problema* (2nd ed.). LIBROKOM.
Sidorovich, V. (2015). *Mirovaya energeticheskaya revolyutsiya.* Al'pina Publisher.
Silverstein, J. E. (2021). Redefining evolution: Life beyond the limits of neo-Darwinian theory. *Social Evolution & History, 20*(2), 3–26.
Simmons, T. (2013). Avifauna of the final Natufian of Eynan. In O. Bar-Yosef, & F. R. Valla (Eds.), *Natufian foragers in levant: Terminal pleistocene social changes in western Asia. International monographs in prehistory* (pp. 284–292). Archaeological Series 19.
Singer, C. (1941). *A short history of science to the nineteenth century.* Oxford University Press.
Smil, V. (1994). *Energy in world history.* Westview.
Smith, P. E. L. (1976). *Food production and its consequences.* Cumming Publishing Company.
Snooks, G. D. (1997). *The ephemeral civilization.* Routledge.
Spassov, S. (2003). The information technology revolution: A new techno-economic paradigm. In D. Ginev (Ed.), *Bulgarian studies in the philosophy of science. Boston studies in the philosophy of science* (Vol. 236). Springer. https://doi.org/10.1007/978-94-017-0961-3_3
Stearns, P. N. (1993). Interpreting the industrial revolution. In M. Adams (Ed.), *Islamic and European expansion. The forging of a global order* (pp. 199–242). Temple University Press.
Stearns, P. N. (Ed.). (1998). *The industrial revolution in the world history* (2nd ed.). Westview.
Steiger, B. (2007). *Worlds before Our Own.* Anomalist Books.
Sterling, B. (2005). *Budushcheye uzhe nachalos'. Chto zhdet nas v XXI veke.* U-Faktoriya.
Stonier, T. (1983). *The wealth of information: A profile of the post-industrial economy.* Methuen Publishing.
Strickland, A. (2023). Mysterious species buried their dead and carved symbols 100,000 years before humans *CNN* 05.06.2023. https://edition.cnn.com/2023/06/05/world/homo-naledi-burials-carvings-scn/index.html
Stringer, C. B. (1990). The emergence of modern humans. *Scientific American, 263*(6), 68–74.
Stringer, C. B. (2011). *The origin of our species.* Allen Lane.
Sylvester, E., & Klotz, L. C. (1983). *The gene age: Genetic engineering and the next industrial revolution.* Scribner.
Tanno, K., Willcox, G., Muhesen, S., Nishiaki, Y., Kanjo, Y., & Akazawa, T. (2013). Preliminary results from analysis of charred plant remains from a burnt Natufian building at Dederiyeh cave in northwest Syria. In O. Bar-Yosef & F. R. Valla (Eds.), *Natufian foragers in Levant: Terminal Pleistocene social changes in western Asia. International monographs in prehistory* (pp. 83–87). Archaeological Series 19.
Tanton, J. H. (1994). End of the migration epoch? Time for a new paradigm. *The Social Contract, 4*(3), 162–173.
Tattersall, I. (2008). *The world from its beginnings to 4000 BCE.* Oxford University Press.
Tattersall, I. (2012). *Masters of the planet: Seeking the origins of human singularity.* Palgrave Macmillan.
Thomlinson, R. (1975). *Demographic problems: Controversy over population control* (2nd ed.). Dickenson.
Thurow, L. C. (1996). *The future of capitalism.* Nicholas Brealey Publishing.
Toffler, A. (1980). *The third wave.* Morrow.

Toffler, A. (1985). *The adaptive corporation*. McGraw-Hill Book Co.
Toffler, A. (1990). *Powershift: Knowledge, wealth, and violence at the edge of the 21st century*. Bantam Books.
Toffler, A., & Toffler, H. (1995). *Creating a new civilization*. Turner Publishing.
Touraine, A. (1974). *The post-industrial society. Tomorrow's social history: Classes, conflicts and culture in the programmed society*. N.Y.
Touraine, A. (1983). *From exchange to communication: The birth of programming society. The human use of human ideas. The discoveries project and eco-technology*. Pergamon Press, Ltd.
Toynbee, A. (1927) [1884]. *Lectures on the industrial revolution of the eighteenth century in England: Popular addresses, notes, and other fragments*. Rivingtons.
Toynbee, A. (1956) [1884]. *The industrial revolution*. Beacon Press.
Turnbaugh, W. A., Nelson, H., Jurmain, R., & Kilgore L. (1993). *Understanding physical anthropology and archaeology* (5th ed.). West Publishing Company.
Turner, J. H., & Machalek, R. (2018). *The new evolutionary sociology: Recent and revitalized theoretical and methodological approaches*. Routledge, Taylor & Francis Group.
Tylecote, R. F. (1976). The technique and development of early copper smelting. In R. F. Tylecote (Ed.), *A history of metallurgy* (pp. 5–9). The Metals Society.
Urlanis, B. T. (1941). *Rost naseleniya v Evrope*. OGIZ-Gospolitizdat.
Valentey, D. I. (red.). (1974). *Marksistsko-leninskaya teoriya narodonaseleniya*. 2-ye izd. Mysl'.
Vandam, R., Mušič, B., & Medarić, I. (2019). Contextualizing Kuruçay Höyük: Assessing the unexplored late chalcolithic landscape near the beginning of early social complexity in SW turkey. *Journal of Field Archaeology, 44*(1), 1–18.
Vega-Centeno, R. (2010). Cerro Lampay: Architectural design and human interaction in the north central coast of Peru. *Latin American Antiquity, 21*(2), 115–145.
Velichko, A. A. (1989). Sootnosheniye izmeneniy klimata v vysokikh i nizkikh shirotakh Zemli v pozdnem pleystotsene i golotsene. In A. A. Velichko, Y. Y. Gurtovaya, & M. A. Faustova (Eds.), *Paleoklimaty i oledeneniya v pleystotsene* (pp. 5–19). Nauka.
Vinogradov, I. V. (2000). Novoye tsarstvo v Yegipte i pozdniy Yegipet. In V. A. Yakobson (Ed.), *Istoriya Vostoka*. Vol. 1. *Vostok v drevnosti* (pp. 370–432). Vostochnaya literatura.
Virginsky, V. S., & Khoteenko, V. F. (1993). *Ocherki istorii nauki i tekhniki s drevneyshikh vremen do serediny XV veka*. Prosveshcheniye.
Vishnevsky, A. G. (2005). *Izbrannyye demograficheskiye trudy. V dvukh tomakh Tom I. Demograficheskaya teoriya i demograficheskaya istoriya*. Nauka.
von Weizsäcker, E. U., & Wijkman, A. (2018) *Come on! Capitalism, short-termism, population and the destruction of the planet. A report to the Club of Rome*. Springer.
Vries, J. D., & Woude, A. V. D. (1997). *The first modern economy. Success, failure, and perseverance of the Dutch economy, 1500–1815*. Cambridge University Press.
Vries, P. (2013). *Escaping poverty: The origins of modern economic growth*. W&R Unipress.
Wallerstein, I. (1974). *The modern world-system* (Vol. 1). Academic Press.
Wallerstein, I. (1987). *The modern world-system* (Vol. 3). Academic Press.
Wells, P. S. (2011). The iron age. In S. Milisauskas (Ed.), *European prehistory. Interdisciplinary contributions to archaeology* (pp. 461–464). Springer.
Wenke, R. J. (1990). *Patterns in prehistory*. Oxford University Press.
White, L. Jr. (1978). *Medieval religion and technology: Collected essays*. University of California Press.
White, T. D., Asfaw, B., DeGusta, D., Gillbert, H., Richards, G. D., Suwa, G., & Howell, F. C. (2003). Pleistocene *Homo sapiens* from middle Awash, Ethiopia. *Nature, 423*, 742–747.
Wijkman, A., & Skånberg, K. (2017). *The circular economy and benefits for society. Jobs and climate clear winners in an economy based on renewable energy and resource efficiency. A report to the Club of Rome*. Club of Rome.
Wilkins, J., Schoville, B. J., Pickering, R., Gliganic, L., Collins, B., Brown, K. S., Hatton, A., et al. (2021). Innovative homo sapiens behaviours 105,000 years ago in a wetter Kalahari. *Nature, 592*(7853), 248–252.

Wilkinson, D. (2022). The power configurations of the central civilization/world system in the eighth century. *Social Evolution & History, 21*(1), 33–64.
Wong, R. B. (1997). *China transformed: Historical change and the limits of the European experience.* Cornell University Press.
Wymer, J. (1982). *The paleolithic age.* Croom Helm.
Xie, C., & Cheng, E. (2016). Deepening in division of labor: An analysis of five great social divisions of labor and intra-sector division of labor. 马克思主义研究, (12).
Yanin, V. L. (\Ed.). (2006) *Arkheologiya.* MGU.
Yarygin, V. N., Vasil'yeva, V. I., Volkov, I. N., Sinel'shchikova, V. V. (1999). *Biologiya.* Vysshaya shkola.
Yasamanov, N. A. (1985). *Drevniye klimaty Zemli.* Gidrometeoizdat.
Yasrebitskaya, A. L. (1993). Material'naya kul'tura i obraz zhizni v Yevrope na iskhode srednevekov'ya. In L. T. Mil'skaya & V. I. Rutenburg (Eds.), *Istoriya Yevropy.* In 8 vols. Vol. 3 *Ot srednevekov'ya k Novomu vremeni* (pp. 16–40). Nauka.
Young, L. D. (2022). Testing tilly: Does war really make states? *Social Evolution & History, 21*(1), 175–199.
Yusoff, K. (2015). Anthropogenesis: Origins and endings in the Anthropocene. Theory. *Culture and Society, 33*(2). https://journals.sagepub.com/doi/abs/10.1177/0263276415581021
Zablotska, Y. (1989). *Istoriya Blizhnego Vostoka v drevnosti (ot pervykh poseleniy do persidskogo zavoyevaniya).* Nauka.
Zeder, M. A., & Hesse, B. (2000). The initial domestication of goats (*Capra hircus*) in the Zagros mountains 10,000 years ago. *Science, 287*(5461), 2254–2257. https://doi.org/10.1126/science.287.5461.2254
Zhdanko, A. V. (1999). Pis'mo v redaktsiyu. Zametki o pervobytnoy istorii (sovremennyye dannyye paleoantropologii i paleoarkheologii o vozniknovenii Homo sapiens). *Filosofiya i Obshchestvo, 5*, 175–177.
Zimmer, C. (2003). Great mysteries of human evolution. *Discover, 24*(9), 34–44.
Zinkina, J., Christian, D., Grinin, L., Ilyin, I., Andreev, A., Aleshkovski, I., Shulgin, S., & Korotayev, A. (2019). *Big history of globalization. The emergence of a global world system.* Springer.
Zinkina, J., Ilyin, I., & Korotayev, A. (2017). The early stages of globalization evolution: Networks of diffusion and exchange of domesticates, technologies, and luxury goods. *Social Evolution & History, 16*(1), 69–85.
Zubov, A. A. (1963). *Chelovek zaselyayet svoyu planetu.* Gheografiya.

# Cybernetic Revolution and Self-managing Systems

# 3

**Abstract**

This chapter is dedicated to the consideration of the Cybernetic Revolution that is the last of the greatest technological revolutions in all history following the Agrarian and Industrial ones. It is a major transition from the industrial production to the production and service sector based on the implementation of self-regulating systems. The first phase of this revolution started in the 1950 and 1960s. Between the 2030 and 2070s, the final phase of this revolution will lead to a new level of self-operating control, namely the level of self-regulating/managing systems. They are systems that, by means of the embedded programs and a number of intelligent components, can regulate themselves to operate independently with no human intervention. These systems will become the major part of technological process during the forthcoming phase of the Cybernetic Revolution named the epoch of self-regulating/managing systems. But these systems are not only technical devices but a wider range of systems and control processes of a biological, physiological, techno-biological, social and other nature, with a high level of self-regulation, which are and will be implemented in various areas (including medicine, genetic engineering, robotics, social relations). Special sections discuss the differences between self-regulating systems and artificial intelligence.

## 3.1 What This Chapter Is About

This chapter is dedicated to the consideration of the Cybernetic Revolution, its phases and main features as well as the characteristics of self-regulating systems and their correlation with AI. According to our concept, **the Cybernetic Revolution** is the last of the greatest production (technological) revolutions in all history



L. Grinin et al., *Cybernetic Revolution and Global Aging*, World-Systems Evolution and Global Futures, https://doi.org/10.1007/978-3-031-56764-3_3

following the Agrarian and Industrial ones. It is a major transition from the industrial production to the production and service sector based on the implementation of self-regulating/self-managing systems.

The first phase of this revolution started in the 1950 and 1960s and brought the development of powerful information technologies, new materials and energy sources, as well as a diffusion of automation. Between the 2030 and 2070s, the final phase of this revolution will lead to a new level of self-operating control in technical systems, namely the level of self-regulating systems.

This chapter explains why the Cybernetic Revolution is called Cybernetic. How can it influence future life? What kind of technological developments can we expect in the next half of this century? And so on.

Self-regulating/self-managing systems are systems that, by means of the embedded programs and a number of smart components, can regulate themselves, responding intelligently to the feedback from the environment as well as to operate and make decisions or suggest the best solutions independently in a wide range of variations or cases of changing situations with minimal or no human intervention. Of utmost importance is the fact that self-regulating/self-manging systems will become the major part of technological process. That is the reason why we denote the final (forthcoming) phase of the Cybernetic Revolution as the epoch of self-regulating systems.

This chapter gives a classification of self-regulating/self-managing systems of different levels, from smart and intelligent systems with elements or limited programs of self-regulation to systems with artificial intelligence that allow making decisions within a fairly wide range of possible (re)actions (self-driving cars of the future; housekeeping robots; a system that independently selects plant varieties suitable for particular regions, etc.). By self-regulating/self-managing systems we mean not only machines and other technical devices but a wider range of systems and control processes, including those of a biological, physiological, techno-biological, social and other nature, with a high level of self-regulation, which are and will be implemented in various areas (including medicine, genetic engineering, robotics, social relations).

Special sections discuss the differences between self-regulating/self-managing systems and artificial intelligence. We list the most important characteristics of the Cybernetic Revolution. These features are closely interconnected and mutually reinforcing. The dominant directions of the final phase of the Cybernetic Revolution are considered.

One of the central concepts of our monograph is the abbreviation MANBRIC, which, according to our forecasts, is an acronym for the names of the seven most active new areas of technological development, which will eventually constitute a powerful system, even a convergence. They are respectively: Medicine, Additive technologies, Nanotechnologies, Biotechnologies, Robotics, Information technologies (Artificial Intelligence), Cognitive technologies. It is this complex that will become the core of future technological development.

## 3.2 Introductory Notes. What Is the Cybernetic Revolution and Why is it Cybernetic?

As defined in Chap. 2, *the Cybernetic Revolution* is the latest of the three most important technological revolutions in history following the Agrarian and Industrial ones. *The Cybernetic Revolution is a great breakthrough from industrial production to the production and services based on the implementation of self-regulating and self-managing systems*[1] (see Grinin, 2006; Grinin & Grinin, 2015a, 2015b, 2016; Grinin et al., 2017a, 2017b, 2020, 2021). The first phase of this revolution started in the 1950 and 1960s and brought the development of powerful information technologies, new materials and energy sources, as well as a diffusion of automation. Between the 2030 and 2070s, the final phase of this revolution will lead to a new level of self-operating control in technical systems, namely, to the level of self-regulating/managing systems. We will witness not only the diffusion of smart and intelligent and truly independent (without human control) operating systems, but also the proliferation of self-regulating systems beyond the purely technical sphere to include also biological, medical, techno-biological, ecological and even social domains. The possibility to organize highly sophisticated self-regulating and self-managing systems in all spheres of production, human activity, health, environment and life will provide new opportunities for controlling some biological, physiological, genetic, chemical, physical and other natural processes that are currently beyond our control.

Thus, the Cybernetic Revolution, which has been unfolding over the last seven decades, will continue for another 60–70 years or so; so its total duration since the 1950s, according to our estimates, will be 130–140 years. We are likely to witness its new, final phase very soon. However, the dramatic changes will bring not only undeniable achievements (as we point out in Chap. 1), but also problems and challenges that humanity has never faced before. At present, some of these changes are already deeply felt. In this regard, it is crucial to develop a deeper understanding of this revolution and its main trends and characteristics. We will return to this a little later; for now, we will focus only on its most important features.

We suppose that *the most important characteristic of the final phase of the Cybernetic Revolution will be a wide implementation of self-regulation* ***to the point of self-management in various systems***. As we said above, by *self-regulating systems we mean not only machines and other technical* devices, *but a wider range of systems and control processes, including those of a biological, physiological, techno-biological, social and other nature, with a high level of self-regulation and self-management which are and will be implemented in various fields (including medicine, biotechnology, genetic engineering, robotics, social relations, etc.).* In the

[1] We borrowed the term from computer science, from its theory of control. The idea is the same as we point out for all systems—to reduce operating costs by increasing automation, ideally to by allowing systems to be self-managing without any human intervention (Diao et al., 2005).

following, we will discuss self-regulating systems in detail, along with examples, current and future implementations, their main characteristics, etc.

Our analysis of such systems relies on the ideas from an academic field called *cybernetics* which studies the control of various systems with complex regulation (see, *e.g.*, Wiener, 1948; Ashby, 1956; Beer, 1959, 1994; Foerster & Zopf, 1962; Umpleby & Dent, 1999; Tesler, 2004; McClelland & Fararo, 2006; Dupuy, 2009; Wieser & Slunecko, 2014).

The emergence of cybernetics as a scientific discipline in the 1940 and 1950s and the development of automation are closely related phenomena, since cybernetics became a scientific response to advances in automation. Cybernetics can be defined as the study of general laws and patterns of receiving, storing and processing information in complex systems. Its main principles are quite suitable for the description of self-regulation. In any case, the concepts of regulation and information are considered of utmost importance in cybernetics. Since the beginning of the Cybernetic Revolution, technology has been intertwined with information processing, and complex administrative systems have become of particular importance. Regulation and self-regulation (as the highest form of regulation) within systems are also major categories in Cybernetics. Hence, since the forthcoming phase will be the epoch of self-regulating systems, we denote it as the Cybernetic Revolution.

The analysis of the Cybernetic Revolution in the present and following chapters will be related to the general theoretical aspects of this production revolution as well as to particular forecasts for certain directions (however we will discuss forecasts elsewhere in this monograph). In Chap. 3 we will speak about self-regulating and self-managing systems as the most important characteristic and outcome of the Cybernetic Revolution. We will show how smart technologies fit into the logic of this revolution. Then, we will consider the main features of the Cybernetic Revolution, which will be manifested to varying degrees in all technologies.

In Chap. 4, we will study the main directions of the Cybernetic Revolution. Among numerous technological directions of the future, we identify seven of the most leading ones (medicine, additive technologies, nanotechnologies, biotechnologies, robotics, information, including AI and cognitive technologies). These trends are emerging and in the future they will eventually combine into a single complex, which we call the MANBRIC convergence. In Part 3 (in Chaps. 8, 9 and 10) we will present a detailed analysis of these seven and some other technological directions and make some predictions about their development over the next four to five decades.

## 3.3 The Difference Between Self-regulating and Self-managing Systems

As we have already mentioned above, there are different kinds of self-regulating systems, so it is worth speaking about a large class of such systems. However, within this class, it is necessary to single out two sub-classes. Such a subdivision makes a lot of sense, because it is far from always possible to use only the term "self-regulating" systems as a general term. The problem with the term "self-regulating systems" is that it is not always appropriate to use it both for cases when systems are able to regulate complex processes to support a certain regime and functions (e.g., artificial organs as kidneys, etc., or self-purification of waste water) as well as for systems that can not only regulate something but also manage (rule, govern) themselves, e.g., such future technological systems as fully autonomous driverless cars or fully functional robots that can move, talk, make decisions, react to dangers, protect people and so on. Both of these types can be grouped into a broad class of new systems. However, we want to divide this class into two sub-classes. The first are the self-regulating systems proper, which are capable of regulating complex processes to support a particular regime and functions. The second are what we want to call self-managing systems, which can control, manage themselves, and even develop themselves (self-learning, self-evolution, etc.). Below we discuss this in detail.

In the present chapter we only briefly consider the leading directions of the final phase of the Cybernetic Revolution.

## 3.4 Self-regulating Systems

### 3.4.1 Automation and Self-regulation

The Cybernetic Revolution began in the mid-twentieth century as a revolution in which automated and autonomous systems began to be widely implemented.

By the end of the twentieth century, automation had penetrated all spheres of human life. Automation implies that a machine is equipped with automatic control devices that allow it to "monitor" the results of its work and make "decisions" what to do next. Thus, the machine acquires the ability to independently perform a long chain of complex operations. Thus, it no longer needs an operator. A human's role is reduced to the design and production of a machine, and to the maintenance of its functioning (Lilley, 1966).

As we have already mentioned, in the course of modernization and especially in the final phase of the Cybernetic Revolution, the introduction of a new type of systems, i.e., of self-regulating ones, is underway. As we will see, their main feature is not just autonomous operation, but also the ability to make decisions independently and, basing on these decisions, to change their functional parameters in variable conditions. In other words, in such systems, the decision-making center is configured in such a way that human intervention and direct control are not required to make a more or less complex decision within the framework of changing tasks, environment and emerging various challenges. Self-regulating systems, with a maximally developed potential to make a complex decision, can be called **self-managing systems**.

Thus, automation is the least developed level of self-regulation since automatic systems operate most of the time and perform most of their functions in a stand-alone mode. They require only a little control and reconfiguration. But on the other hand, the process automation can only operate under stable conditions, while the means of protecting these conditions function within strictly defined parameters of change (e.g., accident or its threat). Thus, it turns out that automation, which was considered the major process at the dawn of the Cybernetic Revolution (e.g., Philipson, 1962; Bernal, 1965; Lilley, 1966, 1976; Benson & Lloyd, 1983), actually turned out to be only the starting point of this transition to self-regulation.

Self-regulation implies autonomous operation under highly variable and even unstable conditions, to which a system is able to adapt, and also making independent decisions based on a set of priorities and parameters. Simply put, the difference between automation and self-regulation/self-management is connected with who makes a decision in changing situations: we or a machine.

However, at present, we have not yet achieved a full-scale and fully functional self-regulation of systems (except in a few particular cases). We believe that we will advance toward this level of self-regulation during the final phase of the Cybernetic Revolution. But in any case, one may point out that there are a number of levels of self-regulating systems of varying complexity between automation and self-management of technical systems.

For example, let us consider a navigator. There are usually several ways to reach destination, but since navigators can take directions from any point, they are supposed to be able to guide a driver to the destination from any place. The number of variations is usually very large. Here the degree of self-regulation can be considered high, although the device itself is not very complicated.

A navigator which seemed to be a magic two decades ago is now considered a common device that can be installed on any smartphone (which itself seemed to be unimaginable a short time ago). But the example of a navigator gives us a good reason to speculate on some characteristics of self-regulating systems.

The navigator is considered to be a self-regulating system, firstly, because it operates under constantly changing conditions and adapts instantly to changing circumstances. Secondly, it adapts to changing conditions in order to achieve a certain goal (to guide us to a certain point). And this condition is the *moment of target setting*, which is embedded in a self-regulating system and the latter strictly adheres to under all circumstances, thus constituting the core, the basis of its functioning, and *this is fundamentally important for self-regulating systems*. Thirdly, the navigator actively imposes its decisions. And if our knowledge of the route is less than that of this self-regulating system, then the latter becomes dominant in the driver-navigator system.

So, we can conclude that decision-making is not a single action, but a whole sequence linked to different circumstances and factors, as well as to characteristics of both a system and an individual (experience, character, urgency and speed of response, knowledge, competencies, mood, etc.). Thus, *self-regulating systems can be classified according to the degree of their influence on decision-making, up to the full acceptance and unconditional decisions.* So systems, like a navigator, can become universal nodes for more advanced self-regulating and self-managing systems.[2] Further we will give such a gradation in terms of the complexity of self-regulating systems.

Another important characteristic of the most complex self-regulating systems is the ability to learn, the presence of flexible programs that can change under the influence of changes in the environment or tasks. And they can also be classified according to the level of complexity of these functions and parameters.

### 3.4.2 Self-regulating Systems and Artificial Intelligence

Already today there are many self-regulating systems that are able to navigate in the environment, reconfigure themselves and find solutions depending on the situation. These are, for example, unmanned aerial vehicles and self-driving cars, homing missiles and artificial earth satellites. Another good example is life-support systems (such as medical ventilation apparatus or artificial hearts). They can regulate a number of parameters, choose the most appropriate mode and detect critical situations. Nano or biochips implanted in our bodies are also examples of self-regulation.

Actually, self-regulating systems have been actively operating in the non-cash financial system for a long time, (the ATM can be considered as a system of the kind). Also, special programs have been already available for a long time that can determine the value of shares and other securities, respond to changes in their value, buy and sell them automatically, make thousands of transactions per day and fix profits. And these are just a few examples of self-regulating systems that already work. But in most cases, they are technical or informational (like industrial robots or computer programs). And today, far from all of them can be called purely self-regulating (not to say self-managing) since the role of a human is still very considerable.

Of course, a large number of self-regulating systems have been created in the form of robots or software. However, as mentioned above, *during the final phase of the Cybernetic Revolution, many self-regulating systems of different from these types will emerges* in such spheres as **biotechnology**, **bionics**, **physiology**, **medicine**, **agriculture**, **the environment**, **social relations**, etc. The number of such systems,

[2] Similarly, applied mechanics had developed over centuries, constantly combining numerous units and systems (for details see Grinin & Grinin, 2015b; Bogolyubov, 1988; Lilley 1966).

as well as their complexity and autonomous character, will dramatically increase. Additionally, they will significantly reduce energy and resource consumption.

Along with self-regulating systems that can navigate in the environment, reconfigure themselves and find solutions depending on the situation (what drones or trading systems on stock exchanges are able to do today), more complex systems will emerge that can adapt to specific contexts, reconfigure their functionality and learn within the framework of the opportunities built into their flexible program, and control people's behavior and minds. These could be, for example, online health monitoring systems. If some of the body's functions are working abnormally, they can take necessary actions, for instance, call an ambulance. Possible examples here are also systems for curing diseased cells by making independent decisions about the degree of their deviation; robots capable of taking care for the bed-bound; artificial organs capable of replacing natural ones to a certain extent; emergency response systems as a techno-social system, etc. In Chaps. 6, 8 and 15 we will describe in detail our vision of such a very complex and sophisticated self-managing system, which we call the "medical-biotechnological environment".[3]

As a result, the opportunity to control various natural, social and production processes without direct human intervention (which is currently impossible or considerably limited) will increase. Social and administrative relations will be regulated by social-technical self-managing systems, which will form a basis for the establishment of an electronic state (e-state; for details see Chap. 14). Human everyday life itself will be increasingly organized by such self-regulating systems (e.g., through the above-described health monitoring, probably the control of day regimes, the regulation or recommendation of load limits, the prevention of illegal actions, etc.) (for more details on this, as well as on the negative aspects of such regulation, see below and in Chaps. 8 and 15).

*Thus, self-regulating* and self-managing *systems are systems that by means of the embedded programs and intelligent components can regulate themselves, respond intelligently to the feedback from the environment and operate and make decisions*

[3] Here is the short essence of a detailed explanation of this system. A medical-biotechnological environment will be a system for health monitoring, control and possible help, when treatment and support of the organism will be performed mostly by autonomous systems, which will be able to function regularly and constantly. In particular, there will be a constant scanning of the organism including certain organs, functions, biometric parameters, etc. This will be a good example of a self-managing system making a complex decision autonomously. At the same time, these systems will be able to monitor significant fluctuations of indicators and make recommendations on short- and long-term living habits, as well as try to prevent deviations from the norm and deviations in behavior. Such a system, having perceived disturbing signals in the mood of the observed persons, can transfer the communication with them directly to a specialist (of course, by recording the conversation and analyzing it via neuro-computer methods or other analysis). That is, the technologies will allow the system to carry out some actions under certain circumstances, for example, to give an injection, artificial respiration, provide first aid or something like that. Or to give instructions to other people (systems) to perform such actions, as well as to contact urgent medical-technological centers.

*or suggest best solutions independently in a wide range of variations with minimal or no human intervention. They are capable of making operational decisions under changing conditions, selecting optimal regimes for a given environment and achieving certain goals and tasks, according to their priority* (Grinin & Grinin, 2015a, 2015b; Grinin & Grinin, 2015b, 2016; Grinin et al., 2017a, 2017b, 2020, 2021; Grinin et al., 2016).

This definition shows the enormous complexity of such self-regulating systems and, respectively, the difficult path toward their creation. The principle of self-regulation requires a constant interaction between the environment and the internal state of the objects to which the activity of the system and its own components is directed.[4] Such systems should also take into account possible changes in tasks and priorities in the selection of a goal, etc. Similarly to *living beings* (biological systems), self-regulating technical and biotechnological systems have a control subsystem (a kind of brain), peripheral subsystems for various purposes (a kind of sensory organs and subsystems that carry out necessary movements and shifts in a complex space) and a system of connecting them through direct communication and feedback. It is clear that central "organs" of a self-regulating system should intervene in the case of unexpected deviations from the predetermined parameters or when an important resetting of parameters is needed. So it is necessary to provide a signal of changing parameters and a message requesting the permission for some changes, as well as a number of possible options.

The above-mentioned complexity of self-regulating and even more so of self-managing systems inevitably leads to the question of whether artificial intelligence can be considered a self-managing system. And it also raises the question of whether it makes sense to speak about AI instead of self-regulating systems. First of all, we should emphasize that in any case AI does not cover the whole range of self-regulating systems. Let us point out that this refers not only to technical but also to biological, medical, compound (e.g., bio-technical, social-technical) and some other types of self-regulating systems that will become widespread in the final phase of the Cybernetic Revolution.

It is no less important to note that *AI and self-regulating/managing systems are related but not synonymous concepts, and are actually concepts defining different classes of objects*. We will return to the issue of AI, including General AI and new AI chats, in Chap. 10, and here we will make only a few brief comments.

Firstly, although the concept of AI is used for a long time, it still remains rather vague and ambiguous, and this fact hampers its scientific application.

Secondly, one should realize that AI refers to technical systems, while the class of self-regulating and self-managing systems, as we have pointed out above, is much broader. For example, silencing or activating genes, implanting artificial organs, etc., by no means relate to AI.

---

[4] By the way, the interrelation between the environment and the 'decision' that a system makes on the basis of changes in the environment, is also the most important idea in cybernetics.

Thirdly, even in technical systems (machines), AI is not a self-regulating system but a generalized concept for technologies that help numerous smart and intelligent and self-regulating systems to function[5]. These are different categories, just like the principle of using electric current and main components versus general approaches and a wide variety of electric machines based on this principle.

In addition, AI is also a theory and/or a field of study that was founded as an academic discipline in 1956 (Crevier, 1993). In this case, AI is a kind of a theory of machines, or more precisely, the theory of intelligent machines. Therefore, it is quite logical that AI is sometimes defined as the science and technology of designing intelligent machines, especially intelligent computer programs.[6]

As we generally understand it, AI is a special universal technology (and a practical field of the theory implementation), that is an (almost) indispensable part of self-regulating systems, just like an electric motor in electric machines; or as an internal combustion engine (ICE) in cars, tractors and other machines. Hundreds and thousands of different machines are based on AI. [7] But just as an internal combustion engine or an electric motor does not determine the principle of operation of all machines, but only of their large class, so AI does not determine the functions and principle of operation of all self-regulating systems, but only of their large class (in many cases of this class AI can be the central and managing part).

However, with respect to self-regulating and self-managing systems, we should always keep in mind that they deal with information of a different nature. At the same time, artificial intelligence will contribute to the advance toward unification and transformation of one type of information into another more convenient for us (in particular, in the form of ICT information). Thus, for example, genetic information, the electrical impulses of the brain and other types of information are already being translated and can be translated in the future into a universal form of information. Thus, with respect to self-regulating systems, we can speak of the role of information as the most important substance and resource, while AI is a technological component at all stages of the process, which generally enables systems and information of different types to be used as subsystems of more complex self-regulating systems.

Thus, information and AI as a technology for its analysis and transformation become an almost essential component of many technologies and systems, while

---

[5] Some analytics regard intelligent as more advanced than smart (e.g., Burian, 2021). In some respects it is true. So, in this chapter we are in a line with such an idea. However, with respect to smart houses, smart cities, etc., this can be not suitable.

[6] AI is sometimes defined as the ability of machines to perform mental (creative) tasks that were previously considered as human tasks.

[7] This definition, introduced by computer scientists Avron Barr and Edward Feigenbaum in the early 1980s, separates rather well the technology/science from the systems that use it: "Artificial Intelligence (AI) is the part of computer science concerned with designing intelligent computer systems, that is, systems that exhibit the characteristics we associate with intelligence in human behaviour—understanding language, learning, reasoning, solving problems, and so on" (Barr & Feigenbaum, 1981: 3).

information units play a subordinate role, and serve to provide more diverse features (for the analysis of different aspects of AI see, e.g., Abi Younes et al., 2020; Eden et al., 2012; Graglia & Von Huelsen, 2020; Gregg, 2018; Hengstler et al., 2016; Liu et al., 2020; Michalski et al., 2013; Montes & Goertzel, 2019; Plebe & Perconti, 2020; Russell et al., 2003).[8]

### 3.4.3 From Smart Technologies to Self-managing Systems

Nowadays, there is a lot of talk about a number of so-called "smart" technologies and things that respond appropriately and quite flexibly to external impacts. It makes sense to show their connection with self-regulating systems. The prefix "smart" is a collective and unsettled term which is added to many devices that can perform actions which were previously controlled by humans.

Some smart things cannot by any means be attributed to self-regulating/self-managing systems (although this does not diminish their distinction). A simple but very illustrative example here is a mattress or pillow that takes (or remembers) the shape of the user's body (or head); another example is chameleon sunglasses that change the intensity of sun protection depending on the brightness of the sunlight. However, these technologies as well as some automatic systems like auto-open doors, auto-switch light, *etc.,* use only some elements of self-regulation. Somewhat more complex are a number of smart things based on nanotechnologies, such as, for instance, a window that can regulate the lighting in a room, etc. (see Chap. 10). More complex smart objects approach the level of self-regulation. They can have a response amplitude within a predetermined range. For example, a temperature controller (connected to Wi-Fi) that can remember a person's climatic preferences has a rather small amplitude of preferences. In fact, there is a sort of continuum of gradations from simple smart things to truly self-regulating systems. However, in most cases one can distinguish between smart technologies and self-regulating systems.

Within self-regulating systems, the processes of identification, memorization, learning and selection of a necessary mode within a spectrum of variants should operate on a much larger scale. In self-regulating systems, the number of variations is more or less unlimited; such systems are capable of selecting an appropriate model in any combination within the large framework of multiple parameters.

As it is said above, intelligent things, devices, programs, machines, etc., in most cases are more advanced.

---

[8] It is important to point out that due to this rapid development, artificial intelligence technology poses a major challenge for humanity (Eden et al., 2012). In recent years, communication technologies, data analysis and surveillance technology have considerably advanced. In 2023 a new rise of intellectual possibility of AI is beginning (see Chap. 10). As a result, the problem has become even more acute. A number of works have already been devoted to the analysis of various aspects of this problem in the present and the future (e.g., Westin, 1966; Ashman et al., 2014; Cecere et al., 2015; Moustaka et al., 2019; Schwartz, 1999; Solove, 2008; Brammer et al., 2020; Alharbi 2020).

Although there are already a large number of both smart and intelligent technologies and self-regulating/self-managing systems, they are not yet sufficient for a future breakthrough. For the real breakthrough, that will open the epoch of self-regulating systems that will come during the final phase of the Cybernetic Revolution, will require a much greater variety of such things than is currently available. A greater number of smart objects and simple self-regulating technologies will be also needed to provide just the right number of building blocks, elements, units and means to launch an innovation breakthrough. At present, in order to trigger the transition to the final phase of the Cybernetic Revolution, the technological prerequisites and innovations are increasingly accumulated along with the creation of ready-made units, variants, blocks, technologies, languages, forms of use, etc., which will become larger elements for the development of new cybernetic (self-regulating) technologies, opening the door to the final phase of the Cybernetic Revolution.[9] In other words, different directions for future technologies of self-regulating systems are formed, which, if combined, can produce more complicated and perfect already self-managing systems and develop a more universal way for their engineering. We believe that smart and intelligent systems will play a significant role in creating the necessary units and elements for complex self-regulating and self-managing systems. In this context, the process appears as an evolution *from "smart" to self-managing systems.*

To illustrate this point, let us use an analogy from the history of technology. In the sixteenth to eighteenth centuries, several dozen basic mechanisms and units were designed, which became the basis (important and even universal units and parts) for the creation of many hundreds and thousands of machines. Just as the creation of universal mechanisms exponentially expanded the existence of various kinds of machines based on these mechanisms and units (although their number remained not very large), so with the creation of various "units", that is, models of self-control, adaptability and flexibility to the environment (appearing in various "smart" machines, systems and materials), the number of self-regulating systems will grow. In other words, the number of "units" and "mechanisms" of self-regulation will not be too large, but the variety of combinations and joint actions in systems will be almost unlimited. Let us continue our analogy with the Industrial Revolution.

A mechanism is a device for transmitting and transforming motion (Bogolyubov, 1988: 63). Thus, a machine can consist of one or more mechanisms. In a similar way, we may say that smart and intelligent systems are devices for transmitting a special kind of information (such as data from the environment, internal and external sources as well as instructions, orders, tasks and goals) in order to carry out necessary reactions and actions of operated systems without human participation in certain situations or various environments. One should

[9] We define this process as the method of evolutionary block assembly (Grinin et al., 2008; Grinin, 2017; Grinin & Grinin, 2019). It is characteristic of the search for evolutionary directions, but also of technological development.

note that such transfer of "information for action" can be carried out in a variety of ways, depending on the peculiarities of an intelligent system: by means of intelligent components, by using of physico-chemical and biological structural and functional features of systems; or by combining them.

Similarly to the connection between machines and mechanisms, a self-regulating system may consist of one, but more often of several or many smart and intelligent systems. This supports a rapid growth in the number of different machines or systems. Thus, by the beginning of the nineteenth century, there were about 200 mechanisms, while at the end of the third quarter of the twentieth century there were about five thousand mechanisms (and many more machines). In other words, their number increased by a factor of 25 from the beginning of the nineteenth century. This is the possible scale of growth of those "units" that will help self-regulating and self-managing systems function in the future (their number will also be very large).

At the same time, self-regulating systems of a universal type will be also developed alongside with such specialized systems. This is a common developmental path in the history of technologies: first, the creation of specialized technologies, then their improvement to a level at which they can be widely applied in other industries, and as a result, the transformation of certain technologies into universal ones, that are part of more complex technologies. We assume that the development of self-regulating systems will follow the same pattern: first, the creation of specialized self-regulating systems, then of universal self-regulating systems, which in the future can become units for various systems.

Thus, as for the "smarter" materials, devices, technologies, etc., the faster the most universal ones will appear, the more combinations of their integration into more complex systems will appear, and the more confident will be the advance toward the creation of self-regulating systems, which in this context will be systems involving various "smart" technologies and AI, united by a flexible control system.

New approaches will succeed in solving many problems of a purely technical nature that were previously insolvable. It is well-known that one of the main assumptions of the reliability theory is that a *technical system has a limit of design complexity*. That is why a simple device is more durable than a complex one. And when using, for example, mechanical parts or electrical connections, there comes a point where the system becomes too expensive and unstable. This was the case with the diode and triode lamps, originally used in computers and other electronics. With the advent of semiconductors, many problems were solved, and the limits of reliability were considerably extended. The elements of intelligent systems push us toward the creation of self-regulating machines of great complexity, but the production robots already face reliability problems.

## 3.5 Classification of Self-regulating Systems

Thus, a self-regulating system is a system in which the main task is not simply to maintain certain parameters in order to preserve the working capacity (less complex systems are also capable of this), but to select parameters for solving problems and, if necessary, to modify the task itself in accordance with changing circumstances. *The ability to select appropriate decisions from a whole range of options is the most important characteristic of self-regulating and especially self-managing systems. Other possible characteristics are: the ability to self-learning; the presence of not rigid but flexible programs that can be modified under the impact of a changing environment or changing tasks.* Probably, the degree of freedom and the range of possible actions in the decision-making process, as well as the degree of self-learning and modification of programs will determine the complexity of self-regulating systems.

From a technical point of view, self-regulating systems differ from systems with elements of controllability and self-regulation, namely: (a) in the complexity of a control subsystem; (b) in the number and sensitivity of direct and feedback links between the control subsystem and peripheral organs; (c) in the ability to perform various movements and (d) in the ability to learn and self-learn. That is, there is a difference between a higher biological organism with a brain and a complete set of sensory organs, and a lower one with only a primitive nervous system. The complexity of control and communication greatly increases the number of combinations in which the system can be involved and affects the relevance of the response to changing situations. Currently, much work is done to improve systems (in particular, robots) in terms of their orientation in space, the selection of an appropriate response to different obstacles and changes in the environment, the ability to choose a solution, etc.

Now let us present our classification, which is, of course, far from perfect, but which allows us to understand the differences between more complex self-regulating systems and simpler ones.

*Classification of self-regulating systems:*

1. Smart systems with elements or a limited degree of self-regulation and autonomy (here there are also a number of levels of complexity).
2. More advanced intelligent systems.
3. Automatic complex systems with a pre-installed program and an independent controller, which are unable and not intended to go beyond the loaded program (so their interaction with the environment is limited). Examples: CNC machines, first-generation industrial robots.
4. Automatic complex systems with limited choice of actions and limited communication with the user (vending machines for a large number of products), sometimes they are parts of a more complex system (like credit card machines; also highly complex control systems).

5. Self-regulating systems capable of navigating in the environment, adapting and finding solutions depending on the situation (navigators, robots with limited functions, etc.; trading systems on the stock exchanges).
6. Complex systems that can navigate in space and reach targets, but are still controlled by humans (drones, reconnaissance aircraft, self-driving or unmanned cars with the driver sitting nearby).
7. More complex systems that can adapt to specific situations, rearrange their functionality, learn within the framework of the possibilities built into a flexible program, control human behavior and conditions (e.g., systems for continuous online monitoring of health status; a system for curing diseased cells by making independent decisions about the degree of their deviation; a robot capable of taking care of lying patients; artificial organs that can to some extent replace natural ones; genetic systems that allow the control or switching off certain genes). Or something more mundane—a smartphone as your personal secretary and assistant. When your smartphone sets an alarm according to your average sleep time. Or it makes an appointment with your doctor or a hairdresser because you missed the visit for a while. Or, just to be polite, it sends a nice message to your friends or family on their birthday. Here you are not involved in decision-making.
8. Artificial intelligence systems capable of making decisions within a fairly wide framework and without any human control and activity (future self-driving cars; housekeeping robots; a system that independently selects plant varieties suitable for particular regions, the complex medical-biotechnological environment above-described).

Systems number 6 and 7 belong to a higher level of self-regulation. So we can call them as self-managing systems. They will strongly change all spheres in the future. However, it is worth seriously thinking now about how to limit the negative sides of these implementations, if we don't want to be slaves to such self-regulation.

Thus, self-managing systems have the highest level of complexity, while self-regulating systems can be at different levels of complexity. However, the difference between self-regulating and self-managing systems does not always depend on their complexity. The difference often depends on the sphere of application of such systems. Technical systems at their highest level of complexity would often be self-managing systems. But medical, genetic, biotecnological systems would often be self-regulating systems. Future universal robots that can move, speak, make decisions, react to dangers, protect people and so on, as well as fully self-driving cars, are self-managing systems. Not to mention robots that will be able to improve themselves and adapt to new conditions (see about them and other examples in Chap. 10). But no less complex systems that enhance immunity, regenerate tissues, artificial organs, nanorobots that can deliver mini doses of drugs, self-cleaning systems in cells and in the whole organism, such as self-cleaning mechanisms for cellular debris, DNA repair mechanisms, etc., would usually be self-regulating systems, because their tasks are simply to regulate in the necessary order (see about them and other examples in Chaps. 8, 9, 10 and 11).

Thus, biological, physiological, genetic, many biotechnological (e.g., self-purification of wastewater), nanotechnological and other natural systems that are being introduced or will be introduced during the Cybernetic Revolution could hardly be called "self-managing systems", since they are generally designed to regulate a certain level of health, norms, etc. That is, they remain self-regulating systems. But AI-based technical systems, as well as SSSs, can rather be called self-managing systems.

## 3.6 The Main Features of the Cybernetic Revolution

### 3.6.1 Characteristics and Trends

Below we identify the most important characteristics of the Cybernetic Revolution. They can be observed today, but the large-scale implementation and maturity of these technologies will take place only in the future. These features are closely interrelated and mutually supportive (for more details see Grinin and Grinin, 2013a, 2013b, 2015a, 2016). For a better understanding of the formulations presented below, one should take into account what has already been discussed—in particular, such points as the increasing role of information; the proliferation of intelligent and autonomously working units and systems; the fact that self-regulating systems gradually become the main part of the technological process; as well as the characteristics of these systems and the increasing connection between technological systems and the environment.

**The most important characteristics and trends of the Cybernetic Revolution are as follows:**

1. The increasing volume of information, its growing complexity and complicating analysis of systems (including the ability of systems to communicate and interact independently).
2. Continuous development of the system of regulation and self-regulation; increasing number of self-regulating systems of different types and nature.
3. Mass use of artificial materials with new properties especially in bio- and nanotechnologies, bionic (techno-biological) and medical fields[10].
4. Application and control of systems and processes of various kinds including living matter, analogues of biological and physiological substances and new levels of organization of matter.
5. Transition to control of deeper and more fundamental processes and levels (down to atomic level, using nanoparticles as building blocks).

[10] The process is similar to the one observed in chemistry over the last hundred years, which has produced tens of thousands of artificial materials.

6. Miniaturization and micronization[11]—the trend toward decreasing size of particles, mechanisms, electronic devices, implants, etc.
7. Resource and energy saving in all areas.
8. Individualization and/or personalization as one of the most important technological trends.
9. Control of human behavior and activity.

**The characteristics of technologies of the Cybernetic Revolution:**

1. The transformation and analysis of information as an essential part of technologies and one of the main resources.
2. We observe the integration of machinery, equipment and hardware with technology (know-how and process knowledge) into a unified technical and technological system.[12]
3. Self-regulating systems will become the major component of technological processes. For this reason, the final (forthcoming) phase of the Cybernetic Revolution should be called *the epoch of self-regulating systems.*
4. Widespread use of AI elements and units in various devices and self-regulating systems.
5. Universalization, i.e., the emergence of more and more multifunctional devices (today it is a computer, a mobile phone, a virtual voice assistant; in the future it will be a universal robot, a unified bio-medical-techno-system [see Chap. 6], etc.).
6. The increasing interconnection between technological systems and the environment.
7. The ability of materials and technologies to adapt to different objectives and tasks (smart materials and technologies) and the radical increase in the ability of systems to select optimal *regimes in the context of particular goals and tasks.*
8. A large-scale synthesis of materials and characteristics of systems of different nature (*e.g.*, of animate and inanimate nature).
9. Implementation of intelligent technologies and a trend toward humanization of their functions (common language, voice, gesture interfaces, appearance, *etc.*).
10. *Different directions and branches of development should generate a system cluster of innovations*[13].

[11] See, e.g., Wang (2010).
[12] During the Industrial epoch, these elements existed separately: technologies were designed on paper or in the minds of engineers. At present, technology helps not only to design but also to manage the creation of new technologies. This facilitates the transition to the age of self-regulating systems.
[13] Different directions of the MANBRIC are already developing toward their convergence and the creation of a system cluster of innovations (for details see Chap. 4, in particular see Fig. 1. Relative global dynamics of patent grants for various technologies). In the future the effect of such clustering will be immense.

### 3.6.2 Some Clarifications on the Characteristics of the Cybernetic Revolution

**Information as an integral part of the systems operation**. It is important to note that with respect to the Cybernetic Revolution, we speak not only about specific information technologies, which have separated as a sector already in its initial phase, and even not only about the development of AI. As we have already pointed out, information is the most powerful resource, the most important component, an attribute of the functioning of any complex system, especially of self-regulating systems. Information, as we have said, is the most important subject of research in cybernetics; thus, information is one of the main components of the Cybernetic Revolution. At the moment, information, like energy, is becoming an integral part of every operating system. At the same time, similarly to energy, information exists in many different forms. It is obvious that genetic information, information coming from the brain in the form of electrical signals, the transmission of information in computers by means of electrons, or verbal information, represent its significantly different forms. But the technological process is moving in the direction that allows one type of information to be transformed into another without loss of meaning through its universal intermediate forms. One may expect that during the final phase of the Cybernetic Revolution we will make significant progress along this path of transformation (but we hope, not at the suggestion of Kurzweil's followers, who speak about direct connection between the computer and the brain).

Thus, obtaining, converting into more convenient or universal forms, as well as the analysis, transformation, further transmission and use of information for analysis, performance of complex intellectual tasks and management is a function not only of specialized information systems, but also in general of all technologies of the Scientific-Cybernetic Production Principle (similarly as electrical technologies began to be applied in the vast majority of modern devices).

**Communication with the external environment**. The importance of the connection between technologies and the external environment is constantly growing. The level, density and complexity of feedback with the environment are increasing as well. The peripheral devices play an increasingly important role in devices, machines, etc. As information and response to it are the key characteristics, technologies and devices become "smart" and start to react and get feedback. As a result, there is an increasing number and level of technologies that have "sense organs" and information analyzers. Such devices and technologies will become the leading characteristic of production and life in the final phase of the Cybernetic Revolution. We expect that future technologies will use an increasing number of parameters that define relations with the external environment and the level and type of analysis. For example, there is a great perspective in the development of robotic effectors, which serve as the organs of the robot: vision, touch, hearing, speech interface and even smell (see in Chap. 10).

We can also argue that the developmental vector is directed toward the creation of anti-entropy technologies that will reduce the loss of important information.

**Personalization and Individualization**. Individualization/personalization has become one of the leading production and business strategies. This is largely due to the economic trend resulting from the consumption by households and individuals (this consumption is, according to many economists, the most important factor for the GDP growth). In this context, the winners are those manufacturers and technologies that are better able to meet individual needs.[14] In fact, the advance toward individualization is supported by the general vector of information technologies, since the user has more and more opportunities to be unique on the Internet, so that a computer, phone or any other gadget, as well as its content, become purely personal. Moreover, many computer programs can now be adapted to individual needs and tastes, so that many versions of the same programs coexist.

Increasingly diverse technical systems tend to remember the needs, tastes and passions of their owners. However, here, major threats to privacy and human rights have already become as real as ever due to the increasing abuse on the part of Internet service providers and other recipients of information. This is a very serious problem to deal with for many decades to come. We will speak about it later.

In the future, self-regulating systems will adapt to changing conditions or user-defined tasks (which is also a future threat). Individualization will be actively manifested in the growing demand for corrective medicine and plastic surgery, and this trend is also emerging in the realms of pharmacology and clinical medicine.

Individualization becomes both a technology and a means of identifying a person by means of the unique biometric and other biological characteristics, and this is already applied and will be used more actively in socio-technical self-regulating systems (for control, identification, search, etc.; for details see Chap. 14). In the future, with the development of new biomedical technologies, the role of individualization will dramatically increase, including in the pharmaceutical sector, and there will be a general advance from mass production of things to adjusting them to individual use. The range of such services can be expanded incredibly. For example, the transition from mass fashion to fashion for small groups or even to individual fashion will be quite possible after the development of artificial fashion designer and tailor technologies using 3D technologies (by the way, this is one of the many examples of professions that may change or disappear). So, the profession of a consultant will become very popular, but at the same time there will be developed technologies to substitute them.

**Saving resources, energy and labor in all spheres**. The general idea of the Cybernetic Revolution is closely related to a constant and comprehensive saving of energy, resources (including labor) and materials, which will begin to be

[14] In the 1950s to 1980s, the creation of computers in the USSR followed the old trajectory, that is, they were considered only as machines that should be used in relevant institutions or in production. At the same time, in the USA, since the 1970s, the trajectory of information technologies has taken a different path of creation of a personal computer (as previously of a personal car), which immediately expanded the market enormously and at the same time increased the power of new information technologies, which began to be used by tens of millions of people instead of tens of thousands of specialists.

implemented due to the mass development of self-regulating systems of a fundamentally new level. In fact, without a breakthrough in savings there can be no growth in the standard of living of the population of the planet (which will continue to rise at least until the end of the twenty-first century) (according to most forecasts, see, e.g., Alkema et al., 2011; Raftery et al., 2012; Podlazov, 2017; Lutz et al., 2018; Korotayev, 2020; UN Population Division, 2012, 2019, 2022a, 2022b; Vollset et al., 2020; Wittgenstein Centre, 2022; Korotayev et al., 2023).

Saving labor was the leitmotif of the previous Industrial Revolution. The Cybernetic Revolution continues this trend. Yet, instead of simple mechanical labor, a large-scale saving of complex and intellectual labor begins, involving sphere after sphere. Today, information and communication technologies together with smart machines, help to perform a large number of operations that previously required the enormous mental effort of millions of people: from calculations and mathematical processing of data to complex financial transactions, from translating foreign words to distance learning, from photography to design, from taking tests to making diagnoses. New AI Chats promise even more (see Chap. 10). Miniaturization, the development of new materials and efficient energy sources, advances in control systems, improved technological precision (e.g., localization of the area of influence using laser or nanotechnologies) and many other advances have provided a general course toward saving resources and energy in many processes (which has also been greatly promoted by rising raw materials and energy prices, as well as environmental policies). However, there is still a lot of waste of resources, and its examples are around us. However, we believe that the completion of the Cybernetic Revolution will significantly advance humanity on this path of saving especially due to biotechnology and nanotechnology, as we will see in Chaps. 9 and 10 (here it is also worth mentioning the well-known predictions of this kind made almost a quarter of a century ago by Weizsacker and Lovins, 1998). It is worth remembering that the need to save resources and the problem of their depletion have been raised almost since the very beginning of the Cybernetic Revolution (and even much earlier). However, one can speak about a scientifically substantiated and widespread forecast (albeit too pessimistic, inertial and contradicted by life) from the first reports to the Club of Rome in the early 1970s (see, e.g., Meadows et al., 1972; Mesarović and Pestel, 1974; Tinbergen et al., 1976; Laszlo, 1977; Pestel, 1989; see also: Gvishiani 1997; Meadows et al., 1972, 1999; Weizsacker and Lovins, 1998). For our analysis of the ideas of the Club of Rome, see Grinin and Grinin, 2021; Sadovnichy et al., 2023). Overall, the world consumption of energy and materials is still growing in absolute terms. But in terms of a single car or a light bulb, the consumption is reduced quite noticeably. We believe that the increasing controllability, self-regulation and self-management, the ability to choose flexible modes, work according to individual programs, the transition to nanotechnology, etc., will reinforce this trend. About the optimization of the economic consumption model as a result of the completion of the Cybernetic Revolution, see Chap. 13.

**Miniaturization and microminiaturization**. Miniaturization, i.e., a constant reduction in the size of units, devices, machines, artificial organs, etc., has been observed since the very beginning of the Cybernetic Revolution (and was noticeable even before it). However, a particularly active development began with the advent of the transistor. The invention of the integrated circuit became a kind of a threshold when miniaturization turned into microminiaturization.[15] The latter was also associated with a radical expansion of control opportunities in the microworld, in particular, with the use of ever new types of energy to influence microparticles (as well as cells and genes), and this level can become fundamental in technologies. Nanotechnology gives amazing examples of microminiaturization rates (see Chap. 10). In this sense, nanotechnology has also a very promising future.

**Universalization**. Universalization is one of the most important and even surprising characteristics and trends of the Cybernetic Revolution. Universalization, i.e., the emergence of more and more multifunctional devices (today it is a computer, a mobile phone, a help desk robot and 3D printers; in the future—a universal robot, a unified medical-biotechnological environment [see above and Chaps. 6, 8, 11, 15], etc.; in the future, perhaps, perhaps, a lot can be done with the help of the nanoassembly, creating various nanostructures and substances (see Chap. 10). Universalization is a vector that sharply changes the previous direction toward professionalization and specialization of technology (see below). It consists in the fact that the vector of technological changes goes in to the direction of gathering the maximum possible operations in specific technologies, making them as broad as possible, constantly combining various existing autonomous machines and mechanisms into a single complex. This also applies to the universalization of competences by means of of self-regulating and self-managing systems, when, with the help of devices and software, an individual can combine dozens of competences, each of which was previously only feasible for a professional or an advanced amateur (we will discuss this further in Chap. 4).

This vector of technological change will incorporate the maximum possible operations from different technologies and will make them widely used, and for this purpose it will constantly combine various existing and independently working techniques and mechanisms. Technologies and devices become more and more functional, incorporating previously independent technologies as subsystems. An example is smartphones, which today integrate at least twenty functions (from the former typewriter to the video camera, from the pencil to the secretary, from the tape recorder to the television), as well as the functions of the computer.

And what about a car? It has so many functions now! It is a house with climate-control, transport with guide, a mini-station for communication and a concert hall at the same time. This universalization has given rise to the do-it-yourself skills we have already mentioned, which have led to the disappearance of a number of professions. At the same time, to continue with this example, cars feed into different professions, by providing, even during their normal use, a huge amount of data that is produced and sent to different "users"—GPS-tracking is just one, and probably

[15] For some aspects of both directions, see Danko et al., 1959; Peercy, 2000.

the best known example. Part of this is that miniaturization and microminiaturization mean that an increasing number of simpler devices are integrated into various more complex devices. There are three main ways to illustrate this development in an easily comprehensible way: (i) Moore's law states that the number of transistors on an integrated circuit doubles every two years; (ii) today, a car is equipped with more electronics than was needed for the first manned space flight; (iii) a smartphone or tablet today has more capacity than a large computer—an electronic data processing machine as it was then called—had in the early 1990s, less than half a century ago.

Processes of universalization are observed in not only electronic devices. We can trace them in the very promising fields of creating multifunctional robots or mini-laboratories which also become increasingly multifunctional. One of the latest trends in universalization is 3D printers, which will probably compete with computers in terms of scale and scope of application. In Chap. 10 we will show that 3D printers are becoming truly multifunctional. As it will be seen, the opportunities provided by such printers are exceptionally far-reaching: from construction to cooking, from the home workshop to museums, from car manufacturing to robotics, from medicine to children's toys, from training models to design. These machines are actively used in such sectors as aircraft construction and rocket engineering to produce individual details, for example, support stands for an aircraft engine (see, e.g., Turichin, 2015; Lipson & Kurman, 2013; Ford, 2015).

In general, the Industrial Production Principle of the eighteenth–twentieth centuries developed specialization, making professionalization evermore narrow, and creating highly specialized technical devices for it. For example, there are dozens of different types of cars, many hundreds of types of tools and so on.[16] It should be noted that universalization was not clearly evident in the initial phase of the Cybernetic Revolution. On the contrary, specialization seemed to be increasing. However, the tendency came to the fore during the modernization phase of the Cybernetic Revolution and has been increasing ever since.

We believe that, first of all, this tendency has not come to maturity yet but it is likely to step forward in the final phase of the Cybernetic Revolution. Secondly, it will lead to the emergence of complex self-managing systems which, being multifunctional, will include a number of subsystems and technologies, but whose main element will be the management system capable of self-regulation and self-management. Thirdly, this trend will respectively manifest itself in different areas. For example, we have already described health monitoring systems that will fulfill many functions. Besides, multifunctional robots are also most likely to appear. Fourthly, this trend will encourage the emergence of closely interconnected complex technologies.

So, we have explained every characteristic, feature and trend of the Cybernetic Revolution. In the following chapters (especially in Chaps. 8, 9, 10, and 11) we

[16] However, universal technical things were also invented during this period (e.g., electric or internal combustion engines).

will return to them again and again, discussing every characteristic and trend in detail as it applies to different technological branches.

All these above-analyzed areas are already closely interlinked and their further development will lead to their mutual support. For example, resource and energy saving can be achieved through choosing optimal modes by autonomous systems that fulfill specific objectives and tasks and vice versa, the choice of an efficient mode will depend on the level of energy and material consumption, and the consumer's budget. Or, the opportunities of self-regulation will make it possible to select a particular decision for the variety of individual tasks, orders and requests (e.g., with 3D printers and choosing an individual program as the optimal one; see Chap. 10). The increasing self-regulation of systems leads to the combination of systems of different nature (e.g., technical and biological, social and biological, technical and social, etc.). And the creation of such symbiotic systems increases the level and scale of controllability. Miniaturization enhances individualization and resource saving, and the course toward individualization and saving requires even more miniaturization. And it is this kind of positive feedback that we trace in many cases in the following chapters.

**Acknowledgment** This chapter has been prepared with support by the Russian Science Foundation (Project **№ 23-11-00160**).

## References

Abi Younes, G., Ayoubi, C., Ballester, O., Cristelli, G., de Rassenfosse, G., Foray, D., et al. (2020). COVID-19: Insights from innovation economists. *Science and Public Policy*, (paaa036). https://doi.org/10.1093/scipol/scaa028

Alharbi, F. S. (2020). Dealing with data breaches amidst changes in technology. *International Journal of Computer Science and Security (IJCSS), 14*(3), 108–115.

Alkema, L., Raftery, A. E., Gerland, P., Clark, S. J., Pelletier, F., Buettner, T., & Heilig, G. K. (2011). Probabilistic projections of the total fertility rate for all countries. *Demography, 48*(3), 815–839. https://doi.org/10.1007/s13524-011-0040-5

Ashby, R. (1956). *An introduction to cybernetics*. Chapman and Hall.

Ashman, H., Brailsford, T., Cristea, A. I., Sheng, Q. Z., Stewart, C., Toms, E. G., & Wade, V. (2014). The ethical and social implications of personalization technologies for e-learning. *Information & Management, 51*(6), 819–832. https://doi.org/10.1016/j.im.2014.04.003

Barr, A., & Feigenbaum, E. A. (eds.). (1981). *The handbook of artificial intelligence*. Heuristech Press/William Kaufmann.

Benson, I., & Lloyd, J. (1983). *New technology and industrial change: the impact of the scientific-technical revolution on labour and industry*. Kogan Page.

Bernal, J. D. (1965). *Science in History* (3rd ed.). Hawthorn Books.

Bogolyubov, A. N. (1988). *Tvoreniya ruk chelovecheskikh. Yestestvennaya istoriya mashin*. M.: Znaniye.

Brammer, S., Branicki, L., & Linnenluecke, M. (2020). COVID-19, societalization and the future of business in society. *Science of the Total Environment, 34*(4), 2–7. https://doi.org/10.5465/amp.2019.0053

Burian, J. (2021). Smart versus intelligent: It's all in the silos. *IndustryWeek,* 12.03.2021. https://www.industryweek.com/technology-and-iiot/automation/article/21157742/smart-vs-intelligent-its-all-in-the-silos

Cecere, G., Le Guel, F., & Soulié, N. (2015). Perceived Internet privacy concerns on social networks in Europe. *Technological Forecasting and Social Change, 96*, 277–287. https://doi.org/10.1016/j.techfore.2015.01.021
Wittgenstein Centre. (2022). *Wittgenstein centre human capital data explorer*. Wittgenstein Centre. http://dataexplorer.wittgensteincentre.org/wcde-v2/
Crevier, D. (1993). *AI: The tumultuous search for artificial intelligence*. BasicBooks.
Danko, S. F., Doxey, W. L., & McNaul, J. P. (1959). The micro-module: a logical approach to microminiaturization. *Proceedings of the IRE, 47*(5), 894–903.
Diao, Y., Hellerstein, J. L., & Parekh S. et al. (2005). Self-managing systems: a control theory foundation. In *Proceedings of the 12th IEEE international conference and workshops on the engineering of computer-based systems (ECBS'05)*. http://www.cs.columbia.edu/~rg2023/pubs/EASe05.pdf
Dupuy, J. (2009). *On the origins of cognitive science. The mechanization of the mind*. MIT Press.
Eden, A. H., Moor, J. H., Soraker, J. H., & Steinhart, E. (Eds.). (2012). *Singularity hypotheses: A scientific and philosophical assessment*. Springer. https://doi.org/10.1007/978-3-642-32560-1
Ford, M. (2015). *Rise of the robots: technology and the threat of a jobless future*. Basic Books.
Graglia, M. A. V., & Von Huelsen, P. G. (2020). The Sixth Wave of Innovation: Artificial Intelligence and the Impacts on Employment. *Journal on Innovation and Sustainability RISUS, 11*(1), 3–17. https://doi.org/10.23925/2179-3565.2020v11i1p3-17
Gregg, B. (2018). The coming political: challenges of artificial intelligence. *Digital Culture & Society*, 4(1), 157–180. https://doi.org/10.25969/mediarep/13530
Grinin, L., & Grinin, A. (2015). Kiberneticheskaya revolyutsiya i shestoy tekhno-logicheskiy uklad. *Istoricheskaya psikhologiya i sotsiologiya istorii, 8*(1), 172–197.
Grinin, A. L., & Grinin, L. E. (2015). The cybernetic revolution and historical process. *Social Evolution and History, 14*(1), 125–184.
Grinin, L. E., & Grinin, A. L. (2016). *The cybernetic revolution and the forthcoming epoch of self-regulating systems*. Moscow Branch of the Uchitel Publishing House.
Grinin, L. E., & Grinin, A. L. (2019). Technological dimension of big history and the cybernetic revolution. In L. E. Grinin & A. V. Korotayev (Eds.), *History & Mathematics: Big History Aspects* (pp. 250–277). Uchitel Publishing House.
Grinin, L. E., Korotayev, A. V., & Tausch, A. (2016). *Economic cycles, crises, and the global periphery*. Springer. https://doi.org/10.1007/978-3-319-41262-7
Grinin, L. E., Grinin, A. L., & Korotayev, A. (2017). Forthcoming kondratieff wave, cybernetic revolution, and global ageing. *Technological Forecasting & Social Change, 115*, 52–68.
Grinin, L., Grinin, A., & Korotayev, A., et al. (2017). The MANBRIC-technologies in the forthcoming technological revolution. In T. Devezas (Ed.), *Industry 4,0—entrepreneur-ship and structural change in the new digital landscape: what is coming on along with the fourth industrial revolution* (pp. 243–261). Springer. https://doi.org/10.1007/978-3-319-49604-7_13
Grinin, L., Grinin, A., & Korotayev, A. (2020). A quantitative analysis of worldwide long-term technology growth: From 40,000 BCE to the early 22nd century. *Technological Forecasting and Social Change, 155*, 119955. https://doi.org/10.1016/j.techfore.2020.119955
Grinin, L., Grinin, A., & Korotayev, A. (2021). COVID-19 pandemic as a trigger for the acceleration of the cybernetic revolution, transition from E-government to E-state, and change in social relations. *Technological Forecasting & Social Change, 175*, 121348. https://doi.org/10.1016/j.techfore.2021.121348
Grinin, L. E., & Grinin, A. L. (2013a). Macroevolution of Technology. In L. E. Grinin, & A. V. Korotayev (eds.), *Evolution: development within big history, evolutionary and world-system paradigms* (pp. 143–178). 'Uchitel' Publishing House.
Grinin, L. E., Grinin, A. L. (2013b). Global technological transformations. In L. E. Grinin, I. V. Ilyin, & A. V. Korotayev (eds.), *Globalistics and globalization studies: theories, research & teaching* (pp. 98–128). Uchitel.
Grinin, L. E., & Grinin, A. L. (2015b). *Ot rubil donanorobotov. Mirnaputikepokhesamoupravlyaemykh sistem (istoriya tekhnologiy i opisanie ikh budushchego)*. Moscow Publishing House 'Uchitel'.

Grinin, A. L., & Grinin L. E. (2021). Analyzing the global problems of the twenty-first century. a review and forecast based on the report to the club of Rome 'Come On!'. *Journal of Globalization Studies, 12* (2), 181–195.
Grinin, L. E., Markov, A. V., & Korotayev, A. V. (2008). *Makroevolyutsiya v zhivoy prirode i obshchestve*. LKI/URSS.
Grinin, L. (2006) *Proizvoditel'nyye sily i istoricheskiy protsess*. URSS.
Grinin, L. E. (2017). *Bolshaya istoriya razvitiya mira: istoriya i evolyutsiya Solnechnoy sistemy*. Moskovskaya redaktsiya iz-datel'stva 'Uchitel'.
Gvishiani, D. M. (ed.). (1997). *Rimskiy klub. Istoriya sozdaniya, izbrannyye doklady i vystupleniya, ofitsial'nyye materialy*. URSS.
Hengstler, M., Enkel, E., & Duelli, S. (2016). Applied artificial intelligence and trust—The case of autonomous vehicles and medical assistance devices. *Technological Forecasting and Social Change, 105*, 105–120. https://doi.org/10.1016/j.techfore.2015.12.014
Korotayev, A., Malkov, S., & Musieva, J. (2023). Demography. Toward optimization of demographic processes. In V. Sadovnichy et al. (Eds.), *Reconsidering the limits to growth. A report to the Russian Association of the Club of Rome* (pp. 97–116). Springer. https://doi.org/10.1007/978-3-031-34999-7_6
Korotayev, A. (2020). The 21st century Singularity in the Big History perspective. A re-analysis. In A. Korotayev, & D. LePoire (Eds.), *The 21st century Singularity and global futures. A Big History perspective* (pp. 19–75). Springer. https://doi.org/10.1007/978-3-030-33730-8_2
Laszlo, E. (Ed.). (1977). *Goals for mankind. A report to the club of Rome on the new horizons of global community*. Dutton.
Lilley, S. (1966). *Men, machines and history: the story of tools and machines in relation to social progress*. International Publishers.
Lilley, S. (1976). Technological progress and the industrial revolution, 1700–1914. In C. M. Cipolla (Ed.), *The industrial revolution, 1700–1914. The Fontana economic history of Europe* (pp. 187–254). The Harvest Press.
Lipson, H., & Kurman, M. (2013). *Fabricated: The new world of 3D printing*. John Wiley & Sons.
Liu, J., Chang, H., Forrest, J.Y.-L., & Yang, B. (2020). Influence of artificial intelligence on technological innovation: Evidence from the panel data of china's manufacturing sectors. *Technological Forecasting and Social Change, 158*, 1–12. https://doi.org/10.1016/j.techfore.2020.120142
Lutz, W., Goujon, A., Kc, S., Stonawski, M., & Stilianakis, N. (2018). *Demographic and human capital scenarios for the 21st century: 2018 assessment for 201 countries*. Publications Office of the European Union.
McClelland, K. A., & Fararo, T. J. (Eds.). (2006). *Purpose, meaning, and action: control systems theories in sociology*. Palgrave Macmillan.
Meadows, D., Meadows, D., Randers, J., & Behrens, W. W. (1972). *The limits to growth*. Potomac Associates.
Meadows, D., Meadows, D., & Randers J. (1999). Za predelami dopustimogo: global'naya katastrofa ili stabil'noye budushcheye. In V. L. Inozemtsev (ed.), *Novaya postindustrial'naya volna na Zapade*. Academia.
Mesarović M. D., & Pestel E. (1974). *Mankind at the turning point: the second report to the club of Rome*. IIASA.
Michalski, R. S., Carbonell, J. G., & Mitchell, T. M. (2013). *Machine Learning*. Springer.
Montes, G. A., & Goertzel, B. (2019). Distributed, decentralized, and democratized artificial intelligence. *Technological Forecasting and Social Change, 141*, 354–358. https://doi.org/10.1016/j.techfore.2018.11.010
Moustaka, V., Theodosiou, Z., Vakali, A., Kounoudes, A., & Anthopoulos, L. G. (2019). Enhancing social networking in smart cities: Privacy and security borderlines. *Technological Forecasting and Social Change, 142*, 285–300. https://doi.org/10.1016/j.techfore.2018.10.026
Peercy, P. S. (2000). The drive to miniaturization. *Nature, 406*(6799), 1023–1026.
Pestel, E. (1989). *Beyond the limit to growth: a report to the club of Rome*. http://www.comitatoscientifico.org/temiSD/oltreilpil/docs/PESTELThe LimitstoGrowth.pdf

Philipson, M. (Ed.). (1962). *Automation: implications for the future*. Vintage.

Plebe, A., Perconti, P. (2020). Plurality: The End of Singularity? In A. V. Korotayev, D. J. LePoire, (eds.), *The 21st century singularity and global futures* (pp. 163–184). Springer. https://doi.org/10.1007/978-3-030-33730-8_8

Podlazov, A. A. (2017). Teoriya global'nogo demograficheskogo protsessa. *Vestnik Rossiyskoy akademii nauk, 87*(3), 256–266. https://doi.org/10.1134/S1019331617030054

Raftery, A. E., Li, N., Ševčíková, H., Gerland, P., & Heilig, G. K. (2012). Bayesian probabilistic population projections for all countries. *Proceedings of the National Academy of Sciences, 109*(35), 13915–13921. https://doi.org/10.1073/pnas.1211452109

Sadovnichy, V., Akaev, A., Ilyin, I., Malkov, S., Grinin, L., Sayamov, Y., & Korotayev, A. (2023). Introduction. Hoping for the future. In V. Sadovnichy et al. (Eds.), *Reconsidering the limits to growth. a report to the Russian association of the Club of Rome* (pp. 1–14). Springer. https://doi.org/10.1007/978-3-031-34999-7_1

Russell, S. J., Norvig, P., Canny, J. F., Malik, J. M., & Edwards, D. D. (2003). Artificial intelligence: a modern approach (Vol. 2). Prentice hall Upper Saddle River.

Schwartz, P. M. (1999). Internet privacy and the state. *Connecticut Law Review, 32*, 815–829.

Solove, D. J. (2008). *Understanding privacy*. Harvard University Press.

St, Beer. (1959). *Cybernetics and management*. English Universities Press.

St, Beer. (1994). *Decision and control: the meaning of operational research and management cybernetics*. John Wiley & Sons Ltd.

Tesler, G. S. (2004). *Novaya kibernetika*. Logos.

Tinbergen, J., Dolman, A. J. (Eds.). (1976). *Reshaping the international order: a report to the club of Rome*. Dutton.

Turichin, G. A. (2015). *Additivnyye tekhnologii v sovremennom proizvodstve*. Report at international workshop *'Bazisnyye tekhnologii pervoy poloviny XXI v. (strukturno-tsiklicheskiy analiz)'*. Sankt-Petersburg, October 1–2.

Umpleby, S. A., & Dent, E. B. (1999). The origins and purposes of several traditions in systems theory and cybernetics. *Cybernetics and Systems, 30*, 79–103.

UN Population Division. (2012). *World population prospects: The 2010 revision*. United Nations.

UN Population Division. (2019). *World population prospects 2019:* volume I: *Comprehensive tables*. United Nations.

UN Population Division. (2022a). *United Nations population division database*. United Nations. http://www.un.org/esa/population

UN Population Division. (2022b). *World population prospects 2022*. United Nations.

Vollset, S. E., Goren, E., Yuan, C. W., Cao, J., Smith, A. E., Hsiao, T., Murray, C. J. L., et al. (2020). Fertility, mortality, migration, and population scenarios for 195 countries and territories from 2017 to 2100: a forecasting analysis for the global burden of disease study. *The Lancet, 396*, 1285–306. https://doi.org/10.1016/S0140-6736(20)30677-2

von Foerster, H., & Zopf, G. W. (1962). Principles of self-organization. In M. C. Yovits & S. Cameron (Eds.), *Self-Organization* (pp. 31–50). Pergamon Press.

Wang, V. X. (Ed.). (2010). *Encyclopedia of information communication technologies and adult education integration*. IGI Global.

von Weizsacker, E. U., & Lovins, L. H. (1998). *Factor Four. Doubling wealth, Halving resource use—The New Report to the Club of Rome*. Taylor & Francis.

Westin, A. F. (1966). Science, privacy, and freedom: Issues and proposals for the 1970s. Part I. The current impact of surveillance on privacy. *Columbia Law Review, 66*(6), 1003–1050.

Wiener, N. (1948). *Cybernetics, or control and communication in the animal and the machine*. MIT Press.

Wieser, M., Slunecko, Th. (2014). Cybernetics, Radical behaviorism, Cultural Psychology, Cultural-historical Psychology. In T. Teo (Ed.), *Encyclopedia of critical psychology*. Springer.

# Cybernetic Revolution and MANBRIC Technologies. When, How and Where Will the Forthcoming Breakthrough Start?

4

**Abstract**

According to the authors' concept, the Cybernetic Revolution is the last of the major production (technological) revolutions in all history following the Agrarian and Industrial Revolutions. It is a major transition, from the Industrial Production Principle to production and service provision based on the implementation of self-regulating/self-managing systems. The first phase of this revolution began in the 1950s and 1960s and brought the development of powerful information technologies. Between the 2030s and the 2070s, the final phase of this revolution will lead to a new level of self-operated control, namely, to the level of self-regulating/self-managing systems, which can operate with no human intervention. Grinin et al. have described these systems in the previous chapter. In the present chapter, the authors study the main directions of the final phase of the Cybernetic Revolution, which will form a peculiar and closely related cluster of innovative directions for the development of technologies. This complex Grinin et al. call the MANBRIC complex/convergence. This is an acronym that includes medicine, additive (3D printers), nano- and biotechnologies, robotics, IT, and cognitive sciences., The authors show that a number of reasons medicine will be the first sphere to start the final phase of the Cybernetic Revolution.

## 4.1 What This Chapter is About

According to our vision, the Cybernetic Revolution is the last of the major production (technological) revolutions in all history following the Agrarian and Industrial Revolutions. It is a major transition, from the Industrial Production Principle to production and service provision based on the implementation of self-regulating/self-managing systems. The first phase of this revolution began in the 1950s and

 
L. Grinin et al., *Cybernetic Revolution and Global Aging*, World-Systems Evolution and Global Futures, https://doi.org/10.1007/978-3-031-56764-3_4

1960s and brought the development of powerful information technologies, new materials and energy sources, as well as the diffusion of automation.

Between the 2030s and the 2070s, the final phase of this revolution will lead to a new level of self-operated control in technical systems, namely, to the level of self-regulating/self-managing systems, which can operate and take decisions or propose the best solutions independently in a wide range of variations or cases of changing situation with minimal or no human intervention. We have described these systems in the previous chapter. In the present chapter, we study the main directions of the final phase of the Cybernetic Revolution, which will form a peculiar and closely related cluster of innovative directions for the development of technologies.

There are different views on the technologies that will become of utmost importance in the coming decades and that will form this cluster. We propose our own vision. We identify seven leading technological directions/branches/fields of the future. These directions are emerging and, in the future, they will eventually converge into a single complex, which we call the MANBRIC complex/convergence. This is an acronym that includes medicine, additive (3D printers), nano- and biotechnologies, robotics, IT, and cognitive sciences (note that the order of the letters in the acronym does not reflect the relative importance of the areas that make up the complex).

At present, these technological fields closely interact and reinforce each other and will do so increasingly in the future. Further, we analyze each branch of the MANBRIC convergence, with some projections of their development in the 2030s to 2070s. In addition, we examine in detail the answers to important questions, namely:

(1) Which of the directions will be integral in this complex?
(2) In which field will the new great breakthrough or the final phase of the Cybernetic Revolution start? Which one will be the first?

We explore different directions such as green energy, robotics, pilotless (self-driving) cars. We show that a number of reasons make us answer to the last two questions that medicine will be the first sphere to start the final phase of the Cybernetic Revolution because among other areas, medicine meets these criteria to the greatest extent. In particular, the medical sphere has unique opportunities to combine the above-mentioned technologies into a single complex.

Even today, all the areas of MANBRIC are very actively used in modern medicine, but in the future, the relationship of all these areas to medicine will increase noticeably, and somewhere fundamentally. In addition, the aging of society, which is now developing rapidly, creates an acute need for labor-saving technologies and gives a powerful impetus to the field of medicine. Progress in the latter will help to extend the working age and improve the health-related (biological) quality of life.

It has made the issues of health and medicine the subject of interest to literally the whole society and a huge part of humanity, and has placed medicine and biotechnology at the forefront of technological development. The leading role of medicine in the final phase of the Cybernetic Revolution (at least in the 2030–2040/50s; after which other areas may come to the fore) is also explained by the resulting economic and social consequences, as well as a number of other factors, to which the pandemic has been added since 2020.

Then we show why the breakthrough will occur in the 2030s/2040s and how the development of self-regulating and self-managing systems will involve the most diverse areas of production, services and life. The chapter discusses many other subjects, in particular, the issue of the very close correlation between the final phase of the Cybernetic Revolution and the so-called sixth technological paradigm.

## 4.2 The Main Directions of the Cybernetic Revolution. The MANBRIC Convergence

### 4.2.1 The Final Phase of the Cybernetic Revolution and the Sixth Technological Paradigm

**About the new innovation cluster**. In the next few decades, human life and activity will change dramatically. Sooner or later, the coming technological wave, that is, the final phase of the Cybernetic Revolution, will undoubtedly affect almost every area of technology and life, penetrating even into the biological foundations of the functioning of the human body. Biotechnology will bring dramatic improvements in the environmental spheres. The interaction between society and nature is also likely to advance on a planetary scale, including the expanding opportunities for influencing the climate. Thus, as we have already pointed out, this will provide additional opportunities for purposeful influence and even control over various natural (perhaps, even global natural), social, production and other processes; so we assume that many of these processes will come under human control (which is now impossible or extremely limited). However, the self-regulating and self-managing systems will reduce constant and direct human intervention and ensure control through self-regulation processes.

Nevertheless, although as a result of the final phase of the Cybernetic Revolution there will affect almost every area of technology and life, a narrower cluster of technologies will eventually become the core of technological changes, largely determining the development of other technologies. In the theory of long-term economic cycles, those clusters of advanced industries that tend to dominate for several decades are called technological paradigms. Thus, in the present chapter, we will use as synonyms the concepts of a new technological paradigm and of a cluster of innovations, i.e., the MANBRIC convergence, of the final phase of the Cybernetic Revolution (explained below).

**Long economic cycles and technological paradigms**. According to the theory of long cycles (or Kondratieff waves), which we have discussed in detail elsewhere, every new cycle, lasting about 50–60 years, gives rise to a new technological (or techno-economic) paradigm (or mode). In the present monograph, we have no opportunity to dwell on this important idea, as well as on the characteristics of long cycles.[1] However, we should point out our conclusions, which are directly related to the present monograph. The fifth long cycle is now coming to an end; respectively, five technological paradigms can be defined in history.[2] The Cybernetic Revolution corresponds to: the fourth wave (the late 1940s–the early 1980s): car manufacturing, artificial materials, electronics; the fifth wave (early 1980s–2020s): microelectronics, personal computers and mobile phones.[3] The sixth long cycle, or wave, is expected to start in the late 2020s and early 2030s. Respectively, the emerging sixth technological paradigm will clearly show up. But for the purposes of our monograph, we should note that this will be just the period when the final phase of the Cybernetic Revolution should start. So, these two processes will merge and thus strengthen the expected technological wave (see Grinin & Grinin, 2014, 2016; Grinin & Korotayev, 2015; Grinin et al., 2016a, 2016b, 2017a). This is one of the reasons why we believe that technological progress may accelerate in the 2030s and 2040s (see also Chap. 12; Grinin et al., 2016, 2017a, 2017b, 2020, 2021a). We think that the sixth technological paradigm will shape the basis of the technological convergence of the final phase of the Cybernetic Revolution (ibid.). A few remarks on how one paradigm (one technological wave) evolves into another could be useful.

[1] Here we refer the reader to a number of our works (Grinin & Korotayev, 2010c, 2014; Grinin et al., 2010; Korotayev & Tsirel, 2010; Korotayev et al., 2011, 2020; Korotayev & Grinin, 2012; Grinin, 2020a, 2020b; Grinin & Grinin, 2015a, 2015b, 2016, 2022; Grinin et al., 2016, 2017a, 2017b, 2020, 2021a; Akaev, 2017).

[2] Technological systems and/or techno-economic paradigms are complex sets of major technological and industrial directions of a given period, which are currently the most advanced and determine the economy (about them see Mensch, 1979; Kleinknecht, 1981, 1987; Dickson, 1983; Dosi, 1984; Freeman, 1987; Tylecote, 1992; Glazyev, 1993; Mayevsky, 1997; Modelski & Thompson, 1996; Modelski, 2001, 2006; Yakovets, 2001; Freeman & Louçã, 2001; Ayres, 2006; Kleinknecht & van der Panne, 2006; Dator, 2006; Hirooka, 2006; Papenhausen, 2008; see also Lazurenko, 1992; Glazyev, 2009; Polterovich, 2009; Perez, 2002).

[3] The full scheme of technological paradigms looks as follows:

– the first wave (1780–the late 1840s): textile industry;

– the second wave (the late 1840s–the 1890s): railway lines, coal, steel;

– the third wave (1890–the end of the 1940s): electricity, chemical industry and heavy engineering;

– the fourth wave (the end of 1940s–the beginning of the 1980s): auto-mobile manufacturing, manmade materials, electronics;

– the fifth wave (early 1980s–-2020s): microelectronics, personal computers, biotechnologies; for the sixth wave see below. Of course, there were many other branches and technologies within each paradigm, but these are the most powerful ones. For example, as we will see in Chap. 9, biotechnology has been developing very quickly.

**General ideas about how technological structures are formed**
A new technological paradigm usually evolves according to the following pattern:

(1) the existing technologies gain some fundamentally new opportunities and levels of complexity (innovations);
(2) technologies that played a minor role within the previous paradigm, come to the fore, naturally accompanied by their qualitative development;
(3) breakthrough systems emerge.[4]

As for the first point, this is a natural enough process but because of its smoothness, it is often only recognized as a milestone in retrospect. At present, we can only observe but not fully perceive such a qualitative development of ICT technologies toward a more complex level of artificial intelligence (which will be discussed in more detail below and in Chap. 10).

As for the second point, we argue that the importance of the technology, which remains peripheral until a certain moment, can increase rapidly in different periods; sometimes such sectors are experimental or narrow systems, applied only in certain areas (e.g., in the military sphere).[5] The leading directions of technological paradigms have many spheres of application and gradually penetrate into new areas. It is also important that fundamentally new systems usually emerge after a new technological paradigm is fully formed as a primary system. Thus, in the sixth technological paradigm, computers that operate on a different basis than today (say, a quantum computer) could become such fundamentally new systems. Or it could be new energy sources (such as hydrogen, fusion energy) or some new technological ways of storing electricity in large quantities, which will give a huge impetus to the development of power generation, including green energy, and electrical engineering based on self-regulating systems. However, the prospects for these innovations in power generation, as well as for quantum computing seem rather vague today. In any case, it is difficult to imagine possible fundamentally new systems, since they can be very unusual and therefore difficult to predict.

One should also take into account that during the formation of a technological paradigm, a whole cluster of new technologies usually emerges, but hardly all of them become leading technologies. Some would develop as secondary and/or narrow sectors, etc.[6] However, even among these, some may take the main roles

[4] For example, in the first and second technological paradigms these were new (steam) transport systems (Grinin, 2022d, 2022e); in the fifth paradigm they were intelligent systems (Grinin, 2022c).

[5] Thus, the first computers appeared in the 1940s, when the third paradigm (Grinin, 2022f) was ending and the fourth was emerging, and their role was generally minor. They started to play the leading role only in the fifth paradigm (Grinin, 2022c). Electric transport appeared along with combustion-engine vehicles, but only during the last decade we observe efforts to make it the leading type of transport. Perhaps, this transition will occur within the sixth technological paradigm, but probably close to its end, that is, around the 2060s (see Chap. 10 for details).

[6] Thus, high-speed trains, which have been developing since at least the 1960s (i.e., since the beginning of the Cybernetic Revolution) and which constitute a truly innovative cluster, although they

at the end of the technological paradigm or during the subsequent paradigm. For example, electric motors for transport appeared almost simultaneously with automobiles in the 1880s and 1890s; nevertheless, the former became the basis, while the latter became an essential but not the leading part of the third mode, and the leading ones of the fourth mode (see Grinin, 2022b). Finally, let us point out that while the prospects for the development of information, digital and communication technologies are great, still they are unlikely to provide future technological breakthroughs and will only become an integral part of other technologies.

**The cluster of new technologies in the final phase of the Cybernetic Revolution.** As already mentioned, the final phase of the Cybernetic Revolution will form a peculiar and closely related cluster of innovative directions for the development of technologies (i.e., the sixth technological paradigm). However, one can hardly define exactly what this cluster will be. For our study it is very important to identify the most important combination of industries (a) whose development will exert a huge influence on the course of the Cybernetic Revolution in the 2030s–2060s; and (b) which will eventually form a highly interconnected system (actually a convergence) of industries that will support and reinforce each other. Then (c) within this set of directions for the next period (at the beginning of the final phase of the Cybernetic Revolution (2030–2040)), the core and integral element will be defined. It will come to the fore the first, then later other areas can and will come to the fore. So the "breakthroughs" in some areas will inevitably cause various innovations to "catch up" and form a system that will ultimately provide opportunities for the completion of the Cybernetic Revolution and also strongly influence innovations in social life.

The unfolding new technological wave will inevitably reduce the importance of many industries that currently play a significant role. So, the chemical industry, while remaining important, may lose its role and transfer the primacy to the biochemical industry; heavy engineering will probably transfer its role to the less material-intensive but "smarter" machines, including robots. Mass production will probably start to decline, while some of its functions will be transferred to additive technologies (3D printers-machines) that produce unique things for individual users (this is called personalization) or to small-scale production; carbon energy will give way to green energy, and so on.

### 4.2.2 A New Cluster of Innovative Technologies—The MANBRIC-Convergence

**Different views on the combination of breakthrough innovative directions.** There are different views on the technologies that will become of utmost importance in the coming decades and that will make up the sixth technological paradigm.

play an important role in a number of countries (Japan, China, France, etc.), have not generally become a very significant sector.

Thus, particular hopes are pinned on the biotechnological revolution and a huge or even dominant role of bioeconomy in the sixth technological paradigm (Sylvester et al., 1983; Fukuyama, 2002; Nefiodow, 1996; see also Nefiodow & Nefiodow, 2014a, 2014b; see also El-Chichakli et al., 2016; Jakimowicz & Rzeczkowski, 2019; Wilenius & Casti, 2015; Vega-González, 2017; Peters et al., 2021; see also Chap. 9).[7] As a result of the rapid development of nanotechnology, the sixth technological paradigm becomes associated with a widely used concept of the NBIC-technology/convergence (see Lynch, 2004; Roco & Baibringe, 2003; Bainbridge & Roco, 2005; Dator, 2006; Kovalchuk, 2011; Akayev, 2012).[8] There are also some researchers (e.g., Jotterand, 2008) who attribute this role to a different set of technological directions—the GRAIN (Genomics, Robotics, Artificial Intelligence, and Nanotechnology) and other combinations can also be found (e.g., Šmihula, 2010). In Chap. 2 we also mention the ideas of the so-called Green Revolution (e.g., Fücks, 2013).[9]

**The MANBRIC convergence: seven leading directions**. We have considered various views on the combination of breakthrough innovation directions that could "lead" the beginning of a new technological wave (i.e., the final phase of the Cybernetic Revolution). Our studies also convincingly prove that, firstly, this set will be larger, and secondly, it is extremely important that the vast majority of researchers do not consider the most important part of the new complex, namely, medicine (see Grinin & Grinin, 2014, 2016).

Thus, in the final phase of the Cybernetic Revolution the most important will be the set of developmental directions which we call **MANBRIC technologies**, which include *medicine, additive* (3D printers), *nano- and biotechnologies, robotics, IT*, and *cognitive sciences*.[10] In our opinion, they will form a systemic cluster of innovations and become the main drivers of the final phase of the Cybernetic Revolution (see Fig. 4.1).

*Note*: It is worth paying attention to the fact that pharmaceutical technologies, which were in a common cluster with medical, bio- and IT technologies, have started to lag behind since 2013. The explanation for this innovation slowdown is related to a rather strong and almost systemic crisis in the pharmaceutical industry, which is associated with a strong over-regulation of this industry, long time for the introduction of new drugs on the market and huge costs for their development and

[7] However, some researchers believe that the sixth paradigm has already emerged, while others (including us) believe that it will only be formed in the near future).

[8] NBIC-Nano-Bio-Info-Cogno.

[9] There are many different approaches; for example, there is an idea about the breakthrough role of the so-called urban innovation systems (Batty, 2018).

[10] The order of the letters in the acronym does not reflect our understanding of the relative importance of the areas that make up the complex. For example, biotechnologies will be more important than nanotechnologies, let alone additive manufacturing. The order is simply for ease of pronunciation. However, it is important to bear in mind that the relative importance of these directions may change over different time periods. So it is possible that nanotechnology will be already more important than biotechnology by 2070.

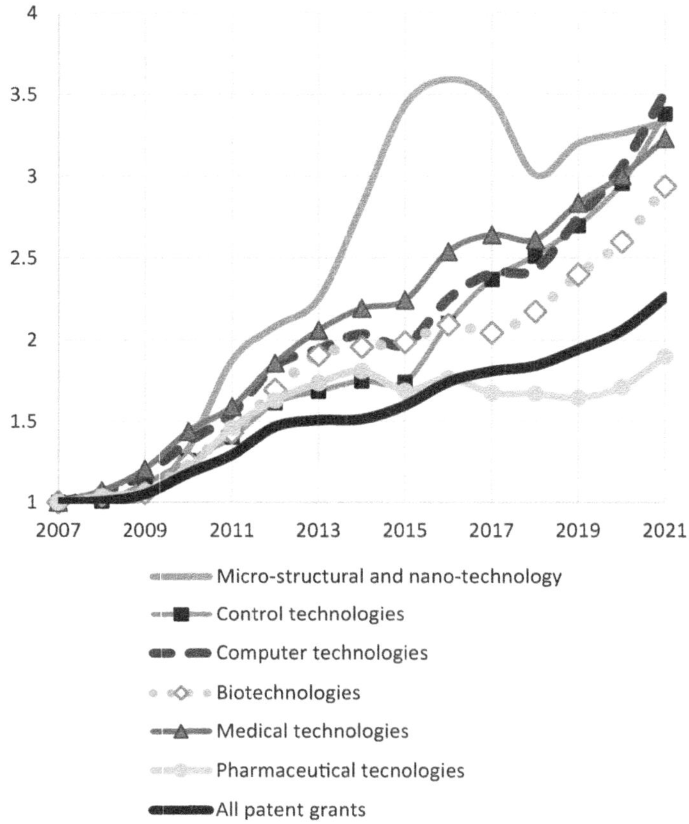


**Fig. 4.1** Relative global dynamics of patent grants for different technologies, 2007–2021, 1 = 2007 level.[11] *Data source* WIPO (2023)

testing, which forced many promising medium-sized companies out of the competition. The huge costs also lead to the mass production and introduction of drugs, which greatly reduces their medical efficacy. With such a system, the pharmaceutical industry cannot turn to a more promising path—small-scale production and, in the future, personalization of medicines (for more on the pharmaceutical crisis,

[11] It seems necessary to say a few words about the control technologies mentioned in this figure. What is control technology? The term control technology stands for devices that control, regulate, monitor, collect data, communicate and diagnose. Along with sensors and actuators, controllers are a central element of automation technology. According to Control Technology (2023), control technologies should come from a single source to ensure the consistent performance required for efficient automation. They are also combined with control software. The integrated control program covers a wide range of motion functions: from single-axis positioning and the electronic cam to robotics (Control Technology, 2023).

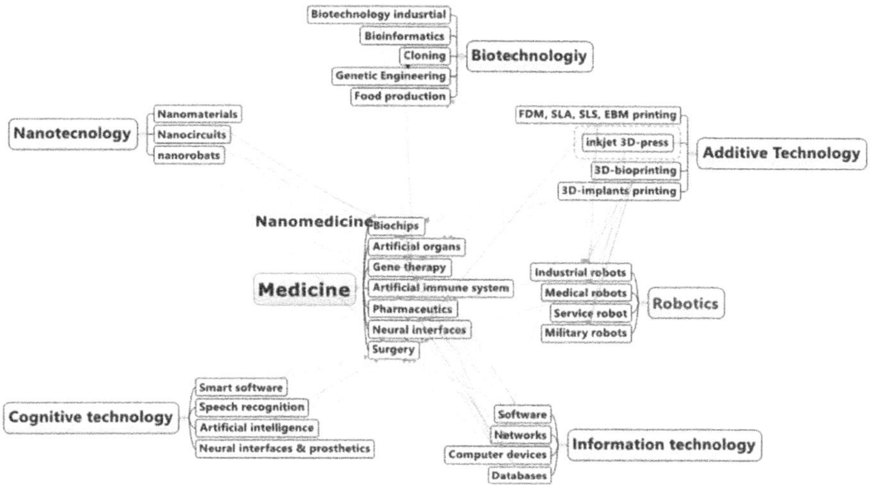


**Fig. 4.2** Interconnections between various MANBRIC technologies (*Source* Grinin et al., 2017b)

see Chaps. 8 and 9). On micro-structural and nanotechnologies see also Fig. 4.1 Chap. 10.

Together they will form an advanced system of self-regulating production—the MANBRIC technologies. Medicine will be a core and integral element of the MANBRIC complex and closely combine bio- and information technologies, robotics, as well as nano- and cognitive technologies and actively develop the implementation of additive technologies in the medical field. We will discuss this below (see also Fig. 4.2).[12]

The development of the MANBRIC complex can be examined via the analysis of patent applications in the fields of medicine, pharmaceuticals, nano- and biotechnologies and, to some extent, control technology (Figs. 4.1 and 4.3), which also show converging growth rates (2017a; Grinin et al., 2016a, 2016b).

### 4.2.3 On Some Technologies of the MANBRIC Convergence

As we have already said, the development of this complex will lead to integration, interaction and synergetic effects in many systems. And it is worth identifying these directions.

[12] In the Introduction, we gave the example of a bionic eye. And this is one of many examples of how medicine can combine innovative technologies (3D printers, ICT, biotechnologies, nanotechnologies, etc.) into a single complex (into a single self-regulating system) for a higher priority goals (both from a humanitarian and economic point of view).

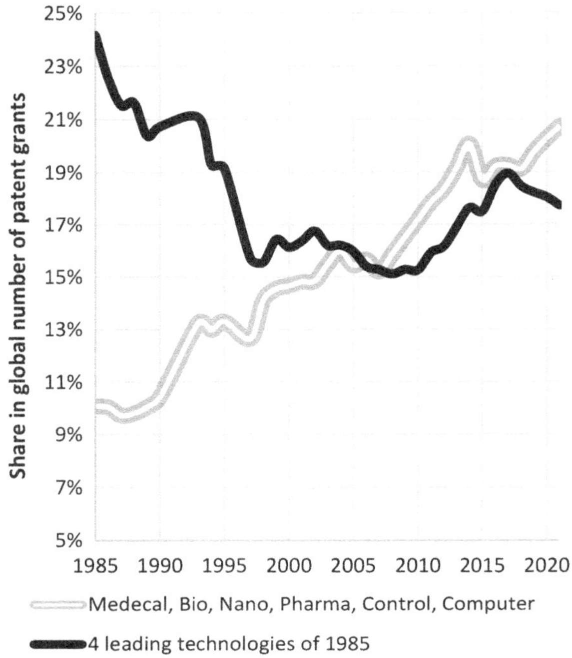


**Fig. 4.3** Dynamics of the combined share of the four most patented technologies worldwide in 1980–1990 (electrical machinery, measuring instruments, machine tools and other special machines) with those in 2000–2021 (medical, pharmaceutical, computer, control, nano and biotechnologies) in the period 1985–2021. *Source* WIPO IP Statistics Data Center 2023

**Medicine**. As mentioned above, medicine will be the central and integral element of the MANBRIC complex. At the same time, in connection with the societal aging process, medicine will increasingly shift to the field of geriatrics, the fight against age-related diseases and syndromes, and technologies for slowing down aging and rejuvenation (see Chaps. 8 and 11).[13]

Along with medicine, related areas will also actively develop: biotechnology and genetics, nanotechnology, robotics (now applied mainly in surgery, but in the future it will develop in many areas of medicine and care), cognitive technologies (to help the disabled and the elderly to compensate for their physical disabilities). The general breakthrough vector can be described as a rapid growth of opportunities to correct or even modify the human biological nature. In other words, our ability to interfere with the human body and its mental processes, and perhaps with its genome, will increase[14]; we will probably have more opportunities for targeted treatments and surgeries instead of modern surgical ones; it may also become possible to grow individual biological tissues, organs or their parts and elements for

[13] Geriatrics is a separate branch of gerontology, which deals with the study, prevention and treatment of diseases of old age. It aims at supporting health by preventing and treating diseases and disability in older adults.

[14] In the Introduction we emphasized that at the same time it will lead to growing worries and threats.

regeneration and rehabilitation of the body, as well as non-biological analogues of biological tissues (e.g., organs, receptors), etc. Additive technologies will be also essential (see Chap. 10).

**Biotechnologies**. Biotechnologies have been rapidly developing for decades. Some researchers write about a biotechnological revolution as the basis for the future radical change (see above). At the same time, biotechnologies are developing in several directions: "red" biotechnologies (pharmaceuticals and medicine), "green" biotechnologies (related to agriculture and environmental improvement) and "white" biotechnologies (widely used in the industrial sphere including in a wide range of production processes based on biocatalysis and fermentation as well as in biofuel production). Biotechnologies are the most important part of pharmaceuticals and genetic engineering, and their role in feeding the world's growing population is considerable (and it will increase in the future). They will help not only to increase productivity, control pandemics, etc., but also to improve production of seafood (so-called aquaculture associated with "blue" biotechnology). We also believe that biotechnologies, together with nanotechnologies, will play (and are already playing) an important role in the functioning of zero-waste production and the self-cleaning and remediation of polluted territories and waters. For more details, information and predictions on biotechnologies see Chap. 9.

**Nanotechnologies**. In the 2000s, many predictions were made about the upcoming breakthrough of nanotechnologies. However, they do not play an important role yet. We believe that the rise of nanotechnology may occur in the 2050s at the earliest. Actually, these technologies will allow research at the molecular and even atomic level, which will make it possible to create a huge number of new materials with unique properties and thus, contribute to the introduction of self-control in many systems, including physiological ones, and also make great progress in terms of saving energy and resources. At the moment, it is worth pointing out the increasing role of nanotechnologies as a branch of medicine and biotechnologies, in particular, in production of vaccines. For more details, information and forecasts on nanotechnology see Chap. 10.

**Robotics** will, of course, not only be developed in medicine, although the nursing robots will probably be the most common type of robots. We believe that robots will be very widespread, replacing millions of people in the service sector, including restaurants and cafes, hotels, cleaning, registration, shops, etc. Self-managing drones (including those delivering goods) and self-driving cars will also spread. The development of robotics may be driven to a significant extent by the increasing demand for them for military purposes (and, probably, in space and in hazardous industries). Military technologies have always paved the way for innovation. Many self-regulating and self-managing systems will appear in the army, while the drones, combat robots, rescue robots (miners, etc.), unmanned vehicles, etc., will become more and more intelligent and widespread. For more details, information and predictions on Robotics see Chap. 10.

**Information Technologies**. In recent years, ITC technologies have clearly developed to a higher level, which can be called the level of artificial intelligence (AI). Although this concept is rather vague, and AI has been used for a long time, now its capabilities have significantly increased thanks to various new technologies (as already well-known as neural networks, remote contact, linguistic systems, facial recognition systems, movement control, determination of individual preferences, machine learning as the latest, including ChatGPT). Moreover, most recently, the AI has been increasingly directed toward influencing social, control and even administrative relations (see Chap. 14; see also Grinin et al., 2021a). We believe that the sixth technological paradigm will allow the ICT in the form of artificial intelligence to mature and become an almost integral part of a new type of system. Moreover, it will become possible to complicate or simplify these systems so that they will be, metaphorically speaking, a kind of block system that can be assembled in a "configuration" required for a particular task, just as today we can increase the memory of a computer or upgrade the graphics card.[15] For more details, information and predictions on AI see Chap. 10.

**Additive technologies** are new universal technologies that make it possible to create a variety of objects using special devices (the 3D printers) and different materials. During the manufacturing process, the material is deposited, layer by layer, to form an object according to a three-dimensional computer model (this process is often, but not quite accurately, called printing, but it is actually a layering method). Such technologies are used in metallurgy and construction, mechanical engineering (to make parts), at home and in workshops, in museums (to make copies of unique exhibits), in medicine (to make analogues of tissues, cartilage and bones, organs, etc.). We suppose that 3D printers will become of particular importance in medicine. We believe that the mass distribution of such technologies will occur in the 2040s and 2050s, helping to dramatically expand opportunities for limited and small-scale production, which is often unprofitable today. At the same time, such technologies can be combined with others, for example, with robots. For example, scientists at the University of Oslo in Norway use 3D printing to create self-learning robots that have their own 3D printers and can repair their own or others' breakdowns by making the desired parts (Yesakova, 2014). For more details, information and predictions on additive technology see Chap. 10.

**Cognitive technologies**. Neural interfaces can become one of the most breakthrough areas of cognitive science and the Cybernetic Revolution in general. Neurointerfaces, or brain-computer interfaces, are technologies that enable interaction between the human nervous system and external devices. The fundamental achievement of cognitive sciences is the opportunities to control artificial organs using brain signals, just as healthy people can control their limbs and body parts.

---

[15] In Chap. 3 we dwelled in detail on the idea that AI is a special technology and not a production system, that is, the concepts of self-regulating and self-managing systems and AI lie in somewhat different dimensions, although they are related.

This holds the prospect that, over time, hundreds of millions of disabled people will be able to compensate for much of their physical limitations with the help of such devices, which will relieve many people from the need to help the disabled. In the future, neural interfaces will probably be used not only by disabled people but also by healthy people in everyday activities, although there are many well-founded fears of misuse on the part of large digital companies and state authorities. In general, the achievements of cognitive science are already applied and will be increasingly applied in various fields, from medicine to robotics, which are moving toward the development of self-regulating systems. However, repeat, the development of this direction faces serious technical, social and moral obstacles.

Therefore, we believe that cognitive technologies will actively show their value in the late phases of the Cybernetic Revolution. It is worth adding that the military and intelligence services are actively exploring the opportunities of cognitive science and technologies in order to use them for their own purposes. This also poses some problems. The discoveries and achievements of cognitive science have been a major trigger for the development of AI technologies (e.g., Neuromorphic computing networks). For more details, information and predictions on cognitive technology see Chap. 10.

## 4.3 The Future Dynamics of the Cybernetic Revolution in the Light of the Logic of Production Revolution

First of all, it should be remembered that the Cybernetic Revolution is one of the largest production (technological) revolutions in all of history; it is the last of such revolution after the Agrarian and Industrial ones. The concept of production revolutions should help us to make some important forecasts about the Cybernetic Revolution with certain backgrounds.

Above we have identified the cluster of the leading innovative technologies that will emerge as a result of the final phase of the Cybernetic Revolution. However, we have not brought to light the very important questions which we formulated at the beginning of this chapter. They are: (1) what direction will be central and integral to this complex? We assume that the development of these technological fields, although closely related, is not and will not be simultaneous. Consequently, the Cybernetic Revolution will *begin not on a large scale, but in a certain narrow sphere, in a certain area.* Hence the question: (2) in which sphere will the final phase of the Cybernetic Revolution begin? Without a clear argumentation of our conclusions on these questions it will be difficult to understand our technological forecasts.

Previously, we have already given a tentative answer to the first question, namely that medicine will be an integral part of the MANBRIC complex. We can also give a preliminary answer to the second question, and again, it is medicine that will be the area where, according to our assumptions, the final phase of the Cybernetic revolution will begin. However, both questions are very important, and therefore require a more comprehensive answer and detailed argumentation, which we will present in the remaining sections of this chapter.

For a more detailed understanding of how, where and why a breakthrough to a new technological wave (the final phase of the Cybernetic Revolution) will begin, as well as the reasoning behind our calculations of the beginning of this breakthrough in the 2030s, we will need to describe the logic and algorithms of the Cybernetic Revolution, which are largely based on the analysis of previous production revolutions—the Agrarian and Industrial ones—and the theory of production revolutions as a whole (to which Chap. 2 was devoted). Therefore, the following section, although somewhat academic, is very important for understanding our theory and conclusions; it also clarifies a lot in relation to Chap. 12.

### 4.3.1 The Analysis of the Phases of Production Revolution as a Means of Forecasting the Future Dynamics of the Cybernetic Revolution

The importance of the theory of production revolutions and production principles lies in the fact that they not only allow a more detailed and fruitful description of the production and technological evolution but also provide scientific means *to forecast the unfolding of the Cybernetic Revolution and of the Scientific-Cybernetic Production Principle.* Our forecast is based on the identified regularities in the phases of production revolutions. This section defines these regularities and the way they can be used in forecasting.

Let us recall that the fundamental idea of the proposed concept of production revolutions is that, *within each production revolution, each of its three phases plays a functionally similar role, while the ratio between the duration of the phases within the framework of each cycle remains approximately the same.* Thus, on the basis of the regularities identified within the Agrarian and Industrial Revolutions, one can make the following assumptions:

- *first*, on the duration of the modernization (intermediate) phase of the Cybernetic Revolution;
- *second*, on the beginning and approximate duration of the final phase of the revolution;
- *third*, on the sectors and directions that will be affected by the technological breakthrough.

Therefore, the theory of production revolutions provides a methodological approach to ground our predictions about the future technological shifts within the Cybernetic Revolution. Let us remind the reader that the initial phase of the Cybernetic Revolution has already been completed (it lasted from 1950 to the early 1990s) and the modernization phase is about two-thirds of the way through its development (it started in the 1990s and expected to end in the 2030s). So we can compare the forecasts of the theory for each phase of production revolution

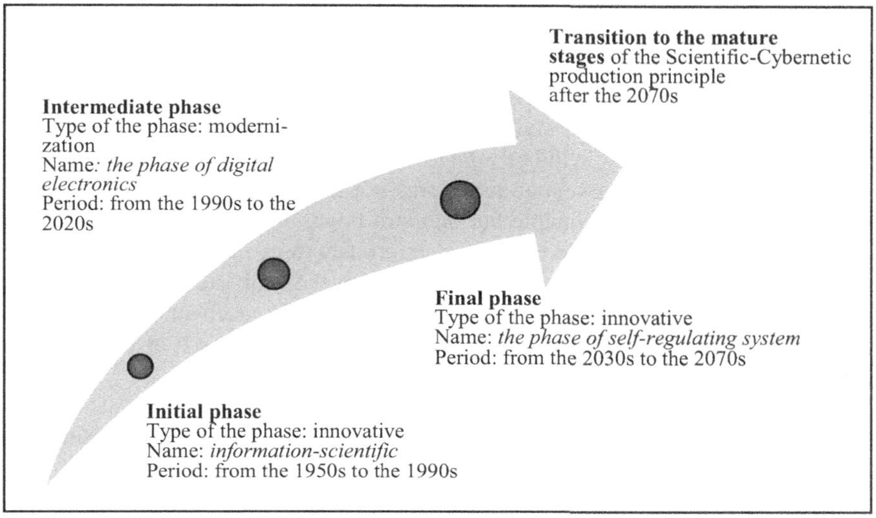


**Fig. 4.4** Phases of the cybernetic revolution

with the current reality, and we can also deduce the role that technologies will play in the final phase of the Cybernetic Revolution (see Fig. 4.4).

To give a better explanation to such a methodology, we formulate a number of functional and processual relations between the initial and final phases of a production revolution, between the initial and modernization phases, and between the latter and final phases of the production revolutions. Knowing the algorithm of how the processes manifested in the initial phase of the production revolution can be transformed during the intermediate and final phases, we provide forecasts of the development of the Cybernetic Revolution for the coming decades, proceeding from the study of its initial and uncompleted intermediate phase (for the mathematical calculations see Chap. 12, Tables 12.3 and 12.4).

### 4.3.2 The Peculiarities of the Initial Phase: The Non-systemic Tendencies Merge into a System While the New Ones Develop

The *initial phase* of the production revolution is characterized by the following peculiarities:

1. *A number of trends and innovations that were non-systemic within the previous production principle acquire a systemic character.* The non-systemic character means that within the previous production principle these phenomena did not play a decisive role and did not result from its main characteristics, whereas within a new production principle the role of these characteristics increases

significantly. This can be exemplified by automatization, which was developed to a certain extent in the industrial production long before the Cybernetic Revolution. One of the main characteristics of the Industrial Production Principle is that production is carried out by machines operated by humans who use their senses, strength and skills. At the same time, some operations are carried out automatically, without human intervention. However, the automation of processes was not essential and it was not a necessary characteristic of the Industrial Production Principle but an extra bonus. In the early twentieth century, automatization started to develop vigorously (e.g., in electrical engineering to prevent accidents, in engines for convenient control, etc.). Nevertheless, it was not decisively important as it was not generally used for the automation of technological processes (for more details see Chap. 2; see also Grinin and Grinin, 2015a, 2015b).

Therefore, at this stage, automation can be regarded as a over-development of such an essential phenomenon as mechanization. Moreover, in the first half of the twentieth century, automation was not the leading trend of the Industrial Production Principle. On the contrary, the leading role was played by the processes of the latest division of labor including the wide distribution of assembly-flow production (a constantly intensifying division of labor is an essential and transparent characteristic of the Industrial Production Principle, strikingly manifested already in manufactories). The development of automation in the second half of the twentieth century is quite a different matter. It became the most important characteristic of the Scientific-Cybernetic Production Principle (in its initial stages) and found new forms of application and implementation in releasing human costs in the control of processes (especially in Information and Communications Technologies [ICT]). As we pointed out in Chap. 3, automation can be considered as the first stage of self-regulation, i.e., the basic quality of self-regulating systems.

Thus, *the non-systemic elements of the previous period are developed to the highest degree during the initial phase of production revolution.* Thus, automation followed mechanization (see, e.g., Lilley, 1966; Philipson, 1962; Bernal, 1965), similar as the chemistry of synthetic materials was sequential to organic chemistry (Domingues et al., 2023; Morgan, 1981; Zvorykin et al., 1962); and just as the Green Revolution in agriculture followed agronomy (Evenson & Gollin, 2003; Thirtle et al., 2003; Renkow & Byerlee, 2010; Ecker et al., 2011). The development of radio and television technologies continued the earlier trend of new means of information transfer. Such continuity can conceal the intensity of the transition from one epoch to another. Thus, it is not surprising that in the 1950s the rise of scientific and technological innovation was initially considered as a continuation of the Industrial Revolution, though it was the beginning of the new Cybernetic

Revolution[16] However, it is important for this section to stress that this super-development had some qualitative characteristics. So we will describe them below.

2. *The former non-systemic characteristics, together with the newly emerging ones merge into a unified system, representing a new production principle.* Automation, chemical synthesis of materials, a powerful development of non-computer electronics and means of communication, the emergence of various engines, a general transition to a new type of energy and fuel, breakthroughs in selection and plant protection, the dismissal of dozens of millions of workers previously employed in agriculture and industry and their transition to the service sector; a number of new directions emerging in technologies, informatics and science—all this creates a principally new situation in the economy and also evidences the start of a new production revolution, namely, of the Cybernetic one.
3. *An important factor with a strong synergetic effect is a temporal density (a cluster pattern) of the emergence and development of a number of directions which are more or less typical of a new production revolution and principle.* Such directions in the 1950s and 1960s were the nuclear power industry, space exploration and the use of space frequencies for communication and other purposes, the deep-sea exploration, information and computer technologies, multiplying equipment, laser technologies and other areas (e.g., genetics, medicine and biotechnology).
4. However, *these innovative spheres can have different destinies*: some of them will experience a peculiar and important development in the second half of the initial phase and in the intermediate phase; while other trends will develop less intensively. Some may become dead-ends (at least, temporally). Thus, the nuclear energy industry currently faces severe constraints due to security concerns and environmental problems, the hopes for mastering thermonuclear energy have fallen short of expectations, while deep-sea exploration (except for the shelf sea) still remains exotic.[17] At the same time, the development of ICT has become the leading trend (see Chap. 10).
5. *The change of the leading sector in the course of the production revolution.* The leading role of some peculiar characteristics and sectors of a new production principle becomes especially obvious at the end of its initial phase or during the modernization phase (as in the case with ICT). These sectors need some time to reach maturity and acquire a systemic character. *Thus, during these first two phases of the production revolution, the leading branches and sectors constantly*

[16] In the previous chapters we have pointed out that this Revolution has been called by different names (in particular, the scientific and technological revolution [Bernal, 1965]). Anyway, in the 1960s, it became completely clear that it was a new technological revolution. While during that period it was still quite acceptable to speak of a new Industrial Revolution (see, e.g., Sylvester & Klotz, 1983), increasing the number of Industrial Revolutions after each technological wave up to three or four (Rifkin, 2011; Schwab, 2016) seems counter-conceptual in our view.

[17] However, the recent energy crisis has contributed to a new rise of nuclear energy. Also, there has been some advance in experiments with thermonuclear energy.

*change while some new sectors are created.*[18] **One of the branches of a new production principle starts to dominate over the others for quite a long period of time (from the end of the initial phase to the end of the intermediate phase)**. This branch becomes the key phenomenon of the production revolution and its driving force, while *its role as a driving force diminishes later*. For example, the wool industry (the most important branch in the initial phase of the Industrial Revolution) became insignificant in the final phase when it was replaced by the cotton industry (for details see Chap. 2; Grinin & Korotayev, 2015; Grinin & Grinin, 2015).[19] The AI continues to be the most widespread technology due to huge development during the initial and modernization phases. It is a very innovative field and we expect new innovations in it. However, even though it may sound contrary to common expectations, one may assume that *in the final phase of the Cybernetic Revolution the ICTs/AI will hardly remain the most important sphere.* AI will obviously be a very important component of many self-regulating and self-managing systems, as we mentioned in Chap. 3, but it will hardly be a dominant sphere. So, in short, our idea is that the initial breakthrough that will launch the final phase of the Cybernetic Revolution will not be the AI, but another direction, which, however, will be based on the achievements of AI as well as other fields of MANBRIC complex.

Later, however, during the final phase of the Cybernetic Revolution (around the 2040–2060s), we can expect a new qualitative breakthrough in ICT. Perhaps new-type (quantum) computers will be invented. But there will also be other possibilities. For example, it can be assumed that sooner or later serious changes will inevitably occur in programming. At present, this process is labor-intensive and slow. It will most likely develop in the direction of simplification and robotization of some elements of programming and especially in the implementation of programs. In other words, machine programming will largely replace human programmers and the trend toward 'self-programming' will advance. For our forecasts about AI and robotics see Chap. 10.

6. *Already during the initial phase of a production revolution, prototype sectors emerge that will become the leading ones in its final phase. However, in the initial phase they do not play the leading role* (see below for more details on future leading sectors).

[18] It is worth remembering that a production principle consists of six phases, while a production revolution consists of three phases, which correspond to the first three phases of a production principle. For details see Chap. 2.

[19] Animal husbandry, which developed during the modernization phase of the Agrarian Revolution did not become the leading direction of the final phase of this revolution (see Chap. 2).

### 4.3.3 Characteristics of the Modernization (Intermediate) Phase: Accumulation of Innovations and Search for a Breakthrough Point

1. *The growing scale of already existing tendencies and the formation of new ones.* On the one hand, in this phase many processes develop (to a varying degree) which appear in the initial phase of the production revolution. On the other hand, in the modernization phase we can trace the roots of those forms that will become leading in the final phase of the production revolution. Therefore, it is important to distinguish between the trends which have already matured and those that are still in the process of formation, and to understand which of them will develop and which of them will become less important, become stable or later decline.
2. *The expansion of development. The need for profound social and political changes.* The spread of new technologies is especially evident in the first half of the modernization phase. In its second half, this expansion saturates to a certain extent and slows down, thus increasing innovation activities. There appears an anticipation of something important. But the crucial element for the formation of a new system is still missing. Moreover, this gap can be manifested not only in the absence of basic technological innovations, but also in the absence of a social context for their implementation. One of the most important characteristics of the modernization phase is that *during this period some profound changes or even breakthroughs in social and political relations should occur.*

For example, with respect to the Industrial Revolution, its modernization phase was the period between the seventeenth and eighteenth centuries. It was the time of social revolutions in England, the Netherlands, the USA and France that changed the world. It was also the time of changes in the world politics: The Thirty Years' War (1618–1648) and the subsequent Peace of Westphalia laid the foundations of international relations for a long time (on the concept of sovereignty and the history of changes in the world order see Grinin, 2009a, 2012a, 2016; Grinin et al., 2016). With respect of the Cybernetic Revolution, its modernization phase is a period from the early to mid-1990s up to the present. Many political and social breakthroughs have been occurred during this period. For a long time it was globalization that has transformed the World System. After the Great Recession, which was itself a powerful transformation in many respects, the period which we denoted as the epoch of new coalitions, that is, an increase in the number of situational and special-purpose alliances and relationships (Grinin, 2009a, 2012a; Grinin & Korotayev, 2010b, 2015; Grinin et al., 2021b), changed significantly. Then came COVID-19 with great disruption and change,[20] and finally the war

[20] See, e.g., Chumakov and Yurchenko (2021), Grinin (2020a, 2020b), Hussain (2021), Irshad (2020), Jasiński and Bąkowska (2020), Kakkar (2021), Seifabadi and Dolatabadi (2022), Widdowson (2021).

in Ukraine, which halted globalization and divided the World System. The new coalitions mentioned above are beginning to blossom.

However, whereas in the seventeenth century a political revolution took place in one country—Britain—today one may speak about an analogue of a political revolution unfolding within the whole World System, that is, dramatic changes are taking place in the geopolitical sphere and world order, as well as in political scene in many countries, including the USA. Elsewhere we have analyzed in detail this process which we denote as the reconfiguration of the World-System. It has been underway since about 2009, that is, since the beginning of the Arab Spring (see Grinin, 2009b, 2012a, 2022a; Grinin & Korotayev, 2010b, 2012, 2015, 2016, 2020; Grinin et al., 2019; Grinin & Grinin, 2015a, 2015b). And it is related to the fact that the economic component of globalization is considerably ahead of its political component, which causes (a) increasing tensions, conflicts, including wars and turbulence in the World System in general; (b) the weakening and destruction of the old world order;[21] (c) an increase in the number of revolutions and revolutionary events; (d) in general, the rollback of economic globalization along with the processes of deglobalization. We believe that these processes will continue for at least the next 8–10 years, and their weakening and related transformations will facilitate the transition to the start of the final phase of the Cybernetic Revolution.[22]

3. *The idea of a decisive component.* During the modernization phase, one can observe the accumulation of opportunities and improvements that will contribute to the start of the final phase of a production revolution. All components should be ready for this start. However, we emphasize that innovations can form a new system only after the key component emerges. At the same time the reconstruction of the relationships within the production system in general will be considerable.
4. *The emergence of a decisive innovation in the new field.* Based on our analysis of production revolutions, we can conclude that the decisive innovation will hardly appear within the most important economic sector.[23] Moreover, certain

[21] The discussions of the inevitable decline of the American power began as early as the 1970–1980s (see, e.g., Vogel, 1979; Kennedy, 1987). Since the 1990s, one observes an increasing number of political forecasts predicting an imminent decline of the American supremacy and simultaneous coming to the fore of Asian economies (Attali, 1991; Colson & Eckerd, 1991; Arrighi, 1994; Frank, 1997; Buchanan, 2002; Kupchan, 2002; Todd, 2003; Wallerstein, 2003; Mandelbaum, 2005; NIC, 2008, 2012; see also: Grinin, 2010; Grinin & Korotayev, 2010a, 2010b, 2015; 2020; Zhdanov & Korotayev, 2022).

[22] About these processes see Grinin (2010, 2012b, 2012c, 2022a), Grinin and Korotayev (2011, 2012, 2016, 2020), Grinin et al., (2016a, 2016b), Grinin et al., (2017a, 2017b, 2017c, 2021b).

[23] We can see some evidence in the previous production revolutions. Irrigated agriculture was not the most important sector of agriculture in pre-state barbarian societies, but then irrigated agriculture came to the fore and became the basis for the first civilizations; similarly, the cotton industry was not the most important industrial sector in the first half of the eighteenth century, but then it became the leading industrial sector.

conditions, including high commercial profitability and attractiveness should be present in this field, which will ensure a steady demand over a long period of time. Nevertheless, the emergence of the breakthrough innovation can remain underestimated for some time.

The decisive innovation for the start of the final phase of the Cybernetic Revolution may emerge in some fields of bio-nano-medicine or in another new branch of medicine.[24] It may be a series of innovations that will transform the growing number of innovations into a qualitatively new system. It is quite possible that such a breakthrough will be associated with the invention of successful methods to fight cancer, since this disease is very different from other diseases and requires solutions at the genetic level as well as the application of fundamentally new technologies.

### 4.3.4 The Peculiarities of the Final Phase

1. *The main characteristics of the production revolution reach their maturity.* All the basic characteristics of the final phase of the revolution can be found in its initial phase, albeit in an undifferentiated, incomplete or undeveloped state. These characteristics of the future system show up appear in the intermediate phase, when the production principle takes a relatively complete, though undeveloped, form.

*Thus, one may infer the main characteristics of the Cybernetic Revolution on the basis of the analysis of the initial and intermediate phases, focusing on their features and developmental dynamics.* This analysis allows defining the most important characteristics of the Cybernetic Revolution, mentioned in Chap. 3, including the saving of resources, miniaturization, individualization, a wider implementation of artificial and smart materials, etc. These characteristics are already present today, but they will be absolutely dominant in the coming epoch.

2. *Some of the numerous directions that develop during the initial phase will inevitably become the leading directions in the final phase.* At the same time, they play a less significant role in the initial phase. Thus, while in the final phase of the Industrial Revolution the main focus was on mechanisms, machines and the replacement of manual labor by machines, in its initial phase machines were only one part of this new direction. At the beginning of the Industrial Revolution, technical innovations (the replacement of manual labor by machines) were less important and the main factor was the intensified labor division. If one considers the Agrarian Revolution, the leading direction of manual (hoe)

[24] The COVID-19 pandemic has seen one such breakthrough with a new type of vaccine (see Chap. 12 for details), although so far they have not yet played the role they were intended to.

agriculture was the use of fertile lands with the help of manual labor (e.g., with the help of a sharp stick or stone hoe). Soil fertility was achieved naturally or by burning plants. As for irrigation technologies, in the initial phase of the Agrarian Revolution they were not so widespread and were determined by the local environment. However, in the final phase they came to the fore and remained leading ones throughout the period when the Craft-Agrarian Principle dominated in production.

Therefore, the leading sector of the final phase of the Cybernetic Revolution has already been formed, but it is one of those sectors that do not yet play a decisive role in the economy. In our opinion, *medicine will play the leading role in the unfolding of the final phase of the Cybernetic Revolution.*

3. *The mutual integration of innovative sectors is complete after the formation of the decisive innovations.* This process intensifies especially during the final phase of a production revolution. Innovations are mutually integrated and form a fundamentally new system. This was the case with the invention of the power spinning loom in the 1760s (which was later continuously improved). Before that, for two decades, other important directions (steam engines, steam power, new types of machines, principles of management in large enterprises, the establishment of the institution of inventions and various technological innovations) had prepared the formation of the fundamentally new sector of cotton mills. This resulted in a cumulative effect of the rapid development of lacking innovations in the field of cotton carding (i.e., the separation of cotton fibers), painting, printing, etc.

*Thus, the breakthroughs in medicine and related technologies will cause the 'catching up' and merging of different innovations into a system that could accelerate the Cybernetic Revolution.* The MANBRIC convergence will develop intensively.

4. One should *distinguish between the breakthrough area and the essence of a new production system.* The breakthrough area only launches profound transformations. The production revolution will fully acquire its logic and 'sense' or 'essence' only later, when the transformations become profound and extended. However, one can try to guess this 'meaning', 'sense' and 'essence' on the basis of the processes occurring during the initial and intermediate phases of production revolution.

## 4.4 Determining the Future Breakthrough Sector—Why Medicine?

### 4.4.1 Characteristics of the Future Breakthrough Sector

As we have shown above, at the beginning of the final phase of a production revolution, one of the numerous directions that previously played a minor role in the economy becomes a breakthrough area.

The analysis of the actual development of production revolutions also suggests the following characteristics of the future sector:

- the commodity produced in this sector should be of primary necessity. For example, cereals during the Agrarian Revolution and cotton during the Industrial Revolution were basic necessities;
- the direction of development of the sector should be in line with the mainstream trends and urgent problems of society (irrigated agriculture could support and increase the sudden exponential population growth; the cotton industry met the needs of increasing urbanization and used the surplus labor force created in the agricultural sector);
- the sector can influence and integrate a significant number of spheres (e.g., during the Agrarian Revolution the irrigation facilities required joint actions in society; and during the Industrial Revolution the transition to machines and steam engines in the cotton industry led to a rapid economic growth, not least due to the reconstruction of transport routes and the expansion of trade);
- technological conservatism in this sector is relatively low;
- the breakthrough sector should provide high profits and rely on steady demand, otherwise it will fail to attract large investments. In addition, new technologies in this sector that emerge in the advanced society should be borrowed by other societies without any obstacles (e.g., government bans, etc.);
- the sector must have a great potential for productivity growth, and the need for productivity growth must remain high for a long time in order to stimulate innovation and investment.

### 4.4.2 A Possible Breakthrough Area

Let us apply these conclusions to the Cybernetic Revolution. It is obvious that the future breakthrough sector of the final phase of this revolution should already have emerged. But which of the existing ones have the characteristics mentioned? Let us consider several possible "candidates" for leading innovations. We will start with energy.

#### 4.4.2.1 Power Generation

The conversion of the entire energy sector to 'green' basis (i.e., to the use of low-carbon energy) will certainly affect the whole economy and, in many respects, transport, as well as promote various innovations, including forms of economic organization. However, unlike the transition to steam energy or electricity, it will hardly have a dramatic and overwhelming impact on the constructive changes in machines, mechanisms and technologies. The issue of the need to switch to "green" energy is determined not by the higher productivity of these technologies and not by greater comfort, but by the problems associated with threats of climate deterioration and natural and humanitarian disasters. Similar situations have occurred many times in history, for example, when soils deteriorated (due to salinization, erosion), aridization, etc. Societies would change in such situations, but such changes do not usually lead to technological revolutions. However, the latter were somehow connected with climate change.[25] But there is no clear correlation here (the climate dramatically changed much more often than technological revolutions would occur); so climate change can only be one of the triggers of a technological revolution if the conditions for it are already basically developed.

On the one hand, the desired financial flows and investments in green energy are simply enormous, up to 5–10 trillion (!) dollars per year for the next 30 years (Durden, 2021a, 2021b). And from this point of view, this area is even ahead of investment in medicine (although one can hardly believe that such investment can be found). But on the other hand, it is not an industry that integrates all innovative areas, nor does it transform everything. In addition, the technological prospects here are vague, and it is also unclear when new energy technologies superior to the carbon ones will be developed. In any case, green technologies will be a very important part of the final phase of the Cybernetic Revolution.

However, with respect to the use of hydrogen as an energy source, not to mention the thermonuclear fusion for electricity production, we can say that if they were to succeed, their impact would obviously be greater than that of traditional green technologies. For example, hydrogen as a fuel could replace the fuel produced from oil, and thermonuclear fusion could dramatically reduce the cost of electricity production and significantly increase productivity. However, while wind and solar energy are already actively used, the thermonuclear energy is still only a dream, and hydrogen energy is still in the experimental stage. And the prospects for their industrial use are vague. This is even more the case for the so-called "green" hydrogen, whose production costs currently far exceed its economic effect (Akaev & Davydova, 2023).[26]

**On possible variants of a new technological cluster with respect to green energy.** Any predictions are always rather hypothetical since reality can bring dramatic

---

[25] For example, the cooling of the fourteenth and then seventeenth centuries increased the demand for warm woolen clothing, thus contributing to the increased production of wool and woolen fabrics and supporting the spread of manufacturing in England. And this was a powerful impetus for the development of the Industrial Revolution there as well.

[26] However, "green" hydrogen combined with green electricity, produced during periods of low electricity demand, when it is almost never used, will be much cheaper. We are talking about hypothetical projects to accumulate the low-demand green electricity mentioned above to produce hydrogen.

changes. Therefore, the components of the defined set of directions should be elaborated. In the last few decades in connection with the course toward climate protection and for environmental reasons, green energy has been actively developing along with the production of electric vehicles. At present, huge efforts are being made in this direction. So it makes sense to replace the abbreviation MANBRIC with MENBRIC, that is, to replace additive technologies with energy technologies. Actually, the developmental rates of new energy and additive technologies are incomparable. However, green energy technologies are still very far from the level of efficiency needed to become a driver of changes that will lead to the integration of all the other innovative areas of the new technological wave (and it is not yet clear how they can influence, for example, medicine or additive technologies) without concentrated and very costly efforts. On the other hand, we think that additive technologies have a rather challenging future, as they can greatly facilitate the delivery of goods (as many products will be manufactured on-the-site and not delivered, similar to how paper products are now much less transported but printed according to the layout provided). And in this sense, additive technologies are also green. So perhaps it is more reasonable to increase the number of areas included in the complex to eight and to call it MANEBRIC?

#### 4.4.2.2 Robotics

We believe that robotics could become the breakthrough area if robots are created that can perform different functions in the service sector, that is, universal robots. It is not without reason that future scientific and technological progress is linked to achievements in the field of robotics. Currently, robotics is widely used and developing rapidly (see, e.g., Makarov & Topcheev, 2003; Katz et al., 2005; Gates, 2007; Dirican, 2015; Nistor & Rusu, 2019; Soto et al., 2020). Nevertheless, one may say that robotics is unlikely to become the breakthrough direction, given the current volume of investment in this field, which grows slowly. However, it will play a very important role in the final phase of the Cybernetic Revolution and should achieve outstanding results, albeit somewhat later, perhaps in the middle of the final phase of the Cybernetic Revolution (for more details see Chap. 10).

As we pointed out in Chap. 3 and above, the development of robotics is strongly stimulated by military needs, since it can enable the use of unmanned military technologies and greatly reduce both manpower requirements and casualties. We think that robotics will be in great demand in the course of the Cybernetic Revolution. Its achievements can significantly and *most noticeably increase labor productivity*. There are particularly great opportunities in the service sector, which requires relatively low skills (hotels, cafes, cleaning, reception, delivery, gas stations, shop sales, etc.) but where the bulk of workers is concentrated.[27]

[27] Already today, there are certain functions in some service sectors where robots have replaced humans, for example, in identifying which items on store shelves are sold out and need to be replenished.

To conclude this section, we may say that the development of robotics in general will be supported by labor shortages, especially in the service sector, and a possible increase in wages in this sector. Both of these trends are currently observed in the USA, so it is likely that the development and implementation of robots will accelerate.

#### 4.4.2.3 Self-driving Cars

We consider the development of self-driving cars as a vivid sign that the future technological breakthrough will be associated with self-regulating/self-managing systems. However, the reliability of such unmanned systems (as well as robots that are in direct and communicative contact with humans) will be a challenge. Any incident can delay their implementation for a long time, as we will see in Chap. 10; nevertheless, there are ambitious plans to start the mass production of such vehicles in the near future (see, e.g., Vesti.ru, 2022). It will be many years before these unmanned vehicles are really widespread. Even electric vehicles are still an exception rather than a rule. There are also many legal and social barriers to the widespread introduction of self-driving vehicles. Thus, we consider the emergence of robotic vehicles as a precursor to the coming final phase of the Cybernetic Revolution, but not its beginning. Finally, truly unmanned systems that do not require a human being at all (whereas today a human should still be highly qualified) are still a distant prospect. Yet, unmanned vehicles could be introduced quite quickly in some situations: on certain routes, in desert areas, as a guided part of a convoy of trucks, etc.

There is another obstacle on the way to the mass introduction of driverless vehicles. While green energy creates many new jobs, robotic vehicles (self-driving cars) on the contrary threaten to put many millions of skilled workers, perhaps the largest section of the working class today, out of work. Naturally, the job losses will provoke active social resistance. On the other hand, today there is a severe shortage of truck drivers in the USA, Great Britain and other countries. If this trend continues, the introduction of self-driving trucks in the future may become easier. However, the efficiency of electric trucks is another matter. And self-driving diesel trucks are a more complex combination. For more details on driverless vehicles see Chap. 10.

### 4.4.3 Medicine as a Field of the First Technological Breakthrough

**Why is medicine to become the breakthrough sphere?** Above we described the criteria that (according to the theory of production revolutions) can be used to predict the future breakthrough industry at the beginning of the final phase of the Cybernetic Revolution. Accordingly, this will be the area that most closely (albeit incompletely) fulfills these criteria. Among other fields, medicine meets these criteria to the greatest extent. Let us consider in detail why medicine is to become the breakthrough field.

(a) Medicine is unique because it stimulates constant activity in the field of new high technologies.
(b) The medical field offers unique opportunities to combine the above-mentioned technologies into a single complex. Many fields (including but not limited to biotechnologies, nanotechnologies, robotics, use of the latest ICT and various devices, cognitive technologies, synthesis of new materials) will be integrated in this field (i.e., MANBRIC technologies, see Fig. 4.1).
(c) There are far fewer social, cultural or structural obstacles to the implementation of these technologies in medicine than in other fields (the same situation with respect to the obstacles to adoption of innovations is observed in the country of invention as well as in many other countries).
(d) The commercial prospects and profits of new technologies in this field are enormous, as people are willing to pay for them. Moreover, the healthcare market will grow steadily and the face of aging. So medicine, as the future breakthrough area, will be highly attractive in commercial terms.
(e) Medicine (both government and privately funded) has produced a significant impact influence on GDP. Taking into consideration the faster GDP growth rates of expected in developing countries and the rapid emergence of a middle class there, it is reasonable to assume that healthcare spending will generally increase significantly. For example, in Germany, 22% of the workforce is employed in the health sector, compared with only 2.3% in the automotive industry (Nefiodow & Nefiodow, 2014a). In the USA the figure is almost 15% (Stiegler et al., 2021: 9), which is comparable with total employment in the whole of the US industry as a whole (20%, see World Bank, 2023).
(f) Medicine can alleviate the shortage of labor force which is exacerbated by a declining young population and a growing number of elderly people.

**Why will the breakthrough occur in the 2030s–2040s?** First, the proportions identified between the phases of production revolutions make it possible to predict the beginning of the final phase around the 2030–2040s (for more details see Chap. 12). But, of course, these reasons alone are absolutely not enough. Second, the aging problem will become very acute by this time, and only medicine can mitigate the consequences of aging. As we will see in Chap. 6, by 2030 aging in developed countries could reach dangerous levels, with the proportion of people aged 60+ likely to approach 30%. We believe that by the 2030s, the trends that are already clearly emerging—the increasing impact of global aging on economic slowdown, fiscal and pension problems and labor shortages—will become very acute and tangible. Third, unique opportunities for medical breakthroughs may emerge in the 2030s or a little later. Fourth, the rapid growth of the global middle class and the level of education of the population, especially in developing countries (NIC, 2012), are expected by the 2030s. In addition, poverty and illiteracy will be reduced in peripheral countries. As a result, the focus will shift from eliminating the most intolerable conditions to the problems of raising living standards, health care, etc.

Fifth, after 2020, the pandemic (and the threat of new pandemics) has been added to these factors, making health and medical issues of interest to almost the entire society and a huge part of humanity, and placing medicine and biotechnology at the forefront of technological and scientific development (see Chap. 14; Grinin et al., 2021a).

In general, medicine has huge potential and favorable circumstances for its development.

Thus, the final phase of the Cybernetic Revolution will start in a particular field, most probably in medicine. However, when speaking about medicine, we should keep in mind that it is a very heterogeneous sphere in terms of possible revolutionary transformations. On the one hand, we understand medicine in a broad sense not only as a sphere (system of special organization, institute, etc.) of medical treatment, but also as a system of healthcare service and even as a health-related sphere of life. On the other hand, we regard medicine as an integral part of the MANBRIC convergence, i.e., because of its potential to involve (and it already actively involves) a great number of other scientific-technological branches. However, not all branches of medicine are able to play such an integral role. Here we speak about a kind of new medicine, its most innovative and often most advanced branches (such as biomedicine, nanomedicine, etc.). And when we speak of medicine as a peculiar sector of breakthroughs, we do not mean medicine in general, but only its various innovative branches and directions, which are particularly interconnected with biotechnologies and other innovative fields. We uncover this point in the next section.

## 4.5 Why Will Breakthrough not Occur in All Branches of Medicine, But Only in Its One or Few Innovative Fields?

It is worth remembering that the Industrial Revolution began in a rather narrow field of cotton textile production and was connected with the solution of quite specific problems: first, the elimination of the gap between spinning and weaving, and then, after increasing the productivity of weavers, the search for the ways to mechanize spinning. However, the solutions to these narrow tasks caused an explosion of innovations mediated by many important elements of machine production (including abundant mechanisms, primitive steam engines, the high volume of coal production, etc.), which gave an impulse to the development of the Industrial Revolution (see Grinin & Grinin, 2015a, 2015b; Grinin & Korotayev, 2015).

Similarly and on the basis of the general vector of scientific achievements, we assume that the final phase of the Cybernetic Revolution will start in one or a few new branches of medicine with its associated field. Some of them have already emerged (e.g., biomedicine or nanomedicine) and will develop as a result of the borrowing of other innovative technologies into medicine.

As for other branches of medicine, the revolutionary changes will begin there later. Moreover, some branches of medicine would fail to transform due to their conservatism. Thus, more radical reforms will occur in these fields in the future.

**Conclusion**. Medicine will therefore be the first sector to enter the final phase of the Cybernetic Revolution, but, later the development of self-regulating and self-managing systems will involve the most diverse sectors of production, services and life. Thus, in the next few decades, we may witness huge technological changes and an acceleration of technological progress, which should overcome the slowdown observed since the 1980s and 1990s. As a result of the processes associated with the final phase of the Cybernetic Revolution, an innovation cluster—the MANBRIC convergence—will emerge and become widespread. Its common characteristics will be a widespread development of self-regulating/self-managing systems and a new phase of saving human labor in its intellectual form as it will also radically change the service sector, which today employs the most part of labor force in developed countries.

This will also lead to a widespread of socio-technical self-regulating systems, which will significantly affect socio-political structures and the state. This will be discussed in Chap. 14 of the monograph.

**Acknowledgements** This research has been supported by the Russian Science Foundation (Project No. **23-11-00160**).

## References

Akayev A. A. (2012). Matematicheskiye osnovy innovatsionno-tsiklicheskoy teorii ekonomicheskogo razvitiya Shumpetera – Kondrat'yeva. In A. A. Akayev, et al. (Eds.), *Kondrat'yevskiye volny: Aspekty i perspektivy* (pp. 314–341). Uchitel.

Akaev, A., & Davydova, O. (2023). Climate and energy. Scenarios of energy transition and global temperature changes based on current technologies and trends. In V. Sadovnichy, et al. (Eds.), *Reconsidering the limits to growth. A report to the Russian Association of the Club of Rome* (pp. 53–70). Springer. https://doi.org/10.1007/978-3-031-34999-7_4

Akaev, A., Korotayev, A., Issaev, L., & Zinkina, J. (2017). Technological development and protest waves: Arab spring as a trigger of the global phase transition? *Technological Forecasting and Social Change, 116*, 316–321. https://doi.org/10.1016/j.techfore.2016.08.009

Arrighi, G. (1994). *The long twentieth century: Money, power, and the origins of our times*. Verso.

Attali, J. (1991). *Millennium: Winners and losers in the coming world order*. Times Books.

Ayres, R. U. (2006). Did the fifth K-wave begin in 1990–92? Has it been aborted by globalization? In T. C. Devezas (Ed.), *Kondratieff waves, warfare and world security* (pp. 57–71). IOS Press.

Bainbridge, M. S., & Roco, M. C. (2005). *Managing nano-bio-info-cogno innovations: Converging technologies in society*. Springer.

Batty, M. (2018). *Inventing future cities*. MIT Press.

Bernal, J. D. (1965). *Science in history* (3rd ed.). Hawthorn Books.

Buchanan, P. J. (2002). *The death of the west: How dying populations and immigrant invasions imperil our country and civilization*. St. Martin's Griffin.

Chumakov, A. N., & Yurchenko, P. S. (2021). Global Pandemic in terms of objective causes and subjective speculations. *Journal of Globalization Studies, 12*(2), 157–168.
Colson, C., & Eckerd, J. (1991). *Why America doesn't Work?* Word Pub Group.
Control Technology. (2023). *Control technology*. https://www.sew-eurodrive.com.au/products/control_technology.html
Dator, J. (2006). Alternative futures for K-waves. In T. C. Devezas (Ed.), *Kondratieff waves, warfare and world security* (pp. 311–317). IOS Press.
Dickson, D. (1983). Technology and cycles of boom and bust. *Science, 219*(4587), 933–936.
Dirican, C. (2015). The impacts of robotics, artificial intelligence on business and economics. *Procedia—Social and Behavioral Sciences, 195*(July), 564–73. https://doi.org/10.1016/J.SBSPRO.2015.06.13
Domingues, J. M., Miranda, C. S., Homem, N. C., Felgueiras, H. P., & Antunes, J. C. (2023). Nanoparticle synthesis and their integration into polymer-based fibers for biomedical applications. *Biomedicines, 11*(7), 1–50.
Dosi, G. (1984). *Technical change and industrial transformation*. St. Martin's Press.
Durden, T. (2021a). *Here is the hidden $150 trillion agenda behind the "Crusade" against climate change*. https://www.zerohedge.com/markets/here-hidden-150-trillion-agenda-behind-crusade-against-climate-change
Durden, T. (2021b). *One bank reveals the dismal truth about the $150 trillion crusade against climate change*. https://www.zerohedge.com/energy/one-bank-reveals-dismal-truth-about-150-trillion-crusade-against-climate-change
Ecker, O., Breisinger, C., & Pauw, K. (2011). *Growth is good, but not good enough to improve nutrition*. International Food Policy Research Institute.
El-Chichakli, B., Von Braun, J., Lang, C., Barben, D., & Philp, J. (2016). Five cornerstones of a global bioeconomy. *Nature, 535*, 221–223.
Evenson, R. E., & Gollin, D. (2003). Assessing the impact of the green revolution, 1960 to 2000. *Science, 300*, 758–762.
Frank, A. G. (1997). Asia comes full circle—with China as the 'Middle Kingdom.' *Humboldt Journal of Social Relations, 76*(2), 7–20.
Freeman, C. (1987). Technical innovation, diffusion, and long cycles of economic development. In T. Vasko (Ed.), *The long-wave debate* (pp. 295–309). Springer.
Freeman, C., & Louçã, F. (2001). *As time goes by: From the industrial revolutions to the information revolution*. Oxford University Press.
Fücks, R. (2013). *Intelligent wachsen. Die grune revolution*. GmbH & Co: Carl Hanser Verlag.
Fukuyama, F. (2002). *Our post-human future: Consequences of the bio-technology revolution*. Farrar, Straus, and Giroux.
Gates, B. (2007). Mekhanicheskoye budushcheye: Microsoft predskazyvayet revolyutsiyu v sfere robototekhniki. *V Mire Nauki, 5*, 36–43.
Glazyev, S. Y. (1993). *Teoriya dolgosrochnogo tekhniko-ekonomicheskogo razvitiya*. VlaDar.
Glazyev, S. Y. (2009). Mirovoy ekonomicheskiy krizis kak protsess smeny tekhnologicheskikh ukladov. *Voprosy ekonomiki, 3*, 26–32.
Grinin, A. L., & Grinin, L. E. (2015). The Cybernetic revolution and historical process. *Social Evolution and History, 14*(1), 125–184.
Grinin, L. E. (2009a). Globalization and the transformation of national sovereignty. Auckland. In J. Sheffield (Ed.), *Systemic development: Local solutions in a global environment* (pp. 47–53). ISCE Publishing: Goodyear.
Grinin, L. (2009b). Privedet li global'niy krizis k global'nym izmeneniyam? *Vek Globalizatsii, 2*, 117–140.
Grinin, L. (2010). Which global transformations would the global crisis lead to? *Age of Globalization, 2*, 31–52.
Grinin, L. (2012a). New foundations of international system or why do states lose their sovereignty in the age of globalization? *Journal of Globalization Studies, 3*(1), 3–38.

Grinin, L. (2012b). Arabskaya vesna i rekonfiguratsiya Mir-Sistemy. In A. Korotayev, J. Zinkina, & A. Khodunov (Eds.), *Sistemnyy monitoring global'nykh i regional'nykh riskov* (pp. 188–223). URSS, LIBROKOM.

Grinin, L. (2012c). Rekonfiguratsiya mira, ili nastupayuschaya epokha novykh koalitsiy. *Istoriya i Sovremennost, 2*, 3–27.

Grinin, L. E. (2016). The new world order and philosophy. In A. N. Chumakov & W. C. Gay (Eds.), *Between past orthodoxies and the future of globalization. Contemporary philosophical problems* (pp. 143–156). Brill: Rodopi. http://www.brill.com/products/book/between-past-orthodoxies-and-future-globalization

Grinin, L. (2020a). How can COVID-19 change geopolitics and economy? *Journal of Globalization Studies, 11*(2), 121–134.

Grinin, L. (2020). Kondratieff waves, technological modes, and the theory of production revolutions. In L. Grinin & A. Korotayev (Eds.), *History & mathematics: Investigating past and future.* Yearbook (pp. 30–79). 'Uchitel' Publishing House.

Grinin, L. (2022a). Revolutions of the 21st century as a factor of the world system reconfiguration. In J. A. Goldstone, L. Grinin & A. Korotayev (Eds.), *Handbook of revolutions in the 21st century: The new waves of revolutions, and the causes and effects of disruptive political change* (pp. 973–996). Springer. https://doi.org/10.1007/978-3-030-86468-2_38

Grinin, L. (2022b). Chetvertyy tekhnologicheskiy uklad (konets 1940-kh – nachalo 1980-kh gg.): rastsvet i zakat promyshlennogo kapitalizma. In L. Y. Grinin & A. V. Korotayev (Eds.), *Kondrat'yevskiye volny: tekhnologicheskiye i ekonomicheskiye aspekty: yezhegodnik.* 'Uchitel' Publishing House.

Grinin, L. (2022c). Pyatyy tekhnologicheskiy uklad (1980-ye gg. – nastoyashcheye vremya). In L. Y. Grinin & A. V. Korotayev (Eds.), *Kondrat'yevskiye volny: tekhnologicheskiye i ekonomicheskiye aspekty: yezhegodnik.* 'Uchitel' Publishing House. In press.

Grinin, L. (2022d). The first technological paradigm. In L. E. Grinin & A. V. Korotayev (Eds.), *Kondratieff waves: Processes, cycles, triggers, and technological paradigms.* 'Uchitel' Publishing House. In press.

Grinin, L. (2022e). The second technological paradigm. In L. E. Grinin & A. V. Korotayev (Eds.), *Kondratieff waves: Processes, cycles, triggers, and technological paradigms.* 'Uchitel' Publishing House. In press.

Grinin, L. (2022f). The third technological paradigm. In L. E. Grinin, & A. V. Korotayev (Eds.), *Kondratieff waves: Processes, cycles, triggers, and technological paradigms.* 'Uchitel' Publishing House. In press.

Grinin, L. E., & Grinin, A. L. (2014). The sixth Kondratieff wave and the cybernetic revolution. In L. E. Grinin, T. Devezas, & A. V. Korotayev (Eds.), *Kondratieff waves. Juglar – Kuznets – Kondratieff* (pp. 354–378). Uchitel.

Grinin, L. E., & Grinin A. L. (2015). *Ot rubil donanorobotov. Mirnaputikepokhesamoupravlyaemykh sistem (istoriya tekhnologiy i opisanie ikh budushchego).* Moscow Publishing House 'Uchitel'.

Grinin, L. E., & Grinin, A. L. (2016). *The cybernetic revolution and the forthcoming epoch of self-regulating systems.* Moscow Branch of the Uchitel Publishing House.

Grinin, L. E., & Grinin, A. L. (2022). The dynamics of Kondratieff waves in the light of the theory of production revolutions. In L. Grinin & A. Korotayev (Eds.), *History & mathematics: Historical and technological dynamics: Factors, cycles, trends* (pp. 201–255). 'Uchitel' Publishing House.

Grinin, L. E., Grinin, A. L., & Korotayev, A. (2017a). Forthcoming Kondratieff wave, cybernetic revolution, and global ageing. *Technological Forecasting & Social Change, 115*, 52–68.

Grinin, L., Grinin, A., & Korotayev, A. (2017b). The MANBRIC-technologies in the forthcoming technological revolution. In T. Devezas, et al. (Eds.), *Industry 4,0—entrepreneur-ship and structural change in the new digital landscape: What is coming on along with the fourth industrial revolution* (pp. 243–261). Springer.

Grinin, L., Grinin, A., & Korotayev, A. (2020). A quantitative analysis of worldwide long-term technology growth: From 40,000 BCE to the early 22nd century. *Technological Forecasting and Social Change, 155*(June), 119955. https://doi.org/10.1016/j.techfore.2020.119955

Grinin, L., Grinin, A., & Korotayev, A. (2021a). COVID-19 pandemic as a trigger for the acceleration of the cybernetic revolution, transition from E-Government to E-State, and change in social relations. *Technological Forecasting & Social Change*.https://doi.org/10.1016/j.techfore.2021.121348

Grinin, L., Grinin, A., & Korotayev, A. (2021b). Global trends and forecasts of the 21st century. *World Futures, 77*(5), 335–370. https://doi.org/10.1080/02604027.2021.1949939

Grinin, L. E., Ilyin, I. V., & Andreev, A. I. (2016). World order in the past, present, and future. *Social Evolution and History, 15*(1), 58–84.

Grinin, L. E., & Korotayev, A. V. (2010a). Will the global crisis lead to global transformations? 1. The global financial system: Pros and cons. *Journal of Globalization Studies, 1*(1), 70–89.

Grinin, L. E., & Korotayev, A. V. (2010b). Will the global crisis lead to global transformations? 2. The coming epoch of new coalitions. *Journal of Globalization Studies, 1*(2), 166–183.

Grinin, L. E., & Korotayev, A. V. (2010c). *Global'nyy krizis v retrospektive. Kratkaya istoriya poyemov i krizisov ot Likurga do Alana Grinspena*. M.: Knizhnyy dom LIBROKOM.

Grinin, L., & Korotayev, A. (2011). The coming epoch of new coalitions: Possible scenarios of the 674 near future. *World Futures, 67*(8), 531–563.

Grinin, L., & Korotayev, A. (2012). Does 'Arab Spring' mean the beginning of world system reconfiguration? *World Futures, 68*(7), 471–505. https://doi.org/10.1080/02604027.2012.697836

Grinin, L. E., & Korotayev, A. V. (2014). Interaction between Kondratieff waves and Juglar cycles. In L. E. Grinin, T. C. Devezas, & A. V. Korotayev (Eds.), *Kondratieff waves. Juglar – Kuznets – Kondratieff. Yearbook* (pp. 25–95). 'Uchitel' Publishing House.

Grinin, L., & Korotayev, A. V. (2015). *Great Divergence and Great Convergence*. Springer International Publishing.

Grinin, L., & Korotayev, A. (2016). MENA region and the possible beginning of world system reconfiguration. In M. Erdogdu & B. Christiansen (Eds.), *Comparative political and economic perspectives on the MENA region* (pp. 28–58). Information Science Reference. https://doi.org/10.4018/978-1-4666-9601-3.ch002

Grinin, L., & Korotayev, A. (2020) Seven weaknesses of the U.S., Donald trump, and the future of American hegemony. *World Futures*. https://doi.org/10.1080/02604027.2020.1801309

Grinin, L., Korotayev, A., Issaev, L., Shishkina, A., Ivanov, E., & Meshcherina, K. (2017). World order transformation and sociopolitical destabilization. National Research University Higher School of Economics. Series: International Relations, WP BRP 29/IR/2017.

Grinin, L., Korotayev, A., & Malkov, S. (2010). A mathematical model of Juglar cycles and the current global crisis. In L. Grinin, P. Herrmann, A. Korotayev, & A. Tausch (Eds.), *History & mathematics. Processes and models of global dynamics* (pp. 138–187). Uchitel.

Grinin, L. E., Korotayev, A. V., & Tausch A. (2016). *Economic cycles, crises, and the global periphery*. Springer.

Grinin, L., Korotayev, A., & Tausch, A. (2019). *Islamism, Arab spring, and the future of democracy. World system and world values perspectives*. Springer. https://doi.org/10.1007/978-3-319-91077-2

Hirooka, M. (2006). *Innovation dynamism and economic growth. A nonlinear perspective*. Edward Elgar.

Hussain, M. (2021). The changing dynamics of liberal international order: Covid-19 and its global implications. *Journal of Globalization Studies, 12*(2), 143–156.

Irshad, S. M. (2020). COVID-19 and global 'At-Risk Community': From benefit-sharing to risk-sharing of economic crisis. *Journal of Globalization Studies, 11*(2), 135–145.

Jakimowicz, A., & Rzeczkowski, D. (2019). Firm ownership and size versus innovation activities over the business cycle: Near-zero inertia as a sign of the transition from the fifth to the sixth Kondratieff wave. *Oeconomia Copernicana, 10*(4), 689–741.

Jasiński, A. M., & Bąkowska, A. M. (2020). Globalization and life satisfaction of the polish population. Before and during the COVID-19 pandemic. *Journal of Globalization Studies, 11*(2), 146–155.

Jotterand, F. (2008). *Emerging conceptual, ethical and policy issues in bionanotechnology*. Springer.

Kakkar, S. (2021). The outbreak of sars-cov-2 (Covid-19): Impact on international tourism. *Journal of Globalization Studies, 12*(1), 145–158.

Katz, R. D., Rosson, G. D., Taylor, J. A., & Singh, N. K. (2005). Robotics in microsurgery: Use of a surgical robot to perform a free flap in a pig. *Microsurgery, 25*(7), 566–569. https://doi.org/10.1002/micr.20160

Kennedy, P. (1987). *The rise and fall of great powers 1500–2000*. Random House.

Kleinknecht, A. (1981). Innovation, accumulation, and crisis: Waves in economic development? *Review, 4*(4), 683–711.

Kleinknecht, A. (1987). *Innovation patterns in crisis and prosperity: Schumpeter's long cycle reconsidered*. Macmillan.

Kleinknecht, A., & van der Panne, G. (2006). Who was right? Kuznets in 1930 or Schumpeter in 1939? In T. C. Devezas (Ed.), *Kondratieff waves, warfare and world security* (pp. 118–127). IOS Press.

Korotayev, A., Bilyuga, S., & Shishkina, A. (2020). Which countries generate Kondratieff waves in global GDP growth rate dynamics in the contemporary world? *Journal of Globalization Studies, 11*(1), 33–63. https://doi.org/10.30884/jogs/2020.01.03

Korotayev, A. V., & Grinin, L. E. (2012). Kondratieff waves in the world system perspective. In L. E. Grinin, T. C. Devezas, & A. V. Korotayev (Eds.), *Kondratieff waves. Dimensions and prospects at the dawn of the 21st century* (pp. 23–64). Uchitel.

Korotayev, A., & Tsirel, S. (2010). A spectral analysis of world GDP dynamics: Kondratieff waves, kuznets swings, Juglar and Kitchin cycles in global economic development, and the 2008–2009 economic crisis. *Structure and Dynamics, 4*(1), 3–57. https://doi.org/10.5070/SD941003306

Korotayev, A., Zinkina, J., & Bogevolnov, J. (2011). Kondratieff waves in global invention activity (1900–2008). *Technological Forecasting and Social Change, 78*(7), 1280–1284. https://doi.org/10.1016/j.techfore.2011.02.011

Kovalchuk, M. V. (2011). Konvergentsiya nauk i tekhnologiy – proryv v budushcheye. *Rossiiskie nanotekhnologii, 3*, 134–158. http://www.nrcki.ru/files/nbik01.pdf

Kupchan, C. A. (2002). *The end of the American era*. Knopf.

Lazurenko, S. (1992). Problemy dolgovremennykh kolebaniy ekonomicheskoy dinamiki. *Voprosy ekonomiki, 10*, 69–75.

Lilley, S. (1966). *Men, machines and history: The story of tools and machines in relation to social progress*. International Publishers.

Lynch, Z. (2004). Neurotechnology and society 2010–2060. *Annals of the New York Academy of Sciences, 1031*, 229–233.

Makarov, I. M., & Topcheev, Y. I. (2003). *Robototekhnika. Istoriya i perspektivy*. Nauka.

Mandelbaum, M. (2005). *The case for goliath: How America acts as the world's government in the twenty-first century*. Public Affairs.

Mayevsky, V. I. (1997). *Vvedeniye v evolyutsionnuyu makroekonomiku*. Yaponiya segodnya.

Mensch, G. (1979). *Stalemate in technology—innovations overcome the depression*. Ballinger.

Modelski, G. (2001). What causes K-waves? *Technological Forecasting and Social Change, 68*, 75–80.

Modelski, G. (2006). Global political evolution, long cycles, and K-waves. In T. C. Devezas (Ed.), *Kondratieff waves, Warfare and world security* (pp. 293–302). IOS Press.

Modelski, G., & Thompson, W. R. (1996). *Leading sectors and world politics: The coevolution of global politics and economics*. University of South Carolina Press.

Morgan P. W. (1981). Brief History of Fibers from Synthetic Polymers. *Journal of Macromolecular Science: Part A – Chemistry, 15*(6), 1113–1131. https://www.tandfonline.com/doi/abs/https://doi.org/10.1080/00222338108066456

Nefiodow, L. (1996). *Der sechste Kondratieff. Wege zur Produktivität und Vollbeschäftigung im Zeitalter der Information*. 1 Auflage/Edition. Rhein-Sieg-Verlag: Sankt Augustin.
Nefiodow, L., & Nefiodow, S. (2014a). *The sixth Kondratieff. The new long wave of the world economy*. Sankt Augustin.
Nefiodow, L., & Nefiodow, S. (2014b). The sixth Kondratieff. The growth engine of the 21st Century. In L. E. Grinin, T. Devezas, & A. V. Korotayev (Eds.), *Kondratieff waves. Juglar – Kuznets – Kondratieff* (pp. 326–353). Uchitel.
NIC—National Intelligence Council. (2008). *Global trends 2025: a transformed world*. National Intelligence Council.
NIC—National Intelligence Council. (2012). *Global trends 2030: Alternative worlds*. National Intelligence Council. www.dni.gov/nic/globaltrends
Nistor, M. T., & Rusu, A. G. (2019). Chapter 3. Nanorobots with applications in medicine. In: C. Vasile (Ed.), *Polymeric nanomaterials in nanotherapeutics* (pp. 123–49). Micro and Nano Technologies. Elsevier. https://doi.org/10.1016/B978-0-12-813932-5.00003-0
Papenhausen, C. (2008). Causal mechanisms of long waves. *Futures, 40*, 788–794.
Peters, M., Jandrić, P., & Hayes, S. (2021). Biodigital technologies and the bioeconomy: The global new green deal? *Educational Philosophy and Theory*, 1–12.https://doi.org/10.1080/00131857.2020.1861938
Perez, C. (2002). *Technological revolutions and financial capital: The dynamics of bubbles and golden ages*. Elgar.
Philipson, M. (Ed.). (1962). *Automation: Implications for the future*. Vintage.
Polterovich, V. (2009). Gipoteza ob innovatsionnoy pauze i strategiya modernizatsii. *Voprosy Ekonomiki, 6*, 4–23.
Renkow, M., & Byerlee, D. (2010). The impacts of CGIAR research: A review of recent evidence. *Food Policy, 35*, 391–402.
Rifkin, J. (2011). *The third industrial revolution. How lateral power is transforming energy, the economy, and the world*. Palgrave Macmillan.
Roco, M., & Bainbrige, W. S. (2003). *Converging technologies for improving human performance. Nanotechnology, biotechnology, information technology and cognitive science*. Springer Science Business Media.
Schwab, K. (2016). *The fourth industrial revolution*. World Economic Forum.
Seifabadi, M. S., & Dolatabadi, A. B. (2022). Covid-19, globalization and strengthening authoritarianism in the middle east. *Journal of Globalization Studies, 13*(1), 135–157.
Šmihula, D. (2010). Waves of technological innovations and the end of the information revolution. *Journal of Economics and International Finance, 2*(4), 58–67.
Soto, F., Wang, J., Ahmed, R., & Demirci, U. (2020). Medical micro/nanorobots in precision medicine. *Advanced Science, 7*(21), 2002203. https://doi.org/10.1002/advs.202002203
Stiegler, K., Martiniano, R., & Forte, G. (2021). *Health care employment projections, 2019–2029: An analysis of bureau of labor statistics projections by setting and by occupation*.
Sylvester, E., & Klotz, L. C. (1983). *The gene age: Genetic engineering and the next industrial revolution*. Scribner.
Thirtle, C., Lin, L., & Piesse, J. (2003). The impact of research-led agricultural productivity growth on poverty reduction in Africa, Asia and Latin America. *World Development, 31*, 1959–1975.
Todd, E. (2003). *After the empire: The breakdown of American order*. Columbia University Press.
Tylecote, A. (1992). *The long wave in the world economy*. Routledge.
Vega-González, L. R. (2017). Using patents and innovation strings to anticipate the next Kondratieff long waves. *International Journal of Technology Intelligence and Planning, 11*(3), 212.
Vesti.ru. (2022). *London massovo vyvedet bespilotniki na dorogi v 2025 godu*. https://www.vesti.ru/auto/article/2899808
Vogel, E. F. (1979). *Japan as number one: Lessons for America*. Harward University Press.
Wallerstein, I. (2003). *The decline of American power. The U.S. in a chaotic world*. New York, London.

Widdowson, M. (2021). From Covid-19 to zero-gravity: Complex crises and production revolutions. *Journal of Globalization Studies, 12*(1), 117–144.

Wilenius, M., & Casti, J. (2015). Seizing the x-events. The sixth K-wave and the shocks that may upend it. *Technological Forecasting and Social Change, 94*, 335–349.

WIPO = World Intellectual Property Organization. (2023). *World intellectual property organization database*. https://www3.wipo.int/ipstats/IpsStatsResultvalue

World Bank. (2023). *World development indicators*. World Bank. https://data.worldbank.org/indicator

Yakovets, Y. V. (2001). *Naslediye N. D. Kondrat'yeva: vzglyad iz XXI veka*. MFK.

Yesakova, P. (2014). V Norvegii raspechatali samoobuchayushchikhsya robotov. Look at Me, November 13. http://www.lookatme.ru/mag/live/experience-news/209343-3d-robot

Zhdanov, A., & Korotayev, A. (2022). Factors of deconsolidation of a liberal democratic regime: The case of the United States. *Cliodynamics, 13*, 1–34. https://doi.org/10.21237/C7clio13054207

Zvorykin, A. A., Osmova, N. I., Chernyshov, V. I., & Shuhardin, S. V. (1962). *Istoriya tekhniki*. Izdatel'stvo sotsial'no-ekonomicheskoy literatury.

# Part II
# Demographic Transformations and Global Aging

In this part, Grinin et al. explore the close relationship between the production (technological) revolutions, which were considered in Part 1, and demographic transformations in the historical process. They also study in detail the complex relationships between the development of the Cybernetic Revolution and the Cybernetic Production Principle, on the one hand, and the process of global aging on the other hand. Throughout history, technological change has been among the most fundamental factors influencing demographic growth and sociocultural development (and just this connection distinguishes the reproduction of humans from any other living beings). However, the demographic dimension of the historical process has not been sufficiently studied. Yet, it is an extremely important aspect, explaining essential features of historical processes; in addition, it also provides a basis for explaining current processes, as well as for forecasting.

Grinin et al. (Chap. 5) consider the dynamics of the ratio between the major technological breakthroughs in history—the Agrarian, Industrial and Cybernetic Revolutions (discussed in Chap. 2)—and major demographic transformations in the historical process. While the authors rely on the conclusions drawn from the analysis of long-term demographic trends, our focus is on the demographic transformations associated with the Cybernetic Revolution. Among these transformations, global aging appears an increasingly important process affecting all spheres of life. In Part 1, Grinin et al. have argued that the Cybernetic Revolution will produce a huge impact on the health-related quality of life and provide opportunities to modify human biological nature. Of no less importance is the fact that the increasing life expectancy and global aging over the next few decades will increasingly affect individual societies and the World System in general as well as further changes in the type of population reproduction.

The analysis of the process of global aging process, the long-term forecasts of the demographic development and its impact on changes in society, in particular its connection with technological transformations, and possible scenarios for the future development of an aging society gives the authors the opportunity to show the close and deep relationship between both, individual areas of technology (medicine, biotechnology, robotics, etc.) with the aging process, and the mutual influence of aging and technological progress in general, as well as the transformation of the economic model, political and social structures and institutions.

Production (or technological) revolutions opened up gigantic ecological niches that made it possible to make qualitative breakthroughs in population growth. In other words, production revolutions involved major demographic transformations. Subsequently, the rate of population growth was increasing hyperbolically. During the Paleolithic, the population grew by an average of 3% per millennium (see Chap. 5 for more detail), whereas in the 1960s there were years when the population of the Earth grew by almost 3% per year. However, afterwards the trend started being reversed and now the population growth rate is only slightly over 1% per year. What has happened?

The Cybernetic Revolution will have a huge impact on the quality of human biological life and on the ability to modify human biological nature itself. No less important is the fact that in the process of increasing life expectancy and global aging, the very type of population reproduction is changing very significantly.

Then (in Chap. 6) Grinin et al. consider in more detail the complex relationship between the development of the Cybernetic Revolution and the Cybernetic Production Principle, on the one hand, and the process of global aging, which will increasingly affect individual societies and the World System as a whole, further changing the type of population reproduction (TPR), on the other. The authors then show (in Chap. 7) the impact of the Cybernetic Revolution on changes in the type of population reproduction, the diverse consequences (both positive and negative) of global aging and the contradictions that the process of global aging and the technological transformations that contribute to it will inevitably cause in the future society. They will also analyze different forms of adaptation to aging that must be developed so that frictions and tensions in society do not impede its development; in addition, Grinin et al. outline the main features of a new type of population reproduction, which, in their opinion, will take shape by the end of this century. This new type of population reproduction will characterize future societies, which Grinin et al. have called cybernetic (it will be discussed in the Conclusion, i.e., Chap. 15).

# Demographic Transformations in the Historical Process

5


## Abstract

In this chapter, Grinin et al. provide an analysis of the demographic development of humankind in its close relationship with technological development, including the trends of the twenty-first century. The analysis of the demographic component, the demographic dimension of the historical process, has not been sufficiently researched. Meanwhile, it is an extremely important aspect that can not only explain essential parts of the development, but also provide a basis for explaining current processes and forecasting our futures. Grinin et al. consider the dynamics of the correlation between the major technological breakthroughs in history—the Agrarian, Industrial and Cybernetic Revolutions (discussed in Chap. 2)—and the major demographic transformations in the historical process. A special place is given to the analysis of the demographic transition in the context of the demographic transformations of the last seven to eight decades. This allows them to reveal the essential features of modern demographic processes, the role of the societal aging process, which will increasingly influence the life of society. Grinin et al. also make predictions about future demographic transformations associated with the development of the societal aging process and the formation of a new type of population reproduction up to the end of this century.


## What This Chapter is About

In this chapter, we provide an analysis of the demographic development of humankind in its close relationship with technological development, including the trends of the twenty-first century. We consider the dynamics of the correlation between the major technological breakthroughs in history—the Agrarian, Industrial and Cybernetic Revolutions (discussed in Chap. 2)—and the major




L. Grinin et al., *Cybernetic Revolution and Global Aging*, World-Systems Evolution and Global Futures, https://doi.org/10.1007/978-3-031-56764-3_5

demographic transformations in the historical process. This allows us to reveal the essential features of modern demographic processes, the role of the societal aging process in them, and to offer forecasts up to the end of this century.

The demographic structure of modern society is very different from that of fifty or seventy years ago. However, deep demographic trends are very long term. Therefore, in order to understand modern demographic processes, it is necessary to make a deep analysis of them, in fact, a demographic survey of the entire historical process. Modern society, which has completed the so-called demographic transition, shows the following pattern of population reproduction: low birth rate, low mortality, low, almost zero (or even negative) natural increase. For the purpose of this monograph, it is important to emphasize that we live in a rapidly aging society. This process will increasingly influence the life of society and its technologies.

The analysis of the demographic component, the demographic dimension of the historical process, has not been sufficiently researched. Meanwhile, it is an extremely important aspect that can not only explain essential parts of the development, but also provide a basis for explaining current processes and forecasting our futures. It is especially important to note that, throughout history, technological change has been one of the most essential factors influencing demographic growth and socio-cultural development. The close correlation between technological revolutions and radical demographic changes was established quite long ago; in fact, it is this connection that distinguishes human reproduction from the reproduction of any other living being.

However, the relationship between demography and production is far from simple and linear; and furthermore, the correlation between the demographic and productive components of the life of societies has persisted throughout the historical process, yet it has changed significantly. This chapter aims to provide a theoretical framework for the relationship between the development of production and technology, on the one hand, and demographic transformations, on the other, during the historical process, and also to describe all the major demographic transformations during human history.

The chapter also reflects on important issues that affect population growth and its limits. We describe the historical types of population reproduction and the causes of their change. All this allows us to offer a forecast about the vectors and main features of the coming demographic transformation in the twenty-first century.

It is shown that as a result of a production revolution and the development of the cycle of a production principle as a whole, there is a change in the type of population reproduction, and together they give a powerful impetus to the qualitative reorganization of the entire social structure and social relations. We conclude that every production revolution is followed by a fundamental demographic change. Conversely, population growth causes such changes in society and intersocial relations, communication and the diffusion of innovations that it significantly accelerates technological development. The chapter analyzes the relationship between each production revolution and each production principle, on the one hand, and demographic transformations and the type of population reproduction, on the other hand, not only in a theoretical model, but in connection with the specific course of the historical process and the quantitative data currently available to science.

All these ideas about demographic trends from the Upper Paleolithic to the late twenty-first century have also been summarized in a convenient tabular form.

A special place is given to the analysis of the demographic transition in the context of the demographic transformations of the last seven to eight decades; in addition, the connection between the ongoing Cybernetic Revolution and global aging is revealed. Predictions are made about future demographic transformations associated with the development of the societal aging process and the formation of a new type of population reproduction.

## 5.1 The Correlation Between Technological and Demographic Changes: General Ideas

In this chapter, we will consider the dynamics of the ratio between three production (technological) revolutions—the Agrarian, Industrial and Cybernetic (which were discussed in Chap. 2)—and the major demographic transformations in the historical process. This will provide a better understanding of Chaps. 6 and 7. In Chap. 6, we will consider in detail the complex relationship between the development of the Cybernetic Revolution and the Cybernetic Production Principle, on the one hand, and the process of global aging, which will increasingly affect individual societies and the World System in general over the several coming decades, as well as the further change of the type of population reproduction. In Chap. 7, (a) we show the impact of the Cybernetic Revolution on changes in the type of population reproduction (TPR), the diverse consequences (both positive and negative) of global aging and the contradictions in future society which will inevitably result from global aging and the technological transformations that support it; (b) we also analyze various forms of adaptation to aging—there elaboration makes it possible to mitigate frictions and tensions that could impede societal development; (c) we outline the main features of a new type of population reproduction, which, in our opinion, should take shape by the end of the present century. This new TPR will characterize the future society, which we denote as cybernetic (we will discuss this in detail in Chap. 15, the Conclusion).

For a better understanding of the purposes of this chapter let us remind the reader that we singled out four main technological stages of the world's productive forces, which we call production principles, they are: (1) Hunter-Gatherer; (2) Craft-Agrarian; (3) Trade-Industrial; (4) Scientific-Cybernetic. The change in production principles is associated with production revolutions (=technological revolutions, so we will use these terms as synonyms). We consider the beginning of such revolutions as the beginning of a new production principle. We single out three production (or technological) revolutions that mark the beginning of new production principles. They are: (1) the Agrarian Revolution; (2) the Industrial

Revolution and (3) the Cybernetic Revolution. We divide each production principle into six phases and each production revolution into three phases. It is important to note that these three phases of a production revolution are at the same time the first three phases of a production principle. The other three phases of a production principle are the phases of its maturity. For details see Chap. 2.

Thus, for the purposes of the present chapter, it is important to realize that *one of the essential characteristics (and consequences) of each production (technological) revolution and all the more so of production principles is their huge impact on demography. In their turn, the demographic changes ultimately create conditions for the transition to a new phase of the production revolution* (Grinin & Grinin, 2023; Grinin et al., 2023). This chapter is devoted to examining this correlation. The subject of the chapter is all the more significant since socio-historical studies investigate quite insufficiently the correlation between the development of productivity and population growth. Sometimes this leads to serious theoretical misperceptions. We can mention as an example that some respected demographers argue that the *type of population reproduction* established as a result of the first demographic revolution remained unchanged for thousands of years until the eighteenth century (Vishnevsky, 1973, 2005). However, this is not entirely true, and in many respects, it is completely false, as we are going to prove below.

The population growth is undoubtedly the most important driving force of the development of human society.[1] The population growth made it possible to settle in new territories and various natural zones. For many millennia, it has contributed to the process of urbanization and continues to do so; it also complicated the societal structure (in particular, it contributed to the emergence of first small and later large states)[2]; it also led to the division of labor, development of culture and the media, the intensification of production and innovation.

However, it is essentially important that population growth could not and cannot be endless. It has always been limited by various factors and we will discuss some of them below. And since the population growth has always been strictly limited, *on the regional and even more on the world-system scale, only major technological breakthroughs could make it possible to overcome these limits and reach a fundamentally new developmental level.*

[1] And one can trace this already at the earliest stages of the historical process (see e.g., Carneiro, 1970; 1978; Claessen & van de Velde, 1985; Chase-Dunn & Hall, 1997; Fried, 1967a; 1967b; Grinin, 2003, 2004a, 2004b, 2006a, 2007, 2011a, 2012b; Grinin & Grinin, 2015; Grinin & Korotayev, 2009b; Kremer, 1993; Korotayev, 1997, 2003a, 2003b, 2005, 2007, 2020a, 2020b; Service, 1975; Turchin et al., 2021, 2022).

[2] Henri J. M. Claessen denotes the combination of demographic characteristics as a societal format. This concept comprises the size of population, the acceptable demographic pressure on the territory and the spatial distribution of population (Claessen, 2000: 3). This format played an extremely important role in creating the conditions for the formation of complex polities and early states (see also Grinin, 2008, 2009, 2011a).

The main idea of the present chapter can be formulated in the following way: *each production revolution and the further development of a production principle lead to the replacement of the type of population reproduction; and together they give a powerful impetus to a qualitative reorganization of the entire social structure and the social relations*. The logic is as follows. As we have shown in Chap. 2, the transition to a new production principle and its strengthening brings more than just a growing output. It changes the whole economic life. The main point here is that, in world historical terms, *this transition implies the creation not only of a new management system, but of a system that has great potential for the future and can be widely diffused and borrowed*. At the same time, the labor productivity, yield capacity and land productivity can increase many times and even by orders of magnitude as a result of production revolutions. In turn, this creates a greater volume of resources and goods. Finally, this leads not just to population growth, but to a change in the type of population reproduction—which, in turn, changes the societal structure (expands and complicates it). As the production principle evolves, the population density increases, which is one of the important reasons for the start of a new production revolution and the transition to a new production principle.

*Hence, every production revolution is followed by a fundamental change in demographic reproduction and* vice versa. *The population growth causes such changes in society and intersocietal relations, in communication and in the diffusion of innovations which greatly accelerate technological development.*

For example, only a sufficiently large population organized by a new supreme power could carry out the large-scale irrigation of the large river valleys, which led to the transition to the final—very important and productive—phase of the Agrarian Revolution at the end of the 4th/beginning of the 3rd millennium BC (see Chap. 2). In the sixteenth/first half of the seventeenth century, the population growth in England created a significant sector of non-agricultural jobs, which employed a substantial part of the population (they were fully or partially engaged in non-agricultural crafts). In the seventeenth/first half of the eighteenth century, this contributed to a powerful rise in both manufacturing and agriculture and thus, paved the way for the final phase of the Industrial Revolution (associated with the transition to machine production and steam energy). The powerful growth of agricultural production in England became the basis for a demographic revolution, with the population growing from 6 million to 9.4 million between 1740 and 1801 (i.e., by 57% (Davis, 2002), an unprecedented result for Europe; see also below). The demographic revolution in Europe, triggered by the Industrial Revolution, continued for several decades (until the beginning of the twentieth century). The global aging, which has intensified during the ongoing modernization phase of the Cybernetic Revolution, is the impetus for the transition to the final phase of this revolution in the 2030s (see about the influence of the societal aging in Chaps. 4, 6, 7, 12 and 13).

## 5.2 The Largest Demographic Transformations (LDT) and Production Revolutions

**The Concept of the Demographic Revolution in Demography**. The concept of the demographic revolution was introduced by the French demographer Adolphe Landry in 1934 (Landry, 1934), who described the process of a rapid population growth in Europe, followed by a sharp decline in the birth rates after the First World War as a result of reduced fertility due to abortion, contraception and family planning (i.e., the process more commonly referred to today as the demographic transition, which we will discuss later).[3] This concept became widely recognized, witnessing a huge population growth starting in Asian, African and Latin American countries since the 1950s, also known as the population explosion (see e.g. Arab-Ogly, 1978; Ehrlich, 1975; Fischer, 1993; Saavedra, 2019; Vishnevsky, 2005, as well as in more detail below). Sometimes they distinguish three demographic revolutions associated with (1) the transition from the foraging to the food production in the Mesolithic and Neolithic eras (see, e.g., Bocquet-Appel, 2011); (2) the transition to a modern industrial society in the eighteenth to nineteenth centuries and (3) the above-mentioned revolution in developing countries, which was particularly intense in the 1950s–1970s. Sometimes only two demographic revolutions are singled out when the last two are combined into one (e.g., Vishnevsky, 1973, 2005).

However, with the development of demographic science, the term 'demographic revolution' lost its ground (and is now used quite infrequently), since it was substituted for the concept of *demographic transition* (introduced by the American demographer Frank W. Notestein in the 1940s and 1950s [Notestein, 1953; see also Coale, 1983, 2013]). This substitution occurred because the change in the type of population reproduction (TPR) resulting from the Industrial Revolution was characterized by two distinct phases: (1) a decline in mortality with preserved high birth rate and, accordingly, rapid population growth and (2) a decline in the birth rate and a transition to a model of low population growth. So, the notion of demographic transition united both phases. At the same time, the first phase began to be denoted as the demographic revolution, although Landry himself (1934) and other researchers spoke and still speak of two phases (e.g., Vishnevsky, 1973, 2005); also some researchers use both terms, as, for example, Chesnais (1992). As a result, the concepts of demographic revolution and demographic transition are often confused or overlap each other.

In addition, one should point out other important ideas related to the demographic theory, which cause *inconvenience and lack a useful application of the notion of demographic revolutions to the historical process*:

[3] However he had a predecessor who used this term earlier, Léon Rabinowicz [Radzinowicz] (1929). Both Rabinowicz and Warren Thompson (1930) connected the demographic revolution with the industrial one. For details see Shubrtova (2005 [1984]). An alternative term for the demographic revolution was and still is the term 'the vital revolution'.

1. For the period of transition from hunter-gathering to agriculture in the Neolithic, the concept of a demographic revolution was introduced in analogy with the demographic revolution of the industrialization period, but unlike the latter, the former concept has not been elaborated and therefore lacks proper theoretical relevance. In particular, the concept of demographic transition, which is now considered as a process of first a powerful increase and then a decline in natural population growth, could not and cannot be applied to the Neolithic era. Therefore, the description of the ancient demographic revolution in terms and methods of modern demographic processes is not always productive, especially since, as we pointed out in Chap. 2, the concept of the Neolithic Revolution only partially coincides with the process that we denote as the Agrarian Revolution.[4] Perhaps, this is related to the above-mentioned false statements that *the type of population reproduction* was established as a result of the first demographic revolution and remained unchanged for thousands of years until the eighteenth century (Vishnevsky, 1973, 2005). In fact, as we will show, the transition to intensive agriculture launched profound and large-scale demographic changes in societies and gave rise to the Eurasian world-system.
2. The concepts of demographic revolution and demographic transition are confused at the level of the individual countries and at the level of the World System, which in some cases poses a serious challenge to theoretical perception (as we will show below).
3. It has been a long time since the Industrial Revolution of the eighteenth and nineteenth centuries, and at present a new production revolution—the Cybernetic one—is unfolding. As we noted in Chaps. 2 and 4, its initial phase took place in the 1950s–1990s, and its final phase will begin in the next two decades. However, since the demographic theory follows its own logic, it turns out that the Agrarian and Industrial production revolutions caused related demographic revolutions, while the Cybernetic one did not. But this does not mean at all that the previous logic of the historical process (that *each production revolution brings about a change in the type of population reproduction*) is violated in the present case. Instead, the logic of *the analysis* of the two most important components of the historical process is only superfluously violated.

**The Largest Demographic Transformations and the Type of Population Reproduction**. Thus, the problem is the lack of a developed theory of the correlation between major technological and demographic changes. At the same time, as we have observed, the concept of demographic revolution cannot be applied to the description of demographic transformations in the historical process, since it is not properly elaborated and is actually absorbed by the concept of demographic

[4] Let us remind that the concept of the Neolithic Revolution includes only the process of transition from hunting and gathering to primitive agriculture and animal husbandry. Meanwhile, the dramatic changes in agriculture, associated with the transition to irrigation and plough farming, occurred much later. Within our approach, this was the final phase of the Agrarian Revolution.

transition. Any attempt to improve it would only increase the terminological confusion, and also would not allow us to separate the level of individual countries from the world-system level.

Therefore, we will no longer use the term demographic revolution. Instead, we employ the concept of *the largest (world-system) demographic transformations* (abbreviated as *LDT*).

The largest demographic transformation (LDT) *is an important and inevitable phase of change in the type of population reproduction (TPR), as well as in the spatial and social structure of the population of a society at the world-system level.* As we will see in detail below, *the type of population reproduction (TPR) does not change at once but through two phases, each implying the largest demographic transformation. The first phase spreads only across a part of the World System (its core), the second phase then spreads across its periphery, and thus, throughout the whole World System.*

Notes to the Definition:

- The LDT can be associated both with large quantitative changes in the population size and with large qualitative changes in the conditions and parameters of the population's life (see below for more).
- The spatial change is determined by population growth which requires settlement in new or previously sparsely populated territories.[5] There would also emerge larger settlements. Thus, urbanization was known since the 4th millennium BC, and in particular, it became an important part of the demographic changes of the nineteenth and twentieth centuries.
- A vigorous population growth inevitably led to the complication of societies and changes in their political and social structure.[6]
- The LDTs, as a world-system phenomenon, are associated with production revolutions, and even more tightly and directly to their phases and results.

---

[5] At the level of individual societies (sometimes continents), the major demographic transformations may be based on active colonization (the colonial demographic model is typical for North America, Australia and Russia). But colonization is usually the continuation/realization of a demographic transformation in a particular natural area supported by technological advances. Even such colonization as that of the Russian South in the eighteenth and nineteenth centuries is the result of military (technological) revolutions and the victory of civilization over nomads (this also applies to the settlement of the American West to an even greater extent). As for the Americas and Australia, their settlement is a direct result of the initial/modernization phases of the Industrial Revolution (Great Geographical Discoveries).

[6] Thus, the emergence of the state is directly or indirectly linked to the LDT of the Agrarian Revolution, that is, with the accelerated population growth during the transition to intensive agriculture (see Grinin, 2011a). But, on the other hand, the emergence of states able to ensure internal peace also contributed to population growth (see below). The development of a broad and intensive international trade in food and other essential commodities finally brought European countries out of the Malthusian trap in the nineteenth century (see more about this below).

**The Type of Population Reproduction (TPR)**[7] is *a stable pattern of population reproduction, which is formed as a result of the actual technological and socio-cultural level of societal development.* This pattern results from the two largest demographic transformations (see above). It is characterized by fertility rate, mortality, natural increase and the associated life expectancy and its quality. The type of population reproduction can change due to an increase/decrease in the birth rates and/or a decrease/increase in mortality. At the same time, the decrease/increase in mortality may be total, that is, for all population groups (e.g., after overcoming epidemics, famines, or with improvements in health care) or in certain age groups (e.g., among infants or elderly people).

Until quite recently, the change in the TPR was associated with a hugely accelerated population growth in comparison with previous periods, going hand in hand with a growing population density. It was primarily a *quantitative change* that would promote certain qualitative changes.

Let us consider this for the sake of clarity (explanations will be given with the data below). We will see that a relative population growth decreases from one production revolution to another. Therefore, the Agrarian Revolution turned out to be the most powerful of the three production revolutions in terms of influencing relative population growth. We mean that during the Agrarian Revolution, the population growth exceeded the population growth in the pre-agricultural period in relative terms much more than it did as a result of the Industrial and Cybernetic Revolutions. Thus, at the start of the Agrarian Revolution, the world population hardly exceeded 10 million people, while in the fifteenth century, prior to the Industrial Revolution, there were already several hundred million people (400–500 million, see below). This means that the quantitative difference consists of an increase of at least 40 times (see, e.g., Korotayev, 2020a; Livi-Bacci, 2017). By the 1950s—the time of the beginning of the Cybernetic Revolution—there were 2.5 billion people. Consequently, during the period of the Industrial Production Principle (i.e., over 500 years), the world population increased by five to six times. In the seven decades of the Cybernetic Revolution, the world population has tripled. According to forecasts, by the end of the twenty-first century the world population will hardly exceed 10–11 billion people (UN Population Division, 2023), that is, during the period of the Scientific-Cybernetic Production Principle, the population will increase only four times. Thus, *while during the Craft-Agrarian Production Principle the population increased in absolute terms by dozens of times, then during the Industrial and Cybernetic Principles it increased only by a few times.* At the same time, in the 2nd millennium CE, the absolute rate of population growth per unit of time increased from century to century. Thus, *the relevance of quantitative changes in the population during the historical process decreased, while the importance of qualitative changes in people's life would grow. This is directly related to the technological progress, which step by step replaced living labor as the most important part of production.*

[7] We use a modified version of a term originally proposed by Vishnevsky (1973).

We can divide qualitative changes into two closely related types: material improvement (better nutrition, living conditions, sufficient payment for work, etc.) and improvement in the biological quality of life (better medicine, special technologies, special ways of maintaining health, etc.).

Although the improvement in the material quality of life has been observed at least since the Agrarian Revolution (in clothing, housing, heating, in some cases a variety of food, etc.), until quite recently it occurred noticeably slower than the population quantitative growth (and often did not affect considerable segments of the population). Only after the completion of the Industrial Revolution the growth in material well-being became on a par with the population growth and later overtook it.

However, during the Cybernetic Revolution, the change in the *TPR became mostly associated with qualitative changes*. In other words, the latest LDT (LDT-5, "adaptation to aging" according to Table 5.4) resulted in improving the quality of biological life (increasing life expectancy as well as the health-related quality).[8] In Chap. 6 we will dwell in detail on the relationship between the material quality of life and the quality of biological life, and also what exactly the quality of biological life implies. It is important to note that the LDTs occur on a world-system scale. The point is that significant changes in the population size have been observed in certain countries—the local demographic transformations.[9] However, they were associated with local improvements, innovations or peculiarities. The world-system scale is, of course, related to the historical scale of world systems, but in general it shows that this is a change in the type of population reproduction (TPR) in the leading part of the World System or its core. The World System originated as an Afroasian world system in the Middle East with the beginning of the Agrarian Revolution, 9–12 thousand years ago (about its development see Grinin, 2011b, 2012b; Grinin & Korotayev, 2009b, 2012a, 2018; Korotayev & Zinkina, 2017; Korotayev et al., 2006a; Zinkina et al., 2019).

**Different Dynamics of Production Revolutions and of the Largest Demographic Transformations (LDTs).** As we mentioned, one may speak about the law of a correlation between production revolutions and the LDTs. One can even argue that *production revolutions and demographic transformations are in some respect*

[8] Of course, there have been sporadic improvements in the quality of biological life since ancient times (take the example of the transition to cooked food, or the emergence of medicine). But from the seventeenth to eighteenth centuries it was precisely due to technology that it began to grow steadily in Europe, and already from the nineteenth century it became a trend.

[9] Thus, there occurred a huge local transformation in China in the end of the seventeenth to the first half of the nineteenth centuries, when the population more than tripled over two centuries from 130 million in 1650 to 420 million in 1850 (see Ilyushechkin, 1986; Mokir & Fotkh, 2012: 61; 61; see also Bray, 1984: 491–494, 598; Ho Ping-ti, 1956; Korotayev et al., 2006; Perkins, 1969: 38; Shiba, 1970: 50). However, it is mainly driven by social innovations (for details see Korotayev & Khaltourina, 2006; Grinin, 2010: 267–271). See also Korotayev et al. (2006b) about the impressive population growth in China in the eleventh century.

*different phases of the same techno-demographic cycle.*[10] *While the production revolution opens up new ecological niches, the LDT fills these niches.*

The different dynamics of technological and demographic changes also show that, although this correlation is evident, still the largest demographic transformations appear linked not to the same but to different phases of production revolutions. This is due to the acceleration of technological process and of the historical process in general (see examples below).[11] On the one hand, the logic of demographic change implies a change of generations while technological breakthroughs can occur even within the lifetime of one generation (this applies in particular to the Cybernetic Revolution). On the other hand, technological revolutions were much less frequent in ancient times. In fact, thousands of years passed between the initial and final phases of the Agrarian Revolution, that is, tens or even hundreds of generations succeeded.

It is also extremely important to understand that the pace of technological change has accelerated so much over the last two centuries that *the demographic changes associated with them have actually become almost continuous, and yet they do not keep up with the technological change, while somewhere they are ahead of them.* And the growing connectedness of the World System also greatly affects both types of transformations, but in different ways. Therefore, although the logic of the connection between technological and demographic changes is very strong, it requires taking into account the difference in dynamics, which we will show below in the following explanations and tables.[12] Thus, we consider the largest demographic transformations as separate breakthroughs that are associated with certain phases of the production revolution, as well as with the mature phases of the production principle.

**We Distinguish Six Largest World-System Demographic Transformations (LDTs) Within the Historical Process**. They are systematized in Table 5.4 and described in Sect. 5.5. The first three LDTs have already taken place. The fourth LDT is also almost complete, though not yet to the full extent in some developing countries. Today we witness the unfolding of the fifth LDT which is associated with aging. The sixth LDT is hypothetical and may be associated with the completion of the Cybernetic Revolution and the mature phases of the Cybernetic Production Principle. We will speak about the fifth and sixth LDTs in more detail in the following chapters.

---

[10] Even more so if one keeps in mind that people are the most important and integral part of the productive forces. We do not touch this point in the present chapter, but it is still an important issue (see Grinin, 2006a, 2006b, 2011b, 2012a).

[11] Let us remind readers that in Chap. 2 we presented the cycle of production revolutions as consisting of three phases.

[12] Wars and depressions, while stimulating technological breakthroughs, also have a negative impact on the demographic dynamics. And this was particularly evident in the first half of the twentieth century.

## 5.3 The Main Constraints on Population Growth

It is obvious that there are more or less severe limitations to population growth at each stage of the historical process.

**The Hunter-Gatherer Production Principle**

The most important factors limiting the population growth among hunter-gatherers are:

(a) a sufficiently large area necessary to supply an individual (in some Siberian regions the so-called 'hunter's path' was about 300 km long);
(b) the inability to hunt more game animals or harvest more plant food/useful biomass than is possible in a particular area without affecting its reproductive capacity in the future;
(c) the need for fairly large 'neutral' areas between communities;
(d) the scarcity of other (non-food) resources (e.g., in the north it is the scarcity of fuel);
(e) if a settlement is permanent, then the resources are limited; if people have to constantly migrate, then having a large number of children is a huge burden when migrating;
(f) women of the pre-agrarian period had natural limitations associated with a long lactation period (due to the inability to feed infants with other foods that were too coarse for them). Hence the limitation of conceptions and births due to prolonged lactation. There were also other factors that reduced the female fertility during the hunting-gathering period compared to the agrarian period, for example, a low-carbohydrate diet (see, e.g., Bocquet-Appel, 2011; Livi-Bacci, 2017);
(g) high natural infant mortality rate characteristic of a nomadic lifestyle.

**The Craft-Agrarian Production Principle**

In a pastoral society some restrictions were eliminated, but the number of pastoralists directly depended on the size of pastures.

In agricultural societies, population growth was limited by the following factors:

(a) the amount of cultivable land (as well as the need to leave most of the potentially cultivable land fallow in extensive/slash-and-burn agriculture, which was widespread for a very long time) (see, e.g., Boserup, 1965);
(b) the fertility of the land, although the development of technology could increase this fertility;
(c) favorable climate and other available natural resources (forests, rivers, etc.);
(d) the risk of natural disasters and epidemics;
(e) external threats and conflicts with neighbors;

(f) the quality of internal peace and order provided by political power—a well-organized and long-lasting centralized state was the exception rather than the rule during this period.

Let us consider these limitations in more detail.

The population growth depended on the fertility of the soil and the amount of cultivable land, which is sometimes defined as the *carrying capacity of the ecological niche* that people occupy. It is possible to calculate mathematically the maximum possible population in a certain area with certain technologies. But, as a rule, society could hardly even come close to this mathematical limit, since conflicts came into play. We believe that after occupying half of the possible capacity (or even far from reaching this half), it became congested: the settlements would become situated too close to each other and friction arose (Carneiro, 1987; Grinin, 2007, 2011a; Grinin & Korotayev, 2012b; Korotayev et al., 2006b; Turchin & Korotayev, 2006; Zinkina et al., 2016).[13] In addition, waste products polluted large areas: as the population density increased, for example, in cities, contagious diseases were likely to spread and could easily eliminate most part of the inhabitants. If the climate changed, then the capacity of the ecological niche often reduced dramatically, with all the subsequent effects (see, e.g., Lee, 2007; Parker, 2013). Crop failures, which could occur from time to time, also had a strong effect on population growth. The threat of warfare and raiding had considerable effect and made it necessary to leave large territories unoccupied or to use places with better natural protection, such as hills and high banks (which were few).

However, there were exceptions when highly organized power, internal peace and attention to management led to a considerable population growth. China achieved especially great success (see, e.g., Grinin, 2007, 2010; Korotayev & Khaltourina, 2006; Korotayev et al., 2018). In such cases, the population could reach large numbers (tens, and in some cases even hundreds of millions of people).[14] While the population growth generally led to intensification of agriculture (Boserup, 1965), even in these states the population growth was also limited. In this case, the population growth was already limited by environmental opportunities (the amount of land and other resources, as well as the increased risk of epidemics, conflicts, etc.) (see, e.g., Chu & Lee, 1994; Grinin, 2007, 2010; Korotayev & Khaltourina, 2006; Korotayev et al., 2011, 2016a, 2016b; Lee & Zhang, 2010; Nefedov, 2013, 2014b). Sometimes, due to some technological innovations and development of new territories, the ecological niche expanded dramatically and the population increased considerably (as was the case in Sung China in the eleventh century AD, when—thanks to the development of the South, where

[13] Including wars for rich territories. They occurred since very ancient times, including the pre-state peoples. For example, bloody battles took place between the Indians, in particular the Sioux and the Chippewans, for the wild rice fields (Lips, 1954: 105–106).

[14] In addition to China, here we can mention Egypt and Babylonia, some countries of Central Asia and Europe (already at the later stages of the Craft-Agrarian Production Principle).

new varieties of rice and other technological innovations emerged—the population almost doubled from 60 to 100 million (see Kychanov, 1986; Lapina, 2002; Korotayev & Khaltourina, 2006; McNeill, 2013; Zhao & Drechsler, 2018). However, in any case, the agricultural economy had a certain maximum population, since sustainable production growth in such systems is an exception rather than a rule. Sooner or later, the land became scarce. Social tensions would increase in the context of land and resources scarcity, and increasing taxes and injustice (see, e.g., Grinin, 2007, 2010; Goldstone, 2016; Grinin & Korotayev, 2012b; Korotayev, 2017; Korotayev & Komarova, 2004; Nefedov, 2004, 2014a; Usher, 1989).

Thus, the partial overcoming of limitations due to a technological breakthrough led to the emergence of long-term (from several decades to two to three hundred years) demographic cycles: as a result of rapid population growth, a country or region fell into the so-called Malthusian trap.[15] Social tensions arose due to the land scarcity, increasing poverty and considerable stratification. As a result, societies sooner or later entered a period of catastrophes and disasters.[16]

This was all the more probable since the first more active and talented rulers could be replaced by others who were soft and more infantile, prone to wastefulness and short-sighted.[17] The social system, when facing restrictions, would become unstable and collapse under adverse circumstances (wars, invasions and prolonged crop failures). This could lead to socio-demographic catastrophes, while the loss of population as a result of famines, epidemics, invasions and other cataclysms sometimes numbered many millions or even tens of millions of people.

As already mentioned, the long-term ('secular') socio-demographic cycles do not characterize all agrarian societies, but only the highly organized ones, which generally were an exception to the rule and which contributed a lot to the development of civilizations.[18] As for demographic catastrophes in Europe (primarily in Western Europe), they were associated, in particular, with epidemics.[19]

---

[15] About the Malthusian trap see Artzrouni and Komlos (1985), Steinmann and Komlos (1998), Komlos and Artzrouni (1990), Steinmann et al. (1998), Usher (1989), Wood (1998), Kögel and Prskawetz (2001), Grinin and Koroatyev (2009a, 2012b), Grinin et al. (2008a, 2008b, 2009), Nefedov (2014c), Guschina and Malkov (2014), Malthus (1798), Steinmann and Komlos (1998), Korotayev et al. (2011, 2014, 2016a, 2016b), Zinkina and Korotayev (2014a, 2014b), Korotayev and Zinkina (2014, 2015, 2022).

[16] For details see Nefedov (2007), Korotayev (2006, 2017), Korotayev et al. (2007a, 2011, 2016a, 2016b), Grinin (2011a), Grinin and Korotayev (2012b), Usher (1989), Chu and Lee (1994), Komlos and Nefedov (2002), Nefedov (2004, 2013, 2014a, 2014b), Korotayev and Komarova (2004), Korotayev and Khaltourina (2006), Turchin and Korotayev (2006), Turchin and Nefedov (2009), Malkov (2009), Lee and Zhang (2010), Goldstone (2016); see also Borsch (2004, 2005).

[17] This trend of dynastic degradation was brilliantly described by the fourteenth-century Arab sociologist Ibn Khaldūn (1958, 1980, 2008).

[18] On the other hand, in some societies or during certain long periods, under the influence of factors such as under population, favorable climatic fluctuations, the development of new lands, etc., the demographic restrictions would not work.

[19] In fact, the first depopulation was caused by the collapse of the Western Roman Empire and the barbarization of Europe (when the efficiency in agriculture, productivity and the ecological niche

The (total) population of Europe amounted to 38.5 million in the year 1000; 73.5 million in 1340 and 50 million in 1450 (Russel, 1975: 36). In other words, we observe a considerable population growth within 340 years (due to the increase in agricultural, industrial and other production). Later a stark depopulation followed, associated with bloody wars, climatic deterioration, and other disasters, but especially with the plague (Black Death) after 1347—the recovery remained incomplete even a century later. During the three years of the first wave of the epidemic in 1348–1350, mortality reached 25% (!) in some countries (Russel, 1975: 55; see also, e.g., McNeill, 1998; Livi-Bacci, 2017).

**The Trade-Industrial Production Principle**

For a long time (until the beginning of the nineteenth century, and in some places even later), the limits of the previous period, that is of the described above Malthusian type, were still strongly perceived, so the amount of cultivable land generally defined the ecological niche. During the Modern Age, the ecological niche expanded markedly for different reasons such as new land cultivation, colonial expansion, the growth of trade and new crops.[20] But although the expansion was considerable, this niche was still generally determined by the amount of cultivable land.

As a result of the final phase of the Industrial Revolution—and with this the development of science, culture, sanitation and medicine—we find a significant decrease in mortality and, consequently a remarkable population growth. The old restrictions were still in place and sometimes quite acutely, but in general, they were weakened.[21]

Importantly, the limitation for population growth was largely eliminated by the end of the nineteenth century in those societies where population numbers were determined by agricultural productivity: more people could be employed in industrial production and the population could grow significantly due to food imports. This could only happen as a result of a strong expansion of trade in grain and other products within the World System, which meant that *industrial countries escaped the Malthusian trap* (see Grinin et al., 2008b, 2009; Grinin & Korotayev, 2015). Thus, the growth of agricultural production in some countries supported

decreased). Thus, according to some sources, between 500 and 650, the population of Europe, decreased by one and a half times, from 27.5 million to 18 million people (Russel, 1975: 36).

[20] In the sixteenth to eighteenth centuries very valuable and productive agricultural crops plants spread from the New World to the Old World, primarily corn and potatoes which played an important role in the development of agriculture and population growth not only in Europe, but also in Asia and Africa (in the latter batata, or sweet potato, became very important) (see, e.g., Frakema, 2019; Galesi, 2022; Zinkina et al., 2019).

[21] So, while in eighteenth-century Britain there was enough grain for both its own needs and for export, then by the mid-nineteenth century the situation had changed: the country could no longer supply itself with agricultural products. However, the population growth did not stop, and now the growth of grain imports (the result of a powerful development of trade and the improved maritime transport) allowed the population to grow rapidly (e.g., Persson & Sharp, 2015).

the growth of industry and science in other countries. As a result, there has generally been significant population growth, especially in industrial and industrializing countries.

However, escaping from the Malthusian trap did not mean the elimination of other limitations, which became particularly evident against the background of growing population and changing demographic proportions. In particular:

(a) in the twentieth century, famine was still a frequent phenomenon in Russia/USSR, Asia and Africa;
(b) social cataclysms and wars significantly influenced the limits of population growth in many countries during certain periods;
(c) the insufficient (by today's standards) development of medicine and pharmaceuticals, as well as the lack of sanitary and health education, hampered the further reduction of the mortality rate (in other words, the limits of its reduction were outlined). By modern standards, of course, the mortality rate was still high for a relatively young European society;
(d) insufficient (by today's standards) level of the GDP per capita, a strong social and property stratification and absence of social policy, which began to be fully implemented only at the beginning of the scientific-cybernetic principle,—all these factors meant restrictions for the potential increase of life expectancy, as well as the subsistence of large families. These factors caused social upheavals. Wars also appeared a fairly significant limitation;
(e) urbanization became the most important limitation (which was one of the most characteristic features of the type of population reproduction during industrialization). The birth rate in cities (even in pre-industrial times) was always lower than in rural areas, while the mortality rate was often higher. Cities grew due to migration from the countryside. When urbanization reached high rates (see Davis, 2002; Zinkina et al., 2019), the overall birth rate began to fall significantly as a result of both, increased modernization and the effect of a number of other factors (Caldwell et al., 2006; Chesnais, 1992; Reher, 2011).

**The Scientific-Cybernetic Production Principle**

The beginning of the Scientific-Cybernetic Production Principle was marked by a phenomenon known as the population explosion. While during the Paleolithic the population grew by 3% per millennium (see more below), there were years in the 1960s when the global population grew by almost 3% per year!

The inability to feed a rapidly growing population was one of the pressing problems in developing countries. The dramatic population growth only exacerbated this limitation, bringing to life the gloomy Malthusian predictions; the possibility of a transition to zero growth was discussed (Arab-Ogly, 1978; Ehrlich, 1975; Meadows et al., 1972; Peccei, 1969; Spengler, 1976). However, during the period from the 1940s to the 1980s, the so-called *Green Revolution* took place in the developing world, which significantly loosened these limitations (see, e.g., Grinin & Grinin, 2015; Grinin & Korotayev, 2015; Pingali, 2012; Wik et al., 2008).

Between 1960 and 2000, the yields in all developing countries increased by 208% for wheat, 109% for rice, 157% for corn, 78% for potatoes and 36% for cassava (FAO, 2004; Grinin & Korotayev, 2015, 2016; Grinin & Grinin, 2015; Pingali, 2012).

Thus, already the first decades of the Cybernetic Revolution, the previous limitations imposed by the development of production, science, medicine, etc., and by the Green Revolution, were significantly removed. All this contributed to a powerful demographic growth ("the population explosion"). However, along with these achievements new obstacles to population growth emerged.

*Paradoxically, the main constraint on population growth (first in the developed world, and then in many developing countries) was science and medicine, which led to a population explosion, reduced mortality and increased life expectancy.* As soon as contraceptives and many other opportunities to limit births through family planning appeared, the psychology of people (also due to the better and more widespread modern education), especially of women, began to change dramatically toward lower fertility, women having one or two children and even remaining childless (e.g., Caldwell et al., 2006; Korotayev et al., 2006a; Lesthaeghe, 2020).[22] This has been facilitated by targeted policies and propaganda of the idea of family regulation and striving for women's equality.[23]

There are a number of other important constraints on population growth:

(a) The process of aging and the consequent change in the proportion of older people in the population of societies naturally limits total fertility and leads to an increase in the death rate, because, ceteris paribus, in an elderly society the death rate is higher than in a young one;
(b) globalization has led to the recognition of the scarcity of the Earth's resources, in particular, in terms of ecological and climatic limits. Thus, although the Malthusian limits have been reduced, and the opportunities provided by science and technology seem almost unlimited, the constraints of the Earth limits now become the limit to population growth;
(c) the example and authority of developed countries play an important role in demographic policies for many developing countries.

[22] Thus, in the USA the number of childless households (which, of course, includes older people whose children have grown up) rose from 51 to 74% between 1960 and 2020, respectively (Urnov, 2022: 40).

[23] Both the government and business benefited from the decline in the birth rate, because it raised the standard of living and created the so-called demographic dividend and freed up many millions of women for the labor market.

## 5.4 General Demographic Changes During Production Revolutions and Production Principles

### 5.4.1 The Hunter-Gatherer Production Principle

**The Hunter-Gatherer Production Principle** corresponds to the type of population reproduction (TPR), which we denote as the *type of simple reproduction* (it can also be called natural [see Table 5.5]). It is characterized by a lower birth rate than in agrarian societies, since the lifestyles impeded female high fertility and ability to raise many children, and also by a medium–high mortality rate (see above). As a result, the reproductive rate of the population was barely reached, and its very slow growth was largely due to the settlement of new territories (Siberia, America, glacier-free territories in Northern Eurasia, etc. see Chap. 2).

**The Upper/Late Paleolithic (40–12 Thousand Years Ago)**. The assumptions about the total world population during this period are based on calculations of a possible population density, which was extremely low; although, in turn, it varied greatly depending on the local natural environment.[24] The population estimates for the Lower Paleolithic vary greatly from 1 to 10 million people (Kapitza, 2009: 38; Korotayev, 2020a; Livi-Bacci, 2017; Thomlinson, 1975; for details see Korotayev et al., 2007a: 14–15). However, we believe that the real figures were much closer to the larger figure than to the smaller one. Kapitza (2006: 19; 2009: 38) estimates the population at about 10–15 million referring to the later date of 9–10 thousand years ago. This was the very beginning of the Agrarian Revolution and it could already be its contribution to the population growth, but only in some and relatively small areas, so it is rather unlikely that the population could grow one and a half to two times within two or three thousand years only due to this phenomenon. Here an important factor was the warming during the Holocene, when vast areas of land were cleared of ice and settled.

**On the Population Growth Rate**. The world's population generally grew so little that it is only possible to distinguish it on a secular or even millennial scale. According to some data, the Paleolithic population increased by only 3–8% (per millennium!), that is, by 0.003–0.008% per year (Livi-Bacci, 2017: 25, Table 1.2; Valentei, 1974: 236; Vishnevsky, 2005: 30). Of course, in some, particularly favorable, places the population growth could be considerably faster, but on the other

---

[24] According to some data, the average population density in the pre-agrarian era was 0.1 people per 1 sq. km. of land (see, e.g., Childe, 1942: Chap. 2; Andrianov, 1978: 21). At the same time, this density fluctuated considerably. In the most favorable areas, it reached 0.66 persons per 1 sq. km (Childe, 1942: Chap. 2). In less favorable areas, the density was one person per hundreds kilometers. Thus, according to Boris O. Dolgikh, in the seventeenth century among the Yukaghirs, there was one person per 200–300 square kilometres (Andrianov, 1978: 22). A similar situation was observed among the Tungus hunters (see Dolgikh, 1960: 15). And according to David Christian's calculations (Christian, 2004: 198), the average population density in the period 10 thousand years BC was one person per 25 sq. km (that is, even less than according to Andrianov's estimates of one person per 10 sq. km).

hand, many groups often experienced depopulation, and the climate during most of the Upper Paleolithic (40–15 thousand years ago) was quite severe over a large area, characterized by significant fluctuations (e.g., Bruxelles & Jarry, 2011; Clark et al., 2019; see Chap. 2). As for the growth of the world population during the Upper Paleolithic era, the estimates should be considered as hypothetical. However, they can provide some insight. Kapitza (2009: 37–38), citing Lutz, 1994, provides a table that shows that 1.6 million years ago the primordial population was 100 thousand people (Ibid.: 38, Table 1).[25] However, already at the beginning of the Upper Paleolithic era (i.e., 40 thousand years ago), the number of people is estimated at 1–5 million (Ibid.: 38, Table 1), which is quite consistent with the above estimates. Thus, according to these assumptions, the population growth during the Lower Paleolithic was noticeably lower than that of the Cro-Magnons and amounted to only some tenths of a percent per millennium (compare with 3–8% per millennium during the Lower Paleolithic). As for the scale of population growth in absolute terms, according to Sergey Kapitza (2006: 19) "By the onset of the Neolithic 9–10 thousand years ago, the absolute growth rate was already ten thousand times higher than at the beginning of the Stone Age". In general, such growth is not surprising, since the cultural and material level of the Cro-Magnons, including the ability to speak, was significantly higher than that of the archaic humans.

### 5.4.2 The Craft-Agrarian Production Principle

*With the transition from hunting-gathering economy to agriculture, that is, the Agrarian Revolution, the population eventually increased tenfold compared to the pre-agrarian period.* While before the start of the Agrarian Revolution the world population hardly exceeded 10 million people (12 thousand years ago it was 6 million according to Livi-Bacci [2017: 25, Table 1.2]), then prior to the Industrial Revolution in the fifteenth century, it was several hundred million people.

**The type of population reproduction (TPR)** during the Craft-Agrarian Production Principle was fundamentally different from that of the previous period. We find a transition from medium–high hunter-gatherer fertility to very high fertility. Mortality, however, also increased slightly (mainly due to child mortality rates), but not as much as the birth rate, which ensured a noticeable natural population growth (Bocquet-Appel, 2011; Cohen, 1998, 2009; Cohen & Armelagos, 1984; Cohen & Crane-Kramer, 2007; Livi-Bacci, 2017). However, the natural increase still remained rather low. But everything is relative. It was significant, that is by more than an order of magnitude higher than in the hunter-gather society, and yet it was extremely low compared to subsequent periods. Thus, according to Urlanis

[25] But here one should keep in mind that Homo habilis (archanthrope) lived in East Africa 1.6 million years ago and was not a direct ancestor of Homo sapiens. Thus, it would be more correct to calculate the growth of the human population from the first Homo sapiens, who lived in Africa (and not only in the East) 200–300 thousand years ago.

(1941: 91), from 1000 to 1500, the natural increase in Europe between 1000 and 1500 was between 0.1 and 0.2% per year (other researchers define it as less than 0.1%, e.g., 0.07% according to Livi-Bacci, 2017: 25, Table 1.3). At times, the natural development was even negative (e.g., during the period of the so-called Black Death or the Mongol conquests). Therefore, we define this TPR as the *type of slow growth* (it can also be called the socio-natural type) (for its other characteristics, see Table 5.5).

However, over the centuries, the overall population growth was quite remarkable. Although the mortality rate was high among both hunters and farmers, there were significant differences here as well. In contrast to the TPR of the Hunter-Gatherer Production Principle, the adult mortality was lower in some developed agrarian societies (although it increased periodically during epidemics, wars, hardships and famines). The infant mortality rate was high in both types of society, but due to a higher birth rate, the population growth in the craft-agrarian society was noticeably higher than in the hunter-gatherer society. As a result of high infant mortality, the average life expectancy in many craft-agrarian societies was not higher than among hunter-gatherers; the adult life expectancy was often higher in some advanced agrarian societies.[26]

Thus, the TPR changed, in particular, because of the increase in female fertility and in the number of births. What could increase fertility? Farming made it possible to greatly increase the birth rate due to (a) the transition to sedentism, (b) the use of pack and draft animals (especially among the nomads) and (c) the shorter birth intervals made possible by the availability of domestic animal milk and cooking, and also by the use of various foods (and devices to cut them very small) to feed small children, so that a woman could wean a child earlier than in hunter-gatherer societies. In the latter, as we mentioned above, prolonged breastfeeding was also a form of pregnancy prevention (e.g., Bocquet-Appel, 2011; Livi-Bacci, 2017). One should also take into account a certain decline in catastrophic mortality.

As a result, already at the initial phases of the Agrarian Revolution, the population growth accelerated sharply, by 5–13 times, reaching 40% per millennium (from 3–8% in the Paleolithic) (Valentey, 1974: 236; Vishnevsky, 2005: 30–31).

The dynamics of change of the TPR in relation to the stages of Craft-Agrarian Production Principle and the growth of its world-system distribution can be briefly presented in the following way.

The initial and modernization phases of the Agrarian Revolution (i.e., the economy of primitive manual farming) correspond to the first largest demographic transformation (LDT), which we have called the "Neolithic" (see Table 5.4). That is, these periods already corresponded to the transition to a new TPR with an extremely high birth rate. However, a noticeable sustainable population growth could take place only in a few regions with a high land capacity, and therefore,

[26] On demographic changes during the period of the Agrarian Revolution see Kozintsev (1980), Korotayev et al. (2007b: 173–175), Grinin (2011a), Bocquet-Appel (2011), Livi-Bacci (2017).

with an ecological niche capable of supporting millions and tens of millions of people within the World System.[27]

The final phase of the Agrarian Revolution (when irrigation and plough farming had spread) corresponds to the transition to a *much more capacious ecological niche,* capable of feeding tens or even hundreds of millions of people within the World System. This was the second LDT, which we call "civilizational" since its unfolding led to the emergence of states and civilizations (see Table 5.4).

However, we should remember that the completion of the Agrarian Revolution in different regions proceeded in different ways, and its completion took thousands of years (for details see Chap. 2). It is extremely important to note that its particular variants strongly depend on the natural environment of societies and regions. As a result, depending on the model of the completion of the Agrarian Revolution (irrigated or non-irrigated plough agriculture), different demographic models developed. The most important among them were the following:

(1) irrigation with a very large population numbers and high population density (Egypt and Mesopotamia are typical examples here);
(2) non-irrigated plough agriculture with a lower population density (Europe).
(3) mixed models.[28]

The first two models actually formed two cultural and historical developmental trajectories (Eastern, including the Far East) and antique-medieval in Europe, and prepared prerequisites for Europe's breakthrough into the Industrial Revolution. Vitaliy Melyantsev (1996: 77) believes that in the Middle Ages, the total carrying capacity of land in Europe was five times lower than in the Eastern countries (see also Huang, 2002). As a result, it was necessary to cultivate much more land to provide for one person than in the East, which was more difficult with only manual labor.

And this, combined with a number of other factors, made labor through increased mechanization of labor proceed more intensively in Europe than in

[27] There was also a marginal TPR among the nomads, with lower birth rates than among agriculturalists but higher than among hunter-gatherers and apparently less regular mortality.

[28] Here we can mention the following two types: 3) fertile lands with partial irrigation, but mostly unirrigated. This is particularly the case in the Middle East, where a model developed with a higher population density than in Europe, but lower than in China due to more fertile soils; 4) an arid model, with infertile soils and sparse populations, where nomadic animal husbandry was developed. Both models played important roles in history; the nomads were an element that united the World System and was a kind of barbarian periphery (Grinin & Korotayev, 2012a, 2018; Korotayev et al. 2021, 2022). In addition, we can also speak about the African model, where farming was mainly non-irrigated and manual (Grinin, 2011a; Korotayev & Khaltourina, 2006; Korotayev et al., 2016a, 2016b). However, this model played a minor role in the historical process. We can also mention models with a sparse population and difficult conditions for agriculture, where non-agricultural economic types (fishermen, sailors, warriors, etc.) were formed. The examples here are some Scandinavian communities.

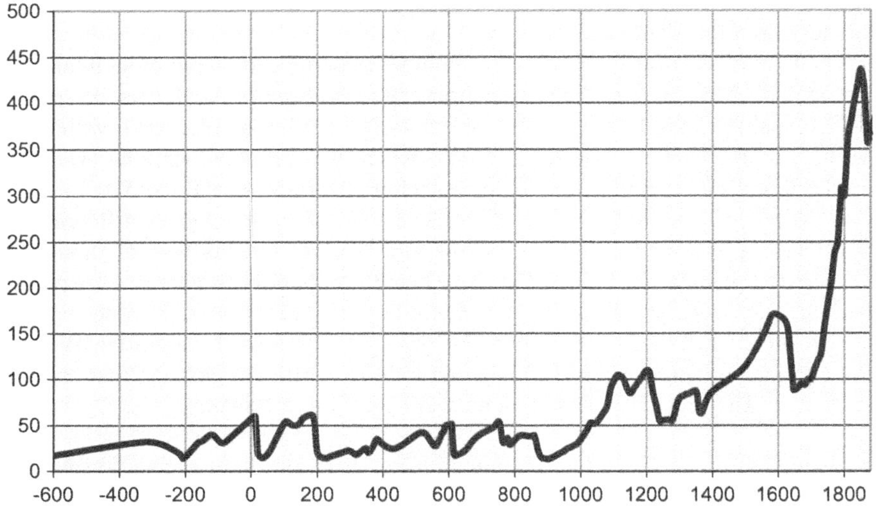


**Fig. 5.1** Demographic dynamics in China (million people) from the seventh century BCE to the mid-nineteenth century CE. *Data Source* Korotayev et al. (2006b)

the East. This would trigger the Industrial Revolution (for details see Grinin & Korotayev, 2015; Grinin, 2010; Grinin & Grinin, 2015).

After the Agrarian Revolution came to an end, the population continued to grow for some time, filling the ecological niche and complicating the geography of settlements, the division of labor, trade and so on. Later, after the easy-to-settle niches had been filled, the population growth became much more difficult and required a systematic organization of society and state, as well as internal peace and order—this had been pointed out earlier. Such conditions were the exception rather than the rule, but when they did appear (as it happened in China), the population grew by leaps and bounds. However, sooner or later, such an organized society fell into the Malthusian trap and its dynamics acquired a cyclical character, with structural and demographic cycles lasting from one hundred to three hundred years long[29] (see, e.g., Fig. 5.1 and Table 5.1).

### 5.4.3 The Trade-Industrial Production Principle

**General Changes**. The correlation between industrialization and the largest demographic changes is widely recognized (Armengaud, 1976; Caldwell et al., 2006; Chesnais, 1992; Dyson, 2010; Grinin & Korotayev, 2015; Grinin & Grinin, 2015; Livi-Bacci, 2017; Minghinton, 1976: 85–89; Mokir & Fotkh, 2012). Indeed, while

[29] See above.

**Table 5.1** Demographic changes during the Craft-Agrarian Production Principle

| Stage/phase | Dates | Techno-social changes | Demographic changes | Limitations | World population at the beginning of the period |
|---|---|---|---|---|---|
| Initial phase of the Agrarian revolution[30] | 12–9 thousand years ago | Development of agriculture and animal husbandry in some areas of the Middle East. Expansion of the ecological niche | Beginning of a new type of population reproduction with an increased birth rate | A very narrow territorial base of a new ecological niche | 8–10 mln[31] |
| Modernization phase of the Agrarian revolution | 9–5.5 thousand years ago | Development and diffusion of agriculture | Formation of a new type of population reproduction with a high birth rate but within a limited ecological niche | The limited ecological niche and intensification of conflicts due to the complexity of societies | 15–30 mln |
| Final phase of the Agrarian revolution | 5.5–3 thousand years ago | Transition to large irrigation systems, plough farming and a drastic expansion of the ecological niche; formation of states and civilizations | Development of a new type of population reproduction with a high birth rate already based on a capacious ecological niche; rapid population growth in some places | A narrow territorial base of a new ecological niche in the valleys of large rivers | 50 mln[32] |
| The last three mature stages of the Craft-Agrarian Production Principle | 3 thousand years ago—1430 | Emergence of political conditions for agricultural intensification; rise of urbanization, crafts and trade; expansion of the Afroeurasian world-system | A powerful development of a new type of population reproduction with a high birth rate based on a capacious ecological niche with considerably expanded territory. Together with the state's impact on the growth of production, internal peace and order, this leads to a relatively rapid population growth | Malthusian trap | about 200–300 mln in the beginning of CE—400 500 mln in the fifteenth century[33] |

30 As we pointed out above and in Chap. 2, the three phases of the Agrarian Revolution are at the same time the first three phases of the Craft-Agrarian Production Principle.

31 10–15 million in the period of 9–10 thousand years ago (Kapitza, 2006: 19; 2009: 38).

32 Valentey (1974: 236); Vishnevsky (2005: 31).

33 The estimates referring to the beginning of our era range from 170 mln (McEvedy & Jones, 1978) to 250/252 (Arab-Ogly, 1978: 88; Livi-Bacci, 2017: 25, Table 1.3; Korotayev, 2020a: 590) and 300–330 million people (Akimov, 1999; Durand, 1977; Scherbakova, 2011), and sometimes even 400 mln (see Korotayev et al., 2007a: 114; Korotayev et al., 2006b).

by the beginning of the Industrial Revolution in the second third of the fifteenth century the World System may have had 400–450 million inhabitants,[34] then at the turn of the eighteenth and nineteenth centuries the world population had more than doubled and exceeded 1 billion people.[35] By 1927, the humanity had reached 2 billion. And in the early 1950s, the world population exceeded 2.5 billion people (Shcherbakova, 2011; UN Population Division, 2023).[36] Thus, over the whole period of the Industrial Production Principle (i.e., within 525 years), the population increased by 5.5–6 times. This is not as remarkable as the relative growth during the Craft-Agrarian Production Principle (see above), but significantly more than the number anticipated for the Scientific-Cybernetic Production Principle (with a maximum fourfold increase; the world population by the end of the present century will be about 10–12 billion people or less [UN Population Division, 2023; Vollset et al., 2020]).

*Thus, in contrast to the technological growth rate, which has accelerated recently, the demographic growth rate is now slowing down from phase to phase.*

**The type of population reproduction (TPR), which gradually emerged during the Industrial Production Principle**, is characterized by a low mortality rate (compared to the previous type) and relatively high, but gradually decreasing, fertility, leading to a high natural increase. The share of young people in the population increases, resulting in the so-called youth bulges.[37] Therefore, we denoted the TPR of Industrial Production Principle as the *type of rapid growth*; it can also be called the *socio-quantitative type* (see Table 5.5). The term "social" means an increase in life expectancy due to improvements in sanitation and medicine and the comforts of urban life. The term "quantitative" is quite appropriate because never before has the population grown so rapidly as in the industrializing and modernizing countries, first in Europe and then in other regions. However, in emphasizing the term "quantitative" we must also point out the increase in the quality of life as a result of the improvements in medicine, comforts of urban life and education.

However, one should distinguish three stages in the historical dynamics of this TPR (if we limit ourselves to only European history). During the first phase

---

[34] The variability ranges from 350 to 390 million for the earlier period of 1400 and from 425 to 540 million estimated for a later period of 1500 (Biraben, 1980; Clark, 1967; Durand, 1974; Korotayev, 2020a: 590; Livi-Bacci 2017: 25, Table 1.3; Maddison, 2003; McEvedy and Jones, 1978; Tanton, 1994); see also Arab-Ogly (1978: 88), Chesnais (1992), Kapitza (2009: 38, Table 1); see also Kapitza et al. (2003: 218) (with reference to Chesnais 1992).

[35] This threshold is dated to 1804, but its estimates vary greatly: from 629 to 961 million people in 1750, and from 813 to 1125 million people in 1800 (see Scherbakova, 2011; e.g., 771 and 954 million, respectively, according to Livi-Bacci [2017: 25, Table 1.3]); see also Korotayev (2020a: 590).

[36] The estimates of the world population in the mid-twentieth century range from 2,400 to 2,557 million people (Scherbakova, 2011; UN Population Division, 2023).

[37] See, e.g., Choucri, 1974; Goldstone, 2016; Urdal, 2004; Korotayev et al., 2011; Korotayev et al. 2014, 2021, 2021b; Goldstone et al. 2012; Grinin 2012c; Grinin & Korotayev, 2012b; Romanov et al., 2021; Korotayev et al., 2023; Sawyer et al., 2022. For an example of such a youth bulge (in South Korea in 1960–1990) see Chap. 15, Fig. 3.

(lasting from the fifteenth to the early nineteenth centuries), the mortality rate decreased (mainly due to the smoothing out of the catastrophic mortality thanks to the development of sanitation technologies [e.g., Armengaud, 1976; Russel, 1975; Urlanis, 1941; Vishnevsky, 2005]) considerably but still not enough; the second stage—from the nineteenth to the early twentieth centuries—a sharp decline in mortality rate along with the maintenance of a high birth rate, the period of the actual demographic revolution in Europe; and the third stage, 1920–1940, when the birth rate in the West gradually declined, reducing the natural growth rate there (e.g., Chesnais, 1992; Landry, 1934; Urlanis, 1941). But in general, until almost the very end of the Industrial Production Principle, the birth rate in Europe considerably surpassed the replacement level. However, then the birth rate began to increase in the Third World (see below).

**The Development in the Sixteenth, Seventeenth, and Eighteenth Centuries.** Although the initial stages of the Industrial Revolution took place between the fifteenth and eighteenth centuries (see Chap. 2; Grinin & Korotayev, 2015; Grinin & Grinin, 2015), the decline in the mortality rate could hardly have been caused by technological advances in manufacturing. This was not immediately evident, because for a long time there were no scientific and technological prerequisites (poor medicine, low level and quality of urbanization and life-saving technologies). Nevertheless, thanks to the growth of production, literacy, culture and science, as well as to the spread of the media and freedom of occupation in a number of countries (primarily in north-western Europe) between the sixteenth and eighteenth centuries, we can trace rather considerable changes that led to the expansion of the ecological niche and a noticeable population growth in a number of countries. In particular, in Britain in the seventeenth and eighteenth centuries, thanks to innovation and productivity growth, the agricultural output began to overtake the population growth, which was rather considerable during certain periods. Thus, over a century and a half (from the sixteenth to the first half of the seventeenth centuries), the population of Britain increased from 2.8–3 million to 5–5.5 million people (e.g., Turchin & Nefedov, 2009; Wrigley et al., 1997), that is, the growth of 66–100%. And the growth rate reached 1% per year. During the second half of the seventeenth century and in the first decades of the eighteenth century, the population in Britain hardly grew.[38]

However, agriculture (along with industry) continued to develop actively, so that the country was able to export large quantities of grain to Europe. Within a century (1700–1800), English agriculture doubled its output (Reisner, 1986: 233). The population growth in the second half of the eighteenth century was impressive, all the more since a decrease in excess mortality in lean years was already noticeable. During the eighteenth century, the country's population grew from 5.5 million to

[38] Among the affecting factors were the Civil War (Goldstone, 2016), emigration to America, plague epidemics, adverse natural and climatic changes (Parker, 2013), very high mortality rate from excessive alcohol consumption (and falling birth rates for the same reason), in particular, the consumption of cheap gin, which grew very rapidly in the first half of the eighteenth century; there were also some other, not entirely clear reasons.

9 million, that is, with a hundred percent increase in food production, the population grew by 64% (Reisner, 1986: 233; see also Mantu, 1937: 127; Trevelyan, 1959: 358; Bairoch, 1971: 47). In the previous periods, a two-thirds increase in population would most likely have caused a socio-demographic catastrophe (see, e.g., Korotayev et al., 2006b).

Let us note that in the sixteenth/early seventeenth century, the population of other European countries, in particular, France and Germany, considerably grew, but in the latter the Thirty Years' War resulted in depopulation (e.g., Parker, 2013).

Since the sixteenth century, one can speak about the beginning of the escape from the limitations of the Malthusian trap (Grinin et al., 2008a, 2008b; 2009; Grinin & Korotayev, 2012b; Korotayev et al., 2011, 2014). However, this escape was a long and difficult process that only ended in the late nineteenth century (see above about this). No wonder that some growth theorists call the period before 1750 the Malthusian era (Clark, 2007; Mokyr & Fotkh, 2012).

We should note that *in the sixteenth to eighteenth centuries two demographic models competed within the World System: the European (technological) and the Chinese/Far East (preserving the former Malthusian pattern).* The latter model is clearly ahead of the former in absolute terms (above we have spoken about the more than threefold demographic growth in China between the seventeenth and nineteenth centuries). But such rivalry remained relevant only until the end of the Industrial Revolution in Britain.

**The Formation of a New TPR in the Eighteenth/Nineteenth Century**. The decreasing mortality rate was noticeable from the end of the eighteenth century and it was associated with the development of science, medicine, education, the rapid spread of knowledge and changes in the state sanitary and medical policies (e.g., Chesnais, 1992; Livi-Bacci, 2017; see also Deane 1965: 259–261).[39] This, together with other factors, contributed to a rather significant acceleration of demographic growth in many European countries in the late eighteenth century, although the discrepancies were significant (Mokir & Fotkh, 2012).[40]

But all these changes did not result in the new largest demographic transformation (LDT) as part of the Industrial Revolution, because the required level of medicine and social relations had not been reached. The largest demographic transformation seems to have been lagging behind, as we pointed out above. The new

[39] It is in this period that we can trace the origins of the current decline in the birth rate, due to the development of medicine (condoms, other forms of contraception and medical recommendations) and the reluctance of women to have children. Thus, according to the French researcher Moheau, French aristocrats in the eighteenth century considered the birth of children to be "the stupidity of all times" (cited in Urlanis, 1941: 226).

[40] Instead of 0.2% annual growth in the sixteenth and seventeenth centuries (which was higher than the average for the previous five centuries, see above), this number tripled in the eighteenth century and the annual growth rate reached 0.6% (Urlanis, 1941: 220). This is an unprecedented result so far. At the same time, the annual natural growth increased throughout the century from 0.19% in the first quarter to 0.71% in the last (Ibid.: 222). This growth was supported by decreasing mortality (Ibid: 222–227) and to some extent by the increasing birth rate, yet not in all countries, but only in some (Britain, Germany) (Ibid: 227).

LDT only emerged as a result of the completed Industrial Revolution and the mature phases of the Industrial Production Principle. In Table 5.4, it is designated as *anti-Malthusian*. However, it could be described as *urban*, since it could hardly have ever taken place without urbanization. As a result, the TPR of the industrial period was formed: reduced mortality with preserved or slowly declining birth rate. And this model was already unfolding during the mature phases of Industrial Production Principle. As a result, the population in Europe grew in an explosive exponential way. The spread of medicine supported the population growth even in not fully industrialized countries, like Russia—this was a harbinger of the future population explosion in the so-called Third World. As a result, in the second half of the nineteenth century, the escape from the Malthusian trap became evident due to the development of transport, global trade, specialization of countries in certain agricultural crops and involvement of huge agricultural lands into circulation (Grinin & Korotayev, 2012b, 2015).

The reduction in mortality rates was observed in different population groups, in particular, due to the elimination of problems such as deadly epidemics and famine. We should particularly note the reduction in infant and female (especially maternal) mortality, and these phenomena intensified compared to the Middle Ages and the Early Modern Period due to the growth in the population of cities, where the necessary sanitary requirements for women in labor had long been disregarded (Chesnais, 1998; Waller, 2003: Chap. 2). This process of the diffusion of necessary sanitation became noticeable from the late seventeenth century, but especially in the eighteenth and nineteenth centuries.

Life expectancy also increased between 1800 and 1900, that is, a period of more than a century. However, Europe and North America were very different from the rest of the world in this respect. While in the World-System core countries (Western Europe and the USA) the life expectancy increased from 35 to 47 years, in the world in general this increase was rather insignificant—from 31 to 33 years (Urnov, 2022: 95). We can add that the spatial distribution of the population changed dramatically due to the rapidly growing urbanization (see Davis, 2002; Zinkina et al., 2019).

**The Transformation of the Type of Rapid Population Growth**. The third period, which started with the First World War, involved mostly only Western Europe and should rather be associated with the subsequent type of population reproduction characteristic of the beginning of the Cybernetic Revolution. Although in the 1920s and 30s the birth rate (together with the natural increase) markedly dropped in a number of Western European countries, these were already the characteristic features of the Scientific-Cybernetic Production Principle, since at the end of the Industrial Production Principle the family planning practices started to be introduced through the primary development of contraception and the mass distribution of medically safe abortion (see Coale, 1983, 2013; Landry, 1934;

Livi-Bacci, 2017; Urlanis, 1941: 368).[41] For more on this period, see a specific section below (Table 5.2).

### 5.4.4 Scientific-Cybernetic Production Principle

First of all, let us make a reservation that in the present section we will speak quite briefly since many ideas will be covered in more detail in Chap. 6.

We denote the type of population reproduction associated with the Cybernetic Revolution as **the type of qualitative changes**, or it can also be called the social-qualitative type. And although the world population has considerably grown during the seventy years of its operation, from 2.5 billion people in 1950 to 8 billion in 2022 (UN Population Division, 2023), that is to say more than tripled, the quantitative growth will not, in our opinion, be the main feature of this TPR. And the main characteristics of the new TPR will be associated with the completion of the so-called demographic transition, that is, the low mortality and low birth rates and near-zero or negative natural increase, high life expectancy and a high proportion of the elderly and old people in its general structure.

What does qualitative growth mean? It means, first, a significant increase in life expectancy; second, ensuring, with the help of medicine, technologies and social programs, the transition to the so-called healthy aging, to a fairly secure, active and fully fledged lifestyle in declining years; third, ensuring adequate care for the elderly and the disabled; fourth, increasingly adapting people with disabilities to an active lifestyle with the help of medicine, technologies and social programs. *Thus, aging and an increase, probably, even a dramatic increase in life expectancy, and an ever more profound and more diverse adaptation of society and the individual to the aging*—these are the main trends of demographic changes within the Cybernetic Production Principle (see in Chaps. 6, 7 and 15).

**General Changes**. The increase in the life expectancy over the last 120 years is impressive. In the first half of the twentieth century, that is, at the end of the final phase of the Industrial Production Principle, the life expectancy grew considerably both in the world—from 33 years in 1905 to 50 years in 1955,—and in the core countries, respectively, from 47 to 67 years (Urnov, 2022: 95; UN Population Division, 2023). In the first decades of the Cybernetic Revolution, that is, in the second half of the twentieth century, life expectancy also increased considerably, but relatively a bit less: from 47 to 67 years worldwide and from 64 to 75 years in the core countries. It continued to increase in the first two decades of the twenty-first century, but at a somewhat slower pace (which is quite natural, since every extra year of life expectancy is now a great victory). But it is important to note that during this period, the general life expectancy in the world grew faster (as

[41] Abortion services were advertised in the newspapers, although they were banned in many countries, along with some contraceptives, but at the same time they were widely sold in shops (Sakevich, 2003).

**Table 5.2** Demographic changes during the Industrial Production Principle

| Stage/phase | Dates | Techno-social changes | Demographic changes | Limitations | World population at the beginning of the period |
|---|---|---|---|---|---|
| Initial phase of the Industrial Revolution | 1430s–1600s | The development of global trade and colonial economy, the development of the first machines and factories, the growth of agricultural productivity in Europe | Some acceleration in population growth, urbanization and the starting escape from the Malthusian trap | Weak development of science, medicine and sanitation | 400–450 mln[42] |
| Modernization phase of the Industrial Revolution | 1600s–1760s | Active growth of primary industry and agriculture in Europe, especially in Britain; development of state policy, education and science. Spread of new types of food in the world (corn, potatoes, sweet potatoes, etc.) | Beginning of a noticeable reduction in mortality rate in Europe, the acceleration of population growth; dramatic population growth in China, following the old development trajectory | Insufficient conditions for the growth of education and population mobility within the estate states | About 500–600 mln[43] |

(continued)

42 See Note 34.

43 According to Akimov (1999) in 1650 there lived 500 mln, and in 1750—795 mln.

**Table 5.2** (continued)

| Stage/phase | Dates | Techno-social changes | Demographic changes | Limitations | World population at the beginning of the period |
|---|---|---|---|---|---|
| Final phase of the Industrial Revolution | 1760–1830s | Creation of the primary machine industry in Britain, the USA and development of new types of transport (railway and steamboats), expanding global trade | Decreasing mortality rate in the world-system core, accelerated population growth, increasing urbanization; the beginning of the formation of the TPR of rapid growth | Insufficient development of industry and medicine in many European countries and very weak development in the world | About 800 mln[44] |
| Mature stages of the Industrial Production Principle | 1830s–1940s | Industrial Revolution, development of urbanization, education, culture and governance spreads across the World System | Continued decline in mortality rate, development of the TPR of rapid growth, preparing for a future population explosion in many colonies and semi-colonies; starting of fertility decline and slowing of population growth in a number of Western countries in the first half of the twentieth century | Still weak expansion of modernization in most countries of the World System; global and social catastrophes of the first half of the twentieth century | about 1.2–1.3 bln[45] |
| Turn of the Industrial and Cybernetic Production Principles<br>An event chronologically refers to the beginning of the Cybernetic Production Principle, but its type of population reproduction refers to the Industrial one[46] | 1940s–1970s | The end of the war, the emergence of the new world order, the liberation of colonies against the background of a powerful scientific and technological rise and the development of medicine | Unprecedented population growth in the developing countries (demographic explosion) due to the penetration of the latest medical technologies. Formation of the TPR of rapid growth throughout the World System | Poor development of agriculture not conforming the demographic rise; poor development of education, culture and urbanization in large parts of the World System | 2.3–2.5 bln[47] |

44 The estimates vary greatly; see Note 35.

45 In 1850, there lived 1,265 mln people according to Akimov (1999) and 1,241 mln according to Livi-Bacci (2017: 25, Table 1.3).

46 However, one should keep in mind that for many developing countries, including African ones, the initial demographic changes (less evident to the Europeans, but very impressive for the time) must be dated to the early twentieth century, and in some regions to the late nineteenth century. See more about it below.

47 World population estimates for the mid-twentieth century range from 2400 to 2557 million people (Scherbakova, 2011; UN Population Division, 2023).

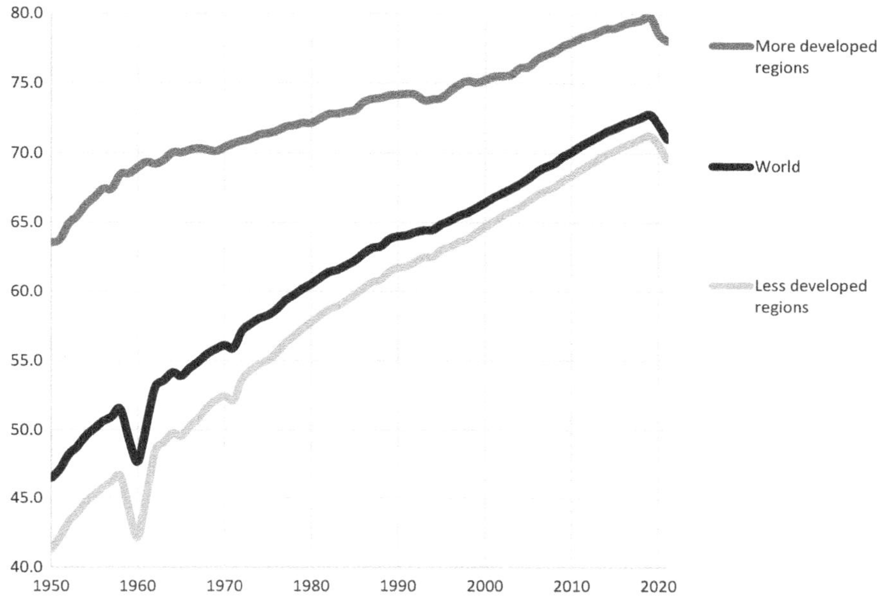


**Fig. 5.2** Life expectancy dynamics in the world, more and less developed regions, years, 1950–2021. *Data Source* UN Population Division, 2023

developing countries caught up with developed ones) from 66.5 years in 2000 to 72.8 years in 2019 (see Fig. 5.2). Respectively, in developed countries it increased from 75.3 to 79.8 years. However, in 2020–2021 there was observed an obvious decline due to the COVID-19 pandemic[48] (see Fig. 5.2).

**On the Demographic Transition**. Most, if not all, demographers consider the so-called demographic transition to be a continuous (or almost continuous) process that usually takes hundred to hundred and fifty years in individual countries.

The demographic transition consists of the following phases: (1) a reduction in the mortality rate while maintaining (or even increasing) a high birth rate, which leads to a high natural increase rate and rapid population growth and (2) a further reduction in mortality rate (albeit at a slower pace) and a rapid reduction in the birth rate, as a result of which the natural increase rate slows down, then becomes close to zero or even negative; the population grows, becomes stable or even starts to decline. At the same time, in the first phase, the share of children and young people in the demographic structure increases and the so-called youth bulges emerge. Meanwhile, the growing share of young people, especially

[48] On the excessive mortality caused by the pandemic (see e.g., Zhigalkin, 2021). In general, on social consequences of the COVID-19 pandemic (see e.g., Chumakov & Yurchenko, 2021; Efremov & Arkhangelsky, 2023; Grinin, 2020; Grinin et al., 2022; Hussain, 2021; Irshad, 2020; Kakkar, 2021; Seifabadi & Dolatabadi, 2022; Widdowson, 2021).

in the context of modernization, increases the risks of social upheavals.[49] In the second phase, the share of the working-age population first increases (due to the decreasing number of children), resulting in the so-called demographic dividend that provides society with significant benefits in economy and standard of living (Bloom & Williamson, 1998; Bloom & Canning, 2008; Bloom et al., 2007; Hawksworth & Cookson, 2008: 7–10; Lee & Mason, 2006, 2011; Barsukov, 2019; Groth et al., 2019; Kotschy et al., 2020; Korotayev et al., 2022, 2022b, 2022c). It was especially evident in Europe in the 1960s–1980s; and in China—in the 1990s–2000s, where it has now largely been overcome. Almost simultaneously, yet with a lag of a few decades, the process of population aging gains momentum—this will take much more time than the exhaustion of the demographic dividend.

If we consider the demographic transition in the world-system context, it has been going on for more than two hundred years by now and will continue for several more decades. Having started in Europe in the nineteenth century (or even in the eighteenth century, if we take France, England and some other countries as a reference), it ended in the First World countries by the end of the twentieth century. From the late nineteenth century, and especially from the mid-twentieth century, the development involved the rest of the countries of the World System. Still, for Tropical Africa and some other countries, this process has not entered its final phase yet.

**The Demographic Transition and the Cybernetic Revolution**. Due to the acceleration of technological progress in the twentieth century, the demographic transition actually took place almost continuously. From the point of view of demography, it is absolutely correct to consider it as a single and continuous process.

However, within the theory of production revolutions and production principles in their relationship with demographic transformations and TPR, it is particularly important to keep in mind that the *first and second phases of the demographic transition must be attributed to different production principles*. Our approach can be formulated in the following statements:

1. The demographic transition could not end during the Industrial Production Principle. There were neither the technological nor, respectively, the scientific and medical prerequisites for it. Industrial capitalism could not economically sustain the increasing number of elderly and unemployed people, and hence, the life expectancy growth would begin to decline.
2. Although the beginning of the demographic transition in European countries corresponds to the last phases of Industrial Production Principle, the social-qualitative type of population reproduction is not typical of the Industrial

[49] See, e.g., Moller (1968), Choucri (1974), Goldstone (2001, 2002, 2016), Urdal (2004, 2006), Korotayev et al. (2011, 2014, 2021, 2021b, 2022, 2022b, 2022c), Goldstone et al. (2012), Weber (2019), Romanov et al. (2021), Sawyer et al. (2022), Grinin (2012c, 2013a, 2013b, 2017, 2022).

Production Principle due to relatively modest living standards, extremely weak social policies and enormous property stratification.
3. The demographic transition in its proper sense (i.e., its second phase) is a phenomenon generally characteristic of the initial phases of the Cybernetic Production Principle.
4. The reduction in the birth rate to the level of population replacement in Europe could be temporary, being a rather unsustainable phenomenon due to the fact that only the first half of the twentieth century appeared to become an unfavorable period (wars, social catastrophes, disasters and especially the Great Depression). Obviously, this affected the desire to have children. Not without reason, when the situation improved in the late 1940s, a baby boom started along with an increase in the birth rate (see below).
5. In our theory we point out that we find in the final stages of any production principle we find a-typical and non-systemic phenomena (Grinin 2006a; Grinin & Grinin, 2015). The dramatic decline in the birth rate in Europe in the 1930s (noted by Landry [1934], Urlanis [1941] and other demographers) was such a non-systemic phenomenon (however, the decline in the birth rate during economically and socially difficult periods is a phenomenon characteristic of the whole human history).
6. The population explosion in the developing countries, which became explicit since the 1950s, refers in its essence to the Industrial Production Principle (when modernization was expanded to the periphery), but it also started in the last phases of Industrial Production Principle—in this sense, it also refers to non-systemic manifestations of the latter.

As the reader will recall, the demographic transition as a single process consists of two main phases: (1) a reduction in the mortality rate while maintaining a high birth rate, which leads to a high natural increase rate and rapid population growth and (2) a further reduction in the mortality rate (albeit at a slower pace) and a rapid reduction in the birth rate, leading to a very slow natural increase rate. Thus, within our theory, *the first phase of the demographic transition should be associated with the Industrial Production Principle; while its second phase should be linked to the initial phases of the Cybernetic Production Principle.*[50] However, a new significant demographic transformation awaits us in the mature stages of the Cybernetic Production Principle (see below).

**The Demographic Changes in the Initial and Modernization Phases of the Cybernetic Revolution (in the 1950s–1990s)**. Since the late 1950s, that is, already at the very beginning of the Cybernetic Revolution, one can quite clearly see the

[50] Such uniting processes can be observed at the junctions of different production principles. For example, the development of kinship systems was already characteristic of the latest stages of the Hunter-Gather Production Principle, but they manifested especially evidently in the first stages of the Craft-Agrarian Production Principle. Monarchical power actively developed within the framework of the Craft-Agrarian Production Principle, but it emerged in the form of an absolute monarchy already in the first phases of the Industrial Revolution.

acceleration of the transition to a new TPR (with low mortality and high birth rate) in developing countries, in fact, in the main part of the World System. This transformation gained momentum in the 1950s and 1960s, then slowed down a little, but still remained very noticeable in the 1970s and 1980s. As a result, as mentioned above, they started to speak about a demographic explosion or demographic revolution in the World System. However, one should note that the acceleration of population growth and the decline in mortality rates in many colonies and semi-colonies actually began already in the late nineteenth/early twentieth century due to the reduction of the catastrophic mortality and the start of the introduction of advanced medicine in some places. As a result, the population of Africa more than doubled over the 60-years period between 1890 and 1950, growing from 93 to 225 million (McEvedy & Jones, 1978: 206; UN Population Division, 2023); while before this period the population had doubled only over the period of several centuries, from 1500 to 1890 when it grew from 46 to 93 million people (McEvedy & Jones, 1978: 206). The beginning of the Cybernetic Revolution, the liberation of colonies and a powerful introduction of advanced medical and sanitation technologies ensured an impressive increase in the World System population, which continued during the modernization phase and will continue, driven by Africa. In fact, this was *the largest demographic transition* of the Industrial Production Principle, but it was shifted in time, coinciding chronologically with the beginning of the Cybernetic Production Principle. In Table 5.4, we denote it as LDT-4, the 'Third World' LDT.

However, the start of the Cybernetic Revolution triggered not only a population explosion in the Third World, but also a rather noticeable increase in the birth rate in the First World, where the pre-war trends of declining birth rates were replaced by a baby boom.[51] Thus, the emerging trend toward decreasing birth rate in the developed countries was interrupted for two decades. Probably, if one imagines that the situation in Europe in the first half of the twentieth century was peaceful, calm and fairly well-off, then the decline in the birth rate in the 1920s, 1930s and 1940s would probably have been slower, and there would have been no baby boom in the 1950s/1960s. The end of the baby boom, although it was quite logical, was clearly accelerated by the massive supply of contraceptives in the 1960s, along with the accompanying ideological justification. We witness revolutionary changes in the ability to control conception and to plan a family. As a result, a drastic reduction in the birth rate began in the developed countries: the second phase of the demographic transition commenced: low birth rate, low mortality rate and low

[51] Thus, between 1925 and 1940, the total fertility rate in the USA fell from 3.13 to 2.06 children per woman. And between 1940 and 1960, it rose from 2.06 to 3.58 children per woman (Chesnais, 1998; UN Population Division, 2023). This boom interrupted the trend toward the *qualitative changes* in TPR for a while, with a small increase. The obvious reasons for this were the new psychology of people who had survived the war and strived to maintain the optimism of life, and this was accompanied by very high growth rates of the GDP and living standards, which was directly opposite to the pre-war situation.

or even negative natural increase; and the modern demographic structure of the population gradually emerged with a trend toward aging.

A different model characterized the so-called Third World in the first decades. In fact, it was a model of the Industrial Production Principle, but (a) it developed in countries with a lower level of development than that observed in the European countries in the nineteenth and early twentieth centuries, since the role of agriculture was more considerable in developing countries than in Europe, and respectively, children were valued more as workers than among the urban dwellers; (b) it employed the achievements of the initial phase of the Cybernetic Revolution in medicine and (c) in societies with traditional ideologies encouraging large families, the fertility rate is higher. But then, in a number of developing countries (however, hardly in all of them), the turn to the second phase of the demographic transition began: a shift toward a sharp decline in the birth rate.

In general, the initial phase of the Cybernetic Revolution coincided with the end of the population explosion in developing countries (even in Africa the birth rate began to decline for some time at a rather rapid pace, but later this decline would considerably slow down [e.g., Zinkina & Korotayev, 2014a]) and the approach to completion of the demographic transition in the first world and the beginning of marked global aging. There was observed a dramatic reduction in mortality rates in all population groups and an increase in life expectancy all over the world.

**The Modernization Phase of the Cybernetic Revolution and the Results of Its First Two Phases (the 1950–2020s)**. The transformation that emerged during the initial phase of the Cybernetic Revolution, associated with the qualitative changes in new TPR (with very low mortality, very low birth rate, low growth and high life expectancy, that is, population aging) intensified during the modernization phase. The demographic dividend is now virtually exhausted in both the first world and less developed countries, including China, where the share of young people has decreased while the proportions of elderly age groups increased. The birth rates have fallen far below even the replacement levels, so in some countries, like Japan and Germany, depopulation has already started (in the latter case, only mitigated by migration t). But depopulation or stabilization of population is traced not only in the first world, but also in less developed countries as well (in the former second world: Russia, Ukraine, Bulgaria, etc.), and even in some developing countries. According to the estimates of the UN Population Division, depopulation has already begun in China, yet here it proceeds at a very slow pace (UN Population Division, 2023). In this regard, Landry (1934) proclaimed that the humanity no longer reproduces itself. Many countries, including India, are actively moving along this path. However, in most African countries the type of the previous phase is still preserved, resulting in a fast population growth in sub-Saharan Africa (Grinin & Korotayev, 2023; Korotayev et al., 2015, 2023a, 2023b; Malkov et al., 2022).

We can *summarize the results of the first two phases of the Cybernetic Revolution.* It is important, since we already observe some important ongoing processes while others are predictive:

(1) the emergence (for the first time in history) of technologies that could effectively control the birth rate;
(2) the fact that this was under the control of women themselves;
(3) the active movement of more and more countries into the second stage of the demographic transition;
(4) the emergence of the type of reproduction with a low birth rate and low mortality rate (TPR qualitative changes or socio-qualitative, see Table 5.5);
(5) the emergence of an aging society (for the first time in history);
(6) the increasing attention is paid to the elderly, the old and the disabled—society is increasingly turning attention to them, many technologies have been already created and are designed to improve the quality of their biological life—however, there are many problems and issues that will exacerbate (and which we will discuss in another chapter);
(7) Trends have clearly emerged that will determine not only demography, but the whole way of life and the development of technologies in the twenty-first century: (a) global population aging and a dramatic change in the demographic structure of the population, with a growing share of old people and a shrinking share of children and youth; (b) the depopulation in a number of countries, and not only in very developed ones; (c) the division of the world's countries into those with a young and those with an old population and (d) an increasing need for migration.

We assume that *the final phase of the Cybernetic Revolution* will start in the 2030s and last until the 2070s or even later. One of the most important issues that will determine the final phase of the Cybernetic Revolution is how quickly African countries will shift to a low birth rate pattern? This will define (a) the total population of the planet and (b) Africa's role in planetary affairs and possibly a planetary crisis if the population growth continues.

However, the most important issue of this period will be population aging, the difficulties associated with it and the search for the ways to adapt to aging (see more about this in Chaps. 6, 7, 11 and 15). At the same time, many countries face severe problems associated with population aging: labor shortages, pension crises and conflicts over the inevitable increase in the working-age limit, various forms of ageism, problems associated with the growth of layers/communities of migrants, as well as geopolitical problems caused by demographic changes. This will be discussed in another chapter. Here we can only add that the scarcity of working-age population will be actively compensated by attracting workers in young countries through remote work.

One way or another, population aging and the fight against aging-associated disease (oncology, dementia, etc.), an increase, probably even a dramatic increase, in life expectancy, and an ever more profound and more diverse adaptation to the aging of society and the individual (including, of course, growing attention to the adaptation of the disabled people)—these are the main trends in the development of demographic processes in the next seven to eight decades, the main demographic transformations of the final phase of the Cybernetic Revolution and the mature phases of the Scientific-Cybernetic Production Principle. Thus, in contrast to the Agrarian and Industrial production revolutions, when the population growth increased dramatically, in the twenty-first century the population growth will gradually become less and less important (world population growth will tend to stabilize, or even decrease to negative values), while the improvement of the quality of life (duration, biological activity, satisfaction with life, degree of relief from diseases and disabilities, etc.) will come to the fore.

**The Mature Stages of the Cybernetic Production Principle**. As we will see in Chaps. 6 and 7, what has been discussed earlier *will lead to the merging of everyday biological life with a totally medical-biotechnological environment. This will require the formation of special public institutions and an increase in their role in adapting to aging.*[52] However, as society adapts with the help of future technologies of the Cybernetic Revolution and new social institutions, gradually solving and experiencing complex internal and international problems and tensions, the features of a new LDT (we denote it as (LDT-6, "techno-bio-medical", see Table 5.4) will become evident. The trend toward increasing life expectancy, yet, with dramatic rollbacks (their harbingers we observed during the COVID-19 pandemic), will continue after the Cybernetic Revolution is complete. However, in order to achieve a radical life extension, it will be necessary to considerably increase the technological component of assistance to the elderly, the disabled, the infirm and other categories needing care and supervision. This is not just about more technologically advanced medicine, where many self-regulating systems will be involved, but about the situation when the *biotechnological environment will become permanent so that life without it may become impossible* (just as today we feel uncomfortable without the Internet and certain gadgets). This, of course, together with the increasing duration and comfort of biological life, will also create many insecurities and problems, which we will discuss in another chapter. *Against the background of the increasing conservatism of an aging society the need for crucial changes will become a new page in history.*

*Thus, after the completion of the Cybernetic Revolution at the mature stages of the Cybernetic Production Principle, we will witness the beginning of the new largest demographic transformation (LDT), associated with the starting accelerated creation of a sustained biotechnological environment and cyborgization,* which we discussed in Chap. 1 (the Introduction) and which we will consider in more detail.

One should note that as society ages and adapts to the aging, different scenarios may become possible in the future: from minor and controlled growth and complete stabilization of the size of the population to a more or less slow but stable depopulation. It is also possible to move toward technologies of (semi-)artificial reproduction, at least one can expect a very wide distribution of improved IVF technologies (already during the final phase of the Cybernetic Revolution).

52 In Chaps. 6 and 7, we will also analyze the major social problems associated with these changes, abuses, the danger of growing of techno-medical power and social tensions. All this can cause both great socio-political difficulties and intensification of internal social conflicts due to the growing ageism.

## 5.5 Conclusion. Major Demographic Transformations and Types of Population Reproduction in the Historical Process (Summary)

**Thus, we identify six largest world-system demographic transformations (LDT) and four types of population reproduction (TPR) within the historical process.** It is worth clarifying the correlation between LDT and TPR, explaining the difference in their numbers. Each production principle corresponds to its own type of population reproduction (see Table 5.4). The type is not formed immediately but in two stages since the technological development within the production principle has a staged development. Therefore, *for each type of population reproduction (TPR), we identify two major demographic transformations* (see Column 4 in Table 5.4). However, the first, i.e., the Hunting-Gathering, Production Principle is not associated with any world-system demographic transformation (LDT), with the possible exception of the Upper Paleolithic or Human Revolution of about 45,000 ybp (which we discussed in Chap. 1, Introduction and Chap. 2). Therefore, the remaining three TPRs (corresponding to Craft-Agrarian, Trade-Industrial and Scientific-Cybernetic Production Principles) quite logically correspond to the six largest demographic transformations. Their characteristics are systematized in Table 5.4 ("The largest world-system demographic transformations". Let us recall that the fifth LDT ("adaptation to aging") is still emerging, while the sixth is very likely but still hypothetical. Also, the fourth TPR is still evolving, although many of its features have already manifested themselves (for details see Chaps. 6 and 7).

The first two leading *world-system demographic transformations* (LDT) are related to the Craft-Agrarian Production Principle (one took place during the first two phases of the Agrarian Revolution, the second began in the final phase of the Agrarian Revolution and ended already in the mature stages of the Craft-Agrarian Production Principle). The third LDT occurred during the Trade-Industrial Production Principle and the fourth demographic transformation took place at the end of the Industrial Production Principle and the start of the Cybernetic Revolution (it involved the developing countries, as we have already pointed out earlier).

Finally, during the first two phases of the Cybernetic Revolution (the 1950s to the 2020s), the fifth LDT has been actively going on and the features of a new type of population reproduction (TPR) are being formed. By the way, we should note that some signs of the next LDT can be retrospectively guessed already at the last stages of the previous production principle. For example, in the first half of the twentieth century it is evident in a number of European countries, where the birth rate approached the replacement level—this is already typical for the Cybernetic Production Principle. Today, too, the signs of the coming demographic transformation can be traced.

And in the final phase of the Cybernetic Revolution, we assume the beginning of the transition to a new LDT-6, which will take place already in the mature phases of the Scientific-Cybernetic Production Principle. So, *for the formation of the future cybernetic society we expect a new (future) world-system demographic transformation, which we denote as* "techno-bio-medical" (see line 6 in Table 5.4).

Also, *the fourth type of population reproduction (TPR) is developing* (see line 4 in Table 5.3), although many of its features can be already identified.

The main characteristics of all four types of population reproduction (TPRs) are described in Table 5.5 ("Production principle and type of population reproduction"). The formation of the TPR took place during the whole production principle and could even partially involve the subsequent production principle (as we can see by the example of the Trade-Industrial and Scientific-Cybernetic Production Principles). *It is very important that the type of population reproduction associated with the Cybernetic Production Principle changes and will change dramatically the previous trajectory of the development of demographic trends*, replacing quantitative growth with qualitative changes in the life of the whole population and every individual. Therefore, an important conclusion can be repeated: in contrast to the rate of technological growth, which accelerates over time from phase to phase (see Grinin et al., 2020), the rate of demographic growth begins to slow down from phase to phase starting from the 1970s.

**Table 5.3** Demographic changes during the Scientific-Cybernetic Production Principle

| Phase | Dates | Techno-social changes | Demographic changes | Limitations | Population at the start of the phase |
|---|---|---|---|---|---|
| Initial phase of the cybernetic revolution | 1950s–1990s | Powerful growth of science, technology and medicine. Major political and social changes in the World System; liberation of the colonies; the rise of social policies | In developing countries, due to the reduction in mortality, a very rapid population growth began; in developed countries, after a period of rising birth rates, there was a transition to a decrease in the birth rate to the replacement level | Huge gap in levels of development between the core and the periphery of the World System; advances in medicine, the rise of female education and rights and economic policy led to declining birth rates in the developed world | 2.5 billion |
| Modernization phase of the cybernetic revolution | 1990s–2030s | Mass diffusion of ICT and AI, growth of education in the developing countries | Transition to TPR of high life expectancy and population aging in most parts of the World System | In terms of reducing the birth rate, traditional and religious ideologies and practices are a limitation | 5.27 billion[53] |
| Final phase of the cybernetic revolution | 2030s–2070s | Emergence of major breakthroughs in medicine; the impact of aging on society; struggle for a new world order | Global aging and adaptation to it; possible completion of the demographic transition in Africa | Traditional and religious ideologies in Africa and parts of Islamic countries | 8.37–8.65 billion (forecast)[54] |
| Mature phases of the Scientific-Cybernetic Production Principle | 2070s–2120s | Growth of conservatism, the final formation of new social institutions associated with aging | Noticeable increase in life expectancy; growth of adaptability to aging and disability due to the mass distribution of techno-biological technologies among the population[55] | Conservatism of the aging society; dangers of excessive techno-medical power; unpredictable dangers | 10–12 billion (according to different forecasts; see also Table 5.6) |

53 Akimov (1999); Shcherbakova (2011).

54 UN Population Division (2023).

55 Among all population groups, but especially among the elderly, disabled and incapacitated (requiring constant care).

**Table 5.4** The largest world-system demographic transformations (LDTs)

| N | Name | Description | Type of population reproduction | Correspondence to the stages of the production principle | Date[56] | Part of the World-System |
|---|---|---|---|---|---|---|
| 1 | Neolithic | Transition from medium–high fertility among hunter-gatherers to a model of high fertility among primitive farmers with a transition from medium–high to high mortality[57]; although generally small, but more or less constant natural increase appeared. But this is a model of high fertility with a limited ecological niche | Slow growth | Initial and modernization phase of the Agrarian revolution | 10–5 thousand years ago | Different regions of Eurasia and North Africa[58] |

(continued)

[56] Dates are given before the beginning of the next world-system demographic transformation while previous world-system demographic transformations are continued in different parts of the World System, coexisting with the new one.

[57] At the same time, the birth rate increased significantly more than the death rate.

[58] = Afroeurasian world system.

**Table 5.4** (continued)

| N | Name | Description | Type of population reproduction | Correspondence to the stages of the production principle | Date | Part of the World-System |
|---|---|---|---|---|---|---|
| 2 | Civilizational | Development of the established model with a high birth rate and the achievement[59] of a somewhat lower mortality rate for a significant part of population on the basis of a capacious ecological niche, which made it possible to increase natural growth and consequently the population growth; as well as improve the quality of life | Slow growth | Final phase of the Agrarian revolution and mature stages of the Craft-Agrarian Production Principle | 5000–500 years ago | In the most part of the Afrasian world-system and partly in American world-systems |
| 3 | Anti-Malthusian | Transition to a model of reduced mortality while maintaining a relatively high birth rate; this resulted in a high natural growth rate | Fast growth | Final phase of the Industrial Revolution and mature stages of Industrial Production Principles | 1700s–1930s | In Europe, North America and several other places |

(continued)

[59] In a number of cases.

**Table 5.4** (continued)

| N | Name | Description | Type of population reproduction | Correspondence to the stages of the production principle | Date | Part of the World-System |
|---|---|---|---|---|---|---|
| 4 | Third World | Transition to a model of considerably reduced mortality while maintaining a high birth rate; all this led to an explosive global population growth | Fast growth | Final stage of the Industrial Production Principle and initial phase of the Cybernetic Revolution | 1890–1980 | In the developing countries |
| 5 | Adaptation to Aging | Transition to a model of an elderly society with relatively low mortality and very low birth rates, minimal growth and an increase in the share of the elderly and old people | Qualitative changes | Initial, modernization and final phase of the Cybernetic Revolution | 1960–2050 | In the majority of the World System countries |
| 6 | Techno-bio-medical | Transition to the model of total technological support for health and quality of biological life with the stabilization (or reduction) of the population, with a very large proportion of the elderly people | Qualitative changes | Final phase of the Cybernetic Revolution, mature stages of the Cybernetic Production Principle and probably beyond it | 2060s–2220s | within the World System |

**Table 5.5** Production principles and types of population reproduction

| *N* | *Production Principle* | *Type of Population Reproduction (TPR)* | *Birth rate* | *Death Rate* | *Natural increase* | *Life expectancy* | *Settlement type* | *Quality of biological life*[60] |
|---|---|---|---|---|---|---|---|---|
| 1 | Hunter-gatherer | Simple reproduction (natural) | Medium–high[61] | Medium–high for all population groups[62] | Very low | Low | No permanent settlements | Low |
| 2 | Craft-agrarian | Slow growth (socio-natural) | Very high[63] | High (especially for children)[64] | Low with a tendency to slow acceleration | Average low; but higher for adults with comparison to hunter-gatherers | Rural settlements; but the number of dwellers is gradually growing | Low (while improving in some aspects) |
| 3 | Trade-industrial | Fast growth (socio-quantitative) | High | Medium–low | High | Medium–high | Cities | Medium |
| 4 | Scientific-cybernetic | Qualitative changes (socio-qualitative) | Low | Low | Low | High and very high | Deurbanization | High |

[60] Includes a certain comfort level, the development of medicine and care services, including for the elderly and the disabled.

[61] Twenty-five–thirty life births per thousand per year.

[62] Twenty-five–thirty deaths per thousand per year.

[63] More than 45 life births per thousand per year.

[64] More than 40 deaths per thousand per year.

**Table 5.6** Dynamics of the global population growth

| Date | World population | Time to grow by 1 billion |
|---|---|---|
| 1801 | 1 billion | >200,000 years |
| 1925 | 2 billion | 124 years |
| 1960 | 3 billion | 35 years |
| 1974 | 4 billion | 14 years |
| 1987 | 5 billion | 13 years |
| 1998 | 6 billion | 11 years |
| 2010 | 7 billion | 12 years |
| 2022 | 8 billion | 12 years |
| 2037 | 9 billion | 15 years |
| 2058 | 10 billion | 21 years |
| ? | 11 billion | ? |

*Data sources* Akimov (1999) (till 1950); UN Population Division (2023) (empirical estimates forv1950–2021; medium projection for 2022–2100)

**Acknowledgements** This chapter is an output of a research project implemented as part of the Basic Research Program at HSE University in 2024 with support by the Russian Science Foundation (Project No. 23-18-00535).

## Appendix

See Table 5.6 and Figs. 5.3, 5.4, 5.5, 5.6 and 5.7.

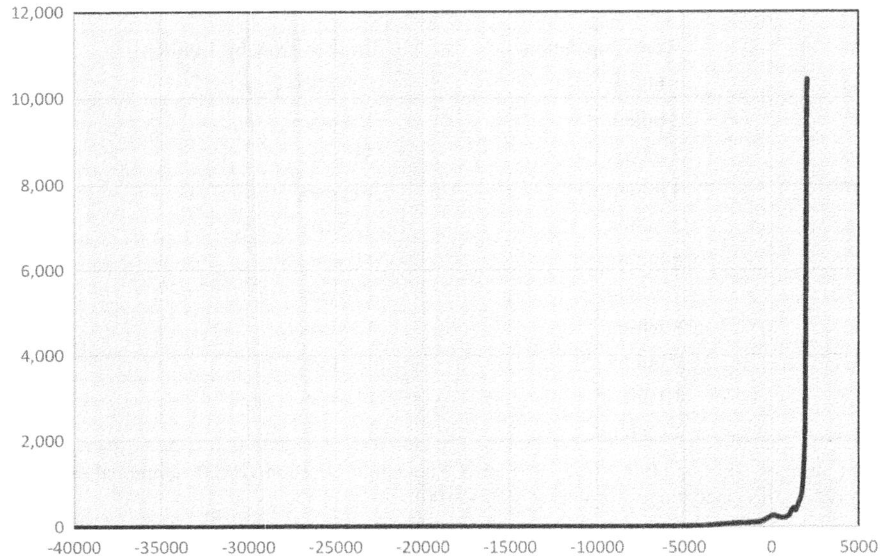


**Fig. 5.3** World population dynamics, millions, 40,000 BCE–2021 CE, with the UNPD medium projection till 2100. *Data sources* Kremer (1993), Livi-Bacci (2017), Korotayev (2020a), UN Population Division (2023)

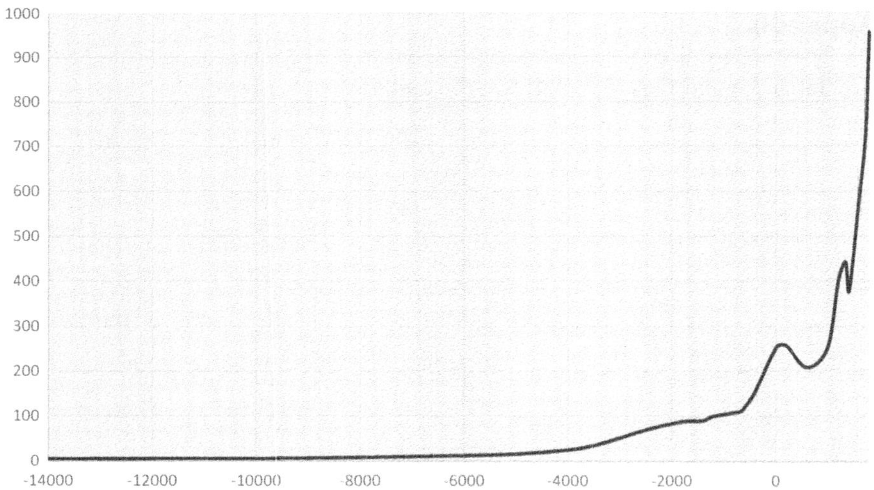


**Fig. 5.4** World population dynamics, millions, 14,000 BCE–1800 CE, natural scale. *Data sources* Livi-Bacci (2017), Korotayev (2020a)

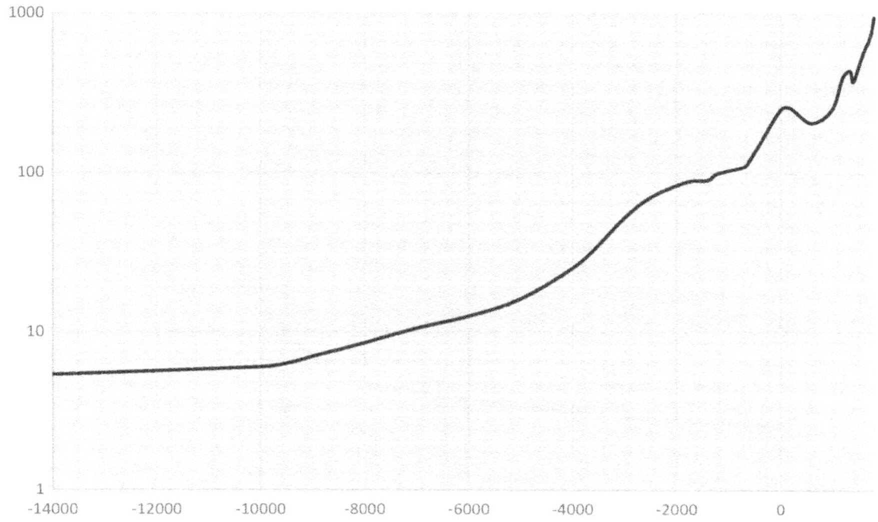


**Fig. 5.5** World population dynamics, millions, 14,000 BCE–1800 CE, logarithmic scale. *Data sources* Kremer (1993), Livi-Bacci (2017), Korotayev (2020a)

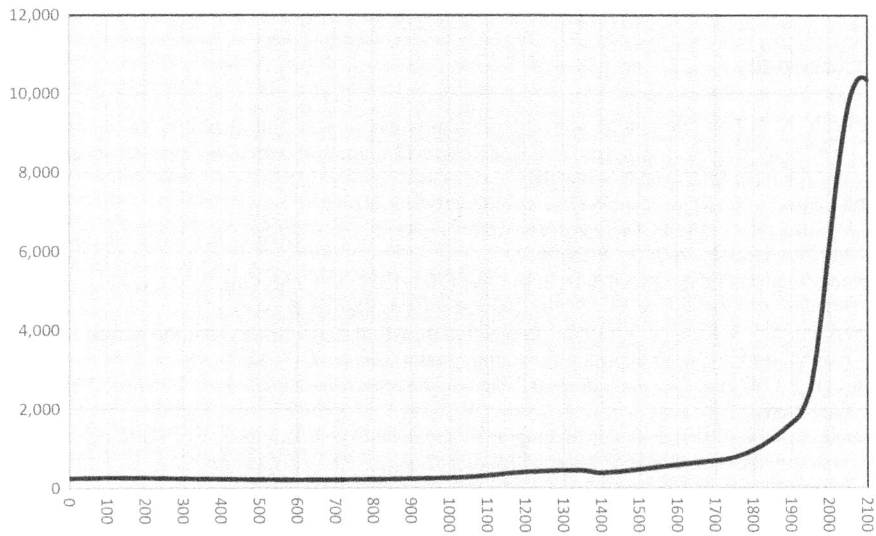


**Fig. 5.6** World population dynamics, millions, 1–2021 CE, with the UNPD medium projection till 2100. *Data sources* Kremer (1993), Livi-Bacci (2017), Korotayev (2020a), UN Population Division (2023)

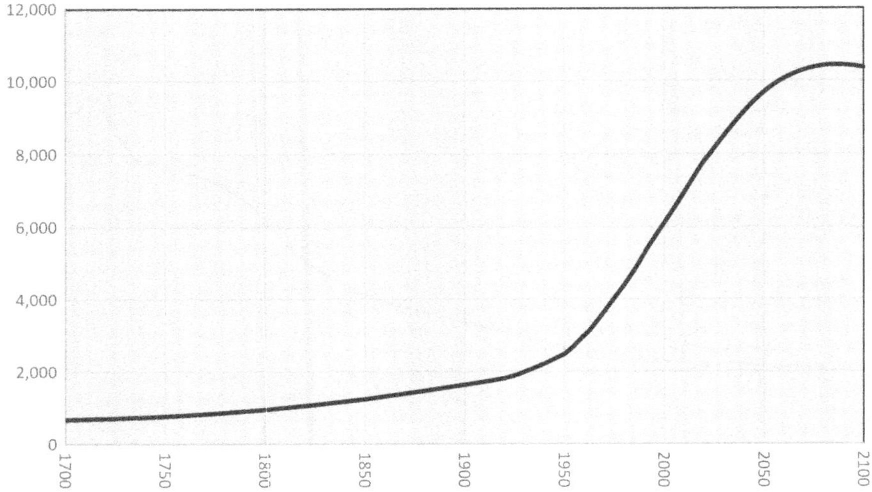


**Fig. 5.7** World population dynamics, millions, 1700–2021 CE, with the UNPD medium projection till 2100. *Data sources* Kremer (1993), Livi-Bacci (2017), Korotayev (2020a), UN Population Division (2023)

## References


Akimov, A. (1999). Na poroge sed'mogo milliarda. Naseleniye i obshchestvo. *Informatsionnyy byulleten' tsentra demografii i ekologii cheloveka instituta narodokhozyaystvennogo prognozirovaniya RAN* № 39, September.

Andrianov, B. V. (1978). *Zemledeliye nashikh predkov*. Nauka.

Arab-Ogly, E. A. (1978). *Demograficheskiye i ekologicheskiye prognozy. Kritika sovremennykh burzhuaznykh kontseptsiy*. Statistika.

Armengaud, A. (1976). Population in Europe 1700–1914. In C. M. Cipolla (ed.), *The industrial revolution, 1700–1914* (pp. 22–76). Harvester Press, Barnes & Noble.

Artzrouni, M., & Komlos, J. (1985). Population growth through history and the escape from the malthusian trap: A homeostatic simulation model. *Genus, 41*(3–4), 21–39.

Bairoch, P. (1971). *Le tiers-monde dans l'impasse. Le démarrage économic du XVIIIe au XXe siècle*. Gallimard.

Barsukov, V. N. (2019). From the demographic dividend to population ageing: World trends in the system-wide transition. *Economic and Social Changes: Facts, Trends, Forecast, 12*(4), 167–182. https://doi.org/10.15838/esc.2019.4.64.11

Biraben, J.-N. (1980). An essay concerning Mankind's evolution. *Population, 4*, 1–13.

Bloom, D. E., & Canning, D. (2008). Global demographic change: Dimensions and economic significance. *Population and Development Review, 34*, 17–51.

Bloom, D. E., & Williamson, J. G. (1998). Demographic transitions and economic miracles in emerging Asia. *World Bank Economic Review, 12*(3), 419–455.

Bloom, D. E., Canning, D., Fink, G., & Finlay, J. (2007). Does age structure forecast economic growth? *International Journal of Forecasting, 23*(4), 569–585. https://doi.org/10.1016/j.ijforecast.2007.07.001

Bocquet-Appel, J. P. (2011). When the world's population took off: The springboard of the neolithic demographic transition. *Science, 333*(6042), 560–561.

Borsch, S. J. (2004). Environment and population: The collapse of large irrigation systems reconsidered. *Comparative Studies of Society and History, 46*(3), 451–468.

Borsch, S. J. (2005). *The black death in Egypt and England*. The American University of Cairo Press.

Boserup, E. (1965). *The conditions of agricultural growth: The economics of agrarian change under population pressure*. Aldine.

Bray F. (1984). Agriculture. In Needham J. (ed.), *Science and Civilization in China*. Vol. 6. Part. 2. *Biology and Biological Technology*. Cambridge, MA: Cambridge University Press.

Bruxelles, L., & Jarry, M. (2011). Climatic conditions, settlement patterns and cultures in the Paleolithic: The example of the Garonne Valley (southwest France). *Journal of Human Evolution, 61*(5), 538–548.

Caldwell, J. C., Caldwell, B. K., Caldwell, P., McDonald, P. F., & Schindlmayr, T. (2006). Demographic transition theory. *Springer*. https://doi.org/10.1007/978-1-4020-4498-4

Carneiro, R. (1970). Theory of the origin of the state. *Science, 169*, 733–738.

Carneiro, R. (1978). Political expansion as an expression of the principle of competitive exclusion. In R. Cohen & E.R. Service (eds.), *Origins of the state* (pp. 205–223). Institute for the Study of Human Issues.

Carneiro, R. (1987). The evolution of complexity in human societies and its mathematical expression. *International Journal of Comparative Sociology, 28*(3–4), 111–128.

Chase-Dunn, Ch., & Hall, T. D. (1997). *Rise and demise: Comparing world-systems*. Westview Press.

Chesnais, J. C. (1992). *The demographic transition: Stages, patterns, and economic implications: A logitudinal study of sixty seven countries the period 1720–1984*. Clarendon Press.

Chesnais, J. C. (1998). Naseleniye SSHA posle 1945 goda. (La population des États-Unis depuis 1945). *Naseleniye i obshchestvo. Informatsionnyy byulleten' tsentra demografii i ekologii cheloveka instituta narodokhozyaystvennogo prognozirovaniya RAN*, August. http://www.demoscope.ru/acrobat/ps29.pdf

Childe, V. G. (1942). *What happened in history*. Penguin Books.

Choucri, N. (1974). *Population dynamics and international violence: Propositions, insights, and evidence*. Lexington Books.

Christian, D. (2004). *Maps of time: An introduction to big history*. University of California Press.

Chu, C. Y. C., & Lee, R. D. (1994). Famine, revolt, and the dynastic cycle: Population dynamics in historic China. *Journal of Population Economics, 7*(4), 351–378.

Chumakov, A. N., & Yurchenko, P. S. (2021). Global pandemic in terms of objective causes and subjective speculations. *Journal of Globalization Studies, 12*(2), 157–168.

Claessen, H. J. M. (2000). Problems, paradoxes, and prospects of evolutionism. In N.N. Kradin, A.V. Korotayev, D.M. Bondarenko, V. de Munck, & P.K. Wason (Eds.), *Alternatives of social evolution* (pp. 1–11). FEB RAS.

Claessen, H. J. M., & van de Velde, P. (1985). The evolution of sociopolitical organization. In H. J. M. Claessen, P. van de Velde, & E. M. Smith (Eds.), *Development and decline. The evolution of sociopolitical organization* (pp. 1–12). Bergin & Garvey.

Clark, C. (1967). *Population growth and land use*. St. Martin's Press.

Clark, G. (2007). *A farewell to alms: A brief economic history of the world*. Princeton University Press.

Clark, G. A., Barton, C. M., & Straus, L. G. (2019). Landscapes, climate change & forager mobility in the Upper Paleolithic of northern Spain. *Quaternary International, 515*, 176–187.

Coale, A. (1983). Frank W. Notestein, 1902–1983. *Population Index, 49*, 3–12.

Coale, A. (2013). K 30-letiyu so dnya smerti Frenka Noutstayna. *Demoskop*, 575–576, November 18–December 1. http://www.demoscope.ru/weekly/2013/0575/nauka01.php

Cohen, M. N. (1998). Were early agriculturalists less healthy than food collectors? In C. R. Ember, M. Ember, & P. N. Peregrine (Eds.), *Research frontiers in anthropology* (pp. 61–83). Prentice Hall.

Cohen, M. N. (2009). Introduction: Rethinking the origins of agriculture. *Current Anthropology, 50*(5), 591–595. https://doi.org/10.1086/603548
Cohen, M. N., & Armelagos, G. J. (Eds.). (1984). *Paleopathology at the origins of agriculture*. Academic Press.
Cohen, M. N., & Crane-Kramer, G. (Eds.). (2007). *Bioarchaeological interpretations of the human past: Local, regional, and global perspectives series*. University Press of Florida.
Davis, S. (2002). *Urbanizatsiya kak spontannyy poryadok*. https://old.inliberty.ru/library/457-urbanizaciya-kak-spontannyy-poryadok
Deane, P. (1965). *The first industrial revolution*. University Press.
Dolghikh, B. O. (1960). *Rodovoy i plemennoy sostav narodov Sibiri v XVII v*. Izd-vo AN SSSR.
Durand, J. D. (1974). *Historical estimates of world population: An evaluation* (p. 10). University of Pennsylvania.
Durand, J. D. (1977). Historical estimates of world population. An evaluation. *Population and Development Review, 3*(3), 253–296.
Dyson, T. (2010). *Population and development. The demographic transition*. Zed Books.
Ehrlich, P. R. (1975). *The population bomb*. Rivercity Press.
Efremov, I., & Arkhangelsky, V. (2023). Global fertility changes in the context of the COVID-19 pandemic. *Journal of Globalization Studies, 14*(2), 165–170.
FAO. (2004). *The state of food and agriculture 2003–2004*. Food and Agriculture Organization of the United Nations.
Fischer, G. (1993). The population explosion: Where is it leading? *Population and Environment, 15*(2), 139–153.
Frankema, E. (2019). Africa and the demographic consequences of the columbian exchange. *Asian Review of World Histories, 7*(1–2), 66–79.
Fried, M. H. (1967a). *The evolution of political society. An essay in political anthropology*. Random House.
Fried, M. H. (1967b). On the concepts of 'tribe' and 'tribal society.' In J. Helm (Ed.), *Essays on the problem of tribe* (pp. 3–20). American Ethnological Society.
Galesi, L. (2022). Maize on the move: The diffusion of a tropical cultivar across Europe. *Environment and History*. https://doi.org/10.3197/096734021X16076828553386
Goldstone, J. A. (2001). Demography, environment, and security: An overview. In M. Weiner & S. S. Russell (Eds.), *Demography and national security* (pp. 38–61). Berghahn Books.
Goldstone, J. (2002). Population and security: How demographic change can lead to violent conflict. *Journal of International Affairs, 56*(1), 3–21.
Goldstone, J. A. (2016). *Revolution and rebellion in the early modern world: Population change and state breakdown in England, France, Turkey, and China, 1600–1850*. 2nd ed. Routledge.
Goldstone, J. A., Kaufmann, E. P., & Toft, M. D. (Eds.). (2012). *Political demography: How population changes are reshaping international security and national politics*. Oxford University Press.
Grinin, L. E. (2003). The early state and its analogues. *Social Evolution and History, 2*(1), 131–176.
Grinin, L. E. (2004a). Democracy and early state. *Social Evolution and History, 3*(2), 93–149.
Grinin, L. E. (2004b). Early state and democracy. In L.E. Grinin, R.L. Carneiro, D.M. Bondarenko, & A.V. Korotayev (Eds.), *The early state, its alternatives and analogues* (pp. 419–463). Uchitel.
Grinin, L. E. (2006a) *Proizvoditel'nyye sily i istoricheskiy protsess*. 3-ye izd. Uchitel'.
Grinin, L. E. (2006b). Proizvoditel'nyye sily kak sotsioyestestvennaya kategoriya. In E.S. Kulpin (Ed.), *Chelovek i priroda: iz proshlogo v budushcheye* (pp. 200–217).
Grinin, L. E. (2007). Nekotoryye razmyshleniya po povodu prirody zakonov, svyazannykh s demograficheskimi tsiklami (k postanovke problemy opredeleniya obshchikh metodologicheskikh podkhodov k analizu demograficheskikh tsiklov). In P.V. Turchin, L. Ye. Grinin, S. Yu. Malkov, & A.V. Korotayev (Eds.), *Istoriya i matematika: Kontseptual'noye prostranstvo i napravleniya poiska* (pp. 219–246). LKI/ URSS.
Grinin, L. E. (2008). Early state, developed state, mature state: Statehood evolutionary sequence. *Social Evolution & History, 7*(1), 67–81.

Grinin, L. E. (2009). The pathways of politogenesis and models of the early state formation. *Social Evolution and History, 8*(1), 92–132.

Grinin, L. E. (2010). *Gosudarstvo i istoricheskiy protsess: Evolyutsiya gosudarstvennosti: Ot rannego gosudarstva k zrelomu.* 2-ye izd., ispr. LIBROKOM.

Grinin, L. E. (2011a). *Gosudarstvo i istoricheskiy protsess. Epokha formirovaniya gosudarstva. Obshchiy kontekst sotsial'noy evolyutsii pri obrazovanii gosudarstva.* 2-ye izd. LKI.

Grinin, L. E. (2011b). Istoki globalizatsii: Mir-sistemnyy analiz. *Vek globalizatsii, 1*, 80–94.

Grinin, L. E. (2012a). *Ot Konfutsiya do Konta: Stanovleniye teorii, metodologii i filosofii istorii.* LKI.

Grinin, L. E. (2012b). Afroyevraziyskaya mir-sistema i istoki globalizatsii. *Naslediye L. N. Gumileva i sud'by narodov Yevrazii: istoriya, sovremennost', perspektivy: Sbornik statey Mezhdunarodnogo nauchnogo kongressa, posvyashchennogo 100-letiyu so dnya rozhdeniya L. N. Gumileva*, Sankt-Peterburg, 1–3 oktyabrya 2012 g. — SPb.: Izd-vo RGPU im. A. I. Gertsena (pp. 211–224).

Grinin, L. E. (2012c). State and socio-political crises in the process of modernization. *Cliodynamics, 3*(1), 124–157 https://doi.org/10.21237/C7clio3112323

Grinin, L. (2013a). State and socio-political crises in the process of modernization. *Social Evolution & History, 12*(2), 35–76.

Grinin, L. (2013b). Gosudarstvo i krizisy v protsesse modernizatsii. *Filosofiya i Obshchestvo, 3*, 29–59.

Grinin, L. (2017). Rossiyskaya revolyutsiya v svete teorii modernizatsii. *Istoriya i Sovremennost, 2*, 22–57.

Grinin, L. (2020). How can COVID-19 change geopolitics and economy? *Journal of Globalization Studies, 11*(2), 121–134.

Grinin, L. E. (2022). Revolutions and modernization traps. In J.A. Goldstone, L. Grinin, & A. Korotayev (Eds.), *Handbook of revolutions in the 21st century: The new waves of revolutions, and the causes and effects of disruptive political change* (pp. 219–238). Springer. https://doi.org/10.1007/978-3-030-86468-2_8

Grinin, L. E., & Grinin, A. L. (2015). *Ot rubil donanorobotov. Mirnaputikepokhesamoupravlyaemykh sistem (istoriya tekhnologiy i opisanie ikh budushchego).* Moscow Publishing House 'Uchitel'.

Grinin, L. E., & Grinin, A. L. (2023). Demographic transformations in the historical process in the light of technological development: Theoretical approach. *Social Evolution & History, 22*(2), 159–202.

Grinin, L., Grinin, A., & Korotayev, A. (2020). A Quantitative analysis of worldwide long-term technology growth: From 40,000 BCE to the early 22nd century. *Technological Forecasting and Social Change, 155*(June), 119955. https://doi.org/10.1016/j.techfore.2020.119955

Grinin, L., Grinin, A., & Korotayev, A. (2022). COVID-19 pandemic as a trigger for the acceleration of the cybernetic revolution, transition from e-government to e-state, and change in social relations. *Technological Forecasting and Social Change, 175*, 121348. https://doi.org/10.1016/j.techfore.2021.121348

Grinin, L., Grinin, A., & Korotayev, A. (2023). Demographic transformations in the light of technological development: Types of demographic reproduction in the past and in the future. *Social Evolution & History, 22*(2), 159–202.

Grinin, L. E., & Korotayev, A. V. (2009a). *Global'nyy krizis v retrospektive. Kratkaya istoriya pod"yemov i krizisov: ot Likurga do Alana Grinspena.* Librokom/URSS.

Grinin, L. E., & Korotayev, A. V. (2009b). *Sotsial'naya makroevolyutsiya. Genezis i transformatsii Mir-Sistemy.* LIBROKOM.

Grinin, L. E., & Korotayev, A. V. (2012a). The Afroeurasian world-system: Genesis, transformations, characteristics. In S.J. Babones & Ch. Chase-Dunn (Eds.), *Routledge handbook of world-systems analysis* (pp. 30–41). Routledge.

Grinin, L., & Korotayev, A. (2012b). *Tsikly, krizisy, lovushki sovremennoy Mir-Sistemy. Issledovaniye kondrat'yevskikh, zhyuglyarovskikh i vekovykh tsiklov, global'nykh krizisov, mal'tuzianskikh i postmal'tuzianskikh lovushek.* LKI.

Grinin, L. E., & Korotayev, A. V. (2015). *Great divergence and great convergence.* Springer International Publishing.
Grinin, L. E., & Korotayev, A. V. (2016). *Blizhniy Vostok, Indiya i Kitay v globalizatsionnykh protsessakh.* Mosk. red. izd-va Uchitel'.
Grinin, L. E., & Korotayev, A. V. (2018). Origins of globalization in the framework of the Afroeurasian world-system history. In T.D. Hall (Ed.), *comparing globalizations, world-systems evolution and global Futures* (pp. 37–70). Springer. https://doi.org/10.1007/978-3-319-68219-8_3
Grinin, L., & Korotayev, A. (2023). Africa – The continent of the future. Challenges and opportunities. In V. Sadovnichy et al. (Eds.), *Reconsidering the limits to growth. A report to the Russian Association of the Club of Rome* (pp. 225–238). Springer. https://doi.org/10.1007/978-3-031-34999-7_13
Grinin, L. E., Korotayev, A. V., & Malkov, S. Yu. (2008a). Matematicheskiye modeli sotsial'no-demograficheskikh tsiklov i vykhoda iz mal'tuzianskoy lovushki: nekotoryye vozmozhnyye napravleniya dal'neyshego razvitiya. In G.G. Malinetskiy & A.V. Korotayev (Eds.), *Problemy matematicheskoy istorii. Matematicheskoye modelirovaniye istoricheskikh protsessov* (pp. 78–117). Librokom/URSS.
Grinin, L. E., Korotayev, A. V., & Malkov, S. Yu. (2008b). Nekotoryye vozmozhnyye napravleniya dal'neyshego razvitiya matematicheskikh modeley sotsial'no-demograficheskikh tsiklov. In L.I. Borodkin (ed.), *Informatsionnyy byulleten' Assotsiatsii "Istoriya i komp'yuter"* № 35. Materialy XI konferentsii AIK. Dekabr' (pp. 30–31). Azbuka.
Grinin, L. E.., Malkov, S. Yu., Gusev, V. A., & Korotayev, A. V. (2009). Nekotoryye vozmozhnyye napravleniya razvitiya teorii sotsial'no-demograficheskikh tsiklov i matematicheskiye modeli vykhoda iz "mal'tuzianskoy lovushki". In S. Yu. Malkov, L. Ye. Grinin, & A.V. Korotayev (Eds.), *Istoriya i matematika. Protsessy i modeli* (pp. 134–210). Knizhnyy dom "Librokom".
Groth, H., May, J. F., & Turbat, V. (2019). Policies needed to capture a demographic dividend in Sub-Saharan Africa. *Canadian Studies in Population, 46*(1), 61–72. https://doi.org/10.1007/s42650-019-00005-8
Guschina, Yu. A., & Malkov, S. Yu. (2014). K voprosu o modelirovanii demograficheskikh tsiklov i o vykhode iz mal'tuzianskoy lovushki. In L. Ye. Grinin & A. V. Korotayev (Eds.), *Istoriya i matematika: Sotsial'no-ekonomicheskiye protsessy* (pp. 29–42). Uchitel'.
Hawksworth, J., & Cookson, G. (2008). *The world in 2050. Beyond the BRICs: A broader look at emerging market growth prospects.* Pricewaterhouse Coopers.
Ho Ping-ti. (1956). Early-ripening rice in Chinese history. *Economic History Review, 9*, 200–218.
Huang, P. C. C. (2002). Development or involution in eighteenth-century Britain and China? A review of Kenneth Pomeranz's "The great divergence: China, Europe, and the making of the modern world economy." *The Journal of Asian Studies, 61*(2), 501–538.
Hussain, M. (2021). The changing dynamics of liberal international order: Covid-19 and its global implications. *Journal of Globalization Studies, 12*(2), 143–156.
Ibn Khaldūn, & 'Abd al-Rahman. (1958). *The Muqaddimah: An introduction to history*, Bollingen Series (Vol. 43). Pantheon Books.
Ibn Khaldūn, & 'Abd al-Rahman. (1980). Prolegomeny k "Knige pouchitel'nykh primerov...." In A. A. Ignatenko (Ed.), *Ibn-Khal'dun* (pp. 121–155). Mysl'.
Ibn Khaldūn, & 'Abd al-Rahman. (2008). Vvedeniye (al-Mukaddima). In A. V. Smirnova (Ed.), *Istoriko-filosofskiy yezhegodnik* (pp. 187–217). Nauka.
Ilyushechkin, V. P. (1986). *Soslovno-klassovoye obshchestvo v istorii Kitaya (opyt sistemno-strukturnogo analiza).* Nauka.
Irshad, S. M. (2020). COVID-19 and global 'at-risk community': From benefit-sharing to risk-sharing of economic crisis. *Journal of Globalization Studies, 11*(2), 135–145.
Kakkar, S. (2021). The outbreak of sars-cov-2 (Covid-19): Impact on international tourism. *Journal of Globalization Studies, 12*(1), 145–158.
Kapitza, S. P. (2006). Ob uskorenii istoricheskogo vremeni. In L.E. Grinin, A.V. Korotayev, & S. Yu. Malkov (Eds.), *Istoriya i Matematika: Problemy periodizatsii istoricheskikh makroprotsessov* (pp. 12–30). Yeditorial URSS.

Kapitza, S. P. (2009). *Obshchaya teoriya rosta chelovechestva. Kak ros i kuda idet mir cheloveka.* Al'pina non-fikshn.
Kapitza, S. P., Kurdyumov, S. P., & Malinetskiy, G. G. (2003). *Sinergetika i prognozy budushchego.* Izd. 2. Yeditorial URSS.
Komlos, J., & Artzrouni, M. (1990). Mathematical investigations of the escape from the Malthusian trap. *Mathematical Population Studies, 2*, 269–287.
Komlos, J., & Nefedov, S. (2002). A compact macromodel of pre-industrial population growth. *Historical Methods, 35*, 92–94.
Kögel, T., & Prskawetz, A. (2001). Agricultural productivity growth and escape from the Malthusian trap. *Journal of Economic Growth, 6*, 337–357.
Korotayev, A. V. (1997). *Faktory sotsial'noy evolyutsii.* Insitutet vostokovedeniya.
Korotayev, A. V. (2003a). George Peter Murdock i shkola kolichestvennykh kross-kul'turnykh (kholokul'turnykh) issledovaniy. In: *Sotsial'naya evolyutsiya: faktory, nablyudeniya, nablyudeniya* (pp. 478–555). Vost. lit-ra.
Korotayev, A. V. (2003b). *Sotsial'naya evolyutsiya: faktory, zakonomernosti, tendentsii.* Vost. lit-ra.
Korotayev, A. (2005). A compact macromodel of world system evolution. *Journal of World-Systems Research, 11*(1), 79–93. https://doi.org/10.5195/jwsr.2005.401
Korotayev, A. V. (2006). *Dolgosrochnaya politiko-demograficheskaya dinamika Yegipta: Tsikly i tendentsii.* Vost. lit-ra.
Korotayev, A. (2007). Compact mathematical models of world system development, and how they can help us to clarify our understanding of globalization processes. In G. Modelski, T. Devezas, & W. Thompson (Eds.), *Globalization as evolutionary process: Modeling global change* (pp. 133–160). Routledge.
Korotayev, A. (2017). Malthusian cycle. In F. M. Moghaddam (Ed.), *The SAGE encyclopedia of political behavior* (pp. 457–462). SAGE. https://doi.org/10.4135/9781483391144.n208
Korotayev, A. (2020a). How singular is the 21st century Singularity?. In A. Korotayev & D. LePoire (Eds.), The *21st century singularity and global futures. A big history perspective* (pp. 571–595). Springer. https://doi.org/10.1007/978-3-030-33730-8_26
Korotayev, A. (2020b). The 21st century singularity in the big history perspective. A re-analysis. In A. Korotayev, & D. LePoire (Eds.), *The 21st century Singularity and global futures. A big history perspective* (pp. 19–75). Springer. https://doi.org/10.1007/978-3-030-33730-8_2
Korotayev, A., Goldstone, J. A., & Zinkina, J. (2015). Phases of global demographic transition correlate with phases of the great divergence and great convergence. *Technological Forecasting and Social Change, 95*, 163–169. https://doi.org/10.1016/j.techfore.2015.01.017
Korotayev, A., Grinin, L., & Grinin, A. (2021). Mathematical model of interaction between civilization center and tribal periphery: A description. *Social Evolution & History, 20*(2), 50–78. https://doi.org/10.30884/seh/2021.02.03
Korotayev, A., Grinin, L., & Grinin, A. (2022). Mathematical model of interaction between civilization center and tribal periphery: An analysis. *Social Evolution & History, 21*(1), 65–100.https://doi.org/10.30884/seh/2022.01.03
Korotayev, A., & Khaltourina, D. (2006). *Introduction to social macrodynamics: Secular cycles and millennial trends in Africa.* KomKniga/URSS.
Korotayev, A., & Komarova, N. (2004). A new mathematical model of pre-industrial demographic cycle. In M. G. Dmitriev & A. P. Petrov (Eds.), *Mathematical modeling of social and economic dynamics* (pp. 157–163). Russian State Social University.
Korotayev, A., Malkov, A., & Khaltourina, D. (2006a). *Introduction to social macrodynamics: Compact macromodels of the world system growth.* KomKniga/URSS.
Korotayev, A., Malkov, A., & Khaltourina, D. (2006b). *Introduction to social macrodynamics: Secular cycles and Millennial trends.* KomKniga/URSS.
Korotayev, A. V., Komarova, N. L., & Khalturina, D. A. (2007a). *Zakony istorii: Vekovyye tsikly i tysyacheletniye trendy. Demografiya, ekonomika, voyny.* KomKniga.
Korotayev, A. V., Malkov, A. S., & Khalturina, D. A. (2007b). *Zakony istorii: Matematicheskoye mode-lirovaniye razvitiya Mir-Sistemy. Demografiya, ekonomika, kul'tura.* KomKniga/URSS.

Korotayev, A., Malkov, S., & Grinin, L. (2014). A trap at the escape from the trap? Some demographic structural factors of political instability in modernizing social systems. In L.E. Grinin & A.V. Korotayev (Eds.), *History & mathematics: Trends and cycles* (pp. 201–267). 'Uchitel' Publishing House.

Korotayev, A., Malkov, S., & Musieva, J. (2023a). Demography. Toward optimization of demographic processes. In: V. Sadovnichy et al. (Eds.), *Reconsidering the limits to growth. A report to the Russian Association of the Club of Rome* (pp. 97–116). Springer. https://doi.org/10.1007/978-3-031-34999-7_6

Korotayev, A., Romanov, D., Zinkina, J., & Slav, M. (2022b). Urban youth and terrorism: A quantitative analysis (are youth bulges relevant anymore?). *Political Studies Review*. https://doi.org/10.1177/14789299221075908

Korotayev, A., Sawyer, P., Gladyshev, M., Romanov, D., & Shishkina, A. (2021b). Some sociodemographic factors of the intensity of anti-government demonstrations: Youth bulges, urbanization, and protests. *Sociologiceskoe Obozrenie = Russian Sociological Review, 20*(3), 98–128. https://doi.org/10.17323/1728-192x-2021-3-98-128

Korotayev, A., Shulgin, S., Ustyuzhanin, V., Zinkina, J., & Grinin, L. (2023b). Modeling social self-organization and historical dynamics. Africa's futures. In V. Sadovnichy et al. (Eds.), *Reconsidering the limits to growth. A report to the Russian Association of the Club of Rome* (pp. 461–490). Springer. https://doi.org/10.1007/978-3-031-34999-7_20

Korotayev, A., Shulgin, S., Zinkina, J., & Slav, M. (2022c). Estimates of the possible economic effect of the demographic dividend for sub-saharan Africa for the period up to 2036. *Vostok (Oriens)*, 2, 108–123. https://doi.org/10.31857/S086919080019128-7

Korotayev, A., & Zinkina, J. (2014). How to optimize fertility and prevent humanitarian catastrophes in Tropical Africa. *African Studies in Russia, 6*, 94–107.

Korotayev, A., & Zinkina, J. (2015). East Africa in the Malthusian trap? *Journal of Developing Societies, 31*(3), 385–420. https://doi.org/10.1177/0169796X15590322

Korotayev, A., & Zinkina, J. (2017). Systemic boundary issues in the light of mathematical modeling of world-system evolution. *Journal of Globalization Studies, 8*(1), 78–96.

Korotayev, A., & Zinkina, J. (2022). Egypt's 2011 revolution. A demographic structural analysis. In J.A. Goldstone, L. Grinin, & A. Korotayev (Eds.), *Handbook of revolutions in the 21st century: The new waves of revolutions, and the causes and effects of disruptive political change* (pp. 651–683). Springer. https://doi.org/10.1007/978-3-030-86468-2_25

Korotayev, A., Zinkina, J., & Andreev, A. (2016a). Secular cycles and millennial trends. *Cliodynamics, 7*(2), 204–216. https://doi.org/10.21237/C7CLIO7230484

Korotayev, A., Zinkina, J., Goldstone, J., & Shulgin, S. (2016b). Explaining current fertility dynamics in Tropical Africa from an anthropological perspective: A cross-cultural investigation. *Cross-Cultural Research, 50*(3), 251–280. https://doi.org/10.1177/1069397116644158

Korotayev, A., Zinkina, J., Kobzeva, S., Bogevolnov, J., Khaltourina, D., Malkov, A., & Malkov, S. (2011). A trap at the escape from the trap? Demographic-structural factors of political instability in modern Africa and West Asia. *Cliodynamics, 2*(2), 276–303. https://doi.org/10.21237/C7clio22217

Korotayev, A., Zinkina, J., & Zlodeev, D. (2018). Great divergence of the 18th century. *Cliodynamics: THe Journal of Quantitative History and Cultural Evolution, 9*(2), 108–123. https://doi.org/10.21237/C7clio8233403

Kotschy, R., Urtaza, P. S., & Sunde, U. (2020). The demographic dividend is more than an education dividend. *Proceedings of the National Academy of Sciences, 117*(42), 25982–25984. https://doi.org/10.1073/pnas.2012286111

Kozintsev, A. G. (1980). Perekhod k zemledeliyu i ekologiya cheloveka. In N. A. Butinov & A. M. Reshetov (Eds.), *Ranniye zemledel'tsy* (pp. 6–33). Nauka.

Kremer, M. (1993). Population growth and technological change: One million B.C. to 1990. *The Quarterly Journal of Economics, 108*(3), 681–716. https://doi.org/10.2307/2118405

Kychanov, Ye. I. (1986). Osnovnyye kanaly sotsial'noy mobil'nosti v Kitaye pri dinastiyakh Tan i Sun (VII–XII vv.). In G.F. Kim & K.Z. Ashrafyan (Eds.), *Klassy i sosloviya v dokapitalisticheskikh obshchestvakh Azii. Problema sotsial'noy mobil'nosti* (pp. 105–117). Nauka.

Landry, A. (1934). *La révolution démographique.*
Lapina, Z. G. (2002). Kitay v period pravleniya dinastii Sun (960—1279). In: A.V. Meliksetov (Ed.), *Istoriya Kitaya: ucheb* (pp. 199–228). 2-ye izd. Izd-vo MGU; Vysshaya shkola.
Lee, F. (2007). Climatic change and Chinese population growth. *Human Geography, 24*(3), 347–364.
Lee, H. F., & Zhang, D. D. (2010). Changes in climate and secular population cycles in China, 1000 CE to 1911. *Climate Research, 42*(3), 235–246.
Lee, R., & Mason, A. (2006). What is the demographic dividend? *Finance and Development, 43*(3), 16–17.
Lee, R., & Mason, A. (2011). *Population ageing and the generational economy: A global perspective.* Edward Elgar.
Lesthaeghe, R. (2020). The second demographic transition, 1986–2020: Sub-replacement fertility and rising cohabitation – a global update. *Genus, 76*(1), 1–38. https://doi.org/10.1186/s41118-020-00077-4
Lips, Y. U. (1954). *Proiskhozhdeniye veshchey.* Inostrannaya literatura.
Livi-Bacci, M. (2017). *A concise history of world population.* Wiley-Blackwell.
Lutz, W. (1994). *The future population of the world. What can we assume today. Heilig: How many people can the Earth feed?* Earthscan Press, L NASA.
Maddison, A. (2003). *The world economy: Historical statistics.* OECD.
Malkov, S. (2009). *Sotsial'naya samoorganizatsiya i istoricheskiy protsess: Vozmozhnosti matematicheskogo modelirovaniya.* LIBROKOM.
Malkov, S., Korotayev, A., & Davydova, O. (2022). World dynamics as an object of modeling (for the fiftieth anniversary of the first report to the Club of Rome). *Computer Research and Modeling, 14*(6), 1371–1394. https://doi.org/10.20537/2076-7633-2022-14-6-1371-1394
Malthus, T. (1798). *An essay on the principle of population, as it affects the future improvement of society with remarks on the speculations of Mr. Godwin, M. Condorcet, and Other Writers.* St. Paul's Church-yard.
Mantu, P. (1937[1906]). *Promyshlennaya revolyutsiya XVIII stoletiya v Anglii.* Sotsekgiz.
McNeill, W. H. (1998). *Plagues and peoples.* Anchor.
McNeill, W. H. (2013). *The pursuit of power: Technology, armed force, and society since AD 1000.* University of Chicago Press.
McEvedy, C., & Jones, R. (1978). *Atlas of world population history.* Facts on File.
Meadows, D. H., Meadows, D. L., Randers, J., & Behrens, W. W. (1972). *The limits to growth: A report for the Club of Rome's project on the predicament of mankind.* Universe Books.
Melyantsev, V. A. (1996). *Vostok i Zapad vo vtorom tysyacheletii.* MGU.
Minghinton, W. (1976). Patterns of demand 1750–1914. In C.M. Cipolla (Ed.), *The industrial revolution, 1700–1914* (pp. 77–186). Harvester Press, Barnes & Noble.
Mokir, D., & Fotkh, G.-I. (2012). Ekonomicheskiy rost v Yevrope v 1700–1870 gg.: teoriya i fakticheskiye svidetel'stva. *Ekonomicheskaya sotsiologiya, 13*(5), 57–101.
Moller, H. (1968). Youth as a force in the modern world. *Comparative Studies in Society and History, 10*(3), 237–260. https://doi.org/10.2307/177801
Nefedov, S. A. (2004). A model of demographic cycles in a traditional society: The case of Ancient China. *Social Evolution & History, 3*(1), 69–80.
Nefedov S. A. (2007). *Kontseptsiya demograficheskikh tsiklov.* UGGU.
Nefedov, S. A. (2013). Modeling Malthusian dynamics in pre-industrial societies. *Cliodynamics, 4*(2), 229–240. https://doi.org/10.21237/C7clio4221335
Nefedov, S. A. (2014a). Modeling Malthusian dynamics in pre-industrial societies. *Social Evolution & History, 13*(1), 172–184.
Nefedov, S. A. (2014b). Modeling Malthusian dynamics in pre-industrial societies: Mathematical modeling. In L.E. Grinin & A.V. Korotayev (Eds.), *History & mathematics: Trends and cycles* (pp. 190–220). 'Uchitel' Publishing House.
Nefedov, S. A. (2014c). Matematicheskiye modeli mal'tuzianskoy ekonomiki. In L. Ye. Grinin, & A. V. Korotayev (Eds.), *Istoriya i matematika: Aspekty demograficheskikh i sotsial'no-ekonomicheskikh protsessov* (pp. 18–28). Uchitel.

Notestein, F. W. (1953). Economic problems of population change. In *Proceedings of the Eighth International Conference of Agricultural Economists* (pp. 13–31). Oxford University Press.
Parker, G. (2013). *Global crisis: War, climate change, & catastrophe in the seventeenth century*. Yale University Press.
Peccei, A. (1969). *The Chasm Ahead*. Macmillan.
Perkins, D. H. (1969). *Agricultural development in China, 1368–1968*. Aldine.
Persson, K. G., & Sharp, P. (2015). *An economic history of Europe*. Cambridge University Press.
Pingali, P. L. (2012). Green revolution: Impacts, limits, and the path ahead. *Proceedings of the National Academy of Sciences of the United States of America*, 109/31, 12302–12308. https://doi.org/10.1073/pnas.0912953109. http://www.ncbi.nlm.nih.gov/pmc/articles/PMC3411969/#r1
Rabinowicz, L. (1929). *Le problème de la population en France. Précédé d'une histoire générale de la population*. Etude de sociologie de la population. Paris.
Reher, D. (2011). Economic and social implications of the demographic transition. *Population and Development Review, 37*, 11–33. https://doi.org/10.1111/j.1728-4457.2011.00376.x
Reisner, L. I. (1986). Genezis sistemy «nauka – tekhnika – proizvodstvo» v zapadno-yevropeyskom regione v XVI–XVIII vv. i otsenka perspektiv etogo protsessa v vostochnykh obshchestvakh epokhi NTR. In L.I. Reisner & B.I. Slavnyy (Eds.), *Istoricheskiye faktory obshchestvennogo vosproizvodstva v stranakh Vostoka* (pp. 111–148). Nauka, Gl. red. vost. lit-ry.
Romanov, D., Meshcherina, K., & Korotayev, A. (2021). The share of youth in the total population as a factor of intensity of non-violent protests: A quantitative analysis. *Polis. Political Studies, 3*, 166–181. https://doi.org/10.17976/jpps/2021.03.11
Russel, J. G. (1975). Population in Europe 500–1500. In C. M. Cipolla (Ed.), *The fontana economic history of Europe* (pp. 25–70). The Middle Ages. Collins/Fontana Books.
Saavedra, L. T. (2019). *The demographic explosion in the Sahel region: Its governance's new challenge*. Instituto Español de Estudios Estratégicos.
Sakevich, V. (2003). Abort – krivoye zerkalo demograficheskoy politiki. *Demoskop Weekly*, 123–124, August 25–September 7. http://www.demoscope.ru/weekly/2003/0123/analit01.php#3
Sawyer, P., Romanov, D., Slav, M., & Korotayev, A. (2022). Urbanization, the youth, and protest: A cross-national analysis. *Cross-Cultural Research, 56*(2–3), 125–149. https://doi.org/10.1177/10693971211059762
Seifabadi, M. S., & Dolatabadi, A. B. (2022). Covid-19, globalization and strengthening authoritarianism in the middle east. *Journal of Globalization Studies, 13*(1), 135–157.
Service, E. R. (1975). *Origins of the state and civilization. The process of cultural evolution*. Norton.
Shcherbakova, Ye. M. (2011). V oktyabre 2011 goda chislennost' naseleniye mira prevysit 7 milliardov chelovek. *Demoskop Weekly*, 481–482, October 10–23. http://www.demoscope.ru/weekly/2011/0481/barom01.php
Shiba, Y. (1970). *Commerce and society in Sung China*. University of Michigan Center for Chinese Studies.
Shubrtova, A. (2005). [1984]. Teoriya demograficheskoy revolyutsii: k voprosu o proiskhozhdenii Translated from Czech. *Demografiye, 3*, 193–199.
Spengler, J. J. (Ed.). (1976). *Zero population growth: Implications*. Carolina Population Center.
Steinmann, G., & Komlos, J. (1998). Population growth and economic development in the very long run: A simulation model of three revolutions. *Mathematical Social Sciences, 16*, 49–63.
Steinmann, G., Prskawetz, A., & Feichtinger, G. (1998). A model on the escape from the Malthusian trap. *Journal of Population Economics, 11*, 535–550.
Tanton, J. H. (1994). End of the migration epoch? Time for a new paradigm. *The Social Contract, 4*(3), 162–173.
Thomlinson, R. (1975). *Demographic problems: Controversy over population control* (2nd ed.). Dickenson.
Thompson, W. S. (1930). *Population problems*.
Trevelyan, G. M. (1959). *Sotsial'naya istoriya Anglii. Obzor shesti stoletiy ot Chosera do korolevy Viktorii*. Izd-vo in. lit-ry.

Turchin, P., Hoyer, D., Korotayev, A., Kradin, N., Nefedov, S., et al. (2021). Rise of the war machines: Charting the evolution of military technologies from the neolithic to the industrial revolution. *PLoS ONE, 16*(10), e0258161. https://doi.org/10.1371/journal.pone.0258161

Turchin, P., & Korotayev, A. (2006). Population density and warfare: A reconsideration. *Social Evolution and History, 5*(2), 121–158.

Turchin, P., & Nefedov, S. (2009). *Secular cycles*. Princeton University Press.

Turchin, P., Whitehouse, H., Gavrilets, S., Hoyer, D., François, P., Bennett, J. S., Feeney, K. C., Peregrine, P., Feinman, G., Korotayev, A., Kradin, N., Levine, J., Reddish, J., Cioni, E., Wacziarg, R., Mendel-Gleason, G., & Benam, M. (2022). Disentangling the evolutionary drivers of social complexity: A comprehensive test of hypotheses. *Science Advances, 8*(25), 1–18. https://doi.org/10.1126/sciadv.abn3517

UN Population Division. (2023). *United Nations population division database*. United Nations. http://www.un.org/esa/population

Urdal, H. (2004). *The devil in the demographics: The effect of youth bulges on domestic armed conflict, 1950–2000*. World Bank.

Urdal, H. (2006). A clash of generations? Youth bulges and political violence. *International Studies Quarterly, 50*(3), 607–629. https://doi.org/10.1111/j.1468-2478.2006.00416.x

Urlanis, B. T. (1941). *Rost naseleniya v Evrope*. OGIZ-Gospolitizdat.

Urnov, M. Yu. (2022). *Mir-sistema na perelome: Izmeneniye balansa sil mezhdu stranami Yadra i Periferii i nachalo global'nogo krizisa*. LENAND.

Usher, D. (1989). The dynastic cycle and the stationary state. *The American Economic Review, 79*(5), 1031–1044.

Valentey, D. I. (red.). (1974). *Marksistsko-leninskaya teoriya narodonaseleniya*. 2-ye izd. Mysl'.

Vishnevsky, A. G. (1973). *Demograficheskaya Revolyutsiya. Voprosy Filosofii, 2*, 53–64.

Vishnevsky, A. G. (2005). *Izbrannyye demograficheskiye trudy. V dvukh tomakh Tom I. Demograficheskaya teoriya i demograficheskaya istoriya*. Nauka.

Vollset, S. E., Goren, E., Yuan, C. W., Cao, J., Smith, A. E., Hsiao, T., Murray, C. J. L., et al. (2020). Fertility, mortality, migration, and population scenarios for 195 countries and territories from 2017 to 2100: A forecasting analysis for the global burden of disease study. *The Lancet*. https://doi.org/10.1016/S0140-6736(20)30677-2

Weber, H. (2019). Age structure and political violence: A re-assessment of the "youth bulge" hypothesis. *International Interactions, 45*(1), 80–112. https://doi.org/10.1080/03050629.2019.1522310

Waller, M. (2003). *London 1700*. Rusich.

Widdowson, M. (2021). From Covid-19 to zero-gravity: Complex crises and production revolutions. *Journal of Globalization Studies, 12*(1), 117–144.

Wood, J. W. (1998). A theory of preindustrial population dynamics: Demography, economy, and well-being in Malthusian systems. *Current Anthropology, 39*, 99–135.

Wik, M., Pingali, P., & Broca, S. (2008). *Background paper for the world development report 2008: Global agricultural performance: Past trends and future prospects*. World Bank.

Wrigley, E. A., Davis, R. S., Oeppen, J. E., & Schofield, R. S. (1997). *English population history from family reconstitution. 1580–1837*. Cambridge University Press.

Zhao, X., & Drechsler, W. (2018). Wang Anshi's economic reforms: Proto-Keynesian economic policy in Song Dynasty China. *Cambridge Journal of Economics, 42*(5), 1239–1254.

Yu, Z. (2021). *Tayny koronavirusa: izbytochnaya smertnost' v SSHA i Rossii*. https://www.svoboda.org/a/tayny-koronavirusa-izbytochnaya-smertnostj-v-ssha-i-rossii/31466295.html.

Zinkina, J., Christian, D., Grinin, L., Ilyin, I., Andreev, A., Aleshkovski, I., Shulgin, S., & Korotayev, A. (2019). A big history of globalization: The emergence of a global world system. *Springer*. https://doi.org/10.1007/978-3-030-05707-7

Zinkina, J., & Korotayev, A. (2014a). Explosive population growth in tropical Africa: Crucial omission in development forecasts (emerging risks and way out). *World Futures, 70*(4), 271–305. https://doi.org/10.1080/02604027.2014.894868

Zinkina, J., & Korotayev, A. (2014b). Projecting Mozambique's demographic futures. *Journal of Futures Studies, 19*(2), 21–40.

Zinkina, J., Korotayev, A., & Andreev, A. (2016). Circumscription theory of the origins of the state: A cross-cultural re-analysis. *Cliodynamics: the Journal of Quantitative History and Cultural Evolution, 7*(2), 187–203.

# Global Aging and Other Demographic Trends Within Cybernetic Production Principle

6

**Abstract**

The chapter analyzes global aging and other important demographic processes, current and future demographic transformations in the twenty-first century, and their correlation with the technological changes that will occur as a result of the Cybernetic Revolution and further trends. Dramatic changes are to be expected in connection with future demographic transformations. They create a powerful impulse that will lead to a qualitative reformatting of the entire social structure and social relations. Based on this analysis, Grinin et al. make forecasts about the new type of population reproduction (TPR) that will characterize the World System by the end of the twenty-first century. Among the main demographic processes, the authors put particular emphasis on the process of global aging. However, even now the global aging is an extremely important issue and, they suppose, it will become the most crucial problem in the future. The chapter analyzes various socio-political risks associated with aging. These risks and problems are of particular concern given that by the mid-twenty-first century two-thirds of people over the age of 65 will live in middle- and low-income countries. Despite this, the ability of modern society to cope with the increasing risks associated with this process is problematic.

## 6.1 What This Chapter is About

The chapter analyzes global aging and other important demographic processes, current and future demographic transformations in the twenty-first century and their correlation with the technological changes that will occur as a result of the Cybernetic Revolution and further trends. Dramatic changes are to be expected in connection with future demographic transformations. They create a powerful impulse that will lead to a qualitative reformatting of the entire social structure and social relations.



L. Grinin et al., *Cybernetic Revolution and Global Aging*, World-Systems Evolution and Global Futures, https://doi.org/10.1007/978-3-031-56764-3_6

Based on this analysis, we offer forecasts about the new type of population reproduction (TPR) that will characterize the World System by the end of the twenty-first century. The first sections of the chapter analyze two opposing demographic processes that together will greatly affect our demographic future.

The first is a very rapid demographic growth in Sub-Saharan Africa, which may continue for a very long time. In any case, it is Africa that will contribute most to the population growth on the world scale.

The second process is an opposite trend of possible depopulation in the second half of the twenty-first century due to decreasing birth rate and population aging. The latest IHME forecast (Bhattacharjee et al. 2024) suggests that the decline of birth rate might be go even more rapid than it has been assumed in recent years. Depopulation is very possible in the developed countries of Western Europe, Eastern European countries, China and many other Asian and Latin American countries. Worldwide, more than 100 countries are likely to lose some part of their population between 2022 and 2100.

Thus, despite Africa's demographic growth, the world-system's population is likely to start declining by the late twenty-first century. Both global trends imply serious problems. The presence of a large proportion of youth in African states is accompanied by national-ethnic tensions, rapid urbanization and other factors, and this can result in serious upheavals and destabilization processes in Africa.

The elderly states of the West and many Asian countries will confront labor shortages, pension crises, confrontations between age groups, the decline of democracy, etc. Among the main demographic processes, we put particular emphasis on the process of global aging. This is all the more important since, in general, studies on this issue are scarce both in the public sphere and in scientific discourses.

However, even now the global aging is an extremely important issue and, we suppose, it will become the most crucial problem in the future. Already today, the share of the population aged 65+ amounts to 20–28% in 22 countries and regions. By 2050, the population aged 65 and over is projected to make 16 percent worldwide, i.e., more than twice the number of children under five and about the same as the number of children under 12. For the first time in human history, people aged 65 and over will outnumber children.

We consider in detail the complex relation between the development of the Cybernetic Revolution/Production Principle and the process of global aging. The latter—over several of the coming decades—will increasingly influence both individual societies and the World System in general as well as the further change in the type of population reproduction. It is important to foresee that the struggle for very long life expectancy will lead to the creation of very complex systems to support a continuous medical-biotechnological environment for elderly and disabled people.

However, this will not have only positive effects, but will also create problems connected with the limitation of human will and freedom. The chapter analyzes various socio-political risks associated with aging. Among the risks and problems associated with population aging, one finds economic slowdowns, pension crises, problems of caring for the growing number of disabled and frail people, maintaining the financial savings of the elderly, the issue of ageism and the vulnerability of the elderly in times of crisis and pandemics and consequently maintaining life expectancy, as well as a stable democracy and even world order.

These risks and problems are of particular concern given that by the mid-twenty-first century two-thirds of people over the age of 65 will live in middle- and low-income countries. However, though the influence of the aging process on various areas of social activity will be dramatic in most countries, the ability of modern society to cope with the increasing risks associated with this process is problematic.

## 6.2 Change in the Long-Term Demographic Trends by the End of the First Quarter of the Twenty-First Century

**Quantitative and Qualitative Aspects of Demographic Transformations.** In Chap. 5 we argued that throughout the history there has been a strong correlation between production (technological) revolutions and the largest demographic transformations (LDTs). The essence of this correlation consists in technological breakthroughs and technological progress in general which have opened up new ecological niches for society and led to a huge quantitative population growth and some improvement in the quality of life (though not always for all people). In this process, the population growth has been the major result of demographic transformation while the improvements in quality of life through technological and social innovations have been supplementary. But with every LDT, the improvements in the quality of life played an increasingly important role (Grinin & Grinin, 2023; Grinin et al., 2023a, 2023b, 2023c). *As a result, during the initial and modernization phases of the Cybernetic Revolution, the qualitative aspect of the demographic transformations came more noticeable to the fore and strongly influenced the World System.* At the same time, while in the initial phase of the Cybernetic Revolution, a very powerful population growth was observed in all countries of the world, and especially in the developing ones (the population explosion described in Chap. 5), in the modernization phase this growth started to slow down. It almost stopped in developed countries, declined sharply and continues to decline in many developing countries. At the same time, mortality has fallen significantly and life expectancy has considerably increased everywhere. Moreover, it is quite obvious that the population growth in the foreseeable future will stop almost everywhere in the foreseeable future, and in some places the reverse trend (of depopulation) may start. *But there is a prospect of a new powerful breakthrough in terms of improving the quality of life* (see Table 6.1).

As shown in Chap. 5, one may distinguish two domains in improving the quality of life (albeit very closely interrelated): (1) material and cultural improvement (better food, housing, education, working and living conditions, opportunities for

**Table 6.1** Demographic changes during the Scientific-Cybernetic Production Principle

| Phase | Dates | Techno-social changes | Demographic changes | Limitations | Population at the start of the phase |
|---|---|---|---|---|---|
| Initial phase of the Cybernetic Revolution | 1950s–1990s | Powerful growth of science, technology and medicine. Major political and social changes in the World System; liberation of the colonies; the rise of social policies | In developing countries, due to the reduction in mortality, a very rapid population growth began; in developed countries, after a period of rising birth rates, there was a transition to a decline in the birth rates to the replacement level | Huge gap in the levels of development between the core and the periphery of the World System; advances in medicine, the rise of female education and rights, and economic policy led to declining birth rates in the developed world | 2.5 billion |
| Modernization phase of the Cybernetic Revolution | 1990s–2030s | Mass diffusion of ICT and AI, growth of education in the developing countries | Transition to TPR of high life expectancy and population aging in most parts of the World System | In terms of reducing the birth rate, traditional and religious ideologies and practices are a limitation | 5.27 billion[1] |
| Final phase of the Cybernetic Revolution | 2030s–2070s | Emergence of major breakthroughs in medicine; the impact of aging on society; struggle for a new world order | Global aging and adaptation to it; possible completion of the demographic transition in Africa | Traditional and religious ideologies in Africa and parts of Islamic countries | 8.37–8.65 billion (forecast)[2] |
| Mature phases of the scientific-Cybernetic Production Principle | 2070s–2120s | Growth of conservatism, the final formation of new social institutions associated with aging | Noticeable increase in life expectancy; growth of adaptability to aging and disability due to the mass distribution of techno-biological technologies among the population[3] | Conservatism of the aging society; dangers of excessive techno-medical power; unpredictable dangers | 10–12 billion (according to different forecasts) |

[1] Akimov (1999); Shcherbakova (2011).

[2] UN Population Division (2022a).

[3] Among all population groups, but especially among the elderly, disabled and incapacitated (requiring constant care).

bringing up children, etc.) and (2) an improved quality of physical life (or health-related quality of life)[4]: better medicine, reduced all-cause mortality, reduced severe consequences and complications of diseases and injuries, increased life expectancy, health promotion, adequate care for the sick and infirm, providing a more or less normal life for chronic patients with life-threatening diseases, the opportunity to lead a healthy lifestyle and an increasingly fulfilling life in terms of physical and mental activity; adaptation of the disabled, etc. In the course of history, both these aspects have been improved, yet the material and cultural aspects would prevail. However, in the second half of the twentieth century/the beginning of the twenty-first century both domains developed rapidly, and it is even difficult to say which one developed faster.

A very relevant concept here is the Health-Adjusted Life Expectancy, HALE. This concept was proposed by Daniel F. Sullivan in 1971, who also developed the theoretical foundations and methodological approaches for its calculations (Sullivan, 1971). Currently, different versions of this indicator or its derivatives are used, but the principle of calculation remains the same—health-adjusted life expectancy (HALE) is defined as the difference between total life expectancy and the number of years lived in an unhealthy state. There are two main approaches to assessing health status when calculating health-adjusted life expectancy. The first is based on official morbidity statistics and identified patterns of deterioration in health after illness—this approach is used by the World Health Organization (WHO). The second approach, practiced, in particular, by Eurostat, is based on self-assessments of the population obtained in the course of various national and international surveys.

**Limits for Improvements in the Material Domain of Quality of Life**. However, today (in the 2020s), at the threshold of the final phase of the Cybernetic Revolution, the rational limits for increasing the material well-being emerge in the developed countries (in developing countries, this limit is still very far away, but even there they are moving rather quickly in this direction). Of course, the desire to consume more and more can hardly be limited, so even the very high level of consumption achieved by the majority of the population in developed countries may appear insufficient. But it is already clear that there are various limitations to the further aspirations to increase the GDP per capita at all costs. At the same time, the population aging and the lack of a necessary number of young people are not least important limitations in this context. In addition, powerful social, economic and climatic constraints also stand in the way: the social ones are associated with the need to reduce the growing inequality in Western societies; the economic ones are related to the fact that opportunities for economic growth are exhausted and climatic constraints are associated with attempts to reduce the anthropogenic impact on climate change, which leads to growing costs of energy and resources. The pandemic threats have also manifested themselves (see Chap. 14, as well as Grinin, 2021a; Grinin, 2020e; Widdowson, 2021). By the way, they have also greatly reduced both the opportunities and desire for endless travel. There are also

[4] Those concepts are close in their meaning, but not completely synonymous.

political limitations since the political component of globalization lags far behind its economic component, which we have already discussed elsewhere (see Grinin, 2009, 2010, 2012a, 2012b, 2022a, 2022b, 2022c); as a result, globalization has bogged down and even started to noticeably roll back, which was intensified by the split resulting from the military conflict between Russia and Ukraine.[5]

It is no coincidence that attempts to make business more social have been intensifying for several years now. The ideas become widely spread that the growing consumption is already counterproductive and one should think about making consumption more rational (including such basic resources as energy and water), and the pursuit of profits and property is not only an ideal, but is generally wrong, and capitalism should be considerably transformed (Schwab & Malleret, 2020; Schwab & Vanham, 2021; von Weizsäcker & Wijkman, 2018; Wallerstein et al., 2013), or even "cancelled" and "closed down".[6] For some time now, social thinkers have been increasingly arguing that neoliberalism, with its emphasis on private initiative and growth, should be abandoned, while other principles should substitute its rules, since it is wrong that neoliberalism favors competition over solidarity and economic growth over social welfare, and that the main goal of companies can no longer be simply the unbridled pursuit of financial profit; now, they are obliged to serve all their stakeholders, and not just those who own shares.[7]

In addition, the powerful trend is to convince and force politicians and businesses to choose the climate-friendly course (and this clearly leads to a slowdown of growth or even decline in living standards). Searching for a balance between the need for economic development and other imperatives is a very difficult task. Still, we believe that in the coming decades, although the standard of living in the world, especially in developing countries (in particular in the least developed), will continue to grow up to a certain level, while *in the World System as a whole*

[5] The decline in living standards in 2022–2023 in Europe, the USA and many other countries is largely due to the abrupt rollback of globalization that occurred then.

[6] Of course, there is a great danger of going too far and triggering an economic crisis, which would only worsen the situation, as well as the danger of throwing the baby out with the bathwater. We have already written about these issues (see Grinin & Grinin, 2021a, 2021b; Grinin et al., 2022b, 2022c).

[7] The need to change the economic rationality of market capitalism, which has become the imperative of the modern society, in favor of the concept of "dignity" and "culture" (which is more important than "cost/price") is mentioned in several reports to the Club of Rome (Andersen, 2020; Berg, 2019; Bozesan, 2020; Korten, 2015; Kuenkel, 2019; Lovins et al., 2018). Some other analysts, like Collier and Kay (2020) and Collier (2018) express the same ideas. Among the possible solutions the following measures are suggested: to adapt the market system in order to reduce the negative impact of consumerism. The environmental and social costs of consumerism must be reflected in market prices, the development of less resource-intensive business models, production systems must be adapted accordingly, and products must be designed for a longer period of use, their renewal and recycling; facilitating the transition to a post-material society; reducing inequality that drives consumption; changing the perception of consumption as something that makes a used product unusable and breaks down; prosperity and a more modest lifestyle; struggle against selfishness and lack of meaning (Berg, 2019: 198–204).

*the standard of living (understood as consumption in the narrow sense of the word) will not grow so actively as it used to.* But this does not mean that there will be no improvements in the quality of life in general. The trend (with strong fluctuations, of course) is that the improvements in the quality of material life will weaken, while the improvements in the domain of physical and mental health, on the contrary, will increase.[8] And global aging will considerably contribute to this, since as a person gets older, s/he gets more concerned not with a career, material acquisitions and entertainment, but with health, the opportunity to live longer, the limitations of old age and serious illness side effects, which one really wants to reduce.

The standard of living in developed countries may even decline at times. Of course, the per capita income and consumption will depend on labor productivity growth, the GDP growth and on other factors. Indeed, the rapidly growing number of robots, AIs and other self-managing systems capable of performing operations currently performed by humans can largely compensate for the exclusion of workers from the labor market, similar as mechanical looms replaced manual artisans in their time, as cars replaced horses, etc. Therefore, there will hardly be an absolute decline in the standard of living in the long run, but it may occur over a period of several years (now we witness just such a period), which will become an impetus for the fastest introduction of labor-saving technologies.

**Aging and the Trend Toward Improving the Health-Related Quality of Life.** So, *one may expect a very significant, and hopefully in some ways unprecedented growth in health-related quality of life.* And we believe that *this will be the main trend perhaps for the next hundred years.* However, one should keep in mind that, first, this trend will hardly be smooth, and it is likely to be accompanied by fluctuations, crises and setbacks (the COVID-19 has been and still is a notable example[9]); and, secondly, the impact of these changes on various aspects of life of contemporary and especially future cybernetic society will be great and, in many ways, conflictory and painful (see Chaps. 7 and 15 for details as well as for possible opposite trends). This is inevitable in the face of restructuring, new and aggressive technologies and an unknown path forward, the benefits of which are often sought by a small but powerful minority.

We argue that *global aging and total adaptation to the growing numbers of the elderly and disabled people will be central to this trend.* This will create or intensify new contradictions, conflicts, tensions and frictions in society, this will require serious changes, and in many ways transform the whole economic life around these

[8] The process will resemble the change of the balance between quantitative and qualitative growth within demographic transformations described above and in Chap. 5.

[9] On social consequences of the COVID-19 pandemic see, e.g., Chumakov & Yurchenko, 2021; Efremov & Arkhangelsky, 2023; Grinin, 2020e; Grinin et al., 2022a, 2022b, 2022c; Hussain, 2021; Irshad, 2020; Jasiński & Bąkowska, 2020; Kakkar, 2021; Seifabadi & Dolatabadi, 2022; Widdowson, 2021.

processes. Moreover, we believe that *the symbiosis of technological and demographic development will intensify in techno-medical and techno-biological terms* (see below).

**The Lack of Attention to the Impact of Global Aging on Society**. We argue that the process of global aging will remain one of the strongest and most powerful factors influencing the lives and future transformations of society and the World System in general for several decades to come. The present chapter (as well as other chapters) will address different directions and options for this influence.

In general, all population projections are unanimous in anticipating the dramatic global aging in the coming decades (Alkema et al., 2011; Kaneda et al., 2016; Lutz et al., 2018; Raftery et al., 2012; UN Population Division, 2019b, 2022a, 2022b; Vollset et al., 2020; Wittgenstein Center, 2022) (see Fig. 6.1).

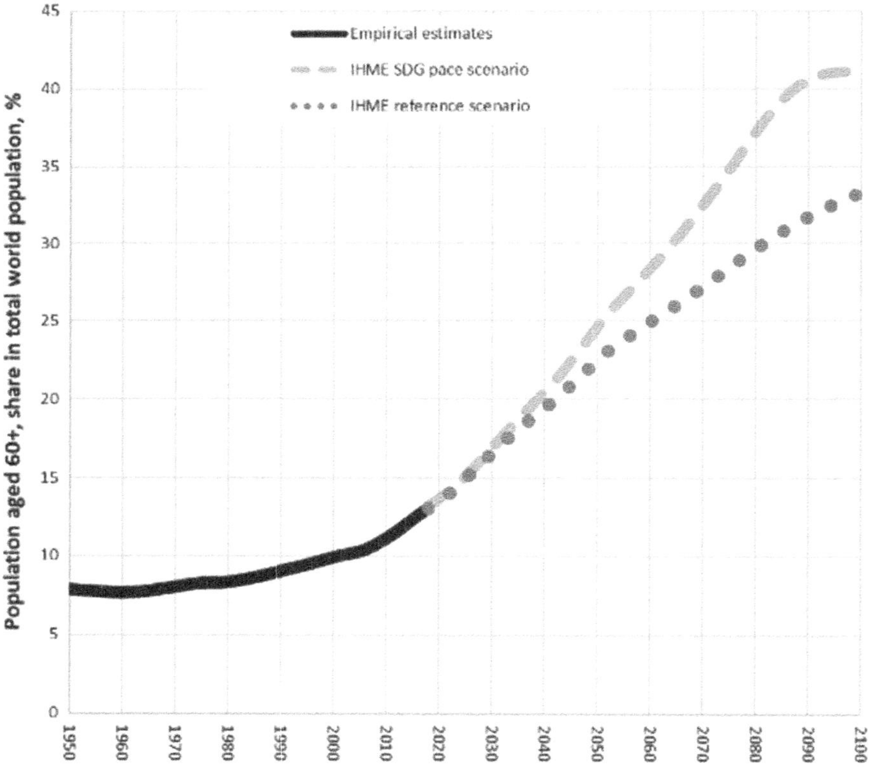


**Fig. 6.1** Population aged 60+, share of total population (%); Institute for Health Metrics and Evaluation at the University of Washington (IHME) empirical estimates for 1950–2020; IHME reference scenario for 2021–2100; IHME Sustainable Development Goals (SDG) pace scenario for 2021–2100. The IHME SDG pace scenario "shows what we can expect population trends to look like if all countries and territories meet the SDGs for education and contraceptive needs by 2030" (Vollset et al., 2020: 1295)

Many studies deal with different aspects of aging and its impact on society, especially on the economy (on this phenomenon, see, e.g., Barsukov, 2019; Bengston, 2018; Bloom & Luca, 2016; Bös & Weizsäcker, 1989; Clark & Spengler, 1980; van Dullemen, 2017; Fichtner, 2018; Fukuyama, 2002; Goldstone, 2015; Goldstone et al., 2015; Hsu & Lo, 2019; Jackson, 2021; Jackson et al., 2008, 2013; Kapelyushnikov, 2019a, 2019b; Lee, 2016; Lee et al., 1988, 2010; Maestas et al., 2016; Medici, 2021; Mitchell & Walker, 2020; Orlická, 2015; Peterson, 1999; Pifer & Bronte, 1986a, 1986b; Popova & Yanik, 2014; Randers et al., 2018; Shcherbakova, 2019; Zimmer, 2016; UN, 2013; Yaşar & Yaşar, 2018). However, unfortunately, even in academic journals and publications one can notice that only little attention is paid to the issue of global aging, especially in terms of its broad and complex impact on all spheres of society.[10] Yet, the UN and related organizations, particularly the WHO, pay quite a lot of attention to the process of global aging, which proceeds at different paces in different countries. And these studies have been conducted since the late 1970s (Kamiya & Hertog, 2020; Medici, 2021; Popova & Yanik, 2014; UN Population Division, 2013, 2015a, 2019a, 2020; WHO, 2017, 2021a, 2021b); many different programs, decisions, declarations, etc., have been adopted in order to set program objectives and goals (e.g., UN Population Division, 2013, 2015a, 2015b, ; WHO, 2017, 2021a, 2021b).

However, even the documents of e international organizations and the European Union do not focus on all aspects of this problem. The analysis of the impact of global aging covers only a few (albeit very important) aspects: pensions, budgetary, legal, gender, ageism and medical in terms of aging-associated disease. However, aging has many more important aspects. In a word, *the level of research into this problem—the problem of the growing influence of aging on all aspects of social life—is clearly disproportionate to its significance.* And this situation of strange passing over in silence has been observed for several decades. As early as in 1999, Peter Peterson argued that the '"graying" of the developed world's population may actually contribute more to reshaping of our collective future' than anything else—at the same time he wondered why, compared to the climate problem, the future manifestations of the process of global aging are so little spoken about (Peterson, 1999: 42). Already ten years before him, Dieter Bös and Robert K. von Weizsäcker had written about the problem of denying the issue of aging (Bös & Weizsäcker, 1989: 345).

[10] In our other works we try to study global aging as an integral problem of the future (Grinin & Korotayev, 2015b, 2016; Grinin et al., 2016, 2017a, 2020); however, a huge amount of work is needed since the phenomenon permeates the entire society and affects all processes.

## 6.3 Africa as the Most Important Puzzle of Demographic Transformations in the Twenty-First Century

Despite the above said, the quantitative growth in the World System will still continue, perhaps until the end of the twenty-first century or even later. Of course, this will not be a 'population explosion' like in 1950–1970 (see Chap. 5); still the world population may increase from current 8 billion people by another 2–3 billion and it is worth stressing that the main population growth will occur in the first half of our century (see Fig. 6.2).

Here, Tropical Africa and several countries in Asia (like Yemen, Afghanistan) may be the main contributors to this growth (see Figs. 6.3, 6.4, 6.5).

To understand the importance of Africa for the future global demography, one should visit the UN 'World Population Prospects' and 'United Nations Population Division Database'. According to these sources (UN Population Division, 2022a, 2022b) the 46 least developed countries (most of which are Sub-Saharan) are among the world's fastest growing countries. Many are projected to double in population. Countries of Sub-Saharan Africa are expected to contribute more than half of the global population increase anticipated through the 2050s. More than half of the projected global population increase up to 2050 will be concentrated in just eight countries; among them, there are four Tropical African countries: the DRC, Ethiopia, Nigeria and Tanzania. Between 2050 and 2100, of the ten countries, which will give the most important contribution to the population increase will be mostly Sub-Saharan countries, while the population of Eastern and South-Eastern Asia, Central and Southern Asia, Latin and Northern Americas and Europe are projected to begin to decline before 2100.

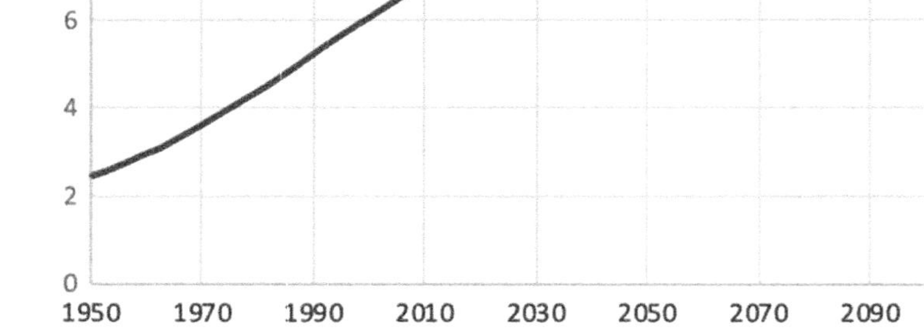


**Fig. 6.2** Dynamics of the world population—empirical estimates (up to 2022) and the medium forecast of the UN Population Division until 2100. *Data source* UN Population Division (2022a)

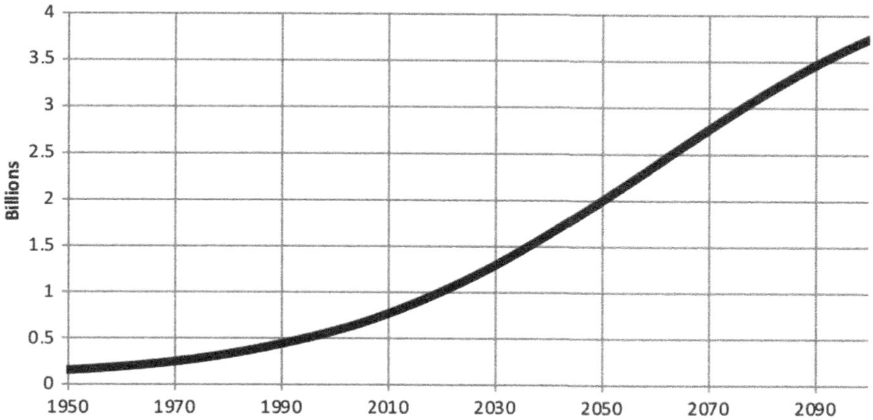


**Fig. 6.3** Dynamics of the population of Tropical Africa for the period 1950–2020 with a projection to 2100 (UN medium projection). *Data source* UN Population Division (2022a)

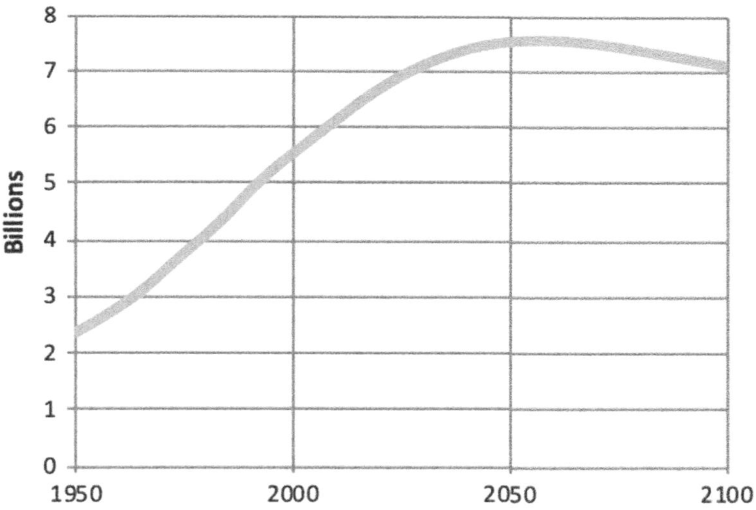


**Fig. 6.4** Dynamics of the population of the rest of the world (i.e., the whole world excluding Tropical Africa) for the period 1950–2020 with a projection to 2100 (UN medium projection). *Data source* UN Population Division (2022a)

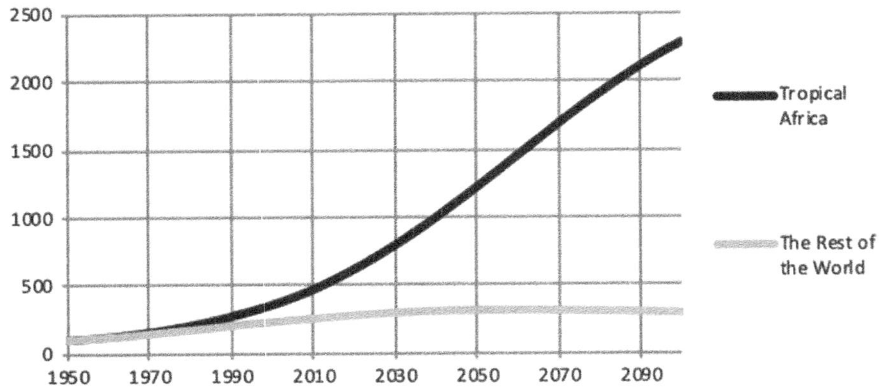


**Fig. 6.5** Relative dynamics of the population of Tropical Africa and the Rest of the World for 1950–2020 projected to 2100 (UN medium projection), 100 = 1950 level. *Data source* UN Population Division (2022a)

Thus, the birth rate is high across Sub-Saharan Africa; however the fluctuations between countries are very considerable.[11] In most countries of this part of the continent, fertility is declining rather slowly[12] (see, e.g., Kebede et al., 2019; Korotayev et al., 2016; May & Rotenberg, 2020; Nzimande & Mugwendere, 2018; Schoumaker, 2019; Zinkina & Korotayev, 2014a;); however, the overall trend—the lower the level of modern education, continuing urbanization, and the higher birth rate in the less developed countries—is quite evident.

One can point to traditional and religious (including Islamic) ideology, and a number of other factors preventing active entry into the second phase of the demographic transition in Tropical Africa (e.g., Grinin, 2020a; Korotayev et al., 2016).[13] There is also no doubt that the type of agriculture (still mostly extensive how agriculture in combination with millions of people engaged in semi-nomadic animal husbandry) greatly contributes to a high birth rate (see Ibidem). On the one hand, the agricultural modernization in Tropical Africa should develop (in some places this is already happening), on the other hand, this situation contradicts the environmental situation in the world, since the increasing environmental pressure caused by modernization can become destructive (Grinin, 2020d). Along with high

[11] Even among the top ten countries in terms of fertility. Thus, in 2021, the birth rate in Niger was 47 children per 1,000 people, and in Mozambique (10th place)—38. And, for example, in Botswana—22 children per 1,000.

[12] This is in contrast to Northern and Southern Africa, where fertility is much lower. However, in almost all countries of Tropical Africa, it is still declining. For example, in Niger in 2005–2010 the total fertility rate was 7.55 children per woman, and currently it is 6.95 (UN Population Division, 2022a).

[13] Islamic ideology also hampers the completion of the demographic transition in a number of Islamic Middle Eastern countries.

population growth, this can also become a factor of pressure on the demographic policy of African countries.

Much will also depend on how deeply, systematically and quickly the green energy transition will occur. For its implementation, in addition to a host of technological and other restraining problems, an important limitation results from the scarcity of many metals and components (like silicon), necessary for green energy, electric vehicles and other equipment. Some African countries (e.g., Chad, Niger, DRC, Zambia, Namibia and Zimbabwe) are rich in ores of manganese, chromites, bauxites, gold, platinoids, cobalt, vanadium, copper, uranium, antimony, beryllium, graphite, titanium, nickel, bismuth, lithium, tantalum, niobium, tin, tungsten—many of which are in great demand in the modern electronics industry and in green energy; in addition, one cannot but mention the reserves of rare earth metals. The need for these minerals can activate foreign capital and accelerate the industrial development of Africa, which will undoubtedly affect the decline in fertility, as well as contribute to the change of Africa's status among the leading geopolitical and economic players with all the pros and cons of their active penetration to this continent (Grinin & Korotayev, 2023; Korotayev et al., 2023).

In Figs. 6.3, 6.4, 6.5, we see that while the total world population, excluding Africa, is projected to start declining from the mid-twenty-first century, the population of Africa will continue to grow, still at a very fast pace. At present, in most countries of Sub-Saharan Africa, fertility is declining rather slowly (Grinin & Korotayev, 2023; Kebede et al., 2019; Korotayev & Zinkina, 2015; Korotayev, Shulgin et al., 2023; Zinkina & Korotayev, 2014a, 2014b). However, as Tropical African countries develop, a reduction in the birth rate is inevitable, but it is not clear how fast this will happen. How fast or slow this process goes will strongly influence the overall growth and the total number of the world population in the twenty-first century and beyond. That is why we stress in the title of this section that Africa is the most important puzzle or intrigue of demographic transformations in the twenty-first century (see Fig. 6.5).

The fast population growth, large youth population in immature states under the situation of national-ethnic tensions, rapid urbanization and other factors threaten Africa with very serious upheavals and destabilization processes.[14] At the same time, the danger of destabilization will grow in direct proportion to the rate of improvement in the quality of life and standard of living, which will inevitably and more or less rapidly increase, given the high standard of living in the outer world for Africa.

The opposite trend, threatening with depopulation especially in the second half of the twenty-first century, is observed in developed countries, China and some Asian, European and Latin American countries.An excessively rapid population

[14] See, for example, Cincotta and Weber (2021); Goldstone (2002); Grinin (2020b, 2020c, 2020d); Korotayev et al., (2011, 2021, 2022a, 2022b, 2023); Korotayev and Zinkina (2022); Mesquida and Weiner (1999); Moller (1968); Romanov et al., (2021); Sawyer et al., (2022); Urdal (2004, 2008, 2012); Yair and Miodownik (2016).

aging, in turn, threatens these countries with a loss of dynamism and authority and furthermore with social problems due to ageism.

Therefore, the demographic problems in the World System should be considered by applying a relational perspective; on the one hand, it is necessary to gradually (but not dramatically) reduce the birth rate in Africa in order to improve the development of countries and reduce their problems, and on the other hand, it is necessary to overcome the limit of depopulation, otherwise we will lose many nations and generate problems of a new type. According to the UN forecasts (UN Population Division, 2022a), *more than 100 countries of the world will face a loss of population between 2022 and 2100.* In other words, *we need to address the need of optimizing demographic processes in terms of quantitative development and with the account of improving the quality of life, which will continue (possibly with noticeable fluctuations) right along.* In 2021, the world birth rate was 16.9 live births per 1,000 people, which is just over 1.2% of the natural increase, or about 2.30–2.35 births per woman (UN Population Division, 2022b; compare with five births per woman in 1950 [ibid.]). Accordingly, a normal population reproduction requires 11 children per 1,000 people (or the total fertility rate of 2.1 births per woman during her lifetime).

It should be noted that the demographic projections of the UN and other organizations are mainly inertial forecasts, made using standard methods, while the options (low, medium and high) differ mainly in terms of projected fertility levels. However, a certain paradigm shift or increase in fertility (at least for some time) in developed countries cannot be ruled out, as trends can change. Of course, developed countries can hardly return to a high birth rate, but it is quite possible to increase the birth rate with the help of social measures (see Arkhangelsky et al., 2015; Gallagher, 2020).

## 6.4 The Final Phase of the Cybernetic Revolution: Aging and Depopulation

**Aging and Demographic Transition on the World-System Scale**. Although we mentioned above that the demographic transition came in a number of countries already to a preliminary completion, it will finally end when the process of population aging reaches its apogee. Thus, population aging is the main characteristic of the last phase of the demographic transition (Bloom & Luca, 2016; Kapelyushnikov, 2019a, 2019b; UN, 2008: vi). Each phase of the latter is associated with a change in the demographic structure of society: (1) a rejuvenation of the population structure due to increasing number of children; (2) then an increase in the number of people of working age and (3) an increase in the number of older people and their share of the population. Each phase of this transition takes several decades.[15]

Thus, although in some countries a zero population growth or even depopulation is already observed due to a considerable decline in the birth rate, not a single country has yet fully completed the demographic transition, not to mention the World System as a whole. These processes will end for many countries in the final phase of the Cybernetic Revolution, while for the World System it will be completed not earlier than in the late twenty-first century. So, during the final phase, most countries of the world will complete the demographic transition and will be affected by aging processes, albeit to varying degrees. However, many African societies will still be younger. Thus the global aging will be the most important process in demographic and social terms. The decline in birth rates will, respectively, finalize the previous trend and increase the degree of aging (due to changes in the demographic structure of society, and, of course, unless the trends change). In Chap. 15, we show what a society in which the demographic transition has been completed will be. We call this society "cybernetic society".

As we have already pointed out and will discuss in the next chapters, this will significantly determine the direction of technological breakthroughs, enhancing the development of medicine and related MANBRIC spheres. For quite a long time there will be a positive feedback: the development of medical and other technologies will increase the life expectancy, and the latter will intensify the search for new medical technologies to support the growing proportion of elderly people and their longevity (taking into account the predicted shortage of people of working age). But closer to the end of the final phase, and even more noticeable at the mature stages of the Scientific-Cybernetic Production Principle, this connection will weaken and then evoke feedback: the intensification of the aging process in societies will weaken the dynamics of development, especially technological development (see Chap. 13 for details; see also Grinin et al., 2020). Thus, aging will become a factor contributing to the unfolding of the final phase of the Cybernetic Revolution, and the most important driver of technological development due to the accumulation of huge financial resources and the need to solve problems associated with labor shortage and health support of the aging population (see also Chap. 4). But during the mature phases of the Scientific-Cybernetic Production Principle, aging will become the factor that will slow down scientific and technological progress.

[15] Within a different approach, they sometimes distinguish four stages here: (1) a significant decrease in the mortality rate with the birth rate remaining the same or slightly falling; (2) a considerable decrease in the birth rate and a further decrease in mortality, yet less considerable than the fall of the birth rate; (3) an increase in the death rate due to demographic aging, and a simultaneous slowdown of the birth rate; (4) an increase in mortality, which approaches and in most cases exceeds the birth rate, completing the process of demographic stabilization, which is often followed by depopulation (see Motrevich, 2000; Chesnais, 1992; Dyson, 2010; Livi-Bacci, 2017; Vollset et al., 2020).

Thus, while during the initial phases of the Cybernetic Revolution the aging process was just one of many processes, during its final phase it will become the most important one (see Grinin et al., 2020). Along with the growing aging population, the adaptation to it also occurs. However, we believe that the development of institutions and mechanisms for *adaptation to aging will be to the greatest extent the result of the mature and final phases of the Scientific-Cybernetic Production Principle*. The analysis of aging and adaptation to it in various aspects will be the subject of this and subsequent chapters.

There are not many studies in which their authors analyze aging in its different dimensions, using classifications and definitions. That is why it is worth quoting some definitions from the book *Aging, Society, and the Life Course* (Morgan & Kunkel, 2016). In Chap. 1, 'Aging and Society' (Ibid.: 1–6), the authors single out and characterize aging in four forms: physical aging, psychological aging, social aging and societal aging.

According to them, **social aging** refers to the ways in which society helps to shape the meanings and experiences of aging. Social aging includes the expectations and assumptions of those around us about how we should behave, what we are like, what we can do and what we should do at different ages. The concept of social aging also refers to the ways in which those expectations influence the opportunities available to us as we grow older. The book applies the concept of social aging to the major dimensions of our social lives, including families, work and health.

**Societal aging** is defined as the demographic, structural and cultural transformations that a society undergoes as the proportion of older adults increases and affects every aspect of social life, from social institutions (such as the family, work, education and politics) to the experiences of aging individuals.

These are good points, however, for the purposes of our study such a classification is not appropriate, because we regard aging as a social and global phenomenon, including both social and societal aging (in the sense of Morgan and Kunkel) and also in medical aspect (see Chaps. 8–11), as a unified complex phenomenon and institutions. It is worth stressing specifically, that for us it is more important to point out **societal aging and global aging**. The former is a complex process connected with the problems, searching for the ways of their solution, transformation of society under the development of aging and so on. The latter is the process and transformation as a result of aging of population at supranational and world-system scales. However, it is often difficult to separate these meanings as well as other aspects and dimensions of aging.

Thus, we think that the division into social and societal aging is not productive for our monograph and consequently, we use the term **societal aging** in a broad sense, uniting social and societal meanings of Morgan and Kunkel as well as some special meanings, connected with public health sphere,[16] etc. We will also use our term **adaptation to societal aging** for such broad complex meanings. If we mean supranational and world-system scale, we will use the term **global aging**.

**The Threat of Depopulation and Its Impact**. So, the world population and its stabilization depend on the trends that will prevail during the final phase of the Cybernetic Revolution and beyond, as well as on the development of these trends. Apart from the issue we have already discussed above,—when will the population growth in Africa start to decline significantly?—the question of whether the depopulation in the rest of the world will continue for a long time becomes even more important. Let us dwell on depopulation in more detail.

First of all, this process, involving more and more countries,[17] means the weakening of a long-term trend of population growth. This growth lasted (albeit with setbacks) for more than ten thousand years and in fact it took place (albeit very slowly) before the start of the Agrarian Revolution (see Chaps. 2 and 5). For a very long time it has been the most important factor, and always remains among the major ones. Its decreasing role, and later the actual elimination of this factor, inevitably leads to the change of the societal economic model.

Processes of depopulation occurred throughout the historical process and could sometimes be associated with technological failures (e.g., the destruction of irrigation systems). However, the most significant cases of depopulation were caused by the destruction of large social and state systems, invasions, especially by nomads and adverse climatic changes. Sometimes all of these misfortunes and catastrophic factors would coincide and trigger a demographic collapse. We mentioned such phenomena in Chap. 5, when speaking about socio-demographic cycles (secular cycles). But these were mostly local depopulations, although in China they could lead sometimes to a reduction in the population by tens of millions of inhabitants. The world-system depopulations were relatively infrequent; the major ones occurred in the second to third centuries in connection with the so-called 'Antonine plague'; in the sixth to seventh centuries in connection with unfavorable climatic shifts and related "Justinian Plague"; in the twelfth century in connection with the Mongol conquests; in the fourteenth century due to unfavorable climatic shifts and related Black Death pandemic; in the seventeenth century in connection with the so-called "Little Ice Age" and the associated new waves of pandemics (see, e.g., Diamond, 1999; Korotayev et al., 1999; McNeill, 1993, 2013; Parker, 2013). But sooner or later the population growth would resume.

In the past, depopulations could have different consequences. They could often cause desolation of territories, a rollback in techno-cultural terms and decline of civilizations. But there were times when depopulations became the starting point

---

[16] Success story of societal aging is in advancing public health and medicine, in reducing the risk of premature death, the control and prevention of diseases and improvement in social and economic development (Chen et al., 2022).

[17] Depopulation is already observed in a number of countries. Between 2022 and 2050, the population is projected to decrease in 61 countries and regions due to a sustained low level of fertility (UN Population Division, 2022b).

for a new economic and cultural breakthrough, since the demographic pressures eased, land hunger ended and agricultural productivity grew. This recurrently happened in medieval and modern China (see, e.g., Korotayev et al., 2006; Lee, 2007; Lee & Zhang, 2010; McNeill, 2013; Grinin, 2010; Grinin & Korotayev, 2012). This was the case in fourteenth-century England after the Black Death (in France, the depopulation was aggravated by the Hundred Years War). The decline in the population contributed to the personal emancipation of peasants, and a sharp increase in wages for workers (wages would achieve this level only several centuries later), which was one of the reasons for the trend toward mechanization (Grinin, 2006; Grinin & Grinin, 2015, 2016; Grinin & Korotayev, 2015a).[18] But one way or another, depopulation would always lead to dramatic social and often political changes.

Thus, in the near future (within 80–100 years), *we can expect a complete change in the trend toward population growth and its replacement by a trend toward a decline of the population for a short or rather long time.* Later, we will talk about alternatives to the depopulation trend. The main mechanism of depopulation consists in a reduced birth rate, but one cannot exclude the possible "black swans" in case of pandemics, wars and other events that can hit an overgrown population and especially the elderly (e.g., the coronavirus pandemic).

Depopulation will certainly affect future societies: it will pose serious financial and economic challenges, in addition to stagnating demand, shrinking markets, etc., with increasing labor shortages and growing financial burdens for maintaining the growing elderly and disabled populations (which we will discuss later). *This will produce a noticeable negative effect.* But at the same time, *this may stimulate technological advancement and labor productivity growth* (as well as opportunities for a rapid movement of labor force from one place to another, from places with excessive labor reserves to places with shortages due to expanding opportunities for remote work). Thus, the technological development will continue to affect the replacement of labor force lost due to depopulation and aging. For some time, one will observe a positive feedback. Furthermore, *depopulation will certainly lead to a considerable restructuring of society (and of the global society as well) and to changes in the political sphere, also in a global perspective.* We will discuss this separately.

As already mentioned, along with depopulation and stabilization of the population, the growth in Tropical Africa will continue. *This will support the formation of the world-system model: the old North—the young South* (see Chap. 7 for more details). At the same time, African countries may come under pressure if their population growth threatens global interests.

**Migration**. As for migration, in recent decades, it has played a relatively small role in economic terms on a world-system scale. However, between 2000 and

[18] Yet, in Russia the depopulation during the Smuta or the Time of Troubles, the period of disorder and great unrest at the beginning of the seventeenth century, on the contrary, led to the enslavement of peasants.

2020, in the high-income countries, the contribution of international migration growth (net inflow of 80.5 million) exceeded the balance of births over deaths (66.2 million) [UN Population Division, 2022b]. For individual countries, both countries with high emigration rates (like Syria, Pakistan, Bangladesh, Nepal, Ukraine, El Salvador, etc.) and countries with high immigration rates (like Turkey, the USA, Jordan, Germany, Saudi Arabia, Russia, Israel, etc.), these movements seem to be quite important. While the political and cultural aspects of migration (often highly conflicting and polarizing host societies) are actively debated, there is not sufficient attention paid to the even greater role of migration in creating channels for the transfer of ideas, technologies and cultures from more developed countries to the less developed ones, since large diasporas from the latter emerge in the former (Grinin et al., 2019). As for the final phase of the Cybernetic Revolution, migration will remain at the current or slightly higher level. But on the other hand, a new type of migration, unprecedented in history, will develop. In other words, there will be a fusion of technological and demographic trends, that is, there will emerge what we would call *virtual migration*, along with the remote working, which is already actively used today (see Grinin et al., 2023a, 2023b, 2023c). This will partly compensate for the shortage of labor resources in developed countries, maintain aging population (due to taxes and deductions) and also improve the living standards in developing countries.

**Disabled People**. In addition to aging and adaptation to it, one should point out another important trend, namely the growing assistance to the disabled and the sick (which, however, is partly covered by the subject of aging). Paradoxical as it may sound, *it is the disabled who will become the vanguard of humanity on the path to the creation of medical-biotechnological environment/systems (and partial cyborgization*). However, this will manifest itself particularly in the mature stages of the Cybernetic Production Principle (see below and in Chap. 7).

## 6.5 Demographic Trends of the Mature Stages of the Scientific-Cybernetic Production Principle

In the last third of the twenty-first century and the early twenty-second century, that is, in the mature phases of the Scientific-Cybernetic Production Principle, the major demographic and related trends will be the following:

1. As already mentioned, there will be a stabilization of population numbers or even global-scale depopulation. The demographic transition will be mainly or almost completed within the framework of the World System, that is, one will observe a transition to the model of an elderly society with low mortality and very low birth rates, minimal growth and increasing share of the elderly and old population (in some societies and regions, this transition will be completed earlier). For more details about such an elderly society, named cybernetic society see Chap. 15.
2. The process of further adaptation to aging at the global, supranational and national levels will continue in socio-economic, institutional and political directions. To a large, and perhaps to a decisive extent, this will be associated with an improved health-related quality of life. Let us remember that we denote the largest demographic transformation (LDT-5) associated with the Cybernetic Revolution, as an *adaptation to aging* (about the latter see Chap. 7). The main adaptation mechanisms will be developed already at the final phase of the Cybernetic Revolution, but the institutionalization of the adaptation to aging will continue later together with the formation of a new LDT—the techno-medico-biological one (LDT-6). A new type of population reproduction will emerge (TPR-4 qualitative changes), which will be associated with the stabilization of population numbers, very high life expectancy, partial cyborgization and with growing involvement of people with disabilities in normal life with the development of the medical-biotechnological environment (MBTE).
3. Possible changes in reproductive technologies and approaches to childbearing.

Let us consider these trends in detail. We should note that *demographic trends will inevitably be closely related to economic ones*,[19] so we will try to show some aspects of this relationship.

**The stabilization of the population size on a global scale** may proceed at different pace, depending on the countries' demographic policies and the possible general agreements on this matter, as well as on many other circumstances or factors that are difficult to predict. We still believe that already during the final phase of the Cybernetic Revolution, the demographic policies of many countries will change in the direction of encouraging childbearing, although it is difficult to predict the outcome, especially if we take into account that it is difficult to

[19] They will also be very closely related to production, technological relationships and trends; we considered them in previous chapters, when discussing self-regulating systems. But they will also be associated with political trends and aspects, which will be discussed in Chap. 7.

restore traditions requiring great personal effort and sacrifice for the well-being of children if they are interrupted even within one generation. For such a policy to be successful, the material impetus should be supported by (a) a change in tax policy in terms of penalizing childless adults, a kind of tax on childlessness which is not discussed today. The introduction of such a tax may also face strong resistance from different groups and movements; and (b) the strengthening of patriotic or religious ideology, when the birth of children is considered as a manifestation of patriotism to the service of one's homeland (in some respects Israel is an example).

Also, many other factors will impede the fertility growth, including what is now called the short-term political vision (horizon), when politicians ignore long-term trends and projects and, paradoxically, the possible labor shortage, which will increase if the birth rate grows and some women will get involved in child care. But, despite the above said, we still believe that the chances are rather high that depopulation will at least slow down, while in the African and some other countries the decline of birth rates will accelerate, and provide a huge and growing demographic dividend (on the role of demographic dividends in different types of countries see Barsukov, 2019; Korotayev et al., 2022a, 2022b). The advances of new approaches to childbearing may also contribute here (see below). That is why we believe that Africa's demographic development will follow the medium or low scenarios, probably closer to the medium one.

**Further Adaptation to Aging and Improvements in Health-Related Quality of Life**. As we mentioned at the beginning of the chapter, the following changes can be expected in the quality of life.

On the one hand, as mentioned above, the improved quality of the material domain of life, that is, of the level of consumption, is likely to stabilize. It will continue to grow in the developing countries (especially in the less developed ones) while in the developed countries and even in the World System in general, this growth will be weak and perhaps even decline somewhat. This may happen for the following reasons: (a) the population growth in the World System as a whole will stabilize (or even begin to decline) and (b) the proportion of elderly people will increase dramatically and this group usually has stable or even falling incomes. Moreover, as we will discuss below, the increasing share of older people in the population may lead to a decline in the share of investments in GDP and a decrease in the rate of personal savings; the growth rate of GDP per capita is likely to decrease as well (Akaev et al., 2023; Aksoy et al., 2015; Grinin et al., 2021b; Kapelyushnikov, 2019b: 43, Table 6.1; Khaltourina & Korotayev, 2007; Korotayev, 2020; Korotayev et al., 2020; Malkov et al., 2022). In addition, older people hardly need a large amount of real estate and expensive durable goods (this is precisely the reason for the above-mentioned decline in investments).

On the other hand, *one may expect a really significant, and in some respects unprecedented improvements in health-related quality of life. As we argued, this will be the main trend for the next hundred years.* The process will start already during the final phase of the Cybernetic Revolution (in fact, it already started some time ago); however, it will acquire a peculiar scale and quality during the mature phases of the Cybernetic Production Principle. As we pointed out above, the impact of these changes on various aspects of life of the future society will be enormous, and they will also create or intensify new contradictions, conflicts, tensions and frictions in society; this will require serious transformations and, in many respects, transform the whole economic life related to these processes. In Chap. 7, we will discuss various scenarios associated with adaptation to aging.

One way or another, the increasing expenditures on the care, monitoring and treatment of the elderly, along with the creation of the necessary social, technological and psychological environment for them will inevitably become more and more significant both in financial and economic and in social terms. The same applies to disabled people. And this is not surprising, since we deal with a large part of the population (the elderly and the disabled), which, in addition, will constantly grow. *In other words, the funds will be redistributed between different aspects of the quality of life: the material domain is likely to receive less funding while the health-related domain of quality of life will get more.*

Let us make a few remarks on some other characteristics associated with a change in the direction of improvements in the quality of life.

**Some Twenty-First-Century Trends in Distribution and Consumption**. The process of aging, adaptation to it and improvements in health-related quality of life will require significant changes in the whole economic cycle from production to consumption. Let us consider this in a bit more detail.

**Distribution**. The shift from improving the quality of life in the narrow–material–sense to improving the health-related quality of life will be closely related to the changing age structure of society, i.e., to the process of aging and the attempts to adapt to it. Thus, the society should significantly change the distribution precisely in the direction of increasing the share of benefits, funds and resources allocated to the elderly, old and disabled people. Adaptation to aging and disability, creation of an appropriate techno-biological environment, health monitoring, treatment, care, etc.—all these aspects are very expensive. And although the transition to this trend has actually started, it still proceeds only with difficulties. Any change in distribution is a conflictual and complicated matter fraught with dramatic fluctuations. And we will talk about this in the paragraph on the pension system. However, the matter is not limited to the pension system, but concerns the general revision of the ideology of public spending, which should be eventually manifested in institutionalized ageism (which will be discussed in Chaps. 7 and 15).

**Taxes**. Changes in the tax system are more than likely, but this is a separate issue that we will not touch on here. Note that the ideas of ESG (Environmental + Social + Governance) and of the so-called stakeholders are already actively introduced into financial and economic practice, thus making businesses becoming responsible for the well-being of territories up to the whole planet (Schwab & Vanham, 2021)—this shows one of the possible directions for the transformation of the tax system: from clear rules and mechanisms to indefinite responsibility. Today it mostly refers to the climate change and the problems of the socially disadvantaged people but in the future cybernetic society it can be used as a mechanism for obtaining funds to help the elderly and infirm (see Chap. 15).

However, while preserving great importance in the economy and distribution system, *the reciprocity* will be combined with an increasing role of non-market mechanisms, which will be associated with reduced consumerism. Today there is much discussion on the need to reduce consumerism. But so far it is more the matter of ideology and someone's interests than real opportunities (especially in developing countries). While this method of encouraging sales has to date no substitute, trends of reducing consumerism will play a larger role in the future (see below).[20]

We will talk about the reduction of consumerism due to global aging in due course (see Chap. 13 for details). Note that the development of societies within the framework of the unfolding Cybernetic Production Principle toward higher degree of planning will also contribute to reducing consumerism (due to the spread of self-managing systems [Grinin & Grinin, 2019; Grinin, 2021a, 2021b]). An increasing role of planning also brings more restrictions for people, and here compromises are required, since conflicts are inevitable. Market mechanisms will shrink due to increasing planning and a growing share of public goods in the overall system of distribution and consumption, and other factors, in particular, due to the possible guaranteed income in society. But it is extremely dangerous to make haste of burying capitalism, market and related mechanisms; their importance will be preserved for a long time and is unlikely to disappear at any stage (Grinin et al., 2022a, 2022b, 2022c, 2023a, 2023b, 2023c).

**Consumption**. We can state the following. The role of human beings in the overall *production-consumption system* will change. It will decline in production (due to the reduced number and proportion of the working population and substitution of humans with self-regulating and self-managing systems) whereas the role of people will grow in the sphere of consumption. We have been observing this process for quite a long time, but in the future, due to the growing number of adult disabled people and elderly, this phenomenon will intensify.[21] As a result, let us repeat, it is necessary that the system moves toward a greater predictability and ordering of consumption (among other things, through a modified system of distribution and reduction of consumerism). Today, they speak much about reducing consumption because of the climate change. Such a reduction may become a reality for some periods due to a change in people's behavior (although this is highly improbable since the economic system is still firmly tied to the growth of consumption). The decrease in consumption may occur due to economic crises and problems, which we currently observe.

[20] Let us note that such a sequence of changes is typical for many transformations: at first, the ideology considerably runs ahead, and it also creates certain awareness. The main danger is to rush at introduction of the relations for which society is not ready yet. But another danger—to ignore promising ideas—is also great.

[21] In this context, the ideas about so−called unconditional basic income, while seeming fantastic at present, still reflect to the full extent the described trend, as well as the frightening predictions how robots and AI will force people out of production.

**Accumulation**. On the one hand, in order to support an ever-increasing number of elderly people, the society needs to increase savings for the future. It is precisely the scarcity of savings that poses one of the main threats to the pension system of an aging society (along with the decline in the available labor force and the lack of mechanisms to secure savings). On the other hand, the system develops in such a way that relations gradually make people save less at the expense of non-market and social support systems (this is driven by aging). Already today, the ideology of rejecting possession in favor of renting (the so-called 'sharing') has started to quickly spread. However, currently it has rather ideological and political roots, reflecting the interests of certain power strata. Hence, the desire for a rapid implementation of this trend seems extremely dangerous. Yet, in the future, these processes may intensify, supported by global aging, since from a certain age people are less inclined to save.

## 6.6 Aging, the Future Medical-Biotechnological Environment and New Largest Demographic Transformation

**What Do We Mean by Improvements in Health-Related Quality of Life?** Let us remember that we denote the type of population reproduction of the Cybernetic Production Principle as the *type of qualitative changes or social-qualitative* (TPR-4, see Chap. 5). And this makes it necessary to say a few words about the health-related quality of life. Of course, the health-related quality of life is in many ways integral to (a) the general quality of life; (b) the ecological and other environments and local climate; (c) the efficiency of material and civil rights and freedoms, including the free disposal of property and freedom of choice, also with respect to one's health; (d) the self-esteem even in the situation when a person cannot take care of himself and (e) the general moral and psychological security and climate in society.

**The Basis of Health-Related Quality of Life Includes but Is Not Reduced to Its Medical and Physiological Characteristics**. In biomedical studies, the quality of life (QoL) is defined as an integral characteristic of a patient's physical, psychological, emotional and social functioning based on his/her subjective perception (Bowling, 1996; Novik & Ionova, 2007; Solovieva & Naimushina, 2011; Spilker, 1996; Staquet, 1998). Here it is important to emphasize that this is an *integral characteristic*, which, in addition, is based on the *subjective perception of a patient*. However, we are considering a slightly different aspect, since we are talking about the health-related quality of life (a) of all healthy and sick people, of people of all ages, although we are of primarily interested in a group of a certain age; (b) we speak about an objective characteristic, which, according to the law of large numbers in sociology, is made up by subjective opinions of a huge number of people; (c) in medicine, the quality of life is determined through a focus on diagnosis, especially for different diseases (Novik & Ionova, 2007), while we speak about the generalized quality of life not focused on medical aspects.

The health-related quality of life includes: (1) the minimum functioning of all basic physiological functions and sensory organs; (1a) the minimum full and enjoying/satisfying performance of all physiological functions (food, sleep, rest, walks, etc.)[22]; (2) the minimum mental health (memory, mind, psyche, the ability to clearly understand the situation, etc.); (3) the ability to maintain competencies, work and/or be socially active (including obtaining and comprehending the necessary information about the life of society) and to serve the society as long as possible; (4) the ability to learn and acquire new knowledge, competencies, etc.; (5) the ability to be minimally mobile both in the place of residence and far beyond its borders, including traveling; (6) the ability to take care of oneself to the minimum sufficient measure until death or as long as possible; (7) statutory free medical assistance in case of both existing and possible new diseases; (8) the ability to feel as comfortable as possible without losing self-esteem, even if a person cannot fully function physically or mentally and needs partial or full care.

Of course, each point here implies many rather complicated and expensive matters. And of course, progress in any sphere is the great achievement of a society while the path to improvement is thorny, contradictory and difficult.

The improvements in health-related quality of life will actually be the most important component of adaptation to aging. But in addition, this adaptation extremely needs: (1) legal means to regulate the above-mentioned issues and relationships related to ageism, and what is of utmost importance, to regulate the use of technologies for the adaptation to aging and disability, whose significance will increase dramatically[23]; (2) the creation of various institutions that could help the elderly and old cohorts to better fit the new realities, from micro-groups, including the ones consisting of or oriented toward the elderly themselves, to the planning of cities, economic development, etc., with a focus on the elderly age groups and (3) the account for the interests of those age groups in political agenda and international relations. We think that fundamental international documents on this matter will be quite numerous, and interactions between various organizations, including pension, medical and other funds, will be very active. It is possible that political parties of a new type will appear, which will focus on certain age groups. However, one may reveal not only advantages but also dangers. Yet, one way or another, the institutions and mechanisms for adaptation to aging will be created, which we will discuss in more detail in the next chapter.

**Technological System and Techno-Medico-Biological Demographic Transformation**. The major point without which an advanced adaptation to aging (and improvements in health-related quality of life) seems impossible, of course,

---

[22] Satisfaction is also measured by various indices that have been actively employed for decades and that will become much more accurate, and numerous and will be widely used (especially if there emerge self-regulating systems related to condition monitoring), for example, the index of physical activity, the index of psychological functioning (see, e.g. Sjöland et al., 1997).

[23] A lot of legal procedures are already implemented: from requirements to include disabled people in the workforce to creating facilities for them in the streets and airports. But when it comes to adaptation to aging, there is much less progress observed.

is a breakthrough in medical technologies, solutions for care for the sick and infirm, monitoring, control, forecasting, etc. (see below). As a result, the techno-medical and biosocial mechanisms of adaptation to aging will be developed and this will contribute to the emergence of a new LDT-6 (defined as the techno-medico-biological one, see Chap. 5).

We denote the future largest world-system demographic transformation (LDT-6) as a *techno-medico-biological* one. It implies a transition to the model of total technological support for health-related quality of life, while stabilizing (or reducing) the population, with a very large share of the elderly people. And this will be connected with the started acceleration of the processes of creating a continuous MBTE (medical-biotechnological environment) and cyborgization, which is discussed in the Introduction.

Let us emphasize that a significant increase in the technological component of care for the elderly, the disabled, the infirm and other needy people is the ultimate condition for achieving a substantial extension of human lifespan. This is not just a matter of technologically advanced medicine, involving numerous self-regulating systems, but the situation when the *biotech environment will be more persistent. In other words, without such an environment, life may even become impossible*, comparable with today, when it is impossible to imagine ourselves without the Internet and gadgets. Just as hundreds of millions of people cannot live without constant medication, so in the future a huge part of the population will be unable to stay outside the MBTE for a long period of time. In addition to increasing longevity and physical comfort, this will, of course, create many threats and problems (one of which may be cohabitation, coexistence with robots and, in general, with self-regulating systems, most of them with AI). We will address the problems and dangers in Chap. 7.

Thus, this system will imply not only a breakthrough in medical technologies, but a powerful breakthrough in care technologies, continuous health monitoring (not only of the elderly and the disabled), control, telemedicine, psychological support, prediction, prescription and emergency care (see below). Psychological support (both therapeutic and emergency intervention), also likely to be provided in most cases by the robotic systems, will become essential (e.g., in the case of a suicide threat; see more in Chap. 10). In addition, it will be necessary to create a system for continuous communication so that people would not feel lonely (the prototypes already exist: for example, pilot projects are carried out in hotels where robots entertain guests). Again, such a system, having detected alarming signals in the mood of the interlocutors, can transfer the communication with them directly to a specialist (of course, after recording the conversation and analyzing it via computer or other analytical means). It is quite possible that recommendations and instructions on what to do in typical cases of deviations from the norm and abnormal behavior would be pre-installed in such a system. Prediction will make it possible to carry out propaedeutic tests on people in order to detect certain deviations, including mental ones, which threaten others with outbreaks of violence (shootings, etc.). Urgent help is also possible in some critical situations, when delay can threaten the life of the observed person. That is, to perform some actions with the help of technology under certain circumstances, for example, to give an injection, artificial respiration, provide first aid or something similar. Or give instructions to other people (systems) to perform such actions, as well as urgently contact medical technology centers. But on the way to such a system there will be many real problems related to legal, moral and political difficulties, as well as to the legitimate reluctance to be under 4-h supervision and to follow the prescriptions of inanimate systems, as well as to be chipped and the like. We will return to the analysis of such a system in Chaps. 8 and 11 on aging.

Of course, the path to such a medical-biotechnological environmental system cannot be quick, let alone smooth, given the attempts of certain forces to use this trend to their advantage. We believe that this movement will finally lead to an *increasing symbiosis between technological and demographic development in the techno-medical and techno-biological terms.*[24]

**Development of Reproductive Technologies and the Relationships Associated with Childbearing**. First, we believe that the already implemented reproductive technologies will develop significantly, so that greatly improved IVF, etc. can become widespread, and infertility will be better cured; respectively, the chances of having a child will increase. In particular, only half a million children are currently born worldwide with the help of IVF (Denisova, 2021: 109; Fauser, 2019; Sciencedaily, 2018), that is, approximately 0.4% of all children born—a surprisingly/remarkably low number, considering that such technologies have been developed for decades. The main reason is that they are still not very effective: more than 2.5 million IVF cycles are performed worldwide each year (ibid.); that means that only one-fifth of all attempts become successful (WHO, 2023). So far, IVF is very expensive (a single IVF cycle costs $12,400, with the average American couple spending about $ 20,000 [Wendorf, 2022]); however, there are predictions of robotization of all processes. Meanwhile, according to embryologists, today a considerable part of all their work is routine, which in the future may be performed by self-regulating systems (Denisova, 2021: 212). Now 15% of couples of reproductive age worldwide are affected by infertility (WHO 2023). In the 30 years since the world's first IVF (that took place in 1978), more than 8 million babies have been born using IVF (Sciencedaily, 2018). Thus, taking into account that the number of childless couples and women in the world annually grows by 5–10%,

---

[24] Certainly, many problems will emerge requiring solutions in terms of adapting these inventions to societal life, especially with respect to the elderly (see Herrmann, 2014). The COVID-19 pandemic has shown how conflicting any kind of restrictions can be, especially if they are sharp, sudden, not completely transparent and the need is—to some extent/for some—questionable (see Chap. 14; see also Grinin, 2021a). Such a medical-biotechnological environmental system can cause skepticism and even rejection. We see such contradictions today too. Indeed, the system of compulsory and universal health care has been introduced in many countries, but in the USA tens of millions of people cannot or refuse to take out medical insurance or to make payments. It would seem that they do harm to themselves, but it is extremely difficult to force them. Thus, the path to constant monitoring of the life cycle of the elderly, who are accustomed to independence, will cause conflicts and difficulties, requiring patience, compromise and adaptation.

it can be assumed that the number of children born through IVF will increase and may eventually reach 4–10% of all those born. For instance, in Israel almost 5% of all babies are now born through IVF.

Second, it is possible that procedures of freezing eggs and sperm will become widespread, which will de facto extend the childbearing age. Today, there are already many cases when women after menopause have children with the help of IVF and donation; there are also cases of the birth of children conceived from sperm of a dead man. Third, it is possible that the legislation, and respectively, the moral restrictions and the complexity of the relationship between the parties in the sphere of surrogate motherhood will be improved, and that a large number of women from less developed countries will become such mothers for families from the developed world. Fourth, it is possible that medicine will develop toward semi-artificial rearing of children (i.e., completely new technologies will appear), which may help to increase the birth rate. Already today prototypes of such technologies are being created. For example, there is a lot of talk about the artificial uterus (Denisova, 2021) and there are even reports of the first steps being taken in this direction.

## 6.7 Life Expectancy and Its Probable Fluctuations

**During the modernization phase of the Cybernetic Revolution**, the life expectancy in general, and the life expectancy for an individual at the age of 60 in particular, increased for several decades. Thus, by 2015, the global life expectancy of an individual aged 60 was 20 years, which is undoubtedly a significant increase (WHO, 2017: 11). The global life expectancy at birth has increased from 46.5 years in 1950 to 66.5 years in 2000 (i.e., by 20 years), by more than five years between 2000 and 2015 (to 71.8 years), and by another year between 2015 and 2019 (to 72.8 years). Yet, between 2019 and 2021, as a result of the COVID-19 pandemic, it dropped very noticeably—from 72.8 to 71, i.e., by almost two years (UN Population Division, 2022a, 2022b, see Fig. 6.6):

Note that here we speak about the global life expectancy. At the same time, overall life expectancy and life expectancy at older ages grow faster in developing countries than in developed ones, which is not surprising. In fact, between 1950 and 2019, the life expectancy increased by 16 years in developed countries while in developing countries it increased by 30 years (see Fig. 6.6).[25]

Once the level of 78–82 years is reached in developed countries, further growth slows down noticeably. Each additional year is gained with difficulty. COVID-19 hit this growth considerably (see Fig. 6.6), and finally even reduced it, since it was the age group of the elderly and very old people who suffered the most (yet

[25] By the way, it should be noted that the gap between developing and developed countries is also decreasing in relation to the above-mentioned the health-adjusted life expectancy (HALE) index. The most intense growth in HALE can be observed in low-income countries; as wealth increases, growth in health-adjusted life expectancy slows down (see Sindyashkina, 2022 for details).

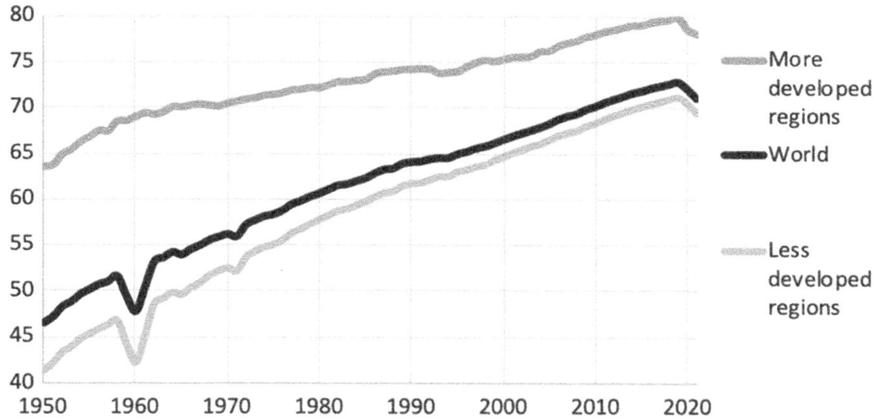


**Fig. 6.6** Life expectancy dynamics in the world, in the more developed regions and in the less developed regions, years, 1950–2021. *Data source* UN Population Division (2022a)

in the USA, this rollback began already before the pandemic [see, e.g., Grinin & Korotayev, 2023; Zhdanov & Korotayev, 2022]). The economic difficulties also produce a negative effect. We hope that within a few years, the life expectancy in developed countries will return to, or even slightly exceed, previous levels. But we do not expect any significant breakthrough in the next decade or more. However, in developing countries the process is likely to continue, although it will hardly develop as fast as before. It is expected that in 2050 the global average life expectancy will be around 77.2 years (UN Population Division, 2022b) mostly in developing countries; but, this seems doubtful.

**During the final phase of the Cybernetic Revolution**, within the four to five decades that this phase will last, we expect significant changes in life expectancy, fundamental breakthroughs in medicine and related fields are expected (see Chaps. 8, 11). In Fig. 6.7 we present the latest UN Population Division's projections.

However, a breakthrough increase in life expectancy will take some time. Moreover, the global-scale results will take even longer. Therefore, according to our forecasts, a noticeable increase in life expectancy may manifest itself no earlier than in the second half of the 2040s in developed countries, and in the 2050s in developing countries. Speaking about a noticeable increase in life expectancy, we mean additional five to ten years for developed countries (to the highest point of 2019, and it will rather be closer to five years), and ten years for developing ones. How can this be achieved, and what are the possible obstacles and variations? First, the advances in the treatment of age-related diseases—oncology, cardiovascular diseases, strokes, respiratory diseases and hypertension—as well as a decrease in the impact on mortality of such diseases as dementia, diabetes,

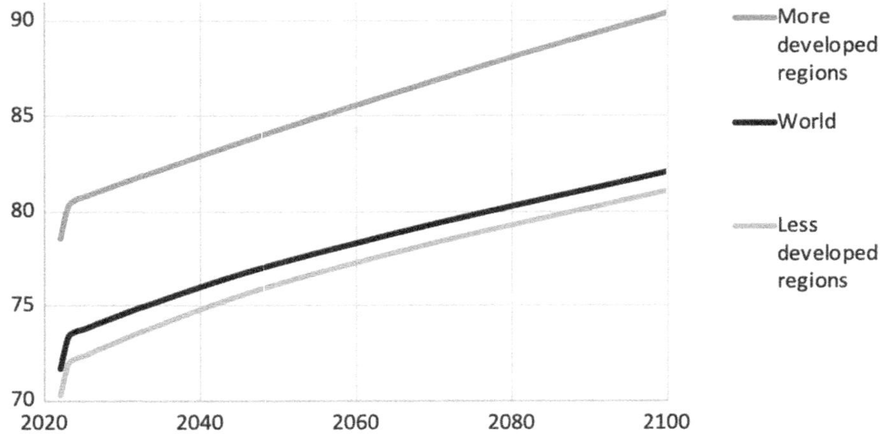


**Fig. 6.7** Life expectancy dynamics in the world, in the more developed regions and in the less developed regions, years, forecast for 2022–2100. *Data source* UN Population Division (2022a)

etc.,[26] can bring about the largest gains in life expectancy (see Chap. 8 for details). In less developed countries, the child mortality is expected to fall, as well as the deaths from malaria, AIDS, intestinal infections and other diseases, whose impact has sharply declined in developed countries. But in connection with the increasing life expectancy, the number of cardiovascular and oncological diseases will rise.

Second, with all the expected achievements in medicine, the life expectancy will not increase indefinitely and may stagnate after a further breakthrough. In addition, the threat of pandemics and other factors (and perhaps, the economic burden of supporting the growing number of old people) will be a very important constraint. It is possible and even likely that the life expectancy will increase with interim setbacks.

Third, the possible increase in life expectancy will depend not only on the achievements in medical and other technologies, but also on the extent to which the adaptation of societies to the aging process will advance and on the success of institutions that promote healthy aging and fullness of life. It is a known fact that the immediate environment, the atmosphere of communication with others, a sense of purpose, different activities, etc., are a powerful factor contributing both to the health conditions and to the joy of life, the desire to fight for it, the will to live, which old people often lack. The Roman wisdom that a healthy mind is in a healthy body (Iuvenal) is generally accepted, while the idea that it is a healthy mind that maintains health and the will to live is less discussed. A stable international order is also essential in this context, since in the face of escalated

[26] The number of people with dementia in the world today is in the tens of millions, and will be in the hundreds of millions in the future, since about eight million new cases of dementia are reported worldwide each year (WHO, 2013: 12).

tension and confrontation, the elderly people's mental state can hardly remain optimal. Thus, we see that the efforts to increase life expectancy in every country are much broader than just medical and social technologies and systems, and they also require global peace (see in this context the work on Social Quality, e.g., van der Maesen et al., 2011).

Fourth, the increase in life expectancy will unfold more in terms of scale than in duration and will affect less developed societies. As a result, by the 2060s, the gap in life expectancy between developed and developing countries will narrow compared with the current period—we disagree with the latest UN Population Division forecast, which completely ignores the effect of the diffusion of technologies of the second demographic transition from economically developed countries to developing ones, which we confidently predict.

**The Mature Stages of the Scientific-Cybernetic Production Principle**. One can expect that as a result of the completion of the Cybernetic Revolution and the development of adaptation to aging, *new levels of life expectancy will be reached, perhaps 10–15 years more than in the pre-Covid period in the developed countries, that is, life expectancy in some countries may exceed 90 or even more years, and it is not out of the question to go closer to the 100 years* (see also Fig. 6.7), although to reach even 90+ is possible only under very favorite conditions (about negative scenarios see Chaps. 7 and 15). Of course, it is still difficult to expect considerable breakthroughs at present. At the same time, the increasing life expectancy will be more and more noticeable as a result of an ever-expanding MBTE (medical-biotechnological environment). However, already today scientists and doctors search for the means to slow down or stop aging, and some directions (which we will discuss in the relevant chapter, Chap. 11) look quite promising. But in any case, the health-related quality of life should be improved in general, although there are fears that this can only be achieved by restricting the freedom of the elderly through technological means. Since depopulation also can hardly go on indefinitely, so at some point the population should either (a) stabilize or (b) move toward a growing population, which will possibly and probably be a controlled process. This may occur due to the development of technologies for artificial childbearing. In this case, the technological and demographic revolutions of the future will merge (into a techno-demographic revolution). One should also keep in mind the role of population in the future world order (i.e., the country's vote, its claims on common resources will depend on the population; or, vice versa, its responsibilities will also depend on it).

**Acknowledgements** This chapter has been prepared with support by the Russian Science Foundation (Project No. **23-11-00160**).

## References


Akaev, A., Malkov, S., Grinin, L., Bilyuga, S., Davydova, O., Grinin, A., Kovaleva, N., Malkov, A., Musieva, J., & Korotayev, A. (2023). Modeling social self-organization and historical dynamics. Modern society and a look into the global futures: Cybernetic W-society. In V. Sadovnichy et al. (Eds.), *Reconsidering the limits to growth. A report to the Russian association of the club of Rome* (pp. 419–460). Springer. https://doi.org/10.1007/978-3-031-34999-7_19

Akimov A. (1999). Na poroge sed'mogo milliarda. Naseleniye i obshchestvo. *Informatsionnyy byulleten' tsentra demografii i ekologii cheloveka instituta narodokhozyaystvennogo prognozirovaniya RAN* № 39, September.

Aksoy, Y., Basso, H. S., Smith, P. R., & Grasl, T. (2015). *Demographic structure and macroeconomic trends.* Bancode de Espana. Documentos de Trabaja. №1628.

Alkema, L., Raftery, A. E., Gerland, P., Clark, S. J., Pelletier, F., Buettner, T., & Heilig, G. K. (2011). Probabilistic projections of the total fertility rate for all countries. *Demography, 48*(3), 815–839. https://doi.org/10.1007/s13524-011-0040-5

Andersen, L. R. (2020). *Bildung—keep growing. A report to the club of Rome.* Nordic Bildung.

Arkhangelsky, V., Bogevolnov, J, Goldstone, J., Khaltourina, D., Korotayev, A., Malkov, A., Novikov, K., Ryazantsev, S., Rybalchenko, S., Shulgin, S., Steck, P., Yuriev, Y., Zinkina, J., & Zvereva, N. (2015). *Critical 10 years. Demographic policies of the Russian Federation: Successes and challenges.* Russian Presidential Academy of National Economy and Public Administration (RANEPA).

Barsukov, V. N. (2019). From the demographic dividend to population ageing: World trends in the system-wide transition. *Economic and Social Changes: Facts, Trends, Forecast, 12*(4), 167–182. https://doi.org/10.15838/esc.2019.4.64.11

Bengtson, V. (Ed.). (2018). *Global ageing and challenges to families.* Routledge.

Berg, C. (2019). *Sustainable action: Overcoming the barriers.* Routledge.

Bhattacharjee, N. V., Schumacher, A. E., Aali, A., Abate, Y. H., Abbasgholizadeh, R., Abbasian, M., ... & Bahri, R. A. (2024). Global fertility in 204 countries and territories, 1950–2021, with forecasts to 2100: a comprehensive demographic analysis for the Global Burden of Disease Study 2021. The Lancet. https://doi.org/10.1016/S0140-6736(24)00550-6

Bloom, D. E., & Luca, D. L. (2016). The global demography of ageing: Facts, explanation, future. *Handbook of Economics of Population Ageing, 1A*, 3–56.

Bös, D., & Von Weizsäcker, R. K. (1989). Economic consequences of an ageing population. *European Economic Review, 33*(2–3), 345–354.

Bowling, A. (1996). *Measuring disease: A review of disease-specific quality of life measurement scales.* Open University Press.

Bozesan, M. (2020). *Integral investing. From profit to prosperity. A report to the club of Rome.* Springer. https://doi.org/10.1007/978-3-030-54016-6.

Chen, C., Maung, K., Koh, J., & Rowe, J. W. (2022). Global differences in adaptation to societal aging: A cross-sectional comparison of 143 countries. https://www.ncbi.nlm.nih.gov/pmc/articles/PMC9766445/

Chesnais, J. C. (1992). *The demographic transition: Stages, patterns, and economic implications.* Clarendon Press.

Chumakov, A. N., & Yurchenko, P. S. (2021). Global pandemic in terms of objective causes and subjective speculations. *Journal of Globalization Studies, 12*(2), 157–168.

Cincotta, R., & Weber, H. (2021). Youthful age structures and the risks of revolutionary and separatist conflicts. In A. Goerres, & P. Vanhuysse (eds.), *Global political demography: Comparative analyses of the politics of population change in all world regions* (pp. 57–92). Palgrave. https://doi.org/10.1007/978-3-030-73065-9_3

Clark, R. L., & Spengler, J. J. (1980). *The economics of individual and population ageing.* Cambridge University Press.

Collier, P., & Kay, J. (2020). *Greed is dead: Politics after individualism.* Penguin Books.

Collier, P. (2018). *the future of capitalism: Facing the new anxieties.* Harper.

Denisova, I. V. (2021). *Sdelay menya tochno. Kak reproduktivnyye tekhnologii menyayut mir.* Individuum.

Diamond, J. (1999). *Guns, germs, and steel: The fates of human societies.* Norton.

Dyson, T. (2010). *Population and development.* Zed Books.

Efremov, I., & Arkhangelsky, V. (2023). Global fertility changes in the context of the COVID-19 pandemic. *Journal of Globalization Studies, 14*(2), 165–170.

Fauser, B. (2019). Towards the global coverage of a united registry of IVF outcomes. *Reprod Biomed Online, 2*(38).

Fichtner, J. J. (2018). Global ageing and public finance. *Business Economics, 53*(2), 72–78.

Fukuyama, F. (2002). *Our post-human future: Consequences of the bio-technology revolution.* Farrar, Straus, and Giroux.

Gallagher, J. (2020). Fertility rate: 'Jaw- dropping' global crash in children being born. https://www.bbc.com/news/health-53409521.

Goldstone, J. A. (2002). Population and security: How demographic change can lead to violent conflict. *Journal of International Affairs, 56*(1), 3–21.

Goldstone, J. A. (2015). Population ageing and global economic growth. In J. A. Goldstone, L. Grinin, & A. Korotayev (Eds.), *History & mathematics: Demography & ageing* (pp. 147–155). Uchitel.

Goldstone, J. A., Grinin, L., & Korotayev, A. (2015). Research into global ageing and its consequences. In J. A. Goldstone, L. Grinin, & A. Korotayev (Eds.), *History & mathematics: Political demography & global ageing* (pp. 5–9). Uchitel.

Grinin, L. (2009). Privedet li global'niy krizis k global'nym izmeneniyam? *Vek Globalizatsii, 2*, 117–140.

Grinin, L. (2010). Which global transformations would the global crisis lead to? *Age of Globalization, 2*, 31–52.

Grinin, L. (2012a). Arabskaya vesna i rekonfiguratsiya Mir-Sistemy. In A. Korotayev, J. Zinkina, & A. Khodunov (Eds.), *Sistemnyy monitoring global'nykh i regional'nykh riskov* (pp. 188–223). URSS, LIBROKOM.

Grinin, L. (2012b). Rekonfiguratsiya mira, ili nastupayuschaya epokha novykh koalitsiy. *Istoriya i Sovremennost, 2*, 3–27.

Grinin, L. (2020b). Dvizheniye k ustoychivym demokraticheskim otnosheniyam i natsional'nomu gosudarstvu kak dlitel'nyy faktor vozmozhnoy destabilizatsii v stranakh afraziyskoy makrozony nestabil'nosti. In L. Grinin, A. Korotayev, & D. Bykanova (Eds.), *Sistemnyy monitoring global'nykh i regional'nykh riskov* (pp. 694–724). Uchitel.

Grinin, L. (2020c). Dvizheniye obshchestv k zrelym sotsial'no-politicheskim i ekonomicheskim otnosheniyam kak dlitel'nyy faktor vozmozhnoy destabilizatsii v stranakh afraziyskoy makrozony nestabil'nosti. In L. Grinin, A. Korotayev, & D. Bykanova (Eds.), *Sistemnyy monitoring global'nykh i regional'nykh riskov* (pp. 666–693). Uchitel.

Grinin, L. (2020d). Radikal'nyy islamizm, religioznyy faktor i destabilizatsiya v stranakh afraziyskoy makrozony nestabil'nosti. In L. Grinin, A. Korotayev, & D. Bykanova (Eds.), *Sistemnyy monitoring global'nykh i regional'nykh riskov* (pp. 810–828). Uchitel.

Grinin, L. (2020e). How can COVID-19 change geopolitics and economy? *Journal of Globalization Studies, 11*(2), 119–131.

Grinin, L. E. (2023a). Accelerating reconfiguration of the world-system in relation to the special military operation in Ukraine and possible future scenarios. Article 1. The impact of the special military operation in Ukraine on the reconfiguration of the world-system. *Vek Globalizatsii, 2*, 50–72. https://doi.org/10.30884/vglob/2023.02.06

Grinin, L. E., & Grinin, A. L. (2019). Sovremennyye globalnyye tendentsii i prognozy na XXI stoletiye. *Istoriya i Sovremennos, t, 4*, 3–35.

Grinin, L. E., & Grinin, A. L. (2021a). Razmyshleniya ob ekonomicheskom roste i budushchem. Stat'ya pervaya. Globalizm vs rost VVP i «zakat Zapada». *Filosofiya i Obshchestvo, 3*, 5–34. https://doi.org/10.30884/jfio/2021.03.01

Grinin, L. E., & Grinin, A. L. (2021b). Razmyshleniya ob ekonomicheskom roste i budushchem. Stat'ya vto-raya. Globalizm, neosotsializm, spaseniye klimata i ekonomicheskiy rost. *Filosofiya i Obshchestvo, 4*, 5–38. https://doi.org/10.30884/jfio/2021.04.01

Grinin, L. E., & Grinin, A. L. (2023). Demographic transformations in the historical process in the light of technological development: theoretical approach. *Social Evolution & History, 22*(2), 159–202.

Grinin, L., & Korotayev, A. (2011). The coming epoch of new coalitions: Possible scenarios of the near future. *World Futures, 67*(8), 531–563.

Grinin, L., & Korotayev, A. (2012). Does 'Arab Spring' mean the beginning of world system reconfiguration? *World Futures, 68*(7), 471–505. https://doi.org/10.1080/02604027.2012.697836

Grinin, L., & Korotayev, A. (2015a). *Great divergence and great convergence. A global perspective.* Springer International Publishing.

Grinin, L., & Korotayev, A. (2016). Global population ageing, the sixth Kondratieff wave, and the global financial system. *Journal of Globalization Studies, 7*(2), 11–31.

Grinin, L., Korotayev, A., & Grinin, A. (2016). Global'noye stareniye naseleniya, shestoy kondrat'yevskiy tsikl i kiberneticheskaya revolyutsiya. *Problemy Teorii i Praktiki Upravleniya, 7*, 54–59.

Grinin, L., Grinin, A., & Korotayev, A. (2017a). Forthcoming Kondratieff wave, cybernetic revolution, and global ageing. *Technological Forecasting and Social Change, 115*, 52–68.

Grinin, L. E., Korotayev, A. V., & Tausch, A. (2019). *Islamism, Arab spring, and the future of democracy*. Springer.

Grinin, L., Grinin, A., & Korotayev, A. (2020). A quantitative analysis of worldwide long-term technology growth: From 40,000 BCE to the early 22nd century. *Technological Forecasting and Social Change, 155*, 119955. https://doi.org/10.1016/j.techfore.2020.119955

Grinin, L., Grinin, A., & Korotayev, A. (2021b). Global trends and forecasts of the 21st century. *World Futures, 77*(5), 335–370. https://doi.org/10.1080/02604027.2021.1949939

Grinin, L., Grinin, A., & Korotayev, A. (2022a). COVID-19 pandemic as a trigger for the acceleration of the cybernetic revolution, transition from E-government to E-State, and change in social relations. *Technological Forecasting and Social Change, 175*, 1–15. https://doi.org/10.1016/j.techfore.2021.121348

Grinin, L., Grinin, A., & Korotayev, A. (2023a). Demographic transformations in the light of technological development: Types of demographic reproduction in the past and in the future. *Social Evolution & History, 22*(2), 159–202.

Grinin, L. E., & Grinin A. L. (2015). *Ot rubil do nanorobotov. Mir na puti k epokhe samoupravlyaemykh sistem (istoriya tekhnologiy i opisanie ikh budushchego).* Uchitel.

Grinin, L. E., Grinin, A. L. (2016). *The cybernetic revolution and the forthcoming epoch of self-regulating systems.* Moscow Branch of 'Uchitel' Publishing House.

Grinin, L. E., & Korotayev, A. V. (2010a). Will the global crisis lead to global transformations? 2. The coming epoch of new coalitions. *Journal of Globalization Studies, 1*(2), 166–183.

Grinin, L. E., & Korotayev, A. V. (2010b). *Global'nyy krizis v retrospektive. Kratkaya istoriya poyemov i krizisov ot Likurga do Alana Grinspena.* M.: Knizhnyy dom LIBROKOM.

Grinin, L., & Korotayev, A. (2015b). Global'noye stareniye naseleniya, shestoy tekhnologicheskiy uklad i mirovaya finansovaya Sistema. In Grinin, L. E. & Korotayev, A. V. (eds.), *Istoriya i matematika: futurologicheskiye i metodologicheskiye aspekty: yezhegodnik* (pp. 31–56). Uchitel.

Grinin, L., & Korotayev, A. (2023). Africa—The continent of the future. Challenges and opportunities. In V. Sadovnichy et al. (Eds.), *Reconsidering the limits to growth. A report to the Russian association of the club of Rome* (pp. 225–238). Springer. https://doi.org/10.1007/978-3-031-34999-7_13

Grinin, L., Korotayev, A., Issaev, L., Shishkina, A., Ivanov, E., & Meshcherina, K. (2017), *World order transformation and sociopolitical destabilization.* National Research University Higher School of Economics. Series: International Relations, WP BRP 29/IR/2017.

Grinin, L. I., Korotayev, A. V., & Malkov, S. Y. (2022). Developed and developing countries: towards the common target with different speeds. *Russian Journal of Economics and Law, 16*(2), 229–244. https://doi.org/10.21202/2782-2923.2022.2.229-244

Grinin, L., Malkov, S., Grinin, A., & Korotayev, A. (2022). Will capitalism die? Reflections on the capitalism of the past, present and future. *Sotsiologicheskiy Zhurnal = Sociological Journal, 28*(2), 100–130. https://doi.org/10.19181/socjour.2022.28.2.8988

Grinin, L., Grinin, A., & Malkov, S. (2023). Socio-political transformations. A difficult path to cybernetic society. In V. Sadovnichy et al. (Eds.), *Reconsidering the limits to growth. A report to the Russian Association of the Club of Rome* (pp. 169–189). Springer. https://doi.org/10.1007/978-3-031-34999-7_10

Grinin, L., Malkov, S., & Korotayev, A. (2023). High income and low income countries. Toward a common goal at different speeds. In V. Sadovnichy et al. (Eds.), *Reconsidering the limits to growth. A report to the Russian Association of the Club of Rome* (pp. 207–224). Springer. https://doi.org/10.1007/978-3-031-34999-7_12

Grinin, L. E. (2006). *Proizvoditel'nye sily i istoricheskiy protsess*. 3rd edn. KomKniga.

Grinin, L. (2020a). Demograficheskiye protsessy kak bazovyy i dlitel'nyy faktor vozmozhnoy destabilizatsii v stranakh afraziyskoy makrozony nestabil'nosti. In L. Grinin, A. Korotayev, & D. Bykanova (Eds.), *Sistemnyy monitoring global'nykh i regional'nykh riskov* (pp. 181–199). Uchitel.

Grinin, L. (2022a). On revolutionary situations, stages of revolution, and some other aspects of the theory of revolution. In J. A. Goldstone, L. Grinin, & A. Korotayev (Eds.), *Handbook of revolutions in the 21st century: The new waves of revolutions, and the causes and effects of disruptive political change* (pp. 69–104). Springer. https://doi.org/10.1007/978-3-030-86468-2_3

Grinin, L. (2022b). On revolutionary waves since the 16th century. In: J. A. Goldstone, L. Grinin, & A. Korotayev (Eds.), *Handbook of revolutions in the 21st century: The new waves of revolutions, and the causes and effects of disruptive political change* (pp. 389–411). Springer. https://doi.org/10.1007/978-3-030-86468-2_13

Grinin, L. (2022c). Revolutions and historical process. In J. A. Goldstone, L. Grinin, & A. Korotayev (Eds.), *Handbook of revolutions in the 21st century: The new waves of revolutions, and the causes and effects of disruptive political change* (pp. 139–171). Springer. https://doi.org/10.1007/978-3-030-86468-2_5

Grinin, L. E. (2023b) Accelerating reconfiguration of the World-System in Relation to the Special Military Operation in Ukraine and possible future scenarios. Article 2. Possible scenarios for the Reconfiguration of the World-System and world order. *Vek globalizatsii, 3*(47). https://doi.org/10.30884/vglob/2023.03.07

Herrmann, P. (2014). *Social policy—production rather than distribution. A rights-based approach.* EHV Academicpress.

Hsu, Y. H., & Lo, H. C. (2019). The impacts of population ageing on saving, capital formation, and economic growth. *American Journal of Industrial and Business Management, 9*(12), 2231–2246. https://doi.org/10.4236/ajibm.2019.912148

Hussain, M. (2021). The changing dynamics of liberal international order: Covid-19 and its global implications. *Journal of Globalization Studies, 12*(2), 143–156.

Irshad, S. M. (2020). COVID-19 and global 'At-Risk Community': From benefit-sharing to risk-sharing of economic crisis. *Journal of Globalization Studies, 11*(2), 135–145.

Jackson, R., Howe, N., Strauss, R., & Nakasima, K. (2008). *The greying of the great powers. Demography and geopolitics in the 21st century*. Center for Strategic & International Studies.

Jackson, R., Macaranas, R., & Peter, T. (2013). *U.S. development policy in an ageing world. New challenges and new priorities for a new demographic era.* A Research Report from Project on U.S. Leadership. Lanham etc.: CSIS (Center for Strategic and International Studies).

Jackson, R. (2021). *The shape of things to come. The macro challenges of population ageing.* Concord Coalition & the Global Ageing Institute.

Jasiński, A. M., & Bąkowska, A. M. (2020). Globalization and life satisfaction of the polish population. Before and during the COVID-19 pandemic. *Journal of Globalization Studies, 11*(2), 146–155.

Kakkar, S. (2021). The outbreak of sars-cov-2 (Covid-19): Impact on international tourism. *Journal of Globalization Studies, 12*(1), 145–158.
Kamiya, Y., & Hertog, S. (2020). *Measuring household and living arrangements of older persons around the world.* United Nations, Department of Economics and Social Affairs, Population Division.
Kaneda, T., Greenbaum, C., & Patierno, K. (2016). *World population data sheet.* Population Reference Bureau. https://www.prb.org/2016-world-population-data-sheet/
Kapelyushnikov, R. (2019a). Fenomen stareniya naseleniya: Ekonomicheskiye effekty. *Ekonomicheskaya Politika, 14*(2), 8–63.
Kapelyushnikov, R. (2019b). *Fenomen stareniya naseleniya: ekonomicheskiye effekty.* Institut ekonomiki RAN.
Kebede, E., Goujon, A., & Lutz, W. (2019). Stalls in Africa's fertility decline partly result from disruptions in female education. *Proceedings of the National Academy of Sciences, 116*(8), 2891–2896. https://doi.org/10.1073/pnas.1717288l1
Khaltourina, D., & Korotayev, A. (2007). A modified version of a compact mathematical model of the world system economic, demographic, and cultural development. In M. G. Dmitriev, A. P. Petrov, & N. P. Tretyakov (Eds.), *Mathematical modeling of social and economic dynamics* (pp. 274–277). RUDN.
Korotayev, A., & Zinkina, J. (2015). East Africa in the Malthusian trap? *Journal of Developing Societies, 31*(3), 385–420. https://doi.org/10.1177/0169796X15590322
Korotayev, A., Klimenko, V., & Proussakov, D. (1999). Origins of Islam: Political-anthropological and environmental context. *Acta Orientalia Hung., 52*, 243–276.
Korotayev, A., Zinkina, J., Goldstone, J., & Shulgin, S. (2016). Explaining current fertility dynamics in Tropical Africa from an anthropological perspective: A cross-cultural investigation. *Cross-Cultural Research, 50*(3), 251–280. https://doi.org/10.1177/1069397116644158
Korotayev, A., Romanov, D., Zinkina, J., & Slav, M. (2023b). Urban youth and terrorism: A quantitative analysis (are youth bulges relevant anymore?). *Political Studies Review, 21*(3), 550–574. https://doi.org/10.1177/14789299221075908
Korotayev, A., & Zinkina, J. (2022). Egypt's 2011 revolution. A demographic structural analysis. In J. A. Goldstone, L. Grinin, & A. Korotayev (Eds.), *Handbook of revolutions in the 21st century: The new waves of revolutions, and the causes and effects of disruptive political change* (pp. 651–683). Springer. https://doi.org/10.1007/978-3-030-86468-2_25
Korotayev, A., Malkov, A., & Khaltourina, D. (2006). *Introduction to social macrodynamics: Compact macromodels of the world system growth.* KomKniga/URSS.
Korotayev, A., Zinkina, J., Kobzeva, S., Bogevolnov, J., Khaltourina, D., Malkov, A., & Malkov, S. (2011). A trap at the escape from the trap? Demographic-structural factors of political instability in modern Africa and West Asia. *Cliodynamics, 2*(2), 276–303. https://doi.org/10.21237/C7clio22217
Korotayev, A., Bilyuga, S., & Shishkina, A. (2020). Which countries generate Kondratieff waves in global GDP growth rate dynamics in the contemporary world? *Journal of Globalization Studies, 11*(1), 33–63. https://doi.org/10.30884/jogs/2020.01.03
Korotayev, A., Sawyer, P., Gladyshev, M., Romanov, D., Shishkina, A. (2021). Some sociodemographic factors of the intensity of anti-government demonstrations: Youth Bulges, urbanization, and protests. *Sociologiceskoe Obozrenie = Russian Sociological Review, 20*(3), 98–128. https://doi.org/10.17323/1728-192x-2021-3-98-128
Korotayev, A., Shulgin, S., Zinkina, J., & Slav, M. (2022). Estimates of the possible economic effect of the demographic dividend for sub-saharan Africa for the period up to 2036. *Vostok (Oriens), 2*, 108–123. https://doi.org/10.31857/S086919080019128-7
Korotayev, A., Shulgin, S., Ustyuzhanin, V., Zinkina, J., & Grinin, L. (2023a). Modeling social self-organization and historical dynamics. Africa's futures. In V. Sadovnichy et al. (eds.), *Reconsidering the limits to growth. A report to the Russian association of the club of Rome* (pp. 461–490). Springer. https://doi.org/10.1007/978-3-031-34999-7_20
Korotayev, A. V. (2020). The 21st century singularity in the big history perspective. A re-analysis. In: Korotayev, A. V., & LePoire, D. (eds). The 21st century singularity and global futures. A big

history perspective (pp. 19–75). Springer, Cham. https://doi.org/10.1007/978-3-030-33730-8_2
Korten, D. C. (2015). *Change the story, change the future. A report to the club of Rome*. Berrett-Koehler Publishers.
Kuenkel, P. (2019). *Stewarding sustainability transformations*. Springer. https://doi.org/10.1007/978-3-030-03691-1
Lee, F. (2007). Climatic change and Chinese population growth. *Human Geography, 24*(3), 347–364.
Lee, F., & Zhang, D. D. (2010). Changes in climate and secular population cycles in China, 1000 CE to 1911. *Climate Research, 42*(3), 235–246.
Lee, R. D., Arthur, W. B., & Rodgers, G. (Eds.). (1988). *Economics of changing age distributions in developed countries*. Oxford University Press.
Lee, R. D., Mason, A., & Cotlear, D. (2010). *Some economic consequences of global ageing. A discussion note for the world bank*. World Bank.
Lee R. D. (2016). *Macroeconomics, ageing, and growth*. Cambridge (Mass.): NBER. NBER Working Paper № 22310.
Livi-Bacci, M. (2017). *A concise history of world population*. Wiley-Blackwell.
Lovins, L. H., Wallis, S., Wijkman, A., & Fullerton, J. (2018). *A finer future. Creating an economy in service to life. A report to the club of Rome*. New Society Publishers.
Lutz, W., Goujon, A., Kc, S., Stonawski, M., & Stilianakis, N. (2018). *Demographic and human capital scenarios for the 21st century: 2018 assessment for 201 countries*. Publications Office of the European Union.
Maestas, N., Mullen, K. J., & Powell, D. (2016). *The effect of population ageing in economic growth, the labor force and productivity*. Cambridge (Mass): NBER. NBER Working paper № 22452.
Malkov, S., Korotayev, A., & Davydova, O. (2022). World dynamics as an object of modeling (for the fiftieth anniversary of the first report to the Club of Rome). *Computer Research and Modeling, 14*(6), 1371–1394. https://doi.org/10.20537/2076-7633-2022-14-6-1371-1394
May, J. F., & Rotenberg, S. (2020). A call for better integrated policies to accelerate the fertility decline in Sub-Saharan Africa. *Studies in Family Planning, 51*(2), 193–204.
McNeill, W. H. (1993). *Plagues and peoples* (2nd ed.). Monticello.
McNeill, W. H. (2013). *The pursuit of power: Technology, armed force, and society since AD 1000*. University of Chicago Press.
Medici, A. C. (2021). *Health sector challenges and policies in the context of ageing populations*. United Nations, Department of Economics and Social Affairs, Population Division.
Mesquida, C. G., & Weiner, N. I. (1999). Male age composition and severity of conflicts. *Politics and the Life Sciences, 18*, 113–117. https://doi.org/10.1017/S0730938400021158
Mitchell, E., & Walker, R. (2020). Global ageing: Successes, challenges and opportunities. *British Journal of Hospital Medicine, 81*(2), 1–9. https://doi.org/10.12968/hmed.2019.0377
Moller, H. (1968). Youth as a force in the modern world. *Comparative Studies in Society and History, 10*(3), 237–260. https://doi.org/10.2307/177801
Morgan, L. A., & Kunkel, S. R. (2016). Aging, society, and the life course, 5th edn. New York, Springer. https://connect.springerpub.com/content/book/978-0-8261-2173-8
Motrevich, V. P. (2000). *Istoricheskaya demografiya Rossii*. Yekaterinburg: Izd-vo Ural, un-ta.
Novik, A. A., & Ionova, T. I. (2007). *Rukovodstvo po issledovaniyu kachestva zhizni v meditsine*. 2-ye izd. M.: Olma Mediagrupp.
Nzimande, N., & Mugwendere, T. (2018). Stalls in Zimbabwe fertility: Exploring determinants of recent fertility transition. *Southern African Journal of Demography, 18*(1), 59–110.
Orlická, E. (2015). Impact of population ageing and elderly poverty on macroeconomic aggregates. *Procedia Economics and Finance, 30*, 598–605.
Parker, J. (2013). *Global crisis: War, climate change and catastrophe in the seventeenth century*. Yale University Press.
Peterson, P. G. (1999). Gray down: The global ageing crisis. *Foreign Affairs, 78*(1), 42–55.

Pifer, A., & Bronte, L. (1986a). Introduction: Squaring the Pyramid. In A. Pifer & L. Bronte (Eds.), *Our ageing society: Paradox and promise*. W.W. Norton.

Pifer, A., & Bronte, L. (Eds.). (1986b). *Our ageing society: Paradox and promise*. W.W.Norton.

Popova, S. M., & Yanik, A. A. (2014). Problemy global'nogo stareniya naseleniya: Analiz dokumentov i strategii OON. *Mezhdunarodnoye Pravo i Mezhdunarodnyye Organizatsii, 3*, 429–443.

Raftery, A. E., Li, N., Ševčíková, H., Gerland, P., & Heilig, G. K. (2012). Bayesian probabilistic population projections for all countries. *Proceedings of the National Academy of Sciences, 109*(35), 13915–13921. https://doi.org/10.1073/pnas.1211452109

Randers, J., Rockström, J., Stoknes, P. E., Golüke, U., Collste, D., & Cornell, S. (2018). *Transformation is feasible. A report to the club of Rome*. Stockholm Resilience Centre.

Romanov, D., Meshcherina, K., & Korotayev, A. (2021). The share of youth in the total population as a factor of intensity of non-violent protests: A quantitative analysis. *Polis-Politicheskie Issledovaniya, 3*, 166–181. https://doi.org/10.17976/jpps/2021.03.11

Sawyer, P., Romanov, D., Slav, M., & Korotayev, A. (2022). Urbanization, the youth, and protest: A cross-national analysis. *Cross-Cultural Research, 56*(2–3), 125–149. https://doi.org/10.1177/10693971211059762

Schoumaker, B. (2019). Stalls in Fertility transitions in sub-Saharan Africa: Revisiting the evidence. *Studies in Family Planning, 50*(3), 257–278. https://doi.org/10.1111/sifp.12098

Schwab, K., & Malleret, T. (2020). COVID-19: *The Great Reset*. N. p.: Forum Publishing. http://reparti.free.fr/schwab2020.pdf

Schwab, K., & Vanham, P. (2021). *Stakeholder capitalism*. John Wiley & Sons.

Sciencedaily. (2018). More than 8 million babies born from IVF since the world's first in 1978. https://www.sciencedaily.com/releases/2018/07/180703084127.htm

Seifabadi, M. S., & Dolatabadi, A. B. (2022). Covid-19, globalization and strengthening authoritarianism in the Middle East. *Journal of Globalization Studies, 13*(1), 135–157.

Shcherbakova, Y. M. (2011). V oktyabre 2011 goda chislennost' naseleniye mira prevysit 7 milliardov chelovek. *Demoskop Weekly*, 481–482, October 10–23. http://www.demoscope.ru/weekly/2011/0481/barom01.php

Shcherbakova, Y. (2019). Stareniye naseleniya mira po otsenkam OON 2019 goda. *Demoskop Weekly*, 837–838. http://demoscope.ru/weekly/2019/0837/barom01.php

Sindyashkina, E. N. (2022). Healthy life expectancy in the context of the united nations decade of healthy ageing (analysis and forecasting). *IMEMO Journal, 1*, 40–53.

Sjöland, H., Caidahl, K., Wiklund, I., Haglid, M., Hartford, M., Karlson, B. W., Karlsson, T., & Herlitz, J. (1997). Impact of coronary artery bypass grafting on various aspects of quality of life. *European Journal of Cardio-Thoracic Surgery, 12*(4), 612–619. https://doi.org/10.1016/s1010-7940(97)00216-9

Solovieva, S. V., & Naimushina, A. G. (2011). Kachestvo zhizni kak mediko-biologicheskaya kharakteristika sostoyaniya zdorov'ya zhiteley severa tyumenskoy oblasti i goroda Tyumeni. *Fundamental'nyye Issledovaniya, 6*, 162–165.

Spilker, B. (Ed.). (1996). *Quality of life and Pharmacoeconomics in clinical trials* (2nd ed.). Lippincott-Raven.

Staquet, M. J. (1998). *Quality of life assessment in clinical trials*. Oxford University Press.

Sullivan, D. F. (1971). A single index of mortality and morbidity. *HSMHA Health Reports, 86*(4), 347–354. https://doi.org/10.2307/4594169

Transition from E-Government to E-State, and Change in Social Relations. Technological Forecasting & Social Change. https://doi.org/10.1016/j.techfore.2021.121348

UN Population Division. (2013). *World population ageing 2013*. United Nations, Department of Economics and Social Affairs, Population Division.

UN Population Division. (2015a). *World population ageing 2015*. United Nations, Department of Economics and Social Affairs, Population Division.

UN Population Division (2015b). *World population prospects 2015*. United Nations, Department of Economics and Social Affairs, Population Division.

UN Population Division. (2019a). *World population ageing 2019*. United Nations, Department of Economics and Social Affairs, Population Division.

UN Population Division. (2019b). *World population prospects 2019*. United Nations, Department of Economics and Social Affairs, Population Division.

UN Population Division. (2020). *World population ageing 2020*. United Nations, Department of Economics and Social Affairs, Population Division.

UN Population Division. (2022a). *United nations population division database*. United Nations, Department of Economics and Social Affairs, Population Division. http://www.un.org/esa/population

UN Population Division. (2022b). *World population prospects 2022*. United Nations, Department of Economics and Social Affairs, Population Division.

UN. (2008). *World economic and social survey 2007: Development in an ageing world*. Department of Economic and Social Affairs.

Urdal, H. (2008). Population, resources, and political violence: A subnational study of India, 1956–2002. *Journal of Conflict Resolution, 52*(4), 590–617. https://doi.org/10.1177/0022002708316741

Urdal, H. (2012). Youth bulges and violence. In J. A. Goldstone, E. P. Kaufmann, & M. D. Toft (Eds.), *Political demography: How population changes are reshaping international security and national politics* (pp. 117–132). Oxford University Press.

Urdal, H. (2004). *The devil in the demographics: The effect of youth bulges on domestic armed conflict, 1950–2000*. World Bank.

van der Maesen, L. J. G., & Walker, A (eds.). (2011). *Social quality. From theory to indicators*. Palgrave Macmillan.

van Dullemen, C. E. (2017). The politics of ageing, global challenges and national solutions. *Journal of Ageing and Neuropsychology. 01*(2017), 1–4. https://doi.org/10.29011/JANP-104.100004

Vollset, S. E., Goren, E., Yuan, C.-W., Cao, J., Smith, A. E., Hsiao, T., Murray, C. J. L., et al. (2020). Fertility, mortality, migration, and population scenarios for 195 countries and territories from 2017 to 2100: A forecasting analysis for the Global Burden of Disease Study. *The Lancet*. https://doi.org/10.1016/S0140-6736(20)30677-2

von Weizsäcker, E. U., & Wijkman, A. (2018). *Come on!* Springer.

Wallerstein, I., Collins, R., Mann, M., Derluguian, G., & Calhoun, C. (2013). *Does capitalism have a future?* Oxford University Press.

Wendorf, M. (2022). *Concept unveiled for the world's first artificial Womb Facility*. https://scienceandstuff.com/ectolife-artificial-wombs/

WHO—World Health Organization. (2013). *Dementia: A public health priority*. https://www.who.int/publications/i/item/dementia-a-public-health-priority

WHO—World Health Organization. (2017). *Global strategy and action plan on ageing and health*. World Health Organization. https://www.who.int/publications/i/item/9789241513500

WHO—World Health Organization. (2021b). *Ageism is a global challenge*. https://www.who.int/news/item/18-03-2021-ageism-is-a-global-challenge-un

WHO—World Health Organization. (2023). *Infertility fact sheet*. https://www.who.int/health-topics/infertility#tab=tab_1

WHO—World Health Organization. (2021a). *Ageing and health*. https://www.who.int/news-room/fact-sheets/detail/ageing-and-health

Widdowson, M. (2021). From Covid-19 to zero-gravity: Complex crises and production revolutions. *Journal of Globalization Studies, 12*(1), 117–144.

Wittgenstein Centre. (2022). *Wittgenstein centre human capital data explorer*. http://dataexplorer.wittgensteincentre.org/wcde-v2/

Yair, O., & Miodownik, D. (2016). Youth bulge and civil war: Why a country's share of young adults explain only non-ethnic wars. *Conflict Management and Peace Science, 33*(1), 25–44. https://doi.org/10.1177/0738894214544613

Yaşar, E., & Yaşar, M. (2018). Global wealth and ageing population. In R. Yilmaz & G. Löschnigg (Eds.), *Studies on Balkan and near eastern social sciences* (vol. 2, pp. 47–59). Peter Lang.

Zhdanov, A., & Korotayev, A. (2022). Factors of deconsolidation of a liberal democratic regime: The case of the United States. *Cliodynamics, 13*, 1–34. https://doi.org/10.21237/C7clio13054207

Zimmer, Z. (2016). *Global ageing in the twenty-first century: Challenges*. Routledge.

Zinkina, J., & Korotayev, A. (2014a). Explosive population growth in tropical Africa: Crucial omission in development forecasts (emerging risks and way out). *World Futures, 70*(4), 271–305. https://doi.org/10.1080/02604027.2014.894868

Zinkina, J., & Korotayev, A. (2014b). Projecting Mozambique's demographic futures. *Journal of Futures Studies, 19*(2), 21–40.

# Global Aging, Adaptation to It and Future Demographic Transformations

7

**Abstract**

This chapter provides a systemic analysis of the development of the process of population and society aging during the twenty-first century. It considers achievements, challenges and conflicts that await us and the world as a whole in connection with the formation of an elderly society and the intensification of the global aging process, also the mutual influence between global aging and technological change. Grinin et al. hope that the study will be all the more relevant and important since the studies of the problem of population aging are limited. However, global aging is already an extremely important problem and it will become the most crucial problem in the future. This means that insufficient attention will result in societal tensions, crises and intergenerational conflicts as well as political and social instability. The chapter analyzes the challenges that will become even worse in the future, including a possible decline in democracy etc. Grinin et al. reveal the need for society to adapt to societal aging through the transformation of a number of its institutions. The authors outline the main features of a new type of population reproduction and new type of society, which will emerge by the end of this century.

**What this Chapter is about**

This chapter provides a systemic analysis of the development of the process of population and society aging during the twenty-first century. It considers achievements, challenges and conflicts that await us and the world as a whole in connection with the formation of an elderly society, the intensification of the global aging process and the stabilization of the population.

We hope that the study will be all the more relevant and important since not only systematic research on these topics and issues is absent, but also more generally, the studies of the problem of population aging are limited in scientific



L. Grinin et al., *Cybernetic Revolution and Global Aging*, World-Systems Evolution and Global Futures, https://doi.org/10.1007/978-3-031-56764-3_7

discourse. However, global aging is already an extremely important problem and it will become the most crucial problem in the future. This means that insufficient attention will result in societal tensions, crises and intergenerational conflicts as well as political and social instability.

The chapter examines the mutual influence between global aging and technological change, and their combined impact on the transformation of society in the twenty-first century century. We also analyze the challenges that society is already facing and that will become even worse in the future: the struggle for resources in society due to the growing number of elderly and sick people, increasing conservatism and a possible slowdown in the technological progress, the aggravation of intergenerational contradictions and conflicts, a possible decline in democracy due to population aging, ageism, a sharp increase in the number of people suffering from dementia and other incurable agerelated diseases in less developed countries, and the growing number of deaths by euthanasia in the developed countries, to name but a few.

We reveal the need for society to adapt to societal age through the transformation of a number of its institutions and they also analyze different ways and forms of adaptation to aging that will need to be elaborated in the future in order to mitigate friction and tensions that could impede societal development. A large section is devoted to various scenarios for the development of future society in the context of its aging.

In this chapter we show the impact of the Cybernetic Revolution on development of technologies, which can be stimulated precisely by the process of global aging and the need to increase life expectancy. But on the other hand, this increased longevity will create new serious problems, tensions and contradictions in an increasingly greying and technological society.

Comprehensive forecasts are offered about the future of society. In particular, we outline the main features of a new type of population reproduction, which, in our opinion, will emerge by the end of this century. We argue that a new demographic and technological system will promote the formation of a new societal type, which can be denoted as the cybernetic society. It will be an extremely high-tech socium with a stable population size, a very large proportion of elderly peo-ple and total technological support for their health and quality of life (which will be achieved by creating a continuously expanding medical-biotechnological environment).

## 7.1 Current Global Aging and Projecting Problems into the Future

So, as we have seen in the previous chapters, aging is already an extremely important process, yet in the future it will become actually a crucial process and factor, determining much of the twenty-first century (and at least the first decades of the twenty-second century). In other words, it is the most important process for the

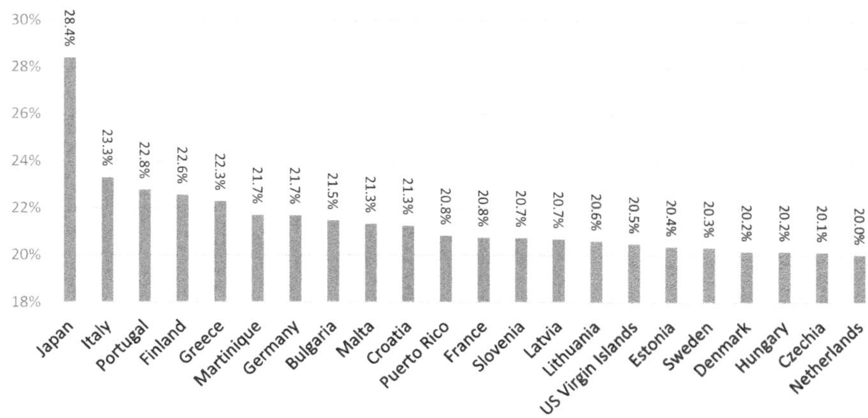


**Fig. 7.1** Countries and territories of the world with the share of the population aged 65+ >20% in 2020. *Source* authors' calculations on the basis of data provided by the UN Population Division database (2022a)

next hundred years or more, provided, of course, that there is no large backslide in terms of life expectancy.

Today, the share of the population aged 65+ is already 20% or more in 22 countries and territories, mostly in Southern and Eastern Europe, but Japan is the leader with 28.4% (see Fig. 7.1; see also Kamiya & Hertog, 2020; Medici, 2021; Shcherbakova, 2019).

In several countries, the share of the population aged 80+ has already exceeded 6%, with Japan again leading the way with 10.2%[1] (see Fig. 7.2).

The number of people aged 60+ already exceeds the number of children under the age of five years (He et al., 2016: 3; WHO, 2021b). By 2050, the share of people aged 65 and over is projected to amount to 16% worldwide which is more than twice the number of children under five and about the same as the number of children under age 12 (He et al., 2016: 3, Tables 2, 3, P.5; UN Population Division, 2022b). For the first time in human history, people aged 65 and over will outnumber children under the age of 5. This unique demographic phenomenon of the "crossing" is unprecedented. Against this backdrop, it is surprising that only relatively little attention is paid to the phenomenon of global aging.

However, as we stated in Chap. 6, the UN and related organizations, especially the WHO, pay quite a lot of attention to the process of global aging, which is progressing at different rates across countries. This work has been actively carried out since the late 1970s (Kamiya & Hertog, 2020; Medici, 2021; Popova & Yanik, 2014; UN Population Division, 2013, 2015, 2017, 2019a, 2020; WHO, 2017a, 2021a, 2021b). Many different programs, decisions, declarations, etc. have been

[1] An even higher percentage (16.6%) is found in Monaco (UN Population Division, 2022a), but this is, of course, a very peculiar country.

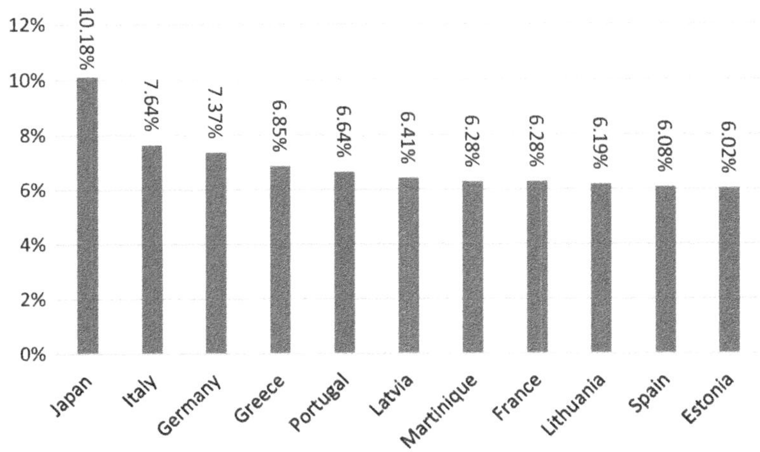


**Fig. 7.2** Countries and territories of the world with the share of the population aged 80+ >6% in 2021. *Source* authors' calculations on the basis of data provided by the UN Population Division database (2022a)

adopted. In general, a set of well-defined problems, tasks and attitudes contribute to a more active and humane inclusion of elderly people in social processes and also help to ensure that aging is not only perceived as a growing burden on society but also as an opportunity to organize the societal development in a certain direction, using the assets of people of mature and old age—their unique experience and life wisdom (see also Grinin et al., 2023b). In general, human rights approaches are being developed, along with ways to combat the bullying of the elderly people and to free them from a sense of doom, the end of life and the feeling that they are a burden to everyone, including themselves (e.g., Kamiya & Hertog, 2020; UN Population Division, 2019a, 2020, 2022b; WHO, 2017a, 2021b). Much has been done to compile national data on aging and to adapt strategic documents at the national level, etc. Nevertheless, one should admit that the main approaches of international organizations hide several delayed risks associated with a change in the hierarchy of goals and a shift in focus toward solving an ever-increasing number of current and local tasks at the expense of strategic ones (Popova & Yanik, 2014). But, of course, international organizations are unable to solve all problems, let alone work out a common strategy, since they are bound by a complex consensus of opinions. Every society at each stage of its development must start to actively integrate this issue into everyday life in order to accelerate the processes of adaptation to aging and the institutionalization of the changing age structure of society. We will speak about this a lot more.[2]

[2] It is worth noting that the UN has also started to talk about adapting to aging. In particular, they emphasize that greying countries should take steps to adapt public programs to the growing proportions of the seniors, including social security and pension systems and by establishing universal

Global aging is an extremely important process that has a widespread effect today and will have an even greater impact in the future. However, global aging and especially its considerable influence on all spheres of society is given unjustifiably little attention.

There are many aspects associated with global aging, and their number will increase even more, and these problems are likely to become more acute, at least for a number of decades.

### 7.1.1 Aspects of Global Aging in Societies and the World System

This influence of aging is multifaceted, since it is generally recognized to have serious macroeconomic consequences.[3]

Although the impact of the accelerated aging of the labor force can be positive in certain aspects (due to their greater experience and matching with jobs [Kapelyushnikov, 2019b: 23, etc.]), in general, the aging of the labor force impedes both the introduction of innovations and the retraining of the labor force, as well as opportunities for increasing labor productivity.[4] It also reduces mobility of the labor force (Kapelyushnikov, 2019b: 23). A negative impact of aging on economic growth rates caused by its diverse impact on society can be observed, in particular, due to a sharp reduction in the number of working hours (Goldstone, 2015; Bloom & Luca, 2016; Kapelyushnikov, 2019b: 42; see also Grinin et al., 2020). According to some studies, a one-percent increase in the share of the elderly (60+) in the total population tends to reduce the per capita GDP growth rate, the share of

---

health care and long-term care systems (UN Population Division, 2022b). However, looking ahead, we note that the logic of thinking here is still outdated: the main role is assigned to the state, which, however, is not fully prepared for this mission. In our opinion, it is necessary to take a broader and deeper approach in order to build institutions for adapting to aging of society. See more on this below.

[3] E.g., Aldwin and Gilmer (2013); Atchley (2004); Barsukov (2019); Binstock (1991a, 1991b); Bengtson et al., (2009); Bloom and Luca (2016); Coleman and Rowthorn (2015); Fichtner (2018); Fukuyama (2002); Galor and Weil (2000); Goldstone (2015); Goldstone et al., (2015); de Grey and Rae (2008); Grinin and Grinin (2015, 2016); Grinin and Korotayev (2010, 2015b, 2015c, 2016a, 2016b); Grinin et al., (2017a, 2017b); Haas (2015); Harper (2006); Hsu and Lo (2019); Jackson et al., (2008, 2013); Jackson (2021); Kapelyushnikov (2019a, 2019b); Lee and Mason (2011); Maestas et al., (2016); Morgan and Kunkel (2007); Orlická (2015); Powell and Khan (2013); Prettner (2013); Popova and Yanik (2014); Park and Shin (2015); Tsirel (2008); UNCTAD (2012); WHO (2017a, 2021a); Zimmer (2016).

[4] According to some reports, labor productivity declines as a result of population aging. Maestas et al. (2016) find out that for 10% increase in the share of older workers in the labour force, labor productivity falls by an average of 3–5% in all age groups. The reasons for this are not entirely clear. Maestas et al. (2016) believe that it may be due to the earlier exit of the most productive workers from the labor market (Kapelyushnikov, 2019b: 29). But, perhaps, the general increase in conservatism affects younger groups as well. However, there is also other evidence on the productivity of older workers (e.g. Burtless, 2013).

investment in GDP, the rate of personal savings and the number of hours worked by −0.14; −0.20; −0.59; and −0.89 percentage points, respectively; subsequently the inflation rate increases by 0.17% (Aksoy et al., 2015).

The strongest impact is, of course, associated with a sharp reduction in labor supply. One may agree that at present and in the foreseeable future this is the main challenge that economies face in the situation of an aging population (Kapelyushnikov, 2019a, 2019b). However, there can be different responses to a powerful challenge. According to our estimates, the unfolding Cybernetic Revolution will respond, although not immediately, with an increase in labor productivity (due to the widespread introduction of self-regulating and self-managing systems), which should also boost the economic growth rate (see Grinin et al., 2016b, 2017a, 2017b, 2020; Grinin & Grinin, 2016). Here we should note that although the final phase of the Cybernetic Revolution should begin in the field of medicine and related areas, and this can already increase labor productivity, for example, by reducing the number of people involved in caring for the sick and elderly, still *the most noticeable increase in labor productivity will begin when the Cybernetic Revolution spreads to new areas, including productive* ones. In particular, there are great opportunities here for the service sector with relatively low qualifications (hotels, cafes, gas stations, etc.).

In general, significant labor productivity growth will become absolutely essential. One cannot but agree that an aging society needs this growth in order to rearrange infrastructure, modernize the economy and maintain high social standards (although there are some nuances, which we discussed in Chap. 6). High productivity becomes even more important as the economically active population shrinks and, consequently, the number of people who need social services grows (Fuecks, 2013). We believe that the growth in the situation of a declining working population can only be achieved through automation, and in particular, through a *move towards self-regulating and self-managing systems* in order to reduce the number of people involved in control and other functions. In the context of an aging society and reduced labor resources, the most important direction will, of course, be the release of people from the service sector, especially in terms of caring for the elderly and the helpless.

We believe that this will be resolved in the course of the Cybernetic Revolution; however, this path is full of inevitable difficulties, crises and conflicts. It will also inevitably raise the question of a comprehensive solution to the problem of the limits of the independence of self-regulating systems (which we observe with respect to AI [see Chap. 10; Grinin et al., 2021]).

**The Coming Pension Crisis as the Main Challenge to the Welfare State**. Strangely enough, the issue of the looming pension crisis as the main challenge to the welfare state is not widely discussed. However, the crisis is coming, and in some cases it is already detected. It will become particularly acute in case of major collapses in the securities markets in which pension funds are invested, as well as possible sovereign defaults. In fact, funds invested in the sovereign debt of developed countries can already be considered spent. As economic growth is weak, public debt is growing steadily and rapidly, inflation is high and the impending debt crisis may become very acute, the pension system is becoming a hostage of the general economic situation. There is an opinion that modern economies are trapped in institutional terms precisely because of their pension systems. And it is the existence of solidarity pension systems and not the phenomenon of population aging as such, that has served as a source of numerous intractable problems (Kapelyushnikov, 2019b: 45–46). We agree with this, but only partially. First of all, modern states have fallen into a trap because they have consumed demographic dividends, spread public debt to unprecedented levels and ignored demographic problems for a long time (and generally still do) (e.g., by not encouraging fertility but, on the contrary, encouraging childlessness). Finally, they drove their economies into stagnation. In addition, the policy of virtually constant and huge money emissions, combined with near-zero or even negative interest rates (see Efimovich, 2020; Grinin, 2021), has led to the devaluation of pension savings and high inflation. Of course, solidarity pension systems are worse than the funded ones, but if there is a state default on debts (and this is quite possible), then formal savings may turn out to be a fiction similar to the right to a pension under the solidarity system. So, it appears that the restructuring of pension systems is largely blocked (Kapelyushnikov, 2019b: 46; about our approaches to possible changes see Grinin & Korotayev, 2013, 2015a, 2015b).

In general, the main problem of the growing pension crisis is, of course, the decline in the proportion and number of the working-age population. And this problem will not only persist but also get worse.[5] However, most countries in the world have hardly prepared themselves for this obstacle, although it has been problematized for many decades (e.g., Bös & Weizsäcker, 1989; Peterson, 1999).

Meanwhile, the global population of people over 60 is growing faster than any of the younger age groups (see above). In two decades, they will make up about a third of the total adult population in developed countries (Konstantinov, 2021). How will pensions be provided as the working population shrinks? We have already touched on this issue in Chap. 4 and will discuss it also in other chapters. Of course, they will have to raise the retirement age, but there are obviously limits. Moreover, it is a very complicated and painful process, fraught with socio-political and age-related conflicts. In addition, it tends to lag behind demand, in other words, the increase in the retirement age lags behind the increase in life expectancy. The estimates for 47 countries for the period from 1965 to 2005 revealed a negative

[5] There is one pension fund for stuff of several international organizations. This is the United Nations Joint Staff Pension Fund (UNJSPF). In addition to the UN Secretariat, 22 international organizations are members of the Fund. Significantly, even in this unique pension fund, the amount of pension payments exceeds the number of contributions collected. The source of the shortage of funds is income from the investment of the accumulated reserves, and, under some circumstances, the reserves themselves (Chetirkin, 2014: 15, 18). However, for many pension funds, the situation of funds shortage gradually becomes the rule. At the same time, the day is coming when more and more funds will begin to incur increasing losses due to the reducing contributions. Moreover, the inflated exchange and debt bubbles, which bring income to these funds, may burst.

correlation between these variables (−0.21) (Bloom & Luca, 2016; Kapelyushnikov, 2019b: 50). It is clear that it will be difficult to increase the retirement age. Democracy can hardly allow ignoring the interests of such a huge part of the voters. All the more so, since demographic forecasts indicate that the population of many countries will decline by 2050, especially in the second half of the twenty-first century (e.g., UN Population Division, 2022a; Vollset et al., 2020; Wittgenstein Centre, 2022; see Chap. 6).

There are some ways to alleviate the problem, but the current trends often go in the opposite direction. For example, the IMF recommends many countries not to pay pensions to working pensioners and this contradicts the problem (we agree that the negative aspects of the solidarity pension system manifest quite vividly here).

Different alternatives are possible, yet all of them to some extent may lead to new threatening phenomena; the worst of which is that society would lose interest in increasing life expectancy in order to reduce the number of old people. Although this scenario seems unlikely at the moment, nevertheless, there is a frightening and gloomy trend of increasing cases of so cold mercy killing or euthanasia (see below). However, taking into account that life expectancy will continue to grow, with possible rollbacks even in developed countries (and even more in developing countries), the issue of directing the Cybernetic Revolution toward efficiency in medicine and related areas (including healthy lifestyles) is becoming increasingly relevant. Modern technologies provide increasing opportunities for remote work in a growing number of professions so that for some time the young population of Africa and other countries will be able to work in the market of developed countries, which will allow the latter to collect certain amounts for pensions for the elderly. Although here there are also obvious limitations, it seems to be a rather perspective way (see Grinin & Korotayev, 2010; Grinin et al., 2023a, 2023b, 2023c, 2023d).

**Human capital and investment in it**. Investment in human capital has become one of the most important trends in economic and social development during the Cybernetic Revolution. This concept comprises the growth of knowledge and skills, the promotion of talents, the selection of the most capable to solve certain tasks, reskilling and retraining, the required standard of living, the level of medical services and health in general, job satisfaction and many other aspects (see, e.g., Goldin, 2016). The compiled Human Development Index (HDI) becomes very important for measuring the business and other types of environments in a country (see, e.g., UNDP, 2022). Developed (and even developing) countries compete with each other in terms of living standards and the rights of citizens. Most of the population in developed countries has a high standard of living and some confidence in the future. The HDI has become an important indicator of the level of social development, reflecting, in addition to GDP per capita, characteristics such as the level of education, culture, health care, the health status of the population and its life expectancy (see, e.g., UNDP, 2022: 272–276). Today investment in human capital is considered as one of the most important economic indicators and production factors. By the way, such trends can also be viewed as a change in the structure of society's productive forces, in which the role of the human component has grown compared to the Industrial Production Principle (in particular, one can trace this in IT companies, where wages are the main expense, while the capitalization of such companies can be very high).

These, as well as many other factors, have made the life of society much more stable than it used to be, although it gave rise to a lot of problems associated with social dependency.

Of course, this is part of what we call the economic type of alienation (see below for details). However, the process of global aging significantly changes the developmental trend of human capital and investment in it.

One can argue that along with negative effects, population aging also brings some benefits in terms of investment in human capital. In particular, population aging provides opportunities and incentives for more active investment in the human capital of new generations, as the declining number of children allows families and the state to invest more in each child (Kapelyushnikov, 2019b: 24–25).If small cohorts of workers have high levels of human capital, then living standards can rise despite an unfavorable age structure of the population (Kapelyushnikov, 2019b: 25; Lee & Mason, 2010). One can only agree to a relatively small extent. For a long time to come, society and the economy will need huge numbers of people who are either relatively low-skilled or whose services, though skilled, cannot generate high value (e.g., teachers or policemen). Even with the introduction of robots and other self-regulating systems, most of this need will remain. Without a sufficient number of workers, the problem of maintaining a high level is difficult to solve, at least until the end of the Cybernetic Revolution—this will take several decades. Furthermore, in the absence of competition, small groups of highly qualified specialists with high human capital become very demanding and selfish. In addition, as Ronald Lee (2016) rightly points out, it is doubtful whether the positive effect associated with the activation of human capital will turn sustainable. Personally, we believe that this trend will hardly be sustainable and long-lasting due to the extremely rapid changes in technologies that will be associated with the unfolding Cybernetic Revolution.

**Aging and the Economic Type of Alienation**. Against this background, we see the need to change the direction of investment in human capital, to gradually increase investment in the elderly and, in every possible way, to encourage retired people to work as long as possible. This will require new methods of retraining elderly people who are capable of learning, as well as incentives and appropriate laws, institutions and mechanisms[6]—the system of investment in human capital

[6] For example, quotas for pensioners, tax breaks (taking into account that people have already accumulated pensions), and mandatory shorter working hours, longer vacations, etc. Of course, business will adapt to this far from immediately, but the path of global aging is very long, and we must move on. Perhaps, it would make sense to set up companies with predominant participation of pensioners and tax advantages or other benefits.

will be constantly improved and adapted to the age structure. This will be a further development of what we call the *economic type of alienation.*[7] What does it involve? How does it differ from the semi-economic type of alienation characteristic of the Industrial Production Principle? The main difference lies in the fact that modern society provides many social guarantees and types of insurance against various problems and misfortunes that can happen to a person (unemployment, accidents, disability, etc.), as well as developed pension and social security systems. On the one hand, society has a huge amount of resources to support the weak. Therefore, each person, enjoying various freedoms granted simply because she or he is a citizen or resident of a certain country, can enjoy social guarantees and rights (to education, health care, protection, assistance in old age or disability, etc.). However, on the other hand, the economic type of alienation is an opportunity to better demonstrate one's skills, qualifications and talents and to receive many times more than the social minimum. Consequently, the coercive factor of labor becomes weaker, while the incentives to activate people's talents and energy increase (for more details see Grinin, 2006; Grinin & Grinin, 2015). The scarcity of labor contributes significantly to the development of the economic type of alienation, because the pressure of unemployment decreases considerably, and respectively, the rigid forms of coercion to work weaken. If some kind of guaranteed income is introduced, this will create rather serious difficulties in the labor market, which in turn will push this type of alienation to maturity.

The development of this type of alienation will be considerably affected by aging for many reasons, including the guaranteed income the elderly people receive in the form of pensions and savings. But in general, one must realize that social division within the Cybernetic Production Principle will be increasingly associated with gender and age differences, as well as other non-class differences (we will return to this issue below). Let us first recall that since the very origin of the Cybernetic Production Principle (in 1950–1970), the female and youth movements, as well as movements related to gender identity (LGBT), have become more active—we are dealing with a lasting phenomenon. The struggle for the rights of women and various minorities has also been going on. The so-called youth bulges, which we have already discussed, became a real challenge; however, now they are replaced by "the elderly bulges" (see, e.g., Fig. 7.3).

---

[7] Each production principle is based on a peculiar set of technologies, which ultimately creates a specific level, a complex, a type of relationship, whose main characteristics manifest in every society that uses the appropriate technology, despite considerable variations in economic and social relations from society to society. We denoted the main characteristics of such a complex of relations corresponding to a production principle as the type of alienation of labor and the individual, using Hegel's and Marx's term of "alienation" (Grinin, 2006).

The type of alienation changes as the production principle changes. Thus, the Hunting-Gathering Production Principle corresponded to the pre-economic type of alienation (because there was no an economy yet); the craft-agrarian associated with the non-economic (i.e. mostly violent) type; the Trade-Industrial—with semi-economic type (free labor relations, but without any social safeguards); and respectively, the Scientific-Cybernetic—with economic, which is still developing (for more on types of alienation, see in Grinin, 2006; Grinin & Grinin, 2015).

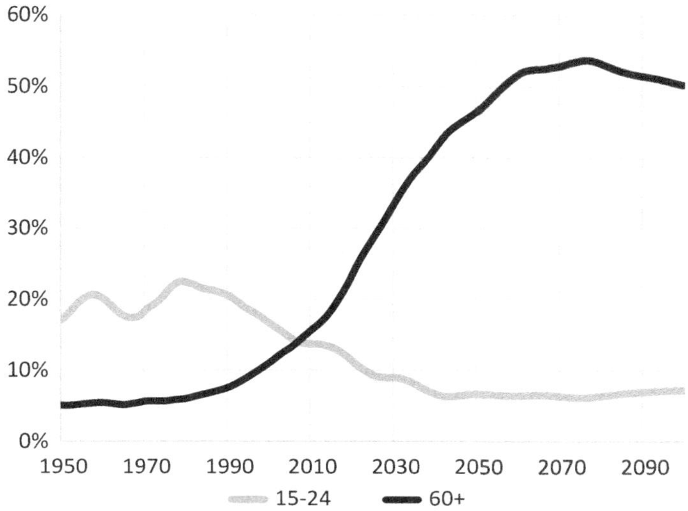


**Fig. 7.3** "The youth bulge" and "the elderly bulge" in South Korea: the dynamics of the share of the population aged 15–24 and 60+ in the total population, 1950–2021 with the UN medium projection till 2100—"South Korean Cross". *Data source* authors' calculations on the basis of materials provided by the UN Population Division database

Thus, while in the 1960s and 1980s, the main problems were the youth riots and the relationship between the generations of fathers and children, now the elderly are coming to the fore, and there is a problem with the relationships between fathers and children, on the one hand, and grandfathers, on the other. We argue that the role of generations in the social division of society will increase, and respectively, the views of different generations should be fully taken into account in various relations, institutions and, possibly, laws. We will return to this issue below.

**Rising Social and Medical Costs**. It is well known that, as life expectancy increases, the number of cases of age-related diseases that are more difficult and costly to treat is also increasing; it has to be mentioned as well that treatment is not always or everywhere available and requires great efforts on the part of relatives, others and the state.[8] This means not least that many people are excluded from productive activity—both those who need care and those who provide it. This can be well illustrated by the group of diseases known collectively as dementia, the most common of which is Alzheimer's disease, which is rapidly growing

[8] For example, spending on medicare in the U.S. has increased about ninefold, from $37 billion in 1980 to $336 billion in 2005, for different reason, among which aging was very important (see Centers for Disease Control, 2007: 5).

in scale and scope in many developing countries. Dementia is a group of symptoms associated with deterioration in cognitive function. Unfortunately, it is not yet completely clear what exactly provokes this pathology. Among chronic diseases, dementia occupies a special place, being one of the main causes of disability and need for care (Dua et al., 2012: 68).

The forecasts are very disappointing. According to the WHO (2017b), the number of people with dementia may triple by 2050.[9] At the time of the press release (WHO, 2017b), there were 50 million patients worldwide, respectively, by 2050 there will be more than 150 million, and possibly much more. At the same time, the cost of treatment and care in 2017 was 16,000 dollars per patient and in total the world expenditures amounted to 818 billion dollars, that is equaling more than 1% (!) of the world GDP. However, not all costs are direct. The total cost includes direct medical expenses, social assistance and informal assistance (loss of income for caregivers). By 2030, these costs are expected to increase by almost two and a half times to $2 trillion, undermining social and economic development and imposing a heavy burden on health and social services, including long-term care systems (see also, e.g., Carrion et al., 2018; Hartman et al., 2018). Naturally, we may hardly expect that global GDP will grow by 2.5 times by 2030; therefore, this money and these losses will lead to cuts in other areas of spending, leaving other needs unmet. Up to 10 million new cases of dementia are reported each year, two-thirds of them in low- and middle-income countries (WHO, 2017b).

Caring for the sick, who require constant supervision and care, is always a heavy cross to bear. But in the case of dementia, the burden is even heavier because the patients are not immobile and are physically quite capable, while mentally they often behave inappropriately and therefore can be in danger. Therefore, when a patient is diagnosed with dementia, it is sometimes said that not one but two patients appear at once in a family, meaning the family member who is the carer and is forced to provide continuous care for the patient. The fact is that family members caring for people with dementia become extremely vulnerable to affective disorders such as major depression and anxiety disorders as a result of nervous stress (Dua et al., 2012: 71–74). When analyzing the problem, the physical and emotional overload of those who care for sick relatives should also be taken into account. At the same time, people with dementia often do not consider themselves to be ill and behave restlessly when they arrive at the hospital, and the staff often do not know how to provide emotional comfort (Petty et al., 2020). It is also important to pay attention to gender and economic inequalities that affect

---

[9] For example, some data and forecasts from OECD (2021) on people with dementia per 1,000 population look as follows:

South Africa: 2.6 (in 2021); 4.9 (in 2050);
India: 5.3 (in 2021); 11.8 (in 2050);
Indonesia 5.7 (in 2021); 13.8 (in 2050);
Mexico 7.7 (in 2021); 19.0 (in 2050);
China 8.2 (in 2021); 24.5 (in 2050).

care (WHO, 2017a: 6; on social policies for elderly people in different countries, see, e.g., Asadzadeh et al., 2022).

Of course, there are many other issues associated with aging and disability.

Returning to the issue of costs, we would like to quote the opinion of the WHO (2017a: 16)—although we disagree with it, it is worth mentioning. According to them, the assumption that the growing needs of an aging population will lead to unsustainable increases in healthcare costs is incorrect, since the situation is intricate. Although older age is commonly associated with increasing healthcare needs, the relationship between the two, the use of healthcare provisions and the related expenditure, is variable. In some high-income countries, the per capita health spending actually decreases significantly after the age of 70 (while spending on long-term institutional care increases). As more people live longer, providing them with the conditions to lead long and healthy lives may actually reduce the pressure of rising healthcare costs. Although more evidence is needed, it is doubtful to predict future healthcare expenditure based on age distribution. The policy should address the issues of financing long-term care and developing innovations in home and community care. In addition, the care sector also offers significant job opportunities (WHO, 2017a: 16).

Perhaps, there is some truth in this approach, but for the foreseeable future (until there are dramatic breakthroughs in medicine and general health care), increasing life expectancy will require ever-increasing expenditures.[10] Of course, they can and should be optimized, but we can hardly avoid the problem.[11] According to the World Bank, global health spending in 2017 already amounted to 12.82 trillion 2017-international dollars PPP (World Bank, 2022). The European Commission estimates that public spending on long-term care for the elderly in the Organization for Economic Co-operation and Development (OECD) countries will increase from 1.6% of the GDP in 2015 to 2.8% in 2060. In the USA, the elderly care market is expected to grow from $438.5 billion in 2017 to $549.7 billion in 2024 (Damien, 2021). These and other data confirm the accelerating growth of spending on health and care for the older generation. The technological breakthroughs may provide a way out of this situation, but many years will pass between the currently growing needs and the solution.

**Care for the Weak in Families and Its Economic Aspects**. Providing care for vulnerable people at home poses a serious problem in economic terms (e.g.,

[10] According to Levy et al. (2020), the *additional* expanses on treating a range of diseases alone amount to US$63 billion per year. This makes $1 out of every $7 spent on high-cost diseases for all Americans over the age of 60 for one year. The authors attribute these additional costs to ageism, or discrimination against older people, which we will discuss in a separate section. The total expenditures on high-cost illness for Americans aged 60+ are almost $450 billion. Let us emphasis that the data refer to 2013, and the costs are already much larger today. And this is far from the total cost of medical care for the older generation in the United States.

[11] It is also necessary to encourage private efforts in this field in every possible way. Already today, about two out of five people in the world rely partly on their own savings, which on average provide about 30% of their income in old age (Damien, 2021).

Kamiya & Hertog, 2020; UN Population Division, 2019a, 2020). Most people who need care stay with their families. This is true in all countries, including the developed ones. For example, in OECD countries, about 10% of adults are involved in home care (OECD, 2011). These are huge numbers! In particular, in the USA, at least 43.5 million people aged 18 and over provide unpaid care to an elderly person who is a family member or a friend (this equals 19% of adults in America). The majority of caregivers are women (67%) with an average age of 50 and providing nearly 20 h of care per week (NAC & AARP, 2009, 2020). In 2012, women aged 50–65 made up the majority of people in Japan leaving or changing jobs in order to care for their parents (WHO, 2017a: 14).

Of course, the majority of care is provided by women: what used to be the burden of caring for children has been partly replaced by the burden of caring for the elderly and dependent relatives. It is mostly women of a young and relatively young age who are likely to care for children, while the care for the vulnerable is provided by women of a mature and elderly age. We can agree with the World Health Organization that, in addition to the general losses associated with the non-participation of a huge number of women in labor force, there are also significant social, political, environmental and economic inequalities that place a separate burden on women in these countries, and need to be thought over (WHO, 2017a: 14).

Even a partial solution to the problem of caring for the infirm in families with the help of self-regulating technologies will, among many other things, eliminate to a large extent the labor shortage caused by the population aging.

### 7.1.2 Modern Concepts and Institutions Related to Aging

**The Concept of Healthy Aging**. Several important UN and WHO policy documents have been elaborated (e.g., Healthy Aging [UN Population Division, 2015]). They can help in the future and will serve as a road map (such documents become a guide for action for decades).[12]

"Healthy aging" is defined as the process of developing and maintaining functional capacities that ensure well-being in old age. The concept of well-being brings together all the components and elements of life that are valuable to people (UN Population Division, 2015). Healthy aging is a concept that shows how public institutions can provide opportunities for people to live their lives and experience everything that is of value to them. At the same time, a person's environment and interactions with that environment become of utmost importance. The fundamental point is that every older person should have an opportunity to lead an active lifestyle in a favorable environment, even if this person suffers from many diseases or from decreasing individual health. The theory underlying healthy aging

[12] Also, since the 1960s, various concepts of successful aging have been developed, that is, life satisfaction in old age (Yevseeva, 2020).

aims to achieve overall well-being and reflects an eco-social model based on the social determinants of health and the accumulation of positive or negative aspects throughout life. Efforts to improve healthy aging require the action of multi-stage and multifaceted schemes, including biological, social, economic and environmental determinants, as well as influences at home, in communities, in cities, countries and regions and around the world (WHO, 2017a: 3). An important point of this concept is the emphasis on what can be done to help older people enjoy their human rights, including the right to use open, affordable, acceptable and good quality healthcare facilities, items and services (WHO, 2017a: 9). One of the determinants of healthy aging and well-being is taking responsibility and striving for dignity, such as striving to take care of oneself when and where possible (ibid.). The concept of healthy aging corresponds to the HALE (the healthy life expectancy) index, which we discuss in Chap. 6. In particular, the self-perceived health status, which is taken into account by Eurostat, is of particular importance for the promotion of healthy aging. Although subjective, it is of course essential for assessing the quality of life and determining the need for health care, social services and long-term care (for details see Sindyashkina, 2022).[13]

Thus, there is a kind of Declaration of the Rights of the Elderly. Like a declaration of human rights, it will serve as a certain guideline for the creation of institutions. The multidimensional character highlighted in the WHO document is important, but every step in this direction will trigger societal changes. The movement toward more fundamental institutions will have a strong structural impact on the whole of society.

However, one should realize that (a) compared with the past, the population aging has accelerated significantly and will continue to do so for a long time; (b) the number of older people (if taken by total number) will increase especially rapidly in middle- and low-developed countries, which are very poorly prepared for this challenge; (c) in general, the world is obviously not sufficiently prepared for the aging societies, while; (d) all countries face a great challenge in preparing health and social care systems to perform at their best during this new demographic transition (WHO 2021b).

**Modern institutions for helping the elderly people**. First of all, one can agree with the conclusion that the established institutions of support for the elderly were created in the early to mid-twentieth century in a completely different demographic and economic situation, without taking into account the long-term problems that they can generate. While in the past the main channel of support for the elderly was provided by families, now it has become the state (Kapelyushnikov, 2019b: 45). However, one cannot fully agree with Karen N. Eggleston and Victor Fuchs, who suggest that the problem of population aging is mostly a consequence of inert and outdated institutions and policies and not of the demographic transition as such (Eggleston & Fuchs, 2012). New institutions could mitigate this problem

[13] For details on the HALE index, see Murray and Lopez (2013); Murray et al., (2002, 2012); Sindyashkina (2022); Wang et al., (2012).

while more competent and patriotic demographic policies could keep the birth rate at a higher level, but the general trend is still related to technological changes that increase life expectancy and provide unprecedented opportunities for family planning.

Institutions play a significant role in social life as they are integral systems that mediate between different spheres of life. In particular, they mediate the links between population aging, on the one hand, and the economy, social and legal spheres, on the other. As one can rightfully note, pension, tax and healthcare systems, labor market and capital market institutions should adapt to the situation with a high and continuously growing proportion of elderly people (Kapelyushnikov, 2019b: 45), but also legal and social trends, ideology and others. Meanwhile, ideology is still far behind in this respect.

In modern society, there are actually few institutions established to help the elderly adapt in society, to support "healthy aging" (see above) or "dignified aging" for those who cannot take care of themselves. The first is the pension system. However, in many countries it is far from being developed; moreover, a number of countries have a so-called solidarity system, in which the size of the pension depends little on the contributions made by the individual himself. As a result, in times of financial crises and inflation, a pension is set by the state, but it may be far from covering a normal standard of living. Even in societies where pensions depend mostly on contributions, current and future pensioners remain under siege from the financial well-being of a state, and the growing number of pensioners makes the default on financial obligations increasingly real (on the problems of pension systems see Bloom & Luca, 2016; Grinin & Korotayev, 2010, 2016a; Kapelyushnikov, 2019a, 2019b; Randers et al., 2018; von Weizsäcker & Wijkman, 2018).

Secondly, there are healthcare systems. Although health care covers the whole society, in some countries there is no free health care. But even there (as in the USA) we find special insurance programs for the elderly. We have already spoken about the increasing cost of medical services due to aging. One can also point to certain assistance and preferences for disabled people in public places where certain conditions are required: wheelchairs, ramps, etc. (and among the disabled, a very large, if not predominant, part are the elderly).

Thirdly, these are institutions, created purposefully or mainly for the elderly. These are social workers who supervise older people and provide them with assistance at home. Also, there are homes for the elderly. As for the latter, they are most common in developed countries, but even there a relatively small part of the elderly live in them permanently. For example, according to the US National Center for Health Statistics (2022), the total number of residents in US nursing homes in 2015 was 1.3 million, while the total number of Americans aged 70 and over at that time was 31.1 million (UN Population Division, 2022a). However, a larger percentage goes to such homes to die. According to statistics, more than

3% of elderly people spend the last part of their lives there (Polonskaya, 2019).[14] Yet, one should note that this last part is often very short. So, according to some estimates, more than 50% of people in nursing homes in the USA die within the first six months (Hellwig, 2018; see also, e.g., España et al., 2021).

Most elderly people are reluctant to leave their homes for a public institution, where they have to live according to a schedule and under supervision. Another quite obvious reason why few elderly people live in such homes is the high cost: not every pensioner and not every family is able to find 60 or more thousand dollars a year to support their elderly relative (in most states the payment is 60 thousand, and this is already the minimum fee—in more comfortable homes it is much higher). Even where provision for pensioners is well established, as in Israel, and pensioners can pay a significant part of the fee from their pension, the cost of living in such houses is rarely less than 20–30 thousand dollars a year, and often more.

The ratio between the monthly pension and the cost of living in nursing homes for the elderly and disabled is 41% in the USA, 51.8% in France, 36.8% in Japan and 60.1% in the UK (Polonskaya, 2019). In other words, if an older person does not have large savings, he/she is unable to live in such a home on a pension (it gives only 40–60% of the required amount), so the elderly people sell their houses and property to pay for living.

The number of older people in nursing homes is generally low. There are different estimates that vary widely. But the costs are very high. In France, more than 600,000 permanent residents and about 400,000 employees work in medical and social institutions for elderly people with disabilities (EHPAD) (cf. Asadzadeh et al., 2022; Yakushev et al., 2020: 3). Of course, in nursing homes, where people are more capable, the staff is smaller, but it is also large. In the USA, 1.3 million people live in nursing homes, another 800,000 live in residences with care services, and nearly 3 million people are employed in social services (Yakushev et al., 2020: 3). Thus, as the number of the elderly and of those unable to care for themselves (see above on dementia) increases, the expanses will become overwhelming for society (both for citizens and for the state) (on the ever-increasing costs of long-term care and the bleak projections for the USA, see, e.g., Dhue & Epperson, 2022; Gleckman, 2022; see also Assaulova, 2020). The same is true, although to a lesser extent, for social workers. In addition, during the COVID-19 pandemic, especially in the first months, the nursing homes showed their vulnerability (e.g., España et al., 2021). Thus, by May 2020, the share of the older category of population in the total number of deaths from coronavirus was 27% in the USA and 32% in Israel. According to the London School of Economics, at the beginning of May 2020, the highest percentage of deaths due to the virus in total mortality

[14] If we consider that people aged 65+ made up 17% of the total population of the United States in 2020 (Jackson, 2021: 3), that is, around 57 million people, then 3% of them will make 1 million 700 people.

was observed in nursing homes in Canada (62%), Ireland (60%), Norway (60%), France (51%) and Belgium (53%) (Yakushev et al., 2020: 2, 3–4, 6).

Thus, nursing homes are unlikely to become the main institution associated with global aging in the future (and the COVID-19 pandemic[15] has confirmed the serious danger to their residents; see above). They are, in fact, still a legacy of previous ideas about charity for the needy.

*Special retirement communities.* Among the currently emerging institutions, we can also mention special retirement communities, although so far they are only common in a few countries. These are mainly Anglo-Saxon countries: the USA, Australia, New Zealand and the UK. They are also organized in Western Europe but are less common there. In the USA, these are full-functioning settlements or even cities with a well-arranged infrastructure specifically designed for the elderly. The main requirement for living in such a city is to be over 55.

There are several types of such settlements. There are even communities of active seniors who travel in mobile homes. Pensioners, who cannot rely on the support of their relatives, often organize their own settlements, where they help each other solve everyday problems. Two types are most common: senior settlements with permanent care and settlements of 'the Sun city' type (Dutton, 2017; Lyubimov, 2013). The latter are special cities and towns where people live mostly normal lives, yet, there may be nursing homes on their territory. This movement originated with the foundation of Sun City in Arizona in the early 1960s. Characteristically, Sun City provides all kinds of infrastructure and employment centers available for retirees. Currently, this Arizona village consists of about 24,000 households; there are about 40,000 retirees who create thousands of jobs. Today there are about 500 such communities in the USA, and they are all similar to the Sun City (there is large enough literature on age-friendly cities and communities around the world, see e.g., Fitzgerald & Caro, 2014).

In total, perhaps, about 3 million people (6–7% of the total number of pensioners) currently live in the settlements for people aged 55 and over in the USA. The initiative has been taken up in other countries. By the end of 2008, there were about 80 villages in the UK, about 350 in New Zealand, and 800 in Australia. In Western Europe, retirement villages are less common, although they exist in Germany, Austria and Denmark. The statistics show that the number of residents does not vary much from country to country and amounts to *5–6% of the total number of retirees* (Asadzadeh et al., 2022; Kuznetsov, 2012).

Residents of a continuing care retirement community (CCRC), can live independently or with the necessary assistance from nurses, depending on their health status, as well as receive medical services of different scales and scopes. The popularity of such centers is constantly increasing. In 2010, more than 2,000

[15] On social consequences of the COVID-19 pandemic see, e.g., Chumakov and Yurchenko (2021); Efremov and Arkhangelsky (2023); Grinin (2020); Grinin et al., (2022b, 2022a); Hussain (2021); Irshad (2020); Jasiński and Bąkowska (2020); Kakkar (2021); Seifabadi and Dolatabadi (2022); Widdowson (2021).

CCRC-type retirement settlements were registered in the USA, with a population of about 650,000 people. In the early 2000s, this number was twice as low (Lyubimov, 2013).

In one way or another, it is important that *such settlements help to develop new institutions, new relationships, and new forms of involvement*. For example, in some settlements of the Sun City type, much work is done by pensioners themselves. They often work voluntarily (i.e., for free), but this does not make them less willing to use their skills for the benefit of the community. For example, residents clean the streets, care for the landscape, and even serve as sheriffs.

In addition, the number of inhabitants is quite noticeable, which makes one scrutinize this kind of social movement and shows that businesses can find a large market for investment in such areas. But, of course, in order to live in well-maintained settlements, one must be able to buy rather expensive property there.

We do not rule out the prospects of such an institution. Such settlements contribute to the formation of necessary institutions and the involvement of older people in active life, although age segregation also has negative sides, including the separation of the older population from the mainstream of life and the strengthening of their conservatism (on the latter, see below).

**Deurbanization**. In Chap. 5, Table 5, we have indicated "deurbanization" in the column for the main types of settlements for the Cybernetic Production Principle. By this, we mean that the process of population decline in the central regions of large and leading cities, which has been going on for quite a long time in many developed countries,[16] will enter a new phase. This may happen due to the acceleration of aging and the need to create a more appropriate and friendly form of community and settlements for the elderly than the megacities with their noise, the fast pace of life, and far from suitable environment. Due to the fact that a huge percentage of the elderly especially the old population will no longer work, as well as in connection with the development of remote work technologies and the spread of autonomous energy production (solar and wind), the need to live in huge cities will decrease and consequently the development will reverse and cities are becoming smaller again. In addition, one of the directions for the development of self-regulating and self-managing systems (already indicated today) is the development of so-called smart cities.

We doubt that smart cities will be megacities. On the contrary, they are likely to be cities with a small or medium-sized populations, where life will be much better arranged. We hope that the interests of the aging part of the population will be taken into account, supporting the emergence *of institutionalized ageism* (see below). It is possible that instead of special towns and settlements for retirees, special suburbs, quarters, microdistricts, etc. will be more common. This could

---

[16] Especially in the form of the so-called suburbanization—the outflow from city centers to their periphery (suburbs). The study of the process of deurbanization started from Brion Barry's work 'Urbanization and counterurbanization' (Berry, 1976).

eliminate the problem of complete exclusion from life. But, of course, the very process of institutionalizing the obligatory consideration of the interests of this population group in planning and their implementation will take time and will not be straightforward and easy. In addition, the adjustment of large cities to new standards will be conflictual and expensive. The creation of a new type of settlement will also be expensive. But it is important that this open up a huge market for investment and GDP growth.

However, we must keep in mind that only just over half of the world's population lives in cities today. So the process of urbanization will continue in developing countries, and quite actively, especially given the considerable population growth in dozens of African and Asian countries (UN Population Division, 2019c). Accordingly, megacities will emerge and grow there. However the global trends toward deurbanization will start to somehow influence urbanization in these countries.

**Euthanasia as a grim omen of the aging process**. We must end this section on a gloomy note. Recently, the use of euthanasia has been expanding. Euthanasia is the voluntary termination of life with the assistance of a doctor, subject to the legally defined procedure for expressing consent and ending life. Euthanasia is subdivided into active euthanasia, that is, voluntary, and passive euthanasia, that is, the withdrawal of life support from patients who are unconscious or otherwise incapable of giving consent. Consent is given by relatives, guardians or other responsible persons. Active euthanasia often includes medically assisted suicide (the provision of lethal drugs at the patient's request). For example, in Belgium, you can buy a lethal "medicine" with a "prescription" for about 70 euros and commit suicide. It is true that, in a pharisaic desire to somewhat cover up the ugliness of this procedure, modern medical organizations have ceased to consider involuntary euthanasia and have sanctioned suicide as euthanasia (Rich & Hida, 2023).

Despite the seriousness and, objectively speaking, the cruelty of such an institution, which is very important from many points of view, little is said about euthanasia in general, in any case not enough. In our opinion, however, this problem is more serious than that of abortion, since it is also symbolic, as a clearly unattractive end to life.

It is of interest to us in the framework of this study because euthanasia is increasingly becoming the lot of the elderly.[17] Voluntary death is chosen by many thousands of older people every year. Of course, these are, first of all, very sick people, often terminally ill, with practically no chance of recovery. But since most of these people are at a very advanced age, we can say that, unfortunately, euthanasia has become one of the modern institutions associated with aging.[18]

---

[17] Unfortunately, however, this procedure is extended in some countries, for example, in Belgium for sick children (Canada is close to this).

[18] The average age of euthanasia recipients in Canada, for example, is 76.3 years, and the most common underlying medical condition cited as a reason for requesting euthanasia is cancer.

It is particularly surprising that euthanasia is widespread not in poor countries, but, on the contrary, in the richest ones: Belgium, the Netherlands, Switzerland, Canada and some others. Yet these countries pride themselves on their humanity, and condemn societies that retain the death penalty. And the number of people who choose lethal injection is constantly growing. For example, in Canada, the number of people choosing euthanasia has increased 13-fold in six years since 2016. The number of deaths by euthanasia reached more than 10,000 in 2021 (Health Canada, 2022; Cheng, 2022). Based on these figures, the Euthanasia Prevention Coalition, a campaign group, estimated that Medical Assistance in Dying (MAiD) cases rose from 10,064 in 2021 to some 13,500 in 2022 (Reinl, 2023). So, this is a one-year increase of 34% from 2021 to 2022. And this number is expected to continue to grow in 2023 and subsequent years would.[19] We suppose that among the excess deaths in Canada, deaths due to euthanasia account for a significant share of all excess deaths.[20] This accounts for 3.3% of all deaths in 2021 and 4.1% in 2022 (Billot, 2023). This is not a token proportion of mortality. In Quebec, 7% of all deaths were state-assisted suicides (Reinl, 2023). We agree with Michael Cook (2023) that Canada's euthanasia law is a human rights disaster. An even higher percentage of euthanasia deaths among all deaths is found in the Netherlands. Euthanasia accounted for 5.1%—or around one in 20 deaths; in 2022 their total number was 8,700. The total number of assisted deaths increased by 13.7% between 2021 and 2022 (Séveno, 2023).

They often justify this legalized murder on the grounds of human freedom of choice. Indeed, there are a number of people who have the opportunity to continue to exist, but who are determined to end their lives because of pain or go this way for other important reasons. However, we fear that the bulk of those who choose euthanasia do not do so entirely voluntarily. Moreover, it is fundamentally important that the majority, probably the absolute majority of those who have made such a decision are old people who cannot pay for their treatment, who have insufficient insurance options, and whose relatives cannot pay. These are often people who lose their will, and who are relatively easy to persuade to make such a decision. If a person is under the pressure of illness or poverty, if he or she is perceived in one way or another as a burden, if doctors persuade him or her to agree in a state of a not-always-clear mind, it is not surprising that he or she can be persuaded. Such legal permission to sanction, in fact, the murder of parents or relatives, which is increasingly being used by children and relatives of the sick and elderly (and, in some cases by parents in relation to their sick children), does not

[19] In Belgium, the number of deaths by euthanasia is also constantly growing, although not as rapidly as in Canada. Nevertheless, between 2003 and 2022, the number of euthanasia deaths increased by more than 8.5 times, from 342 to 2,966 people, and by more than 10% between 2021 and 2022 (CNE.news, 2023; Statista.com, 2023).

[20] Preliminary data show that from the end of March 2020 to the end of August 2022, there were an estimated 53,741 excess deaths in Canada, 7.6% more deaths than expected in the absence of a pandemic. During this period, at least 42,215 deaths were directly attributed to COVID-19. https://www150.statcan.gc.ca/n1/daily-quotidien/230112/dq230112c-eng.htm.

contribute in any way to the humanization of society, and respect for the elderly; it will lead to serious moral and social consequences. The easier a society treats death, the more problems it will have. And if children choose a lethal injection for their parents (and the children's attitudes to this have a huge impact on the decision to die voluntarily), then the prospects for successful aging in society are far from being as good as it seemed.

It is disturbing that such an institution develops in seemingly advanced societies, which should be a model for developing countries. In fact, "it's on a perilous road to mass euthanasia and ever-more pressure on the sick, disabled and poor to end their lives prematurely" (Reinl, 2023). This would be a bad example; it would greatly impede the normal development of the global aging process. Is modern society really moving toward the state of some archaic societies where gerontocide was practiced? Will the development of the global aging process really end with an increase in the number of people voluntarily departing from life? We hope that this will not happen on a large scale, but an unattractive scenario cannot be completely ruled out.

As with any moral issue, one can argue about the admissibility or inadmissibility of euthanasia. However, there are several important points that speak to the deep hypocrisy of societies that allow euthanasia. First, suicide is condemned in any society, and those who enforce another to commit suicide or even assist in the suicide of another, can be punished. In additions, they provide medical assistance to those who commit suicide unsuccessfully. In the case of euthanasia, the state authorizes both suicide and direct assistance to it. What is the difference? Essentially, the difference is that society is unwilling to bear financial losses. This seems to be a clue. But these are the richest societies. There are many centers of psychological help for people who want to commit suicide, they are helped to gain the strength to live and in the case of euthanasia, on the contrary, they are persuaded to die…

Secondly, all countries that have allowed euthanasia have banned the death penalty in their country. It turns out that taking the life of convicted murderers, villains, maniacs, is extremely dangerous to society, is inhumane, they must be supported at the expense of the budget, and taking the last years and months from the elderly is humane. At the same time, European countries condemn those (like China) that apply the death penalty. Is this not hypocrisy?

Thirdly, as mentioned above, the majority of those who choose euthanasia are people who are unable to continue treatment because of poverty. They try to persuade them to agree to a voluntary death. But every life is sacred. What is particularly dangerous is that euthanasia is extended to other groups of people, especially sick children. The latter situation resembles the condemned practice of eugenics, and this eugenics program would be admired by any of the world's past genocidal maniacs (Carter, 2023; see also Tilley, 2023). Now Canada is planning to legalize euthanasia for drug addicts with no other illness in March 2024 (Tilley, 2023).

Fourthly, the physicians who advise and carry out euthanasia have forgotten the Hippocratic Oath. The Hippocratic Oath clearly states: "I will not give a lethal drug to anyone if I am asked, nor will I advise such a plan." And yet, it happens all the time.

Finally, fifthly, it is alarming that the protocol for obtaining consent to voluntary death is rapidly changing in the direction of simplifying this very serious and responsible procedure, the decision-making time is shortened, the number of witnesses is reduced, the procedure for the mentally ill is simplified, etc. All this is a wake-up call that condemning old people to death is becoming more and more commonplace.

Euthanasia, as an evolving institution, poses a threat to aging society. If you do not give birth to children and kill the elderly, then where will such a society go? Yes, euthanasia also shows that the aging of society is far from being a soft process, it is a process that profoundly transforms society. But we hope that euthanasia will not become the main path for aging societies (for more details on the ethical problems associated with euthanasia see, e.g., Bélanger et al., 2019; Bellens et al., 2020; Devos, 2021; Gibson et al., 2022; Keown, 2018; Pesut et al., 2020; Raus et al., 2021).

## 7.2 Aging in the Future

### 7.2.1 Aging and Its Complex Impact During the Final Phase of the Cybernetic Revolution and the Mature Stages of the Cybernetic Production Principle

**How and Why Global Aging Will Affect Almost all Spheres of Life and Society in the Future**. Of course, it is extremely important to analyze how far this process will go and, above all, to determine the proportion of the elderly and old people in the demographic structure of society. Below, we make some predictions on this point. However, in reality, they may deviate in one direction or another. Aging affects strongly (a) economic indicators (employment, consumption and its structure, growth rates, investment, etc.); (b) political dimensions (e.g., whether older people will be the leading electoral stratum; what share of budgets will be allocated to them); (c) social parameters (to what extent social programs will be oriented toward them; how much the professional structure will be adapted to them); and of course, (d) the medical system. We have already mentioned the pension system above. In the following, we will discuss various scenarios. In any case, the social as well as the political structure will experience an increasing trend in population aging. The impact of aging on democracy will be discussed below. As we have already pointed out, in developed societies the elderly start to replace children in terms of the care needed. But caring for the elderly requires more money and effort from society, despite the fact that the main concerns fall on the family and relatives so far.

We should focus on such important institutions as (1) the preservation of older people's savings and property; and (2) social lifts, which may be hindered by ageist tendencies in society (we will consider them in a separate paragraph). We will discuss conservatism inherent to an older society separately in Sect. 7 of this chapter and in Chap. 13 (see also Jackson et al., 2008: 188), but this characteristic will affect many spheres. Here one may note that both the media and their ideological orientation in terms of condemning or approving conservatism can change significantly.

At present, it becomes obvious that while in technological terms the developmental vector has already been fully determined toward extension of life expectancy, it is not at all clear from the political, financial and economic perspectives whether modern society is prepared to solve the problems associated with increasing lifespan. To what extent, for example, is it prepared to simply preserve the savings and assets of older people, and not just theirs, of course? How secure will they be in a situation of unprecedented growth in public debt and the increasing power of bankers and financiers? Will they burn out in financial crises, defaults and inflation, or in the emerging (with the help of technology) manipulations and prohibitions (e.g., there are attempts to ban cash); zero and negative interest rates (Grinin, 2021), as a result of which pension money may depreciate? We have touched a little on this issue above, but in general, it is a subject for a separate and complex study. It is sad that this issue is not in the mainstream at all, and still, it is fundamental to understanding the future of aging generations.

With the threat of possible permanent pandemics, the increasing influence of the media, AI technology and many other factors, what will be the position of older people? Will society ensure a real increase in life expectancy and in the quality of biological (health-related) life? Or will we face a serious social crisis, a backslide (the rehearsal of which we have recently witnessed during the pandemic when the elderly turned out to be the most vulnerable group [e.g., España et al., 2021; UN Population Division, 2020: 17–26; Yakushev et al., 2020]), which we will have to overcome in order to advance the process of increasing life expectancy and to improve the quality of biological life?

It is interesting to note that while the initial and modernization phases of the Cybernetic Revolution generated technologies mainly for young people (up to 40 years of age), with the trend toward the elderly intensifying, the final phase should be much more technologically oriented toward serving the elderly (e.g., robotic technologies for the care of the infirm, not to mention medical technologies [see Chap. 10]). This will be an important aspect of the above-mentioned adaptation to aging.

As for the world-system dimension associated with the process of global aging, the opposite between older and younger societies will be indicated, while the link between them can be strengthened in the situation of mutual dependence associated with the age structure, complementarity (need for finance and labor force), or it can become a confrontation within the World System or a model of *the old North—the young South* (see Chap. 6; we will discuss these scenarios later). Western aid and investment in the training of the workforce in developing societies is possible, but there may also be conflicts between them. In the future, as young societies get older, the situation will change even more. Therefore, we would like to believe that over the next thirty years, some institutional solutions with respect to age people and their problems will be developed, otherwise the Cybernetic Revolution may stagnate, while the internal tensions in the aging society will increase (although this tension may take different forms).

### 7.2.2 Some Forecasts

The population aging projections, even the inertial ones, are quite impressive. The growth rate of the proportion of people 60 + is several times higher than the overall growth rate of the world's population (UN Population Division, 2013, 2022a, 2022b). In the early 1980s, there were no 'aging countries' in the world, i.e. countries where older people would consume more than young people. In 2010, there were already 23 "aging countries" and by 2040 there will be 89 (UN Population Division, 2022b; UNFPA, 2012). According to the most recent UN medium projection, by 2100, 2.47 billion people, more than a quarter of the world population, will be over 65 and only 2.30 billion will be under 20 (UN Population Division, 2013, 2022a, 2022b). The number of people over 80 will increase sixfold, from about 140 million in 2020 to more than 860 million by the end of the twenty-first century (Gallagher, 2020); thus, the number of the 80-year-olds should increase by almost two orders of magnitude by 2100 compared to 1950! (see Fig. 7.4).

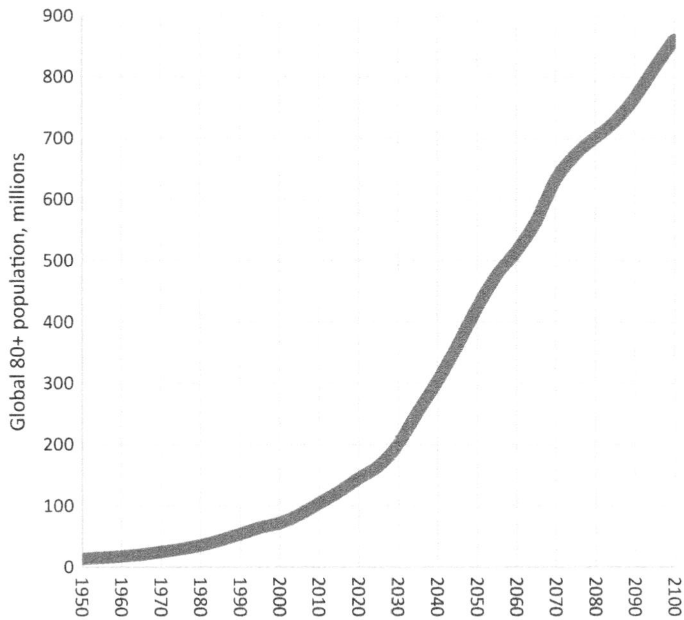


**Fig. 7.4** Global population aged 80+, millions; UN Population Division empirical estimates for 1950–2021 and the UN Population Division medium forecast scenario for 2022–2100. *Data source* authors' calculations on the basis of data provided by the UN Population Division database (2022a)

Although 2100 may seem a long way off, one should clearly understand that the growth rate of the elderly population increases as the median age rises and the pace of aging continues to accelerate (Kamaiya & Hertog, 2020; Medici, 2021; Shcherbakova, 2019; UN Population Division, 2019a, 2020, 2022a, 2022b; Zimmer, 2016). Between 2020 and 2030, the share of people aged over 60 in the global population will increase by 34% (UN Population Division, 2022a, 2022b; WHO, 2021b). By the 2050s, drastic changes in the path of global and national aging will occur. Thus, the population aged 90 years and over will increase by a factor of 3.6 by 2050 (from 22.8 million people in 2021 up to 82.9 (UN Population Division, 2022a, 2022b; cp. Shcherbakova, 2019), see Fig. 7.5.

Between 2000 and 2050, the proportion of the world population aged 60 and over will more than double from about 9.9% to 22%. The absolute number of people aged 60 and over is projected to increase from 1.1 billion in 2021 to 1.4 billion by 2030 and 2.1 billion by 2050, and may reach 3.1 billion in 2100 (UN Population Division, 2022a, 2022b; cf. WHO, 2017a: 11, 2021b). The number of people aged 80 years and over is expected to triple between 2021 and 2050, to 458 million (UN Population Division, 2022a, 2022b; cf. WHO, 2021b).

By 2050, about 34% of the population aged 60 and over will live in Europe, and 25% will reside in Latin America and Asia. Although Africa has the youngest

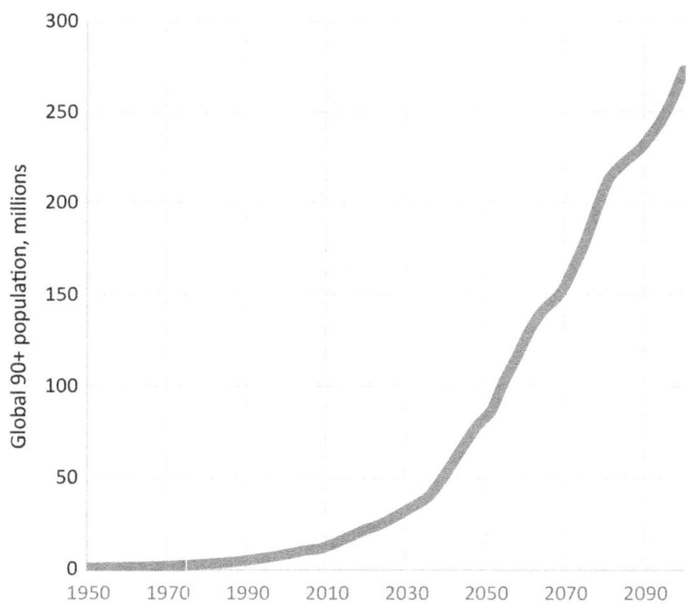


**Fig. 7.5** Global population aged 90+, millions; UN Population Division empirical estimates for 1950–2021 and the UN Population Division medium forecast scenario for 2022–2100. *Data source* authors' calculations based on data from the UN Population Division database (2022a)

population of any major region, it is also projected to age rapidly, with the percentage of people aged 60 years and over rising from the current five percent (in 2022) to nine percent by 2050. Keep in mind that older people will be increasingly concentrated in less economically developed regions (UN Population Division, 2015, 2022a, 2022b; WHO, 2017a: 11). By 2050, about 65% of older people over 60 years of age will live in low- and middle-income countries (WHO, 2021b). This forecast is particularly alarming in terms of improving the quality of biological life (or health-related quality of life) of the elderly and old people, taking into account the underdeveloped pension system in many of these countries.

Accordingly, the problem of pensions and social security for the elderly may become extremely acute. As has already been mentioned, the number of people aged 60 years and over (1,109 million) already in 2022, almost twice exceeded the number of children under five years of age (663 million) (UN Population Division, 2022a, 2022b; cf. WHO, 2021b). It is a somewhat frightening prospect that by 2050 there will be almost twice as many people aged 60 and over (2,132 million) than teenagers and young people in the 15–24 age group (1,299 million) (UN Population Division, 2022a, 2022b; cf. WHO, 2021b).

Already by 2030, fundamental transformations will occur. In 2030, the proportion of the population aged 65 and over will be 20% or more already in 59 countries/territories of the world (cf. Shcherbakova, 2019; UN Population Division, 2022a, 2022b).[21] By 2030, people aged 60 and over will make up one-sixth (16.6%) of the world population (UN Population Division, 2022a, 2022b; cf. WHO, 2021a).

Such forecasts should make politicians think twice and start changing policies. However, many people think that the situation is not that serious and so they do not understand the urgency of the decision to be taken and the complexity of the measures that could, if not reverse the trend, but at least slow it down (about this see, e.g., in Gallagher, 2020).

### 7.2.3 Ageism

**Ageism in the Modern Sense**. Ageism is usually referred to as stereotypes of thinking or behavior that often result in discrimination against people because of the stereotypes about their age. Ageism is usually interpreted as a negative factor,

[21] In Japan, the share of people aged over 60 has already reached 36% (UN Population Division, 2022a, 2022b; cf. WHO, 2021b). It is significant that this was forecasted quite a long time ago, at least as early as in 2008 (Jackson et al., 2008). But the trend has not changed, as the recent forecasts confirm it. We are afraid that the same authors' prediction that the proportion of older people in Japanese society will reach 40% by 2050 (Jackson et al., 2008) may also come true; in fact, according to the UN's latest medium projection, the proportion in Japan in 2050 will be 44%, whereas, according to their lowest fertility projection, it will be even 47% (UN Population Division, 2022a, 2022b). And, according to one of the IHME forecasts (see above), even at the global level this share may reach 40% by 2090 (Vollset et al., 2020).

as a stable stereotype or as an established form of age discrimination. It manifests itself in various ways: from interpersonal and everyday relationships to actions at the level of societal institutions. For example, ageism can manifest itself in a bias based on prevailing ideas that people over a certain age (60+) or, conversely, too young, are undesirable to employ. Or in rather rough ideas about young and middle-aged people in relation to the elderly: 'they have lived their time, why do they need travel, houses, money, etc.' Or, on the contrary, in relation to the young: "You have everything ahead of you, you have nothing to rush, you will have time to buy everything, go everywhere, etc."; "he is still young, how can he become in charge? let someone older do it, etc." Ageism can also be a kind of *self-ageism*, when people evaluate their position according to established stereotypes: "I'm already old, I can hardly get married, it's time for me to die, I don't need it; nobody needs me" etc.

Levy et al., 2020 point to three manifestations of ageism in the healthcare system: "there are three discrete predictors of ageism: age discrimination, defined as detrimental treatment of older people; negative age stereotypes, defined as the negative beliefs of older persons about older people in general; and negative self-perceptions of aging, defined as older people's negative beliefs about their own aging" (Levy et al., 2020: 175).[22]

Ageism in this sense is very common not only in medicine, but also in everyday life, in various situations. It is even believed that one in two people in the world adheres to ageist views (WHO, 2021a). Yet, here one should bear in mind that if such views become so widespread, then they do make sense. Therefore, it is perhaps much more reasonable to channel and institutionalize ageism than to try to eliminate it; not least because it would mean tackling only the extreme and most negatively affecting manifestations of ageism. We will return to this in more detail.

Ageism, in its extreme and harshest forms, causes much damage. This is expressed in various aspects. Firstly, ageism makes it impossible to use very large 'forces' of the elderly, who put a cross on themselves or give up, as well as clip the wings of the young. Secondly, the manifestations of ageism, especially with respect to the sick, elderly and vulnerable people, lead to deterioration in their health, which, in addition to their moral suffering and that of their relatives, results in very high costs, which we will consider later. Thirdly, the harsh manifestations of ageism in relation to social lifts and careers for young and middle-aged people lead to stagnation in society,[23] and in relation to aged workers—to difficulties in transfer of experience and knowledge. Below we will talk about the problems of rising gerontocracy and a slowing down of social mobility in society. Fourth, in

---

[22] An example of the first point: why did you come, you should think about death; why try there, he/she will die anyway, etc. An example of the second point: what is there already to wait for, you need to get used to getting sick; what do you want, age, everyone gets sick at that age. An example of the third point: nothing to wait for, you will die soon.

[23] This is especially evident in the army and paramilitary structures, where career progress in peacetime depends on time in service.

extreme cases, it can lead to direct conflicts between generations. In the 1960s and 1970s, the youth revolts shook Western societies. In the future, declining fertility and the dominance of older cohorts (as researchers have pointed out, e.g., Fukuyama, 2002) may lead to intergenerational conflicts. As Fücks rightly points out, there is a huge potential for intergenerational conflicts related to the distribution of the numbers projected for aging and depopulation (Fuecks, 2013: 42). It is also important to understand that the manifestations of ageism differ in different countries and cultures, depending on living standards and social insurance, belonging to a particular social stratum, religiosity, the strength of traditions, the state of the economy, demographic trends and much more.

**Ageism as a Complex and Ambiguous Social Phenomenon**. As already mentioned, ageism is not a superficial phenomenon. It has strong foundations in society, since these ideas make people (a) relate their own desires and interests to their age status; (b) relate other people's possibilities to their age status; and (c) facilitate communication and subordination in society. It is one thing for relatively young children to give up on elderly parents as hopeless and to try to deprive them of the right to make decisions, dispose of their property, etc. Here ageism with an emphasis on the young's selfishness is obvious. But it is another thing when children do not take into account the age of their parents (or grandparents) and their stamina, believing that they can "pull" like young people and help them constantly. Here, just the ageist ideas that older people need more peace and rest, would be very helpful. It is one thing for older parents (grandparents) to keep young people on a short leash, believing that they know better what young people need; it is another thing when people of age, being still quite capable, completely refuse to help a young family with children who can barely make ends meet (justifying this by saying that only after retirement they can live their own lives). So, here the ageist ideas that the elders should help the young would be appropriate. It is one thing to be independent when children grow up, but it is another thing when they send their older parents to homes for elderly people with light hearts, even if it is comfortable.

In other words, it is important to optimize ideas about the roles and relationships between people of different ages and take into account psychological and personal traits, as well as people's actual abilities at a given age and their aspirations. In addition, stereotypes are useful, but it is dangerous that they lag far behind the changes. Since the average life expectancy has increased significantly, the features that used to be typical for people 60 years old, who were considered old by others and by themselves, no longer apply to people who are today 60 years of age and over—most people feel much younger than their ancestors did at the same age 50–100 years ago (see, e.g., Lutz et al., 2008; Sanderson & Scherbov, 2007, 2010, 2013, 2015). The age range has shifted. And it will continue to shift in the future.

The change in the perception of age is especially noticeable in the way people dress, in the use of cosmetics, in hairstyles, in a word, in their appearance, where the line between young and old has almost disappeared. In addition, cosmetic surgery and medicine, in general, have begun to play a huge role, especially with the improved possibilities for rejuvenating surgery and, going hand in hand, in some areas, the lower thresholds of availability.

Thus, ageism can vary significantly; this is a very ambiguous concept, by far not limited to negative aspects. When relatives (or society as a whole) take care of an old sick person, realizing that he/she can no longer do much on his/her own, but deserves a good attitude—this is ageism, but a positive and humane one. When the elderly help the young, because the latter cannot cope with problems during a period of economic turmoil, this is ageism, but a kind and noble one. When children give way to the elderly in transport, can this be considered a bad ageism? In other words, ageism often appears positive, yet we treat it as a valid norm, while in other cases, in negative situations, we get angry. Thus, ageism, like any other phenomenon (e.g., social inequality), cannot be assessed only as positive or negative. This phenomenon (or institution, if institutionalized ageism takes shape) has both positive and negative sides. The question is how to balance them, so that there is neither a strong infringement of some age cohorts nor the substitution of general human rights by age rights (e.g., the claim that the elderly are always right; or "all the best is for the children"; "the young are always dear to us"). Selfishness is inherent in people and their groups, including age (gender) groups.[24]

**Optimization and Institutionalization of Ageism as a Vector for the Development of the Future Aging Society**. From the above said, it follows that an institutionalized balance of interests is needed and that the balance should be mobile and take into account the changes in the age composition of the population.

On the one hand, a gerontocratic regime should in no case be established (i.e., a situation when a society becomes dominated by the elderly not only in terms of numbers but in terms of their influence over all most important matters, their occupancy of positions and privileges). This will dramatically reduce the opportunities for social lifts in society and increase welfare dependency. Many priorities in society will also change, putting maintenance of the level of health, peace and medicine in the first place, rather than the development of a person of varied abilities (including sports and an active lifestyle). On the other hand, the elderly and the old should not become the object of change, a powerless part of society, whose retirement age is constantly being raised, whose pensions are being cut and who face various restrictions and humiliations. Thus, it would be good if there were political forces representing these cohorts. If women have achieved much in terms of equal rights with men (and in fact, in some cases, they have more rights) by fighting for these rights, then one may expect that older cohorts could also fight for their rights.

So, it is impossible to solve the problem of ageism simply by asking older people to increase their self-esteem, by asking others to change their attitudes toward

[24] From ethnology and history we know examples of both: the elimination of the elderly, that is geronticide, among many peoples; and gerontocracy—among the Australian aborigines, ancient Spartans, and many early farmers (see, e.g., Artemova, 2003; Artemova & Korotayev, 2003; Adegbindin, 2011; Maxwell & Silverman, 1989).

them; or by trying to awaken in respect for other ages in stronger cohorts. One must realize that (a) society will face this problem for a long time due to global aging, and we will have to learn to live with it; (b) the problem of older cohorts will increase, so it is more appropriate to institutionalize some issues related to age inequality than to try to ensure equal rights for all, which may turn out to be unrealistic. In general, one can assume a rather difficult and conflicting path of resetting and adaptation to aging (see below for details). Accordingly, the problem of ageism, which is still acute today, will manifest itself in ever-new aspects and directions and will change society considerably. We think that *ageism will go through several stages* in its development, some of which are quite contradictory. But the empirically worked out laws, rules and institutions will gradually lead to the emergence of a system that will take into account the interests of different age groups and establish their rights, duties and restrictions, up to quotas in political representation and taking positions (as observed today with respect to quotas for women or people of color). In particular, *ageism may gradually be institutionalized in a certain and systematic consideration (separation) of the interests of different age groups in settlements, interests, possibly political representation,* etc. This develops the above-mentioned idea that the economic type of alienation and social division in the Cybernetic Production Principle will increasingly be associated with age and gender differences.

**In our view, the main directions of institutionalization will be related to**:

(a) de jure or de facto different age-specific rights. Of course, such a difference in rights does exist today, firstly, with respect to minors and the rights of people reaching a certain age to receive a pension. Otherwise, legally adult people are formally equal in rights. However, we believe that in the future the number of rights or restrictions associated with age will increase greatly. This will not necessarily be expressed in law, but may take place in the form of stable public perceptions, legal precedents and the like. For example, in most countries constitutions or constitutional laws declare equal rights for women and men in all spheres. But in practice, as a result of the prevailing legal practice in family, marriage and sexual relations, women may in some respects have more rights than men (e.g., in respect of rights to children in case of divorce, child-support payments, early retirement in some countries, etc.; see also above about quotas).[25] Here we speak about this without any value judgment, just to show how the institutionalization of rights and restrictions (duties) in relation to age division can take place. The restrictions imposed on the elderly (65+) by COVID-19 also show how severe restrictions can be imposed on

[25] However, with regard to equal pay for equal work, although women have achieved a great deal, and "overdid it" somewhere, but on the whole, the advantages still remain with men. It is fair to say that the concept of popular psychology, that it is the man who should be able to provide financial support for his family (bring home the bacon), prevails.

people according to their age. Accordingly, there can be a lot of combinations of rights, restrictions and obligations, depending on the society and its cultural and historical characteristics, social position, economic situation and much more;

(b) the self-organization of age groups of citizens in various forms, ranging from associations and mutual aid societies to world organizations; from the provision of mutual assistance to the lobbying of interests;

(c) the structuring of the electorate according to age categories and, accordingly, the formation of political and legal programs;

(d) the strengthening of the ideological orientation in the discussion of the rights and duties of age groups (including all kinds of campaign, as we see in the feminist movement);

(e) the inclusion of the age aspect in all kinds of planning and activities (e.g., in the planning of buildings, in legal acts and laws, in requirements for employers, landlords). Here, to some extent, the institutionalization of disabled people today can serve as a guideline for developing institutionalization of old age tomorrow[26];

(f) the expansion of judicial and legal activity to protect the rights of people in old age.

**Ageism and characteristics of the future generations**. We should point out that due to increasing life expectancy, the psychological and other gaps between the generations may increase, which can complicate the problem of ageism and lead to confrontation between generations. Given that the elderly (60+) will be the largest cohort in terms of numbers, the confrontation may intensify, but not between fathers and children, but between children and grandchildren, on the one hand, and grandfathers, on the other. It is worth thinking about what the leading (age) generations will be, what their defining features will be. So, by 2050, people born between the 1970s and 1990s will play an increasingly important role. These are mostly the so-called Millennials (Generation Y). One must note that in developed countries, these are hardly the most active generations (in other countries, of course, they may have different characteristics). By 2100, the generations born between the 2020s and 2050s will play a huge role, yet, these are already small generations, accustomed to technological and technical progress, which they see as automatic (in fact, the above-mentioned generation is also inclined to such an attitude), accustomed to relying on it completely. They will hardly fight for progress vigorously, but they will not stand up against it (and this is an advantage). For them, globalization will mean much more than for the previous generation. Thus, the background and prerequisites for a slowdown in development (see Chap. 13; Grinin et al., 2020) are already formed. On the other hand, we mentioned above that there will also be a need for innovation due to the need to adapt to aging and

[26] Here, to a certain extent, the institutionalization of people with disabilities can serve as a guide.

longer life expectancy. Therefore, one can hope that we will also see a reversal of this trend as well.

### 7.2.4 On Some Probable Features and Challenges for the Future Aging Society

In this paragraph, we summarize some of the characteristics and possible problems, and trends of the future society, partly repeating what has already been discussed.

**Aging as a Characteristic of Society and the Changing Perceptions of Age**. If we assume that the trend of increasing life expectancy will continue (albeit with fluctuations) and the number of age cohorts will increase, then one should pay attention to the following factors.

(1) *Graduality*. Aging will proceed gradually and somewhat imperceptible so that society can and most likely will adapt to it.
(2) *The elderly bulge*. If we take the Japanese society as a model, which already has almost 30% of the population aged 60–65 years and over, then by 2100 there will be 45–50% of people 60+ (see Fig. 7.6).

This means that these people will make up the majority of the adult population, given that the share of children and young people will decrease. The number of people aged over 80 years old will reach a very high level—20%, that is, their number will correspond to the current generation of people aged 65+. In graphs, this will look like an elderly bulge (similar to a youth bulge) (see Figs. 7.3 and 7.6 above).

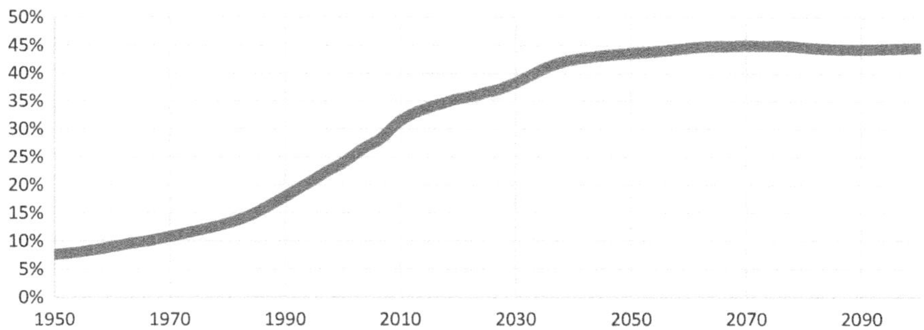


**Fig. 7.6** "Elderly bulge" in Japan: dynamics of the share of the population 60+ in total population, 1950–2021 with the UN medium projection till 2100. *Data source* authors' own calculations based on materials provided by the UN Population Division database

(3) *The changing perception of age.* One can assume that a shift will occur in self-perception of aging toward feeling 15 to 20 years younger than the chronological age (see, e.g., Lutz et al., 2008; Sanderson & Scherbov, 2007, 2010, 2013, 2015). A similar shift has already taken place in society over the past 50 years and it was caused by the dramatic increase in life expectancy. And some manifestations of ageism are partly related to this, when the public opinion maintains a stereotype that 55–65 years old should be classified as old age, while the older people, reaching this age, after retirement have leisure time and are full of energy and vitality, and they object because only now they start to enjoy life to the full, with traveling, arranging various meetings, events, other forms of activity, up to active sports and significant physical exercise. This means that the age of 60–65 years will not be considered as old, and even not as very mature, but as middle age (as today's 40–45 years), and the age of 65–80 years—as today 60–70 years.

(4) Raising the retirement age. Due to the factors described above, the retirement age should be between 70 and 75 years, and it would also be right to equalize the retirement age for men and women. Today it turns out that in many countries women live longer and retire earlier than men. However, this manifestation of sexism is hardly discussed.

(5) The importance of medicine. Today, medical solutions for rejuvenating the appearance of the body make a huge contribution to changing the perception of age; however, medicine will play an even greater role in the future, when it will gain opportunities to really rejuvenate the body at least slightly through new technologies (stem cells are just one of many), which we consider in the next chapters (Chaps. 8, 9 and 11). The age at which children are born will be postponed by 10–15 years due to new technologies (some of which are already available today), for example, freezing eggs and sperm to be used many years later (for details see Chap. 6). Perhaps, this will become common practice (just to be on the safe side). In general, a large part of medicine will focus on rejuvenation and plastic surgery (which will become as commonplace and regular as a holiday in a health resort), so it will be more difficult to understand people's age visually (hence the growing role of biometrics). All this will also be discussed in the chapters below.

**More conservatism and less dynamism**. Just as society will gradually become accustomed to aging, the general trend toward weakening activity will be realized also gradually. This is to affect the following points to a greater or lesser extent.

(6) *Increased conservatism, reduced activity in various forms and innovation.* Because young people are not affected strongly by their life experiences, they have a different view of many things, and are generally more inclined to change. One cannot say the same about the elderly. Of course, a strong need can also force older people to change, but one way or another, the growing number of older people leads to more conservative behavior (see Chap. 13 for details). The increasing share of older people will result in an increase in

the number of workers, while the activity of a very large part of workers and their creative potential is likely to be reduced compared to present day and even more so compared to the period of 20–30 years ago (see Chap. 6).

(7) *Development of the habits and style of youth among the elderly*. Of course, we do not exclude the emergence of opposite trends, but the probability of their emergence is not so high, and these trends will not be able to completely extinguish the indicated inclination toward conservatism. For example, sports events for older people will be held and become more common than today (similar to sports for the disabled). But, of course, this will not replace sport for the young, just as the Paralympic competitions cannot replace the Olympic Games. In other words, in connection with aging and adaptation to it, some forms of young people's activities will spread to mature and older people. And this, along with other factors and the above-mentioned opportunities provided by medicine, will contribute to the development of the above-described trend (point 3—the changing perception of age) and also to a gradual change in social psychology in the perception of what normally corresponds to a certain age (as we remember, this is a very important basis for self-ageism and ageism). But, of course, even with the medicine of the future, the differences between the young and the old will remain great, and accordingly, the general trend will be toward a decline in the vigor of society. So, the task will be to mitigate this trend as much as possible, while at the same time adapting to it.

(8) *The intensification of the aging process* will lead to a corresponding increase in the number of disabled people in society, since the older people are, the greater the proportion of them will be disabled. Thus, in 2012, more than 46% of people in the world aged 60 and over had a disability. More than 350 million older people have a moderate or severe disability (UN Population Division, 2020; UNFPA, 2012). The issue of disability and the issue of aging can therefore, to some extent, become one. It is important to note, that at present, both legislation and other measures and institutions in society have advanced more in relation to disabled people.[27] Therefore, one may assume that society's adaptation to aging will also develop on this basis, including aging with disability.[28] The number of disabled people is very high. As mentioned earlier, by 2017, there were 50 million patients in the world, and by 2050 there will be more than 150 million, and possibly much more (WHO, 2017b).

(9) *The strengthening trend toward gerontocracy* (Rich & Hida, 2023). This may be due to the fact that the number of mature and older people will increase. It is obvious that this will reduce social mobility and increase dependency. Gerontocracy among the elite is especially dangerous. Look at the current US elite. And compare John F. Kennedy, elected at 43, with Joe Biden, elected at

[27] This is largely thanks to the UN and its Convention on the Rights of Persons with Disabilities adopted in 2006, as well as the activities of international organizations of persons with disabilities.

[28] The so-called minimum standards for the elderly and disabled are already being worked out (Age and Disability Capacity Program (ADCAP) 2015, in particular, it was noted there that in 2015, 15% of world's population live with some degree of disability on Earth).

78. And many other leading politicians are a match for Biden. The dominance of the elderly in politics and other key areas will contribute in particular to the stagnation of society and its conservatism.

**Other socio-demographic changes and problems in society**. Aging will affect many relationships and make different trajectories possible.

(10) *Bullying older people*. As already mentioned, it is possible that events will develop toward bullying older people and reducing their standard of living. Yet, this is a path to a dead end, since an overreach in this context will hardly contribute to the increase in life expectancy. However, it is quite possible that the growing number of older people will lead to the adoption of various tough laws. In particular, these may imply a permanent increase in the retirement age and the imposition of age limits for holding various positions. This is already observed in one way or another, but not systematically, e.g., the age of officers, presidents (if it is regulated by constitutions); sometimes of professors, etc. However, such lists can be greatly extended. All this may become a source of conflict and should be optimized.

(11) *The weakening of the youth's vigor*. Due to the shortage of young people, their positions may in some cases become exceptional, which will weaken their motivation and energy (thus, a decrease in vigor may also occur in the young cohorts).[29] But in any case, youth and vigor will be valued, and they will be clearly missed. There will appear small strata of young people, who are spoiled from childhood[30] and are less adapted to the struggle of life; so they will be even more infantile and will need more time to grow up than Generation Z. Of course, other opposite trends are also possible, but it is unclear yet where they may appear until it becomes possible to somehow support the birth rate.

(12) *Difficulties with security forces*. Recruiting into the army and police may be very difficult due to the lack of young people unless technologies help and people are partly replaced by robots. By the way, the war in Ukraine has already shown a shortage of human resources, with the result that both the age of conscription and the age of officers and reservists are rising.

(13) *Difficulties in the economy due to labor shortages*. The self-regulating systems can also play an important role here, especially robots, which will partially replace humans in mass professions (see Chap. 10). However, it is also worth adding that the production system itself will change significantly during the Scientific-Cybernetic Production Principle. The pattern of the division of

[29] The decline in the birth rate in Russia in the 1990s led to low numbers of applicants in the 2010s and 2020s. As a result, the requirements for applicants and students in the absence of competition have greatly decreased. There is a similar situation in recruiting in the U.S. army.

[30] The fewer children there are, the more "spoilt" they are on average; therefore, a decline in the birth rate can lead to the infantilization of generations, especially if more children are born into families where the parents are already beyond the current childbearing age.

labor will change dramatically as well. Technologies will enable intelligent and self-regulating systems to take over many functions, while people will acquire a wide range of skills to help them communicate with these systems. We have clearly observed this trend for several decades, but it should deepen and intensify. As a result, the endless fragmentation of functionality, characteristic of the Industrial Production Principle will come to an end with the advent of more and more professional specialists, and people will be able to become polymaths precisely with the help of self-regulating systems (and their powerful functionality): lawyers, doctors, designers, programmers, manufacturers, cooks, various masters (on 3D-printers), journalists, writers, artists, psychologists, etc., not to mention the professions that can already be mastered even today: photographers, editors, designers, IT specialists, etc. In addition, remote access will allow a significant equalization of opportunities for people from different countries, places and from different backgrounds.

(14) *The problem of caring for people with disabilities.* Despite all the advances in medicine, the number of people needing care will grow. Only the introduction of self-regulating/managing systems and robots can free a part of the population currently occupied with care. As we have seen, there are many such people, especially in families. Due to the weakening of the institution of the family will exacerbate the problems, but we believe that they will be resolved during the Cybernetic Revolution and subsequent periods (see Chap. 10).

(15) *Changes in marital and family relationships.* It is possible that the number of marriages and cohabitations of unequal ages will increase, not only between young women and older men, but also vice versa. At the same time, the number of interracial and interethnic marriages and cohabitations will increase, as the youth will be concentrated in some countries (in Africa, Asia), while money and position will be concentrated in others. At the same time, due to the increasing life expectancy, the number of singles will increase due to divorces, which may increase in aging couples, the death of one of the partners, a decline in the total number of marriages and children and so on.

(16) *The role of age in social stratification and self-organization.* As we have argued above, modern society is somewhat similar to primitive societies with respect to the lines of social division (age, gender and other non-class cleavages). Probably, with an increase in life expectancy, the role of age strata/cohorts will also increase. This will lead to a certain self-organization along the age lines. For example, in some countries or under certain circumstances, there emerge mutual support groups in which 'younger' seniors or stronger people help 'older' seniors or those with reduced abilities. There are many examples of this in low-, middle- and high-income settings, that is in countries with different developmental levels (WHO, 2017a: 9). There are various alternatives to such social self-organization or institutionalization along the age lines,

within the framework of conventional age ranks or inter-rank relations.[31] In general, we believe that social divisions in cybernetic society will increasingly be related to age differences. Interestingly, since the second half of the twentieth century, the role of class (economic) differences in the social struggle has been decreasing, whereas the differences associated with gender, age and other anthropological characteristics have, to a great extent, become more pronounced due to female, youth and other movements (see above).

As we have pointed out above, the process of institutionalizing age categories will intensify. Our idea is that the importance of generations in the social division of the cybernetic society will increase, and, accordingly, the positions of different generations should be clearly taken into account in various relationships, institutions and, possibly, laws. Otherwise, the trends of global aging, the introduction of new technologies and the possibility of changing the biological nature of human beings (see Grinin & Grinin, 2015; Grinin et al., 2017a) will create certain ethical and legal difficulties, which will appear even more serious than today's gender problem. In any case, the age categories will give rise to new social minorities and other problems.[32]

(17) *The business orientation toward generations of mature and elderly people.* Thus, together with aging and adaptation to it, everything will inevitably become age-related: fashion, sports, leisure, clubs, special settlements, classes, specialists, tourism, cosmetology (and of course medicine) and many other things will become specific to the seniors. Many new professions and new jobs will be created in connection with the aging process of society. Accordingly, the economy will be also restructured for the elderly (e.g., Jackson et al., 2008: 127–128), and the entrepreneurial activity among the elderly is more than likely.

### 7.2.5 Scenarios for the Future Older Society

From the trends and options presented so far, it follows that there are various developmental scenarios and their specifics will depend on the characteristics of societies, the balance of power, the pace and direction of technological development, possible pandemics, global trends in politics, economy and ideology, climate change, and many other factors, some of which are still unknown and unpredictable

[31] However, in our opinion, it is desirable to combine various forms of self-organization of the elderly and old people with the environment of younger people so as not to increase the generation gap. The experience of the elderly is also extremely important. In a word, it is necessary to combine self-organization along the lines of age with the principle of connectedness of generations.

[32] We do not rule out the possibility that if age segregation develops in settlements (see above), then social divisions like modern fraternities, that is, territorial associations based on age, as well as all kinds of horizontal ties between such communities, will become important.

hazards. We describe a number of scenarios below. Probably, all of them may come true in the future (in the form of certain phenomena and trends); however, it is of utmost importance which will be the leading one.

**Transition to a New Type of Population Reproduction**. Before analyzing the scenarios, it is worth reminding that all of them would have taken place on the basis of a new type of population reproduction. After the completion of the Cybernetic Revolution at the mature stages of the Scientific-Cybernetic Production Principle, we will witness the beginning of the new largest demographic transformation (LDT), associated with the accelerated creation of a sustainable medical-biotechnological environment. As a result, the characteristics of a new type of population reproduction (TPR), called social-qualitative type of population reproduction will become evident (see Chap. 5, Table 5).[33] This will be a transition to the model of total technological support of health and biological quality of life with the stabilization (or reduction) of the population, with a very large proportion of elderly people. In order to achieve a radical life extension, it is necessary to considerably increase the technological component of assistance to the elderly, the disabled, the infirm and other categories needing care and supervision. This is not just about more technologically advanced medicine, where many self-regulating systems will be involved, but about the situation when the medical-biotechnological environment will become permanent, making life almost impossible without it (just as today we feel uncomfortable without the Internet and certain gadgets). This, of course, together with the increasing duration and comfort of biological life, will also create many uncertainties and problems, which we will discuss in another chapter. Against the background of the increasing conservatism of an aging society the necessity of decisive changes will become a new page in history.

One should note that, as society ages and adapts to aging, different scenarios may become possible in the future: from minor and controlled growth and complete stabilization of the population, to a more or less slow but stable depopulation. It is also possible to move toward technologies of (semi)-artificial reproduction, at least we can expect a very wide distribution of improved IVF technologies (already during the final phase of the Cybernetic Revolution).

**Scenario one: The conflict between generations and the crisis of democracy**. As already mentioned, society's life and mode of operation during adaptation to aging and transition to a new type of population reproduction should change dramatically. Some researchers reasonably point out that the political assessment of such a large-scale adjustment should come from an aging electorate, whose interests and preferences may be radically different from those of the younger voters (Bloom et al., 2011). We also argued above that in a democratic system it is impossible (at least for a long time) to ignore the interests of a huge part of society, that is, of the older population. But it could also mean a course toward gerontocracy. The

[33] See also Grinin and Grinin (2023); Grinin et al., (2023c).

researchers began to think about this at least twenty to twenty-five years ago. For example, Peterson wrote that global aging may even undermine the foundations of democracy itself (Peterson, 1999; see also Bös & Weizsäcker, 1989). Taking into account the shift toward direct democracy due to new technologies and the development of socio-technical self-regulating systems (see Chap. 14 on their use for administrative regulation, as well as Grinin et al., 2021), the role of the older generation may additionally grow, so that its political influence may lead to a generation gap.[34] The conflict may intensify for various reasons: distribution, budget tensions, foreign policy, certain laws, etc. In the conflict of generations, the crisis of democracy is inevitable, since it will become factional anyway.

The problem of ideology can also be very serious, but in this respect, much depends on future trends, including those related to climate. However, one should keep in mind that future political development under democracy can be affected by such trends as the voluntary segregation of the population on the basis of age (which we have already discussed above), and in this case, the old ideas and ideologies are preserved. The results of such a struggle may be different, but in any case, these conflicts will be traumatic for society.

**Scenario two: Peace and order**. People will strive to ensure that life continues according to the established (diverse and not boring) order, and they will resist anything that disrupts it. Hence, there could be a move toward something like authoritarian democracy or even authoritarianism. There will also be widespread conformism, an inactive life position (more precisely, active only when defending a certain line) and an uncreative approach. Intergenerational conflicts will be absent or weak. The youth either follows a conformist trend or is subordinated. At the same time, the economic development is weak. Jackson et al. (2008) note that psychologically, the older society will become more conservative in its outlook and possibly more risk-averse in its electoral and leadership behavior. The elderly dominance of the electorate will tend to lock in current public spending commitments at the expense of new properties.

**Scenario three: Sustainable development**. Accustomed to constant and even rather dynamic development, the generation that grew up during the final phase of the Cybernetic Revolution will believe that development is necessary, yet it should be regulated (which is quite consistent with the idea of self-regulating systems). Hence, a combination of sustainability and development will be sought. The developmental directions are planned (somewhat similar to the socialist planning). This would be quite a good option, although it may prevent revolutionary changes (about the balance between the need for development and the growing conservatism in elderly society see Chap. 13).

---

[34] Here one should take into account that the future older generation is quite sophisticated in technologies today, so that their participation in this kind of democracy can be very active, in contrast to the fact that the current older generation is far from being always friendly with technologies, and therefore, is less active in virtual reality.

**Scenario four: Homeostasis**. Due to the growing conservative tendencies and the threat of climate change, all changes and growth are strictly controlled, there is almost no development and everything is regulated by technologies.

**Scenario five: Militant ageism**. Within the trend to strengthen the position and authority of the older generation, the opportunities for development of the younger generation may weaken. In the absence of social lifts, the path to prosperity may then become quite narrow. This may be similar to the career progression of military personnel in peacetime, which depends largely on time in service.[35] Hence, a society may appear calm and smooth on the surface, while internal tensions may grow due to discontent with unrealized opportunities.

**Scenario six: The rise of techno-medical power**. As we have already noted, the transition to a demographic transformation associated with super-long life expectancy will ultimately emphasize all kinds of technologies that are solely capable of maintaining the health of an aging population. At the same time, a very significant part of diagnostics, monitoring, and various types of treatment will be transferred to the patients' homes and residences, etc. (This trend is already evident today). On the one hand, there will be a decentralization of the hospital system and of communication between doctors and patients (doctors will be largely replaced by electronic consultants, while patients themselves, as mentioned above, will have opportunities to access a kind of medical education). On the other hand, there will be an extremely high concentration of data, their processing and necessary actions, up to direct orders, blocking, etc. In fact, as we assume, a kind of medical-biotechnological environment will eventually emerge, in which many people will have to stay permanently or for a long time (see Chaps. 6 and 8). This will actually mean constant control, observation, data transmission to the data center, implementation of routine procedures, probably communication between an observed individual and the above-mentioned environment, with possible instructions, advice and recommendations; transmission of information in case of violation of the regime or instructions to the observed; it is also possible to provide emergency assistance, as well as an urgent call for help, etc. All this leads directly to a sharp increase in the power of technocrats and doctors. In particular, when we speak of technocrats, we mean programmers or some new specialists who will design responses to a very complicated flow of information. At the same time, it is important to understand that once you have adopted some settings, it will be very difficult to cancel them later, no less difficult than the law, and perhaps even more difficult. The position of medical doctors will also change—the personal power of specialists will be replaced by a kind of corporate power, tied to complex technologies and imperatives that cannot be questioned. Possible pandemics or other threats to health and to the given life expectancy will only strengthen this power, as well as the possible segregation of the elderly. As we

---

[35] Like in Pushkin's *Eugene Onegin*: "…who, in good time, would calmly get fortune, and dignity, and glory".

have pointed out above, society will have to go along a difficult path of optimizing such influence by creating some kind of counterbalances to such impersonal power, perhaps, some supervisory or other bodies that can control the situation.

However, one should keep in mind that if the elderly are locked up and/or put under the constant control of the AI, this may negatively affect their perception of the world, and, consequently, their health. Taken together, the negative aspects may outweigh the benefits. Therefore, finding a way for the elderly to feel comfortable in a good and stimulating atmosphere is an extremely important task. Accordingly, the psychological dimension should be enhanced, as well as the aspect of different ways of communication. We should add that the geriatric technologies, of course, together with increased duration and comfort of biological life (health-related quality of life), will also pose many challenges (including possible cohabitation and coexistence with robots and, in general, with self-regulating systems; see also Grinin et al., 2021).

### 7.2.6 Scenarios for the Future Aging World System

Let us consider how global aging will affect the changes in the global order and the balance of power in the world, as well as the geopolitical balance. The scenarios will depend on the ratio between nationalism and globalism (globalization). And aging will affect them in turn.

*To a large, and even to a decisive extent, the formation of a new balance of power and world order can be affected by the struggle between young, growing societies and older, stagnating ones.* This is the idea of the aging North and the younger South (Bengtson, 2018; Fichtner, 2018; Goldstone, 2015; Goldstone et al., 2015; Haas, 2015; Powell & Khan, 2013; Zimmer, 2016). In this context, we will consider several possible options and scenarios. But first, let us point out that, as we have written (; Grinin, 2016; Grinin & Korotayev, 2020; Grinin et al., 2016a, 2016b), a new world order may emerge in the 2050s. But in connection with global aging, the division of the world into young and old societies, the struggle for a new world order will inevitably begin, of course, already in the context of globalization, which may well experience a new upsurge.

**Scenario one: The senility and decline of the developed world and the emergence of new players on the scene**. Researchers note that population aging in the USA and in its traditionally allied countries will reduce their ability to provide national and global security, and respectively, developing countries will change the future global security landscape. And this, together with population decline and possible GDP contraction in many developed countries, will greatly affect the decline of the developed world (Jackson et al., 2008; see also Jackson et al., 2013; Jackson, 2021; Grinin & Korotayev, 2015c, 2023). In particular, Jackson et al. (2008) note that in tandem with the shrink of the population and GDP the global influence of the developed world will likely decline. The aging developed countries will face a chronic shortage of young-adult manpower which will pose challenges both for their economies and their security forces.

On the whole, one can agree with this, as well as with the fact that in the situation of a shortage of young people, the older societies will face not only economic but also security problems (ibid.).[36] Yet, unmanned military (and industrial) technologies (see Chap. 10) can help to mitigate the imbalance between developed and currently developing countries to some extent, but, by no means completely. Such a struggle can lead to various outcomes: from increased confrontation and stopped globalization to the emergence of some new foundations of the world order.

**Scenario two: Young countries give dynamism to the World System**. It is possible that, if the foundations of the world order secure some of the advantages of the rising countries, this can temporarily give dynamism to the whole World System. By the end of the century, however, this dynamism will be exhausted.

**Scenario three: Global conservatism**. However, if the world order is strengthened with developed countries at the top, conservatism may start to spread to younger countries (such ideas are already articulated today: no need for growth, etc.; for details see Chaps. 1 and 13). At the same time, as noted by Jackson et al. (2008), an aging developed world may struggle to remain culturally attractive and politically relevant to younger societies.

**Scenario four: Activity despite aging as a national policy**. However, the struggle for a new world order in the situation of depopulation and aging in the developed world could lead to a certain rise in the aging countries of the developed world. Older countries will be forced to become more active and increase their birth rates, which could give an impetus to new technologies related to reproduction.

**Scenario five: Quasi-colonialism (ageist globalization)**. Capitals and corporations are increasingly involving young and mature generations in less developed countries, raising their level while thus providing for the older generation in their own countries (see Grinin & Korotayev, 2010). This will clearly enhance globalization. There will emerge zones of influence related to language, traditions and geopolitics based on remote work (see in particular Grinin et al., 2022a, 2022b, 2023a, 2023b, 2023c, 2023d). The increasing importance of content management systems will allow for less physical presence so that more remote foreign labor will be involved. This will undoubtedly be facilitated by advances in translation systems. Lingua Google (or something like that) will appear in the world as an active part of contacts. But today, only a few experts realize that 'the distribution of the working-age populations will be crucial to whether humanity prospers or withers' (Gallagher, 2020).

We emphasize that these are all scenarios for the next 50 years. By the end of the century, the situation will change significantly, since the older countries adapt

---

[36] In particular, with respect to possible wars. Smaller family size may make the public less willing to risk scarce youth in war (Jackson et al., 2008). See also above.

to aging, and the younger ones will not be so young at all. A cybernetic society will be established (see below and Chap. 15). In this situation, other scenarios are possible: the consensus of an aging world or the dominance of new actors.

## 7.3 Conclusion

To conclude, the future society will face complex and, in many ways, new challenges. It will be difficult to find a suitable long-term solution that will fit all and last forever. On the one hand, many problems simply cannot be solved without powerful technological development. On the other hand, it will inevitably be necessary to address radically the issues of the limits to the independence of self-managing systems (as we observe in relation to AI in Chaps. 4 and 10). On the one hand, there are already guidelines for healthy aging (see above) that emphasize the importance of striving for dignity, such as the desire to take care of oneself, whenever it is possible (WHO, 2017a: 9). The aspiration for dignity is very important for good health. However, on the other hand, the increasing number of older people dramatically increases the burden on society and younger generations and may lead to pressure on the elderly by decreasing their standard of living. Also, technological solutions (inevitably) start to limit people's dignity and increase intervention in their privacy. It is extremely difficult to find a balance, not to mention that such control may unprecedentedly increase medical power and even lead to a kind of terror. This is important in the context of the future development of medical-biotechnological environment that will carry out a cycle of health control from observation to prescription (see above). For more details see Chaps. 6 and 8.

The need for investment, in particular, the investment in human capital also appears controversial. On the one hand, the population aging and the increase in the median age will inevitably require a revision of the HDI, but on the other hand, investing in older people is a risky business. Thus, new approaches need to be developed that would become attractive to both the state and business.

We believe that as a result of the largely symbiotic development of global aging and society's adaptation to it, as well as the institutionalization of old age, on the one hand, and a powerful development of self-regulating/managing systems, a new type of society—cybernetic one—will emerge.

***Cybernetic society*** *is the society that will take shape (and is already taking shape) as a result of the completion of the Cybernetic Revolution and the mature phases of the Scientific-Cybernetic Production Principle. It will be (a) super-technologized; (b) socio-technologically regulated at all levels; (c) a society where, as a result of the aging process, the division into age categories will become much more socially significant than it is today (i.e., age will become a much more important social marker than it is today).* We will discuss this in more detail in the Conclusion (Chap. 15) of the monograph.

**Acknowledgements** This chapter is the output of a research project implemented as part of the Basic Research Program at HSE University in 2024 with support by the Russian Science Foundation (Project No. 23-18-00535).

## Appendix

### How Could the Global Aging Affect the Global Value System?[37]

We believe that taking the general information on the basic issues of human evolution into account is necessary for understanding our reasoning concerning the evolutionary perspectives of aging for the global value systems. On the basis of the relevant literature,[38] we hypothesized that a shift toward prosocial values should be attributed to the age effect rather than the cohort effect and thus should be expected to accompany global aging in the coming decades. Following Carlo and Sandman, we define prosocial values as "beliefs esteemed by an individual or group that promote concern and care for the welfare of others" (Carlo & Sandman, 2012: 2709). Values of care and cooperation are frequently subsumed under prosocial values—thus, co-operators and altruists are classified as prosocial individuals (van Lange & Liebrand, 1989: 209); Chatterjee et al. state that prosocial values can be expressed "by a high importance to care for the welfare of close others" (Chatterjee et al., 2017: 842), features of highly prosocial societies include "empathic tendencies and care-giving values" (Hilser, 2021: 2), people with prosocial motives are rendered to be more cooperative (Giacomantonio et al., 2010: 824) etc.

To test these hypotheses, we use the data from the long-term global sociological project "World Values Survey" (WVS) from the first to the seventh wave of the WVS. We used a longitudinal data file (World Values Survey, 2020). We considered multivariate models in which the variables describing values depend on age and cohort and are controlled for gender, education level (numerical value from 1 to 8), marital status (categorical variable), labor market position/employment (categorical variable), income level (numerical variable) (on the relevance of this control see, e.g., Zinkina et al., 2022). We also controlled for fixed effects in

[37] This section reproduces in an amended form some parts of our earlier articles (Korotayev et al., 2021, 2023) prepared in collaboration with Marina Butovskaya, Sergey Shulgin, and Julia Zinkina.
[38] Abraham and Feldman (2018); Agwanda and Harris (2009); Aiello and Dean (1990); Akhter-Khan (2020); Barnes and Magdalena (2020); Bernhard et al., (2006); Bowles (2008, 2009); Burkart et al., (2017); Chang (2018); Chapman et al., (2021); Chisholm (2017); Chisholm et al., (2016); Choi and Bowles (2007); Coall and Hertwig (2010); Coall et al., (2018); Dahlberg (1981); Darwin (1872); DeVore and Lee (1968); De Dreu et al., (2014); Doğan et al., (2018); Euler and Weitzel (1996); García and van den Bergh (2011); Del Giudice and Belsky (2011); Hagmann (2007); Handley and Mathew (2020); Hilbrand et al. (2017); Hrdy (2005, 2009); Kaplan et al., (2000); Khorasani and Lee (2020); Kim et al., (2014); Konner (2018); Lovejoy (1981, 2009); Mattison et al., (2018); Olowolafe et al., (2020); Püschel et al., (2021); Raghanti et al., (2018); Richerson and Boyd (2005); Robson et al., (2006); Rodrigues (2018); Sear (2018); Travis and Saa (2021); Wiessner (2019); Willführ et al., (2018); Zhu and Chang (2020); Zhu et al., (2020, 2021).

all country-waves using the corresponding dummy variables. To address cross-country differences, we used groups of regions (Latin America, Eastern Europe, MENA and others) where we expected a similar relationship at the individual level between the independent variable of interest (age & cohort) and our dependent variables, which represent different individual values (Korotayev et al., 2021).

Those tests yielded the following results:

We have differentiated between age and cohort effects. It has long been observed that some values tend to change with age; for example, older people tend to be more religious than younger people (Argue et al., 1999; Deaton, 2009; Levin & Taylor, 1997; Moberg, 1953; Riley & Foner, 1968). However, it remains still disputable whether this fact should be attributed to the age effect per se or to the cohort effect. We have attempted to distinguish between these two effects using a multifactor model applied to data from the World Values Survey (1981–2020), and to find values for which the age effect strongly outweighs the cohort effect. We found the following groups of values to change the most as respondents get older.

*Religiosity.* In all six subsamples, older respondents are significantly more likely to report frequent attendance at religious services (F028). Older respondents are also significantly more likely to assert that God is very important in their lives (F063) in the OECD, Eastern Europe and East and Southeast Asia. Older respondents are also more likely to state that religion is very important in their lives (A006) in the OECD, Latin America and Eastern Europe. They are also more likely to report being active members of a church/religious organization (A098) in OECD and Sub-Saharan Africa.

*Prosocial values.* In all of the subsamples, older respondents are significantly more likely to assert that:

- avoiding a fare on public transport (F115) is never justifiable (significant in OECD, Eastern Europe, Latin America, Middle East and North Africa, Sub-Saharan Africa; statistically insignificant but in the predicted direction in East and Southeast Asia);
- cheating on taxes (F116) is never justifiable (significant in OECD, Eastern Europe, Latin America, Middle East and North Africa, East and Southeast Asia; marginally significant with one-tailed significance test in Sub-Saharan Africa); and
- accepting a bribe (F117) is never justifiable (significant in OECD, Eastern Europe, Latin America, Middle East and North Africa; marginally significant with one-tailed significance test in East and Southeast Asia; statistically insignificant but in the predicted direction in Sub-Saharan Africa).

*Conservation versus Openness.* In all the six subsamples of the countries, older respondents tend to incline more to values of Conservation (Security, Tradition, Conformity) as opposed to Openness (significant in OECD, Eastern Europe and East and Southeast Asia; significant with one-tailed significance test in Latin America; marginally significant with one-tailed significance test in Sub-Saharan

Africa; statistically insignificant but in the predicted direction in the MENA region).

*Self-Transcendence/care for the others vs. Self-Enhancement/self-empowerment.* Our analysis of these composite variables reveals that in all subsamples older respondents tend to support values of Care for the Others (Benevolence, Universalism) as opposed to Self-Enhancement. In OECD countries this coefficient changes with age by 0.0943, i.e. for each increase in age group, other things being equal (controlling for gender, income, education, etc.), the person shifts toward Self-Transcendence from Self-Enhancement by 0.0943; in Eastern Europe by 0.0609; in Latin America by 0.0493; in East & South-East Asia by 0.0533. It is significant in OECD, Eastern Europe, East and Southeast Asia, Middle East and North Africa; significant with a one-tailed significance test in Latin America; marginally significant with a one-tailed significance test in Sub-Saharan Africa).

Thus, older respondents tend to support prosocial values and values of care for the others more than younger respondents, and, as our analysis demonstrates, this finding should be directly attributed to the age factor rather than the cohort factor. In other words, support for prosocial values and values of care for others is not related to the particular values of the cohorts of people who are currently older and can be expected to increase with age as the currently younger cohorts also enter their senior age. Thus, our findings are consistent with the evidence presented recently by other authors showing that the general increase in prosocial motivations with age (including charitable donations or volunteering for the common good) has been purely altruistic in nature[39] (Mayr & Freund, 2020; Zhu & Chang, 2020; Zhu et al., 2020, 2021). Prosocial behavior has been shifting fundamentally with aging, and older adults have a reduced self-favoring bias in their decision-making (Lockwood et al., 2020). Moreover, along with the general increase in prosocial behavior with aging, antisocial behavior and violence significantly decrease across the lifespan worldwide (Hiraiwa-Hasegawa, 2005; Liberman, 2008). Values related to conservatism and cooperation seem to underlie some important strategies of the older generation. Hence, formerly evolved as a basic evolutionary adaptation aimed at maintaining the stability of society and favorable conditions for the survival of offspring in small-scale societies in the Paleolithic, this adaptation continues to

[39] The current research has investigated the relationship between age and prosocial values, which is independent of the cohort effects. However, expressed prosocial values are not necessarily equivalent to prosocial motivations. It is important to distinguish between people's moral motivations (which lead to their own moral behaviors) and their moral values (which lead to their judgments about the moral behaviors of others). They can be contradictory to each other, leading to the phenomenon of moral hypocrisy (Batson et al., 1999). This is an important point to take into account, but, in the context of the current research focusing on global value shift, both moral motivations and moral values are important, because both one's own moral behavior (guided by moral motivations) and the condemnation of others' immoral behaviors are important in defining the increase in prosociality and, eventually, in assessing the potential macro-social implications of the value shifts viewed in this chapter.

provide obvious benefits, despite the intensive processes of technological (information) development and the transformations in the field of employment and the organization of social life. The stability of such an adaptation testifies to its exceptional importance for the survival of humankind in the past and present. Our results may provide additional support for the hypothesis of dual evolutionary foundations of political ideology (Claessens et al., 2020; Duckitt & Sibley, 2009; Feldman & Johnston, 2014).

Given our findings, with the general tendency of the world's population to aging, we might expect an increase in the emphasis on the social conservatism dimension, along with an increase in adherence to prosocial values, even when there is no immediate reward for doing so. This can be viewed as an increase in strong reciprocity, which Bowles and Gintis define as follows: "Where genetically unrelated members of a group benefit from mutual adherence to a social norm, agents may obey the norm and punish its violators, even when this behavior cannot be justified in terms of self-regarding, outcome-oriented preferences" (Bowles & Gintis, 1998: 1). "Global aging" seems to invoke mostly negative connotations. However, our analysis suggests that in terms of value orientations, global aging may have some positive effects by increasing support for prosocial values at the global level.

**Acknowledgements** This chapter is an output of a research project implemented as part of the Basic Research Program at HSE University in 2024 with support from the Russian Science Foundation (Project No. 23-18-00535).

## References

Abraham, E., & Feldman, R. (2018). The neurobiology of human allomaternal care; implications for fathering, coparenting, and children's social development. *Physiology & Behavior, 193*, 25–34.

Adegbindin, O. (2011). The problem of gerontocracy in Africa: The Yorùbá perspective as illustrated in the Ifá corpus. *Human Affairs, 21*(4), 454–469.

Agwanda, T., & Harris, G. (2009). People-to-people peacemaking and peacebuilding: A review of the work of the New Sudan Council of Churches. *African Security Studies, 18*(2), 42–52.

Aiello, L., & Dean, C. (1990). *An introduction to human evolutionary anatomy*. Elsevier Academic Press.

Akhter-Khan, S. C. (2020). Providing care is self-care: Towards valuing older people's care provision in global economies. *The Gerontologist, 20*(20), 1–9.

Aksoy, Y., Basso, H. S., & Smith, P. R. (2015). *Demographic structure and macroeconomic trends* (p. 1528). Banco de Espana. Documentos de Trabaja. No.

Aldwin C., & Gilmer D. (2013). *Health, illness and optimal aging: Biological and psychosocial perspectives*. Springer.

Argue, A., Johnson, D. R., & White, L. K. (1999). Age and religiosity: Evidence from a three-wave panel analysis. *Journal for the Scientific Study of Religion, 38*(3), 423–435.

Artemova, O. (2003). Monopolization of knowledge, social inequality, and female status: A cross-cultural study. *Cross-Cultural Research, 37*(1), 62–80. https://doi.org/10.1177/1069397102238922

Artemova, O., & Korotayev, A. V. (2003). Monopolization of information and female status: A cross-cultural test. *Cross-Cultural Research, 37*(1), 81–86. https://doi.org/10.1177/1069397102238922

Asadzadeh, M., Maher, A., Jafari, M., Mohammadzadeh, K. A., & Hosseini, S. M. (2022). A review study of the providing elderly care services in different countries. *Journal of Family Medicine and Primary Care, 11*(2), 458–465.

Assaulova, D. (2020). Pochemu SSHA zhdet krizis sistemy dolgosrochnogo ukhoda za pozhilymi. *Forbes.ru*, https://www.forbes.ru/forbeslife/445333-pocemu-ssa-zdet-krizis-sistemy-dolgosrocnogo-uhoda-za-pozilymi

Atchley R. C. (2004). *Social forces and aging*. Wadsworth/Thompson Learning.

Barnes, B. E., & Magdalena, F. V. (2020). Traditional peacemaking processes among indigenous populations in the Northern and Southern Philippines. In S. P. Stobbe (Ed.), *Conflict resolution in Asia: Mediation and other cultural models* (pp. 63–82). Lexington Books.

Barsukov, V. N. (2019). From the demographic dividend to population ageing: World trends in the system-wide transition. *Economic and Social Changes: Facts, Trends, Forecast, 12*(4), 167–182. https://doi.org/10.15838/esc.2019.4.64.11

Batson, C. D., Thompson, E. R., Seuferling, G., Whitney, H., & Strongman, J. A. (1999). Moral hypocrisy: Appearing moral to oneself without being so. *Journal of Personality and Social Psychology, 77*(3), 525–537.

Bélanger, E., Towers, A., Wright, D. K., Chen, Y., Tradounsky, G., & Macdonald, M. E. (2019). Of dilemmas and tensions: A qualitative study of palliative care physicians' positions regarding voluntary active euthanasia in Quebec. *Canada. Journal of Medical Ethics, 45*(1), 48–53.

Bellens, M., Debien, E., Claessens, F., Gastmans, C., & Dierckx de Casterlé, B. (2020). It is still intense and not unambiguous. Nurses' experiences in the euthanasia care process 15 years after legalisation. *Journal of Clinical Nursing, 29*(3–4), 492–502.

Bengtson, V. (Ed.). (2018). *Global ageing and challenges to families*. Routledge.

Bengtson, V. L., Gans, D., Putney, N. M., & Silverstein, M. (2009). *Handbook of theories of aging*. Springer.

Bernhard, H., Fischbacher, U., & Fehr, E. (2006). Parochial altruism in humans. *Nature, 442*, 912–915.

Berry, B. J. L. (1976). *Urbanization and counterurbanization*. Sage Publications.

Billot J. (2023). *Euthanasia responsible for 4.1% of all Canadian deaths in 2022*. https://unherd.com/thepost/euthanasia-responsible-for-4-1-of-all-canadian-deaths-in-2022/

Binstock, R. H. (1991a). Aging, politics, and public policy. In B. B. Hess & E. W. Markson (Eds.), *Growing old in America* (pp. 325–340). Transaction.

Binstock, R. H. (1991b). From the great society to the aging society: 25 years of the older Americans Act. *Generations, 15*, 311–318.

Bloom D. E., Canning D., & Fink G. (2011). *Implications of population aging for economic growth*. NBER Working Paper. No 16705.

Bloom, D. E., & Luca, D.L. (2016). *The global demography of ageing: Facts, explanation, future*. Institute for the Study of Labor.

Bös, D., & Von Weizsäcker, R. K. (1989). Economic consequences of an ageing population. *European Economic Review, 33*(2–3), 345–354.

Bowles, S. (2008). Conflict: Altruism's midwife. *Nature, 456*, 326–327.

Bowles, S. (2009). Did warfare among ancestral hunter-gatherers affect the evolution of human social behaviors? *Science, 324*, 1293–1298.

Bowles, S., & Gintis, H. (1998). *The evolution of strong reciprocity*. SFI Working Paper 1998-08-073. https://sfi-edu.s3.amazonaws.com/sfi-edu/production/uploads/sfi-com/dev/uploads/filer/a3/cf/a3cf6a69-5ff8-4eb8-8f26-235b5405d4cb/98-08-073.pdf

Burkart, J. M., Van Schaik, C., & Griesser, M. (2017). Looking for unity in diversity: Human cooperative childcare in comparative perspective. *Proceedings of the Royal Society b: Biological Sciences, 284*(1869), 20171184.

Burtless G. (2013). *Is an aging workforce less productive?* Boston, MA. The Sloan Center on Work & Retirement, Boston College Retrieved from http://agingandwork.bc.edu/blog/is-an-aging-workforce-less-productive

Carlo, G., & Sandman, D. (2012). Prosocial learning in adolescence: The mediating role of prosocial values. In N. M. Seel (Eds.), *Encyclopedia of the sciences of learning* (pp. 2709–2711). Springer. https://doi.org/10.1007/978-1-4419-1428-6_1730

Carrion, C., Folkvord, F., Anastasiadou, D., & Aymerich, M. (2018). Cognitive therapy for dementia patients: A systematic review. *Dementia and Geriatric Cognitive Disorders, 46*(1–2), 1–26.

Carter C. A. (2023). *Euthanasia is out of control in Canada. The country moves toward making "medical assistance in dying" even more widespread.* https://wng.org/opinions/euthanasia-is-out-of-control-in-canada-1690889071

Centers for Disease Control and Prevention and The Merck Company Foundation. (2007). *The state of aging and health in America 2007.* The Merck Company Foundation. chrome-extension://efaidnbmnnnibpcajpcglclefindmkaj/https://www.cdc.gov/aging/pdf/saha_2007.pdf

Chang, L. (2018). Life history theory. In M. H. Bornstein (Ed.), *The SAGE encyclopedia of lifespan human development* (pp. 1277–1281). SAGE publications.

Chapman, S. N., Lahdenperä, M., Pettay, J. E., Lynch, R. F., & Lummaa, V. (2021). Offspring fertility and grandchild survival enhanced by maternal grandmothers in a pre-industrial human society. *Scientific Reports, 11*, 3652.

Chatterjee, M. B., Baumann, N., & Koole, S. L. (2017). Feeling better when someone is alike: Poor emotion regulators profit from pro-social values and priming for similarities with close others. *Journal of Personality, 85*(6), 841–851.

Cheng, M. (2022). *'Disturbing': Experts troubled by Canada's euthanasia laws.* Associated Press.

Chetyrkin, Y. (2014). Pensionnyy fond mezhdunarodnykh organizatsiy. *Mirovaya ekonomika i mezhdunarodnyye otnosheniya,* 5, 15–20. DOI: https://doi.org/10.20542/0131-2227-2014-5-15-20.

Chisholm, J. S. (2017). How attachment gave rise to culture. In H. Keller & K. A. Bard (Eds.), *The cultural nature of attachment: Contextualizing relationships and development* (pp. 273–299). MIT Press.

Chisholm, J. S., Coall, D. A., & Atkinson, L. (2016). Mother–infant cultural group selection. *Behavioral & Brain Sciences, 39*, e35.

Choi, J. K., & Bowles, S. (2007). The coevolution of parochial altruism and war. *Science, 318*, 636–640.

Chumakov, A. N., & Yurchenko, P. S. (2021). Global pandemic in terms of objective causes and subjective speculations. *Journal of Globalization Studies, 12*(2), 157–168.

Claessens, S., Fischer, K., Chaudhuri, A., Sibley, C. G., & Atkinson, Q. D. (2020). The dual evolutionary foundations of political ideology. *Nature Human Behaviour, 4*, 336–345.

CNE.news. (2023). *Number of Belgian euthanasia deaths increases.* https://cne.news/article/2718-number-of-belgian-euthanasia-deaths-increases

Coall, D. A., & Hertwig, R. (2010). Grandparental investment: Past, present and future. *Behavioral and Brain Sciences, 33*, 1–59.

Coall, D. A., Hilbrand, S., Sear, R., & Hertwig, R. (2018). Interdisciplinary perspectives on grandparental investment: A journey towards causality. *Contemporary Social Science, 13*(2), 159–174.

Coleman, D., & Rowthorn, R. (2015). Population decline—making the best of inevitable destiny? In J. A. Goldstone, L. E. Grinin, & A. V. Korotayev (Eds.), *History & mathematics: Political demography and global aging* (pp. 26–41). Uchitel Publishing House.

Cook, M. (2023). Has Canada finally realized that its euthanasia law is a human rights disaster? *Mercatornet* 23 Jan 2023. https://mercatornet.com/has-canada-finally-realized-that-its-euthanasia-law-is-a-human-rights-disaster/82749/

Dahlberg, F. (Ed). (1981). *Woman the gatherer.* Yale University Press.

Damien, Y. (2021). *Kakiye vozmozhnosti dlya investitsiy otkryvayet stareniye mirovogo naseleniya. Pyat' klyuchevykh segmentov dlya vlozheniy v «serebryanuyu» ekonomiku.* https://www.vedomosti.ru/economics/blogs/2019/05/14/801316-vozmozhnosti-investitsii-starenie-naseleniya
Darwin, C. (1872/1965). *The expression of the emotions in man and animals.* University of Chicago.
De Dreu, C. K. W., Balliet, D., & Halevy, N. (2014). Parochial cooperation in humans: Forms and functions of self-sacrifice in intergroup conflict. In A. Elliot (Ed.), *Advances in motivation science. Advances in motivation science* (Vol. 1, pp. 1–47). Elsevier.
De Grey, A., & Rae, M. (2008). *Ending aging: The rejuvenation breakthroughs that could reverse human aging in our lifetime.* Martin's Press.
Deaton, A. (2009). *Aging, religion, and health.* No. w15271. National Bureau of Economic Research. https://www.nber.org/papers/w15271
Del Giudice, M., & Belsky, J. (2011). The development of life history strategies: Toward a multi-stage theory. In D. M. Buss & P. H. Hawley (Eds.), *The evolution of personality and individual differences* (pp. 154–176). Oxford University Press.
DeVore, I., & Lee, R. B. (Eds). (1968). *Man the hunter.* Aldine Publishing Company.
Devos, T. (2021). *Euthanasia: Searching for the full story.* Springer.
Dhue, S., & Epperson, S. (2022). Inflation drives long-term care costs even higher. Here's how planning ahead can help families afford it. *CNBC* 30 Aug 2022. https://www.cnbc.com/2022/08/30/planning-ahead-can-help-families-afford-long-term-care-amid-inflation.html
Doğan, G., Glowacki, L., & Rusch, H. (2018). Spoils division rules shape aggression between natural groups. *Nature Human Behaviour, 2*(5), 322–326.
Dua, T. et al. (2012). *Dementia: A public health priority.* World Health Organization.
Duckitt, J., & Sibley, C. G. (2009). A dual-process motivational model of ideology, politics, and prejudice. *Psychological Inquiry, 20*, 98–109.
Dutton, P. V. (2017). From sanatoriums to sun city. The creation of 'Active Retirement Living' in Arizona. *Le Mouvement Social, 258*(1), 85–107.
Efimovich, A. L. (2020). Negative rates and other financial technologies in modern economic reality on the world-system scale. *Journal of Globalization Studies, 11*(2), 63–76.
Efremov, I., & Arkhangelsky, V. (2023). Global fertility changes in the context of the COVID-19 pandemic. *Journal of Globalization Studies, 14*(2), 165–170.
Eggleston, K. N., & Fuchs, V. R. (2012). The new demographic transition: Most gains in life expectancy now realixed late in life. *Journal of Economic Perspective, 26*(3), 137–156.
España, P. P., Bilbao, A., García-Gutiérrez, S., Lafuente, I., Anton-Ladislao, A., Villanueva, A., Uranga, A., Legarreta, M.J., Aguirre, U., & Quintana, J. M. (2021). Predictors of mortality of COVID-19 in the general population and nursing homes. *Internal and emergency medicine, 16*(6), 1487–1496.
Euler, H. A., & Weitzel, B. (1996). Discriminative grandparental solicitude as reproductive strategy. *Human Nature, 7*, 39–59.
Feldman, S., & Johnston, C. (2014). Understanding the determinants of political ideology: Implications of structural complexity. *Political Psychology, 35*, 337–358.
Fichtner, J. J. (2018). Global ageing and public finance. *Business Economics, 53*(2), 72–78.
Fitzgerald, K. G., & Caro, F. G. (2014). An overview of age-friendly cities and communities around the world. *Journal of Aging & Social Policy, 26*, 1–18.
Fuecks, R. (2013). *Intelligent Wachsen.* Carl Hanser Verlag.
Fukuyama, F. (2002). *Our post-human future: Consequences of the bio-technology revolution.* Farrar, Straus, and Giroux.
Gallagher, J. (2020). Fertility rate: 'Jaw-dropping' global crash in children being born. *BBC* 15 July 2020. https://www.bbc.com/news/health-53409521
Galor, O., & Weil, D. N. (2000). Population, technology, and growth: From Malthusian stagnation to the demographic transition and beyond. *American Economic Review, 90*(4), 806–828.
García, J., & van den Bergh, J. C. J. M. (2011). Evolution of parochial altruism by multilevel selection. *Evolution and Human Behavior, 32*, 277–287.

Giacomantonio, M., De Dreu, C. K., Shalvi, S., Sligte, D., & Leder, S. (2010). Psychological distance boosts value-behavior correspondence in ultimatum bargaining and integrative negotiation. *Journal of Experimental Social Psychology, 46*(5), 824–829.

Gibson, J. A., Boerstler, J. J., Chase, J., & Smith, C. (2022). Withdrawing life-sustaining treatments v. euthanasia: The ethical difference. *Nursing, 52*(7), 59–61.

Gleckman, H. (2022). How health care costs eat into the incomes of older adults. *Forbes* 22 Sept 2022. https://www.forbes.com/sites/howardgleckman/2022/09/22/how-health-care-costs-eat-into-the-incomes-of-older-adults/?sh=376d7cf86e41

Goldin, C. (2016). 'Human capital. In C. Diebolt & M. Haupert (Eds.), *Handbook of cliometrics* (pp. 55–86). Springer Verlag.

Goldstone, J. A. (2015). Population ageing and global economic growth. In J. A. Goldstone, L. Grinin, & A. Korotayev (Eds.), *History & mathematics: Demography & ageing* (pp. 147–155). Uchitel.

Goldstone, J. A., Grinin, L. E., & Korotayev, A. V. (2015). Introduction, research into global aging and its consequences. In J. A. Goldstone, L. E. Grinin, & A. V. Korotayev (Eds.), *History & mathematics: Political demography and global aging* (pp. 5–9). Uchitel Publishing House.

Grinin, L. (2006). *Proizvoditel'nyye sily i istoricheskiy protsess*. Uchitel Pubishing House.

Grinin, L. E. (2016). The new world order and philosophy. In A. N. Chumakov & W. C. Gay (Eds.), *Between past orthodoxies and the future of globalization. Contemporary philosophical problems* (pp. 143–156). Rodopi. http://www.brill.com/products/book/between-past-orthodoxies-and-future-globalization

Grinin, L. (2020). How can COVID-19 change geopolitics and economy? *Journal of Globalization Studies, 11*(2), 121–134.

Grinin, L. (2021). Otritsatel'nyye stavki i drugiye noveyshiye finansovyye tekhnologii. *Obshchestvo i Ekonomika, 2*, 18–30.

Grinin, L., & Grinin, A. (2015). *Ot rubil do nanorobotov. Mir na puti k epokhe samoupravlyayemykh sistem (istoriya tekhnologiy i opisaniye ikh budushchego)*. Uchitel.

Grinin, L., & Grinin, A. (2016). *The cybernetic revolution and the forthcoming epoch of self-regulating systems*. Uchitel.

Grinin, L. E., & Grinin, A. L. (2023). Demographic transformations in the historical process in the light of technological development: Theoretical approach. *Social Evolution & History, 22*(2), 159–202.

Grinin, L., Grinin, A., & Korotayev, A. (2017a). Forthcoming Kondratieff wave, cybernetic revolution, and global ageing. *Technological Forecasting and Social Change, 115*, 52–68.

Grinin, L., Grinin, A., & Korotayev, A. (2017b). The MANBRIC-technologies in the forthcoming technological revolution. In T. Devezas, J. Leitão, & A. Sarygulov (Eds.), *Industry 4.0—entrepreneurship and structural change in the new digital landscape: What is coming on along with the fourth industrial revolution* (pp. 243–261). Springer. https://doi.org/10.1007/978-3-319-49604-7_13

Grinin, L., Grinin, A., & Korotayev, A. (2020). A quantitative analysis of worldwide long-term technology growth: From 40,000 BCE to the early 22nd century. *Technological Forecasting and Social Change, 155*, 1–19. https://doi.org/10.1016/j.techfore.2020.119955

Grinin, L., Grinin, A., & Korotayev, A. (2021). Global trends and forecasts of the 21st century. *World Futures, 77*(5), 335–370. https://doi.org/10.1080/02604027.2021.1949939

Grinin, L., Grinin, A., & Korotayev, A. (2022a). COVID-19 pandemic as a trigger for the acceleration of the cybernetic revolution, transition from E-government to E-state, and change in social relations. *Technological Forecasting and Social Change, 175*, 1–15. https://doi.org/10.1016/j.techfore.2021.121348

Grinin, L., Grinin, A., & Korotayev, A. (2023a). Future political change. Toward a more efficient world order. In V. Sadovnichy et al. (Eds.), *Reconsidering the limits to growth. A report to the Russian Association of the club of Rome* (pp. 191–206). Springer. https://doi.org/10.1007/978-3-031-34999-7_11

Grinin, L., Grinin, A., & Korotayev, A. (2023b). Global aging and our futures. *World Futures, 79*(5), 536–556. https://doi.org/10.1080/02604027.2023.2204791

Grinin, L., Grinin, A., & Korotayev, A. (2023c). Demographic transformations in the light of technological development: Types of demographic reproduction in the past and in the future. *Social Evolution & History, 22*(2), 159–202.

Grinin, L. E., Ilyin, I. V., & Andreev, A. I. (2016a). World order in the past, present, and future. *Social Evolution and History, 15*(1), 58–84.

Grinin, L. E., & Korotayev, A. V. (2010). Will the global crisis lead to global transformations. 1. The global financial system: Pros and Cons. *Journal of Globalization Studies, 1*(2), 70–89.

Grinin, L., & Korotayev, A. (2013). Will the global crisis lead to global transformations? In J. Sheffield, A. Korotayev, & L. Grinin (Eds.), *Globalization: Yesterday, today, and tomorrow* (pp. 366–410). Emergent Publications.

Grinin, L. E., & Korotayev, A. V. (2015a). Global population ageing, the sixth Kondratieff wave, and the global financial system. In J. Goldstone, L. E. Grinin, & A. V. Korotayev (Eds.), *History & mathematics: Political demography and global ageing* (pp. 81–106). Uchitel.

Grinin, L., & Korotayev, A. (2015b). Global'noye stareniye naseleniya, shestoy tekhnologicheskiy uklad i mirovaya finansovaya Sistema. In L. Grinin & A. Korotayev (Eds.), *Istoriya i matematika: Futurologicheskiye i metodologicheskiye aspekty* (pp. 31–56). Uchitel.

Grinin, L., & Korotayev, A. (2015c). *Great divergence and great convergence, A global perspective.* Springer.

Grinin, L., & Korotayev, A. (2016a). Global population ageing, the sixth Kondratieff wave, and the global financial system. *Journal of Globalization Studies, 7*(2), 11–31.

Grinin, L. E., & Korotayev, A. V. (2016b). Introduction. Global evolution and global ageing. In L. E. Grinin & A. V. Korotayev (Eds.), *Evolution and big history: Dimensions, trends, and forecasts* (pp. 5–17). 'Uchitel' Publishing House.

Grinin, L., Korotayev, A., & Grinin, A. (2016b). Global'noye stareniye naseleniya, shestoy kondrat'yevskiy tsikl i kiberneticheskaya revolyutsiya. *Problemy Teorii i Praktiki Upravleniya, 7*, 54–59.

Grinin, L., & Korotayev, A. (2020). Seven weaknesses of the U.S., Donald Trump, and the future of American hegemony. *World Futures*. https://doi.org/10.1080/02604027.2020.1801309

Grinin, L., & Korotayev, A. (2023). Africa—the continent of the future. Challenges and opportunities. In V. Sadovnichy et al. (Eds.), *Reconsidering the limits to growth. A report to the Russian Association of the club of Rome* (pp. 225–238). Springer. https://doi.org/10.1007/978-3-031-34999-7_13

Grinin, L., Malkov, S., Grinin, A., & Korotayev, A. (2022). Will capitalism die? Reflections on the capitalism of the past, present and future. *Sotsiologicheskiy Zhurnal = Sociological Journal, 28*(2), 100–130. https://doi.org/10.19181/socjour.2022.28.2.8988

Grinin, L., Malkov, S., & Korotayev, A. (2023). High income and low income countries. Toward a common goal at different speeds. In V. Sadovnichy et al. (Eds.), *Reconsidering the limits to growth. A report to the Russian association of the club of Rome* (pp. 207–224). Springer. https://doi.org/10.1007/978-3-031-34999-7_12

Haas, M. L. (2015). Population aging and the future of the great powers. In J. A. Goldstone, L. E. Grinin & A. V. Korotayev AV (Eds.), *History & mathematics: Political demography & global aging* (pp. 133–146). 'Uchitel' Publishing House.

Hagmann, T. (2007). Bringing the sultan back in: Elders as peacemakers in Ethiopia's Somali region. In L. Buur & H. M. Kyed (Eds.), *State recognition and democratization in Sub-Saharan Africa* (pp. 31–51). Palgrave Macmillan.

Handley, C., & Mathew, S. (2020). Human large-scale cooperation as a product of competition between cultural groups. *Nature Communications,* 11(702). https://doi.org/10.1038/s41467-020-14416.

Harper, S. (2006). Addressing the implications of global aging. *Journal of Population Research, 23*(2), 205–223.

Hartman, Y. A., Karssemeijer, E. G., van Diepen, L. A., Rikkert, M. G. O., & Thijssen, D. H. (2018). Dementia patients are more sedentary and less physically active than age-and sex-matched cognitively healthy older adults. *Dementia and Geriatric Cognitive Disorders, 46*(1–2), 81–89.

He, W., Goodkind, D., & Kowal, P. (2016). *An aging world: 2015, international population reports*. U.S. Government Printing Office. http://www.census.gov/library/publications/2016/demo/P95-16-1.html

Health Canada. (2022). *Third annual report on medical assistance in dying in Canada*. Health Canada.

Hellwig T. (2018). *Vse o zhizni v SSHA*. https://allusalife.ru/%D0%B0/novosti-saita/doma-prestarelyih-v-ssha

Hilbrand, S., Coall, D. A., Meyer, A. H., Gerstorf, D., & Hertwig, R. (2017). A prospective study of associations among helping, health, and longevity. *Social Science & Medicine, 187*, 109–117.

Hilser, H. H. B. (2021). *Collective stewardship and pathways to change: Understanding pro-social values, connectedness to nature and empathic capacity to cultivate ecocentrism in rural communities of North Sulawesi, Indonesia*. Thesis for the degree of Doctor of Philosophy in Human Geography, The University of Exeter.

Hiraiwa-Hasegawa, M. (2005). Homicide by men in Japan, and its relationship to age, resources and risk taking. *Evolution and Human Behavior, 26*(4), 332–343.

Hrdy, S. B. (2005). Evolutionary context of human development: the cooperative breeding model. In C. S. Carter, L. Ahnert, K. E. Grossmann, S. B. Hrdy, M. E. Lamb, S. W. Porges & N. Sachser (Eds.), *Attachment and bonding: A new synthesis*. 92nd Dahlem Workshop Report (pp 9–32). MIT.

Hrdy, S. B. (2009). *Mothers and others: The evolutionary origins of mutual understanding*. Harvard University Press.

Hsu, Y. H., & Lo, H. C. (2019). The impacts of population ageing on saving, capital formation, and economic growth. *American Journal of Industrial and Business Management, 9*(12), 2231–2246. https://doi.org/10.4236/ajibm.2019.912148

Hussain, M. (2021). The changing dynamics of liberal international order: Covid-19 and its global implications. *Journal of Globalization Studies, 12*(2), 143–156.

Irshad, S. M. (2020). COVID-19 and global 'At-Risk Community': From benefit-sharing to risk-sharing of economic crisis. *Journal of Globalization Studies, 11*(2), 135–145.

Jasiński, A. M., & Bąkowska, A. M. (2020). Globalization and life satisfaction of the polish population. Before and during the COVID-19 Pandemic. *Journal of Globalization Studies, 11*(2), 146–155.

Jackson, R. (2021). *The shape of things to come. The macro challenges of population ageing*. Concord Coalition & the Global Ageing Institute.

Jackson R., Howe N., Strauss R., & Nakasima K. (2008). *The greying of the great powers. demography and geopolitics in the 21st century*. Center for Strategic & International Studies.

Jackson R., Macaranas R., & Peter T. (2013). *U.S. development policy in an ageing world. New challenges and new priorities for a new demographic era. A research report from project on U.S. leadership*. Center for Strategic and International Studies.

Kakkar, S. (2021). The outbreak of sars-cov-2 (Covid-19): Impact on international tourism. *Journal of Globalization Studies, 12*(1), 145–158.

Kamiya, Y., & Hertog, S. (2020). *Measuring household and living arrangements of older persons around the world*. United Nations, Department of Economics and Social Affairs, Population Division.

Kapelyushnikov, R. (2019a). Fenomen stareniya naseleniya: Ekonomicheskiye effekty. *Ekonomicheskaya Politika, 14*(2), 8–63.

Kapelyushnikov, R. (2019b). *Fenomen stareniya naseleniya: ekonomicheskiye effekty*. Institut ekonomiki RAN.

Kaplan, H., Hill, K., Lancaster, J., & Hurtado, A. M. (2000). A theory of human life history evolution: Diet, intelligence, and longevity. *Evolutionary Anthropology: Issues, News, and Reviews, 9*(4), 156–185.

Keown, J. (2018). *Euthanasia, ethics and public policy: An argument against legalisation*. Cambridge University Press.

Khorasani, D. G., & Lee, S. H. (2020). Women in human evolution Redux. In C. Willermet & S. H. Lee (Eds.), *Evaluating evidence in biological anthropology: The strange and the familiar* (pp. 11–34). Cambridge University Press.

Kim, P. S., McQueen, J. S., Coxworth, J. E., & Hawkes, K. (2014). Grandmothering drives the evolution of longevity in a probabilistic model. *Journal of Theoretical Biology,* 353, 84e94.

Konner, M. (2018). Nonmaternal care: A half-century of research. *Physiology & Behavior, 193,* 179–186.

Konstantinov A. (2021). Kak zamedlit' stareniye. https://expert.ru/expert/2021/37/kak-zamedlit-stareniye/?mindbox-click-id=7b3b7ede-5674-42b9-90dd-60a6af4a084f&utm_source=mindbox&utm_medium=email&utm_campaign=LuchshieStati2021-NEaktivnaya

Korotayev, A., Butovskaya, M., Shulgin, S., & Zinkina, J. (2021). Impact of population ageing on the global value system. *Vek Globalizatsii, 4,* 69–80. https://doi.org/10.30884/vglob/2021.04.05

Korotayev, A., Butovskaya, M., Shulgin, S., & Zinkina, J. (2023). How can the global ageing affect the global value system? An evolutionary perspective. *Social Evolution & History, 22*(1), 57–76. https://doi.org/10.30884/seh/2023.01.03

Kuznetsov A. (2012). Poselki dlya pensionerov. https://xn----dtbhaacat8bfloi8h.xn--p1ai/social-settlements

Lee R. (2016). *Macroeconomics, aging, and growth.* Cambridge (Mass.). NBER Working Paper № 22310.

Lee, R., & Mason, A. (2010). Some macroeconomic aspects of global population aging. *Demography, 47*(supplement), S151–S172.

Lee, R., & Mason, A. (2011). *Population ageing and the generational economy: A global perspective.* Edward Elgar.

Levin, J. S., & Taylor, R. J. (1997). Age differences in patterns and correlates of the frequency of prayer. *The Gerontologist, 37*(1), 75–88.

Levy, B. R., Slade, M. D., Chang, E.-S., Kannoth, S., Shi-Yi, W. (2020). Ageism amplifies cost and prevalence of health conditions. *The Gerontologist, 60*(1), 174–181. https://doi.org/10.1093/geront/gny131

Liberman, A. (Ed.). (2008). *The long view of crime: A synthesis of longitudinal research.* Springer.

Lockwood, P., Abdurahman, A., Gabay, A. S., Drew, D., Tamm, M., Husain, M., & Apps, M. A. (2020). Ageing increases prosocial motivation for effort. *Preprint at.* https://doi.org/10.31234/osf.io/8c5ra

Lovejoy, C. O. (1981). The origin of man. *Science, 211*(4480), 341–350.

Lovejoy, C. O. (2009). Reexamining human origins in light of Ardipithecus ramidus. *Science,* 326(5949), 74–74e8.

Lutz, W., Sanderson, W., & Scherbov, S. (2008). The coming acceleration of global population ageing. *Nature, 451*(7179), 716–719.

Lyubimov, M. (2013). *Pensionnyye poseleniya v SSHA: gde dostoyno vstretit' starost'.* https://tranio.ru/articles/pensionnye_poseleniya_v_ssha_gde_dostoyno_vstretit_starost/

Maestas, N., Mullen, K. J., & Powell, D. (2016). *The effect of population ageing on economic growth, the labor force and productivity.* NBER.

Mattison, S., Moya, C., Reynolds, A., & Towner, M. C. (2018). Evolutionary demography of age at last birth: Integrating approaches from human behavioural ecology and cultural evolution. *Philosophical Transactions of the Royal Society b: Biological Sciences, 373*(1743), 20170060.

Maxwell, R. J., & Silverman, P. (1989). Gerontocide. In R. Bolton (Ed.), *The content of culture: Constants and variants* (pp. 511–523). Human Relations Area Files.

Mayr, U., & Freund, A. M. (2020). Do we become more prosocial as we age, and if so, why? *Current Directions in Psychological Science, 29*(3), 248–254.

Medici, A. C. (2021). *Health sector challenges and policies in the context of ageing populations.* United Nations, Department of Economics and Social Affairs, Population Division.

Moberg, D. O. (1953). Church membership and personal adjustment in old age. *Journal of Gerontology, 8*(2), 207–211.

Morgan, L. A., & Kunkel, S. R. (2007). *Aging, society, and the life course.* Springer Publishing Company.

Murray, C. J. L., & Lopez, A. D. (2013). Measuring the global burden of disease. *The New England Journal of Medicine, 369*, 448–457. https://doi.org/10.1056/NEJMra1201534

Murray, C. J. L., Ezzati, M., Flaxman, A. D., et al. (2012). GBD 2010: Design, definitions, and metrics. *Lancet, 380*, 2063–2066. https://doi.org/10.1016/S0140-6736(12)61899-6

Murray, C. J. L., Salomon, J. A., Colin, D. M., & Lopez, A. D. (Eds.). (2002). *Summary measures of population health: Concepts, ethics*. World Health Organization.

NAC & AARP. (2009). *Caregiving in the U.S.: A focused look at those caring for someone the 50 or older*. https://assets.aarp.org/rgcenter/il/caregiving_09.pdf

NAC & AARP. (2020). *Caregiving in the U.S. 2020: A focused look at family caregivers of adults age 50*. https://www.caregiving.org/wp-content/uploads/2021/05/AARP1340_RR_Car egiving50Plus_508.pdf

OECD. (2011). *Help wanted? Providing and paying for long-term care*. https://www.oecd.org/els/health-systems/help-wanted-9789264097759-en.htm

OECD. (2021). *Health at a glance 2021: OECD indicators*. OECD Publishing. https://doi.org/10.1787/ae3016b9-en

Olowolafe, O. V., Olonade, O. Y., & Adetunde, C. O. (2020). A contextual analysis of three major tribes' traditional conflict resolution mechanisms in pre-literate Nigerian society. *Applied Research Journal of Humanities and Social Sciences, 3*(3), 1–13.

Orlická, E. (2015). Impact of population ageing and elderly poverty on macroeconomic aggregates. *Procedia Economics and Finance, 30*, 598–605.

Park, D., & Shin, K. (2015). Impact of population ageing on Asia's future growth. In J. A. Goldstone, L. Grinin, & A. Korotayev (Eds.), *History & mathematics: Political demography & global ageing* (pp. 107–132). Uchitel.

Pesut, B., Greig, M., Thorne, S., Storch, J., Burgess, M., Tishelman, C., Chambaere, K., & Janke, R. (2020). Nursing and euthanasia: A narrative review of the nursing ethics literature. *Nursing ethics, 27*(1), 152–167.

Peterson, P. G. (1999). Gray down: The global ageing crisis. *Foreign Affairs, 78*(1), 42–55.

Petty, S., Dening, T., Griffiths, A., & Coleston, D. M. (2020). Meeting the emotional needs of hospital patients with dementia: A freelisting study with ward staff. *The Gerontologist, 60*(1), 155–164. https://doi.org/10.1093/geront/gny151

Polonskaya, N. L. (2019). Doma prestarelykh v raznykh stranakh. https://dom-prestarelih.kiev.ua/blog/doma-prestarelyh-v-raznyh-stranah

Popova, S., & Yanik, A. (2014). Problemy global'nogo stareniya naseleniya: Analiz dokumentov i strategii OON. *Mezhdunarodnoye Pravo i Mezhdunarodnyye Organizatsii, 3*, 429–443.

Powell, J. L., & Khan, H. T. (2013). Aging and globalization: A global analysis. *Journal of Globalization Studies, 4*(1), 137–146.

Prettner, K. (2013). Population aging and endogenous economic growth. *Journal of Population Economics, 26*(2), 811–834.

Püschel, H. P., Bertrand, O. C., O'reilly, J. E., Bobe, R., & Püschel, T. A. (2021). Divergence-time estimates for hominins provide insight into encephalization and body mass trends in human evolution. *Nature Ecology & Evolution, 5*(6), 808–819.

Raghanti, M. A., Edler, M. K., Stephenson, A. R., Munger, E. L., Jacobs, B., Hof, P. R., Sherwood, C. C., Holloway, R. L., & Lovejoy, C. O. (2018). A neurochemical hypothesis for the origin of hominids. *Proceedings of the National Academy of Sciences, 115*(6), E1108–E1116.

Randers, J., Rockström, J., Stoknes, P. E., Golüke, U., Collste, D., & Cornell, S. (2018). *Transformation is feasible. A report to the club of Rome*. Stockholm Resilience Centre.

Raus, K., Vanderhaegen, B., & Sterckx, S. (2021). Euthanasia in Belgium: Shortcomings of the law and its application and of the monitoring of practice. *The Journal of Medicine and Philosophy: A Forum for Bioethics and Philosophy of Medicine, 46*(1), 80–107.

Reinl J. (2023). *Canada is on track for another record-busting year of euthanasia deaths, with a 35% jump to some 13,500 state-sanctioned suicides in 2022, a DailyMail.com projection shows*. https://www.dailymail.co.uk/news/article-12166823/Canada-track-record-busting-year-euthanasia-deaths-35-jump-13-500-cases.html

Rich M., & Hida H. (2023). *A yale professor suggested mass suicide for old people in Japan. What did he mean?* https://www.nytimes.com/2023/02/12/world/asia/japan-elderly-mass-suicide.html

Richerson, P. J., & Boyd, R. (2005). *Not by genes alone: How culture transformed human evolution.* University of Chicago Press.

Riley, M. W., & Foner, A. (Eds.). (1968). *Aging and society: An inventory of research findings.* Russell Sage Foundation.

Robson, S. L., van Schaik, C. P., & Hawkes, K. (2006). The derived features of human life history. In K. Hawkes & R. L. Paine (Eds.), *The evolution of human life history* (pp. 17–44). School of American Research Press.

Rodrigues, A. M. M. (2018). Demography, life history and the evolution of age-dependent social behaviour. *Journal of Evolutionary Biology, 31*(9), 1340–1353.

Sanderson, W. C., & Scherbov, S. (2007). A new perspective on population aging. *Demographic Research, 16*, 27–58.

Sanderson, W. C., & Scherbov, S. (2010). Remeasuring aging. *Science, 329*(5997), 1287–1288.

Sanderson, W. C., & Scherbov, S. (2013). The characteristics approach to the measurement of population aging. *Population and Development Review, 39*(4), 673–685.

Sanderson, W. C., & Scherbov, S. (2015). Are we overly dependent on conventional dependency ratios? *Population and Development Review, 41*(4), 687–708.

Sear, R. (2018). Family and fertility: Does kin help influence women's fertility, and how does this vary worldwide? *Population Horizons, 14*(1), 18–34.

Seifabadi, M. S., & Dolatabadi, A. B. (2022). Covid-19, globalization and strengthening authoritarianism in the middle east. *Journal of Globalization Studies, 13*(1), 135–157.

Séveno, V. (2023). Euthanasia accounted for 1 in 20 deaths in the Netherlands in 2022. *I Am Expat* 13 April 2023. https://www.iamexpat.nl/expat-info/dutch-expat-news/euthanasia-accounted-1-20-deaths-netherlands-2022

Shcherbakova, Y. (2019). *Stareniye naseleniya mira po otsenkam OON 2019 goda* (pp. 837–838). Demoskop Weekly. http://demoscope.ru/weekly/2019/0837/barom01.php

Sindyashkina, E. N. (2022). Healthy life expectancy in the context of the United Nations decade of healthy aging. *Analysis and Forecasting. IMEMO Journal, 1*, 40–53.

Statista.com. (2023). *Number of registered euthanasia instances in Belgium from 2002 to 2021.* https://www.statista.com/statistics/1098051/number-of-euthanasia-instances-registered-in-belgium/

Tilley, C. (2023). *Canada to legalize euthanasia for DRUG ADDICTS with no other illness in March.* https://www.dailymail.co.uk/health/article-12651003/Canada-legalize-medically-assisted-dying-addicted-drugs.html

Travis, C., & Saa, W. (2021). Restorative peacebuilding in Liberia: Traditional practices of mourning and reconciliation. In R. Nicholson (Ed.), *Natural healing as conflict resolution* (pp. 100–125). IGI Global.

Tsirel, S. V. (2008). Zametki ob istoricheskom vremeni i putyah istoricheskoj evolyucii. In L. E. Grinin, A. V. Korotayev, & S. Y. Malkov (Eds.), *Istoriya i matematika: Modeli i teorii* (pp. 246–278). LKI/URSS.

UN Population Division. (2013). *World population ageing 2013.* United Nations, Department of Economics and Social Affairs, Population Division.

UN Population Division. (2015). *World population ageing 2015.* United Nations, Department of Economics and Social Affairs, Population Division.

UN Population Division. (2017). *World population ageing 2017.* United Nations, Department of Economics and Social Affairs, Population Division.

UN Population Division. (2019a). *World population ageing 2019.* United Nations, Department of Economics and Social Affairs, Population Division.

UN Population Division (2019b). *World population prospects 2019.* United Nations, Department of Economics and Social Affairs, Population Division.

UN Population Division (2019c). *World urbanization prospects 2019.* United Nations, Department of Economics and Social Affairs, Population Division.

UN Population Division (2020). *World population ageing 2020*. United Nations, Department of Economics and Social Affairs, Population Division.
UN Population Division (2022a). *United Nations population division database*. United Nations, Department of Economics and Social Affairs, Population Division. http://www.un.org/esa/population.
UN Population Division (2022b). *World population prospects 2022*. United Nations, Department of Economics and Social Affairs, Population Division.
UNCTAD, (2012). *Doklad o mirovykh investitsiyakh, 2012 god. Konferentsiya Organizatsii ob"yedinennykh natsiy po torgovle i razvitiyu*. https://unctad.org/system/files/official-document/wir2012overview_ru.pdf
UNDP (United Nations Development Programme). (2022). *Human development report 2021–22: Uncertain times, unsettled lives: Shaping our future in a transforming world*.
UNFPA. (2012) *Ageing in the twenty-first century: A celebration and a challenge*. UNFPA and HelpAge International. https://www.unfpa.org/publications/ageing-twenty-first-century
US National Center for Health Statistics. (2022). National center for health statistics database. https://www.cdc.gov/nchs/fastats/nursing-home-care.htm
Van Lange, P. A., & Liebrand, W. B. (1989). On perceiving morality and potency: Social values and the effects of person perception in a give-some dilemma. *European Journal of Personality, 3*(3), 209–225.
Vollset, S. E., Goren, E., Yuan, C.-W., Cao, J., Smith, A. E., Hsiao, T., Bisignano, C., Azhar, G. S., Castro, E., Chalek, J., Dolgert, A. J., & Murray, C. J. L. (2020). Fertility, mortality, migration, and population scenarios for 195 countries and territories from 2017 to 2100: A forecasting analysis for the global burden of disease study. *The Lancet, 396*(10258), 17–23. https://doi.org/10.1016/S0140-6736(20)30677-2
Wang, H., Dwyer-Lindgren, L., Lofgren, K. T., et al. (2012). Age-specific and sex-specific mortality in 187 countries, 1970–2010: A systematic analysis for the global burden of disease study 2010. *Lancet, 380*, 2071–2094. https://doi.org/10.1016/S0140-6736(12)61719-X
von Weizsäcker, E. U., & Wijkman, A. (2018). *Come on!* Springer.
WHO—World Health Organization. (2017a). *Global strategy and action plan on ageing and health*. World Health Organization.
WHO—World Health Organization. (2017b). *Dementia: Number of people affected to triple in next 30 years*. https://www.who.int/news/item/07-12-2017-dementia-number-of-people-affected-to-triple-in-next-30-years
WHO—World Health Organization. (2021a). *Ageing and health*. https://www.who.int/news-room/fact-sheets/detail/ageing-and-health
WHO—World Health Organization. (2021b). *Ageism is a global challenge*. https://www.who.int/news/item/18-03-2021-ageism-is-a-global-challenge-un
Widdowson, M. (2021). From Covid-19 to zero-gravity: Complex crises and production revolutions. *Journal of Globalization Studies, 12*(1), 117–144.
Wiessner, P. (2019). Collective action for war and for peace. *Current Anthropology, 60*(2), 224–244.
Willführ, K. P., Johow, J., & Voland, E. (2018). When the mother-in-law is just as good—differential mortality of reproductive females by family network composition. *PLoS ONE, 13*(3), e0193252.
Wittgenstein Centre. (2022). *Wittgenstein centre human capital data explorer*. http://dataexplorer.wittgensteincentre.org/wcde-v2/
World Bank. (2022). *World development indicators online*. World Bank. http://data.worldbank.org/indicator

WVS—World Values Survey. (2020). *Longitudinal data file, 1981–2020* (file: WVS_TimeSeries_R_v1_2.rds). Institute for Comparative Survey Research. https://www.worldvaluessurvey.org/

Yakushev, Y., Sinyavskaya, O., & Voron, O. (2020). *COVID-19 i smertnost' v domakh prestarelykh v mire i v Rossii.* Institut sotsial'noy politiki NIU VSHE. https://www.hse.ru/data/2020/05/15/1546185037/%D0%98%D0%A1%D0%9F%20%D0%92%D0%A8%D0%AD_%D0%A1ovid-19%20%D0%B8%20%D1%81%D0%BC%D0%B5%D1%80%D1%82%D0%BD%D0%BE%D1%81%D1%82%D1%8C%20%D0%B2%20%D0%B4%D0%BE%D0%BC%D0%B0%D1%85%20%D0%BF%D1%80..%D1%85_

Yevseeva, Y. V. (2020). Teoriya uspeshnogo stareniya: sovremennyye issledovaniya: Vvedeniye k tematicheskomu razdelu. *Sotsial'nyye i gumanitarnyye nauki. Otechestvennaya i zarubezhnaya literatura.* Series 11.

Zhu, N., & Chang, L. (2020). An evolutionary life history explanation of sexism and gender inequality. *Personality and Individual Differences, 157*, 109806.

Zhu, N., Lu, H. J., & Chang, L. (2020). Understanding the diversity and fluidity of human morality through evolutionary psychology. *Evolutionary Behavioral Sciences, 14*(4), 404–409.

Zhu, N., Lu, H. J., & Chang, L. (2021). Trust as social investment: A life-history model of environmental effects on ingroup and outgroup trust. *Personality and Individual Differences, 168*, 110303.

Zimmer, Z. (2016). *Global ageing in the twenty-first century: Challenges.* Routledge.

Zinkina, J., Ustiuzhanin, V., & Shulgin, S. (2022). Social macroevolution and changes in the human value systems. How has modernization affected human values? *Social Evolution & History, 21*(2), 91–105.

# Part III

# MANBRIC-Technologies in the Forthcoming Epoch of Self-regulating Systems (2030s–2090s)

The present Part 3 of the monograph 'The Cybernetic Revolution and Global Aging' is devoted to forecasts about the final phase of the Cybernetic Revolution. For some fields, the forecasts are provided till the end of the twenty-first century. In Part 3, Grinin et al. present a detailed analysis of the future development of a whole range of technologies. These are breakthroughs in medicine, additive (3D printing), nano- and biotechnologies, robotics (including drones and driverless vehicles), information (AI) and cognitive technologies, which we group under the designation of MANBRIC technologies (MANBRIC is an acronym). Taken together, these areas will support the development at the final phase of the Cybernetic Revolution, which can also be called the epoch of self-regulating/managing systems. These are systems that can operate without human control. This will provide groundbreaking opportunities and may also create new unprecedented problems.

As the authors have shown in Chap. 4, medicine will be the first sphere to launch the final phase of the Cybernetic Revolution. Of course, medicine will hardly remain the only area of breakthrough, the breakthrough will come from the convergence of the whole MANBRIC complex. Nevertheless, in many respects medicine will be the key integral element of this convergence. That is why, while trying to characterize every MANBRIC branch, Grinin and co-authors pay special attention to medicine (Chap. 8 is devoted to it). Further, the authors show how the aging of the world's population can lead to reorganization of medicine, so that the care for the elderly is at the forefront of innovative changes in medicine (Chap. 11 is devoted to this topic).

Biotechnologies are particularly closely related to medicine and are discussed in Chap. 9. The other MANBRIC technologies are discussed in Chap. 10. The authors try to describe the characteristics and prospects of each field, as well as the interrelationships between these different technologies. Thus, such seemingly unrelated areas as the development of AI and cognitive technologies appear unexpectedly close. Nanotechnologies are diffused in almost all of the MANBRIC sectors, and, together with medicine and biotechnology, will perform miracles in the creation of artificial organs and tissues, as well as in pharmaceuticals, etc. Additive technologies also appear unexpectedly universal, capable of facilitating the development of all areas of life. Robotics promises to provide universal solutions, from caring for the disabled to replacing humans in dangerous jobs. The prospects for AI seem almost unlimited and will influence every aspect of life. Grinin et al. aspire to show that the MANBRIC convergence is not a fantasy, but a real and powerful trend. The authors also show how the above-described characteristics of the Cybernetic Revolution (microtization, saving energy and resources, transition to the control over deeper and more fundamental processes and levels, etc.) will manifest themselves in the future (although they already show up). The chapters seek to explain the diversity of future self-regulating/self-managing systems in general and also in the specific areas analyzed in these chapters.

# Medicine and the Cybernetic Revolution: On the Way to Control Over the Human Body

8

**Abstract**

In this chapter, the authors will focus on current and, especially, future technological transformations in medicine during the final phase of the Cybernetic Revolution. Medicine is and will be a branch around which different technologies (AI, robots, 3d-printing, bio, nano and cognitive technologies) combine. The present chapter provides more examples of such interrelationships denoted by the authors as MANBRIC-complex. Grinin et al. argue that there is a wide range of prospects for the development of medicine in the twenty-first century. The general vector of breakthroughs makes it possible to expand our ability to modify the human body and genome. These changes will probably include the following: the expansion of the opportunities to perform minimally invasive operations instead of the current surgical ones; the cultivation of specific biological materials, body parts and other elements for the regeneration and rehabilitation of an organism, as well as artificial analogues of biological materials (tissues, bodies, receptors, etc.) and so on. Along with the achievements that have been and will be made in medicine, the authors will also identify a number of serious problems and challenges, such as the increasing medicalization of humans and the current stagnation in the pharmaceutical industry.

## 8.1 What This Chapter is About

In this chapter, we will focus on current and, especially, future technological transformations in medicine during the final phase of the Cybernetic Revolution. These transformations will greatly enhance the medical establishment's ability to affect our biological nature as well as life expectancy and health-related quality of life. In the present chapter, we often use the terms 'medicine' and 'health care' as synonyms. Their aim is to describe the emergence of a new medical system organized



L. Grinin et al., *Cybernetic Revolution and Global Aging*, World-Systems Evolution and Global Futures, https://doi.org/10.1007/978-3-031-56764-3_8

around a complex of applied technologies. For example, one already observes a strong interconnectedness between different related areas.

There is a particularly close relationship between the seven leading technological directions of the future, which we identify in Chap. 4 as the MANBRIC convergence. This acronym includes medicine, additive technologies (3D printing), nano- and biotechnologies, robotics, IT and cognitive sciences. There are many new directions in which these fields are expanding. We expect that in the future, they will eventually converge into a single system. We expect this convergence to be driven largely by the medical field due to its unique features and ability to combine the above-mentioned technologies.

We have already witnessed the development of this process: the use of robots in surgery and patient care, information technologies applied for remote medical treatment, neural interfaces for the treatment of mental illness and brain research; gene therapy and engineering, nanotechnologies for developing effective means of providing artificial immunity, biochips for monitoring organisms, new materials for growing artificial organs and many other innovations are becoming a new and powerful sector of the economy. In the present chapter we provide more examples of such interrelationships.

In Chap. 4 we have also pointed out that the final phase of the Cybernetic Revolution will begin somewhere at the intersection of medicine with many other technologies.

So, today medicine is a crucial economic sector, and will become even more powerful in the future. Along with the achievements that have been and will be made in medicine, we will also identify a number of serious problems and challenges, such as the increasing medicalization of humans and the current stagnation in the pharmaceutical industry.

We will argue that there is a wide range of prospects for the development of medicine in the twenty-first century. The general vector of breakthroughs encompasses the rapidly expanding opportunities for modifying our biological nature. In other words, it may become possible to expand our ability to modify the human body and genome. These changes will probably include the following: the expansion of the opportunities to perform minimally invasive operations instead of the current surgical ones; the cultivation of specific biological materials, body parts and other elements for the regeneration and rehabilitation of an organism, as well as artificial analogues of biological materials (tissues, bodies, receptors, etc.) and so on.

Undoubtedly, the fight against incurable (or fatal) diseases and the fight against aging and its consequences will become the most important trends. These trends are tightly integrated and will naturally merge, since the older the society, the more patients, especially the elderly, will be diagnosed with incurable diseases. These trends will also be closely associated with the direction of new rehabilitation procedures for the disabled.

As society ages, their numbers will continue to grow. We also believe that emerging trends like implantation, reproductive medicine, gene therapy and aesthetic medicine will remain among the most promising dimensions of innovative change in medicine.

In the present chapter, we devote considerable attention to our forecasts. According to them, by the end of the twenty-first century, the development of self-regulating systems in medicine, together with increasing life expectancy and the general aging of society will lead to the creation of a super-complex system that we can define as a special medical-biotechnological environment (MBTE). Moreover, this environment will become permanent.

This means that life outside it may become almost impossible, just as we can hardly imagine our life today without the Internet and gadgets. Just as many hundreds of millions of people cannot live without lifelong medication, so in the future a huge part of the population will not be able to stay outside this MBTE for long. This system implies not only a breakthrough in medical technologies, but also a powerful breakthrough in care technologies, continuous monitoring of people's condition (not only of the elderly and disabled), remote use of medicines, psychological support, prediction of future diseases, new prescriptive medicines and more efficient and effective emergency care. Of course, along with the increase in the length and comfort of biological life, there will emerge many dangers and problems, which we also discuss in the present chapter.

## 8.2 Medicine in the Initial and Modernization Phases of the Cybernetic Revolution

In the last decades of the nineteenth—first half of the twentieth century, the development of medicine was particularly effective in those areas of health care that eliminated various infectious diseases (e.g., cholera, yellow fever, typhoid, tetanus, polio, whooping cough, syphilis, measles, malaria, diphtheria, etc.); medicines to eliminate inflammation of the joints, treat blood poisoning and reduce child and maternal mortality and issues related to underdeveloped obstetrics.

Having generally solved these problems in developed countries, medicine already takes on new tasks during the Cybernetic Revolution: the fight against age-related diseases, rehabilitation and improving the quality of life; the development of contraception, reproductive, aesthetic and other fields of medicine. In fact, a number of new areas of medicine emerged during the initial phase of the Cybernetic Revolution (from the 1950s to the 1990s). At the same time, the areas that had emerged earlier, including physical tools such as lasers, radiation, ultrasound, etc. continued to develop. Electronics boosted the development of medicine.

During the modernization phases of the Cybernetic Revolution (from the 1990s to the present day), progress in medical science has allowed a steady reduction in infant mortality, and also the mortality from diseases of mature and old age would also considerably reduced. Since the 1980s and 1990s, significant progress has been made in the ongoing fight against the most common causes of death:

heart attacks, strokes, cancer (although to a lesser extent), etc. As a result, life expectancy has increased significantly. Aging and improving the health-related quality of life has become an increasingly important concern in medicine (see Chap. 6). Considerable progress has been made in many new diagnostic methods and technologies, such as computer tomography, nuclear magnetic resonance introscopy, X-ray photography (Mirsky, 2010: 19), infertility treatment and psychiatry. Major developments have been made in the field of organ transplants, artificial organs, surgery, etc. The modernization phase facilitated the differentiation of medicine that had started many years earlier and also supported the emergence of new branches such as geriatrics (see Chap. 11), nanomedicine, biomedicine, stem cell research and generative medicine (see Strategy… 2013; Wagner et al., 2006; Minger, 2006).

## 8.3 Medicine on the Threshold of a Major Breakthrough: Achievements, Problems, Objectives and Prospects

According to our estimates, the start of the final phase of the Cybernetic Revolution will occur in the 2030s or in the 2040s, mainly in medicine. On the eve of a future breakthrough, it seems reasonable to summarize, analyze and systematize a number of achievements, problems and prospects (which, as mentioned, will begin in the field of medicine). The distinction between achievements, problems and prospects is rather conditional since many phenomena and processes may occur simultaneously.

### 8.3.1 Achievements

**General Characteristics of the Achievements**. We have already spoken about the achievements in terms of reduced mortality rates in all population groups, significantly increased life expectancy and improved quality of biological life. In Chaps. 6 and 7, we have pointed out that in the 1960s to 1980s, medicine provided increasing opportunities for family planning. In Chap. 11, we consider the new direction of medicine associated with aging—geriatrics. Here we should point out that the increasing availability and quality of medical services in a growing number of countries is undeniable. This is no coincidence, as it is related to major social transformations, especially with the growing in the number of hospitals and health care personnel. It is not surprising that medicine has become the leading sector in terms of the GDP in developed countries, and its share of the GDP continues to grow.

At the present stage of development, most new technologies are based on the improvement of earlier discoveries and inventions. Numerous improvements and modifications are constantly being made, which contribute greatly to the popularization, cheapening and distribution of technologies. One should also mention the automation of various processes and the synthesis of various new methods. At the same time, innovations in different fields of medicine are being introduced at a rapid pace, and what seemed novel yesterday and was only discussed in specialized journals, now becomes an established practice. As we have pointed out in Chap. 4, the efficient implementation and integration are characteristic of the modernization phase of the Cybernetic Revolution. In general, although technologies have been introduced into medicine for many decades, its modern phase can rightly be called the *technological period*.

Particularly noteworthy are achievements in the treatment of mental disorders and neurological diseases, sleep disorders, eye diseases, the cure of stomach ulcers, etc. Significant progress has also been made in immunology, with the production of new vaccines reaching absolutely unprecedented levels. It is impossible not to mention that today's medicine is highly computerized, especially in the field of diagnostics and various automatic control systems. And one of the most important sources for the success of modern medicine is the close relationship with MANBRIC technologies with ICT/AI, robotics, biotechnologies (including genetic engineering), additive, nano- and cognitive technologies, etc. (which we will discuss in more detail below and in the next chapters). The MANBRIC integration not only explains significant achievements, but also serves as the basis for future major breakthroughs. In this section we have decided to focus only on just a few aspects that characterize the achievements in medicine.

**Gene Therapy** is a separate branch of modern medicine. A major contribution to its development has been made by the Human Genome Project, which aimed at determining the sequence of human DNA (Brown, 2000; Stein, 2004; Gibbs, 2020; Yu & Hu, 2021). The leading countries continue to spend billions of dollars on gene therapy research. However, it is a long way from determining the structure of the genome to understanding its functions, and this scientific discipline is still at the very onset of its development.

**Surgery**. Surgery has been known since ancient times.[1] In modern times, it has always been one of the most advanced branches of medicine. Below we will briefly describe the most striking achievements in modern surgery.[2] They are of interest because they describe practical techniques.

*Cardiac Surgery*. Today, the heart can be temporarily stopped while the doctors correct different defects (replace or repair cardiac valve, patch the ventricular septal defect, perform coronary artery bypass graft surgery to improve the myocardial blood supply), and then they can restart it. Today, such surgeries are performed everywhere, even in extremely difficult situations, for example, a heart surgery on

---

[1] The recent archaeological discovery made in a cave in Borneo, Indonesia, completely changes our understanding of the Stone Age: the scientists found an amputee who lived 31,000 years ago. Meanwhile, just two hundred years ago, many people who had limbs amputated died either from blood loss and shock, or from infection (Prillaman 2022).

[2] When preparing the paragraph on surgery, we used data from Petrov 2014.

a baby in the womb (CBS, 2017; Cleveland Clinic, 2021). Well-established technical support systems are in place to perform such surgeries. For instance, if the heart stops beating, the heart–lung machine both pumps blood and oxygenates it.

*Vascular surgery and microsurgery*. The development of optical technologies and the use of special microsurgical instruments made it possible to reconstruct the thinnest blood and lymphatic vessels and to repair nerves. It became possible to reattach ("replant") a limb or part of a limb amputated as a result of an accident, with full restoration of its functions. It has also become possible to graft skin or take a part of an organ (e.g., the intestine) and use it as a material to connect vessels with arteries and veins in the necessary organ.

*Brain surgery* is one of the most complex disciplines. The advanced magnetic resonance imaging makes it possible to create functional models of the working brain. The tracking system allows the surgeon to operate with the utmost precision, and to determine the functional structures and boundaries of a tumor. Modern 3D endoscopes provide a three-dimensional imaging and a depth of field that makes the image absolutely realistic. Carcinogenic tissue is cut into fragments and removed using a sophisticated ultrasonic aspirator, thereby reducing the overheating, which is characteristic of other surgical techniques.

**Transplantation Surgery** is certainly a surgical discipline, but for the purposes of our work we will treat it as a separate branch. Transplantation is one of the major breakthroughs in the control of the human body and a path to high-level biotechnical systems. At present, such surgeries are already performed, and, for example, hearts, lungs, livers, pancreases and kidneys can be transplanted. However, human donor organs are scarce, so the black market in human kidneys is growing, while people who distribute donor organs without special arrangements are brought to court as criminals around the world. And here we see how the achievements of medical science appear integral with problems.

**The Development of Previously Narrow Sectors of Medicine**. **The Interaction with Other Technologies**. In addition to the above-mentioned, a number of other specialized disciplines are developing, for example, drug addiction treatment and gender reassignment, as well as fields such as sports medicine, aviation medicine, bioethics and so on (Leigh et al., 2010; Smith, 1979).

Aesthetic and cosmetic medicine is currently developing vigorously and its main task is to correct defects or improve attractiveness by changing the shape of various parts of the face and body (wrinkles, rejuvenation, facelift, liposuction, breast augmentation, abdominoplasty, hair transplantation etc.). The global market for cosmetic surgery and medical aesthetics is worth more than $15 billion (Zhokhova, 2011; Brand Essence, 2020; ISAPS, 2021: 7). But it will be many times larger in the future, we will return to this topic below.

**Pharmaceuticals**. In the first fifty years of the Cybernetic Revolution (since the 1950s), a large number of new drugs were produced and made available. This trend has not stopped. The volume of pharmaceuticals consumed is steadily increasing every year (Demirel & Mazzucato, 2008; Kondratieff, 2011); for example, the total number of medical prescriptions dispensed in the U.S. from 2009 to 2021 grew from 3.9 million to 6.4 million (Statista, 2022a). The sector remains highly profitable. Over the past 15 years, the revenue from the global pharmaceutical market has more than doubled, reaching US$ 1.4 trillion. This is a significant increase from 2001, when the market was valued at just US$ 390 billion (Statista, 2022b). Some researchers consider such a growth as one of the main motivations for medical development (Demirel & Mazzucato, 2008; Williams, 2014). Of course, there are serious challenges here which we will discuss below.

### 8.3.2 The Problems

**General Characteristics of the Problems**. The problems of modern medicine are numerous and multifaceted. We have systematized them as follows:

(1) problems related to purely medical opportunities, for example (a) viral diseases; (b) incurable diseases; (c) the difficulties of dealing with new diseases, such as AIDS or COVID; (d) the rapid spread of such diseases of old age and modern civilization as diabetes, dementia (including Alzheimer's disease), apoplexy, etc., which we discuss in Chap. 7; (e) infertility problems.
(2) Problems of rehabilitation of patients with serious diseases (especially cancer and cardiovascular diseases) and disabled people, including incomplete recovery of functions, severe side effects, actual disability of many apparently cured patients, etc. All this significantly reduces the quality of biological life.
(3) Problems related to affordable health care and its level of development. The differences between developed and developing countries in this respect are enormous. For example, the developed countries spend 10–20 times more per capita on health care than the developing countries (OECD, 2022). In the latter, problems that have long been solved in developed countries (infectious diseases, fever, etc.) are still relevant. However, in some developed countries, medicine appears to be unavailable to a significant part of the population, due to overestimation of the cost of medical services and incomplete coverage by health insurance.
(4) Problems with the status of medicine in society, when, on the one hand, it may be insufficiently supported by the state—for example, by keeping the financial status of doctors rather low—and, on the other hand, one observes that the authority of medicine is expanding along with the process of medicalization (see about it below). The COVID-19 pandemic has vividly demonstrated how and to what extent the influence of medical authorities can increase and even surpass the political and economic power.[3]

[3] On social consequences of the COVID-19 pandemic see, e.g., Chumakov & Yurchenko, 2021; Efremov & Arkhangelsky, 2023; Grinin, 2020; Grinin et al., 2022; Hussain, 2021; Irshad, 2020; Jasiński & Bąkowska, 2020; Kakkar, 2021; Seifabadi & Dolatabadi, 2022; Widdowson, 2021.

(5) There are also problems of bullying of elderly people and a certain medical ageism, which we have discussed in Chap. 7.
(6) The insufficient individualization in medicine, when the same treatment protocols are applied to many millions of patients; even though doctors are supposed to treat each patient individually. Similar problems exist in pharmacology.
(7) Increasing adverse secondary effects of medicines and drugs, including a huge number of deaths from overdoses of prescribed opioids.

The potential solutions to some of these problems are fairly clear. With growing prosperity and higher literacy rates in developing countries, the situation with available healthcare will obviously improve. Some problems, like incurable diseases or the development of personalized medicines, are issues for the future. But some problems, like medicalization, are intensifying and are not yet in the spotlight.

**The Fight Against Currently Incurable Diseases** is the most important direction in medicine. With a rapidly aging population, the potential danger of age-related diseases will increase. According to the WHO, the leading cause of death is ischemic heart disease (16%), with stroke and chronic obstructive pulmonary disease in second and third place, accounting for about 11% and 6% of all deaths respectively (WHO, 2020, 2023a, 2023b; see also Centers for Disease Control, 2007: 4). Cancer mortality in developed countries has reached the same level as coronary heart disease mortality (Sung et al., 2021). The present tendency is that with growing life expectancy cancer diseases take first place among diseases. But today, many types of cancer have become noticeably less deadly, thanks to new methods, drugs and early diagnosis (NIH, 2021; Simon, 2020). It turns out that the fight against cancer simultaneously demonstrates the achievements of modern medicine, and also its huge problem. So far, the situation with cancer mortality has not changed dramatically, and no radical cure or technology has been discovered. It is quite possible that cancer will not be conquered even by the 2040s. The situation is exacerbated by the fact that, as the population ages, the incidence of cancer increases significantly. It is clear that cancer treatment requires considerable changes. If we defeat this disease, there will be a strong impetus for a breakthrough in medicine, its transition to a completely new level, as well as to a substantial increase in life expectancy. We will talk about possibilities for defeating cancer in the future in another section (see also Chap. 10).

Every year many patients with previously incurable diseases literally come back-from-the-dead thanks to new medical knowledge and procedures. However, people with cardiovascular and other serious diseases do not always return to a full life, in fact, this happens in a minority of cases. Very often they become disabled, and this number is growing rapidly in the world.[4] The number of people with paralysis is growing. For example, in the USA nearly 5.4 million people live with

[4] According to the WHO, more than 1.3 billion people worldwide, or about 16 percent of the world's population, have some form of disability (WHO 2023a, 2023b; World Bank 2011).

paralysis, with the leading causes of paralysis being stroke, spinal cord injury, MS and cerebral palsy (Armour et al., 2016). Modern methods of cancer treatment (in particular, chemotherapy and radiotherapy) greatly affect the quality of biological and psychological life of those who recover.

**Medicalization and the Threat of Growing Techno-Medical Power**. Note that the growing importance of medicine can be easily traced in the phenomenon of medicalization (Blasco-Fontecilla, 2014), that is, the definition of various forms of deviant or unusual behavior as diseases. Many aspects of human behavior (e.g., attention deficit hyperactivity disorder (ADHD) in children) that have never been related to medicine, are being described in medical terms and require medical observation and intervention (see Yudin, 2008; Moynihan et al., 2008). On the one hand, this activity has a positive effect since it draws attention to the problems that were previously attributed only to criminal inclinations, bad behavior, the natural course of events, or were simply ignored. And as society's ability to solve problems increases, some phenomena move from personal to public problems, and those considered dangerous to society can be decriminalized. On the other hand, there are financial, corporate and other private interests that can exacerbate the problem. These interests may seek funding from budgets or insurance companies, taking resources away from those who are more in need and from society in general. The result of such activities may be small, or even simply negative, or it may create problems where there were no problems before (e.g., Blasco-Fontecilla, 2014). Both extremes can often be combined. Here we may give the examples of very ineffective and expensive treatment of drug addiction, and the prescription of opiates, opioids and other narcotic drugs for acute and chronic pain, especially in the USA; this leads to a huge number of overdose deaths.[5] Yet, it also generates large profits.

Thus, the trend toward medicalization is actively supported by medical research and practicing corporations, big pharma and other interested stakeholders, including insurance companies (see, e.g., Fukuyama, 2002: Chap. 3). In addition, people "suffering" from newly discovered diseases are eager to receive appropriate justifications for their weaknesses and vices, as well as to gain some benefits and privileges. We also observe that the practice of changing certain individual characteristics (from the shape of the nose to the choice of gender), which is actively promoted by medical companies and corporations, is becoming more and more widespread. Thus, *the social and political role of medicine is growing rapidly.* And this is probably the most important aspect of medicalization, which unlike other aspects, is hardly discussed.

In general, the role and real power of the medical authorities, which actively rely on the power of the state, has grown steadily, especially in the twentieth

[5] The opioid overdose deaths in the USA increased to 75,673 in the 12-months ending in April 2021, from 56,064 the year before (NCHS, 2021). The opioid overdose deaths continue to rise rapidly.

century, when the concept of medicalization was fixed at the end of the century (Conrad, 1992, 2005; Illich, 1975). For a long time, there has been an active campaign to recognize many vaccines as compulsory for use, which causes controversy. We have all witnessed how forced vaccination has literally electrified and split societies.

During the pandemic, it became obvious that the uncontrolled growth of the role of medicine and the pharmaceutical industry actually leads to the emergence of a medical-pharmaceutical complex, which is practically more effective than the military-industrial complex and which merges with the state using its power. As a result, the power of doctors increases by an order of magnitude. This power is largely monopolized and, since it is impossible to control the whole population in the old way, leads to the accelerated formation of an electronic state (which we will discuss in Chap. 14). These and other innovations will allow the formation (and considerable control) of really powerful and large forces, huge financial and political interests and they may become the arena of serious social conflicts that can affect the whole society (see Chap. 9).

We should note that modern technological solutions inevitably start to limit people's rights and dignity, and also increase the interference into human privacy. And with the development of technologies, this problem will considerably exacerbate. It seems extremely difficult to find an optimum here, not to mention the fact that on the basis of such control, medical power and even a kind of terror can grow in an unprecedented way. This is very important in the context of the future development of the medical-biotechnological environment that will carry out the cycle of health control from observation to prescription (which will be discussed below, and also in Chaps. 6 and 7). We believe that the problem of medicalization and increasing medical-pharmacological and techno-medical power will become very acute in the future. Therefore, we need special restrictive measures already today, which could optimize the use of medicine and reduce the opportunities for arbitrariness and manipulation by state-corporations.

**Systemic Problems in the Pharmaceutical Industry and the Search for Their Solutions**. As we have already mentioned, the pharmaceutical industry has made considerable progress. However, along with this progress, the number of serious systemic problems in the pharmaceutical industry is increasing. In fact, one can even speak about a crisis. In particular, the number of officially approved medicines has stagnated in the recent decades (Saigitov, 2015; Martyushev-Poklad, 2015; Arnum, 2022). On the other hand, the number of required clinical trials for drugs steadily increases (Clinical Research Fastrack, 2020; Woollett, 2012).

Firstly, the development of new effective drugs and formulations has become a time-consuming and resource-intensive process (drug discovery and development takes 10 to 17 years and costs from 1.5 to 3 billion dollars). Secondly, the process is further complicated by increasing governmental control.

Third, the *considerable cost of developing new medicines also requires a huge market for their distribution*. But new drugs may turn out to be rather ineffective and so they do not help many patients, while the harm from improperly chosen drugs is enormous. There is also a tendency for people to take too many and often unnecessary medicines. This in turn reduces the organism's natural ability to resist disease, weakens the immune system and creates a need for new drugs. Meanwhile, such a large amount of medication is by no means always needed, not to mention the fact that it always has side effects.[6]

So, in general, we are dealing with a kind of dead end in pharmaceutical progress. One of the main reasons for this is the fact that modern pharmacology makes little use of such an important direction of the Cybernetic Revolution as individualization/personalization. The growth of pharmaceutical production is connected with unification, which at the same time reduces the efficiency, since even well-known diseases can take different courses. The general organization of pharmaceutical production, monopolization and megalomania, also interfere. We agree with the researchers (e.g., Wasden & Williams, 2012) who consider personalization as a way out of the crisis (see below in another section). The slowdown in innovation in the pharmaceutical industry has clearly manifested in several aspects, in particular, in the fact that the patent grant rate in this sector has been declining since 2013 (see Fig. 1 in Chap. 4). Whereas it was previously comparable to the development rate in medicine and biotechnology and exceeded the average patent grant rate in general, now it is significantly lower than the average rate. Since 2020, due to the COVID-19 pandemic, the patent grant rate in pharmaceuticals has been increasing. However, we believe that this is a fluctuation rather than a trend.

However, one should note that during the pandemic, one could observe a sharp turn in the lifting of restrictions on the introduction of new medicines, in particular in the case of coronavirus vaccines. They were approved in great haste and introduced into the widest (unprecedented before) use, actually, without any long and large-scale experimental group tests. Of course, this circumvention of the law led to numerous side effects and often quite serious consequences (which will be perceived for many years to come). But at least it may be a blessing in disguise, as it will probably push the development toward working out a more optimal pharmacological legislation, which will positively affect future medicine.

[6] It is obvious that the mass-market drugs have a significant disadvantage: their effectiveness decreases and they help only some patients, whose number can range between 30 to 50 percent (Martyushev-Poklad 2015). Medication errors also cause serious side effects. For example, according to some data (probably, overestimated), the prescription errors caused more than 100,000 deaths per year in the late 1990s and early 2000s (Null et al., 2004: Table 1). Medical errors are quite costly to society. For example, medical billing errors cost Americans $210 billion annually. Approximately 12 million Americans are misdiagnosed every year. And it is estimated that 7,000 to 9,000 patients die every year from medication errors (Medical Error Statistics 2022).

### 8.3.3 Some Prospects and Developmental Vectors

**General Characteristics**. A wide range of prospects for the development of medicine in the twenty-first century are currently being outlined. It is clear that these promising areas will turn out to be important, while others will not. It is difficult to make predictions in this context, yet we are confident about some forecasts, in particular, that the fight against incurable diseases (or those with high mortality) and **the fight against aging and its consequences will be the most important trends**, and that these two trends will naturally merge (in fact, this is already noticeable today, but will become even more obvious in the future; for details see Chap. 11). The direction of rehabilitation of the disabled will also be very close to these two directions. Even today, the proportion of disabled people among the elderly and old is particularly high, but it will become even higher as society ages. The consequences of incurable diseases with a high risk of death, will unfortunately also lead to disability. We believe that today implantation, reproductive medicine, gene therapy and aesthetic medicine have manifested and will remain among the most promising branches.

But in this section, we can only talk about a few points, and we will present the system of tasks in the next paragraph.

**Fighting Viral Diseases and Pandemics**. Virology has achieved great success, but unfortunately, our knowledge of viruses is still poor (Evans & Shapiro, 1997; Lwoff et al., 1962; Solé & Elena, 2018). Despite great efforts and some successes, annual epidemics of influenza and other respiratory diseases continue to trouble the population, affect economies and divert doctors, claiming millions of lives and causing complications for tens of millions of people every year. Undoubtedly, the work to reduce or minimize these epidemics is very important and promising. Until 2020, AIDS was considered the most dangerous viral disease, which claimed many lives and destroyed many destinies, and the final victory over it is still a matter of the future. However, the coronavirus pandemic became even more widespread and deadly. It also revealed the most important challenge for medicine—the fight against possible new pandemics. Let us recall that humankind has not experienced pandemics for almost a hundred years, since the Spanish flu.

**Remote Medical Care or Telemedicine** is one of the many prerequisites for the start of the final phase of the Cybernetic Revolution in the 2030s. Telemedicine has been developing last 10–15 years, especially during the pandemic period. It means that the quality of medical care services will not depend so much on the qualifications of the medical staff of a particular medical care unit. So, it can become a means of equalizing the conditions for patients. See below for details.

**Reducing the Cost to the Body. Fight Against Aging and for Increasing Life Expectancy**. Both these trends are of utmost importance for medicine. Making medical interventions less invasive, the drug doses—minimal, consequences—mild, and pain—less insensitive, etc.—these trends have been observed in medicine for at least two centuries. And they will intensify along with the development of relevant technologies (for more on nanotechnology see Chap. 10).

Improving the accuracy of treatment is a very important direction that can turn the treatment of diseases into a controllable process, not just occasionally but constantly.

The *fight against aging, together with the treatment of specific manifestations of aging and related problems, will become one of the major tasks of medicine.* This emerging prospect will be developed and implemented throughout this century (for more details see Chap. 11). Various anti-aging and rejuvenation technologies are being discussed (see *ibid.* and below). Technologies alone are not enough, especially when continuous monitoring and prevention are needed. In order to maintain increasing life expectancy and its high biological quality, a specific medical-biotechnological environment will begin to form, in which an increasing part of the population will spend an ever longer time (see Chaps. 6 and 7). On the basis of such processes, a new type of society will begin to form—Cybernetic Society (see Chap. 15, Conclusion).

**Diagnostics, Monitoring Health and Drug Dosing**. Even today, the line between medical diagnosis and treatment becomes more and more blurred. During the final phase of the Cybernetic Revolution, there will be a breakthrough in this area of medical care. Biosensors are the key components of devices that can provide the development of self-regulating systems (see below and in the next chapter). They will be associated with the creation of systems for monitoring health and supporting an organism; the treatment will also be carried out mainly by autonomous systems which will be able to function regularly and continuously. There are also very promising opportunities to dramatically reduce the doses of drugs using nano- and microtechnologies (we will talk about them in Chap. 10). In addition, the use of nano-elements to deliver drugs to target diseased cells will make it possible to replace some surgery with conservative treatment.

**The Breakthrough Vector in Medicine and Related Branches** can be defined as a rapid increase in the *opportunities for correcting or even modifying human biological nature.* It will also become possible to *radically expand the opportunities to dramatically increase life expectancy and improve people's physiological abilities and health-related quality of life (HRQoL).*[7] Medicine will also develop toward the prevention and prophylaxis of diseases, including the development of diagnostics. It is widely believed that the medicine of the future will be preventive medicine. Of course, medicine will hardly be able to prevent all problems in advance, since every human organism reacts in a different way. But medicine will certainly be able to diagnose diseases at a much earlier stage, and improvements in prevention will play an increasingly important role in increasing life expectancy. We believe

[7] As we discussed in Chap. 7, the characteristics of a healthy life are also expressed in the Healthy Life Expectancy (HALE) index, which (according to the WHO methodology) takes into account years of life lost due to premature mortality and years lived with poor state of health (Disability Adjusted Life Years, DALY) (see Murray et al., 2002, 2012; Murray & Lopez, 2013; Wang et al., 2012; Sindyashkina, 2022).

that the medicine of the future will certainly develop toward constant monitoring and rapid response to deviations.

### 8.3.4 Tasks and Directions

Let us formulate the main tasks that will help us to better understand the next section on the future directions of medicine. Of course, some of these tasks are already quite relevant today, but we believe that their importance in the hierarchy of tasks will increase significantly.

1. To increase life expectancy and improve health-related quality of life.
2. To reduce limitations for the disabled and to manage chronic diseases.
3. Victory over incurable diseases, including incurable diseases of old age (such as dementia, Parkinson's disease, etc.)
4. To improve the effectiveness and personalization of treatment.
5. Advances in the control of the human body, including mental and psychological processes.
6. Prevention and (permanent) monitoring of health status.
7. Progress toward genome control.
8. Organ transplantation.
9. To reduce the invasiveness of surgery and mitigate the adverse effects of medicines and procedures, etc.
10. To increase opportunities to suppress epidemic outbreaks of viral colds (flu, COVID, etc.)
11. To reduce the influence of the human factor in medicine and increase the productivity of doctors' work via the introduction of self-regulating/managing systems and merging medicine with other MANBRIC technologies.
12. To increase available medical services (the number of patients, the frequency of their requests for help, the wide range of services).

We will consider some of these goals below.

## 8.4 Major Technologies and Forecasts Concerning The Future of Medicine

We base our forecasts on the following provisions:

(1) the importance of the above-described directions in relation to the current urgent problems of medicine, which should / can be ultimately solved in the future. For example, it is impossible to reduce the influence of the human factor (i.e., to *reduce the workload of medical staff of all qualification levels—from junior staff to senior doctors—and thus, to increase the availability of medicine*)

without robotization (see Chap. 10) and without the introduction of other self-regulating systems (including the above-mentioned monitoring of the patients' condition)[8];
(2) the already manifested trends;
(3) the general developmental trends. In particular, the increasingly close symbiosis between medicine and other MANBRIC fields, and the emergence of related fields.

### 8.4.1 Improving the Precision of Treatment

Already today, surgeons make extensive use of special flexible instruments which allow them to reach the most inaccessible parts of the body without making large incisions in the abdomen. During surgery, they also use endoscopes and video cameras which transmit a high-power image to the monitor. To solve the problem of human hand tremor, special devices and installations are used to replace the surgeon's hand. These allow the surgeon to control the smallest movements of the instrument (including laser or ultrasound). Undoubtedly, the development of robotics and other self-regulating systems, combined with such methods such as laser surgery, will allow making treatment to be minimally invasive. In addition, the microminiaturization of all devices and instruments will make the diagnosis of disease more accurate. This will allow less aggressive treatment and the inevitable incisions will become less and less noticeable. Moreover, in many cases they will no longer be needed, as it will be possible to use nanorobots, biochips and other technological advances to act inside the body, through the vascular system or by other means. Non-traumatic surgery already takes advantage of natural orifices. For example, the first transvaginal cholecystectomy—removal of the gallbladder was performed through a woman's natural opening orifice (Marescaux et al., 2007). There are also substances that can mark cancer cells and effectively remove them surgically (Gongalsky et al., 2016).

Another critical direction is to reduce the harm associated with the use of drugs and medical procedures—an important direction that can transform the treatment of diseases into a controllable process. One method that will become precise is drug delivery to target cells. Here nanotubes, which we will consider in the section on nanotechnologies (in Chap. 10), are likely to play a key role. Other methods include the correction of genetic disorders, changes in the technology for less invasive and less harmful surgical procedures; positive regulation of the immune system, etc.

[8] This will be discussed further in Chap. 10; in particular, we will speak about robots that can replace junior medical staff in caring for the sick, old and infirm.

### 8.4.2 Surgery: Moving Toward Roboticized Surgery

It is obvious that surgical technologies will change significantly in the direction of (a) remote medicine; (b) less traumatic surgery; and (c) robotic surgery. The first two have already been described above, so let us focus on the last direction.

Surgical robots are a rapidly developing sector with many advantages. Surgical robots can be classified into different functional categories: assisting surgical robots, telesurgical instruments, navigation systems, robots for precise positioning and robots for specific surgical tasks (Taylor, 1997). In 2010, the world's first pancreas transplant was performed entirely using the da Vinci surgical system.[9] From time to time, there appears information about organ transplants performed using this surgical system (e.g., a kidney was transplanted in Dubai in 2021). However, this trend has recently slowed down (Kelly, 2020) due to the fact that technologies are not developing as fast as we need, and because of the high cost of surgical robots.[10] On the down side, there are also a significant number of cases of injuries and deaths following robotic surgery. Doctors from Rush University Medical Center, the University of Illinois and the Massachusetts Institute of Technology have provided data showing a large increase in injury cases and fatalities after operations performed by a robot. Fatalities and other problems ranged from 13.3 cases per 100,000 surgical procedures in 2004 to 50 cases in 2012 (Pinkerton, 2013). The FDA recorded a dramatic increase in deaths from robotic surgeries in 2013 (from 32 to 58 in 2013 according to Alemzadeh et al., 2016).

However, every year progress is also being made toward creating safer robotic technologies. As these technologies are more convenient, efficient and more cost-effective, they will inevitably become more widespread and more actively used in future surgery. One may expect that more and more surgeries will be performed in the future without the active participation of a surgeon. For some details about the robotic surgery see in Chap. 10.

[9] It is a large machine which is equipped with flexible 'hands'—manipulators with a set of surgical tools. A very small incision is made in the patient; therefore the operations are less painful and patients need a shorter period for recovery. Robots can use all the latest tele-video systems which allow doctors to see the operation clearly, magnified and in color. The doctor watches the monitor and controls the robot, sitting in the other part of the surgery (in the future he could also be in any other city or even country), the assistant watches the robot and the patient. For the purpose of watching the surgery process in full detail, the HD 3D screens are installed. Robot-assisted surgeries are becoming very popular, for example, the medical companies in the USA use the billboards to attract more clients to these painless fast procedures (Pinkerton 2013).

[10] The cost of robotic surgery is higher than traditional surgery, ranging from US$ 30,000 to $50,000 in the USA. The Da Vinci price starts at $1 million (Pinkerton 2013) to $2 million, and robotic surgery generally costs from $3,000 to $6,000 more than traditional laparoscopic surgery (Crew 2020).

### 8.4.3 Transplantation and Artificial Regeneration

As we discussed above, there is a shortage of human donor organs. The solution to the problem of organ shortage is to search in different directions. These include the following:

1. Using part of a donor organ, growing a new organ using stem cells, including the patient's own (e.g., doctors have successfully transplanted an ear made from a patient's own living cells [Businesswire, 2022]).
2. Growing an organ for a patient, who has had part of it removed by surgery.[11]
3. The possibility of xenotransplantation (transplantation of animal organs into humans).[12]
4. The development of biological material substitutes, i.e., bionic tissues, parts and whole organs using technologies such as 3D printers, which actively come into use now, for example, to produce artificial bones (Saijo et al., 2011; Xu et al., 2014). This direction seems to be very promising. So there are growing opportunities for tissue engineering (we will discuss this in the next Chapter). Laboratories use healthy skin and cartilage cells to replace damaged bone or cartilage.[13] A special hydrogel has been invented for 3D bioprinting that combines nano- and biotechnology (Agate et al., 2020).

These technologies are promising with respect to the development of cell therapy and tissue regeneration. Scientists already design various artificial organs: skin, retina, trachea, blood vessels, heart, ear, eye, limbs, liver, lungs, pancreas, bladder and ovaries. At the same time, doctors have made good progress in growing (and replacing with artificial analogues) flat organs, such as cartilage and skin cells. However, the situation is more complicated with three-dimensional organs and tissues (including blood vessels and hollow organs, such as the ureter and bladder, uterus and vagina), although some progress has been made here too.

One of the most serious problems with the implantation of new tissues and organs is the outcome and the compatibility with new tissues. We forecast that the eventual discovery of how to 'deceive' the mechanism of immune suppression of foreign cells will be the major breakthrough in the field of organs and tissues regeneration and transplantation (see above). Some steps have already been taken in this direction. For instance, one can point to the opportunity to control processes

---

[11] Russian scientists are developing a new method for treating liver cirrhosis. It involves placing a cellular scaffold made of a special porous material—air gel—in place of the removed diseased part of the organ. It will trigger cell regeneration and help the body repair damaged tissue (Gritsenko & Maleva, 2023).

[12] Recently, the first person received a gene-edited pig heart (Bush 2022), which unfortunately failed (Bajaj 2022).

[13] Once a sufficient number of cells have been grown, they are implanted in the developed materials, which are based on polysaccharides and special substrates which control their growth. The growth conditions of the cells in these structures are very similar to their natural environment.

by affecting the key elements of the initial subsystem of the human organism, in this case switching off the most vigilant systems associated with immune protection (just as anesthesia switches off the perception of pain during surgery). The important event was when Japanese scientists discovered a way to reprogram the functions of cells. For example, skin cells were reprogrammed and substituted the damaged cells of an eye. Such cells are not rejected, so this direction is exceptionally promising (Kostina, 2013).

In order to grow an organ, a biodegradable scaffold material is needed, and there are certain difficulties here. The created scaffold materials can be either natural—for example, from the same organ of a person—or synthetic, consisting of polymers or ceramics (Moskalev, 2018). And here we encounter a complex problem on the way to organ replacement and tissue regeneration– the need to learn how to cultivate cells of different types outside the body and in large quantities. An important factor is the selection of growth factors that allow the cells to mature into the desired cell type for a given organ. In particular, the complexity of creating hollow organs lies not only in the shape, but also in the need to use bioprinters to create layers of cells of different types. Here, artificial biocompatible materials that can survive in the patient's body without causing rejection could help. Thus, there is hope that in the future, it will become possible to grow solid organs, such as hearts, livers, or kidneys that are capable of performing basic functions. These organs have a more complex structure and consist of several types of cells. To create them, a base of donor organs is used. In order to prevent rejection, the original organs are cleansed of the donor's own cells, leaving only the connective tissue scaffold, which is treated with the patient's stem cells grown outside the body. As they grow in a bioreactor, the stem cells form a full-fledged organ from an empty scaffold, genetically identical to the recipient's body (Moskalev, 2018).

One may speak about far-reaching prospects for the so-called artificial spinal cords that is implants that can cure paralysis and give people the ability to move. Clinical trials of such devices are already underway (Hussein, 2022; see also Chap. 10).

Cosmetologists have started injecting human cadaver tissues into their clients. We can agree with the opinion that it is equivalent to a medical transplant (Landman, 2023).

### 8.4.4 Therapy

Here, the general control of the state of health is of the utmost importance, together with the development of diagnostics, including the use of the Bio-Nanochip Technologies and other biomolecular self-regulating systems. We have discussed above the treatment of incurable diseases in the other branches of medicine: endocrinology (diabetes mellitus), cardiology, oncology and psychiatry, etc. Let us consider the opportunities for fighting cancer and two other topics.

**The prospects for fighting cancer** can be outlined in several directions. In particular, it can be a combination of methods that rely on analytics based on the study of big data and AI. For example, the development of a precise selection and delivery of drugs targeted at cancer cells. Some medical centers (e.g., the Comprehensive Cancer Center at Johns Hopkins University in Baltimore and the Medical Center of Cambridge University) have already modeled a strategy by which specific medicines could target and control metastatic cancer cells. AI can play a major role in disease detection. Currently, AI can detect skin cancer better than a doctor (ESMO, 2018). For instance, other technologies have been developed, such as the self-administered Human Papillomavirus Test, which may help prevent cervical cancer (Palmisano et al., 2003). A very promising option is to train the patient's immune system to recognize and destroy cancer cells. One approach would be to identify immune cells in a tumor, grow them in the laboratory and then infuse them back into the patient—a technique called adoptive cell transfer. We already have the ability to select immune cells that recognize any cancer antigen. This strategy would not be cheap, with each course of therapy costing between US $10,000 and $100,000. But, if researchers could develop a therapy that cured metastatic cancer, the whole cost–benefit analysis and the health economic models would change dramatically (Willyard, 2016).

One of the most promising immunotherapies is CAR T-cell therapy. CAR stands for "chimeric antigen receptor" and T-cells are a type of immune cell that kills virally infected, damaged, or cancerous cells. This method has already saved many lives, including children's. At just six years old, Emily Whitehead faced a life-threatening relapse of acute lymphoblastic leukemia, the most common cancer in children. Emily's cancer had resisted more than 16 months of chemotherapy treatments when her parents were told that her cancer had relapsed and that she would not survive. Her parents enrolled their daughter in a clinical trial of a new immunotherapy treatment designed to turn Emily's immune system into a powerful weapon against cancer. The CAR T-cell therapy worked, and Emily's cancer went into complete remission immediately. In June 2012, at the age of 7, Emily was discharged from the hospital. She became so popular that many doctors have her photo on their walls (Mukherjee, 2022).

Our immune system has special killers—T-cells—that are supposed to target and eliminate cancer cells. But Emily's T-cells were unable to recognize the cancer cells. To remedy this, scientists extracted her T-cells, reengineered them with a special receptor (CAR), and then put these highly trained T-cells back into Emily. In general, this highly promising method is just another, very elegant and sophisticated, way of boosting our own immune system, which has all the tools it needs to fight and defeat cancer.[14]

There are some hopes for genetic treatment by tracking tumor DNA circulating in the blood (Yong, 2014). Researchers at Mirati Therapeutics in San

[14] It proves once again the point which was made 200 years ago by the founder of vaccines, Edward Jenner, 'The deviation of man from the state in which he was originally placed by nature seems to have proved to him a prolific source of diseases.'.

Diego, California have identified a molecule that can stop some of the deadliest cancers, including pancreatic and colorectal cancer. The molecule, called MRTX1133, blocks the activity of a common mutant protein in a once 'incurable' cancer; it causes a reduction in human pancreatic tumors when grown in mice and transplanted into patient (Tang & Kang, 2022).

**Fighting epidemics and pandemics and boosting immunity**. One of the urgent problems facing modern medicine, as we said above, is the inability to prevent, or at least minimize the spread of, the annual epidemics of the viral colds. And accordingly, one of the tasks of future medicine will be to learn how to suppress such outbreaks. Here different directions are possible. We have recently witnessed the greatest, unfortunately not successful, experiment in the history of medicine. These were attempts to use innovative vaccines to increase collective and individual immunity and the body's resistance to the COVID-19 virus. However, these methods failed, and to a certain extent even weakened the immunity of many people. Therefore, the way to increase people's immunity to viral epidemics by methods other than mass vaccination, which does not have a great effect, seems to be optimal. This will provide many other advantages for solving problems both in medicine and society. As we will see below, research into boosting our immunity also moves us along the path of the personalization of medicine.

Let us consider the opportunities for using artificial antibodies and the increasing opportunities of using the individual's own immune system. There will never be a universal drug against all diseases, but strengthening the immune system is one of the universal directions that can change this situation and help in the fight against various diseases.

The human immune system has a special tool—antibodies. Antibodies are molecules synthesized to fight cells of foreign origin, called antigens. The damage caused by antigens usually leads to the destruction of foreign cells and to recovery. Antibodies are produced specifically for each antigen. They are produced by special immune cells called lymphocytes, which accumulate and circulate in the blood throughout life. So everyone has their own protective system based on the 'history of diseases'.

Scientists have repeatedly tried to produce artificial antibodies. Various methods have been used, the most widespread was to isolate antibodies from animal blood, but unfortunately the degree of purification remained low. In 1970, Nobel laureates Cesar Milstein and Georges Köhler found a method to produce a specific type of antibody called monoclonal antibodies. By injecting the antigens into a mouse and isolating the antibodies from its spleen, the scientists managed to obtain separate antibodies that could be cloned by making multiple copies of them. However, such cells could only exist for a short time. By hybridizing them with cancer cells, long-lived self-cloning antibodies were produced—hybridomas. Nowadays, much medical research is focused on the production of antibodies by other means (Schirhagl et al., 2012) and also the creation of chemoreceptors (Dickert et al., 2001). Antibodies are already widely used in pregnancy tests, in the diagnosis of many diseases and in laboratory experiments.

Artificial antibodies have low production costs, reagent reproducibility and increased affinity, as well as detection of fungal toxin in grain samples. Synthetic antibodies are currently the fastest-growing class of therapeutics (Frenzel et al., 2013; Pile, 2013) and can significantly strengthen the trend toward individualization of medicine. Some other immune-boosting technologies are discussed in the next section.

*We suppose that in the final phase of the Cybernetic Revolution there will be considerable progress in the creation of artificial antibodies and their acceptance by the organism.* There is no doubt that progress in this field will lead to a breakthrough in medicine. The creation of artificial antibodies will play an important role in the prevention and treatment of many serious diseases; they will prevent the rejection of transplanted organs, etc. This will make it easier to control the course of a disease and, if possible, to suppress and defeat it. The progress toward the creation and acceptance level of artificial antibodies will mean a significant increase in the *opportunities to control processes that were previously inaccessible to controllable interference and the emergence of self-regulating systems to regulate such interference.*

### 8.4.5 Cell and Gene Therapy

A huge achievement of twentieth-century medicine and biotechnology was the ability to work at the cellular level. This level will be one of the most important in the medicine of the future. It clearly reflects one of the characteristics of the Cybernetic Revolution, namely, the transition to control of deeper and more fundamental processes and levels. It is also connected with other characteristics mentioned earlier: miniaturization and microtization. The use of micro- and nanotechnologies will allow better performance at the submolecular or even atomic level, which will allow solutions to a wide range of medical problems and, in general, greatly expand our opportunities. For example, we have shown that in order to understand how our immunity works, one must be able to go down to the cellular and subcellular levels.

**The use of cellular and intracellular mechanisms in future medicine and the fight against aging**. As mentioned above, a very promising trend is to be able to "teach" our immune system in order to make it more effective. One possibility is to teach the body to detect any pathogen with artificial, monoclonal, or polyclonal antibodies. This has already improved in the treatment of various cardiovascular, cancer, respiratory, hematological, autoimmune and infectious diseases (Singh et al., 2018). It is difficult to overestimate the potential of this technology.

Another approach may be to control the behavior of cells in critical situations, for example, by inducing the self-destruction of infected or uncontrolled cells. The process of self-destruction is called *apoptosis*. It is one of the most promising methods to defeat serious diseases, including cancer. The research into this field has been carried out from the 1960s. It shows that some cells often die in strict compliance with the predetermined plan. For example, the embryo of the microscopic worm nematode consists of 1,090 cells before hatching, but later some of them die and only 959 cells remain in the adult worm organism (Raff, 1998; Ridley, 1996). The mechanism of apoptosis is related to the activity of signaling molecules and special receptors, which receive the signal and launch the processes of morphological and biochemical changes, resulting in cell death. An opportunity to trigger the self-destruction of the disease-causing cells can make the fight against diseases controllable. It also allows for rapid recovery without the long rehabilitation period required after surgery, chemotherapy, or radiotherapy. Conversely, it is also possible to switch off the mechanism of cellular self-destruction in order to save an organism from certain diseases and, probably, to control the process of aging (see Chap. 11). We suppose that the Cybernetic Revolution will enable medicine to make progress in this direction and to control it. That is, it will be possible to induce the death of unwanted cells and at the same time block the mechanism of the death of necessary cells.

*Like the monoclonal antibodies and the artificial immune system, they will become powerful self-regulating systems.*

It is very difficult to find specific physiological mechanisms which determine the aging process, but it finally became tangible after the invention of the genetic structure of special cell bodies necessary for cell division—telomeres. Telomeres are considered to be one of the reasons why cells age and die when an organism reaches a certain age, although scientists have been hotly debating this issue since the last century (Slagboom et al., 1994; we will talk about telomeres and other aging control functions in more detail in Chap. 11).

**Gene therapy**. There are two approaches to gene therapy: *fetal gene therapy,* when foreign DNA is introduced into the zygote (fertilized egg) or a germ at an early stage of development; thus, it is expected that the introduced material will be inherited. The second approach is *somatic gene therapy,* when the genetic material is introduced only into somatic (i.e., non-germinal) cells and it is not transferred to sex cells.

Another approach is to activate the body's own genes in order to completely or partially overcome the impact of the mutated gene. The striking example of such an approach is the use of hydroxyurea to activate fetal hemoglobin synthesis in patients with sickle-cell anemia and thalassemia. Thus, in November 2017 in California (USA) the world's first procedure to 'edit' the genome of an adult directly inside his body took place. The patient was a man with mucopolysaccharidosis type II (Hunter syndrome) (Marchione, 2017).

Gene therapy can become an example of the individualization of technologies and targeted processes. On the basis of the genetic data, the most appropriate treatment will be adapted for individual patients, and, if it is necessary, the defective genes will be corrected. In addition, the activation of necessary genes and gene silencing (if required) are also possible. Gene therapy can be used to choose the appearance of a future child (eye color, skin color, etc.). In the future it may be possible for babies to be born almost to order, these will be 'perfect babies'

(Fukuyama, 2002 citing McGee, 1997).[15] In other words, that means that parents will select desirable traits in a child before it is born. So it is not impossible that the geneticists will find 'the genes' for such traits as nobility, aggression or self-esteem and even intelligence, and thus create an 'improved' baby. Such genetic improvement will remind face and body improvement through plastic surgery. In other words, it will be impossible to make a genius or a champion out of any child, but it is not excluded that it will be possible to improve his potential—just as it is possible at present to improve the sports and intellectual potential by means of educational technologies and other conditions.

We agree with Francis Fukuyama, who believes that the future achievements of the 'biotechnology revolution' should be accepted with great prudence (Fukuyama, 2002). Moreover, it begins to move from the category of assumptions to the category of real possibilities. In particular, in January 2019, the Chinese authorities confirmed the birth of the world's first genetically modified humans in Shenzhen. The biologist who performed it, He Jiankui, who presented his research at the genome-editing summit in Hong Kong in late 2018, was arrested and imprisoned (Harney, 2018; Browne, 2022).

Presumably, gene therapy will be used in sports medicine, as its inherent potential is no longer sufficient to achieve the best results in top-level sports. It can also become a new tool in the attempts of pharmaceutical companies to avoid the control of anti-doping committees.[16] In addition, cell therapy, gene therapy and molecular therapy (as well as immunity enhancement) can in the future become effective means of combating aging and rejuvenating the body (see Chap. 11).[17]

### 8.4.6 Changes in Human Reproductive Capacity

This is a particularly important field of medicine. Although the number of incurable diseases causing infertility is decreasing, the number of childless couples and infertile men and women in the world is increasing every year. Nevertheless, the only opportunity for these patients is in vitro fertilization. In addition, due to the development of medicine, there is an increasing number of women who want to have children after their reproductive age (e.g., Annegret Raunigk, a 65-year-old woman from Germany gave birth to quadruplets [McKie, 2015]). We believe that

[15] It is difficult to say how 'perfect' they will be and what kind of problems will as a result of these technologies. For example, the ability to predict a baby's gender has resulted in gender imbalance in China. As a result, there is a disproportionate number of boys.

[16] On the achievement of superhuman abilities through genetic engineering technologies, see also the conclusion to this chapter.

[17] In addition to these anti-aging technologies, we also highlight: 1) drugs (geroprotectors, "drugs for old age"); 2) technologies of regeneration, substitution/cyborgization (organ replacement, cultivation of organs and tissues; transplantation of intestinal biota); 3) genetic engineering technologies and a number of others. In Chap. 11, we describe in detail these and other emerging technologies for combating aging, rejuvenating the body, and their prospects.

already existing reproductive technologies will develop significantly, so that highly improved IVF, etc. can become widespread, infertility will be better curable.[18] Secondly, robotization of all processes is predicted, since, according to embryologists, now the lion's share of all their work is routine, which in the future will be performed by self-regulating systems (Denisova, 2021: 212). Thirdly, it is possible that procedures of eggs and sperm freezing will become widespread, which will de facto extend the childbearing age. Today, there are already many cases when women after menopause have children with the help of IVF and donation; there are also cases of the birth of children conceived from the sperm of a dead man (Njagi et al., 2023).

At the same time, the development of synthetic insemination (i.e., without one of the parents) is not ruled out. This is how scientists at the University of Cambridge in the UK succeeded in creating a mouse embryo from mouse stem cells, which then developed into a heart and brain, without using either egg or sperm (Amadei et al., 2022). This achievement can become the basis for the development of new treatments for infertility.

One should mention the technologies for growing an embryo outside the woman's body. The transplantation of reproductive organs becomes possible. Scientists are developing an artificial womb that can then be transplanted into a woman with a damaged womb or even into a man, which will radically change the concept of sex (McKie, 2002) and raise new ethical problems (Romanis, 2018; Simonstein, 2006; Werner & Mercurio, 2021). The first patent for a full-fledged artificial uterus was filed in 1955. The device consisted of a reservoir of amniotic fluid, a synthetic umbilical cord, blood pumps, an artificial kidney and a water heater. This version was not complex enough and therefore did not work properly, but after an unsuccessful first attempt, the technology was significantly improved. Artificial womb experiments have been successfully conducted in several places, for example, in Italy, where an artificial womb was grown and transplanted into a woman. Scientists from Suzhou, China, have developed technology that can control the process of embryo development in the laboratory. An 'artificial womb' or 'long-term gestation device' could theoretically mean that a woman no longer has to bear children herself (Chen, 2022). Moreover, technology already allows for interventions in the child's body, even before birth (MacKimm, 2022).

Thus, it is likely that medicine will begin to move toward semi-artificial rearing of children (i.e., completely new technologies will be developed), which may contribute to an increase in the birth rate. In our opinion as a result of the final phase of the Cybernetic Revolution the number of experiments with artificial fertilization will increase and the growth of embryo outside the woman's body will become a reality.

[18] Only half a million children worldwide are born with the help of IVF (Fauser 2019; Denisova 2021: 109; Sciencedaily 2018), which is about 0.4% of all children born. See also Chap. 6.

### 8.4.7 Aesthetic and Corrective Medicine

As already mentioned, aesthetic and cosmetic medicine is developing rapidly. The development of these fields is linked to an increase in the standard of living as people spend more and more money on personal care, and, accordingly, these industries become highly profitable. Many millions of plastic surgeries are performed all over the world; however, according to the International Society of Aesthetic Plastic Surgery (ISAPS), the leader in per capita number is not even the USA (they rank the sixth), but South Korea with 13.5 cosmetic surgeries per 1,000 people. At the same time, a surgery to give girls a slightly more Western appearance is common in East Asian countries. The global market for cosmetic and aesthetic medicine is growing rapidly. For example, the global growth rate between 2020 and 2021 alone shows a leap from 6 to 16% depending on the type of procedures; with abdominoplasty increasing the fastest (ISAPS, 2021: 7).

One of the major achievements of plastic surgery is facial transplantation. The first full face transplant was performed in France in 2005 on a woman who had been mauled by her dog (Barret et al., 2011). Another well-known operation was performed in March 2012, when doctors at the University of Maryland Medical Center gave a 37-year-old man a new face, including jaw, teeth and tongue. In a decade, seven patients worldwide received faces coming from human donors. Five patients died from complications (Siemionow, 2017), but there have also been a few surprisingly positive results.

One may assume that aesthetic medicine will continue to grow over the next decades, due to the technological development and the widespread use of already proven technologies (e.g., various types of facelift, liposuction, body shaping, etc.), as well as the rising standard of living in developing countries. The wealthier a society, the more money people spend on their health and beauty. Taking into account the growth of the world's middle class, this direction and all types of aesthetic medicine are expected to develop rapidly.

Although the most important task of corrective medicine will remain as it is today, the correction of congenital and acquired defects of the body and appearance. It is likely that, as new technologies are created based on the achievements of surgery and genetic engineering, the use of these procedures will become much more widespread and possibly even to some extent prescribed by one's lifestyle. As is the case today, a significant number of successful people, are willing to pay considerable sums for both aesthetic and dental procedures. The growth of aesthetic and corrective medical technologies can cause serious psychological problems, including issues of an individual's self-identification.

## 8.5 Medicine in the Context of Self-regulating Systems and Characteristics of the Cybernetic Revolution

### 8.5.1 Increasing Patient Competencies and Changes in the Medical Profession

The development of medical technologies can significantly change the work of doctors and their interactions with patients, and not only in terms of their partial replacement by robots or other self-regulating systems. There are many other aspects as well. In particular, the development of intelligent technologies that promote self-measurement of medical parameters and self-diagnosis, will allow patients to perform a number of medical functions by themselves. The analysis of important indicators (such as blood pressure, blood sugar, pregnancy tests, urinalysis) allows self-diagnosis, etc. Many self-diagnostic devices can be easily carried out today without the presence of a doctor. Through AI, we can increasingly focus on self-monitoring the most important human bio-indicators. This capacity is a harbinger for the transition to a system of continuous health monitoring. We expect that people will probably live in a holistic medical and biotechnological environment in the future (see below).

In the future, diagnostics are likely to be transferred to mobile devices with biochips which do not require the involvement of specialists. In the USA, a device has already been invented that can quickly diagnose a heart attack (Ren et al., 2021) and provide first aid in the case of cardiac arrest. AI has already been entrusted to detect thyroid cancer (Konnikova et al., 2021); and it is also able to identify rare genetic diseases by facial features (Hsieh et al., 2022). So, these technologies allow a number of doctor's functions to be performed by patients. The latest types of chats, like Chat-GPT-4, are to a certain extent able to make diagnoses based on the symptoms described by a person. Perhaps, in the future, self-managing systems will issue prescriptions instead of doctors.

In addition, the development of telemedicine, the creation of a system of continuous health monitoring, robotization and so on will probably reduce the number of face-to-face contacts between a doctor and a patient (and the number of doctors in the future as well). At present, centers of best practice (e.g., Centers of Excellence), have been built where the leading doctors perform surgery and consult their colleagues online (Elrod & Fortenberry, 2017). Of course, one may face considerable disadvantages here, but the expansion of available medical services has unfortunately meant a loss of personal communication (similar as the expansion of commerce has almost eliminated the previously existing rather nice habit of friendly contacts between buyers and sellers in small shops).

Thus, the doctor's job in its current form may lose some of its current attributes. At present such a metamorphosis is taking place in the service sector (such as photo service, typesetting and page makeup, design, selection of interior, purchase of tourist vouchers, choice of routes, various types of consultation, guides, etc.). Of course, the profession of a doctor will continue to exist, but by the end of the final phase of the Cybernetic Revolution the number of doctors will decrease considerably and perhaps also their income. We mentioned above that saving the direct labor of physicians is one of the leading tasks for the future of medicine. An increase in the number of doctors due to a necessity would make a technological breakthrough more problematic, since it would dramatically increase the training requirements and increase costs.

### 8.5.2 The Medical and Biotechnological Environment as a Self-regulating Supersystem

One of the revolutionary changes during the final phase of the Cybernetic Revolution will be the emergence of health monitoring systems, in which the treatment and support of the organism will largely be carried out by autonomous systems that will be able to function regularly and constantly (we have already discussed these systems in Chaps. 6 and 7). Of course, such a system will hardly emerge immediately, but will gradually combine with already existing self-regulating and intelligent systems to form a super-complex self-regulating system. Biosensors and biochips will become essential elements of such a system.

Biosensors and biochips are electronic registering devices which use biological material (see Chap. 9 for details). Currently, they are actively used in medicine for diagnostics (e.g., of autoimmune diseases [Yang et al., 2019], blood tests [Chi et al., 2022], and of course in COVID-19 tests [Jiang et al., 2022]), as well as in many other areas, in particular for DNA detection (Toren et al., 2020). An example of biosensors used at home is the glucometer, a device used to determine the glucose concentration in blood. Biosensors combined in biochips are also used to measure physical activity. They are applied in production to measure various parameters: the proportions of mixture, concentration of toxins, toxic gasses, etc. They develop biosensors and nanorobots that can monitor the spread of viruses in the blood online (Cavalcanti et al., 2008).

For athletes, biosensors are already tools for controlling their physiological indicators to calculate the level of physical activity, and probably their capabilities will increase. During surgeries, biosensors can monitor the necessary parameters and prompt the surgeon to take further action.

One can easily imagine that in the future biosensors will become an integral part of human life, fulfilling the function of a constant scanner of the organism or of certain organs, and even transmitting the related information to medical centers in case of potential threats or serious deterioration of health. At the same time, intelligent computer systems will be able to monitor significant fluctuations in indicators and give recommendations on short- and long-term lifestyle habits.

So, we can forecast that in the future a system of almost total monitoring and control will be created, not only for patients, but also for different age groups (primarily the elderly, of course), as well as for people in different risk groups. This system will reach its full potential in the last third of the twentieth century. It is extremely important that this future will come gradually, step by step, as has happened with AI and ICT, and that it will unfold imperceptibly to some extent.

Speaking about future mature systems that are directed toward continuous monitoring, we do not mean a more technologically advanced medical service, nor even a more complex self-regulating monitoring system. Instead, we argue that *a specific medical, biological and technological environment (MBTE) will emerge.* Moreover, and this is a very important point, this environment will become permanent. This means that life without it may already become impossible, just as we cannot imagine life without the Internet and gadgets. Just as many hundreds of millions of people cannot live without constant use of drugs, so in the future a huge proportion of the population will not be able to live outside this MBTE for a long time.

First, such an environment will be created in medical institutions. The mini-systems described above can be integrated into a large system which monitors a large number of people, for example in medical centers, therapeutic facilities, hotels, etc. We have already mentioned the decreasing number of hospitals, partly due to online remote monitoring, which significantly reduces the need for patients to visit or stay in hospitals. One can imagine that such systems will be able to detect potentially dangerous situations and quickly respond to critical situations.

Then this environment will spread to individuals who will already be well prepared for it through the diffusion of smart devices that constantly monitor their bodies and biorhythms. Millions upon millions of people will live within this environment. Gradually, it will expand and, apparently, merge with the system of the so-called smart houses. The programs that make personalized recommendations for individuals will become a reality. Built-in sensors will control and regulate all vital processes, suggesting the time and dosage of medication, the time of physical activity, noting the required exercises, which vary according to personal needs, and recommending the most appropriate diet, and so on.

This system, based on new technologies, will lead to breakthroughs in many different areas of health: medicine, care, the continuous monitoring of people's condition (not only of the elderly and disabled), remote control of health, remote medicine, psychological support, prediction, prescriptions and emergency care (see below). Psychological counseling (both therapeutic and urgent) will also be provided by robotic systems. Such care will become essential (e.g., in the case of a suicide threat). For instance, a system that detects disturbing signals in the mood of the individuals it is observing, will be able to transfer communication with them directly to a specialist (of course, by recording the conversation and analyzing it using neuro-computer methods or other analysis). Such a system is likely to have pre-programmed recommendations and instructions on what to do in typical cases of deviations from the norm and behavioral deviations. Forecasting will allow propaedeutic testing of people to identify certain deviations, including mental ones that threaten others with outbreaks of violence (shooting, etc.). Urgent help can be provided in critical situations, such as when a delay can threaten the life of the observed individual. Technologies will enable the vast interconnected health system to carry out specific actions under certain circumstances. These may include giving an injection, providing artificial respiration, providing first aid and so on. Or to give instructions to other people (systems) to perform such actions, as well as to contact medical-technical centers urgently.

Along with the increased duration and comfort of biological life, there are likely to emerge many dangers and challenges (including cohabitation, coexistence with robots and with self-regulating systems in general, most of which will have annoying AI features). Humankind may face complicated ethical and legal problems with respect to such monitoring, since there is always the danger that the surveillance 'Big Brother' will take advantage of this situation. Of course, people's freedom of action will be restricted, because sometimes it is more difficult to resist machines than human wishes. At the same time, certain health imperatives will emerge. In fact, everyone will have their own electronic nurse (just as the children of the prosperous citizens of ancient Greece had teachers from among the slaves, and the children of nobility of landowners had teachers from among the servants). These healthcare systems may be particularly important for monitoring children and caring for sick people who are confined to their homes. If there emerge some relatively cheap multifunctional robots that can flexibly react to changes, people's lives will become much more comfortable (but in that case their independence will be reduced).

In any case, the development of such systems will face numerous real problems related to legal, moral and political difficulties, as well as the natural unwillingness to be under constant surveillance 24 h a day and to follow the prescriptions of soulless systems.

### 8.5.3 Some Characteristics of the Cybernetic Revolution and Medicine

Above, we have paid much attention to such characteristics of the Cybernetic Revolution as individualization, the choice of optimal regimes in the context of certain goals and tasks, the transition to the control over major and more fundamental processes and levels (for more details see Chap. 3). Now we will continue the analysis of its other characteristics with respect to their implementation in medicine.

The integrative role of medicine will become evident in the MANBRIC convergence. One observes an increasingly close symbiosis between medicine itself and other MANBRIC fields, as can be seen from the examples (in particular, the technologies for the creation of artificial tissues and organs). As a result, a new field of medicine (which could be closely related to some other innovative technologies) is emerging, which will then gradually affect other fields and bring about dramatic achievements. For example, the creation of an artificial cornea by the scientists at Stanford University (USA) was made possible by joint research in the field of chemistry, nanotechnologies, biology and medicine. One of the most distinctive features of modern medicine is its 'bio-related trends', which imply a wide use of approaches based on molecular and cellular biology methods. This trend is implemented, in particular, in biomedicine and nanomedicine (see Strategy... 2013; Wagner et al., 2006). This trend started to particularly intensively advance with the development of AI (Goecks et al., 2020). The drug delivery systems will also change dramatically. Nanotechnology, in particular, the nanotubes, will probably play a key role here. We will talk more about this symbiosis below in Chaps. 9 and 10.

**Further development of self-regulating systems**. As we pointed out in the previous section, the diffusion of self-regulating systems in medicine and related fields will intensify. Self-regulation/self-management is one of the most important characteristics of the Cybernetic Revolution. We also anticipate the creation of self-regulating super-systems, in particular, the future MBTE (medical-biotechnological environment) which was described in detail above.

Self-regulation and controllability of systems are manifested in many branches of medicine (e.g., the creation of artificial immunity, regulation of the cell self-destruction, reproductive technologies, artificial organs, as well as the use of sensors, robots, AI programs, various intelligent systems, etc.). It is one of the most important directions that will be realized in the 2030–2060s. Another manifestation of self-regulation will be the technological and automated control of the processes of human organs (through necessary albuminous compounds, cells, antibodies, activation of the immune system, etc.). In other words, treatment will become more targeted.[19]

**Miniaturization. Economy, Optimization of Resource Consumption and Other Characteristics**. Medicine moves toward increasing *miniaturization*. There are two meanings of the term 'miniaturization' in relation to medicine. The first refers to a trend toward constantly decreasing the size of instruments, down to micro- or even nano-size (Peercy, 2000). The second meaning identifies a trend toward a reduction of the zone of medical intervention on the human organism. This second meaning implies that the contact during surgery is focused only on the targeted epidermal layers. For example, some laser eye surgeries with the use of lasers aim at removing tissues only a few microns thick. Such operations do not require any subsequent rehabilitation.

The optimization of resource consumption will be reflected in the economy of medicines, thanks to the targeted delivery and minimized interference with the organism. There will be less need for hospital treatment, as operations will become more targeted. Consequently, the rehabilitation period will become minimal. More people will be treated at home, as the development of telemedicine is rather probable when doctors will be able to monitor the patient's indices online and issue the necessary prescriptions remotely. The cost of medical treatment should fall dramatically. Saving money (as well as resources) is one of the most important directions for a healthy economy.

---

[19] One of today's optogenetic technologies provides a good example and a general idea of how this can work. The essence of the technology is that a fragment of DNA which encodes specific membrane proteins is integrated into the genome. These light-activated proteins (from the light source implanted in the brain tissue or through transosseous luminescence) can create an ion flow inside the cell and thus lead to its activation (Saigitov, 2015).

The achievements in medicine will make a significant contribution to *the optimization of resource consumption*: first, by increasing life expectancy and second, by improving human health and productivity. As we pointed out in Chap. 6, we expect a breakthrough in life expectancy to occur in the 2060s and 2070s. Perhaps, it will increase up to ten years compared to life expectancy in the pre-COVID period in the developed countries; accordingly, the life expectancy in the developing countries will increase for no less a period, but rather longer than in the developed ones. The global increase in life expectancy and quality of life will result in huge savings of resources, yet, not in the usual economic sense, but in the sense that we may be able to significantly enhance what is most valuable to people, their health and years of life; these are completely unique and non-renewable resources.

The increase in life expectancy and, above all, the improvements in maintaining the quality of life and individuals' activity for as long as possible, in the context of the above-mentioned characteristics of the Cybernetic Revolution, require the further development of self-regulating/managing systems, individualization, the selection of optimal regimes, enormous energy savings (including emotional ones), unique experiences and a new perception of the world. One can also note that there will be an opportunity to preserve the experiences of previous generations as well as the personal contact between long-livers and their descendants. It becomes of particular importance in a situation of a rapid technological development, which may lead to a rejection of the experience and knowledge accumulated by previous generations.

## 8.6 Conclusion. Future Medicine and the Path to Cyborgization

As we pointed out in Chap. 1 (Introduction), it is obvious that many achievements in medicine will impel our civilization to a situation where more and more humans will become more or less extent cyborgs. Thus, we are following the path of development of self-regulating systems connected with medicine, which will be constituted by the elements of different origin: natural biological and artificial (in two senses: artificial biological and artificial by using other materials). All that we have discussed about artificial organs and tissues will contribute to the breakthrough in the field of production of absolutely new materials, which will expand the implementation of non-biological elements in the human body. Thus, the Cybernetic Revolution is closely associated with the process that can be called cyborgization. People with disabilities can benefit from the development of medicine and cyborgization as they will be able to significantly compensate for their disadvantages (see also Grinin & Grinin, 2020, 2021, 2022).

Neural interfaces provide new opportunities for the partial cyborgization of the disabled people. Neural interfaces are the technologies that enable interaction between an individual's nervous system and external devices (Brain-Computer

Interface). They implement the interaction between the brain (or spinal cord, nervous system) and AI through computer systems (or other devices) that can be realized via electrode contact with the skin of the head or via electrodes implanted into the brain. And with their help, it will become possible to achieve certain progress in the treatment of diseases associated with the degeneration of brain tissues or deviations in their functioning, as well as in the adaptation of the disabled or paralyzed people who are unable to move. Therefore, one may quite reasonably argue that neural interfaces could become one of the most breakthrough areas of the Cybernetic Revolution (see about them in Chap. 10, in the section on cognitive technologies).

The neural interfaces are already widely implemented. There are neural interfaces that even allow prosthetic devices to be moved by brain signals. At present we have devices that allow paralyzed people to speak, write and even use computers. For example, the famous astrophysicist Stephen Hawking used such a system. The neurosurgeons at the University of Pittsburgh School of Medicine performed a miracle when they implanted a chip in the brain Tim Hemmes. Being paralyzed, he can move a bionic prosthesis with his mind. The prosthesis has a special computer that directs neural impulses from the brain to the desired action (Pylyshyn, 2003). Global media actively discussed the news that Italian and Swedish surgeons had fitted the electrical prosthesis to 22-year-old drummer Robin Ekestam, who had lost his arm to cancer (Lepuschitz, 2010). The neural interfaces and cognitive technologies are advancing at a rapid pace (see Chap. 10).

Thus, we should be aware of the fact that these technological achievements actually do not only mean the formation of a new direction in medicine, but also a step toward the *cyborgization* of a human being and the creation of transcybernetic systems (i.e., the systems that combine elements of different nature). Of course, this can give rise to a certain and quite justified anxiety. On the other hand, expanding the opportunities for not only a long but also an active life is hardly possible without significant support for the sensory organs and other parts of the body that weaken as a result of aging and other causes. After all, contact lenses, artificial teeth, dental fillings, bones, hearing aids, artificial blood vessels, mitral valves, etc. enable hundreds of millions of people to live and work while they are still human. The same is true with respect to more complex systems and functions.

However, we consider it more of a fantasy that the human body will one day be completely replaced by non-biological material, except for the brain or the organs which support the senses (see Introduction; the well-known ideas about such a future for humankind are presented by Kurzweil, 1999, 2005; Schneider, 2009; see also in Rybalkina, 2005: 333). This will never come true. People who propose such solutions, for example, to replace supposedly less durable and comfortable biological material by technological inventions (e.g., the replacement of haematocytes by billions of nanorobots, etc.) in their forecasts try to use the outdated logic that was widespread several decades ago in science fiction or horror stories: the replacement of biological organisms with technical ones. The modern logic of scientific and technological progress, including the latest achievements in bioengineering, shows a movement toward the synthesis of biological forms and technical solutions into a unified system. Technical achievements can hardly replace the biological mechanisms which have been selected over many millions of years. On the contrary, we should follow the path of 'repair', improvement, the development of self-regulation and support of biological mechanisms by some technical solutions.

The human brain is very closely connected with the body and sensory organs, and most of its functions are based on the control of the body, which does not imply an efficient functioning outside its biological basis. The opportunities provided by science and medicine to replace worn organs will increase, but the biological foundations of a human being will always exist and should prevail. If it is possible to help the human body by different means, including methods of activation of immune system, opportunities of genetic repair, finding methods of blocking or slowing down the aging process, etc., it is much more reasonable to preserve the human biological foundation. As we have mentioned in Chap. 1, it is more probable that the process of cyborgization will always remain "supplementary" to the biological component of the organism, capable of significantly improving the quality of life and prolonging it. In any case, in the nearest decades quite radical breakthroughs are possible with respect to cyborgization, but nevertheless cyborgization is unlikely to go too far (Grinin & Grinin, 2020, 2021, 2022).

Let us conclude this chapter with a paragraph on the issue which to a certain extent continues the theme of cyborgization, namely, the improvement of individuals' natural abilities by means of medical (especially genetic) and other technologies. We should note that, at present, all these technologies are aimed at restoring the lost functions of the individual. This does not exclude the future possibility that this direction will provide an opportunity to move toward the improvement of natural, intellectual and physical abilities beyond the natural limits. However, in fact this may hardly happen before the end of the twenty-first century. Probably, the process will be similar to the process in the field of plastic surgery, which started out as a way of repairing damaged tissues, but then became a beauty industry. Nevertheless, the desire to become a superman may be fraught with unknown dangers, and it should be treated with caution, as some qualities may cause the deterioration of others. In addition, the growing number of technological means already noticeably increases both the level of human perception of the world and the scale of human influence on it.

In this context, it is worth mentioning the work (Lagasse & Levin, 2023) which places the focus of future medicine (biomedicine in symbiosis with ICT and AI) on launching mechanisms of tissue and organ regeneration. Although the authors' view is very simplistic (the complexity of the body does not allow it to be controlled from the outside without huge problems), it is much more preferable than the ideas of transhumanists and other supporters of total cyborgization.

Living bodies are "machines" not in the naive mechanistic sense, which deprives them of obvious intelligence and free will, but in the sense that a "machine" is something that can be rationally understood, repaired, improved and, if necessary, recreated.

Possessing tissues and organs with innate collective intelligence opens a new path for biomedicine—regenerative medicine. Regenerative medicine can use natural problem-solving molecular and cellular networks to stimulate complex organ repair and tissue reprogramming. Bioelectric networks formed by somatic cells provide a very convenient interface for harnessing the collective intelligence of cells and tissues.

In their new paper, Levine and Lagasse (2023):

- formulate the concept of transformational regenerative medicine using tissue intelligence;
- offer an option for accessing the wisdom of the body to regulate regeneration based on the integration of neurobiology and behavioral science;
- propose the use of evolutionary bioelectricity as an interface to native high-level anatomical control systems.

**Acknowledgements** This chapter has been prepared with support by the Russian Science Foundation (Project No. **23-11-00160**).

## References

Agate, S., Argyropoulos, D. S., Jameel, H., Lucia, L., & Pal, L. (2020). 3D Photoinduced spatiotemporal resolution of cellulose-based hydrogels for fabrication of biomedical devices. *ACS Applied Bio Materials, 3*(8), 5007–19.

Alemzadeh, H., Raman, J., Leveson, N., Kalbarczyk, Z., & Iyer, R. K. (2016). Adverse events in robotic surgery: A retrospective study of 14 Years of FDA data. *PLoS ONE, 11*(4), e0151470. https://doi.org/10.1371/journal.pone.0151470

Amadei, G., Handford, Ch. E., Qiu, Ch., De Jonghe, J., Greenfeld, H., Tran, M., & Martin, B. K. et al. (2022). Synthetic embryos complete gastrulation to neurulation and organogenesis. *Nature,* August 25, 1–3. https://doi.org/10.1038/s41586-022-05246-3.

Armour, B. S., Courtney-Long, E. A., Fox, M. H., Fredine, H., & Cahill, A. (2016). Prevalence and causes of paralysis—United States, 2013. *American Journal of Public Health, 106*(10), 1855–1857. https://doi.org/10.2105/AJPH.2016.303270

Arnum, P. Van. (2022). New drug approvals in 2021: The numbers and trends. *DCAT Value Chain Insights.* https://www.dcatvci.org/features/new-drug-approvals-in-2021-the-numbers-and-trends/.

Bajaj, S. (2022). Why did the first human patient to receive a pig heart transplant die? *Smithsonian Magazine.* https://www.smithsonianmag.com/science-nature/why-exactly-did-the-first-human-patient-to-receive-a-pig-heart-die-180980361/.

Barret, J. P., Gavaldà, J., Bueno, J., Nuvials, X., Pont, T., Masnou, N., Colomina, M. J., Serracanta, J., Arno, A., & Huguet, P. (2011). Full face transplant: The first case report. *Annals of Surgery, 254*(2), 252–256.

Blasco-Fontecilla, H. (2014). Medicalization, wish-fulfilling medicine, and disease mongering: toward a brave new world? *Revista Clinica Espanola, 214*(2), 104–107.

BrandEssence. (2020). *Medical aesthetics market size, share, scope, trends, and industry growth analysis by 2027.* Report Analysis 2021–2027. Brand Essence® Market Research and Consulting. https://brandessenceresearch.com/medical-device/medical-aesthetics-market.

Brown, K. (2000). The human genome business today. *Scientific American, 282*(1), 50–55.

Browne, Ed. (2022). The Story of He Jiankui, Now Out of Jail After Editing DNA of Unborn Babies. *Newsweek*, April 11.https://www.newsweek.com/he-jiankui-out-jail-editing-genome-dna-unborn-babies-crispr-1696846.

Bush, E. (2022). Pig heart transplants in humans show signs of success. *NBC News*, https://www.nbcnews.com/science/science-news/pig-heart-transplants-humans-show-signs-success-rcna37672.

Businesswire. (2022). 3DBio therapeutics and the microtia-congenital ear deformity institute conduct human ear reconstruction using 3D-bioprinted living tissue implant in a first-in-human clinical trial. *Businesswire*, June 2. https://www.businesswire.com/news/home/20220602005051/en/3DBio-Therapeutics-and-the-Microtia-Congenital-Ear-Deformity-Institute-Conduct-Human-Ear-Reconstruction-Using-3D-Bioprinted-Living-Tissue-Implant-in-a-First-in-Human-Clinical-Trial.

Cavalcanti, A., Shirinzadeh, B., Zhang, M., & Kretly, L. C. (2008). Nanorobot hardware architecture for medical defense. *Sensors, 8*(5), 2932–2958.

CBS. (2017). *Exclusive: Doctors perform heart surgery on baby still inside womb to remove tumor.* https://www.cbsnews.com/philadelphia/news/chop-doctors-fetal-heart-surgery/.

Centers for Disease Control and Prevention and The Merck Company Foundation. (2007). *The state of aging and health in America 2007*. Whitehouse Station, NJ: The Merck Company Foundation. chrome-extension: //efaidnbmnnnibpcajpcglclefindmkaj/https://www.cdc.gov/aging/pdf/saha_2007.pdf

Chen, S. (2022). Chinese scientists create AI nanny to look after embryos in artificial womb. *South China Morning Post*, https://www.scmp.com/news/china/science/article/3165325/chinese-scientists-create-ai-nanny-look-after-babies-artificial.

Chi, J., Wu, Y., Qin, F., Su, M., Cheng, N., Zhang, J., & Li, C. et al. (2022). All-printed point-of-care immunosensing biochip for one drop blood diagnostics. *Lab on a Chip*, 22, 16, August 9, 3008–14. https://doi.org/10.1039/D2LC00385F.

Chumakov, A. N., & Yurchenko, P. S. (2021). Global pandemic in terms of objective causes and subjective speculations. *Journal of Globalization Studies, 12*(2), 157–168.

Cleveland Clinic. (2021). Surgery on baby in womb removes life-threatening tumor on heart. *Cleveland Clinic*. https://my.clevelandclinic.org/patient-stories/534-surgery-on-baby-in-womb-removes-life-threatening-tumor-on-heart.

Clinical Research Fastrack. (2020). Growth of clinical trials. *Clinical Research Fastrack*. https://clinicalresearchfastrack.com/growth-of-clinical-trials/.

Conrad, P. (1992). Medicalization and social control. *Annual Review of Sociology*, 209–32.

Conrad, P. (2005). The shifting engines of medicalization. *Journal of Health and Social Behavior, 46*(1), 3–14.

Crew, B. (2020). Worth the cost? A closer look at the Da Vinci Robot's impact on prostate cancer surgery. *Nature, 580*(7804), S5–7. https://doi.org/10.1038/d41586-020-01037-w.

Demirel, P., & Mazzucato, M. (2008). The evolution of firm growth dynamics in the us pharmaceutical industry: Is 'Structure' in the growth process related to size and location dynamics? *IKD Working Paper, 38*(9), 1–28.

Denisova, I. V. (2021). *Sdelay menya tochno. Kak reproduktivnyye tekhnologii menyayut mir*. Individuum.

Dickert, F. L., Hayden, O., & Halikias, K. P. (2001). Synthetic receptors as sensor coatings for molecules and living cells. *The Analyst, 126*, 766–771.

Efremov, I., & Arkhangelsky, V. (2023). Global fertility changes in the context of the COVID-19 Pandemic. *Journal of Globalization Studies, 14*(2), 165–170.

Elrod, J. K., & Fortenberry, J. L. (2017). Centers of excellence in healthcare institutions: What they are and how to assemble them. *BMC Health Services Research, 17*(1), 15–24.

ESMO. (2018). *Man against machine: Artificial intelligence is better than dermatologists at diagnosing skin cancer*. Annals of Oncology Press Release. https://www.esmo.org/newsroom/press-releases/artificial-intelligence-skin-cancer-diagnosis.

Evans, H., & Shapiro, M. (1997). Viruses. In: Manual of Techniques in Insect Pathology (pp. 17–53). Elsevier.

Fauser, B. (2019). Towards the global coverage of a united registry of IVF outcomes. *Reprod Biomed Online* № 2(38).

Frenzel, A., Hust, M., & Schirrmann, Th. (2013). Expression of recombinant antibodies. *Frontiers in Immunology, 4*, 217.

Fukuyama F. (2002). *Our post-human future: consequences of the bio-technology revolution*. Farrar, Straus, and Giroux, New York.

Gibbs, R. A. (2020). The human genome project changed everything. *Nature Reviews Genetics, 21*(10), 575–576.

Goecks, J., Jalili, V., Heiser, L. M., & Gray, J. W. (2020). How machine learning will transform biomedicine. *Cell, 181*(1), 92–101.

Gongalsky, M. B., Osminkina, L. A., Pereira, A., Manankov, A. A., Fedorenko, A. A., Vasiliev, A. N., Solovyev, V. V., et al. (2016). Laser-synthesized oxide-passivated bright si quantum dots for bioimaging. *Scientific Reports, 6*(1), 24732. https://doi.org/10.1038/srep24732

Grinin, L. (2020). How can COVID-19 change geopolitics and economy? *Journal of Globalization Studies, 11*(2), 121–134.

Grinin, A., & Grinin, L. (2020). Crossing the threshold of cyborgization. *Journal of Big History, 4*(3), 54–65.

Grinin, A. L., & Grinin, L. E. (2021) Are we on the threshold of cyborgization? In L. E. Grinin, I. V. Ilyin, & A. V. Korotayev (eds.), *Globalistics and Globalization Studies: Current and Future Trends in the Big History Perspective* (pp. 355–368). Volgograd: 'Uchitel' Publishing House.

Grinin, A. L., & Grinin, L. E. (2022). Cyborgization: to Be or not to Be? In L. E. , Grinin & A. V. Korotayev (eds.), Evolution: Trajectories of Social Evolution (pp. 187–202). Volgograd: 'Uchitel' Publishing House.

Grinin, L., Grinin, A., & Korotayev, A. (2022). COVID-19 pandemic as a trigger for the acceleration of the cybernetic revolution, transition from E-Government to E-State, and change in social relations. *Technological Forecasting and Social Change, 175*, 1–15. https://doi.org/10.1016/j.techfore.2021.121348

Gritsenko, D., & Maleva, Y. (2023). Kletochnoye vremya: rossiyskiye uchenyye izobreli «protez» dlya pecheni. Kak zastavit' sobstvennyye kletki organizma vosstanovit' bol'noy organ.. *Izvestiya* 09.08.2023. https://iz.ru/1556127/denis-gritcenko-iuliia-maleva/kletochnoe-vremia-rossiiskie-uchenye-izobreli-protez-dlia-pecheni

Harney, A. (2018). China orders investigation after scientist claims first gene-edited babies. *Reuters*, sec. Healthcare & Pharma. https://www.reuters.com/article/us-health-china-babies-genes-idUSKCN1NV19T.

Hsieh, T.-Ch, Bar-Haim, A., Moosa, Sh., Ehmke, N., Gripp, K. W., Pantel, J. T., & Danyel, M. et al. (2022). GestaltMatcher facilitates rare disease matching using facial phenotype descriptors. *Nature Genetics, 54*(3), 349–57. https://doi.org/10.1038/s41588-021-01010-x

Hussain, M. (2021). The changing dynamics of liberal international order: Covid-19 and its global implications. *Journal of Globalization Studies, 12*(2), 143–156.

Hussein, S. (2022). Spinal implant enables paralyzed man with severed spine to walk again: ScienceAlert. *ScienceAlert*. https://www.sciencealert.com/implant-allows-man-with-severed-spine-to-walk-again.

Illich, I. (1975). The medicalization of life. *Journal of Medical Ethics, 1*(2), 73–77.

Irshad, S. M. (2020). COVID-19 and Global 'At-Risk Community': From benefit-sharing to risk-sharing of economic crisis. *Journal of Globalization Studies, 11*(2), 135–145.

ISAPS. (2021). Aesthetic plastic surgery national databank. *STATISTICS* 2020–2021. https://cdn.theaestheticsociety.org/media/statistics/2021-TheAestheticSocietyStatistics.pdf

Jasiński, A. M., & Bąkowska, A. M. (2020). Globalization and life satisfaction of the polish population. Before and during the COVID-19 Pandemic. *Journal of Globalization Studies, 11*(2), 146–155.

Jiang, F., Xiao, Z., Wang, T., Wang, J., Bie, L., Saleh, L., Frey, K., Zhang, L., & Wang, J. (2022). Rapid and sensitive multiplex detection of COVID-19 antigens and antibody using electrochemical immunosensor-/aptasensor-enabled biochips. *Chemical Communications, 58*(52), 7285–7288. https://doi.org/10.1039/D2CC01598F

Kakkar, S. (2021). The outbreak of sars-cov-2 (Covid-19): Impact on international tourism. *Journal of Globalization Studies, 12*(1), 145–158.

Kelly, S. (2020). Robotic surgeries surge to 15% of All procedures, despite limited evidence. *MedTech Dive*. https://www.medtechdive.com/news/robotic-surgeries-surge-to-15-of-all-procedures-despite-limited-evidence/570370/.

Kondratieff, V. B. (2011). Globalnaya farmatsevticheskaya promyshlennost. *Perspektivy,* July 18. http://www.perspektivy.info/book/globalnaja_farmacevticheskaja_promyshlennost_2011-07-18.htm.

Konnikova, M. R., Cherkasova, O. P., Nazarov, M. M., Vrazhnov, D. A., Kistenev, Y. V., Titov, S. E., Kopeikina, E. V., Shevchenko, S. P., & Shkurinov, A. P. (2021). Malignant and benign thyroid nodule differentiation through the analysis of blood plasma with terahertz spectroscopy. *Biomedical Optics Express, 12*(2), 1020–1035. https://doi.org/10.1364/BOE.412715

Kostina, G. (2013). Pokoleniye R. *Ekspert*, 63–65.

Kurzweil, R. (1999). *Age of spiritual machines*. Penguin Books.

Kurzweil, R. (2005). *The singularity is near: When humans transcend biology*. New York, NY: Viking Penguin.

Lagasse, E., & Levin, M. (2023). Future medicine: From molecular pathways to the collective intelligence of the body. *Trends in Molecular Medicine, 29*(9), 687–710. https://doi.org/10.1016/j.molmed.2023.06.007

Landman, B. (2023). Human cadaver tissue could be the secret to looking younger. *NYPost* 19.06.2023. https://nypost.com/2023/06/19/human-cadaver-tissue-could-be-the-secret-to-looking-younger/

Leigh, J. P., Tancredi, D., Jerant, A., & Kravitz, R. L. (2010). Physician wages across specialties: Informing the physician reimbursement debate. *Archives of Internal Medicine, 170*(19), 1728–1734.

Lepuschitz, R. (2010). BCIS: The new ontological dimension of cyberspace. *Masaryk UJL & Tech., 4*, 147.

Lwoff, A., Horne R., & Tournier, P. (1962). A system of viruses. In: *Cold Spring Harbor Symposia on Quantitative Biology*, 27, 51–55. Cold Spring Harbor Laboratory Press.

MacKimm, D. (2022). Unborn baby undergoes heart surgery in womb after doctors detect tumor. *ABC4* Utah (blog), https://www.abc4.com/news/unborn-baby-undergoes-heart-surgery-in-womb-after-doctors-detect-tumor/.

Marchione, M. (2017). AP exclusive: US scientists try 1st gene editing in the body. *AP NEWS*, https://apnews.com/article/technology-business-health-genetic-frontiers-science--4ae98919b52e43d8a8960e0e260feb0a.

Marescaux, J., Dallemagne, B., Perretta, S., Wattiez, A., Mutter, D., & Coumaros, D. (2007). Surgery without scars: report of transluminal cholecystectomy in a human being. *Archives of Surgery, 142*(9), 823–826. https://doi.org/10.1001/archsurg.142.9.823

Martyushev-Poklad, A. (2015). Chto proiskhodit s meditsinoy: protokol vskrytiya (1). *Aftershock*. URL: http://aftershock.su/?q=node/300253.

McGee, G. (1997). *The perfect baby: A pragmatic approach to genetics*. Lanham, Md.: Rowman and Littlefield.

McKie, R. (2015). German woman aged 65 gives birth to quadruplets after IVF treatment. *The Guardian,* May 23. URL: http://www.theguardian.com/world/2015/may/23/annegret-raunigk-german-woman-birth-quadruplets-aged-65.

Medical Error Statistics. (2022). Medical error statistics [2020]: Deaths/Year & Malpractice Rates. *My Medical Score*. Accessed September 8, 2022. https://mymedicalscore.com/medical-error-statistics/.

Minger, S. L. (2006). Regenerative medicine. *Regenerative Medicine, 1*(1), 1–2.

Mirsky, M. B. (2010). *Istoriya meditsiny i khirurgii*. GEOTAR-Media.

Moskalev, A. A. (2018). *120 let zhizni—tol'ko nachalo. Kak pobedit' stareniye?* Moskva: Eksmo.

Moynihan, R., Doran, E., & Henry, D. (2008). Disease mongering is now part of the global health debate. *PLoS Medicine, 5*(5), e106. https://doi.org/10.1371/journal.pmed.0050106

Mukherjee, S. (2022). *The song of the cell: An exploration of medicine and the new human*. Simon and Schuster.
Murray, C. J. L., & Lopez, A. D. (2013). Measuring the global burden of disease. *The New England Journal of Medicine, 2013*(369), 448–457. https://doi.org/10.1056/NEJMra1201534
Murray, C. J. L., Salomon, J. A., Colin, D. M., & Lopez, A. D. (Eds.). (2002). *Summary measures of population health: Concepts, ethics*. World Health Organization.
Murray, C. J. L., Ezzati, M., Flaxman, A. D., et al. (2012). GBD 2010: Design, definitions, and metrics. *Lancet, 2012*(380), 2063–2066. https://doi.org/10.1016/S0140-6736(12)61899-6
NIH—National Institutes of Health. (2021). *Annual report to the nation: Rapid decrease in lung cancer and melanoma deaths lead overall continued decline in cancer death rate*. https://www.nih.gov/news-events/news-releases/annual-report-nation-rapid-decrease-lung-cancer-melanoma-deaths-lead-overall-continued-decline-cancer-death-rate.
Njagi, P., Groot, W., Arsenijevic, J., Dyer, S., Mburu, G., & Kiarie, J. (2023). Financial costs of assisted reproductive technology for patients in low-and middle-income countries: a systematic review. *Human Reproduction Open* (2), hoad007.
NCHS—National Center for Health Statistics. (2021). *Drug overdose deaths in the U.S. Top 100,000 Annually for Immediate Release*: *November 17, 2021*. https://www.cdc.gov/nchs/pressroom/nchs_press_releases/2021/20211117.htm
Null, G., Dean, C., Feldman, M., Rasio, D., & Smith, D. (2004). *Death by Medicine*. URL: http://www.whale.to/a/null9.html.
OECD. (2022). *Health expenditure and financing*. https://stats.oecd.org/Index.aspx?DataSetCode=SHA.
Palmisano, M. E., Gaffga, A. M., Daigle, J., Brinkman, J., Mire, K., Lenczyk, K., Martin, D. H., & Hagensee, M. E. (2003). Detection of human papillomavirus DNA in self-administered vaginal swabs as compared to cervical swabs. *International Journal of STD & AIDS, 14*(8), 560–567. https://doi.org/10.1258/095646203767869183
Peercy, P. S. (2000). The drive to miniaturization. *Nature, 406*(6799), 1023–1026.
Petrov S. V. (2014). (ed.) *Obshchaya khirurgiya*. Manual. GEOTAR-Media, Moscow.
Pile, D. (2013). Artificial antibodies. *Nature Photonics, 7*(1), 3–3. https://doi.org/10.1038/nphoton.2012.354
Pinkerton, S. (2013). The pros and cons of robotic surgery. *Wall Street Journal*, November 17. https://online.wsj.com/article/SB10001424052702304655104579163430371597334.html.
Prillaman, Mc. K. (2022). Prehistoric Child's amputation is oldest surgery of its kind. *Nature, 609*(7927), 453–454. https://doi.org/10.1038/d41586-022-02849-8
Pylyshyn, Z. W. (2003). *Seeing and visualizing: It's not what you think*. Mit Press.
Raff, M. (1998). Cell suicide for beginners. *Nature, 396*, 119–122.
Ridley, M. (1996). *The origin of virtue*. New York, NY: Viking.
Ren, X., Ellis, B. W., Ronan, G., Blood, S. R., DeShetler, C., Senapati, S., March, K. L., et al. (2021). A multiplexed ion-exchange membrane-based MiRNA (MIX·miR) detection platform for rapid diagnosis of myocardial infarction. *Lab on a Chip, 21*(20), 3876–3887. https://doi.org/10.1039/D1LC00685A
Romanis, E. Ch. (2018). Artificial womb technology and the frontiers of human reproduction: Conceptual differences and potential implications. *Journal of Medical Ethics, 44*(11), 751–55.
Rybalkina, M. (2005). *Nanotekhnologii dlya vsekh*. Nanotechnology News Network.
Saigitov, R. (2015). Velikaya stagnatsiya meditsiny. *Lenta.ru*, April 11. http://lenta.ru/articles/2015/04/11/biotech/.
Saijo, H., Kanno, Y., Mori, Y., Suzuki, S., Ohkubo, K., Chikazu, D., Yonehara, Y., Chung, U.-I., & Takato, T. (2011). A novel method for designing and fabricating custom-made artificial bones. *International Journal of Oral and Maxillofacial Surgery, 40*(9), 955–960.
Schirhagl, R., Qian, J., & Dickert, F. L. (2012). Immunosensing with artificial antibodies in organic solvents or complex matrices. *Sensors & Actuators: b. Chemical, 173*, 585–590.
Schneider, S. (2009). Science fiction and philosophy: From time travel to superintelligence, 1st Edn Malden, MA: Wiley-Blackwell. ?

Sciencedaily. (2018). *More than 8 million babies born from IVF since the world's first in 1978*. https://www.sciencedaily.com/releases/2018/07/180703084127.htm.

Seifabadi, M. S., & Dolatabadi, A. B. (2022). Covid-19, Globalization and strengthening authoritarianism in the middle east. *Journal of Globalization Studies, 13*(1), 135–157.

Siemionow, M. (2017). The decade of face transplant outcomes. *JOurnal of Materials Science: Materials in Medicine, 28*(5), 1–5.

Simon, S. (2020). *Facts & figures 2020 reports largest one-year drop in cancer mortality*. https://www.cancer.org/latest-news/facts-and-figures-2020.html.

Simonstein, F. (2006). Artificial reproduction technologies (RTs)—all the way to the artificial womb? *Medicine, Health Care and Philosophy, 9*(3), 359–365.

Sindyashkina, E. N. (2022). Healthy life expectancy in the context of the united nations decade of healthy ageing. *Analysis and Forecasting. IMEMO Journal, 2022*(1), 40–53.

Singh, S., Tank, N. K., Dwiwedi, P., Charan, J., Kaur, R., Sidhu, P., & Chugh, V. K. (2018). Monoclonal antibodies: A review. *Current Clinical Pharmacology, 13*(2), 85–99. https://doi.org/10.2174/1574884712666170809124728

Slagboom, P. ES., Droog, S., & Boomsma, D. I. (1994). Genetic determination of telomere size in humans: A twin study of three age groups. *American Journal of Human Genetics, 55*, 876–882.

Smith, M. W. (1979). A guide to the delineation of medical care regions, medical trade areas, and hospital service areas. *Public Health Reports., 94*(3), 248.

Solé, R., & Elena, S. F. (2018). Viruses as complex adaptive systems. In: *Viruses as Complex Adaptive Systems*. Princeton University Press.

Statista. (2022a). Total drug prescriptions dispensed U.S. 2009–2021. *Statista*. https://www.statista.com/statistics/238702/us-total-medical-prescriptions-issued/.

Statista. (2022b). Pharmaceutical market revenue forecast 2012–2022. *Statista*. https://www.statista.com/statistics/817562/revenue-forecast-for-global-pharma-market/.

Stein, L. D. (2004). Human genome: End of the beginning. *Nature, 431*(7011), 915–916.

Strategy of Development. (2013). *Strategiya razvitiya meditsinskoy nauki v Rossiyskoy Federatsii na period do 2025 goda*. http://rosminzdrav.ru/health/62/Strategiya_razvitiya_meditcinskoj_nauki.pdf.

Sung, H., Ferlay, J., Siegel, R. L., Laversanne, M., Soerjomataram, I., Jemal, A., & Bray, F. (2021). Global cancer statistics 2020: GLOBOCAN estimates of incidence and mortality worldwide for 36 cancers in 185 countries. *CA: A Cancer Journal for Clinicians, 71*(3), 209–249. https://doi.org/10.3322/caac.21660.

Tang, D., & Kang, R. (2022). Glimmers of hope for targeting oncogenic KRAS-G12D. *Cancer Gene Therapy*, 1–3.

Taylor, R. H. (1997). Robots as surgical assistants: Where we are, wither we are tending, and how to get there. artificial intelligence in medicine. *Lecture Notes in Computer Science* 1211: 1–11. Springer Berlin Heidelberg. URL: http://link.sprin ger.com/chapter/https://doi.org/10.1007/BFb0029430.

Toren, P., Smolka, M., Haase, A., Palfinger, U., Nees, D., Ruttloff, S., Kuna, L., et al. (2020). High-throughput roll-to-roll production of polymer biochips for multiplexed DNA detection in point-of-care diagnostics. *Lab on a Chip, 20*(22), 4106–4117. https://doi.org/10.1039/D0LC00751J

Wagner, V., Dullaart, A., Bock, A.-K., & Zweck, A. (2006). The emerging nanomedicine landscape. *Nature Biotechnology, 24*(10), 1211–1217.

Wang, H., Dwyer-Lindgren, L., Lofgren, K. T., et al. (2012). Age-specific and sex-specific mortality in 187 countries, 1970–2010: A systematic analysis for the global burden of disease study 2010. *Lancet, 380*, 2071–2094. https://doi.org/10.1016/S0140-6736(12)61719-X

Wasden, Ch., & Williams, B. (2012). *Owning the disease: A new transformational business model for healthcare*. http://pwc.com/ie/pubs/2012_new_transformational_business_model_for_healthcare.pdf.

Werner, K. M., & Mercurio, M. R. (2021). Ethical considerations in the use of artificial womb/placenta technology. *Seminars in Perinatology*, 151521. Elsevier,

WHO. (2020). *The top 10 causes of death*. https://www.who.int/news-room/fact-sheets/detail/the-top-10-causes-of-death.

WHO. (2023a). *Disability factsheet 2023*. WHO. https://www.who.int/news-room/fact-sheets/detail/disability-and-health#:~:text=An%20estimated%201.3%20billion%20people%20%E2%80%93%20or%2016%25%20of%20the%20global,experience%20a%20significant%20disability%20today.

WHO. (2023b). *WHO mortality database*. Geneva: WHO. https://platform.who.int/mortality

Widdowson, M. (2021). From Covid-19 to zero-gravity: Complex crises and production revolutions. *Journal of Globalization Studies, 12*(1), 117–144.

Williams, S. (2014). *3 Pharma giants growing significantly faster than Glax-oSmithKline PLC*. http://www.fool.com/investing/general/2014/12/26/3-pharma-giants-growing-significantly-faster-than.aspx. Date accessed: 2.01.2015.

Willyard, C. (2016). Cancer therapy: An evolved approach. *Scientific American* https://www.scientificamerican.com/article/cancer-therapy-an-evolved-approach/.

Woollett, G. R. (2012). Innovation in Biotechnology: Current and future states. *Clinical Pharmacology and Therapeutics, 91*(1), 17–20.

World Bank. (2011). *New world report shows more than 1 billion people with disabilities face substantial barriers in their daily lives*. https://www.who.int/mediacentre/news/releases/2011/disabilities_20110609/en/.

Xu, N., Ye, X., Wei, D., Zhong, J., Chen, Y., Xu, G., & He, D. (2014). 3D artificial bones for bone repair prepared by computed tomography-guided fused deposition modeling for bone repair. *ACS Applied Materials & Interfaces, 6*(17), 14952–63

Yang, A., Xuan, R., & Murrell, D. F. (2019). A new indirect immunofluorescence BIOCHIP method for the serological diagnosis of bullous pemphigoid: A review of literature. *Australasian Journal of Dermatology, 60*(3), e173–e177. https://doi.org/10.1111/ajd.13034

Yong, Ed. (2014). Cancer biomarkers: Written in blood. *Nature, 511*(7511), 524–26.https://doi.org/10.1038/511524a

Yu, J., & Hu, S. (2021). On the ultimate finishing line of the human genome project. *The Innovation, 2*(3).

Yudin, B. G. (2008). Meditsina i konstruirovaniye cheloveka. *Znanie, ponimanie, umenie,* 1, 12–20. http://cyberleninka.ru/article/n/meditsina-i-konstruirovanie-cheloveka.

Zhokhova, A. (2011). My sdelayem vam krasivo. *Forbes*, June 3. http://m.forbes.ru/article.php?id=69681.

# 9 Biotechnologies in Perspective: Major Breakthroughs, Development of Self-regulating Systems and Possible Social Confrontations

**Abstract**

In this chapter Grinin et al. concentrate on describing the current and future transformations in biotechnology in different spheres and its fusion with medicine and other MANBRIC technologies in terms of their forecasted huge impact on our future. The authors anticipate that in four decades our life will be absolutely impossible without biotechnologies. In particular, they have great hopes that biotechnologies: (a) will make a huge contribution to the victory over incurable and other diseases, including viral ones; (b) will enable a large number of people suffering from genetic abnormalities, dysfunctions of internal organs or tissue damage to return to a more active life with the help of artificial tissues and organs; (c) will greatly contribute to the fight against aging; (d) will considerably mitigate environmental problems, especially in cities, as well as facilitate the solution of the problem of biodiversity; (e) will become the leading sector in tackling the problem of food supply; (f) will significantly change technological and production processes, allowing waste-free production; and (g) will expand our ability to strategically manage natural processes. However, as COVID-19 pandemic has demonstrated, we could expect a great trouble if we will not control biotechnologies.

## 9.1 What This Chapter is About

In this chapter we concentrate on describing the current and future transformations in biotechnology in different spheres and its fusion with medicine and other MANBRIC technologies in terms of their forecasted huge impact on our future. The role of biotechnology has been increasing since the 1940s. And it proves essential at present.

We suggest that particularly breakthrough innovations will probably result from the synthesis of biotechnologies with other MANBRIC technologies (in particular,



L. Grinin et al., *Cybernetic Revolution and Global Aging*, World-Systems Evolution and Global Futures, https://doi.org/10.1007/978-3-031-56764-3_9

with medicine, nanotechnologies and AI), as well as within the synthetic direction of innovation—bionics. The chapter places particular emphasis on the probable breakthroughs in biotechnology and genetic engineering related to medicine (including the opportunities for creating artificial organs and tissues).

One distinguishes four most important fields of biotechnology, which are often classified using color codes. Red biotechnology is a health-related branch, white biotechnology is linked with application of biotechnology to the industrial sector, for example, biocatalysis or fermentation; green biotechnology is related to food and environment-related areas; and blue is biotechnology which uses marine organisms and resources.

The chapter describes each of these areas, as well as bioeconomy, and makes extended projections about the possible future developments of every branch. Much attention is given to other fields, in particular to approaches to improving the environment, the development of aquaculture and agriculture and solutions for the global food problem.

The chapter pays much attention to modern achievements, problems and challenges linked with biotechnology, and also shows possible ways to overcome them and deal with the dangers that may arise as a result of the active implementation of these achievements in the future (based on the experience of the anti-COVID campaign and the fight against GMOs). Biotechnologies will play a crucial role during the final phase of the Cybernetic Revolution (in the 2030s–2070s), especially with respect to synthesis with the medical field and development of pharmaceuticals, they will become an essential part of medical technologies.

However, the role of biotechnology will also be vital in a number of other industries and fields. Already today biotechnologies provide numerous and vivid examples of self-regulating systems, but the development of such systems in the future will become even more notable (e.g., in genetic engineering, self-regulation of artificial physiological systems, production of artificial tissues, etc.). Moreover, in the future, the creation of huge databases of plant varieties and variations will allow the whole process of getting a transgenic plant to proceed without human participation, thus, to become self-regulated. So, one may assume that by the end of the final phase of the Cybernetic Revolution, agricultural biotechnologies will be already developed to the degree that modified products will be able to respond even to the smallest fluctuations of local conditions.

We anticipate that the role of biotechnology will not only increase in the future, but its branches will become extremely important for human life and environment. In particular, we have great hopes that biotechnologies (and bioeconomy in general):

(a) will make a huge contribution to the victory over incurable and other diseases, including viral ones;
(b) will enable a large number of people suffering from genetic abnormalities, dysfunctions of internal organs or tissue damage to return to a more active life with the help of artificial tissues and organs;
(c) will greatly contribute to the fight against aging;

(d) will considerably mitigate environmental problems, especially in cities, as well as facilitate the solution of the problem of biodiversity;
(e) will become the leading sector in tackling the problem of food supply (similar as it solved it to a large extent during the "green revolution" in the initial phase of the Cybernetic Revolution);
(f) will significantly change technological and production processes, allowing waste-free production; and
(g) will expand our ability to strategically manage natural processes.

In a word, we believe that in three or four decades our life will be absolutely impossible without biotechnologies.

## 9.2 Biotechnology. An Introduction

There are several dozen definitions of biotechnology. It is not surprising, since biotechnology is a broad and multifaceted subject, which has been changing with its development. Nevertheless, there are useful official international definitions, for example, by the UN: 'Biotechnology represents a complex sphere of activity. The basis of it is made by the complex of methods giving a chance to the person to purposefully change the structure of … DNA, or genetic material, plants, animals and microorganisms as a result of obtaining useful products and technologies' (UN, 1992: ch. 16).

We interpret *biotechnology as a range of scientific and industrial methods for producing various products by means of using living organisms, biological processes and related self-regulating systems.* It is important that even though biotechnology constantly develops and expands its scope, the genetic engineering has become its major trend and unites many different biotechnological fields (as we see from the definition above).

As we have mentioned in Chap. 4, many scholars consider biotechnology as a breakthrough sector of the nearest future. There are discussions about a coming biotechnology revolution and a paradigm shift and so on (Sylvester & Klotz, 1983; Fukuyama, 2002; Nefiodow, 1996; see also Nefiodow & Nefiodow, 2014a, 2014b; Avise, 2004; Das, 2007; Harris, 2010; Holdrege, 2010; McCabe & McCabe, 2008; Phillips & Su, 2009; Peccoud, 2016; Bröring, 2021; see also Rifkin, 2011, 2019; McCormick & Kautto, 2013; de Besi & McCormick, 2015; Wilenius & Kurki, 2012; Wilenius & Casti, 2015; El-Chichakli et al., 2016). However, as the reader may remember, we believe (see Chaps. 3 and 4) that the final phase of the Cybernetic Revolution will be characterized by a system (or complex) of breakthrough technologies, namely, the MANBRIC-convergence, rather than a single breakthrough technology. And biotechnologies will play a crucial role as part of this system. However, medicine will remain the central and integrating element of this convergence for quite a long time.

Currently, one distinguishes the four most important fields of biotechnologies, which are often classified with color codes.

**Red biotechnology** is a health-related branch, which includes biomedicine, diagnostics, pharmaceuticals and genetics. This direction is also increasingly associated with the quality of life, and its importance is continuously growing along with the intensification of the global aging trend.

**White biotechnology** is currently often linked to the application of biotechnology in the industrial sector, for example, the use of biocatalysis or fermentation. In a certain sense, this is one of the oldest branches of biotechnology; however, it is also considered as the direction with promising prospects for mitigating the problems of pollution and climate change through the large-scale replacement of fossil resources with renewable ones (such as biogas, biofuel), as well as environmentally friendly construction materials or biodegradable chemicals and other recyclable and renewable resources and processes.

**Green biotechnology** is related to food and environment fields and is also closely connected with genetic engineering. The huge database of genetic information offers great potential for designing desirable traits in plants and animals. In some sectors, white and green biotechnologies are so closely related that it is difficult to separate them.

**Blue Biotechnology** is biotechnology which uses aquatic organisms and resources (fish, shellfish, other aquatic animals, sponges, algae etc.) for food and other purposes including water cleaning (e.g., with the help of such technologies as algae-based biopolymers and biodegradable packaging). It is an actively developing branch; however, marine farms could potentially produce several times more food than they do today.

However, since biotechnologies continue to develop, the researchers have different approaches in attributing biotechnologies to one or another branch (but we do not dwell on this issue here; see, e.g. DaSilva, 2004). They also use other colors to differentiate, e.g., yellow (or insect) biotechnology is related to food production, when microorganisms and active genes in insects are used in agriculture, medicine, perfumery etc. (see below).

**Bioeconomy**. Biotechnologies have provided the basis for a fairly powerful industry. Since there are great hopes on white and green biotechnologies with respect to the climate and ecology preservation, they attract growing amounts of investments, and so their production volumes are increasing. It is no coincidence that in the last decade, people have been talking not so much about biotechnologies, but about the bioeconomy, seeing the latter as a direction that can change the developmental paradigm and create the conditions for a system-wide transformation toward sustainability, paving the way for a new era (de Besi & McCormick, 2015; El-Chichakli et al., 2016; McCormick & Kautto, 2013; OECD, 2009; UN, 2020; Wilenius & Casti, 2015; Wilenius & Kurki, 2012). Bioeconomy is considered as one of the major priority areas for the development of the European Union, which even elaborated Strategic goals and related prospects for the development of bioeconomy until 2050 (see EC, 2018, 2019). Undoubtedly, biotechnology and bioeconomy are closely interrelated. However, bioeconomy includes both innovative and already routine production sectors. Meanwhile, speaking about the future, we primarily mean innovative and breakthrough technologies. Therefore, in the present chapter we mainly focus on biotechnologies.

## 9.3 Development of Biotechnology

Although biotechnology is a rather new field, the history of 'traditional' microbiological production goes back to the Stone Age: yeast bread, yogurt, beer, wine and vinegar have been produced since ancient times. The first scientific foundations of biotechnology were laid by Louis Pasteur who discovered fermentation. In the late nineteenth and early twentieth centuries, microbiological knowledge was actively accumulated and found more and more practical applications in many fields. In the late nineteenth and early twentieth centuries, biofertilizers and biological preparations for pest control and against plant diseases were developed. Experiments were also carried out in bioconversion (Glick & Pasternak, 2003; Volova, 1999). In 1917, the Hungarian engineer Károly Ereky introduced the term 'biotechnology'. The World Wars revealed an urgent necessity to organize mass production of cheap drugs, food and vitamins. Thus, the production of penicillin, which started in 1943, became one of the first mass biomedical developments.

In the first decades after World War II a large-scale production of amino acids, unicellular feed proteins (from oil and paper-pulp industry waste) and steroids was organized, and cell culture of animals and plants was mastered. The intact cells of microorganisms began to be widely used for the production of medical substances of the steroid type, and large-scale production of vaccines was organized (Volova, 1999). The production of pharmaceuticals became a successful and also very profitable business, so capital and scientific forces had been attracted to this area.

Thus, biotechnology has been actively developing since the beginning of the Cybernetic Revolution. During the last decades of the initial phase of the Cybernetic Revolution and the transition to its intermediate phase (from the 1970s to the early 1990s) biotechnology became quite a significant industrial sector, making significant contributions to agriculture (both to crop production and to animal husbandry, including veterinary medicine), the food and chemical industries, pharmaceuticals and general medicine. The breakthroughs in biotechnologies are associated, first of all, with achievements in the study of transferring hereditary information. In 1953, James Watson and Francis Crick defined the structure of the DNA molecule. The discovery of an amazing similarity of genomes of almost all living beings has become the basis for genetic manipulations. It laid the foundation for basic opportunities of a purposeful transfer of genes from one organism to another, and opened up enormous prospects. The following years witnessed a prodigious number of genome discoveries. The period from the 1970s to the 1990s was also marked by a wave of advances in molecular biology. As a result, genetic engineering became a powerful branch of biotechnology.

From the 1970s, the qualitatively new level of development of biotechnology within the framework of the Cybernetic Revolution put it on a new basis.

The period from the 1990s to the 2000s was marked by a powerful advance in biotechnology. It became a rapidly growing sector so that many countries started to invest significant resources (Glen et al., 2013). No wonder that today biotechnology is considered as a sector that can become the engine of a new innovative breakthrough.

## 9.4 Biotechnology Today: Achievements, Problems, Goals and Prospects

### 9.4.1 Achievements and Challenges

**Achievements**. The importance of biotechnologies can be proved by the wide use of their achievements in various fields. For example, it is widely applied in food industry as well as in chemical production (in particular, in production of polysaccharides, biodegradable polymers, biocatalysis and also in the creation of new materials, for example, bioplastics), in energetics, agriculture, municipal services (e.g., in waste recycling). Biotechnology is connected with medicine and pharmacology, nanotechnology, cosmetology and the military sector. Biotechnologies become common for individuals who use bio additives and vitamins, special products in a diet, certain types of cosmetics, etc.

In short, numerous achievements are evident. Let us start with green biotechnology. Its success is evidenced by the fact that even in Africa, with its rapidly growing population, crop yields are outpacing population growth. Agriculture could hardly achieve this without biotechnological innovations, including new genetic editing and new breeding techniques.[1] However, the problem of food supply, including a sufficient availability of protein, remains very pressing.

In Chap. 8, we highlighted an explosive growth of drug production, diagnostics and other areas of red biotechnology.[2] Thus, the reduced mortality from incurable diseases and a better quality of life of those who managed to recover from them are provided by the progress of biotechnology. However, in the last decade, the introduction of new drugs and pharmaceutical preparations has slowed down (see below). An exception is the emergency introduction of new types of vaccines against COVID-19 into practice (which, according to many opinions, was unjustifiably urgent).[3] We will talk more about that.

[1] The enormous achievements in breeding different varieties and crops for different soil and climatic conditions are made with the help of genetic engineering. For example, for metal resistance and salt-tolerance in plants (Grinin et al., 2010) the scientists constructed special genes-controllers which launch only under certain conditions (e.g. with a high concentration of salt in the soil).

[2] One of the evidence for this is the fact that more than a hundred different vaccines are currently being developed by dozens of companies all over the world (Corey et al., 2020; Jeyanathan et al., 2020).

[3] On social consequences of the COVID-19 pandemic see, e.g., Chumakov and Yurchenko (2021), Efremov and Arkhangelsky (2023), Grinin (2020), Grinin et al. (2022), Hussain (2021), Irshad

Particularly noteworthy are biotechnology achievements in the field of creating artificial tissues and organs (see below, as well as Chap. 8), when either artificial materials are used or animal biological tissues (organs) are processed and create organs or tissues with artificially grown human cells. There are also advances in genetic engineering and genetic modifications.

Important achievements of biotechnologies (white, green and blue) should be highlighted in the field of biodegradable materials and the removal of pollutants from soil and water.[4] However, the creation of such biodegradable materials is only a small part of the success that biotechnology has achieved in the *synthesis of new materials*, the history of which dates back to the beginning of the Cybernetic Revolution (see, e.g., about the synthesis of Humalog, which is a widely applied synthetic analogue of human insulin [Woollett, 2012]). This is reminiscent of the history of the development of chemistry: at first people learned to produce the known substances, and then to make artificial materials.

Finally, white biotechnology is associated with industrial production of various products, including biofuels whose global production amounts to over 100 million tons (mainly, in the USA, Europe and Brazil). However, the majority of biomass for biofuel (80%) comes from forestry residues (Kopetz, 2013). Meanwhile, the efforts to preserve the ecological balance of the planet and reduce the use of wood can strongly influence this alternative source of energy.

We would like to finish this paragraph, in which we have barely covered all the achievements, by mentioning a direction that has a tremendous potential. We mean the development of biosensors, i.e., miniature devices capable of not only examining, but also monitoring the body over a certain period of time. Due to their size, they can be embedded into the human body and record numerous functions in a way previously inaccessible for doctors. This will greatly facilitate diagnosis and improve the quality of treatment. We agree with the widespread opinion that biosensors, will be one of the most important directions in the development of medicine and, in general in healthcare system (see Chap. 8). We will discuss this in detail below.

**Achievements, Problems, Challenges and Objectives**. Many of the achievements of biotechnology can also be considered as problematic even to the point of dividing societies. Others can be considered as challenges, the resolution of which becomes an important goal for future development (our forecasts of how these problems can be solved in the future are presented below). And another good example here is GMOs. The anti-COVID vaccines may serve as a highly illustrative example.

(2020), Jasiński and Bąkowska (2020), Kakkar (2021), Seifabadi and Dolatabadi (2022), Widdowson (2021).

[4] The range of bioplastic products is already very large. From 2000 to 2008, the world consumption of compostable plastics made of starch, sugar and cellulose increased by 600 percent (Ceresana Research, 2011). However, so far, the production of oil-free plastic amounts only one–two percent of total production of plastics (Leshina, 2012). Despite difficulties, biotechnology brings hope for more nonpolluting and renewable production which in the long term will allow saving resources.

The agricultural ('green') biotechnology, which has already been involved during the initial phase of the Cybernetic Revolution is based on genetic modification (see, e.g., Borlaug, 2001). One of the most widespread and widely discussed methods of genetic modification is the transfer of genes resistant to the glyphosates, most commonly the chemical herbicide Roundup (Richard et al., 2005; Williams et al., 2000).[5] As a result, when treated with Roundup, genetically modified plants remain intact but the weeds are killed. Thus, GMOs allow a significant reduction of expenses and do not require a long process of selection. The importance of glyphosate-based herbicides (such as Roundup) has reached such a level that their complete ban would be catastrophic for the entire agricultural system of the world. It's estimated that there would be an annual loss of $6.76 billion in global farm income and a reduction in global soybean, corn and canola production equal to 18.6 million tons, 3.1 million tons and 1.44 million tons respectively. Also, there would be additional carbon emissions from increased fuel usage and decreased soil carbon sequestration, equivalent to adding 11.77 million cars to the roads (Brookes et al., 2017). However, there is a lot of discussion about the negative impact of glyphosates on human health, and it seems there are no easy ways to resolve this issue (see for example Peillex & Pelletier, 2020).

Despite various restrictions, the overall production of GMO is quickly growing. Since 2010, the area under GMO cultivation in developing countries has exceeded that in developed countries (Clive, 2011). The analysis of the global economic effect of biotech crops shows the growing profit from the reducing production costs (up to 50%) and increasing opportunities to grow larger crops on the same agricultural land. In addition, a significant increase in crop yields could be supported by removing restrictions on the distribution of GMOs, in such a case the world GDP could grow by several hundred billion US dollars. The movement against cloning, GMOs, computer selection, etc. has been already spreading in different countries. Such a reaction is quite natural, legitimate and in many respects useful though it may happen that conservatism could suppress progress. Within the framework of these struggles and clashes, a balanced decision could be found which will become important in the long term and give an impetus to the development.

We need to consider the manifest and potential problems connected with GMOs and the use of nitrogen-containing herbicides associated with them. It is clear that the environmental damage resulting from the widespread use of nitrogen-containing herbicides, such as Monsanto's Roundup, is very great.[6] And its active substance, glyphosate, accumulates not only in soil, water, etc., but also in the human body. No wonder there are over 11,000 lawsuits in the USA alone against the company that makes Roundup herbicide. However, in comparison with the

[5] Roundup was developed in the 1970s by the American company Monsanto for weed control. In the 1990s, the company created genetically modified crops resistant to it—soybeans, corn and cotton. Since then, this nitrogen-containing herbicide has become the most common in the world. In the USA alone, it accounts for half of the market.

[6] However, other types of herbicides are also harmful, the range of which includes several tens of thousands of items, based on more than 1000 chemical compounds (Yagafarova et al., 2013: 153).

problem of hunger or malnutrition they may seem less important. Actually, the main harm comes not from the gene modification itself, but from the direction of this modification, in particular with regard to herbicide resistance. However, there is hope that it is biotechnology that can greatly reduce this harm.[7] Thus, the challenges facing biotechnology and society in general are to prove that GMOs are harmless to humans, or to make them harmless, or to find another way to solve the food problem, taking into account (which is very important) the expected growth in incomes of the population of the currently poor and poorest countries and associated growth in their nutrition needs. But so far, there is no other main path than genetic modification or other biotechnological innovations.[8] There are also socio-economic problems that need to be addressed; in particular, the considerable monopolization of such technologies by a few corporations (such as Monsanto[9]) and the huge dependence not only of farmers but of society as a whole since these agricultural corporations are able to influence society through the media and other means.

Biofuel can also be seen as both an achievement and a challenge. The problem is that consuming a huge amount of crops for biofuel production exacerbates the food shortage.

**Low Drug Efficacy**. We spoke above and also in Chap. 8 about the huge advances in pharmaceuticals (red biotechnology). However, the huge average costs of developing a new drug (several billion dollars) require a broad market in order to pay off these expenditures. As a result, drugs and medications are intended for an excessively wide range of consumers, that is, they are universal. And as a result, they become insufficiently effective.[10] If there was a differentiation of drugs for narrower groups of consumers (taking into account the characteristics of these groups), the drug efficacy could be much higher (but in this case companies would be unable to recoup the costs). So, there are opinions that, the expansion of pharmaceutical production is accompanied by decline in its efficiency, with drugs only suppressing symptoms rather than having a healing effect. The growth of pharmaceutical production is connected with unification which leads to decreasing efficiency since even well-investigated diseases often proceed individually. There is also a problem of the increase in serious side effects, also associated with insufficient effectiveness.

---

[7] In particular, it was found that almost all chemical compounds used as herbicides are to some extent transformed by microorganisms. Currently, a significant number of bacteria, strains of fungi, actinomycetes and algae have been isolated, which degrade these substances into non-toxic compounds. (Yagafarova et al., 2013: 153).

[8] Of course, in relation to African and some other countries there is still a very large reserve—the transition from extensive agriculture, which dominates in large regions (like the Sahel in Africa), to intensive. But this will not solve the food problem completely.

[9] Monsanto acquired more than a dozen seed companies around the world, e.g. in September 2016, Monsanto was bought by Bayer for US$ 66 billion.

[10] Not to mention such problems as bacterial resistance to antibiotics (Alanis, 2005; Spratt, 1994).

The problems of the pharmaceutical industry have been clearly manifested in the weakening of the industry's innovative capacity, particularly in the fact that the rate of patent grants in this industry has been declining since 2013 (see Fig. 9.1). Whereas it used to be comparable to the pace of development in medicine and biotechnology, and exceeded the average rate of patent grants in general, it is now well below the average rate. Since 2020, due to the COVID pandemic, the rate of patent grants in pharmaceuticals has been increasing. Still, we believe that this is not yet a trend, but a fluctuation.

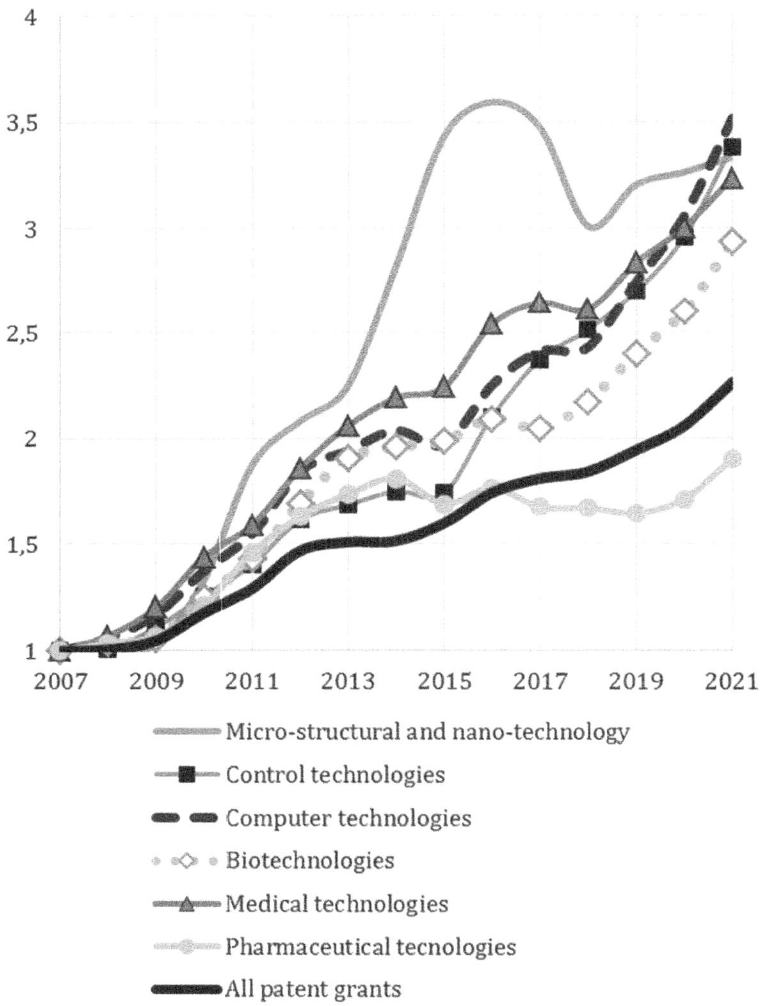


**Fig. 9.1** Relative global dynamics of patent grants for various technologies, 2007–2021, 1 = 2007 level *Data source*: WIPO (2023). (Fig. 9.1 is repeated from Chap. 4)

**The Need to Focus on Personalized Care and Resource Saving**. The effectiveness of medicine and, therefore, of medical care in general, is impossible without an active trend toward individualization/personalization of the drugs needed by patients. This process has already started and, we believe, it is inevitable, since the expanding mass production of medications and drugs leads medical care to a dead end, although it is very profitable for pharmaceutical giants. All this was vividly demonstrated by the anti-COVID vaccines: the mass scale was unprecedented, there was no individualization, and the profits were fantastic. The desire to produce drugs on a mass scale is actually similar to 'covering an area', which has never been effective but has always been resource-intensive. In modern armies, instead of a large amount of conventional ammunition, a relatively small amount of precision-guided munitions is used, and using this comparison, we can say that we need to move to precision-guided drugs and medicines.[11] There are many paths to individualization.

One of them is a new model of pharmaceutical and medical business known as 'Owning the disease: A new transformational business model for healthcare' (Wasden & Williams, 2012: 2). The idea is that a patient receives a full range of services to solve health problems connected with an actual (or potential) disease. In other words, a company supervises a patient over a long period of time, having granted this right in return for considerable financial benefits. Here we mean primarily chronic diseases such as asthma, epilepsy, dementia and hypertension. Since no single company currently offers a complete range of solutions for the treatment of any chronic disease, and given that more than 80% of healthcare costs (in the USA) are for lifelong chronic diseases, the idea is promising. That is why the company which will be able to create a platform for 'mastering the disease', will have many strategic advantages over the competitors. It is clear that we actually mean large biomedical companies.

However, the conservatism of the existing pharmaceutical and medical institutions and the huge financial interests of very influential actors behind them will certainly hamper this transition. So, the task is just to make pharmaceutical companies focus on a particular patient (rather then on mass production); improve both patient outcomes and resource saving, including reducing inconvenience and suffering for patients. Such a turn will create conditions for the emergence of many small biomedical and pharmaceutical companies, since the capture of the market by the largest corporations increasingly becomes an obstacle to development.

Medicalization, that is, the unrestrained prescription of medicines for every occasion and the recognition of all deviations as a disease, began to replace all other forms of improving people's conditions (see Chap. 8). And this leads to a huge waste of resources and harm to patients. In the USA alone, more than

[11] The present high-precision artillery systems are many times more accurate than previous guns dthanksto the AI guidance systems, and therefore, they are both more economical and more effective.

$200 billion is spent every year on unnecessary or misprescribed drugs and procedures; another 300 to 500 billion is spent on treating diseases that could be prevented through changes in lifestyle (Martyushev-Poklad, 2015). These figures may be exaggerated, but in any case they are impressive. There is also evidence (also probably overestimated) that in the late 1990s and early 2000s, more than 100,000 people died every year from improperly selected drugs in the USA (Null et al., 2003: Table 1. Estimated Annual Mortality and Economic Cost of Medical Intervention). In general, iatrogenic factors (associated with unskilled medical interventions) are one of the leading causes of death in economically developed countries (Martyushev-Poklad, 2015). The current situation is probably even more depressing in this respect.[12] Thus, all these problems point to the need to change the organization of health care and the strategies of pharmaceutical companies (see more below). This should (a) increase the effectiveness of drugs; (b) reduce the damage they cause to the organism; and (c) thereby improve the biological quality of life of the sick and the elderly.

### 9.4.2 Prospects and Social Confrontation

The prospects for biotechnologies are exciting. So far they are closely connected with microbiology, and since microorganisms are found almost everywhere, the sphere of application of biotechnologies seems limitless (from space needs to production and processing of mineral resources). Finally, since biotechnologies are closely connected with medicine and nanotechnologies, as well as with food safety, climate saving, etc. they will become one of the main spheres in which the final phase of the Cybernetic Revolution (in the 2030s–2070s) and the consecutive epochs will unfold.

However, there is an essential contradiction which becomes an obstacle on the way to this far-reaching future. The current phase of development of biotechnology is characterized by two major trends: (1) an extensive distribution of new technologies with resulting simultaneous improvements; (2) increasing social tensions and even struggles for necessary changes in some spheres of social life due to the implementation of these technologies. In order for the final phase of a production revolution to begin, the development of technologies during the modernization phase must reach a fairly high level of diversity and 'density'. Taking into account that biotechnologies are innovative branches, any country that wants to become a leader will have to develop them anyway.

On the other hand, such a wide implementation of biotechnologies will, undoubtedly, intensify public, diplomatic and economic resistance to the change

[12] The situation especially worsened during the COVID pandemic. Now we are constantly receiving more and more data on the excess mortality as well as very serious side effects in result of the vaccination. There is also another problem. This is that the overdose deaths from opioids increased to 75,673 in the 12-months period ending April 2021, up from 56,064 the year before (NCHS, 2021). The overdose deaths from opioids is also continuing to increase in 2022 and 2023.

of traditions, national features, real or imagined harm. We have discussed above the opposition to GMO. However, in 2020–2022 we witnessed a huge social confrontation caused by the compulsion to vaccinate against COVID-19 and the unwillingness of a large part of society to obey this compulsion and restrictions. This confirmed our assumptions that already 'in the nearest two decades prior to the start of the Cybernetic Revolution one may expect not only significant achievements in the field of biotechnologies, but also intensification of social struggle and fierce competition between various forces' (Grinin & Grinin, 2015: 254). We argue that this intense struggle is related to two major trends characteristic of the modernization phase of a productive revolution, that is, the period preceding the start of the final phase of the Cybernetic Revolution: (1) a wide range of applied new technologies, going hand in hand with their constant improvement; (2) an increasing public opposition to many changes. Therefore, on the one hand, in the next 15–20 years we will observe a widespread introduction of biotechnologies in our lives; on the other hand, this will undoubtedly intensify social, diplomatic and economic struggles against changes in traditions, national characteristics, real or perceived harm, etc. And while earlier the most acute struggle used to be against GMOs, it has now reached a very high level with respect to vaccines (Strandberg, 2020; Vieten, 2020; Brennan, 2020; Kowalewski, 2020; Carothers & Press, 2020; Armed Conflict Location & Event Data Project, 2020; Louwerse et al., 2021; Bonnevie et al., 2021; for more details see Chap. 14 of the present volume, as well as Grinin et al., 2022). One should note that with regard to COVID-19 vaccines, we observe significant progress, since vaccines have been developed and introduced based on new principles and this, undoubtedly, contributes to the development of red biotechnology (Abi Younes et al., 2020; Brem et al., 2020; Chung et al., 2020; Javaid et al., 2020; Musyuni et al., 2020; Zimmer et al., 2020). But simultaneously this progress also entails a very high price. Unfortunately, this is a widespread phenomenon associated with the introduction of innovations. We suppose that, one way or another, this large-scale experiment in history will lead to further development of biotechnological science. But there is no doubt that the price of such a progress could have been much lower.

Meanwhile, other reasons also come to the fore. Since biotechnology is increasingly associated with the ability to slow down climate change, on the one hand, and on the other hand, it is believed that the struggle to increase productivity and the active use of fertilizers (especially nitrogen) harms the climate, the idea of reducing the use of mineral fertilizers has emerged. However, this came into sharp contradiction with the demands of food security and the interests of farmers. As a result, in 2022, we witnessed a popular revolt in Sri Lanka, where green agronomy led to food shortages (see, e.g., Kostin et al., 2022; Pathirana, 2022), and also the unrest among farmers in the Netherlands and other European countries; in the future it is fraught with increasing confrontation.

## 9.5 Forecasts of Development of Biotechnologies in the Final Phase of the Cybernetic Revolution and Beyond

Now, proceeding from the current trends and the general logic of the development of the Cybernetic Revolution, one may outline *the future developmental milestones for biotechnology* during the final phase of the Cybernetic Revolution (in the 2030s and 2070s) as well as afterward, i.e., during the mature phases of the Scientific-Cybernetic Production Principle up to the end of the twenty-first century.

Of course, one can hardly anticipate the vector and timeline of particular discoveries. However, we suppose that the importance of white, green and blue biotechnologies will increase later than that of red biotechnologies, which, together with medicine, will be at the forefront of the beginning of the Cybernetic Revolution.

Therefore, we will first consider forecasts concerning biotechnologies related to medicine and genetic engineering, aimed at modifying the human body and then forecasts for other bioareas. In both cases, we will relate our predictions to the challenges outlined above, as well as to the characteristics of the Cybernetic Revolution (as described in Chap. 4), especially those which are most manifest in the development of biotechnology.

### 9.5.1 Red Biotechnology, Biomedicine and Gene Engineering

**Red Biotechnology and Biomedicine**. There are many development opportunities associated with the creation of new drugs, vaccines, biomedical materials, diagnostics and many other products. Despite the widespread idea of the biotechnological revolution (see references above), we suppose that in the beginning of the future technological breakthrough it is medicine rather than biotechnology that will play the major and integrating role. However, medicine can hardly make such a breakthrough on its own. It should integrate with some other innovative fields, including biotechnology. Since the latter has already become an important part of innovative medicine and this integration will increase, the term *biomedicine* is more appropriate in this context. It is one of the most promising fields of the future medicine which will allow breakthroughs in the treatment of many diseases, including fatal ones, as well as in the monitoring or regulation of various functions of organisms.

We think that biomedicine has the most promising prospects in studies at the cellular level. We can expect breakthroughs in the development of artificial immunity, the cultivation of artificial tissues, creation of artificial blood and organs, including the use of non-biological components, as well as new possibilities for regeneration. We have discussed this in detail in Chap. 8. However, one should emphasize that these will be purely biotechnological innovations although intended for medical purposes. The importance of biotechnologies will grow in all fields that we identify as emerging, but in the next few decades they will become essential in medicine: the fight against aging and the rejuvenation of the body (about this see Chap. 11). In particular, we should mention the research on stem cells, which is already actively underway, but in the future, the modification of these cells can become a very important method applied in various spheres: in growing tissues and organs, rejuvenation and in cloning (Gurdon & Colman, 1999).

**Gene Therapy and Genetic Engineering**. The above-described technologies are also closely related to gene therapy and engineering. Both these fields are closely connected with cellular biotechnology and can support future breakthroughs in the treatment of various diseases. In particular, the fight against cancer is unthinkable without understanding various processes at the cellular and even subcellular level. Gene and molecular therapy, genetic engineering technologies will play an important role among anti-aging technologies. Gene therapy is the targeted modification of the genetic apparatus of cells in order to acquire new properties in the body or to treat diseases (this can be done using specially designed viral particles, such as the use of mRNA in vaccines against COVID-19). This therapy is currently used to treat rare genetic diseases, but in the future, we think its application will greatly expand. With regard to genetic engineering, there are ideas that it can be used to repair damage to the body at a subcellular level, in particular to neutralize so-called mutant mitochondria. Mitochondria often lose important pieces of their DNA, resulting in their inability to form required proteins. Since the mitochondrial DNA contains very few genes, there are ideas about their possible transfer to a safer place in the cell nucleus in the future (De Grey & Rae, 2007). Finally, in order to perceive opportunities for longevity, it is extremely important to study long-livers in search of their genetic features. One may expect interesting and promising discoveries in this field (more about all these technologies see in Chap. 11).

The research at the cellular level also offers prospects for reproductive technologies. In particular, scientists at the University of Cambridge in the UK managed to create a mouse embryo, which then developed a heart and brain, from mouse stem cells, without using either an egg or sperm (Harrison et al., 2017).

The recent boom in synthetic biology has led some researchers to anticipate far-reaching and promising prospects for the biotech industry, as well as possible new explosive innovations. The work of synthetic biologists today is the prototype of a new generation of cyber-biological systems (Peccoud, 2016: 1).[13] Let us highlight two important predictions that with greater or lesser probability may come true as part of the Cybernetic Revolution and its completion in the 2050s–2070s and later. First, a rapid development of genetic and other databases, together with the automation of a number of processes, gives hope that in the relatively near future, a new approach to the resynthesis of DNA molecules may be cheaper than storing and distributing the existing plasmids. This means that the application of

[13] Peccoud does not give a clear definition, but argues that synthetic biology encodes control algorithms in DNA molecules of organisms that can reproduce, communicate with each other or use complex networks of interactions between hosts and pathogens, prey and predators, etc. These engineered biological systems have an unprecedented level of complexity (Peccoud, 2016: 41).

genetic biology will become much wider.[14] Second, in this sector, the process that was previously observed in the IT and other related fields, that is, DIY (do it yourself) can unfold. Biologists are expected to democratize biological research by bringing it into "their kitchen and their garage", just as the personal computer movement challenged the dominance of government computing in infrastructure in the 1970s. They are excited to challenge the incumbents who dominate the biotechnology industry (Peccoud, 2016: 4).[15] However, it should not be forgotten that the supposed democratization in the area of IT and communications had been in some sense short-lived, establishing new plutocrats (see Freeland, 2012).

It is worth mentioning an original experiment with the application of genetic engineering, which caused heated debate. We mean the so-called three-parents conception. The essence of the technology is that the mother had genetic abnormalities in the mitochondrial DNA (mtDNA), which can be the carrier of numerous dangerous mutations. And it was replaced by the mtDNA of another woman, who can be called the second mother. The child was born in 2017, but it is not yet clear whether the experiment was successful and there will be no deviations in this individual's health (for details of this unique case see (Reardon, 2017). In 2023, a baby with DNA from three people was born in the UK.

As for some absolutely revolutionary breakthroughs in genetic biology and engineering, it is still difficult to predict their outcome. In any case, we doubt that even in the distant future it will become possible to control the human genetic program. It is too difficult and too dangerous with respect to such a highly complex system as the human body. Nevertheless, it is quite possible to significantly influence the gene program, and development in this direction will be quite intensive. But we should be aware of the dangers of such interference, especially if it remains uncontrolled. Even if today the discussions about genetic castes that could arise if private influence on the genetic program of embryos is allowed, seem like a distant fantasy, it is necessary to think already today about ways to prevent this from happening.

**About Some New Technologies**. Among the range of new technologies that should prove themselves, we may mark out the process known as biopatterning, already described in Chap. 8. It involves combining biological cells grown in the laboratory with non-biological material, which forms the basis for structuring artificial tissues. Later, the cells replace this non-biological basis, to deposit active biomolecules on solid surfaces for self-assembly on unconventional substrates. Of course, this opens

---

[14] Already today one may speak about significant achievements in the field of artificial DNA. The Japanese biologists have created the first artificial genomic DNA that can replicate and develop outside the cell. The scientists succeeded in initiating the process of gene expression and extracellular replication in the DNA created in a cell-free system of nucleic acids and proteins. The results of the study were published in the journal ACS Synthetic Biology (Okauchi & Ichihashi, 2021). About whole-genome synthesis, see below.

[15] However, it is worth mentioning that a number of other technologies (e.g., cars and aircraft) also originated from a similar home-made stage (Agamerzyan, 2016: 8).

up enormous possibilities for replacing tissues and, in the future, even parts of organs and entire organs, thus making it possible to solve many medical problems.

Another more complex and understudied process that may have a promising future (since it combines bio- and nanotechnologies) is the so-called biological self-assembly, which is a hybrid method that combines natural molecular tools with synthetic nanoscale constructs (Zhang, 2003). It is also intended to create artificial biological tissues, not at the level of cells, but molecules.

We can also mention the genetic technology of whole-genome synthesis. It is an extension of the above-mentioned booming field of synthetic biology. Researchers use software to design genetic sequences that they produce and then introduce into a microbe, thereby reprogramming it to fulfill a required task. This method is used in biomedicine.

Finally, it is worth mentioning a new type of RNA-based vaccine against COVID-19. Although they have not yet proven their effectiveness, we believe that they open up a path to new technologies that will hardly be smooth (as we have already observed), but that will ultimately be of considerable importance in the future. In general, considerable advancement can be expected in the field of virology in terms of the creation of new drugs, as well as in the application of viruses in other areas (Solé & Elena, 2018; White, 2021). Viruses will no longer be considered only as dangerous pathogens; instead, they will become a powerful and even major weapon of genetic engineering (which is already observed). Being actually pure information particles, viruses will open up new horizons for the transfer of genetic information both within and between species, and even kingdoms, helping to create previously unknown combinations and to introduce them into the genome in a way that is natural and familiar to the body. It is rather probable that just the discoveries in the world of viruses will enable science to truly decipher and master the genetic code. A breakthrough in virology, possibly within the next few decades, can be as important as a breakthrough in understanding the nature of bacteria, both pathogenic and beneficial. The kingdom of viruses is so vast and so diverse, that it is even difficult to imagine the opportunities possibly opening up.[16]

**Biosensors** can be generally defined as devices consisting of a biological recognition system, often called a bioreceptor, and a transducer (Ferrari, 2006). Different biological materials such as enzymes, cells and antibodies are used in biosensors

[16] However, bacteria still remain a very promising area, in particular the intestinal bacteria, which considerably influence human well-being (see more about this in Chap. 11). There also emerged the direction of the so-called postbiotics, i.e., metabolic products of probiotic microorganisms. The technical definition of postbiotics is a 'preparation of inanimate microorganisms and/or their components that confers a health benefit on the host' (Żółkiewicz et al., 2020). This definition was agreed upon by a panel of experts assembled by the International Scientific Association for Probiotics and Prebiotics (ISAPP) in 2019. Closely related are paraprobiotics, which could be defined as 'inactivated microbial cells (non-viable) that provide a health benefit to the consumer.' Postbiotics and paraprobiotics can promote and maintain health. Paraprobiotics also appear to have antioxidant effects, indicating potential applications in the food and pharmaceutical industries (Jang et al., 2018).

(Rusmini et al., 2007; Vo-Dinh et al., 2001). Biosensors are capable to convert biological energy into electrical signals. They are especially useful for analyzes in medicine, for example, to determine metabolites or hormone levels, they also enable control of various parameters in organisms during surgery. At present they are used in many fields including measurement of environmental pollutants.[17] Hundreds and even thousands of biosensors can be combined in biochips. A biochip is a miniature device, actually a real laboratory that can perform thousands of biochemical reactions simultaneously. Biochips help to quickly analyze a large number of biological parameters for different purposes, including diagnosis of cancer, infections and intoxications (Fung et al., 2001; Research & Markets, 2021). The combination of biochips and nanorobots seems rather promising in terms of the online monitoring of the spread of viruses in blood (Cavalcanti et al., 2008). They will also be capable of performing a huge number of different measurements including at the cellular level, which is absolutely unrealizable today. Thus, as we emphasized several times, the development unfolds toward self-regulating systems with biotechnology holding great potential. In particular, biochips will be one of the most important directions in the development of medicine and, in general, not only a system for monitoring the state of the body and health, but a special -medical-biotechnological environment in which people will stay permanently or for a long time (see more on this in Chas. 6 and 8).

**Restructuring of the Pharmaceutical Industry**. As discussed above (and in Chap. 8), reorganization of the whole medical system is a major challenge, but it is impossible without dramatic changes in the pharmaceutical industry. The over-regulated system for approving new drugs appears to be one of the most serious obstacles to the development of a more flexible pharmaceutical industry. It takes many years, sometimes up to 15–20 years, for a new drug to be approved. The reason is obvious: the need to protect patients' health from unsafe, not sufficiently tested drugs. This system makes the development of new drugs so expensive (and requires good government connections) that only very large companies can afford it. As a result, a number of more flexible small and medium-sized companies are deprived of the opportunity to operate in this market. During the Covid-19 pandemic, the procedures for vaccine inspection, testing, etc. were essentially removed. One can hardly describe this experiment as successful since the effectiveness of the vaccines appeared to be far below expectations while the side effects turned out to be serious. However, this shows that the pharmaceutical industry needs to look for alternatives to become more flexible and less monopolized, while maintaining the conditions for the production of approved and high-quality drugs. It is still unknown how to find the solution to this problem but eventually, even

[17] For example, to check the quality of products, in particular the presence of pesticides in them (https://www.rfbr.ru/rffi/ru/science_news/o_2124009). We believe that in the future, devices capable of measuring the degree of naturalness, environmental friendliness of products, the amount of GMOs in them, etc. will go on free sale.

through trial and error, the solution will be found (perhaps, through the development of computer simulation, which will accelerate the process). And it will lead to faster drug production and more effective drugs. We also do not exclude that the production of drugs will pass through the Hegelian triad from the creation of drugs in pharmacies to their mass industrial production, and then back again when based on future innovative technologies and self-regulating systems aimed at individualization, the drugs will be produced in medical institutions and in a new type of pharmacies. Thus, there will be a new symbiosis of medicine and pharmaceuticals. During the Cybernetic Revolution, this path of small-scale production and universalization of skills, including through the use of 3D printers, will be followed by a growing number of new industries. Above, we also presented the forecasts about the transition to the creation of DNA and even of genomes at the level of small private laboratories and even at home.

### 9.5.2 Forecasts for Other Fields of Biotechnology

Below we will consider the development of (1) green biotechnology in the following aspects: (a) environmental protection, primarily the purification of polluted waters and territories; and (b) solutions for sufficient food production; and also (2) blue, (3) yellow and (4) white biotechnologies.

#### 9.5.2.1 Green Biotechnology

**The Solution of Urban and Other Environmental Problems.** Undoubtedly, some important changes will occur in the application of biotechnologies for the solution of environmental problems of water purification, recycling of waste, liquidation of stray animals, etc.[18] The process when biological organisms are used to remove or neutralize environmental pollutants through a metabolic process is called bioremediation. For example, they use different microorganisms (like bacteria, fungi, algae) for water purification.[19] In the future these and similar problems will be solved by the development of self-regulating systems, which will make it possible to solve a number of technological and scientific problems.

Here one may assume that biotechnologies will be applied first of all in urban ecology, because the urban population will increase by 40–50% in the coming decades and, according to the forecast of the UN Population Division (UN Population Division, 2018; see also NIC, 2012), by more than two-thirds by the middle of this century. Thus, just as in the late nineteenth and early twentieth centuries biotechnologies helped people cope with mass infections, in the mid-twenty-first century, the latest biotechnologies may be able to help to solve the most vexing urban problems since at least two-thirds of the population will live in cities.

[18] Perhaps, it will be promoted through introducing genes for sterility or something of that kind.

[19] With their help we also get biogas from waste recovery.

But the need for ecological self-regulating systems arises not only in cities; it must be extended to the cleaning of reservoirs and other ecosystems. The creation of ecological self-regulating systems will considerably reduce expenses and free up huge territories occupied by waste deposits, as well as allow breeding fish in self-cleaning reservoirs. It can be assumed that an important direction will be the creation of self-regulating ecological systems in the resort and recreational areas, which will provide the best conditions for recreation and business.[20]

**Food Production and Genetic Engineering**. Green biotechnology could become important in the future to provide food for the world's growing population, which, according to the forecasts of the UN, will continue to grow for several decades, and possibly until the end of the twenty-first century and beyond (UN Population Division, 2022a, 2022b; for details see Chap. 6). The powerful development of digital DNA data storage and other databases will enable the development of new technologies such as genetic editing and breeding techniques, which will contribute to the growth of food production and reduce its costs.[21]

Thus, while transforming an organism, they will insert not only a separate useful gene (Simon et al., 1983), but a whole set of necessary genes which will function in response to environmental conditions. In the future green biotechnology will use huge databases of encoded genetic information to design desirable plant and animal traits (such as responsiveness to changing environmental conditions). Today the largest genetic database (NCBI) contains all known genes from all genetically studied species and their combinations, including chromosomes and even all the data from the Human Genome project. The database project is absolutely free and available to everyone. This may become extremely important in the case of climate changes. It will allow the selection of the most optimal varieties of seeds and seedlings for a unique combination of weather conditions and territory (a kind of reproduction of evolutionary selection by automatic search in databases). So we suppose that by the 2050–2060s, green biotechnology will help agriculture develop to a level at which the adaptibility of modified products will allow them to respond

[20] One possible challenge is given by the fact, that the application in cities is much easier due to the existing infrastructure, serving a densely populated area; on the other hand, rural areas are especially challenging because of the lack of such infrastructure. While certain problems do not exist to the same extent in these areas (pollution, noise, ...), dealing with them poses additional challenges.

[21] A very good example of a comprehensive solution to the problems of intensifying agriculture and increasing its variability while preserving the environment based on a self-regulating system are special cubes, an agricultural startup of a vertical farm with technology to cultivate high-quality vegetables in a fully controlled environment that can increase production. With the Planty Cube, customers can start a vertical farming business with a smaller number of units and expand as the business grows. The smart hydroponic farm, shaped like a shipping container, is made of stacked rows and shelves of planters on each wall. After seeds are sown through their smart seed capsules, a computerized system controls the environment inside the Planty Cube and helps the plants thrive without constant human attention (CES, 2019). It is also a good example of combining different technologies with biotechnology and expanding the opportunities for growing plants in any region.

even to the smallest fluctuations of local conditions.[22] Moreover, it is quite possible that in the future the whole process of getting a transgenic plant will proceed without human participation, and thus, it will become self-regulating. We can also forecast the emergence of a self-regulating supersystem that will analyze soils in different places and areas. If they create a database that characterizes the optimal chemical and biochemical composition of different soils, then it will be possible to measure their variation in composition from year to year and to recommend what, where and how much fertilizer to use. This will save the work of thousands of agronomists.

The solution to the global protein shortage crisis may come in other ways, perhaps, even through the mass production of artificial foods. Even now the global production of imitation or cultured meat in different countries amounts to about $2 million. Of course, at present its production is too expensive: a gram of lab-grown meat costs hundreds of dollars. However, there is an (over-)optimistic forecast that the global market for cultured meat will reach about $ 2.8 billion (https://www.alliedmarketresearch.com/cultured-meat-market-A06670).[23] It seems that the production of artificial meat would face many obstacles and may appear far from being so energy efficient, as some researchers estimate.[24]

#### 9.5.2.2 Blue Biotechnology

Blue biotechnology could play a huge role. We mean primarily a widespread artificial cultivation of seafood (the so-called aquaculture). Due to the depletion of fish and seafood in the world oceans, a transition is possible similar to the one that took place thousands of years ago from foraging (hunter-gathering) economy to the food production/agriculture (this transition is called the Agrarian Revolution, see Chap. 2). World aquaculture production has grown tremendously: from less than one million tons in the early 1950s to 48.1 million tons in 2005, an average annual growth rate of about 8.8% (Subasinghe et al., 2009; some details about global seafood see in Smith et al., 2010). In 2020, global aquaculture production volume reached 122.6 mln tons (87.5 mln—animals and 35.1 mln—algae; and aquatic animal production is projected to grow by a further 14% by 2030 [FAO, 2022]). The global consumption of aquatic foods has increased at an average

[22] Moreover, it may become possible to order producers or collectors to create varieties of plants for individual greenhouses, hotbeds or plots. Farmers will be able to select individual plants and fodder. Also people will be able to invent a houseplant hybrid suitable for their individual interior and order its production and delivery. Thus, individualization will reach a new level. It is quite possible that similar achievements will be possible with respect to domestic animals: it will be possible to breed animals with peculiar characteristics within separate breeds of animals (or even by individual order).

[23] However, it is reported that Shanghai company CellX launched a pilot plant for the production of artificial meat (RIA Novosti, 2023).

[24] For example, according to Tuomisto and Teixeira de Mattos (2011), the cultured meat can reduce land use by 99%, water use by 96%, and energy consumption by 45%.

annual rate of 3% since 1961, compared to a population growth rate 1.6% (Ibid.). Aquatic foods help supply people with animal proteins in many countries.[25]

Today, the volume of aquaculture production is already approaching 50% of global fish and seafood production (FAO, 2022). However, it is expected that in the future it will become possible to produce several times more food at seafarms (see, e.g., Weizsäcker & Wijkman, 2018; see also Rosenberg, 2017). With huge undeveloped water areas and new technologies, the transition to artificial farming can solve the problem of a growing population. It is particularly important for Africa, where fish consumption is growing slowly (Blow & Leonard, 2007; FAO, 2022: 13). According to the studies from Stanford University (Stanford news, 2021), by 2050 the global demand for seafood will increase by 80%; respectively, one may speak about a "revolution" at the seafood market and the solution to the problem of food supply on a global scale.[26]

#### 9.5.2.3 Yellow Biotechnology

*Prospects for yellow (insect) biotechnology.* We have mentioned above the yellow or insect biotechnology. This plays a minor role today, although it is strongly developing especially in connection with ideas to add insects to the Western people's diet. However, the scope for the future application of the increased knowledge about insects is huge. Yellow biotechnology also involves microorganisms that will be increasingly used to produce a range of new products, including the so-called bacterial polymers. This branch of biotechnologies will also manifest in pharmaceuticals and agriculture. As for the sphere of use of insects, here one can also expect a wide implementation of useful and remarkable discoveries. In particular, the genetically modified insects will probably be used extensively in the wilderness to achieve certain goals. At the same time, we can anticipate the creation of very interesting biological self-regulating systems (of course, with insertions of other components), prototypes of which already exist. Today genetic modification can change a whole population; for example, via the actively applied method of distributing genes via 'decoy' individuals. For example, the infertile mosquitoes were massively released into the wild population. Such an ineffective cross-breeding led to the reduction in a number of insects (Tkachuk et al., 2011; Benedict & Robinson, 2003; Lenharo, 2023, see also the Debug project [https://debug.com/] which

[25] In 2019, aquatic foods globally provided about 17 percent of animal proteins and 7 percent of all proteins. For 3.3 billion people, aquatic foods provide at least 20 percent of the average per capita intake of animal protein. In some countries, like Bangladesh, Indonesia, Ghana and Mozambique aquatic foods contribute half or more of total animal protein intake (FAO, 2022: 17, 18).

[26] However, the use of marine organisms (both animals and plants) is possible in a variety of areas, for example, in breakthrough innovations in medicine, in particular in tissue regeneration; there may also be ideas regarding the fight against age-related diseases, for example, with the help of the so-called immortal jellyfish Turritopsis dohrnii. Although the jellyfish does not age in the same way as humans, its functioning provides insights for new research on cancer, neurodegenerative and cardiovascular diseases, or other pathologies associated with aging (Kuriyama et al., 2021). The study of marine animals can also help in the search for minerals. Thus, it appears that algae do not grow everywhere in the ocean, but only in iron-containing zones (Stoll, 2020).

proposes that 'good bugs' which do not bite and are not infectious will substitute for 'bad bugs'). Such an approach appears much safer than chemical exposure. We believe that in the future the variety of such technologies will increase markedly; they will become commonplace, which will expand the opportunities to influence ecological chains and manage processes.[27] Of course, there is also a great danger of causing serious damage to ecosystems. Flying insects can be used to carry mini cameras for nature observation (Iyer et al., 2020), but also, of course, for reconnaissance activities and military purposes.

In addition, new knowledge about the properties and abilities of insects can be used in the most unexpected areas. Thus, it turned out that ants can sense the cancer. And there are ideas to use them in hospitals as a 'less invasive' diagnostic tool (Morrison, 2022).

#### 9.5.2.4 White Biotechnology

By interacting with other branches, *white biotechnology* promises remarkable achievements in ecology and zero-waste production, particularly in biological recycling, a process in which organisms take up "waste" materials, decompose them and produce new materials from them, using biological processes such as anaerobic digestion or photosynthesis. Bioeconomy recycling also involves so-called biopolymers and biodegradable materials. Various projects are using microorganisms in some new production processes. There is already a whole rapidly growing sector of geo-microbiology, which studies the microorganisms that drive the Earth's biogeochemical cycles, mediating mineral precipitation and dissolution and absorbing and concentrating metals.[28]

### 9.5.3 Some Characteristics of the Cybernetic Revolution in the Future Development of Biotechnology

**Advance on the Way to Self-Regulating Systems**. Biotechnologies provide great opportunities for creating self-regulating systems. As we argued above, almost all of the described new technologies are somehow related to biological self-regulating systems and can be used in the creation and transformation of artificial biological tissue in the patient's body (bioprinting, self-assembly), in the transformation of stem cells into various tissues and organs, including for reproductive technologies, in genetic engineering (modification of individual genes and genomes, special gene-controllers, neutralization of mutant mitochondria), in gene therapy that affect cells and immunity (mRNA vaccines), in the creation

[27] Perhaps we can witness this soon enough. The non-profit World Mosquito Program (WMP) has announced that it will release modified mosquitoes in many of Brazil's urban areas over the next decade. They plan to produce up to five billion bacteria-infected mosquitoes per year (Lenharo, 2023).

[28] There is also an idea to use bacteria to extract palladium contained in scanty amounts in industrial waste (Fücks, 2013).

of artificial DNA and even of new organisms (synthetic biology, cyber-biological systems), in body monitoring (biosensors), in genetically modified bacteria for medical purposes (whole-genome synthesis), in pollution treatment using microorganisms (bioremediation), in the use of genetically modified insects to change ecology or entire population, etc.

Thus, a self-regulating biological system may emerge that has never existed before. We have already observed a prototype of autonomous and self-regulating biological systems (without direct human participation but still controlled by humans) that will be widely and actively used in almost all areas of life in the future.

Biotechnology can considerably contribute to a *breakthrough in resource saving* and help to solve many global issues, in particular, in organizing a recycling system. They also can cheapen the production of medicines and food. Food production can develop in different ways: through increasing yields, through breeding more resistant or efficient varieties, and through reducing seed costs. The breakthrough in the latter field is connected with *totipotency* which is an ability of plants to form a full-fledged organism from a single cell. For example, gene transfer can make a potato variety resistant to the Colorado beetle, or reduce the susceptibility to drought, cold and other stresses (Grinin et al., 2010). The development of cleaning technologies will help both saving processes and the environment.

**Synthesis of New Materials and Products**. The creation of self-regulating systems involving biotechnologies, in particular through genetic manipulations, opens an important direction in the field of new materials with desirable properties (e.g., genetic material). We have mentioned above the creation of artificial biotissues, modified plants and other bio-organisms, the creation of artificial DNA, new organisms and even cultured meat, etc. And the prospects here are enormous, revolutionary and no less important than the opportunities that have opened up for chemistry in the creation of new materials. But the potential dangers along the way are great as well. We must always remember that the coronavirus leaked from a laboratory!

At present, genetic engineering can create not only specific genes but also entire genomes and even chromosomes. The artificial chromosomes can be inserted and thus, add some new genes absent in the organism (Dymond et al., 2011). It potentially allows creating substitutes for the natural raw materials, for example, leather. The respective projects are already developed. For example, the Modern Meadow Company aims at making a revolution in the clothing industry by growing leather and other types of animal skin in laboratories (Zagorski, 2012).

The process of creating bioengineered leather will involve several steps. At first scientists will select millions of cells from donor animals. These can be both cattle or exotic species that are often killed just for their skin. Then these cells will be multiplied in bioreactors. At the following stage, the cells will be mingled into a mass which will be formed in layers using 3D bioprinter. The skin cells will create collagen fibers, while the 'flesh' cells will form a real soft tissue. This process will take several weeks after which soft and fatty tissues can be used for food production. Despite the exoticism and strangeness of the above-described process, it is actually very similar to the process of production of artificial furs, which made it possible to solve the problem of warm clothing.

**Individualization**. We have talked a lot about it above. We can add that biotechnology will allow the development of individualization for people in various directions, including genetic changes in individuals, their children and, quite probably, also their pets—dogs, cats, etc. (in the future, perhaps, even cloning pets), cultivated plants;[29] permanent selection of nutritional supplements and nutraceuticals, etc.

**Synthesis of MANBRIC Technologies with Biotechnology**. Biotechnology provides a good example of convergence with different MANBRIC innovation areas. First of all, let us recall the convergence of biotechnology and medicine into biomedicine. There is a very close, rapidly growing and increasingly complex relationship between biotechnologies, especially genetics, on the one hand, and ICT, artificial intelligence, on the other. This is primarily, the rapid growth of digital DNA data storage, and also the automation of biotechnological processes using AI and ICT at the most different levels (synthetic biology in particular). No wonder there are forecasts (see above) about the rise of cyber-biological systems that combine biological and computer systems. Medical implants are a synthesis of various branches of technologies. Thus, it is quite probable that particularly breakthrough innovations will result from the synthesis of biotechnologies with other MANBRIC technologies, as well as within the framework of a synthetic innovation direction—bionics.

The interrelations within the triangle of bio-, nano- and additive technologies are very promising and already provide interesting results in the field of artificial biological tissues and organs. In the future these interrelations may generate new developmental lines, which will become one of the leading directions (see Chap. 10). Future systems for delivering milli-doses of drugs to the body will combine bio- and nanotechnologies, AI and possibly additive technologies, as well as organic and inorganic materials. Biochips combine achievements of biology, modern microelectronics and nanotechnologies and it seems to be of utmost importance to combine technical and biological elements for the future self-regulating systems.

**Results**. Thus, in the future, the role of biotechnology will not only increase, but its sectors will become extremely important for life and the environment. In particular, there are great hopes that biotechnology (and the bioeconomy in general): (a) will make a major contribution to the victory over incurable and other diseases, including viral diseases; (b) will enable a large number of people suffering from genetic abnormalities, dysfunction of internal organs or tissue damage to return to a more active life with the help of artificial tissues and organs; (c) will make a major contribution to the fight against aging; (d) will considerably mitigate environmental

[29] Individualization in biotechnologies is associated, first of all, with opportunities to change genome and to add new properties to an organism.

problems and facilitate the solution of the problem of biodiversity; (e) will become the leading sector in solving the problem of food supply; (f) will significantly change technological and production processes, creating a waste-free production; and (g) will expand our ability to manage natural processes. In a word, in the future our life will be absolutely impossible without biotechnologies.

**Acknowledgements** This chapter is an output of a research project implemented as part of the Basic Research Program at HSE University in 2024 with support by the Russian Science Foundation (Project No. 23-18-00535).

## References

Abi Younes, G., Ayoubi, C., Ballester, O., Cristelli, G., de Rassenfosse, G., & Foray, D., et al. (2020). COVID-19: Insights from innovation economists. *Science and Public Policy*, (paaa036). https://doi.org/10.1093/scipol/scaa028.

Agamerzyan, I. (2016). Tekhnologicheskoye liderstvo: Vospolzovatsya shansom. In V. Burov (Ed.), *Vyzov 2035* (pp. 8–15). Olimp-Biznes.

Alanis, A. J. (2005). Resistance to antibiotics: Are we in the post-antibiotic era? *Archives of Medical Research, 36*(6), 697–705. https://doi.org/10.1016/j.arcmed.2005.06.009

Armed Conflict Location & Event Data Project. (2020). *COVID-19 and Political Unrest in Northeast India.* http://www.jstor.org/stable/resrep26621.

Avise, J. C. (2004). *The hope, hype, and reality of genetic engineering: Remarkable stories from agriculture, industry, medicine, and the environment.* Oxford University Press.

Benedict, M. Q., & Robinson, A. S. (2003). The first releases of transgenic mosquitoes: An argument for the sterile insect technique. *Trends in Parasitology, 19*(8), 349–355. https://doi.org/10.1016/S1471-4922(03)00144-2

Blow, P., & Leonard, Sh. (2007). A review of cage aquaculture: Sub-Saharan Africa. *FAO Fisheries Technical Paper, 498*, 188–207.

Bonnevie, E., Gallegos-Jeffrey, A., Goldbarg, J., Byrd, B., & Smyser, J. (2021). Quantifying the rise of vaccine opposition on twitter during the COVID-19 Pandemic. *Journal of Communication in Healthcare, 14*(1), 12–19. https://doi.org/10.1080/17538068.2020.1858222

Borlaug, N. E. (2001). "Zelenaya revolyutsiya": vchera, segodnya i zavtra.. *Ekologia i zhisn,* 1. http://www.ecolife.ru/jornal/econ/2001-4-1.shtml.

Brem, A., Viardot, E., & Nylund, P. A. (2020). Implications of the coronavirus (COVID-19) outbreak for innovation: Which technologies will improve our lives? *Technological Forecasting and Social Change, 163*, 2–6. https://doi.org/10.1016/j.techfore.2020.120451

Brennan, E. (2020). *Coronavirus and protest: How Covid-19 has changed the face of American activism.* United States Studies Centre. https://www.ussc.edu.au/analysis/coronavirus-protest-how-covid-19-has-changed-the-face-of-american-activism

Brookes, G., Taheripour, F., & Tyner, W. E. (2017). The contribution of glyphosate to agriculture and potential impact of restrictions on use at the global level. *GM Crops & Food, 8*(4), 216–228. https://doi.org/10.1080/21645698.2017.1390637

Bröring S. (2021). *Moving the BioEconomy from mind to market: what drives technology transfer and innovation adoption?* Presentation at IAMO Forum June 7th 2021.

Carothers, T., & Press, B. (2020). The global rise of anti-lockdown protests—and what to do about it. *World Politics Review*, Benjamin Press. https://www.worldpoliticsreview.com/articles/29137/amid-the-covid-19-pandemic-protest-movements-challenge-lockdowns-worldwide. Accessed November 6, 2020.

Cavalcanti, A., Shirinzadeh, B., Zhang, M., & Kretly, L. C. (2008). Nanorobot hardware architecture for medical defense. *Sensors, 8*(5), 2932–2958.

Ceresana Research. (2011). *Market study: Bioplastics*. http://www.ceresana.com/en/market-studies/plastics/bioplastics/.
CES—Consumer electronics show (2019). *Planty Cube* https://www.ces.tech/articles/2019/planty-cube.aspx.
Chumakov, A. N., & Yurchenko, P. S. (2021). Global pandemic in terms of objective causes and subjective speculations. *Journal of Globalization Studies, 12*(2), 157–168.
Chung, Y. H., Beiss, V., Fiering, S. N., & Steinmetz, N. F. (2020). COVID-19 vaccine frontrunners and their nanotechnology design. *ACS Nano, 14*(10), 12522–12537. https://doi.org/10.1021/acsnano.0c07197
Clive, J. (2011). *Global status of commercialized biotech/GM crops. ISAAA Briefs*. International Service for the Acquisition of Agri-biotech Applications.
Corey, L., Mascola, J. R., Fauci, A. S., & Collins, F. S. (2020). A strategic approach to COVID-19 vaccine R&D. *Science, 368*(6494), 948–950. https://doi.org/10.1126/science.abc5312.
Das, G.G. (2007). Information age to genetic revolution: embodied technology transfer and assimilation—a tale of two technologies. *Technological Forecasting and Social Change, 74*(6), 819–842. https://doi.org/10.1016/j.techfore.2006.07.001.
DaSilva, E. J. (2004). The Colours of biotechnology: Science, development and humankind. *Electronic Journal of Biotechnology, 7*(3). https://www.scielo.cl/scielo.php?script=sci_arttext&pid=S0717-34582004000300001
de Besi, M., & McCormick, K. (2015). Towards a Bioeconomy in Europe: National, regional and industrial strategies. *Sustainability, 7*, 10461–10478.
De Grey, A., & Rae, M. (2007). *Ending aging: The rejuvenation breakthroughs that could reverse human aging in our lifetime*. Martin's Press.
Dymond, J. S., Richardsonm, S. M., Coombesm, C. E., Babatz, T., Muller, H., Annaluru, N., Blake, W. J., et al. (2011). Synthetic chromosome arms function in yeast and generate phenotypic diversity by design. *Nature, 477*(7365), 471–476. https://doi.org/10.1038/nature10403
EC—European Commission. (2018). A sustainable Bioeconomy for Europe: Strengthening the connection between economy, society and the environment: Updated Bioeconomy strategy. Brussels.
EC—European Commission. (2019). *The European green deal*. COM(2019) 640 final.
Efremov, I., & Arkhangelsky, V. (2023). Global fertility changes in the context of the COVID-19 Pandemic. *Journal of Globalization Studies, 14*(2), 165–170.
El-Chichakli, B., Von Braun, J., Lang, C., Barben, D., & Philp, J. (2016). Five cornerstones of a global bioeconomy. *Nature, 535*, 221–223.
FAO. (2022). *In brief to the state of world fisheries and aquaculture 2022. Towards Blue Transformation*. Rome, FAO. https://doi.org/10.4060/cc0463en
Ferrari, M. (Ed.). (2006). *BioMEMS and biomedical nanotechnology*. Springer.
Freeland, C. (2012). *Plutocrats: The rise of the new global super rich and the fall of everyone else*. The Penguin Press.
Fücks, R. (2013). *Intelligent Wachsen. Die Gruene revolution*. GmbH & Co: Carl Hanser Verlag.
Fukuyama, F. (2002). *Our post-human future: Consequences of the bio-technology revolution*. Farrar, Straus, and Giroux.
Fung, E. T., Thulasiraman, V., Weinberger, S. R., & Dalmasso, E. A. (2001). Protein biochips for differential profiling. *Current Opinion in Biotechnology, 12*(1), 65–69.
Glen, T., Giovannetti Jaggi, G., Bialojan, S., Zürcher J., & Hillenbach J. (2013). *Beyond borders*. Biotechnology Industry Report 2013. Ernst & Young.
Glick B. R., & Pasternak, J. J. (2003). *Molecular biotechnology: Principles and applications of recombinant DNA*. American Society Microbiology Press.
Grinin, A. L., Kholodova, V. P., & Kuznetsov, V. V. (2010). Sravnitel'nyy analiz fiziologicheskikh mekhanizmov soleustoychivsti razlichnykh sortov gorchitsy. *Vestnik Rossiyskogo Universiteta Druzhby Narodov, 1*, 27–38.
Grinin, L. (2020). How can COVID-19 change Geopolitics and Economy? *Journal of Globalization Studies, 11*(2), 121–134.

Grinin L. E., & Grinin A. L. (2015). *Ot rubil donanorobotov. Mirnaputikepokhesamoupravlyaemykh sistem (istoriya tekhnologiy i opisanie ikh budushchego)*. Moscow: Moscow Publishing House 'Uchitel'.

Grinin, L., Grinin, A., & Korotayev, A. (2022). COVID-19 pandemic as a trigger for the acceleration of the Cybernetic Revolution, transition from E-Government to E-State, and change in social relations. *Technological Forecasting and Social Change, 175*, 1–15. https://doi.org/10.1016/j.techfore.2021.121348

Gurdon, J. B., & Colman, A. (1999). The future of cloning. *Nature, 402*(6763), 743–746.

Harris, J. (2010). *Enhancing evolution: The ethical case for making better people*. Princeton University Press.

Harrison, S. E., Sozen, B., Christodoulou, N., Kyprianou, C., & Zernicka-Goetz, M. (2017). Assembly of embryonic and extraembryonic stem cells to mimic embryogenesis in vitro. *Science*, 356(6334), eaal1810. https://doi.org/10.1126/science.aal1810.

Holdrege, C. (2010). *Beyond biotechnology: The barren promise of genetic engineering*. University Press of Kentucky.

Hussain, M. (2021). The changing dynamics of liberal international order: Covid-19 and its global implications. *Journal of Globalization Studies, 12*(2), 143–156.

Irshad, S. M. (2020). COVID-19 and global 'At-Risk community': From benefit-sharing to risk-sharing of economic crisis. *Journal of Globalization Studies, 11*(2), 135–145.

Iyer, V. et al. (2020). Airdropping sensor networks from drones and insects. In: *Proceedings of the 26th Annual International Conference on Mobile Computing and Networking*, 1–14.

Jang, H. J., et al. (2018). Antioxidant effects of live and heat-killed probiotic Lactobacillus plantarum Ln1 isolated from kimchi. *Journal of Food Science and Technology, 55*(8), 3174–3180.

Jasiński, A. M., & Bąkowska, A. M. (2020). Globalization and life satisfaction of the polish population. Before and during the COVID-19 Pandemic. *Journal of Globalization Studies, 11*(2), 146–155.

Javaid, M., Haleem, A., Vaishya, R., Bahl, S., Suman, R., & Vaish, A. (2020). Industry 4.0 technologies and their applications in fighting COVID-19 pandemic. *Diabetes & Metabolic Syndrome: Clinical Research & Reviews, 14*(4), 419–422. https://doi.org/10.1016/j.dsx.2020.04.032.

Jeyanathan, M., Afkhami, S., Smaill, F., Miller, M. S., Lichty, B. D., & Xing, Z. (2020). Immunological considerations for COVID-19 vaccine strategies. *Nature Reviews Immunology, 20*(10), 615–632. https://doi.org/10.1038/s41577-020-00434-6

Kakkar, S. (2021). The outbreak of sars-cov-2 (Covid-19): Impact on international tourism. *Journal of Globalization Studies, 12*(1), 145–158.

Kopetz, H. (2013). Renewable resources: Build a biomass energy market. *Nature, 494*(7435), 29–31. https://doi.org/10.1038/494029a

Kostin, M., Grinin, L., & Korotayev, A. (2022). Revolutionary and quasi-revolutionary events of 2022: a brief review. *Sistemnyi Monitoring Globalnyh i Regionalnyh Riskov = Systemic Monitoring of Global and Regional Risks, 13*, 607–635. https://doi.org/10.30884/978-5-7057-6184-5_17

Kowalewski, M. (2020). Street protests in times of COVID-19: adjusting tactics and marching 'as usual. *Social Movement Studies*, 1–8.https://doi.org/10.1080/14742837.2020.1843014

Kuriyama, M., Miyake, H., Toshino, Sh., & Kubota, Sh. (2021). First record of an immortal Jellyfish, Turritopsis Dohrnii (Weismann, 1883) from Tanabe Bay, Wakayama, Japan. *Biogeography, 23*, 62–66.

Lenharo, M. (2023). Massive mosquito factory in Brazil aims to halt dengue. *Nature, 616*, 637–638. https://doi.org/10.1038/d41586-023-01266-9

Leshina A. (2012). Plastiki biologicheskogo proiskhozhdeniya.. *Khimiya i Zhisn—XX vek, 9*, 2–5.

Louwerse, T., Sieberer, U., Tuttnauer, O., & Andeweg, R. B. (2021). Opposition in times of crisis: COVID-19 in Parliamentary debates. *West European Politics, 44*(5–6), 1025–1051. https://doi.org/10.1080/01402382.2021.1886519

Martyushev-Poklad, A. (2015). Chto proiskhodit s meditsinoy: protokol vskrytiya (1). *Aftershock*. http://aftershock.su/?q=node/300253.

McCabe, L. L., & McCabe, E. R. B. (2008). *DNA: Promise and peril.* Berkeley; Los Angeles: University of California Press.

McCormick, K., & Kautto, N. (2013). The bioeconomy in Europe: An overview. *Sustainability, 5*(6), 2589–2608.

Morrison R. (2022). Ants have the ability to sniff out cancer in humans, study reveals, suggesting they could be used in hospitals as a 'less invasive' alternative for diagnosis. *Dailymail.* https://www.dailymail.co.uk/sciencetech/article-10595727/Ants-ability-sniff-cancer-humans-study-reveals.html.

Musyuni, P., Aggarwal, G., Nagpal, M., & Goyal, R. K. (2020). A case study: Analysis of patents on coronaviruses and Covid-19 for technological assessment and future research. *Current Pharmaceutical Design.* https://doi.org/10.2174/1381612826666200720233947

NCHS—National Center for Health Statistics. (2021). *Drug Overdose Deaths in the U.S. Top 100,000 Annually For Immediate Release: November 17.* https://www.cdc.gov/nchs/pressroom/nchs_press_releases/2021/20211117.htm.

Nefiodow L. (1996). *Der sechste Kondratieff. Wege zur Produktivität und Voll-beschäftigung im Zeitalter der Information.* 1 Auflage/Edition. Rhein-Sieg-Verlag: Sankt Augustin.

Nefiodow, L., & Nefiodow, S. (2014a). *The Sixth Kondratieff. The new long wave of the world economy.* Sankt Augustin.

Nefiodow, L., & Nefiodow, S. (2014b). The Sixth Kondratieff. The growth engine of the 21st century. In L. E. Grinin, T. Devezas, & A. V. Korotayev (eds.), *Kondratieff Waves. Juglar—Kuznets—Kondratieff* (pp. 326–353). Volgograd: Uchitel.

NIC (National Intelligence Council). (2012). *Global Trends 2030: Alternative Worlds.* URL: www.dni.gov/nic/globaltrends.

Null, G., Dean, C., & Feldman, M. et al. (2003). Table 1: Estimated annual mortality and economic cost of medical intervention. *Death by Allopathy. Death by Medicine. Extracts.* http://www.whale.to/a/null9.html.

OECD. (2009). *The Bioeconomy to 2030: Designing a policy agenda.* OECD.

Okauchi, H., & Ichihashi, N. (2021). Continuous cell-free replication and evolution of artificial genomic DNA in a compartmentalized gene expression system. *ACS Synthetic Biology, 10*(1), 3507–3517. https://doi.org/10.1021/acssynbio.1c00430

Pathirana, S. (2022). Sri Lanka protesters reject all-party gov't, want Rajapaksas out. *Al Jazeera,* 12.07.2022. https://www.aljazeera.com/news/2022/7/12/sri-lanka-crisis-no-to-all-party-govt-say-protest-leaders.

Peccoud, J. (2016). Synthetic Biology: Fostering the cyber-biological revolution. *Synthetic Biology, 1*(1), 1–7. https://doi.org/10.1093/synbio/ysw001

Peillex, C., & Pelletier, M. (2020). The impact and toxicity of glyphosate and glyphosate-based herbicides on health and immunity. *Journal of Immunotoxicology, 17*(1), 163–174. https://doi.org/10.1080/1547691X.2020.1804492

Phillips, F., & Su, Y. S. (2009). Advances in evolution and genetics: Implications for technology strategy. *Technological Forecasting and Social Change, 76*(5), 597–607. https://doi.org/10.1016/j.techfore.2008.08.006.

Reardon, S. (2017). Genetic details of controversial 'three-parent Baby' revealed. *Nature,* 544(7648).

RIA Novosti. (2023). V Kitaye zapustili pervyy zavod po proizvodstvu iskusstvennogo myasa. *RIA Novosti* 10.08.2023. https://ria.ru/20230810/myaso-1889260120.html.

Richard, S., Moslemi, S., & Sipahutar, H. et al. (2005). Differential effects of glyphosate and roundup on human placental cells and aromatase. *Environmental Health Perspectives, 113*(6), 716–20. https://doi.org/10.1289/ehp.7728. http://www.ncbi.nlm.nih.gov/pmc/articles/PMC1257596/.

Rifkin, J. (2011). *The third industrial revolution. How lateral power is transforming energy, the economy, and the world.* Palgrave MacMillan, London.

Rifkin, J. (2019). *The green new deal: Why the fossil fuel civilization will collapse by 2028, and the bold economic plan to save life on earth.* St. Martin's Press.

Rosenberg A. (2017). *Can farmed fish feed the world sustainably? Aquaculture could hold the key to feeding the world while fighting global warming* December 8. https://www.universityofcalifornia.edu/news/can-farmed-fish-feed-world-sustainably.
Rusmini, F., Zhong, Z., & Feijen, J. (2007). Protein immobilization strategies for protein biochips. *Biomacromolecules, 8*(6), 1775–1789.
Seifabadi, M. S., & Dolatabadi, A. B. (2022). Covid-19, globalization and strengthening authoritarianism in the middle east. *Journal of Globalization Studies, 13*(1), 135–157.
Simon, R., Priefer, U., & Pühler, A. (1983). A broad host range mobilization system for in vivo genetic engineering: Transposon mutagenesis in Gram Negative Bacteria. *Nature Biotechnology, 1*(9), 784–791.
Smith, M. D., Roheim, C. A., Crowder, L. B., Halpern, B. S., Turnipseed, M., Anderson, J. L., Asche, F., et al. (2010). Sustainability and global seafood. *Science, 327*(5967), 784–786. https://doi.org/10.1126/science.1185345
Solé, R., & Elena, S. F. (2018). *Viruses as complex adaptive systems*. Princeton University Press.
Spratt, B. G. (1994). Resistance to antibiotics mediated by target alterations. *Science, 264*(5157), 388–393.
Strandberg, T. (2020). Coronavirus: US and UK governments losing public trust. *The Conversation*. http://theconversation.com/coronavirus-us-and-uk-governments-losing-public-trust-137713. Accessed 4 February 2021
Stanford news. (2021). *Stanford-led study suggests rising and shifting demand for seafood by 2050*. https://news.stanford.edu/2021/09/15/different-seafood-2050/.
Stoll, H. (2020). 30 years of the iron hypothesis of ice ages. *Nature, 578*(7795), 370–371. https://doi.org/10.1038/d41586-020-00393-x
Subasinghe, R., Soto, D., & Jia, J. (2009). Global aquaculture and its role in sustainable development. *Reviews in Aquaculture, 1*(1), 2–9.
Sylvester, E., & Klotz, L. C. (1983). *The Gene Age: Genetic engineering and the next industrial revolution*. Scribner.
Tkachuk, A. P., Kim, M. V., Savitsky, V. Yu., & Savitsky, M. Yu. (2011). Perspektivy ispol'zovaniya transgennykh nasekomykh v programmakh biokontrolya. *Zhurnal obshchey biologii, 72*(2), 93–110.
Tuomisto, H. L., & Teixeira de Mattos, M. J. (2011). Environmental impacts of cultured meat production. *Environmental Science & Technology, 45*(14), 6117–6123.
UN. (1992). *Conference on environment and development*. Agenda 21. Chapter 16. Rio de Janeiro, 3–14 June. https://docs.google.com/gview? url=http://sustainabledevelopment.un.org/content/documents/Agenda21.pdf&embedded=true.
UN. (2020). *United nations. A new social contract for a New Era, the 18th nelson Mandela annual lecture given by united nations' 18th Nelson Mandela Annual Lecture given by the UN Secretary-General António Guterres*. www.un.org/sustainabledevelopment/a-new-social-contract-for-a-new-era/.
UN Population Division. (2018). *World urbanization prospects 2018*. United Nations.
UN Population Division. (2022a). *United nations population division database*. United Nations, Department of Economics and Social Affairs, Population Division. http://www.un.org/esa/population.
UN Population Division. (2022b). *World population prospects 2022*. United Nations, Department of Economics and Social Affairs, Population Division.
Vieten, U. M. (2020). The 'New Normal' and 'Pandemic Populism': The COVID-19 crisis and anti-hygienic mobilisation of the far-right. *Social Sciences, 9*(9), 165. https://doi.org/10.3390/socsci9090165
Vo-Dinh, T., Cullum, B. M., & Stokes, D. L. (2001). Nanosensors and Biochips: Frontiers in biomolecular diagnostics. *Sensors and Actuators b: Chemical, 74*, 2–11.
Volova, T. G. (1999). *Biotechnologiya*. Izdatel'stvo Sibirskogo Otdeleniya RAN Rossiyskoy Akademii Nauk.

Wasden, Ch. L., & Williams, B. S. (2012). *Owning the disease: A new transformational business model for healthcare*. http://download.pwc.com/ie/pubs/2012_new_transformational_business_model_for_healthcare.pdf.
von Weizsäcker, E. U., & Wijkman, A. (2018). *Come on!* Springer.
White, R. A. (2021). The future of virology is synthetic. *Msystems, 6*(4), e00770-e821. https://doi.org/10.1128/mSystems.00770-21
Widdowson, M. (2021). From Covid-19 to zero-gravity: Complex crises and production revolutions. *Journal of Globalization Studies, 12*(1), 117–144.
Wilenius, M., & Casti, J. (2015). Seizing the x-events. The sixth K-wave and the shocks that may upend it. *Technology Forecasting Social Change, 94*, 335–349.
Wilenius, M., & Kurki, S. (2012). *Surfing the sixth wave: Exploring the Next 40 years of global change*. Turku: Finland Futures Research Center.
Williams, G. M., Kroes, R., & Munro, I. C. (2000). Safety evaluation and risk assessment of the herbicide round-up and its active ingredient, glyphosate, for humans. *Regulatory Toxicology and Pharmacology, 31*(2), 117–165. https://doi.org/10.1006/rtph.1999.1371
WIPO = World Intellectual Property Organization. (2023). *World intellectual property organization database*. https://www3.wipo.int/ipstats/IpsStatsResultvalue.
Woollett, R. (2012). Innovation in Biotechnology: Current and future states. *Clinical Pharmacology and Therapeutics, 91*(1), 17–20.
Yagafarova, G. G., Ibragimova, Ch. I., & Mallyabaeva, M. I. (2013). Comparative characteristics of the biological stability of nitrogen containing herbicides. *Bashkirskiy Khimicheskiy Zhurnal, 20*(3), 153–155.
Zagorski, I. (2012). Ne myasom yedinym: kozhanyye kurtki budut vyrashchivat' v laboratorii. *Vesti.ru*. September 20. http://www.vesti.ru/doc.html?id=912084&cid=2161.
Zhang, S. (2003). Fabrication of novel materials through molecular self-assembly. *Nature Biotechnology, 21*, 1171–1178.
Żółkiewicz, J., Marzec, A., Ruszczyński, M., & Feleszko, W. (2020). Postbiotics—a step beyond pre- and probiotics. *Nutrients, 12*(8), 2189.
Zimmer, C., Corum, J., & Wee, S.-L. (2020). Coronavirus vaccine tracker. *The New York Times*, June 10. https://www.nytimes.com/interactive/2020/science/coronavirus-vaccine-tracker.html.

# Nanotechnologies, Robotics, Artificial Intelligence and Other MANBRIC Technologies in the Long-Term Development

# 10

**Abstract**

In this chapter, Grinin et al. present an analysis of the development of a number of technologies: nanotechnologies, robotics, additive and cognitive technologies, ICT and AI, as well as forecasts for their future. Together with medicine, they are integrating in the united MANBRIC complex (this is an acronym made up of the first letters of seven technologies). Each direction has its own future, problems and coordinate system. The authors try to forecast major breakthroughs for each technology. In general, according to Grinin et al., they have a bright future. The chapter analyzes the possible advances in various technologies, highlights the challenges, problems and difficulties along the way and analyzes the possible negative consequences. A comprehensive analysis of these technologies within a single chapter allows Grinin et al. to show that the final phase of the Cybernetic Revolution will not be a wave of diverse innovations, but a complexly integrated and interconnected set of new generation technologies that will create an era of self-regulating/self-managing systems. The chapter opens up to the reader many new perspectives, ideas, facts and problems that humanity will face in the process of the final phase of the Cybernetic Revolution.

## 10.1 What This Chapter is About

In this chapter, we present an analysis of the development of a number of technologies: nanotechnologies, robotics, additive and cognitive technologies, ICT and AI, as well as forecasts for their future. Together with medicine, they are integrating in the united MANBRIC complex (this is an acronym made up of the first letters of seven technologies). Each direction has its own future, problems and coordinate system.



L. Grinin et al., *Cybernetic Revolution and Global Aging*, World-Systems Evolution and Global Futures, https://doi.org/10.1007/978-3-031-56764-3_10

We try to forecast major breakthroughs for each technology. In general, we believe they have a bright future. Nanotechnologies can have a significant impact on saving materials and energy, making major advances in medicine and many other industries and creating materials with amazing properties. Additive technologies promise to change the economy, moving from mass production to small-scale or customized one.

A universal smart robot will be able to alleviate the problems of labor shortage by replacing a nanny, a nurse, a cook and even a sapper, as well as many other professions. Cognitive technologies will enable rehabilitation of many millions of people who are currently paralyzed or otherwise disabled. AI will be an engine of innovation in almost all industries and spheres. One can expect driverless cars and pilotless aircrafts/drones will become widespread in the future.

The chapter analyzes these and other possible advances in various technologies, and also highlights the problems and difficulties along the way, as well as present arguments and reasons why certain forecasts may appear untenable. At the same time, we also analyze the possible negative consequences, including the threat to jobs posed by robots.

The described technologies are different, but there is a very important thread that unites these different technologies.

First, they all provide excellent examples of self-regulating and self-managing systems, that is, systems that can make decisions with minimal human intervention or even autonomously. Self-managing systems are the basis of the future technological paradigm and the main characteristic of the Cybernetic Revolution.

Second, all the technologies described in the chapter belong to the complex of innovative directions MANBRIC, for which we forecast a great future.

Third, using the example of the technologies described below, we can see the convergence of MANBRIC technologies. The MANBRIC clustering in robotics is an excellent example of this, as it combines the maximum number of technological directions. One must pay attention to the mutual influence of technological fields as they impact the direction of development choices. In particular, the chapter shows how AI programmers can learn from cognitive science, how they try to design robotics for the home Internet of robots (giving opportunities to connect robots to a home computer or an operation center). We trace the closest connection between MANBRIC directions in ensuring the functioning of institutions and industries. Thus, the military sphere will be an important driver for the development of these MANBRIC technologies, but of particular interest to the military are robotics, including drones, AI and cognitive technologies.

Fourth, we try to show how different features of the Cybernetic Revolution and the complexity of the functions of self-regulating systems are manifested in each technology. In particular, the functionality of robots can be distinguished as follows: active information processing managing the whole system (its receiving, analysis, distribution, transformation, etc.) by means of flexible interaction with the environment, existing contours of feedback and direct links that fulfill various functions and the collection of information to increase efficiency.

We pay particular attention to the relationship between robotics, medicine and biotechnology. It is shown that robotics and ICT, medicine, bio, nano, additive and cognitive technologies are already quite obviously combined in medical institutions, particularly when working with disabled people. In addition to medicine, AI plays a special role in the integration of MANBRIC. One can speak about a rapid development in the progress of artificial intelligence. We agree that the future is closely associated with the development of artificial intelligence, which will be involved in a very large number of self-regulating/self-managing systems. It is important to define the similarities and differences between self-regulating systems and artificial intelligence.

Described in a single complex, the MANBRIC technologies have different levels of development and will mature at different times, but all are likely to flourish in the future. While AI has already reached a very high development level, with respect to nano and additive technologies including robotics, one can expect an explosive period of development only in the 2050–2060s. One can expect a mass development of new means of transport in the middle or at the end of the final phase of the Cybernetic Revolution (between the late 2040s and 2060s).

However, self-driving electric vehicles can become a powerful source of technological development during the mature phase of the Scientific-Cybernetic Production Principle. Like self-driving cars, drones will sooner or later achieve true autonomy. Due to the great economic effect and convenience of such robots, they will have a very promising future.

A comprehensive analysis of these technologies within a single chapter allows us to show that the final phase of the Cybernetic Revolution will not be a wave of diverse innovations, but a complexly integrated and interconnected set of new generation technologies that will create an era of self-regulating/self-managing systems.

Thus, in this chapter, we solve a twofold problem—on the one hand, to show the logic of and development prospects for each of the technologies described. On the other hand, we try to show how the development of each of these technologies fits into the MANBRIC convergence as a whole.

We appraise how MANBRIC will develop in the future, how it will strengthen it in the future, how it will prepare medical technologies for the initial breakthrough that will launch the final phase of the Cybernetic Revolution and ensure the powerful development of self-regulating systems.

The chapter opens up to the reader many new perspectives, ideas, facts and problems that humanity will face in the process of the final phase of the Cybernetic Revolution.

## 10.2 Nanotechnologies as the Tool for Mastering the Microworld

### 10.2.1 On Nanotechnology: An Introduction

Humankind has been using nanomaterials for a long time. With the current knowledge about nanoparticles one can explain the peculiar properties of the well-known materials created in ancient times such as various enamels, painting materials, damask steel, etc.

Nanotechnology is a widely used concept that can be conventionally defined as *an interdisciplinary field of applied science and technology that develops practical methods and research, analysis and synthesis, and also methods of producing nanomaterials by controlled manipulation of tiny objects down to individual molecules and even atoms*. However, the idea of how to most effectively use nanotechnology for particular problems (primarily in the field of heath) has appeared quite recently. The wide range of topics covered by nanotechnology makes it problematic not only to define it, but also to classify nanoproducts.

The main point of nanotechnology is the control of matter on a scale smaller than 1 μm, usually between 1 and 100 nm (up to 100 nm in one measurement). A nanometer is a billionth of a meter, or $10^{-9}$ m, or a conditional construction of ten hydrogen atoms built in a row.

Why have nanoparticles become so popular? The matter is that in fact this is a very broad new domain which opens up colossal opportunities for mastering the microworld.[1] It is just at this level that the fundamental property of matter is clearly revealed, which is the realization of antipodal properties in its various systems. For example, gold is a conductor at the macro level, but it is an insulator at the nano level. The particles of some substances, which are between 1 and 100 nm in size, have very good catalytic and absorptive properties, while other materials have wonderful optical properties. At the nanoscale, the surface-to-volume ratio changes, and so do the properties of matter. In nature, there are nanosystems that are capable of organizing themselves into special structures and thus acquiring new properties, for example, the biopolymers (proteins, nucleic acids).

The peculiarity of nanoscience consists in the fact that it deals with atoms—compound particles of matter (to repeat: a conditional construction of ten hydrogen atoms built in a row of one nanometer in size). Scientists are now closer to understanding how it will be possible to force atoms to 'build' things. If so, such an approach opens up fantastic opportunities to create new materials with desirable properties. Scientists believe that it would become possible to mechanically move individual atoms using a manipulator (atom-by-atom assembly of objects according to the Nobel laureate Richard Feynman). Some of them think that

[1] Knowledge of the properties of nanoparticles also opens up great opportunities for understanding the properties of matter and its evolution. In particular, all systems contain small amounts of hundreds of undetectable substances, many of which are present in the form of nanoparticles (Melikhov, 2018: 10).

this process will not contradict the known physical laws, although there is quite reasonable skepticism about this (in particular in connection with the known chemical laws, see below). At present, various ingenious techniques are used for such manipulations, but there is no solid solution yet.

### 10.2.2 Nanotechnology as a Result of the Cybernetic Revolution: Origin and Development

The first practical steps in nanotechnologies (as well as the ideological interpretation of the field) were made in the 1950s,[2] whereas the term, according to some scholars, was introduced in 1974 by the Japanese physicist Norio Taniguchi. Nanotechnology emerged as a result of the Cybernetic Revolution. However, for quite a long time it remained in the background of other important branches of science. Practical interest in nanotechnology grew rapidly at the end of the initial phase of the Cybernetic Revolution, in the 1980s, coinciding with the publication of Eric Drexler's books (1987) *Engines of Creation: The Coming Era of Nanotechnology and Nano-Systems: Molecular Machinery, and Computation* (see also Drexler, 1992, 2013). However, the term became widespread when it was picked up by the media. With the start of 'the nanotechnology race', the word 'nano' frequently appeared on television and in print. This meant that nanotechnologies began to be considered as a strategic branch of a future hegemon (along with others: biotechnologies, green energy industry, etc.). Their ultimate task is to win the market of industrial production of new, important and in-demand technologies. The country that succeeds in this will be able to ensure its own economic growth and development for many years to come.

The nanotechnology race began with an initiative by President Bill Clinton in 2000, which became a landmark and launched the competition between countries. Then Japan and Western Europe joined the race, conducting research in nanotechnologies as part of their national programs (see Lane & Kalil, 2007). More and more attention was paid to nanotechnology in China, South Korea and many other countries, including Russia, which had a relatively good starting position in this field (Dementiev, 2008). Today nanotechnology is one of the most intensively developing sectors of the economy.

[2] The prospects of this field were heralded by the Nobel laureate Richard Feynman in his report 'There's Plenty of Room at the Bottom', presented in 1959 at the California Institute of Technology at the annual meeting of the American Physical Society.

### 10.2.3 The Development of Nanotechnologies in the Course of the Cybernetic Revolution

**The initial phase**. The characteristics and opportunities of nanotechnologies correlate with the concept of the Cybernetic Revolution, which is not surprising since they originated within this revolution and besides, will play an increasingly important role in the course of its development.

The initial phase of the Cybernetic Revolution (from the 1950s to the early 1990s) was the formative period for nanotechnology. Roughly speaking, this period began in 1959 when Richard Feynman presented the idea of constructing new materials from nanoparticles, and ended with President Bill Clinton's initiative in 2000 (Callaway, 2020). This period is characterized by a large number of scientific discoveries; however, many of them had no application at the time (e.g., Rybalkina, 2005: 21).

At this stage, the development of nanotechnology was in many ways defined by the creation of probe microscopy. These devices are the eyes and hands of nanotechnologists. In particular, in 1981, the German physicists created the scanning tunneling microscope (STM), which can measure interactions between atoms (Chen et al., 2021). Then, in 1985, the American physicists developed the technology to precisely measure particles of one nanometer in diameter. Since the 2000s, inventors have been moving toward seeing atoms in a microscope. In 2010, researchers at UCLA used a cryo-electron microscope to see the atoms of a virus. Over the last decade, there have been breakthroughs in obtaining clear images down to the individual atoms (e.g., Chen et al., 2021).

**Nanotech euphoria**. The modernization phase (the period of diffusion of innovations) of the Cybernetic Revolution is the period of the emergence of 'modern nanotechnology' (from the 1990s to the 2020/30 s). Nanotechnology has become involved in industrial production, while nanomaterials are penetrating into various fields: engineering, medicine, transport, aerospace, electronics, etc.

The first phase of the development of nanotechnologies (between 2000 and 2005) is associated with the so-called 'passive nanostructures' (incremental nanotechnologies), but generally it involved the production and use of nanodisperse powders. They have been added in order to modify the properties of basic construction materials: metals and alloys, polymers, ceramics, and are also used in cosmetics, pharmaceuticals, etc. Today this is considered to be a primitive generation of nanomaterials which are already widely used in manufacturing and in many products. Only a few nanoprojects have been introduced in high-tech industries.

The broad prospects offered by nanotechnologies, fueled by various interests and the mass media, gave rise to a real euphoria. However, most of the predictions proved wrong or were hardly possible to fulfill.[3] Figure 10.1 clearly shows the

[3] For example, according to the BCC Research (2012), the sales volume of nanotechnology products amounted $11.67 billion in 2009 and could reach $290 billion by 2028 (Emergen Research, 2021). However, there have been even more optimistic forecasts. For example, according to the

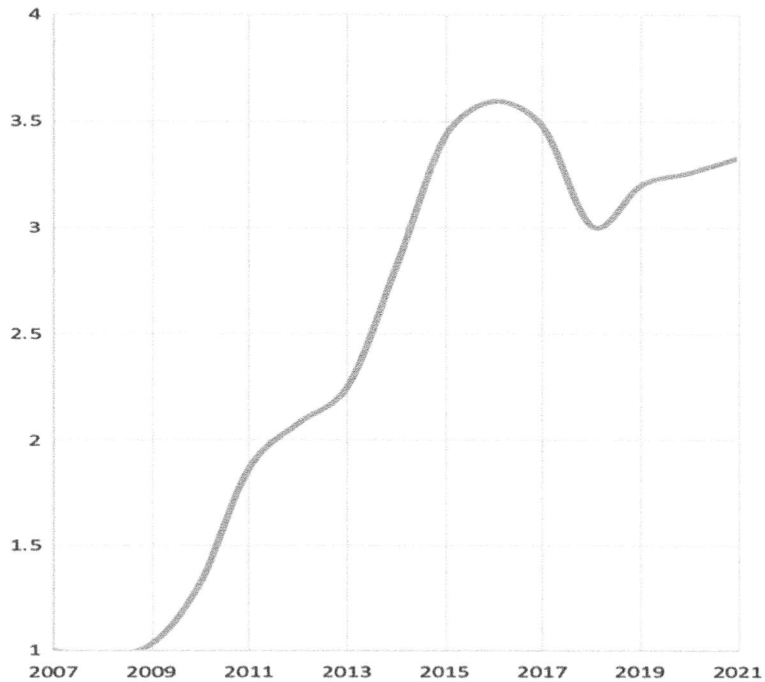


**Fig. 10.1** Relative global dynamics of patent grants in micro-structural and nanotechnologies, 2007–2021, 1 = 2007 level. Data source: WIPO (2023) (see also Fig. 10.1 in Chap. 4)

explosive growth in the number of patent grants in nanotechnologies, which well explains the euphoria of forecasts in this area, as well as the fact that development in the field of patent grants in nanotechnologies has stalled since 2016.

People always want innovations to go faster than they do. At the same time they do not see the obstacles and challenges and do not take into account economic problems. Moreover, such over-optimistic predictions are often driven by a simple desire to attract huge investments. Thus, the growth rates of nanoproduction volumes are far from being as fast as previously predicted.

Some analysts suppose that after 2020 we will enter the era of 'radical nanosystems' in the form of nanorobots and that nanobiotechnological and nanomedical systems will develop rapidly. However, we doubt such rapid development. The theory of production revolutions suggests that despite the numerous innovations that will appear during its modernization phase, they will hardly make a breakthrough to the higher innovative level, while many will remain in low demand. The discoveries that will provide the basis for the breakthrough are being prepared, while

UK Department of Trade and Industry, the demand for nanotechnologies will exceed $1 trillion annually by 2015, and the number of experts in this field will increase to two million people.

the breakthrough itself will come later. In the case of nanotechnology, this will most likely happen between the 2030s and 2050s, i.e., the real take-off of nanotechnology will actually take place a decade or two later than many researchers predict. Nevertheless, the coming decades will see the development of many earlier achievements in various fields. Outstanding innovations will certainly appear during these decades.

### 10.2.4 Forecasts for the Final Phase of the Cybernetic Revolution and Beyond

One can trace all the characteristics of the Cybernetic Revolution in the future development of nanotechnologies: (1) transition to the control of deeper and more fundamental processes and levels (down to the atomic level with the use of nano-particles as building blocks); (2) deep interconnection between directions of MANBRIC convergence (see below); (3) mass invention of technologies based on self-regulating systems (in which nanotechnology will play an essential role); (4) production of new materials with new properties (as we have seen above, there is a great opportunity to achieve this at the nano level); (5) resources and energy saving (e.g., nanomaterials in window glass can drastically reduce energy consumption in homes; nanotechnologies will help to deliver an optimum amount of medicine directly to the damaged area or even to separate cells); (6) nanotechnology will increase the trend toward super-miniaturization, etc.

#### 10.2.4.1 Nanotechnology in Medicine: Great Opportunities

Despite the serious progress of nanotechnologies in electronics and other branches, we believe that the real nanotechnological breakthrough will most likely occur first in medicine, which will give an additional boost to its use in other fields. As a result, the breakthrough in the final phase of the Cybernetic Revolution will be provided by a deep integration of medicine with bio and nanotechnologies, resulting in the vigorous development of bionanomedicine. All these will create important backgrounds for the future era of self-regulating systems in medicine.

One of the future directions in medicine is the development of diagnostic methods that reduce costs. We have already mentioned nanochips (in Chaps. 8 and 9), which will play an important role here. The nanorobots will be able not only to perform medical functions, but also to control individual cell nutrition and excretion of waste products, will be put into practice. Nanorobots can be used to solve a wide range of problems, including the diagnosis and treatment of diseases, the fight against aging, the reconstruction of some parts of the human body, the production of various heavy-duty constructions, etc. (Giri, Maddahi, & Zareinia, 2021; Mallouk & Sen, 2009; Nistor & Rusu, 2019).

We have already mentioned some integration trends in these branches in Chaps. 8 and 9. In general the prospects for such integration are apparent within two to three decades. According to some forecasts, nanobiostructures (i.e., capable of carrying medical nanosensors and even reconfiguring the cells of an organism) will become commonplace. Their active use in diagnostics and as a means of boosting immunity will become an important direction in nanotechnology (Zhang et al., 2022). Already today, some advanced medical organizations apply nanotechnologies to create a biochip that allows a quick diagnosis of a number of dangerous diseases, including tuberculosis (e.g., Gavas et al., 2021). The development of nanotechnology in new materials that mimic new properties, such as those of bone tissue, is promising. The miniaturization of these robotic platforms has led to many applications in precision medicine (Soto et al., 2020). Nanotechnologies are already implemented in surgeries such as nano-neuroknitting to repair severed optic nerves, in implantation of artificial limbs with high precision, cardiac surgery, etc.

Currently, nanotechnologies are not yet used in medicine as mass technologies (see Chap. 8 for details). However, the research, experiments and tests, carried out today, can reveal the future role of nanotechnologies in medicine and give an idea of the main principles of their application, as well as some details of the future successful and breakthrough nanotechnologies.

Here is an interesting example. At the Laboratory of Nanophotonics at Rice University in Houston, Professors Naomi Halas and Peter Nordlander invented a new class of nanoparticles with unique optical properties—nanoshells (Henderson et al., 2022). Twenty times smaller in diameter than red blood cells (erythrocytes), they can move freely in the blood system. Special proteins, antibodies that attack cancer cells, are specifically attached to the surface of the cartridges. A few hours later after their injection, the organism is exposed in infrared light, which the nanoshells convert into thermal energy. This energy destroys cancer cells, while leaving the neighboring healthy cells virtually unharmed.

One of the areas where huge efforts of nanotechnology are concentrated is the fight against cancer (National Institute of Health, 2016). Many new ways to fight cancer are based on nanotechnologies (Aghebati-Maleki et al., 2020; Gavas et al., 2021; National Institute of Health, 2016; Yao et al., 2020). Approved cancer nanomedicines to date are liposomal drug delivery and magnetic fluid hyperthermia based on magnetic iron oxide nanoparticles (Soetaert et al., 2020). For example, the latter is used for carcinoma treatment and is based on heating iron oxide nanoparticles that are introduced into the infected tissue and exposed to a magnetic field, causing particles to heat up and destroy malignant cells (Gavilán et al., 2023).

Researchers are working to develop nanomaterial-based delivery platforms that will reduce the toxicity of chemotherapy drugs and increase their overall efficacy (National Cancer Institute, 2017a).

Nanotechnologies are also being investigated for the delivery of immunotherapy. This includes the use of nanoparticles to deliver drugs, immunostimulatory or immunomodulatory molecules in combination with chemo- or radiotherapy, or as adjuvants to other immunotherapies (National Cancer Institute, 2017b), in particular for the delivery of Herceptin, an effective drug for the treatment of breast cancer.

The value of nanomaterial-based delivery has become apparent for new types of therapeutics such as those using nucleic acids, which are highly unstable and susceptible to degradation. These include DNA- and RNA-based genetic therapeutics. This therapeutics is used in many cases to target 'undruggable' cancer proteins. Additionally, the increased stability of gene therapies delivered by nanocarriers, and often combined with controlled release, has been shown to prolong their effects (National Cancer Institute, 2017b).

It is extremely important that the development of nanotechnologies can radically reduce the side effects, as well as often very serious and sometimes even fatal consequences of medical interventions for patients, including the invasiveness and the damage caused by excessive doses of drugs, including chemotherapy, not to mention the discomfort and pain experienced by patients. And the cost of treatment can be drastically reduced.[4] In Chaps. 8 and 9 we have already discussed the use of nanostructures in new types of vaccines. Despite serious shortcomings and problems, this direction will undoubtedly develop.

Of course, new materials and the acquisition of new properties by materials through nanotechnologies (e.g., nanomaterials with self-cleaning mechanisms to remove bacteria from blood vessels) will play an important role in the development of medicine. This can manifest itself in the creation of artificial biological tissues, artificial immunity, support for damaged tissues, and much more. Some examples have already been given in previous chapters.

#### 10.2.4.2 The Connection with Biotechnologies and Agriculture

Other important directions of nanotechnology include research into nanobiotechnology beyond medicine, particularly, in agriculture. Here, fields such as veterinary medicine and animal husbandry are akin to the field of human medicine. Animals, like humans, need treatment, protection from infectious diseases, enhanced immunity, and the development of feeding technologies. Thus, in a number of cases, similar nanotechnologies can be applied to medicine, veterinary medicine and agriculture. One can mention here the development of controlled protein synthesis technologies to obtain peptides with desirable immunogenic properties. Vector systems for the cloning of immunologically important proteins from pathogens and new generation vaccines with high efficacy and safety are being created. They work on creating nanoparticles for the production of genetically modified proteins, the development of biochips and test systems for biological screening (Persidis, 1998), immune monitoring and the prediction of dangerous and contagious animal

[4] In this respect, one of these technologies, created for personal use, is interesting. An American chemical engineer Mark Davis has developed a new way to fight his wife's cancer. He discovered a special capsule that has a structure similar to sugar and is therefore not rejected or excreted by the body. This capsule contains a drug that can remain in the body for weeks. It searches for a tumor that moves within the blood vessel system. Cancer cells are more acidic than normal healthy cells, and, when they are found, the capsule opens and releases the powerful medicine. His wife with a terminal pancreatic cancer at the stage of metastasis successfully underwent such a treatment (LaPook, 2013).

diseases. Biochip technology is constantly improving and production has become cheaper. This has become even more evident with the wave of development of tests for the COVID-19.

Nanotechnologies are already actively applied in various sectors of agriculture, in particular, in the production of animal feed which allows a considerable reduction in their consumption and makes it more accessible. In crop production, the use of nanopowders with antibacterial components provides increased resistance to adverse weather conditions and increases the productivity of many food crops, such as potatoes, cereals, vegetables, fruit and berries. However, the future use of nanotechnologies in the agriçultural sector holds much greater promises. In particular, nanotechnology can also help to solve the problems of wastewater treatment. Perhaps, more important is the expectation that nanotechnologies and the application of biotechnologies will support significant progress toward the creation of self-regulating systems in agriculture, so that agricultural operations will largely be carried out in an autonomous mode. Many technologies will appear to promote this process. Thanks to their more targeted action and dosage of drugs and substances, these technologies are likely to significantly reduce resource and energy costs and environmental impact. Thus, the introduction of membrane systems for purification, as well as special biocidal coatings and silver-based materials will facilitate and increase the level of livestock managment and the provision of high quality water.

It is expected that the use of nanotechnologies will change land cultivation through the use of nanosensors, nanopesticides and a decentralized water purification system. Nanotechnologies will make it possible to treat plants at the genetic level and to create high-yielding plant varieties that are particularly resistant to unfavorable conditions (Balabanov, 2010). Nanopesticides also can appear less harmful for the environment (Jain, 2004; Kah & Hofmann, 2014). Today there are some innovative ideas that can be further developed in agriculture. In particular, microbial preparations based on associative, endophytic and symbiotic bacteria have emerged. These preparations are designed to produce and transport various enzymes and low-molecular biological active agents (nano-objects) in plants. These can improve the adaptation of plants to unfavorable environmental conditions: pollution by toxic metals, salinization, superacidity, etc.

A fundamental approach is essentially developed for the purpose of obtaining high quality seed material. This approach is as follows: biologically active and phytosanitary components which can increase the adaptation of seeds and plants to real negative environmental conditions are constructed in the form of multifunctional nanochips (Ruban et al., 2011; Voropaeva et al., 2012).

Nanotechnologies can be used in a number of other areas, including being one of the means to preserve and restore the ecological environment.[5]

[5] Chinese scientists have developed a system that can generate electricity through the decomposition of organic matter and removal of organic compounds from waste water. Yanbiao Liu with colleagues have developed a nanotube-based photocatalytic fuel cell that uses solar energy to break down organic compounds in waste water and convert chemical energy into electrical one.

#### 10.2.4.3 New Materials and Their Properties

**Self-regulation and self-organization of nanoparticles. Synthesis of new materials with desired properties**. A close connection between nanotechnology and self-regulation within systems is based on their ability to change processes of self-organization of matter, forcing molecules and atoms to be ordered in a particular spatial and structural way. So, one of the major challenges for nanotechnology is to induce molecules to group and self-organize in the necessary pattern to acquire new properties. The supramolecular branch of chemistry is concerned with this problem and explores interactions that can organize molecules to create new substances and materials. Various processes of self-assembly have already been discovered, such as the electrochemical anodic oxidation (anodizing) of aluminum. However, nanoparticles require largely (or even fundamentally) different technologies for self-assembling. This is a very promising way to create a huge family of materials with desired properties. The creation of new materials with desirable properties is a direct way to make systems work according to predetermined scenarios.

We are still at the beginning of this path. Nevertheless, at present within the field of nanocomposite construction different technologies have been developed that produce substances that, for example, are capable of protective coating, self-cleaning, antibacterial protection, etc. Such nanotechnologies produce striking examples of various self-regulating systems, for example, self-cleaning nanocoatings (i.e., self-cleaning mechanisms to remove bacteria from vessels or self-cleaning nanopolish products for car glass). Nanopolishes modify the surface in such a way that a drop of water slides across it, collecting all the dirt, whereas on a smooth surface, on the contrary, a drop of water slides across it and leaves the dirt on the surface. This is called the 'lotus effect'. The idea is borrowed from nature: the leaves of the lotus plant are covered with the tiniest bulges and cavities of wax so water runs down and washes the dirt away. Fabrics may be developed that give clothing a variety of properties, for example, protection against mosquitoes (Ciera et al., 2019). Nanofibers have great potentials in the medical field, for example, in a wireless garment with embedded textile sensors for simultaneous detection and continuous monitoring of ECG, respiration and physical activity. In the military, wireless communication of a fabric to a central unit allows medics to conduct remote triage of casualties to help them respond faster and safer [Nanotechnology is especially used in sports and following the current trend, sportswear of all types will experience the nanotechnology revolution, in the near future (Ciera et al., 2019; Harifi & Montazer, 2017; Ahn et al., 2006).

But even more important prospects open up in the creation of materials with unique properties for industry. Thus, according to some (possibly exaggerated) reports, the addition of graphene to nanotubes dramatically increases the useful properties of various materials. For example, the addition of nanotubes to tires reduces their weight by a one-third and their braking distance by 45. The addition of nanotubes to aluminium makes the resulting composite material four times stronger and 20 times harder than conventional aluminium (Ocsial, 2017).

There is a great potential for the use of new nanoproperties in electronics in other industries as well. For example, an unusual property of crystals opens up opportunities to study new magnetic phenomena (see Melikhov, 2018; Smith, 2022). Thin films of oxygen, copper and titanium nitride have been obtained. Their electrical resistance is a thousand times lower than that of conventional titanium nitride. This could be a technological breakthrough in the development of next-generation resistors and transistors (Baron et al., 2021).

Many nanotechnologies aim at energy saving and invention of alternative energy sources. So, the trend toward miniaturization of technology not only increases the operating speed of electronic devices and the packing density on the chip (see below), but also reduces their energy consumption. For example, 'smart glass' for rooms can respond to changing lighting and ambient temperature by changing its transparency and heat conduction. There are many different projects that lead to energy and cost savings. To name just one of the utmost importance, a widespread use of electronic paper could prevent deforestation.

#### 10.2.4.4 Nanomechanical Engineering

Nanotechnology has considerable prospects, which could be observed in the development of nanomechanics, nanomechanical engineering and nanorobotics. The components of nanoelectronics, photonics, neuroelectronic interfaces and nanoelectromechanical and other systems are also likely to be developed. There are ideas, that based on the achieved results, we can expect advance in the development of nanosystems capable of self-assembly into three-dimensional networks.

It is also possible to talk about the use of nanotechnology in molecular devices, nuclear design, etc. As we will discuss in the AI section below, the idea of computers that will use nanotechnology to create and store information has been around for some time. There are ideas that such computers can be developed on the basis of nanoelectronics; however, according to Eric Drexler (1987), nanomechanics can also become a basis for storing information. He even proposed mechanical designs for the main components of a nanocomputer—memory cells, logical bytes. From special structures, such as fullerenes, nanotubes, nanocones and others, molecules can be assembled in the form of various nanodetails—gears, rods, bearing details, rotors of molecular turbines, moving parts of manipulators, etc. The assembly of the parts into a mechanical design can be performed by the assemblers (self-assemblers) with the biological macromolecules attached to the details which are capable of selective connection with each other.

Professor James Tour and his colleagues at Texas Rice University have tried to realize some of Drexler's ideas about nanomechanics. In 2005, they created a molecular-mechanical design—the all-molecular four-wheeled nanocar, about 2 nm wide, which consumes light energy. It consisted of about 300 atoms and had a frame and axles. The development and construction of the nanocar took eight years. The scientists even plan to create nanotransport devices and nanotrucks to transport molecules to conveyors in nanofactories (Balabanov, 2010).

Certainly, they are more like toys, than practical research. They remind us of the steam toys similar to the mechanisms created by the Greek mechanic Heronus

Alexandrinus who amazed the public in the first century CE. His devices barely resembled steam engines. But unlike Heronus, who did not even think of a practical use for steam, today's nanotechnologists are preoccupied with practical applications. Therefore, the creation of nanomechanical engineering seems quite real, albeit in the long term. It would most likely happen near the end of this century. The same applies to nanorobotics, although the widespread diffusion of the latter may occur much earlier (see below).

Many of today's futurological predictions about the development of nanomechanics, nanoelectronics, nanoengineering, etc. seem too bold and in some respects even fantastic. However, nanotechnologies will move in this direction one way or another. Anyway, it is obvious that both nanomechanical engineering and nanorobotics will take the development of self-managing systems to a new level, leading to the creation of an industry that will design such systems (similar to how the use of cars promoted their industrial manufacture—mechanical engineering).

#### 10.2.4.5 Self-Assembly: Its Essence and Feasibility

Will self-managing systems be created, such as molecular assemblers, which are something like 3D printers at the molecular level? When we talk about molecular assemblers, we mean devices that can convert a given mass of matter in such a way that it can be reassembled into something else (into whatever we want), given a molecular model (assembly program) for it. This means that the assembler can work in much the same way as genes and ribosomes produce proteins, breaking them down into nucleotides and reassembling them. It was about molecular assemblers that Eric Drexler dreamed of when he described the coming assembler revolution, which he said would be more important than medical technologies, space horizons, advanced computers and new social inventions because it would affect each of these and many other areas.

> Before moving on to a discussion of whether such assemblers are possible, it should be noted, by the way, that a number of future technologies that Drexler was trying to describe and use fit very well with our understanding of self-regulating/managing systems. For example: 'Assembler: A molecular machine that can be programmed to build virtually any molecular structure or device from simpler chemical building blocks. Like a computer-controlled machine shop. Cell Repair Machine: A system that includes nanocomputers and molecular size sensors, as well as tools programmed to repair damage to cells and tissues. Automated engineering: the use of computers to carry out technical developments, in the extreme cases, detailed studies with little or no human assistance to a given general specification. Automated engineering is a specialized form of artificial intelligence' (Drexler, 1987).

There are very strong arguments against such ideas. After all, atoms are not grains of sand, it is impossible to move and sort them just in such a way (because they are bound together not by mechanical, but by electromagnetic and other bonds). In particular, Smalley concluded that, when viewed outside of biological systems, the idea of molecular assembly and molecular manufacturing is in fundamental conflict with the laws of chemistry. In a public debate with Drexler in the pages of scientific journals, he argued that atoms simply cannot be put into place by mechanical means—you have to force them to form the appropriate bonds. But creating molecular machines that can do this is simply impossible (see e.g., Smalley, 2001; Drexler, 2003; see also Ford, 2015).

Nevertheless, it is possible that some kind of assemblers will be created (but this is clearly not a matter of the near future), and on this basis it will be possible to significantly expand our capabilities, as well as to open up a large niche for development and solving a number of important problems. But humanity's technological aspirations have repeatedly encountered the fact that the idea of limitless energy or raw material possibilities (like perpetual motion projects) inevitably encounters very serious obstacles. Therefore, the idea of finding something that leads to unlimited abundance is utopian. Such assemblers will require a lot of energy and resources, and it is obvious that they will not be for free. And still, this will be a very significant step on the endless path to finding the philosopher's stone that can turn everything into anything.

## 10.3 Robotics as a Direction in the Development of Self-regulating/Self-managing Systems

The concept of a robot is very ambiguous since today computer programs, manipulators and other mechanisms, human-like autonomous devices as well as some nano- and microdevices can be called robots. However, despite this diversity, a number of the main characteristics of robots of all types can be defined. It is of particular importance to us that these characteristics coincide to a large extent with the characteristics of the Cybernetic Revolution and its technologies. That is why an ideal robot (which can move, work and solve problems and communicate in a meaningful way depending on the situation) is a good example of a self-regulating/self-managing system. Many definitions also emphasize the aspiration to treat the robot as a self-regulating system. For example, one definition states that a *robot is a device capable of moving independently in space, coping with tasks of picture recognition and analysis, possessing a high degree of mobility, being able to analyze a situation through feedback and also being able to predict situations, relying on its own experience and available information* (definition by Professor Shigeru Vataat, see Nakano, 1988: 26).

The opportunities for the use of robots are undoubtedly vast. In particular, they can help to solve the problem of caring for the growing number of elderly people and the related problem of labor shortages.

### 10.3.1 Robotics at the Modernization Phase of the Cybernetic Revolution

The initial phase of the Cybernetic Revolution convincingly showed a powerful rise of robotics which is an entirely new branch of mechanical and electronic engineering; thus, hardly anyone doubted its promising prospects for development and widespread implementation. Futurologists expected the coming era of universal robots [e.g., Moravec, 1988]) and suggested, that the development of the robot, and robotics as an industry have limitless potential (e.g., Kahn, 1982: 182). However, the development of robotics has been much more modest. The reason is that a big part of the large-scale industry in the Western countries has moved to China and other developing countries since the 1990s. As a result, the share of industry (and the number of workers) in the Western countries has declined, the need to replace workers has significantly decreased, and respectively, the investments in relevant research have been reduced.

In the 1990s, robots continued to develop: their characteristics, software and interface were improved; their control became easier, etc. But the focus was on developing information robots and not industrial robots (see below; we have already spoken about the use of such robotic programs at stock exchanges in Chap. 4). Nevertheless, in the 1990s and 2010s there were (and still are) many optimistic and over-optimistic forecasts on the development of robotics, including for industrial purposes.[6] It was argued that the task of the next decade or two was to robotize everything (Proydakov, 2016). But the reality turned out to be much more prosaic. For example, Terry Gow (Bonev, 2013; World Robotics, 2013), the head of Taiwan's *Foxconn*, the world's largest microelectronics manufacturer, was going to increase the number of robots from 10,000 robots to 300,000 in 2012 and to 1 million by 2014 (Auslander, 2014). But he only managed to fulfill 5% of what he promised.

Such a leap forward in forecasting opportunities is quite a common case. As we have pointed out above, this can also be observed in other innovative industries, including electric vehicles, unmanned vehicles, nanotechnology, 3D printers, etc. This is due to the natural difficulty of forecasting, and is often linked to the PR campaigns of manufacturers who use such forecasts to actively involve investors in a new project.

There are currently about 3 million industrial robots in the world (IFR, 2020). We mean only the most advanced machines with at least three axes of motion and the possibility of free programming. Robots are primarily used for work in sterile conditions (in the electronics and pharmaceutical industries), product assembly and packaging (Smirnova, 2011). Nevertheless, a significant number of robots are still used in car factories. Robots are actively employed in logistic centers and in some other areas. Of course, robots have serious advantages over humans: they work much faster and besides, their work is of a higher quality, etc. However, we suppose that the industrial direction of robotics will hardly make a breakthrough during the initial period of the final phase of the Cybernetic Revolution in the 2030s and beyond. There will be other directions of robotics that could revolutionize this sector (see below).

[6] At the same time, they have been the basis for pessimistic social forecasts that robots will take millions of jobs from society and create huge unemployment. More on this below.

In this regard, it is important to note that the number of robots in everyday life (domestic robots), medicine, logistics, services, entertainment and leisure is growing rapidly (IFR, 2020). The number of military robots is also growing significantly. Household robots, including domestic vacuum cleaners, are the largest group of consumer robots. Nearly 18.5 million units (+6%), worth $4.3 billion, were sold in 2019. It is also clear that the majority of sales in most non-industrial robotics sectors are less complex designs, although the situation is somewhat different in the medical sector. A surprisingly growing number of robots for rehabilitation and non-invasive therapy makes them the largest medical application in terms of units in the medical robotics sector (IFR, 2020, 2022).

The fastest growing service robots are: Autonomous Mobile Robots (AMR), delivery robots, cleaning robots, disinfection robots and medical robots. However, the use of robots in many other areas (promising in perspective, as discussed below) remains at the level of experimentation and enthusiasm. So the reality of robot penetration is far from what was predicted. So far, there is no evidence of the rapid "proliferation" of robots in the workplace, which has been and remains one of the most common predictions since the 1980s (e.g., Ford, 2015, 2021; see more below).

At present, in addition to the first-generation robots (the most numerous), various second-generation robots are developed. Active work is carried out on the design of the third-generation robots with high levels of intelligence, adaptability and orientation.

### 10.3.2 Robots: Classification and Some Forecasts

Currently, we can talk about a large number of different types of robots (we believe that there are at least 20 types). Conditionally, we will divide them into four groups, without pretending to be complete and taking into account the fact that the boundaries between the groups are conditional: (1) already widely used; (2) already used or under development, whose prevalence will occur in the future; (3) largely predictable; (4) specific robots.

**First group**: (1) industrial; (2) information and communications (voice, chat bots, software, etc.); (3) military (spy-robots, drones, mine-clearing, etc.) and rescue robots (this type is not yet widely used). However, there is a movement to ban military robots, but this is unlikely to stop the process; (4) medical (for details see below).

In our opinion, medical, military and rescue robots are particularly promising.

**Second group**: (5) personal/domestic service robots (including domestic helpers, such as robot cooks, security guards, companions, pet care, child education, etc.); (6) social robots (providing various services, including care for the infirm), the so-called assistive robots; (7) professional service robots and logistics robots used in business (cleaners, cooks, waiters, inspection and maintenance robots, receptionists, hotels, couriers, gardeners, loaders, warehouse workers, etc.)[7]; (8) civil servant robots (the name is, of course, conditional). In particular, there are developments (in Israel and South Korea) of robots that can replace border guards and customs officers. They have prospects, especially in tourist countries. They are the first robot guards for prisons. Their task is to detect the aggressive behavior in prisoners and to help the staff in difficult situations. Well, the idea of Robocop has long been popularized. It is difficult to say how relevant the guards will be in the future, but recently the main direction in this direction is to solve the problem with the help of AI programs through video cameras; (9) agricultural robots, in particular already used for milking cows and picking apples (field robots); 10) sexual robots[8]; 11) high-tech toys (pet toys, useful in particular for patients, including those suffering from dementia, and old, infirm people who live with such toys as with real pets); 12) drones and self-driving vehicles (see a separate paragraph on them).

All these types of robots will be more or less in demand in the future, but we believe that service and social robots have the greatest prospects.

**Third group**: (13) planetary robots (designed for mining on the Moon, asteroids, etc.); (14) underwater robots (for algae cultivation and deep-sea mining); (15) biorobots (there are already experimental and exclusive samples), i.e., robots that copy living beings, including humans, with similar appearance and sense organs; (16) robots that are copies of real people.

This is a rather exotic group, whose development will depend on many conditions.

**The fourth group** comprises specific robots, some of them are already in use or under development[9]: (17) nanorobots (see above and below about them); (18) microrobots (see below); (19) neuro-piloted robots linked to the development of biointerfaces. These are robots controlled by signals from the human brain or muscles. They can be used by the disabled and in telepresence work; (20) sports robots of various types, including neuro-piloted ones; (21) small scale flying robots like robobees.[10] As we can see, most of this group are robots, which are especially actively used in medicine or for the adaptation of disabled people, which promises them great prospects in the future.

---

[7] Courier robots have been widely introduced in recent years (delivery drones are also used, but less frequently).

[8] Robots already show considerable results and prospects in this sphere (Yeoman & Mars, 2012). There are also apologists for this direction. However, it is quite possible (and it would be reasonable) that such production can be forbidden. Let us note that feminists have already begun such a campaign (see Griffin, 2015).

[9] This group particularly demonstrates the ambiguity of the above-mentioned concept of a robot.

[10] RoboBee—is a tiny robot capable of partially untethered flight. Its wingspan is 3 cm. It cannot carry a control system or a power source for flight (Wood et al., 2013). It is intended to be used for search and rescue, surveillance and artificial pollination (and, we believe, for military purposes as well).

*From the above, it can be seen that the potential for the development of robotics is enormous, and this (together with the development of the technical capabilities of robots) will mean that in the future more and more new areas, directions and functions will be robotized both in production and in everyday life, in medicine, in private affairs, as well as in the public sphere and public services.*

### 10.3.3 Problems that Robotics Should Solve in Order to Become a Locomotive of Innovative Development

These problems are different in nature; and in general, they show the complexity of the tasks that facing robotics will have to face. And their solution will allow robotics to acquire very significant opportunities, while the services provided by robots will become necessary and common (e.g., Proydakov, 2016).

Firstly, in order to make robots multifunctional and able to process a much larger amount of information than at present, it is necessary to solve technical problems associated with increasing the electrical efficiency and power supply capacity,[11] the performance of the robot's computer-computer systems, and also to improve electromechanics. Secondly, the development of the so-called effectors, which serve as the robot's organs: vision, touch, hearing, speech interface and even smell. In particular, the main problem for vision is the efficient processing of the incoming video stream; for hearing—the recognition of noisy continuous speech (these problems may be partially solved in the process of AI development).

Thirdly, these are problems of the limbs, so to speak. These include mechanics, orientation and tactile perception. It is necessary to teach robots to move at different speeds, on different surfaces, to step over obstacles or avoid them. For the hands, it is necessary to solve (a) problems of mechanics, in particular, the rapid construction of a trajectory of movements and control of the movement of the hands; (b) to create an advanced system of tactile sensors so that the "fingers" could feel the touch of objects, the force of compression, the temperature of the object being grasped, etc. (see more about this in relation to surgical robots).

There is also the problem of creating artificial muscles for robots, which is particularly important for anthropomorphic robots so that they could control facial expressions, lip movements, etc. Artificial muscles could obviously help to solve the problems, and could also be useful in medicine (for bioprostheses, artificial skeletons, etc.).

Solving each of these problems, not to mention solving them in the system, will help create robots that are well oriented, can move freely, can pick up objects of various shapes and weights and put them in the right place, can distinguish sounds and speech, can see, can perceive and distinguish objects and so on. In other

[11] Here, the emerging transition in microelectronics away from the silicon carbide (SiC), for which there is currently no alternative to. Gallium nitride (GaN), which is more efficient in a number of parameters, is likely to help. GaN has already become the leading new material for microelectronics in a wide range of segments (see, e.g., Bondar, 2018).

words, there could appear very intelligent and effective helpers, which could be both universal and specialized with the help of programs. In other words, numerous areas of the service sector, the household, our livelihoods, our contacts, medicine, as well as many areas of production and trade, would change markedly and at some point radically. We believe that we will be able to observe these changes over the next three to four decades during the final phase of the Cybernetic Revolution, but they will be especially evident after the 2040s.

**Robotics in medicine and nursing**. We have already spoken (in Chaps. 4 and 8) about a successful use of robots in medicine and their role in the Cybernetic Revolution. The global medical robotics market is rapidly developing and is expected to grow at a CAGR of 16.67% from 2022 to 2029, reaching over $23.10 billion by 2029 from $5.90 billion in 2020 (Globenewswire, 2019).

Surgical robots are the most widely used and represent a rapidly developing sector (see Chap. 8; Pinkerton, 2013; Beck, 2013). They are expected to triple over the next few years (Globenewswire, 2019). Robot-assisted surgery has many advantages. Already today, robots help perform very delicate operations, especially in eye surgery, which requires the utmost accuracy (Smotrim.ru, 2016). Of course, both forecasts seem overly optimistic, however the expectations show that these are potentially very large markets to profit from. There are certain opportunities for the growth in surgical robotic market in Brazil, India, China and other emerging economies (Globenewswire, 2019). Nevertheless, along with the benefits, surgical robots also bring new problems and fears, since the surgeries involving these systems carry significant risks, including an increase in the number of injuries and deaths in robotic surgeries (Pinkerton, 2013). However, the situation is improving, thanks to technological innovation. In particular, safer systems are being developed (such as the American HeroSurg or the Dexmo Glove, developed by a team of scientists from Europe and China [Media Release, 2016]).[12] These robotic surgery systems have given doctors a sense of touch, a feeling as if they were actually holding the surgical tools, making it much easier to control the robotic surgeon. Moreover, haptic technology would also improve the ability to distinguish cancerous tissues from normal tissue. So medicine and robotics are only at the beginning of the journey; however, this direction is very promising.

Not only mechanical robots but also micro- and nanorobots are introduced into surgical practice. Advances in nanotechnology are particularly suited to the field of neurosurgery. Nanodevices provide the basis for greater control and accuracy, for example, when reconnecting nerves. Oncology is another area where surgical nanorobots could be useful, particularly for imaging tumor margins. The detection and mapping of tumor margins during surgery can be greatly improved by involving nanorobots into tumor excision procedures. The plan is to inject the patient with nanorobots that use chemical sensors trained to detect different concentrations

[12] For example, HeroSurg allows overcoming many of the limitations of existing robotic laparoscopic systems, including collision avoidance capability, modularity and automatic patient/bed adjustment.

of E-cadherin and beta-catenin to find tumor tissue margins and metastatic sites. A cluster of nanorobots will assemble on the tumor and transmit an electromagnetic localization signal to the surgeon for further investigations.

However, the development of robotics, including microrobots, in medicine is by no means limited to surgery. There are forecasts about the creation of similar micro-nano robots made of flexible materials that can change their shape for specific tasks. This, in turn, could have a dramatic effect on how a drug interacts with cells in the body (Binysh et al., 2022).

Currently, robots are actively involved in nursing-care and medical care, including robots for activities that benefit people with disabilities and older adults in the course of their daily lives and potentially with a wide range of functions from dressing to taking medications and playing. In Japan, nurse robots are actively being tested. They help patients to get out of bed, render help to stroke victims in regaining control over their limbs (Khel, 2015) and improve walking capacity (Moucheboeuf et al., 2020). *GeckoSystems* makes robotic nurses and robotic attendants, which, with telepresence capabilities, will allow doctors and nurses to remotely monitor and examine patients as well as change bed linens and give medicine to patients. In our opinion, this is one of the most challenging directions for the use of robots in the future. Many nurse robots are humanoids (Cairns, 2021), which makes them more comfortable for patients (Cairns, 2021). Hospital robots became very useful during the COVID-19 hospital staff shortage.[13] This market is estimated to be worth hundreds of billions of dollars. There are different ideas about the use of robots in medicine as medical transport, sanitation, disinfection and rehabilitation robots (Stanger & Cawley 1996; Crawford, 2016; see also about robots that take a patient's blood pressure using only a simple touch [Kim et al., 2022]).

### 10.3.4 Forecasts for the Development of Robotics in the Final Phase of the Cybernetic Revolution

We believe that in the next few decades robotics will develop more rapidly not in industry but in other areas. One may ask the question, in which areas will robots be most widely used?

The first is medicine and care for the sick, elderly and infirm, where robots of all kinds, from nanorobots to anthropomorphic robots, will be used. There also will be created the self-regulating/self-managing systems that ensure the functioning of physiological processes, as well as robots with different "skills"—assistant robots, doctors, nurses and security guards. We believe that this will be one of the main

[13] On social consequences of the COVID-19 pandemic see, e.g., Chumakov and Yurchenko (2021), Efremov and Arkhangelsky (2023), Grinin (2020), Grinin et al. (2022), Hussain (2021), Irshad (2020), Jasiński and Bąkowska (2020), Kakkar (2021), Seifabadi and Dolatabadi (2022), Widdowson (2021).

directions that will allow at least a partial solution to the problems of care for the elderly.

Secondly, robots will be widely introduced in the service industry. Why? First of all, since it currently involves the major part of the working population in the developed countries, up to 80 percent (Acemoglu & Restrepo, 2017; Turja & Oksanen, 2019). It is rapidly developing in all countries, including China, India and the Third World countries in general. Thus, in economic terms, it is the most promising sector for labor replacement.

Thirdly, the use of robots is correlated to the problem of reduced labor supply due to global aging and lower birth rates around the world. Robotics can significantly alleviate labor shortages, or even largely solve the problem. This will be especially important in low-paid or labor-intensive sectors, from childcare and health care to waiters and cleaners; from security guards to loaders; from secretaries to truckers.[14] We have been observing this process for many decades (as in the case of computer-aided robots), with many occupations becoming extinct or reduced in number. But the problems of labor shortage, however, are only getting worse, even in highly paid jobs like truckers.

We do not consider the threat of structural unemployment here, it exists, and we must be ready for it. Although it is quite serious, in our opinion, it is not an existential threat. One should understand that the process of replacing human labor does not proceed immediately, even if society is technologically ready for it. Any practical (applied) innovation requires a number of conditions, from legal to economic, as well as psychological and others. Therefore, the replacement of human labor will not be rapid, even in certain spheres (and even more so in the economy as a whole) and it will take decades, not years. In addition, this can significantly change the problem of migration. Robots can largely replace migrants.

Fourthly, robots may replace humans in dangerous jobs and activities and in high-paying professions (like lawyers, and doctors), where possible. Here, technological unemployment will only be more painful, because people will lose highly paid jobs.

We believe that these four areas of the introduction of robots into the economy and life will develop most rapidly. And already based on these successes, home robots will develop very actively (for cleaning, cooking, other domestic tasks, managing a household, babysitting, etc.). This will obviously provide a huge market.[15]

As for the universal and anthropomorphic 'smart' robots, it is necessary to recognize that the forecasts about the emergence of such 'smart' technologies in the near future do not yet correspond to the real opportunities. Their time will come much later, in the end of the 2040s—the 2050s or even later. Moreover,

[14] Self-driving cars are also robots, but we will discuss them in a separate paragraph.

[15] However, it is possible, as some researchers believe, that one of the most perspective directions in development of robots will be intimate services (we have already spoken about it above and in the Introduction) if it is not forbidden.

we should not worry that the twenty-first century will be the century of the post-biological world when as a result of the natural selection, robots will displace humans from the pedestal of evolution and develop under the influence of the new post-biological evolution, which can exceed the rates of the biological evolution millions of times (Wadhawan, 2007). We believe such point of view looks more like a scientific fiction than a scientific forecast.

There have been, and still are, different assumptions about the role of robotics in the near future. In 2007, Bill Gates (2007) considered that robotics was approximately in the same position as computers used to be in the 1970s when they founded the Microsoft Company together with Paul Allen, and he apparently anticipated that in the 2030s robotics would become as important as ICT is today. However, we doubt that this prediction will come true by the appointed time. Many companies are working on such developments, but in general, unfortunately, there is not as much business interest in this direction yet, as, for example, in biotechnologies, even though robotics already has a rather long history. Today the total volume of robot production worldwide is rather small. In 2020, 384,000 industrial robots were installed (IFR, 2021); for comparison, in the same year, China alone produced 27,020,000 cars, so the number of industrial robots is absolutely insufficient for an economic takeover.

However, there is no doubt that a bright future is awaiting us due to robotic development. But most likely, its rise will already happen during the apogee of the Cybernetic Revolution on the basis of development of future technologies. So, we assume that certain, though not revolutionary, achievements in this area will occur in the 2020s. In the 2030–2040s, we will see a much more significant increase in robotics, whereas an explosive development of robotics will happen a little later, in the 2050s and 2060s. By this time, as we have said, we can also expect the creation of truly 'intelligent' robots. Hardly all of them will be anthropomorphic, most likely their design will be defined by functions. However, universal robots are also likely to emerge. Perhaps, evolutionary robotics, in which robots are able to improve themselves and adapt to new conditions, may also develop by this time. They will not only be fully self-managing systems, but also self-evolving systems (which are already being developed in the field of AI, see below).

Today, however, many robots are remotely controlled by human operators, and only very few models have the ability to perform some tasks autonomously. These (and other) systems after all are still closer to the operated machines, but they are advancing toward self-regulation and self-managing. People need robots that are not directly operated by a person, but that can work with a high degree of autonomy (see Yuschenko, 2015).

In addition, robots (excluding automated production) work almost individually. In order for robotization to be profound and robots to become ubiquitous, significant progress should be made in the field of creating working and communicating teams of robots of different types. The solution to the problem of robot group interaction is already actively posed today. The development of effective methods of interaction and combining the efforts of many robots is important in military matters, as well as for robots in emergency situations, complex production tasks, etc.[16]The task of providing control over multi-robot systems in the future is already set for robotics. A group of machines is better capable of performing qualitatively different tasks than a single robot (Yuschenko, 2015). So far, there are no breakthrough solutions. But if you look further ahead, then robots should be integrated into various self-regulating and self-managing systems, and this will require interaction between system elements and subsystems. If we talk about socio-technical self-regulating systems (SSS; see Chap. 14 about them), where robots can partially or completely replace certain groups of public servants (policemen, firemen, etc.), then the interaction not only with other robots but also with humans and other STS, is crucial. And this requires new forms of information exchange. Today we are talking about the future creation of special Internet, information services or even social networks for robots to exchange information with each other, which will form the inchoate collective intelligence of robots (Waibel et al., 2011). But we think that in the future such an exchange will be different.

At the end of the paragraph, we can formulate some interesting questions for research: to what extent will the robotization of society help a) to choose a career for self-development, creative, not boring, not routine? b) If robot nannies and assistive robots for people with disabilities can be created, and if enough robots reduce the need for labour and engagement of women into economy, will this be an impetus for women to have more children? (Reproductive technologies can, of course, play an important role here); c) what role can robotization play in the development of the e-government and e-state? (As mentioned above, robots can be integrated into socio-technical self-regulating systems, which helps to lead toward electronic state; see Chap. 14 on the electronic state); d) how can numerous moral problems be solved, including the problem of people's emotional attachment to robots?

## 10.4 Transport and the Cybernetic Revolution

**Self-driving cars**. In many aspects, autonomous cars are similar to robots. Moreover, unmanned vehicles, like robots, clearly demonstrate the very 'idea' of the Cybernetic Revolution as a revolution of self-regulating/managing systems. So we can propose that in the forthcoming decades the breakthrough is most likely to occur in the direction of autonomous traffic and its management. This means that transport in general, vehicles and other transport systems will become driverless and thus self-managing. It is more likely that these will be electric vehicles, although it is possible that driverless cars running on hydrogen or other fuels will become more widespread.

Therefore, since there are many technical limitations and problems in the development of electric cars these problems will slow down driverless electric cars.

[16] For example, the goal of the RoboBee project is to make a fully autonomous swarm of flying robots for different tasks.

In addition, there are many specific obstacles, including legal and organizational (safety, various fears and conservatism) on the way to the mass implementation of self-driving cars. It is impossible to overcome them quickly. Moreover, since people can drive cars themselves and like to do it, few will be willing to pay much more for a robot than for an ordinary car. This, combined with the fear that unmanned vehicles could put millions of people out of work, may hinder their mass adoption. However, from an economic perspective, self-driving vehicles could be more profitable and could dramatically change freight transport as well as taxi services. Such unmanned vehicles can significantly substitute taxi (Khel, 2015) and truck drivers. But certain legal and social difficulties may emerge here.

At present, many companies report the releases of their own models of self-driving cars. However, it is difficult to define the actual number of self-driving cars in use, since for the purposes of their promotion, the manufacturers give very inflated figures of hundreds of thousands, millions and even tens of millions of units, which seems incredible. In addition, there are big differences in what vehicles can be considered unmanned. Most of the so-called self-driving cars are still equipped with autopilots classified as Level 1 and Level 2. Using information about the situation around, the car equipped with the Level 2 autopilot can maintain its course in the lane, and accelerate and decelerate independently. This is done using sensors, detectors and, in some cases, cameras. The driver must keep his hands on the steering wheel of the car and not remove them. There are hundreds of advanced unmanned vehicles with Level-3 autopilot (when the driver does not need to keep his hands on the wheel, but is simply in the car),[17] and given the high cost of maintenance, there are very few cars that can drive without human presence.

It seems that the widespread use of self-driving cars will not happen in the next 10–15 years, but they can be introduced, work and make a profit in areas where cars move along a given route, that is, in cargo transport and maybe in some sectors of public transport, as well as in closed areas (at airports or in industrial zones).[18] In other cases, we expect that there will be a gradual development of autopilots, so cars will move toward self-driving, but will not be able to achieve it for a long time.[19] However, sooner or later such a breakthrough will happen. But there are many technical, organizational and legal problems to be solved;[20] As well as ethical ones, including problems in the vein of Isaac Asimov.

[17] In fact, they are still being tested and there are many restrictions, including speed. In any case, the driver must remain in his seat, be fastened and ready to resume driving at the system's first request. Its activity is monitored by a separate camera in the interior.

[18] This fully applies to self-driving electric locomotives, which are being developed in some countries.

[19] It is worth bearing in mind that the autopilot in aircraft appeared a very long time ago, but there are still no unmanned aircraft.

[20] They are discussed. In particular, the European Parliament proposed to insure the liability of unmanned vehicles. Some countries have already adopted such laws. But legal issues are often resolved very slowly.

For example, Peter Dizikes (2016) describes that research shows inconsistent public opinion on the safety of driverless cars. He informs that in a series of surveys conducted in 2015, the researchers found that people generally take a utilitarian approach to safety ethics. They would prefer autonomous vehicles to minimize casualties in situations of extreme danger. This would mean, for example, that a car with one passenger would veer off the road and crash to avoid a crowd of ten pedestrians. At the same time, the survey's respondents said, they would be much less likely to use a vehicle programmed in this way. In essence, people want driverless cars that are as pedestrian friendly as possible—except for the vehicles they would be riding in. "Most people want to live in a world where cars will minimize casualties", says Iyad Rahwan, an Associate Professor at the MIT Media Lab and co-author of a new paper outlining the study. "But everyone wants their own car to protect them at all costs".

So today the development of various models of such self-managing systems (i.e., driverless cars) should not be considered as the beginning of the final phase of the Cybernetic Revolution but only an important precursor to its inevitable start in the 2030s–2040s. However, later in the middle or at the end of the final phase of the Cybernetic Revolution (between the late 2040s and 2060s) one can expect a mass development of this new means of transport. Moreover, self-driving electric vehicles can become a powerful source of technological development during the mature phase of the Scientific-Cybernetic Production Principle.

**Drones**. Pilotless aircrafts or drones, like robots in general, vary greatly in size, complexity and capabilities (from a small toy to almost an aircraft). As already mentioned, they are particularly actively used in the military sphere in which new and new modifications are constantly appearing. But today drones start to be actively used for peaceful purposes (a fairly common trend in technological development) for checking the power lines or delivering first-aid equipment (they can also be used as messengers), in the agricultural sector and for many other purposes when inexpensive and continuous aerial support or monitoring is required.

Like driverless cars, drones are still semi-self-managing, they are controlled by a person on the ground. But their capabilities are growing, and so are their applications, to the point when they can repair the roof of one's house. In particular, "agro-drones" are expected to be used in agriculture, especially for transporting fertilizers, seeds, etc. This is important given the off-road nature and complexity of delivering to certain areas. There are also plans to transport people, including the development of taxi drones, but so far they cannot be put into practice due to technical and legal problems. A big problem is finding a simple and environmentally friendly power source for drones. Some experts believe that the future belongs to powerful drones powered by lithium ion batteries. However, there is experience of developing hydrogen-powered drones.

Like self-driving cars, drones will sooner or later achieve true autonomy. Due to the great economic effect and convenience of such robots, they have a very promising future.

## 10.5 3D Printing

**Universalization and 3D printing**. Universalization, as we discussed this in Chap. 3, is one of the most important and even surprising characteristics and trends of the Cybernetic Revolution. Universalization is associated with a trend that combines in one direction a whole range of skills and capabilities from other fields and technologies; In other words, universalization, that is, the emergence of more and more multifunctional devices (today it is a computer, a mobile phone, a help desk robot; general AI, see below; in the future—a universal robot, a single medical-biotechnological environment [see Chaps. 6, 8, 11, and 15], etc.; in the future, it is also possible that many things can be done with the help of nanoassemblies, creating various nanostructures and substances). It lies in the fact that the vector of technological changes moves in such a way as to gather together the maximum number of possible operations in specific technologies, to make them as broad as possible, constantly combining various existing autonomous machines and mechanisms into a single complex. This also applies to the universalization of competencies with the help of self-regulating systems, when, with the help of devices and software, an individual can combine dozens of competencies, each of which was previously feasible only for a professional or an advanced amateur (we already talked about this in Chap. 4).

One of the most recent trends in universalization is 3D printing. As we will see, the opportunities provided by such printers are enormous: from construction to cooking, from a home workshop to museums, from motor car engineering to robotics, from medicine to children's toys, from training models to design; from toys to weapons. These machines are actively used in such industries as aircraft construction and rocket engineering to produce individual details, for example, support stands for an aircraft engine (see, e.g., Turichin, 2015). And just because they are used in such fields their development requires considerable investment.

In fact, these printers are actually a universal home workshop or a universal production, construction or factory. And they will acquire new functions and integrate new subsystems in the future. Widespread use of 3D printers can eliminate long technology chains in some production sectors. It will be enough to have a sketch and to make (to 'print', 'fuse') a detail at home or in a 3D printing center. It will also be possible to organize small-scale piece goods production. Engineers could also develop simple food 3D printers that can print, for example, candies, pizza or designer cookies. Futurologists expect a great future for this direction of 3D printing (Lipson & Kurman, 2013; Ford, 2015).

And, we believe that the universalization trend will continue to expand the capabilities of self-regulating systems in the future, particularly for robots that will be able to use the functionality of 3D printers, and, possibly, the future self-driving transport, and other functions.

**3D printing: general characteristics**. 3D printers are machines that can create 3D designs. The history of 3D printers began in the 1980s, the first 3D printers similar to modern ones appeared in the 1990s, and since then the technology has come a long way, both in terms of developing its capabilities and greatly reducing its size and price. Some observers believe that the impetus for the emergence of 3D printing was given by the trend toward the reduction of large-scale production and the expansion of small-scale production that emerged at the end of the last century, as well as the general trend toward individualization of production.

These devices may be of different sizes and can produce a wide range of objects. The most common material is plastic, but some printers can work with metal (with the exception of aluminum), as well as hundreds of other materials, including high-strength composite materials, elastic materials, and even wood (Ford, 2015). With regard to 3D printing, we can also mention the melting of powder agglomerations of both polymeric elements and metal powders (here we can observe a link between printers and nanotechnologies).

Although often referred to as 3D printing (similar to paper-based printers), the term used by experts is "growing" or "building up" (see, e.g., Turichin, 2015). It seems to us to be more comprehensive, as it more accurately reflects the actual process of creating products. Layer by layer, material is built up until a detail or other object is "grown". A distant analogy can be found in the work of the potter's wheel.

3D printing is the process of creating a real three-dimensional solid physical object from a template designed using 3D computer-aided design (CAD) software, and then the 3D printer produces a real product. The basis of 3D printing is additivity, that is the merging (fusing) of materials and the creation of a specific design and construction (such technologies are called **additive**). Fusing layers consists of a series of iterative cycles of creating three-dimensional models, applying a layer of material on the working surface (lift) of the printer, lowering the lift platform by a distance equal to the thickness of one layer and removing waste from the working area surface. The cycles follow each other continuously.

In industrial production (metallurgy and mechanical engineering), a detail is most often produced by subtraction, that is by removing material (although casting and other methods are known), turning and drilling, thus eliminating superfluous material. For example, General Electric intended to use such technologies to reduce the weight of an aircraft engine by about 500 kg (Ford, 2015).

**The potential of 3D printing in medicine**. As already mentioned, the potential applications of additive technologies are very broad. But it is especially important for us to emphasize that these potentialities are vitally important for medicine: in the creation of artificial tissues, parts of organs and the organs themselves, as well as more durable parts of the body, such as bones, including those made from high-strength metals. We have already given a number of examples of this kind of use of bioprinters in Chaps. 8 and 9. It is of fundamental importance that this can be done in hospitals, laboratories, medical centers, etc., that is, applied to specific cases.

**The principle of self-reproduction of the like**. It is very curious that in this technology (as well as in robotics) there is another principle of the Cybernetic Revolution, which may become very important in the future, but which so far is weakly manifested (except in some programming developments) and, perhaps, in some biotechnologies. We mean the principle of self-reproduction of the like.

In 2005, the RepRap community of enthusiasts was founded. The project is based on two ideas: any RepRap printer can print another RepRap printer; all developments of 3D printing device are in the public domain.

In eight years, four generations of RepRap 3D printers have been developed. However, even now, the task of replicating one RepRap device with another is not completed, yet. It is one thing to print plastic parts; it is another to create microelectronics and metal elements of the extruder design (Plotnikov, 2014).

## 10.6 Cognitive Science and Cognitive Technologies

**Cognitive sciences and government interest**. Cognitive science studies the nature of the mental and nervous processes that control movement and many other bodily processes. It is actually a large complex of diverse fields related to intellectual processes, consciousness, knowledge, memory, etc. We refer here to fields such as cognitive neurophysiology, cognitive neuroscience, etc. In recent decades, many discoveries have been made that explain some of the mechanisms and reactions of our brain and mind, including the work of so-called neuromediators. A considerable number of new generation neurostimulating drugs have been created and are actively used in medicine. In general, there has emerged a new branch of pharmacology—neuropharmacology.

In 1924, the German scientist Hans Berger made the first recordings of human brain activity by attaching electrodes to the head (Wolpaw & Wolpaw, 2012). Later electrodes were implanted directly into the human brain.[21]

The key technological achievement has been the development of new brain-scanning technologies (including computed tomography, etc.), which for the first time made it possible to obtain multiple images of the inside of the brain and to obtain direct and indirect data on its functioning. Nowadays, many research organizations are trying to create a database of neuronal cells and their types (according to the latest data, there are 90 billion neurons in the human brain). This will advance the interpretation of the mechanism of operation of visual system by developing a functional classification of different types of neurons in the brain.

In recent decades, we have observed a very rapid growth in research in this field. This is reflected in the rapidly growing global neurotechnology market, which is expected to grow from $10.7 billion in 2020 to $12.82 billion in 2022 (EMR, 2020; PR, 2022). This is due to a number of reasons, including the increased interest from the military, intelligence agencies and companies that are developing artificial

[21] In 2010, there were already about 35,000 people in the world with the electrode implanted in their brains (see Swaab, 2014: 292–294), while in 2019, BBC News in a report with speaking title 'Brain implants used to fight drug addiction in US', claimed that the number was 180,000 (Wakefield, 2019).

intelligence technologies. It is clear that all of them have great opportunities for research funding. As to the military and intelligence services, their interest is quite obvious, as is that of large media and political centers seeking to find new ways of influencing people. As far as IT companies are concerned, they have taken a course in the study of how the brain works in order to try to build new IT technologies by analogy with the principles of the brain. And so-called neural networks and neural learning have become the result of this course. It is significant that these forces work in tandem and have at their disposal hundreds of millions of dollars from public funds (e.g., Cepelewicz, 2016). We will return to this issue in the paragraph on AI. Government-sponsored research is very active in the USA, but also in China, Europe, Korea, Japan, India, Russia and other countries.

**Cognitive technologies and medicine**. Although the research is being intensively driven by the government interests, nevertheless, we believe that the resulting achievements will be manifested in the development of the key areas of medicine.[22] Naturally, cognitive technologies are closely related to bio-, nano- and information technologies. First of all, cognitive science and technology can give breakthrough opportunities for the disabled and people with disabilities by minimizing their problems.[23] Since aging and disability are unfortunately closely linked, the problem of aging and dementia in developed countries also gives a powerful impetus to the development of neuroscience. One promising development from the University Hospital of Lausanne is an artificial spinal cord. The implant sends electrical impulses to the muscles, mimicking the brain, and could one day help people with severe spinal injuries to stand, walk and even play sports. The breakthrough is that the implanted leads are longer and wider, with the electrodes positioned to match the spinal nerve roots, so more muscles can be accessed with this new technology. At the same time, Israeli scientists presented the results of their work to restore the spinal cord function, giving hope to thousands of paralyzed people to get back on their feet (Rowald et al., 2022).

The so-called neuroimplants that help to restore or improve certain functions have great prospects. Various bionic implants are already developed and tested (e.g., bionic eye,[24] artificial (electronic) nose,[25] bionic prosthetic foot (Herr & Grabowski, 2011), knee (Dawley et al., 2013), other bionic or electronic organs. Since the control of these organs requires communication with the brain or nervous system, this significantly complicates and expands the possibilities of cognitive technologies. In particular scientists are already adjusting the functioning of an artificial eye, ear, and even heart by means of neural interfaces.

[22] The usual path for innovation to spread from military-state interests to business applications and dissemination to the general public.

[23] For example, for the first time, a completely paralysed person has been able to form sentences using an implanted device that reads brain signals (Servick, 2022).

[24] First Bionic Eye Implant (Da Cruz et al., 2013).

[25] In the future, a compact neuromorphic 'electronic nose' may be useful for doctors to diagnose diseases, for security services to detect weapons and various substances, etc. (Sandomirskaya, 2021).

One can also observe the beginning of the development of many practical applications that can be implemented from the results of the study of the brain: preventive measures, treatment, rehabilitation, development of the capabilities of operators of complex systems, education and others. Thus, several years ago, the Journal of Neural Engineering published an article about a study by American engineers and neuroscientists who had succeeded in artificially enhancing the capabilities of the human brain and memory by installing neuroimplants (Hampson et al., 2018).

However, the opinion that in the 2030s some nanodevices will be implanted in the human brain and will be able to perform the input and output of the necessary signals from the brain cells and that this may even make learning and getting education much easier is very doubtful. Even if such a cyborgization could be realized in principle, it will happen much later.

**Neural interfaces** or brain-computer interfaces (which we discussed in Chaps. 3 and 8) may become one of the breakthrough directions of cognitive science and the Cybernetic Revolution in general. Let us recall that neural interfaces are technologies that connect the human brain and/or nervous system to external devices. They usually implement the interaction between the brain and the computer system. The fundamental achievement of cognitive sciences is an opportunity to control artificial organs via the brain signals, as healthy people do.

After it was established that the electrical activity of neurons could be used to control robotic manipulators, the study of neural interfaces became even more active (Lebedev & Nicolelis, 2006). Now it has become possible to transmit neuronal signals to devices and thus to operate artificial limbs with natural accuracy (see above).

All kinds of experiments in this direction are conducted in different countries and companies. With the rise of medicalization and the power of digital and pharmaceutical corporations, this kind of development frightens many, and with good reason. Therefore, it would be wise to bring such experiments under the control of society and channel them in a safe direction.

In the future neural interfaces could be used not only in medicine, but also in everyday life, for example to monitor the state of a driver's or an operator's brain and automatically wake them up if they fall asleep. However, there are great risks associated with a wide introduction, because of the strong limitation on human privacy, and the great temptation to use such private information for military, commercial or political purposes (e.g., Chowdhury, 2020). This danger may increase if the future neural interfaces become wireless and can be controlled remotely. Today such chips are as heavy as tiny wires..[26] Musk's company continues active experimenting, and although one can hardly believe their predictions

---

[26] This is how Elon Musk described the Neuralink technology. Measuring four by four millimeters, the chip is claimed to be connected to a thousand "threads" thinner than a human hair, which enter the brain through four holes drilled in the skull. Electrodes embedded in the threads, which are 'sewed' to the chip, can then pick up electrical signals from neurons that indicate activity in the brain, which is then translated into movement or behavior (Chowdhury, 2020). But during the

of early breakthroughs, the process continues. It needs regulation, especially since Musk's company has already moved on to human experimentation.

In general the achievements of cognitive science are already in use and their application will increase even more in the fields that move toward self-regulating systems—from medicine to robotics, from cybernetics to problems of artificial intelligence, and, of course, for military purposes.

However, some serious technical and social difficulties can hamper the development of this direction. Among obstacles one can mention, firstly, the immune rejection. Secondly, many nanostructures, for example, nanotubes, which were predicted to have a bright future, turned out to be very toxic to the human body (Kotov et al., 2009). Thirdly, the implantation of external devices traumatizes the whole organism, despite all serious efforts to reduce the impact (Grill et al., 2009) and causes an inflammatory process called foreign body reaction (Lotti et al., 2017). Neuronal interface implantation triggers acute and subsequent chronic inflammatory responses at the interface with neurons and nerves, damaging surrounding tissues and worsening neural interface functionality (de la Oliva et al., 2018). Electrical impedance at the tissue/device interface also increases as a result of fibrotic tissue formation around the implant (Gunasekera et al., 2015). Moreover, immune cells such as macrophages continue to migrate to the implant site, releasing pro-inflammatory cytokines that perpetuate the immune response and compromise the long-term usability of the neural interfaces (Del Valle et al., 2015; Riva & Micera, 2021). Another problem is the difference in electrical conductivity between biological material and technical device, although there has been made certain progress in solving this problem (Abidian & Martin, 2009). But even if we solve these problems we will still need some powerful software to process the brain signals. At the same time, it is very important to find the means to provide feedback between a device and the human brain, in other words, the brain should not only send a signal, but also receive it from the device. Overcoming these limitations will help to rapidly advance neural interfaces to a new level. However, in order to avoid the mistakes and problems like those caused by uncontrollable spread of computer games (but with consequences on a much larger scale), it is necessary to prevent the misuse of data and the influence on the mind.

At the end of this section, we would like to express the following idea. For several reasons (for details see Grinin & Grinin, 2015; 2016: 147–148; 2021; Grinin et al., 2021; Grinin, 2019; Grinin & Korotayev, 2015), the emergence of essentially new forms of communication (as there have been different types of electronic communication: e-mail, social networks, other Internet-connections, mobile phones,

presentation, Musk showed a larger chip, the size of a large coin, which was implanted in the brain of a pig. This chip has 1024 wires. Characteristically, they are implanted into the brain by a robot. The chip is powered by a battery that lasts for a day—the battery is recharged wirelessly, like a smartphone. Data transfer is the same as via Wi-Fi. The entire implantation procedure takes less than an hour. Behind the outwardly progressive goal, however, lies the desire to work the human brain with AI. At a deeper level, the chip is meant to boost human brainpower so that we can keep pace with a future shaped by artificial general intelligence (Chowdhury, 2020).

etc.) is hardly possible in the coming decades. The development of communication has made great progress in recent decades and has generally even exceeded the level of overall development. Most likely, the revolutionary new forms of mass communication will not appear until the late twenty-first century.[27] However, we suppose and hope that this new form of mass communication will not be the communication via neural interfaces or via the direct implantation of chips in the human brain (so that communication goes from the source directly to the brain). There is still a long way to go. And even if it succeeds, we think, it will hardly be widely implemented in the next few decades (due to ethical, medical and legal constraints). Nevertheless, the threat is not far-fetched, so we can hardly ignore such opportunities and we should think about the restrictions beforehand since this possibility raises concerns.

## 10.7 Artificial Intelligence

### 10.7.1 Development of ICT and Some of Its Dimensions

**Miniaturization is the phenomenon characteristic of the current technological progress**. In the early 2010s, a new 'core' of information technologies was formed based on the transition from microelectronics to nanoelectronics. We are talking about the rapid reduction in the size of processor chips. The speed of size reduction in the manufacturing process is evident from the following data: in the 1970s—3 μm (3,000 nm) (Zilog and Intel); in the 1980s–early 1990s—0.8 μm (800 nm) (Intel and IBM); in the late 1990s–early 2000s—180–130 nm (a number of companies); in the early 2010s—process technology at 45, 32, 28 nm (Intel, etc.). In 2019, Intel released processors based on the 10 nm process technology (the company cannot manage the production of 7 nm chips yet), while the TSMC released chipsets for mobile devices based on the 7 nm process technology (Apple A12, Kirin 980 and Snapdragon 855). At the same time, their production technologies are considerably different: Intel with its 10 nm can fit up to 100 million transistors per square millimeter, while the TSMC, with its 7 nm, can fit only 66 million. Accordingly, the miniaturization processes described above together with the increase in the power of computer technology and the speed of electronic communications have led to the most powerful spread of information and computer technologies.

The AMD Company has announced the imminent creation of 5 nm processors. But is it possible to create a whole new generation of computers with such a reduction in size? It is hardly possible because even under the most optimistic assumptions, such processes will have only a slightly higher productivity (up to 15%), while the shortage of chips will seriously hamper the development of these

[27] However the powerful progress in existing ICT is quite possible within the next three-four decades (see below).

technologies. The bottleneck, where data try to get in, moving between the drive and the CPU, consumes a lot of power and generates much heat and this limits further improvements (Khel, 2015). Now it becomes clear that it is unlikely that it will be possible to create a fundamentally new generation of computers in this way. Hence, there is more and more talk and information about progress in the field of quantum computers, which, if they appear, will not be so soon.

Quite long ago, there appeared an idea of storing data using specific environments (e.g., magnetic, electric, or optical); with the advent of nanotechnologies, it becomes possible to store information, for example, by replacing silicon, the basic material used in the production of semiconductor devices, or by carbon nanotubes.[28] In this case a bit of information can be stored in the form of a few atoms, say 100 atoms. This would reduce the processor size by an order of magnitude and significantly increase their operating speed. The number of transistors in a processor has reached now tens of billions (see above), and, now the aim is to create a processor with trillions of transistors (Moore, 2021). However, there are doubts about the possibility of this technology completely replacing traditional transistors.

**Speed of development**. Since the 1960s, the so-called Moore's Law, named after Intel co-founder Gordon Moore, is better described as the "observation" that processing power doubles every two years or so. This was an empirical observation, but due to its widespread popularization, it formed the idea that the tendency of the computing industry was to prioritize a new semiconductor node to fit that time frame (Campbell, 2021; Tsvetkov, 2017). One way or another, the power of computers grew exponentially, reaching amazing capabilities. However, since about 2007–2010, 'Moore's law' no longer works, since the trends either lag behind or overtake it (Kish, 2002; Zhang, 2022).

However, while the first Moore's law concerns a continuously increasing complexity of microcircuits for a short period of time (two years), then the 'second Moore's law' formulated by Eugene Meyeran in 1998, argues that the cost of factories producing microcircuits increases exponentially with the complexity of the microcircuits produced. And this is what we observe today in the monstrous monopolization of chip production, which has led to a deep crisis and a chip 'war'. In general, such increases in performance and productivity have led to the fact that the level of ICT has approached the limits beyond which a new qualitative breakthrough is already in sight. There are different views on when this will happen, and how it will manifest itself. Many new information technologies have emerged which we will discuss below.

In general, information technology has become ubiquitous; in fact, it is present in almost every sphere, and with the development of AI, as we discussed in Chap. 3, the situation is rapidly moving toward the point when, figuratively speaking, AI will become a kind of block system that can be assembled in a

[28] The transition in microelectronics from silicon carbide to gallium nitride has already been mentioned above, but obviously, this will not be enough for the project described.

'configuration' required for a particular task, just as today we can increase the memory of a computer or upgrade the graphics card.

### 10.7.2 Artificial Intelligence: General Characteristics

Thus, in recent years, the ITC technologies have clearly evolved to a higher level, that can be conditionally described as the level of artificial intelligence (AI). Although this concept is rather vague, and the AI has been used for a long time, today its capabilities have increased significantly thanks to various new technologies (neural networks, remote contact, linguistic systems, facial recognition systems, movement control, determination of individual preferences, machine learning and many others). Moreover, most recently, AI has become increasingly directed toward influencing social control, and even administrative relationships (see Grinin et al., 2021).

In the early 2010s, the concept of four main driving forces in the ICT market was formulated: social networks, mobile solutions, 'cloud computing', and means of processing large amounts of information (Belousov, 2016: 22). So since then, we can speak about a new wave of ever-increasing interest in AI. Previously, three waves have been identified: the first, in 1950–1960 is associated with work on machine translation and game programs; the second (1980s) is the development of expert systems; the third wave started in the late 1990s (Proydakov, 2016: 121). The idea has acquired a completely material reality: whoever owns artificial intelligence owns the world. With the account of the new types of AI (like ChatGTP), this is increasingly taking on a life-changing significance for the world. The AI race is accelerating. The sphere of influence of AI is expanding rapidly. Here is one of the lists of intellectual fields, or branches of AI:

- logical AI;
- search;
- pattern recognition;
- representation;
- inference;
- common sense knowledge and reasoning;
- learning from experience;
- planning;
- epistemology;
- ontology;
- heuristics;
- genetic programming (McCarthy, 2007; see also McCarthy, 1990, 2000; Mitchell, 1997; Shanahan, 1997; Thomason, 2003).

However, this list, which was incomplete from the moment it was drawn up, was quickly supplemented by common areas of activity: conducting dialogues and correspondence on a wide variety of topics, imitating art (painting, composing, poetry, etc.), compiling various texts, selecting necessary materials, etc.

It is believed that the increasing use of AI will lead to the adaptation of technologies in traditional economic sectors along the entire value chain and transform them, leading to the algorithmization of almost all functions, from logistics to company management, from pricing policy to market analysis.

In Chap. 3, we presented some definitions of AI, which are generally quite numerous. It is believed that there are three directions in these definitions: (1) emphasis on the similarity between AI and human intelligence as a result of activity (in particular, the possibility of AI dialogue, which we find today in many programs and in bots); (2) emphasis on the fact that AI programs are capable of learning and self-development (neural networks, deep learning, etc.); (3) definition of AI as a software and hardware complex, capable to some extent of solving problems and making decisions (more often in assisting in decision-making), as well as other intellectual functions.[29] Now it has become relevant to distinguish between the previous view of AI (or Artificial Narrow Intelligence, ANI, Narrow AI) (as a working program that replaces humans in some spheres) and general AI (AGI, Strong AI). This distinction appeared in theory a long time ago. Here is one of the definitions: An artificial general intelligence (AGI) is a hypothetical intelligent agent which can understand or learn any intellectual task that human beings or other animals can. A more precise definition, in our opinion, is that it is capable of performing most of the tasks that a human being can perform (Landgar, 2021; Turing, 1950). The AGI does not exist yet, and there is a fierce debate in the computer industry about how to create it, and whether it can even be created at all (Gates, 2023). But in the context of recent breakthroughs, it turns out that programmers with the help of supercomputers and, a huge team that teaches and "trains" AI, have come much closer to this threshold than it seemed possible until recently (see, e.g., Landgar 2021). Time will tell whether this is the most important technological advance since the graphical user interface, as Gates (2023) argues, or it is still a less breakthrough change. But it is definitely a dramatic change.

**The ability to make decisions**. Researchers (e.g., Zgurovsky & Zaichenko, 2013) often note the following capabilities of AI (compared to the previous levels of ICT):

(1) the presence of a goal or a set of goals of functioning;
(2) the ability to plan its actions and search for solutions to problems;
(3) the ability to learn and adapt behavior at work;
(4) the ability to work in a poorly formalized environment, under conditions of uncertainty, work with fuzzy instructions;

[29] In Chaps. 4 and 9, we also referred to the currently popular definition of Cyber-Physical Systems.

(5) the ability to self-organize and self-develop;
(6) the ability to understand natural language texts;
(7) the ability to generalize and abstract the accumulated information.

So, the AI is really capable of making decisions. And there are a number of areas in which people are already entrusting ICTs and their algorithms to make decisions that are not advisory, but virtually or actually final (e.g., algorithmic trading on financial markets; collecting some legal evidence using video and infrared cameras, whose data are acceptable in court; border control for recognizing the contents of luggage and even individuals; credit decisions based on an algorithmic assessment of creditworthiness through indirect signs, etc.). First of all, these are areas where the amount of information and the necessary speed of its processing are too much for a human to handle. And the number of such spheres will, of course, increase. But still, in most cases, not only today, but also in the future, the AI will mainly support decision-making to one degree or another, or be part of a more sophisticated decision-making complex, embedded in different more complex and technologically integrating domains. In other words, AI should not be associated with self-regulating/managing systems, but should be only an integral part of many of them.

In this context, it is important for us to determine the place of AI in the world of self- regulating/self-managing systems, in which we already live and will continue to live. In Chap. 3, we defined AI as follows: AI is a special universal technology (and a practical field of the theory implementation), which is an (almost) indispensable part of self-regulating systems, just like an electric motor in electric machines; or an internal combustion engine (ICE) in cars, tractors and other machines. Hundreds and thousands of different machines are based on AI. But just as an internal combustion engine or an electric motor does not determine the operating principle of all machines, but only of their large class, so AI does not determine the functions and operating principle of all self-regulating systems, but only of their large class.

**Limits**. The rapid development of AI technologies naturally requires various restrictive measures. They are already lagging behind. However, some things are still being done. For example, in 2021, after several years of preparation, UNESCO adopted recommendations on the ethical aspects of artificial intelligence (UNESCO, 2021). Overall, this is a useful document that states that there is a need to provide a universal framework of values, operating principles and mechanisms to guide states in developing their laws, policies and other documents related to AI, in accordance with international law. However, human rights and fundamental freedoms must be respected, protected and promoted throughout the life cycle of AI-based systems. But, of course, this document remains a mere declaration, which no one intends to follow.

On October 7, 2022, the White House Office of Science and Technology Policy (OSTP) released five guidelines to guide the design, use and implementation of automated systems. The Blueprint for an AI Bill of Rights is a set of five principles and associated practices to help guide the design, use and deployment of automated systems to protect the rights of the American public in the age of artificial intelligence (https://www.whitehouse.gov/ostp/ai-bill-of-rights/what-is-the-blueprint-for-an-ai-bill-of-rights/). But of course, this is completely insufficient, especially since the state structures (including the intelligence services) themselves actively violate these principles and plan to continue to do so. The US government is developing certain measures to control the development of AI, however, due to the great interest of the secret services in this development, such measures may resemble the story when the fox was assigned to guard the henhouse.

In late July 2023, shortly after the advent of ChatGPT and other similar generative neural networks, the UN Security Council held a meeting on the topic "Artificial Intelligence: Opportunities and Risks for International Peace and Security". At the meeting, the UN Secretary General supported the calls by a number of member countries of the world organization for the creation of a new UN body on AI issues, which would help eliminate future threats, as well as establish and implement internationally developed monitoring and control mechanisms. Of course, such a body will be useful, but taking into account that the most powerful states are striving to use generative AI as much as possible for the purposes of control, censorship, instilling the desired ideology, intelligence and military, its ability against the backdrop of secrecy and increasing competition of states in the fastest possible creation of ever new and more powerful AI will be clearly limited. The UN Security Council also announced the first global summit dedicated to the security and regulation of artificial intelligence technology.

In connection with the advent of ChatGPT and others like it, calls have intensified to slow down the race in the field of AI, to impose a kind of moratorium on the further development of AI, at least for a certain period of time. So, in March 2023, an open letter appeared, signed by many famous people, both involved in the development of AI and not involved. They call on all AI labs to immediately pause training of AI systems more powerful than GPT-4 for at least 6 months (Musk et al., 2023; see also Vincent, 2023). In May 2023, Geoffrey Hinton, sometimes referred to as 'The Godfather of A.I.' said he had resigned from Google, where he had worked for more than a decade and become one of the most respected voices in the field, so he could freely speak out about the risks of AI. A part of him, he said, now regrets his life's work. "I console myself with the normal excuse: If I hadn't done it, somebody else would have", Dr. Hinton said during a lengthy interview last week in the dining room of his home in Toronto, a short walk from where he and his students made their breakthrough (Metz, 2023). Google and Alphabet CEO Sundar Pichai said "every product of every company" will be impacted by the quick development of AI, and warned that society needs to prepare for technologies like those that have already been introduced (Elias, 2023). In short, there are many voices for strengthening control over the development of AI. We are afraid that all measures will be declarative, that problems and conflicts in society will increase in order to start a movement to create really working mechanisms to limit the omnipotence of digital companies.

On the other hand, there are powerful forces that would like to install AI not just to maintain their own benefits, but to completely reformat society, in order to bring it under total control and to undermine any possible opposition. It is not without reason that high-tech gurus and Silicon Valley prophets are creating a new universal narrative to legitimize the authority of algorithms and Big Data. As a result, they could then empower Big Brother. This dream of people will give algorithms the authority to make the most important decisions in their lives (for a philosophical essay with description of such ideology see Harari, 2016).

However, it is absolutely necessary to regulate the pace of development of AI. It is worth pointing out an interesting connection with our ideas about the rate of technological progress, which is non-linear and will change dramatically. As we point out in Chap. 12, there has been a slowdown in technological growth in recent decades, starting from the 1970s. But our analysis shows that there are a number of reasons to expect that the global technological growth rate will return to a hyperbolic trajectory for some time in the forthcoming decades, corresponding to the beginning of the final phase of the Cybernetic Revolution (Grinin et al., 2020a, 2020b). Perhaps, the great acceleration in the development of new Large Linguistic Models (or new chats, i.e., new forms of AI) about which many analysts of AI (e.g., IBM, 2023) have been talking after the appearance of ChatGTP can be regarded as a signal of a such a turn from a slowdown in technological progress to its new acceleration.

### 10.7.3 Symbiosis of Cognitive Disciplines and AI

The development of ICT and AI has evolved into a multi-faceted relationship, as they become interdependent on each other. They are in the closest symbiotic relationship of all the technologies in MANBRIC (we have given examples in almost all chapters, including this one). Above we described the combination of neurotechnologies; those for ICT and for AI. However, the connection between AI and cognitive disciplines extends beyond and has other aspects as well. For example, one component of the MANBRIC can determine the direction of another. The fact is that over at least last three decades, it has been the achievements of cognitive sciences that have become one of the most important drivers of AI. We mean the rapidly developing neural network technology (to be more precise, now they are called neuromorphic networks, and they are studied within neuromorphic computing).[30] For the development of AI, it became necessary to understand how the human brain works (e.g., Hawkins & Blakeslee, 2004). The

[30] At the same time, when modelling neural networks, developers try to make them as similar as possible to biological neural networks in terms of information processing. In these networks, information is encoded in the form of intervals between impulses generated by a neuron in response to locally integrated spatial and temporal excitation from impulse signals arriving at its inputs. The technology focuses on a fully hardware implementation (Benderskaya & Tolstov, 2013; Gavrilov & Kangler, 2015).

lack of knowledge about the structure and functioning of the brain, the mechanisms of memory, decision-making, foresight and other intellectual functions of the brain at deep levels has become an obstacle to progress along the path of machine learning and other abilities. As a result, a symbiosis between cognitive sciences and technologies, on the one hand, and programming, on the other hand, began to form, on the way to studying the work of the brain in order to use it for machine learning technologies, as well as the opportunity to influence human consciousness with the help of AI. This holds great promise, so this movement has been actively encouraged and led by government structures. It is believed that the key to future success lies in the collection, accumulation, processing, analysis and use of brain data. The USA has launched large-scale programs to collect and analyze brain data, but they also do it in Europe, China and other countries. The 2016 'Apollo Project of the Brain' was very instructive in this regard, and the US Government committed $100 million to it. This intelligence project aims to 'reverse-engineer' the brain to reveal algorithms that could enable computers to think more like humans. The Intelligence Advanced Research Projects Activity (IARPA), a research organization for the intelligence community modeled after the defense department's famed DARPA, has spent $100 million on a similarly ambitious project. The Machine Intelligence from Cortical Networks program, or MICrONS, aims to reverse-engineer one cubic millimeter of the brain, study the way it makes computations and use those findings to better inform machine learning and artificial intelligence algorithms (Cepelewicz, 2016). To fully understand the unusual and massive task, we can cite the data from the Allen Institute (2020), which states that their task was to do something that had never been done before: cut a one cubic millimeter section of the brain cortex into ~25,000 ultra-thin slices, take ~125,000,000 images of those slices and assemble them into a 3D volume containing ~100,000 cells, 2.5 miles of wiring and 1,000,000,000 synaptic connections.

In a word, the development of AI in many of its directions is based on attempts to imitate biological mechanisms of various kinds.[31] Of course, it is very difficult for AI to reach the capabilities of the brain. The most advanced neural networks today have tens of layers. There are hundreds of thousands or millions of such layers in the human brain. This, of course, limits many opportunities, including the so-called deep learning (Schmidhuber, 2014; Gavrilov & Kangler, 2015). Nevertheless, even the first steps have brought very impressive successes for AI and at the same time make them worrying since they will primarily take advantage of structures, intentions and technologies that are secret and unaccountable. "It's a

[31] As already mentioned, neuromorphic networks are characterized by the ability to model fragments of the biological nervous system and the intellectual properties of the brain; evolutionary computation models natural evolution and formalizes the basic laws of genetics; swarm intelligence models the social behavior of organisms that live in a colony (swarm, flock, etc.); artificial immune systems model the basic principles of biological immune systems; fuzzy systems are based on studies of the interaction of organisms with the environment (Skobtsov, 2008).

substantial investment because we think it's a critical challenge, and [it'll have a] transformative impact for the intelligence community" (Cepelewicz, 2016).

One of the latest examples of sophisticated AI based on neural networks is ChatGPT. Its capabilities are amazing. Obviously, in the relatively near future (within 5–15 years), the ChatGPT will begin to actively replace human intellectual work in writing texts in many areas: journalism, advertising, poetry, even medical and scientific texts, etc. (Biswas, 2023; Lund and Wang, 2023). ChatGPT is one of a group of so-called Large Linguistic Models (LLM) (i.e., new forms of AI).

If these kinds of chatbots become very helpful secretaries and intellectual assistants, this could dramatically facilitate preparatory work in many areas of intellectual activity. But there is a risk, and a significant one, that people will stop double-checking the data and rely on the answers, which will reduce the level of generalization and depth, as well as reliability, not to mention the generalization of personal experience. In general, this is a process that has been going on for a long time, let us take Wikipedia as an example, but now it will accelerate. People will get lazy, and their style will start to follow the style of the chat (which can be filled with whatever sponsors and hosts see fit). This means that AI will start to dictate the style of thinking and presentation.

Thus, we are approaching a new phase in the introduction of new AI capabilities into our intellectual life. This will require a considerable transformation of education, art and science (which will, of course, partially degrade under such pressure), as well as of other areas. It will turn out that a monstrous amount of artificially created intellectual products will diminish its quality.

The danger is that such chats will immediately begin to be used as an indirect powerful ideological and propaganda machine (the more relevant texts with the right bias will cascade into one's memory), increasing the propagative effect), and can also serve as a powerful censorship mechanism. Probably, a battle of ideological chats will begin (just as a current battle for and between websites is ongoing). And the most serious danger of the development of AI (and other technologies) is that we will realize these dangers only when it turns out to be very difficult to change. As Bill Gates notes, "Whatever limitations it [AI] has today will be gone before we know it" (Gates, 2023). Therefore, the regulations should be established in advance.

### 10.7.4 Artificial Intelligence and Virtual Reality

Despite the great progress in the development of ICT and AI, it should be noted that the market for their development has started to grow much more slowly than before. This is quite understandable, as a huge part or even most of humanity has computers, mobile phones, and of course access to the Internet. On the one hand, as we have seen, the new AI technologies are actively supported and funded by states and their structures, which are almost becoming the main customers. And it is here that the projects of gigantic scale and resource consumption are being created. But on the other hand, there are no breakthrough areas that can be compared to the mass character technology development in the previous period. [32] The idea of the Internet of Things has not materialized on the expected scale, the development of self-driving cars has been hampered (see above), etc. In general, this fits well with our concept that during the modernization phase of the Cybernetic Revolution, which we are now at the end of, there is a powerful diffusion of much improved basic technologies, the foundations of which were laid earlier. And the newly emerging technologies no longer seem so innovative and new. At the same time, these improved and highly differentiated technologies are already encompassing the whole society and attempting to reconstruct it completely. In addition, in this phase, the constraints and crises begin to be felt, and the foundations are being laid for a new technological breakthrough.

With this background, digital giants seek to penetrate into a wide range of fields (including medicine, see below), and also create new areas of innovations. For instance, the modification and augmentation of reality is one of the largest such fields. The most concentrated attempt to turn toward artificial reality was the transformation of Facebook into the Meta, though it seems to have failed completely. Some time ago there was news about rather large transactions within the field of artificial reality, in particular, about the "land auction" and the sale of digital paintings (while the artists committed themselves to destroying the originals). The idea that we are on the verge of a new technological revolution, in which business will migrate from the physical to the virtual world, which was expressed several years ago, has not become very visible; however, such a turn is gradually emerging. It is hard to imagine how powerful this shift will be, and so far the financial results have been disappointing.

In any case, the opportunities for changing many things, including the appearance of a person who can already change hair and makeup in the information space, and in principle also change voice and appearance, are constantly growing. It is no coincidence that one of Gartner's recent predictions is that the development of AI toward the creation of a hyper-realistic reality may lead to the situation when people will stop believing their own eyes, so that the transition to a world of zero trust will begin (this also applies to the capabilities of the aforementioned ChatGPT and the like). No one and nothing in this world can be trusted without confirmation in the form of a cryptographic digital signature. It is likely, and even quite possible, that scam bots will join the scammers.[33] This opens up many opportunities for the beneficial use of deception technologies. In particular, if chats are to provide investment advice (and traders are already starting to ask for this

[32] It is likely that a new type of chats will become a breakthrough direction, but they are unlikely to be able to become a saleable product for a wide range of buyers. It is not without reason that at least some of the services offered by Chat-GTP and others like it are free.

[33] For some time now, they have been developing a negotiator bot that can lie and negotiate with humans. According to Quartz, during the training, the system used more than 5.8 thousand real human dialogues during negotiations, collected through an online crowdsourcing platform (Amazon Mechanical Turk. https://www.cnews.ru/news/top/2017-06-15_iskusstvennyj_intellekt_facebook_nauchilsya_vrat).

kind of advice), then it is easy to imagine how it will be possible to manipulate the investors' sentiment.

The IT clones of dead or real people can also complicate the situation. The South Korean company DeepBrain AI proposes a Rememory program to create IT clones. This can be a digital lifetime copy of a person, which is created as a result of a long communication (up to 7 h) which is necessary for the neural network to study a person's character, habits, fix appearance and manner of communication and voice. After death, a digital copy can be created based on photos and videos and a short interview with the person's relatives. Afterward, the relatives of the deceased can have a kind of IT spiritualistic session with him. But the cost of the procedure is high. A digital copy will cost between $ 12,000 and 24,000; each conversation will be to the tune of $ 1200 (Coupeau, 2023).

This way, artificial intelligence can significantly change people's attitudes to life and death, as the deceased person can continue to live among us in a way and communicate through their copy. Their images, thanks to AI, will even be able to learn new facts and news, which will reinforce the illusion of the presence of a living image. Probably, for many, the very fact of losing a loved one may become less tragic. This may lead to additional problems of mental disorders in particularly impressionable people and new psychological and social problems. The prevalence of digital copies will increase, but not rapidly.

In our opinion, it is unlikely that the development of artificial reality technologies will have such a breakthrough direction as mobile phones, since their consumer properties are very specific, suitable only for rich people or those who, as they say, have nothing to do. But in general, of course, such a movement can significantly change some relationships. It is possible that artificial reality will be actively developed (if it is not already developed) by intelligence and military agencies in order to misinform the enemy.

Crypto money is becoming a more serious factor in financial affairs and the state has also started to actively invest and embed itself in this sector. But that is another matter. A change in the nature of money and finance, and with it fundamental changes in the financial sector, including the eradication or even the destruction of modern financial agents (including banks), is occurring. Further, a change in the relationship between financial authorities and the population can lead to unpredictable consequences. With respect to undermining the hegemony of dollar, the cryptocurrencies may one day become the world's equivalent of precious metals, equally accepted in all corners of the globe.

### 10.7.5 AI and Health

As mentioned above, the shrinking profitability and the exhaustion of ICT expansion opportunities means that digital companies are actively penetrating into different spheres. One of the most important areas here is the health sector. In previous chapters and here, we have already touched on the importance of using ICT and AI in medicine and biotechnology, as well as the prospects in the use of nanotechnologies, robots, cognitive technologies and other things in medicine that are inconceivable without AI. We would like to focus on another important aspect. We will talk not only about medicine in the narrow sense of the word as a field of treatment and rehabilitation, but more generally as a field of control and self-control over health, lifestyle and life processes, condition monitoring, prognosis and diagnosis and so on. A wide range of devices are being created—watches, smart wristbands, etc., that monitor the pulse and various biorhythms, human activity (e.g., what distance, at what speed, what trajectory a person walks) and much more.

Moreover, there are increasingly active attempts to use those devices to track and control a person's location, calls, messages, transactions, activities, sleep, driving, etc. They are also working on tracking human emotions and trying to limit their unwanted manifestations. The technology itself is reminiscent of the famous lie detector, because the AI that determines emotions focuses on a person's breathing and heartbeat. But now, instead of sensors on the body, ordinary radio waves are used, and you can use a regular Wi-Fi signal. This means that in the foreseeable future, not only will it be impossible to hide our emotions from an invisible spy hidden on a wireless network, but AI will begin to teach us how to behave. The prototypes are already being developed. 'Master, take it easy, and pull yourself together', says the new wearable device, vibrating or squeezing the wrist. The development will help people who find it difficult to control their emotions (Glyantsev, 2019, 2021).

Such things generally seem quite reasonable and can make life easier and help to maintain health. After all, what can be said against the idea of using software to determine emotional burnout by heart rate? Emotional burnout is a real psychological syndrome caused by constant stress, the scourge of the twenty-first century, according to the authors of the development. Early diagnosis of emotional burnout can help to prevent overwork, stress, depression, nervousness and aggression. The researchers note that this method can also be used in the complex treatment of mental disorders and headaches (Muraya, 2022). Of course, this can be a really good way to monitor yourself if it stays at the level of voluntary use, and the data is not used or 'leaked' to be abused in terms of pushing services and other things. Yes, of course, all sorts of devices and smart things are now voluntary, but we have all seen how voluntariness turns into voluntary-obligatory, and then mandatory things.

In the context of our study, this brief review allows us to confirm our previous conclusions and to draw new ones: (1) there is a clear merging of MANBRIC technologies with medicine as an integral link in its core; (2) there is an obvious shift in the use of AI in the field of medicine and, in general, in the way of life and life of a person related to his health and the quality of biological life; (3) this trend will intensify; [34] (4) thus, the first prerequisites are being created for what we call

[34] It is characteristically that even such a probable future technology as a quantum computer, which will be able to outperform modern computers by orders of magnitude, will, according to futurologists, focus largely on the ability to perform calculations of various processes in the human body,

the medical-biotechnological environment, designed to create a kind of artificial environment for the constant monitoring of human condition, which we expect to be fully operational in a few decades; (5) finally, it is extremely important that this trend should restrict human rights and freedoms as little as possible.

### 10.7.6 Ideas for a New Kind of Computer

**Options for new approaches.** As has already been mentioned, the development of information technology is now on the eve of creating a new type of computer which is fundamentally superior to the today's computers. There are different ideas about the basis of such breakthrough technologies for the computers of the future. The most popular are the predictions about the future of quantum computers. [35] In layman's terms, the fundamental difference between quantum computers and conventional computers lies in the way information is processed. Conventional processors perceive information in a binary system, that is, to put it simply, data takes on values in the form of either one or zero. Quantum computers perform computations in which information can have a value both of one and of zero, the calculations are performed not in ordinary bits, but in what are known as qubits. At the same time, unlike bits, qubits can take on different values at the same time, and as a result, they can perform calculations that a conventional processor is inherently incapable of (Datta et al., 2005; Veldhorst et al., 2015).

There have also been reports of attempts to create a photonic computer. The idea of nanocomputers lives on. As we said above, according to Eric Drexler, it is nanomechanics, not nanoelectronics, that can become such a base. Drexler has developed mechanical designs for the main components of the nanocomputer. Its main components can be pushed in and out of cores that interdependently block each other's movements (Balabanov, 2010; Drexler, 1987, 1992, 2013). There is information about the creation of a so-called single-atom computer, which operates at a temperature close to absolute zero. This is a significant step toward the creation of a quantum computer.[36]

---

including the brain, and, as expected, it is medicine that will be one of its main application areas (see: Bagrayev, 2015).

[35] Dozens of them have already been created, but there are very big problems with errors in the calculations. Although there is information about progress in this direction.

[36] For the first time, it became possible to create a working transistor based on a single atom. For a single-atom transistor to be used in real devices, one atom must be positioned exactly on a silicon chip. According to the nanotechnology journal Nature Nanotechnology (Fuechsle et al., 2012), this is exactly what researchers have achieved. Using a scanning tunnelling microscope (a device that allows researchers to see atoms and manipulate them with precision), they cut a narrow channel through the silicon base. Phosphine gas was then used to place a single phosphorus atom between two electrodes in the desired area. When an electric current passes through such a device, it amplifies and transmits an electric signal, which is the basic principle of operation of any transistor (Tadviser, 2015).

**Prospects**. It is, of course, difficult to predict what type of computer will be the breakthrough computer leading into the future. Sooner or later such a breakthrough will happen. However, the push for a new type of computer could be accelerated by the race to build ever more powerful AI, which requires very large digital resources. However, we do not believe that it will be implemented on a mass scale soon. Let us try to explain why we think this is the case. At the moment, this new type of computer does not have the mass market appeal that the digital giants are accustomed to. The level of development of the most popular computers is quite satisfactory for users (if someone is not personally satisfied, it is only because they have not updated their models). Therefore, it is impossible to replace most of them with more productive ones, except by force. But we still need to create the infrastructure. This is all the more difficult since more and more people prefer to use mobile phones. The latter can still be made more powerful, and they can also be massively replaced only by force. After all, new equipment will obviously be much more expensive (perhaps by orders of magnitude) than the current equipment, which is not exactly cheap. However, even the strong efforts of governments to replace petrol cars with electric vehicles have not yet yielded big results. In other words, these new types of computers will be available only for government and very large companies that would like to replace expensive supercomputers. But this is a limited market. Even if a new type of computer were to emerge within a decade, its development would be limited due to a small or limited market. It would mostly pave the way to become the basis of a new technological paradigm in the future. And taking into account the fact that the need for them is not particularly great even within the military and in the space industries, their technological emergence may be delayed due to insufficient funding. However, if the AI race accelerates, government efforts, either directly or through proxy companies (including leading digital giants), could accelerate the development of ultra-fast computers. In this case, the dangers of uncontrolled use of AI would increase exponentially.

**Models within existing technologies, neuromorphic processors**. There are interesting directions within existing technologies. These include, in particular, neuromorphic processors which are being developed at the intersection of biology, physics, mathematics, computer science and semiconductor production, and which are based on the usual transistors, but with a different architecture that is similar to the structure of biological brain neurons. Similar to biological model, an artificial neuron has one output (axon), the signal from which can go to a large number of inputs of other neurons and can thus change their state. Enthusiasts believe that neuromorphic processors are one of the most promising developments in the field of computing. Today, they are just a new model of programmable computing, but in the near future they are expected not only to speed up the execution of labor-intensive computing tasks on the fly with minimal power consumption, but will also open up new harmonious aspects of the digital lifestyle, as seen in wildlife. Over time, neuromorphic processors have every chance to extend and complement the capabilities of modern processors with new technologies that will allow the computers of the future to operate, adapt and learn using algorithms that resemble the way humans think (Sandomirskaya, 2021; Thakur et al., 2018). We do not rule out the possibility of such a bright future, although there are still doubts as to whether neuromorphic processors will become as popular as the current ones, since they are still intended for a relatively limited range of users.

However, the development of fundamentally new types of processors is underway and it can be assumed that they will be realized some time in the future. But of course, such a basis for AI may increase the dangers of its misuse many times over, so it is necessary to think in advance about the limits of such technologies.

Thus, in this chapter, we have considered a whole group of closely related new technological areas: nanotechnology, robotics (including self-driving vehicles), additive and cognitive technologies and AI. We believe that their development will enhance the development of the MANBRIC convergence as a whole, prepare medical technologies for the initial breakthrough that will launch the final phase of the Cybernetic Revolution and ensure the powerful development of this phase, which we call the phase of self-regulating/self-managing systems.

**Acknowledgements** This chapter has been prepared with support by the Russian Science Foundation (Project No. **23-11-00160**).

## References

Abidian, M. R., & Martin, D. C. (2009). Multifunctional Nanobiomaterials for Neural Interfaces. *Advanced Functional Materials, 19*(4), 573–585.

Acemoglu, D., & Restrepo, P. (2017). Robots and jobs. *NBER Working paper, 23285.*

Aghebati-Maleki, A., Dolati, S., Ahmadi, M., Baghbanzhadeh, A., Asadi, M., Fotouhi, A., Yousefi, M., & Aghebati-Maleki, L. (2020). Nanoparticles and Cancer Therapy: Perspectives for Application of Nanoparticles in the Treatment of Cancers. *Journal of Cellular Physiology, 235*(3), 1962–1972. https://doi.org/10.1002/jcp.29126

Ahn, Y. C., Park, S. K., Kim, G. T., Hwang, Y. J., Lee, C. G., Shin, H. S., & Lee, J. K. (2006). Development of high efficiency nanofilters made of nanofibers. *Current Applied Physics, Nano Korea 2004 Symposium on NT Challenge, 6*(6), 1030–35. https://doi.org/10.1016/j.cap.2005.07.013.

Allen Institute. (2020). *An Automated Pipeline for Understanding how the brain is wired.* https://alleninstitute.org/what-we-do/brain-science/news-press/articles/automated-pipeline-understanding-how-brain-wired.

Auslender D. (2014). IPhone 6 budut sobirat' roboty 'foksboty'. *Hi-News.ru.* URL: http://hi-news.ru/phone/iphone-6-budut-sobirat-roboty-foksboty.html.

Bagrayev, N. T. (2015). *Perspektivy razvitiya kvantovykh vychisleniy. Doklad na Vtorom mezhdunarodnom seminare «Bazisnyye tekhnologii pervoy poloviny XXI veka (strukturno-tsiklicheskiy analiz).* 1–2 oktyabrya. SPb.: Sankt-Peterburgskiy politekhnicheskiy universitet Petra Velikogo.

Balabanov, V. I. (2010). *Nanotekhnologii: pravda i vymysel.* M.: Eksmo'

Baron, F. A., Mikhlin, Yu. L., Molokeev, M. S., Rautskiy, M. V., Tarasov, I. A., Volochaev, M. N., Shanidze, L. V., et al. (2021). Structural, optical, and electronic properties of Cu-Doped TiNxOy grown by ammonothermal atomic layer deposition. *ACS Applied Materials & Interfaces, 13*(27), 32531–32541. https://doi.org/10.1021/acsami.1c08036

BCC Research. (2012). Nanotechnology: A Realistic Market Assessment. *BCC Research*. URL: http://www.bccresearch.com/report/nanotechnology-market-applications-products-nan031e.html.

Beck, M. (2013). Study raises doubts over robotic surgery. *The Wall Street Journal*. February 19. URL: http://www.wsj.com/articles/SB10001424127887323764804578 314182573530720. Date accessed: 22.11. 2014.

Belousov, D. (2016). Smena tekhnologicheskogo uklada: Novyye formaty biznesa, gosudarstva i obshchestva. In V. Burov (Ed.), *Vyzov 2035* (pp. 17–41). Olimp-Biznes.

Benderskaya, Ye. N., Tolstov, A. A. (2013). Tendentsii razvitiya sredstv apparatnoy podderzhki neyrovychisleniy. *Nauchno-tekhnicheskiye vedomosti SPbGPU*, 3 (174), 9–18

Bonev I. (2013). *The truth about Foxconn's Foxbot industrial robots*. February 14. URL: http://coro.etsmtl.ca/blog/?p=73.

Bondar, D. (2018). Nitrid galliya—prem'yer sredi novykh materialov poluprovodnikovoy mikroelektroniki. *Komponenty i Materialy, 4*, 134–137.

Binysh, J.et al. (2022). Active elastocapillarity in soft solids with negative surface tension. *Science Advances*. https://doi.org/10.1126/sciadv.abk3079.

Biswas, S. (2023). ChatGPT and the future of medical writing. *Radiology*, February 2: 223312. https://doi.org/10.1148/radiol.223312.

Cairns R. (2021). *Meet Grace, the ultra-lifelike nurse robot*. https://edition.cnn.com/2021/08/19/asia/grace-hanson-robotics-android-nurse-hnk-spc-intl/index.html.

Callaway, E. (2020). 'It opens up a whole new universe': revolutionary microscopy technique sees individual atoms for first time. *Nature, 582*, 7811, 156–57. https://doi.org/10.1038/d41586-020-01658-1.

Campbell, C. (2021). *Inside the Taiwan Firm That Makes the World's Tech Run*. https://time.com/6102879/semiconductor-chip-shortage-tsmc/.

Cepelewicz, J. (2016). The U.S. Government Launches a $100-Million 'Apollo Project of the Brain. URL: https://www.scientificamerican.com/article/the-u-s-government-launches-a-100-million-apollo-project-of-the-brain/.

Chen, Zh., Jiang, Y., Shao, Y.-T., Holtz, M. E., Odstrčil, M., Guizar-Sicairos, M., Hanke, I., Ganschow, S., Schlom, D. G., & Muller, D. A. (2021). Electron Ptychography achieves atomic-resolution limits set by lattice vibrations. *Science, 372*(6544), 826–31. https://doi.org/10.1126/science.abg2533.

Chowdhury, H. (2020). Elon Musk wants to plug a microchip into your brain: here's why The billionaire entrepreneur is set to unveil a prototype of his brain-computer interface. *Telegraph*, August 27. https://www.telegraph.co.uk/technology/2020/08/27/elon-musk-wants-plug-microchip-brain/.

Chumakov, A. N., & Yurchenko, P. S. (2021). Global pandemic in terms of objective causes and subjective speculations. *Journal of Globalization Studies, 12*(2), 157–168.

Ciera, L., Beladjal, L., Van Landuyt, L., Menger, D., Holdinga, M., Mertens, J., Van Langenhove, L., De Clerk, K., & Gheysens, T. (2019). Electrospinning repellents in polyvinyl alcohol-nanofibres for obtaining mosquito-repelling fabrics. *Royal Society Open Science,* 6(8), 182139. https://doi.org/10.1098/rsos.182139.

Coupeau, N. (2023). *Rememory, The AI that Helps Us Talk to the Dead*. https://greekreporter.com/2023/02/01/rememory-the-artificial-intelligence-that-helps-us-talk-to-the-dead/.

Crawford, M. (2016). 6 Applications for robotics in medicine. *The American Society of Mechanical Engineers*, September 14. https://www.asme.org/topics-resources/content/top-6-robotic-applications-in-medicine.

Da Cruz, L., Coley, B. F., Dorn, J., Merlini, F., Filley, E., Christopher, P., Chen, F. K., Wuyyuru, V., Sahel, J., & Stanga, P. (2013). The Argus II Epiretinal prosthesis system allows letter and word reading and long-term function in patients with profound vision loss. *British Journal of Ophthalmology, 97*(5), 632–636.

Datta, A., Flammia St, T., & Caves, C. M. (2005). Entanglement and the power of one qubit. *Physical Review A, 72*(4), 042316. https://doi.org/10.1103/PhysRevA.72.042316

Dawley, J. A., Fite, K. B., & Fulk, G. D. (2013). EMG control of a bionic knee prosthesis: Exploiting muscle co-contractions for improved locomotor function. *2013 IEEE 13th International Conference on Rehabilitation Robotics (ICORR),* 1–6.

Del Valle, J., De La Oliva, N., Müller, M., Stieglitz, T., & Navarro, X. (2015). Biocompatibility evaluation of parylene C and polyimide as substrates for peripheral nerve interfaces. *Int. IEEE/EMBS Conf. Neural Eng. NER, 2015,* 442–445. https://doi.org/10.1109/NER.2015.7146654

de la Oliva, N., Navarro, X., & del Valle, J. (2018). Time course study of long-term biocompatibility and foreign body reaction to intraneural polyimide-based implants. *Journal of Biomedical Materials Research. Part A, 106*(3), 746–757. https://doi.org/10.1002/jbm.a.36274

Dementyev V. Ye. (2008). *Nanotekhnologicheskaya initsiativa SSHA—opyt politiki tekhnologicheskogo liderstva. Teoriya i praktika institutsional'nykh preobrazovaniy v Rossii.* Moscow: TSEMI RAN.

Dizikes P. (2016). *Driverless cars: Who gets protected? Study shows inconsistent public opinion on safety of driverless cars.* https://news.mit.edu/2016/driverless-cars-safety-issues-0623.

Drexler, K. E. (1987). *Engines of creation: The coming era of nanotechnology.* Anchor Press/Doubleday.

Drexler, K. E. (1992). *Nano-systems: Molecular machinery, manufacturing, and computation.* John Wiley & Sons.

Drexler, K. E. (2003). Drexler and Smalley make the case for and against 'molecular assemblers.' *Chemical & Engineering News, 81*(48), 1.

Drexler, K. E. (2013). *Radical abundance: How a revolution in nanotechnology will change civilization.* Public Affairs.

Efremov, I., & Arkhangelsky, V. (2023). Global fertility changes in the context of the COVID-19 Pandemic. *Journal of Globalization Studies, 14*(2), 165–170.

Elias J. (2023). Google CEO Sundar Pichai warns society to brace for impact of A.I. acceleration, says 'it's not for a company to decide'.

Emergen Research. (2021). *Nanotechnology Market Size to Reach USD 290.93 Billion in 2028* https://www.globenewswire.com/news-release/2021/05/18/2231845/0/en/Nanotechnology-Market-Size-to-Reach-USD-290-93-Billion-in-2028-Advancements-in-Technology-and-Rising-Investment-in-Research-Development-by-Market-Players-is-Driving-Industry-Growth.html

EMR. (2020). Expert market research 2020. *Global Neurotechnology Market Outlook.* https://www.expertmarketresearch.com/reports/neurotechnology-market.

Ford, M. (2015*). Rise of the robots: Technology and the threat of a jobless future.* Basic Books.

Ford, M. (2021) .*Rule of the robots: How artificial intelligence will transform everything.* Basic Books.

Fuechsle, M., Jill A. Miwa, Mahapatra, S., Ryu, H., Lee, S., Warschkow, O., Hollenberg, L. C. L., Klimeck, G., & Simmons, M. Y. (2012). A single-atom transistor. *Nature Nanotechnology, 7*(4), 242–46. https://doi.org/10.1038/nnano.2012.21.

Gates, B. (2007). Mekhanicheskoye budushcheye: Microsoft predskazyvayet revolyutsiyu v sfere robototekhniki. *V Mire Nauki, 5,* 36–43.

Gates, B. (2023). *The Age of AI has begun: Artificial intelligence is as revolutionary as mobile phones and the Internet.* https://www.gatesnotes.com/The-Age-of-AI-Has-Begun.

Gavas, Sh., Quazi S., & Karpiński, T. M. (2021). Nanoparticles for cancer therapy: current progress and challenges. *Nanoscale Research Letters, 16*(1),173. https://doi.org/10.1186/s11671-021-03628-6.

Gavilán, H., Rizzo, G. M. R., Silvestri, N., Mai, B. T., & Pellegrino, T. (2023). Scale-up approach for the preparation of magnetic ferrite nanocubes and other shapes with benchmark performance for magnetic hyperthermia applications. *Nature Protocols,* 1–27. https://doi.org/10.1038/s41596-022-00779-3.

Gavrilov, A. V., & Kangler, V. M. (2015). Neyromorfnyye tekhnologii: sostoyaniye i perspektivy razvitiya. *Materialy VII Vserossiyskoy nauchno-tekhnicheskoy konferentsii 'Robototekhnika i iskusstvennyy intellekt'* RII-2015 (pp. 148–154). Zheleznogorsk.

Giri, G., Maddahi Y., & Zareinia K. (2021). A brief review on challenges in design and development of nanorobots for medical applications. *Applied Sciences, 11*(21), 10385. https://doi.org/10.3390/app112110385.

Globenewswire. (2019). *Surgical Robots Market: 2019 Global Future Scope, Aging Population, Application, Industry Growth, Demand, Region, Competitive Analysis and Forecast 2026.* https://www.globenewswire.com/en/news-release/2019/10/22/1933426/0/en/Surgical-Robots-Market-2019-Global-Future-Scope-Aging-Population-Application-Industry-Growth-Demand-Region-Competitive-Analysis-and-Forecast-2026.html.

Glyantsev, A. (2019). *Novyy braslet schityvayet emotsii vladel'tsa.* URL: https://smotrim.ru/article/1216271.

Glyantsev, A. (2021). *Bolshoy brat: nashi emotsii budut otslezhivat' s pomoshch'yu Wi-Fi.* https://smotrim.ru/article/2519509.

Griffin, A. (2015). Sex robots should be banned, say campaigners, as engineers look to add AI to sex toys. *Independent*, September 15. URL: http://www.independent.co.uk/life-style/gadgets-and-tech/news/sex-robots-should-be-banned-say-campaigners-as-engineers-look-to-add-ai-to-sex-toys-10501622.html.

Grill, W. M., Norman, Sh. E., & Bellamkonda, R. V. (2009). Implanted Neural Interfaces: Biochallenges and Engineered Solutions. *Annual Review of Biomedical Engineering, 11*, 1–24.

Grinin, L. E. (2019). Kondratieff waves, technological modes, and the theory of production revolutions. In L. E. Grinin & A. V. Korotayev (Eds.), *Kondratieff Waves: Spectrum of Opinions* (pp. 95–144). Uchitel' Publishing House.

Grinin, L. (2020). How can COVID-19 change geopolitics and economy? *Journal of Globalization Studies, 11*(2), 121–134.

Grinin, L. E., & Grinin, A. L. (2015). *Ot rubil do nanorobotov. Mir na puti k epokhe samoupravlyaemykh sistem (istoriya tekhnologiy i opisanie ikh budushchego).* Uchitel.

Grinin, L., & Grinin, A. (2016). *The cybernetic revolution and the forthcoming epoch of self-regulating systems.* Moscow branch of Uchitel Publishing House.

Grinin, L., & Grinin, A. (2021). The dynamics of Kondratieff waves in the light of the theory of production revolutions. In: L. E. Grinin, & A. V. Korotayev (eds.), *Kondratieff waves. Historical and Theoretical Aspects* (pp. 59–113). Volgograd: 'Uchitel' Publishing House.

Grinin, L., Grinin, A., & Korotayev, A. (2020a). A quantitative analysis of worldwide long-term technology growth: From 40,000 BCE to the early 22nd century. *Technological Forecasting and Social Change, 155*, 1–15. https://doi.org/10.1016/j.techfore.2020.119955

Grinin, L., Grinin, A., & Korotayev, A. V. (2020b). Dynamics of technological growth rate and the forthcoming singularity. In: A. V. Korotayev, D. LePoire (eds) *The 21st century Singularity and global futures. A Big History perspective* (pp. 287–344). Springer. https://doi.org/10.1007/978-3-030-33730-8_14.

Grinin, L., Grinin, A., & Korotayev, A. (2021). Global Trends and Forecasts of the 21st Century. *World Futures, 77*(5), 335–370. https://doi.org/10.1080/02604027.2021.1949939

Grinin, L., Grinin, A., & Korotayev, A. (2022). COVID-19 Pandemic as a trigger for the acceleration of the cybernetic revolution, transition from E-Government to E-State, and change in social relations. *Technological Forecasting and Social Change, 175*, 1–15. https://doi.org/10.1016/j.techfore.2021.121348

Grinin, L. E., & Korotayev, A. V. (2015). *Great divergence and great convergence: A global perspective.* Springer.

Grinin, L. E., & Korotayev, A. V. (2021). Introduction. Long-wave dynamics research in historical and theoretical aspects. In: L. E. Grinin, & A. V. Korotayev (eds.), *Kondratieff Waves. Historical and Theoretical Aspects* (pp. 4–9). Volgograd: 'Uchitel' Publishing House.

Gunasekera, B., Saxena, T., Bellamkonda, R., & Karumbaiah, L. (2015). Intracortical recording interfaces: Current challenges to chronic recording function. *ACS Chemical Neuroscience, 6*(1), 68–83. https://doi.org/10.1021/cn5002864

Hampson, R. E., Song, D., Robinson, B. S., Fetterhoff, D., Dakos, A. S., Roeder, B. M., She, X., et al. (2018). Developing a hippocampal neural prosthetic to facilitate human memory encoding and recall. *Journal of Neural Engineering, 15*(3), 036014. https://doi.org/10.1088/1741-2552/aaaed7

Harari Y. N. (2016). *Yuval Noah Harari on big data, Google and the end of free will* https://www.ft.com/content/50bb4830-6a4c-11e6-ae5b-a7cc5dd5a28c.

Harifi, T., & Montazer, M. (2017). Application of nanotechnology in sports clothing and flooring for enhanced sport activities, performance, efficiency and comfort: A review. *Journal of Industrial Textiles, 46*(5), 1147–1169.

Hawkins, J., & Blakeslee, S. (2004). *On Intelligence*. Times Books.

Henderson, L., Neumann, O., Kadria-Vili, Ya., Gerislioglu, B., Bankson, J., Nordlander, P., & Halas, N. J. (2022). Plasmonic Gadolinium Oxide Nanomatryoshkas: Bifunctional magnetic resonance imaging enhancers for Photothermal cancer therapy. *PNAS Nexus*, 1(4), pgac140. https://doi.org/10.1093/pnasnexus/pgac140.

Herr, H. M., & Grabowski, A. M. (2011). Bionic ankle-foot prosthesis normalizes walking gait for persons with leg amputation. *Proceedings of the Biological Sciences, 279*(1728), 457–464.

Hussain, M. (2021). The changing dynamics of liberal international order: Covid-19 and its global implications. *Journal of Globalization Studies, 12*(2), 143–156.

IBM. (2023). *Augmented work for an automated, AI-driven world. AI won't replace people—but people who use AI will replace people who don't.* https://www.ibm.com/thought-leadership/institute-business-value/en-us/report/augmented-workforce.

IFR. (2020). *International federation of robots 2020.* IFR Press Conference 24th September 2020. Frankfurt. https://ifr.org/downloads/press2018/Presentation_WR_2020.pdf.

IFR. (2021). *International federation of robots 2021. IFR presents World Robotics 2021 reports.* https://ifr.org/ifr-press-releases/news/robot-sales-rise-again.

IFR. (2022). *International federation of robots 2022. World Robotics 2022—Service Robots report released.* https://ifr.org/searchresults/68416577790a7a730ae7e92f949a704e/.

Irshad, S. M. (2020). COVID-19 and Global 'At-Risk community': From benefit-sharing to risk-sharing of economic crisis. *Journal of Globalization Studies, 11*(2), 135–145.

Jain, K. K. (2004). Applications of biochips: From diagnostics to personalized medicine. *Current Opinion in Drug Discovery & Development, 7*(3), 285–89.

Jasiński, A. M., & Bąkowska, A. M. (2020). Globalization and Life Satisfaction of the Polish Population. Before and during the COVID-19 Pandemic. *Journal of Globalization Studies, 11*(2), 146–155.

Kah, M., & Hofmann, Th. (2014). Nanopesticide Research: Current trends and future priorities. *Environment International,* 63, 224–35. https://doi.org/10.1016/j.envint.2013.11.015.

Kahn, H. (1982). *The coming boom: Economic, political and social.* N. Y.: Simon & Schuster.

Kakkar, S. (2021). The outbreak of sars-cov-2 (Covid-19): Impact on international tourism. *Journal of Globalization Studies, 12*(1), 145–158.

Khel, I. (2015). 10 tekhnologiy, kotorymi dolzhen zapomnit'sya 2015 god. *Hi-News*. URL: http://hi-news.ru/technology/10-texnologij-kotorymi-dolzhen-zapomnitsya-2015-god.html.

Kish, L. B. (2002). End of Moore's Law: Thermal (Noise) death of integration in micro and nano electronics. *Physics Letters, Section A: General, Atomic and Solid State Physics,*https://doi.org/10.1016/S0375-9601(02)01365-8

Kotov, N. A., Winter, J. O., Clements, I. P., Jan, E., Timko, B. P., St, C., Pathak, S., et al. (2009). Nanomaterials for Neural Interfaces. *Advanced Materials, 21*(40), 3970–4004.

Landgar (Shumikhin S.). (2021) *Artificial General Intelligence—the search for the Holy Grail of artificial intelligence.* URL: https://habr.com/ru/users/Landgar/.

Lane, N., & Kalil, T. (2007). *The national nanotechnology initiative: Present at the creation. Issues in Science and Technology.* http://www.issues.org/21.4/lane.html.

LaPook, J. (2013). Engineer develops new way to fight wife's cancer. *CBS evening news*, August 5, 1:42 PM. https://www.cbsnews.com/news/engineer-develops-new-way-to-fight-wifes-cancer/

Lebedev, M. A., & Nicolelis, M. A. L. (2006). Brain-machine Interfaces: Past, Present and Future. *Trends in Neurosciences, 29*(9), 536–546. https://doi.org/10.1016/j.tins.2006.07.004.

Lipson, H., & Kurman, M. (2013). *Fabricated: The new world of 3D printing.*

Lotti, F., Ranieri, F., Vadalà, G., Zollo, L., & Di Pino, G. (2017). Invasive Intraneural interfaces: Foreign body reaction issues. *Frontiers in Neuroscience,* 11. https://doi.org/10.3389/fnins.2017.00497.

Lund, B. D., & Wang, T. (2023). *Chatting about ChatGPT: How may AI and GPT impact academia and libraries?* Library Hi Tech News.

Mallouk, Th. E., & Sen, A. (2009). Powering Nanorobots. *Scientific American, 300*(5), 72–77.

McCarthy, J. (1990). *Artificial intelligence, logic and formalizing common sense.* http://www-formal.stanford.edu/jmc/ailogic.pdf.

McCarthy, J. (2000). *Concepts of logical AI.* http://www-formal.stanford.edu/jmc/concepts-ai.pdf.

McCarthy, J. (2007). *What is artificial intelligence?* Computer Science Department. Stanford University. November 12, http://wwwformal.stanford.edu/jmc/whatisai/whatisai.html.

Media release. (2016). *Deakin builds robotic surgical system with sense of touch.* https://www.deakin.edu.au/about-deakin/news-and-media-releases/articles/deakin-builds-robotic-surgical-system-with-sense-of-touch.

Melikhov, I. V. (2018). *Fiziko-khimicheskaya evolyutsiya tverdogo veshchestva.* BINOM. Laboratoriya znaniy.

Metz, C. (2023). 'The Godfather of A.I.' Leaves Google and Warns of Danger Ahead For half a century, Geoffrey Hinton nurtured the technology at the heart of chatbots like ChatGPT. Now he worries it will cause serious harm. https://www.nytimes.com/2023/05/01/technology/ai-google-chatbot-engineer-quits-hinton.html?searchResultPosition=1.

Mitchell, T. (1997). *Machine learning.* McGraw-Hill.

Moore, K. (2021). *Cerebras' New Monster AI Chip Adds 1.4 Trillion Transistors Shift to 7-nanometer process boosts the second-generation chip's transistor count to a mind boggling 2.6-trillion.* 20 APR 2021. https://spectrum.ieee.org/cerebras-giant-ai-chip-now-has-a-trillions-more-transistors.

Moravec, H. P. (1988). *Mind children.* Harvard University Press.

Moucheboeuf, G., Griffier, R., Gasq, D., Glize, B., Bouyer, L., Dehail, P., & Cassoudesalle, H. (2020). Effects of robotic gait training after stroke: A meta-analysis. *Annals of Physical and Rehabilitation Medicine,* 63(6), 518–34. https://doi.org/10.1016/j.rehab.2020.02.008.

Muraya, O. (2022). *Rossiyskoye prilozheniye opredelit vygoraniye po ritmu serdtsa.* . https://smotrim.ru/article/2812453.

Musk, E. et al. (2023). *Pause giant AI experiments: An open letter. We call on all AI labs to immediately pause for at least 6 months the training of AI systems more powerful than GPT-4.* https://futureoflife.org/open-letter/pause-giant-ai-experiments/

Nakano, E. (1988). *Vvedeniye v robototekhniku.* Mir.

National Cancer Institute. (2017a). *Washington University—Center for Multiple Myeloma Nanotherapy (CNMN).* https://www.cancer.gov/nano/research/alliance/ccne/washington,

National Cancer Institute. (2017b). *Treatment and therapy.* https://www.cancer.gov/nano/cancer-nanotechnology/treatment,

National Institute of Health. (2016). *Nanotechnology to improve early detection and treatment of colorectal cancer.* https://www.nih.gov/research-training/nanotechnology-improve-early-detection-treatment-colorectal-cancer.

Nistor, M. T., & Rusu, A. G. (2019). Chapter 3—Nanorobots with applications in medicine. In C, A. Vasile (ed.), *Polymeric Nanomaterials in Nanotherapeutics* (pp. 123–49). Elsevier, https://doi.org/10.1016/B978-0-12-813932-5.00003-0.

Ocsial. (2017). *Strong and ultra-light aluminium composite with carbon nanotubes.* https://ocsial.com/news/-strong-and-ultra-light-aluminium-composite-with-carbon-nanotubes-/

Persidis, A. (1998). Biochips. *Nature biotechnology, 16*(10), 981–983.

Pinkerton, S. (2013). The pros and cons of robotic surgery. *The Wall Street Journal*, November, 17. http://www.wsj.com/articles/SB10001424052702304655104579163430371597334.

Plotnikov, S. (2014). Tret'ya promyshlennaya revolyutsiya. O 3D-printerakh podrobno i s video. *Ferra.ru,* Aptil 21. http://www.ferra.ru/ru/periphery/review/3d-printers/#.VU9cGCG8PGd.

PR—Precedence Research. (2022). *Healthcare Neurotechnology Market.* https://www.precedenceresearch.com/neurotechnology-market.

Proydakov, E. (2016). Robototekhnika cherez dvadtsat' let. In V. Burov (Ed.), *Vyzov 2035* (pp. 114–131). Olimp-Biznes.

Redolfi Riva, E., & Micera, S. (2021). Progress and challenges of implantable neural interfaces based on nature-derived materials. *Bioelectronic Medicine, 7*(1), 6. https://doi.org/10.1186/s42234-021-00067-7.

Research and Markets. (2021). *Biochips—Technologies, Markets & Companies*. https://www.researchandmarkets.com/reports/4748187/biochips-technologies-markets-and-companies?utm_source=dynamic&utm_medium=BW&utm_code=fd3k8s&utm_campaign=1327398+-+Biochips+Technologies%2c+Companies%2c+Applications+%26+Markets%2c+2028+-+94+Companies+are+Included+Along+with+a+Listing+of+121+Collaborations+Between+Companies&utm_exec=chdo54bwd.

Rowald, A., Komi, S., Demesmaeker, R., Baaklini, E., Hernandez-Charpak, S. D., Paoles, E., Montanaro, H., et al. (2022). Activity-dependent spinal cord neuromodulation rapidly restores trunk and leg motor functions after complete paralysis. *Nature Medicine, 28*(2), 260–271. https://doi.org/10.1038/s41591-021-01663-5

Ruban, I., Voropaeva, N., Istomin, M., & Figovsky, O. (2011). Presowing Nanotreatment of Seeds with Use of Biologically Active Nanochips. *Scientific Israel—Technological Advantages, 13*(4). https://www.elibrary.ru/item.asp?id=35348441.

Rybalkina, M. (2005). *Nanotekhnologii dlya vsekh*. Nanotechnology News Network.

Sandomirskaya, Yu. (2021). Iskusstvennyy intellekt i neyromorfnyye vychisleniya: vtoroye dykhaniye. *Kommersant–Nauka, 47*, 26. https://www.kommersant.ru/doc/5089550.

Schmidhuber, J. (2014). *Deep learning in neural networks: An overview. Technical Report IDSIA-03–14*. URL: http://arxiv.org/pdf/1404.7828.pdf

Seifabadi, M. S., & Dolatabadi, A. B. (2022). Covid-19, globalization and strengthening authoritarianism in the Middle East. *Journal of Globalization Studies, 13*(1), 135–157.

Servick, K. (2022). Brain implant enables man in locked-in state to communicate. *Science, 375*(6587), 1327–1328.

Skobtsov, Yu. A. (2008). *Osnovy evolyutsionnykh vychisleniy*. Donetsk.

Smalley, R. (2001). Of chemistry, love and nanobots. *Scientific American, 285*, 76–77.

Smirnova, Yu. (2011). Roboty vmesto lyudey. *Forbes,* August 4. http://www.forbes.ru/tehno-slideshow/tehnologii/71560-roboty-vmesto-lyudei/slide/1.

Smith, A. (2022). New magnetic phenomenon discovered in nanoscale magnets with immense industrial potential. *Azonano,* April 13. https://www.azonano.com/news.aspx?newsID=38981.

Smotrim.ru. (2016). Uluchshennaya sistema dlya robotizirovannoy khirurgii podarila medikam chuvstvo osyazaniya. *Smotrim.ru,* October 4. https://smotrim.ru/article/1311753.

Soetaert, F., Korangath, P., Serantes, D., Fiering, S., & Ivkov, R. (2020). Cancer therapy with iron oxide nanoparticles: agents of thermal and immune therapies. *Advanced Drug Delivery Reviews, Integration of Hyperthermia and Drug Delivery*, 163–164, 65–83. https://doi.org/10.1016/j.addr.2020.06.025.

Soto, F., Wang, J., Ahmed, R., & Demirci, U. (2020). Medical micro/nanorobots in precision medicine. *Advanced Science, 7*(2), 2002203. https://doi.org/10.1002/advs.202002203

Stanger, C. A., & Cawley, M. F. (1996). Demographics of rehabilitation robotics users. *Technology and Disability, 5*, 125–137.

Swaab, D. (2014). *My—eto nash mozg: Ot matki do Al'tsgeymera*. Ivan Limbakh Publishers.

Tadviser. (2015). *Zakon Mura*. https://www.tadviser.ru/index.php/%D0%A1%D1%82%D0%B0%D1%82%D1%8C%D1%8F:%D0%97%D0%B0%D0%BA%D0%BE%D0%BD_%D0%9C%D1%83%D1%80%D0%B0.

Tae-Ho, K., Bao, C., Chen, Z., Kim, W. S. (2022). 3D printed leech-inspired origami dry electrodes for electrophysiology sensing robots. *Flexible Electronics, 6*(1). https://doi.org/10.1038/s41528-022-00139-x.

Thomason, R. (2003). Logic and artificial intelligence. In E. N. Zalta (ed.), *The Stanford Encyclopedia of Philosophy*. http://plato.stanford.edu/entries/logic-ai/.

Thakur, Ch. S., Lottier, M. J., Cauwenberghs, G., Indiveri, G., Kumar, K., Qiao, N., Schemmel, J., Wang, R., Chicca, E., & Olson, H. J. (2018). Large-scale neuromorphic spiking array processors: A quest to mimic the brain. *Frontiers in Neuroscience, 12*, 891.

Tsvetkov, V. Ya. (2017). Zakon Mura i drugiye. *Mezhdunarodnyy zhurnal prikladnykh i fundamental'nykh issledovaniy*, 1–2, 370–370. URL: https://applied-research.ru/ru/article/view?id=11205.

Turichin, G. A. (2015). *Additivnyye tekhnologii v sovremennom proizvodstve*. Report at the Second International Seminar: "The Basic technologies of the first half of the nineteenth century (structural and cyclic analysis)" Sankt-Petersburg, October 1–2.

Turja, T., & Oksanen, A. (2019). Robot acceptance at work: A multilevel analysis based on 27 EU countries. *International Journal of Social Robotics, 11*(4), 679–689.

Turing, A. (1950). *Computing machinery and intelligence*. Mind.

UNESCO. (2021). *Recommendation on the ethics of artificial intelligence*. UNESCO. https://unesdoc.unesco.org/ark:/48223/pf0000380455.

Veldhorst, M., Yang, C. H., Hwang, J. C. C., Huang, W., Dehollain, J. P., Muhonen, J. T., & Simmons, S. et al. (2015). A two-qubit logic gate in silicon. *Nature, 526*(7573), 410–14. https://doi.org/10.1038/nature15263.

Vincent, J. (2023). *Elon Musk and top AI researchers call for pause on 'giant AI experiments'*. https://www.theverge.com/2023/3/29/23661374/elon-musk-ai-researchers-pause-research-open-letter

Voropaeva, N., Figovsky, O., Ibraliu, A., Shehu, I., Kadiasi, N., Varlamov, V., & Karpachev, V. (2012). Innovative nanotechnology for agriculture. 1. Research of processing presowing seeds by nanochips and study the effects of nanotechnology on growth, development and crop capacity. *Scientific Israel—Technological Advantages, 14*(1). https://www.elibrary.ru/item.asp?id=35405275.

Wadhawan, V. K. (2007). Robots of the future. *Resonance, 12*, 61–78.

Waibel, M., Beetz, M., Civera, J., D'Andrea, R., Elfring, J., Gálvez-López, D., et al. (2011). RoboEarth. *IEEE Robotics & Automation Magazine, 18*(2), 69–82. https://doi.org/10.1109/MRA.2011.941632

Wakefield, J. (2019). Brain implants used to fight drug addiction in US. *BBC News*, November 8, sec. Technology. https://www.bbc.com/news/technology-50347421.

Widdowson, M. (2021). From Covid-19 to zero-gravity: Complex crises and production revolutions. *Journal of Globalization Studies, 12*(1), 117–144.

WIPO = World Intellectual Property Organization. (2023). World Intellectual Property Organization database. https://www3.wipo.int/ipstats/IpsStatsResultvalue.

Wolpaw, J. R., & Wolpaw, E. W. (2012). Brain-computer interfaces: Something new under the sun. In J. R. Wolpaw & E. W. Wolpaw (Eds.), *Brain-Computer Interfaces: Principles and Practice* (pp. 3–12). Oxford University Press.

Wood, R., Nagpal, R., & Wei, G.-Y. (2013). Flight of the Robobees. *Scientific Americanum, 308*, 60–65.

World Robotics. (2013). *Industrial robots*. http://www.worldrobotics.org/.

Yao, Y., Yunxiang, Zh., Lihong, L., Xu, Y., Chen, Q., Wang, Y., Wu, S., Deng, Y., Zhang, J., & Shao, A. (2020). Nanoparticle-based drug delivery in cancer therapy and its role in overcoming drug resistance. *Frontiers in Molecular Biosciences*, 7. https://doi.org/10.3389/fmolb.2020.00193.

Yeoman, I., & Mars, M. (2012). Robots, men and sex tourism. *Futures, 44*(4), 365–371. https://doi.org/10.1016/j.futures.2011.11.004

Yuschenko A. (2015). Roboty ne meneye opasny, chem atomnaya bomba. *Lenta.ru*, May 11. https://lenta.ru/artices/2015/05/11/robot/.

Zgurovskiy, M. Z., & Zaychenko, Yu. P. (2013). *Osnovy vychislitel'nogo intellekta.* Kiyev: Naukova dumka.

Zhang, N. (2022). *Moore's Law Is Dead, Long Live Moore's Law!* ArXiv Preprint ArXiv:2205.15011.

Zhang, Y., Zhang, Y., Han, Y., Gong, X. (2022). Micro/Nanorobots for medical diagnosis and disease treatment. *Micromachines, 13*(5), 648. https://doi.org/10.3390/mi13050648.

# Anti-aging as a Key Challenge for the Medicine of the Future

# 11

**Abstract**

The title of this chapter expresses well not only its main idea, but also formulates one of the key ideas of the whole book. Grinin et al. give many arguments in favor of these ideas, and in the last paragraph of this chapter they summarize them. The authors believe that medicine of the future will be cardinally new kind of medicine because of anti-aging technologies, that is, technologies for slowing down and ennobling aging, as well as stopping and alleviating the problems associated with it. All this will become an integral part of the future of medicine; perhaps even reforming whole system of medicine. This will also mean that more and more attention will be focused on the treatment of age-associated diseases, on the cure or mitigating of the so-called geriatric syndromes (or age-related conditions). The perception of aging in society will also change. Thus, the changing focus of medicine on the problems of aging together with innovations in anti-aging technologies will imply a medical adaptation to aging. The latter will develop gradually up to the late twenty-first century and become an essential part of the common process that Grinin et al. have called adaptation to aging.

## 11.1 What This Chapter is About

The title of this chapter expresses well not only its main idea, but also formulates one of the key ideas of the whole book.

We offer many arguments in favor of these ideas, and in the last paragraph of this chapter we summarize them. Despite the fact that the problems associated with aging receive much attention in the medical field, the issues related to aging do not onstitute a single problem but are scattered over different concerns such as types of diseases, diet, exercise, genetics, etc.



L. Grinin et al., *Cybernetic Revolution and Global Aging*, World-Systems Evolution and Global Futures, https://doi.org/10.1007/978-3-031-56764-3_11

Therefore, today's idea of the medicine of the future, in principle, is that it will be approximately the same as before with the addition of much greater reliance on technology, including AI and robotics. Yet, we believe that there will be cardinally new kind of medicine available because of anti-aging technologies, that is, technologies for slowing down and ennobling aging, as well as stopping and alleviating the problems associated with it, that will become an integral part of the future of medicine; perhaps even reforming the entire system of medicine.

This will also mean that more and more attention will be focused on the treatment of age-associated diseases, on the cure or mitigating of the so-called geriatric syndromes (or age-related conditions). At the same time, a noticeable aspect of medicine will concentrate on rejuvenation, plastic surgery and other anti-aging corrective procedures (which, perhaps, will become commonplace and regular not unlike a vacation in a sanatorium).

The perception of aging in society will also change. Thus, the hanging focus of medicine on the problems of aging together with innovations in anti-aging technologies will imply a medical adaptation to aging. The latter will become an essential part of the common process that we have called adaptation to aging. But, of course, such a transformation will not happen immediately. It will develop gradually up to the late twenty-first century.

## 11.2 Preliminary Remarks. Aging and Future Medicine

Every year the problems of aging and its impact on the body and mind are gaining increasingly more attention in medicine. However, practicing doctors still consider these problems not as a single and increasingly important area of medicine, but as a number of problems, each of them associated with specific types of disease, that, accordingly, should be treated in domains of different specialists. The current basic idea of the future medicine is that it will remain almost the same institution but will be much more technologized. However, we believe that the evolution of medicine will proceed in the direction of a growing role of aging in its practical aspect; thus, the anti-aging technologies will play an increasingly important role. We mean not only the fight against the major age-associated diseases, but also the increasingly widespread implementation of technologies aimed at slowing down aging and, if we may say so, making it more dignified. All of this, as well as technologies and practices for preventing and alleviating various problems, discomforts and inconveniences associated with old age, will become an integral part of the new medicine and, in fact, will form its core direction. Increasing attention will be paid to the treatment of age-associated diseases, but also to the elimination or alleviation of the so-called geriatric syndromes (or age-related conditions). At the same time, a very noticeable part of medicine will deal with rejuvenation, plastic surgery and other anti-aging corrective procedures (which may become commonplace and regular not unlike a vacation in a sanatorium). Of course, this will also change the perception of aging in society (see Chap. 7), facilitating society's adaptation to aging, the most important part of which will be the process of medical adaptation to aging.

This explains why the title of this chapter—Anti-Aging as the Key Challenge for the Medicine of the Future—expresses not only its main idea, but also one of the key ideas of this book.

## 11.3 Aging: Theories, Views and General Approaches

**Aging: A Disease or a Natural Process?** It is only recently that aging has become a major cause of death. We can say that this trend began to intensify during the Cybernetic Revolution. In the absence of global wars and famines and with low infant mortality, various age-associated diseases have become the main problem which causes death of tens of thousands of people every day (Dattani et al. 2018; WHO, 2020).[1]

There are a significant number of age-associated diseases. For example, the Charity Age UK identifies about 15 such diseases (Age UK, 2022). According to the NCOA (The National Council on Aging, USA) the most important are hypertension, arthritis, diabetes (especially type 2), ischemic heart disease, high cholesterol, atherosclerosis, cancer, dementia (including its most common cause, Alzheimer's disease), Parkinson's disease, cataract, etc. (NCOA, 2021).

However, aging itself is not usually an official cause of death and is not recorded statistically. The leading causes of death are cardiovascular, oncological and other diseases. The question of whether it is worth highlighting aging as a common cause is very debatable, and there is no single position on the issue.

On the one hand, there seems to be some truth in the assumptions of the proponents of this idea (Beth, 2019; Gallego, 2016; Musa, 2016)—that the most common serious diseases such as cardiovascular disease, cancer, atherosclerosis, diabetes or dementia—are probably linked by a common onset mechanism that manifests itself at the molecular and/or genetic level in the course of a lifetime. From the age of forty, the risk of these diseases, as well as the risk of dying, doubles approximately every eight years (Harman, 2001).

On the other hand, it is obvious that aging, from a clinical point of view, brings many problems that are apparently unrelated: metabolic slowdown, cells and tissue damage, accumulation of intracellular and other debris, a decrease in the energy, loss of muscle mass, and the accumulation of genetic errors, a weakening or dysfunction of the immune system, organ failure and so on (Harman, 1988; Kanasi et al., 2016; Troen, 2003).

[1] Aging in general is not typical of a wild animal community, since under natural conditions animals are more likely to die before reaching old age. However, it is characteristic that somewhat similar processes are observed among captive animals (Finch et al., 1990; Ricklefs, 2000). Animals, that are protected from the usual causes of earlier death, age and have an increase in diseases of old age.

The older a person gets, the more acute the accumulated problems become and the more negatively they affect the body inflammation, trauma, damaged tissues and organs, genetic disorders. As a result of such problems, particularly of cellular aging, when more and more cells lose their regenerative capacity and degrade, serious disruptions occur. For example, with age, thrombosis occurs in the circulatory system, atherosclerosis develops, and the blood supply to the heart muscles and brain could be disrupted, leading to strokes and heart attacks.

However, although the prospect of including aging in the official list of diseases is very vague, the process of recognizing the need to take into account and treat various age-related health disorders is growing. Thus, the International Classification of Diseases (ICD), the official list of diseases, began to include 'age-related conditions' that refer to conditions associated with aging (WHO, 2022). There is a related concept—geriatric syndromes (or age-related conditions). This term is used to denote clinical conditions in the elderly and senile which go beyond the 'disease' category. Geriatric syndromes are multifactorial conditions that appear in response to a decline in the function of many organs and systems. Despite their heterogeneity, geriatric syndromes have many common features (Inouye et al., 2007; Larrad-Jiménez et al., 2021). They are particularly common among the elderly people and may include the following: dizziness, falls, urinary and fecal incontinence, weakness, mobility problems, insomnia, hearing and vision loss, senile asthenia and a number of others. They are all widespread, multifactorial and associated with increased morbidity and adverse outcomes (Tinetti et al., 1995; Inouye et al., 2007; Dodson & Chaudhry, 2012; see Chukaeva & Larina, 2017 for the practical application of this concept).

**Aging as an important scientific and practical biomedical problem**. There are proponents of the approach that aging is a kind of disease that can be treated and, in principle, cured radically (see, e.g., De Grey & Rae, 2007). Moreover, experts often note that aging according to the ICD criteria has increasingly come to be treated as a disease (Khaltourina et al., 2020; Zhavoronkov & Bhullar, 2015). But more common are theories that aging is inevitable but can be slowed down. The aging of the body is a multifactorial process. There are different opinions about what is the main factor of aging—the natural aging changes in tissues, organs and cells (including the factors that are discussed below), other factors affecting the failing efficiency of physiological processes, or whether this is a programmed process (many scholars, e.g., Moskalev (2018), deny this).

From an evolutionary perspective, there are no processes in the world that would continue indefinitely. This rule is a broad generality of many negative processes that manifest themselves with age. In any system, correction and self-regulation weaken with time (Grinin, 2016a, 2016b; Grinin, 2017). Every organism eventually loses its functionality and collapses. The above multi-level processes may not be directly related, but they are the result of an internal imbalance in a living system, when self-regulatory mechanisms, which normally compensate for malfunctions in the body and maintain homeostasis, break down. Every system, even the most reliable, accumulates errors over time, which sooner or later lead to serious malfunctions and its failure. In the body, the cell DNA is damaged over time and errors accumulate in them. The signals between cells become degraded or distorted, and the molecules that regulate the activity of genes stop working properly. It has even been shown mathematically that it is impossible to reverse aging (Nelson & Masel, 2017).

However, these are general theoretical ideas. But from a practical point of view, it is extremely important that the accumulated medical experience proves that the life of the body can be prolonged, and the quality of its work can be improved if help is provided in time. And every year (even a month) of prolonged life or extended health is a great victory for science, medicine and human genius. But at the same time, it is a problem since it is necessary to at least maintain this level, which requires tireless and great efforts on the part of society and every individual. The increase in mortality during the COVID-19 pandemic particularly among the elderly showed the fragility of achievements in terms of increasing life expectancy and improving the biological quality of life.[2] This aging process, associated with an increase in life expectancy and its biological quality, is rightly described in the well-known UNFPA document (2012) as 'A Celebration and A Challenge', because the achievement of an unprecedented life expectancy is a triumph of medicine, science and social relations. But maintaining such a high level of life expectancy, and even more so increasing it, is a constant and powerful challenge for humanity, both from nature and from negatively affecting social phenomena. In any case, we can agree that in order to slow down or prevent aging, we need to look beyond nature (DiNicolantonio & Fung, 2019).

Today, science and medicine still have very little insight into the nature of aging, its biological mechanisms and causes. On the other hand, interest in this subject is growing and in recent decades there has been a real boom in the study of the causes and in the search for ways to stop, to slow down aging, to improve the biological and social quality of life of aging people and also to find ways to rejuvenate the body, at least partially and temporarily. In general, as Claudio Franceschi (Franceschi & Campisi, 2014; Franceschi et al., 2006), one of the leaders in this field, rightly points out, a lot of disparate facts have already been accumulated, but there are not enough concepts to unify them. Nevertheless, the field of research is expanding and different directions gain impetus. Theories of aging have emerged based on different biological mechanisms, which we will discuss below.

**Basic theories of aging and its causes**. There are a number of causes (and signs) of aging, although there are great differences, both in the number of such signs and in their hierarchy. López-Otín et al. (2013) identify nine major cellular and molecular causes of aging: chronic inflammation, increased genomic instability; telomere attrition; epigenetic alterations; altered intercellular communication; protein homeostasis disorder; stem cell exhaustion; cellular senescence; mitochondrial

---

[2] On social consequences of the COVID-19 pandemic see, e.g., Chumakov & Yurchenko, 2021; Efremov & Arkhangelsky, 2023; Grinin, 2020; Grinin et al., 2022; Hussain, 2021; Irshad, 2020; Jasiński & Bąkowska, 2020; Kakkar, 2021; Seifabadi & Dolatabadi, 2022; Widdowson, 2021.

dysfunction; deregulation of cellular signalling pathways.[3] In this chapter we will discuss most of these phenomena in one way or another.

As indicated above, the causes of aging and ways of slowing it down are being investigated from many directions. In particular, the role of different genes in aging and longevity has been studied (see, e.g., Troen, 2003). More applied studies are now carried out to study rates of aging, biological clocks, biomarkers that allow determining the biological age of different body systems, and they also study specific methods of intervening in individual mechanisms of the body's aging process (Booth & Brunet, 2016; Hughes & Reynolds, 2005).

The theory of oxidative stress has become popular. As was shown already in the last century (Beckman & Ames, 1998; Harraan, 1955), one of the processes of aging is the accumulation of oxidative damage to cells and tissues. Oxidants are substances that take electrons from other substances during a chemical reaction. Molecules and atoms that have lost electrons as a result of such reactions may transform into free radicals, that is, particles containing one or more unpaired electrons on the outer electronic level. These are usually small molecules with an extraordinary reactivity, constantly striving to take the missing electron from some other atom, which then becomes a free radical itself. Thus, a chain reaction is set off that destroys molecules (in particular, DNA, lipids and proteins), and, in general, the cells and tissues of the body. Since the main oxidizing agent in the body is oxygen, free radicals are associated primarily with the so-called reactive oxygen species (ROS), which include oxygen ions, free radicals and peroxides. Oxidants and free radicals can form and accumulate in the body for many reasons: poor diet, bad habits, stress (Halliwell & Gutteridge, 2015). Moreover, there is already considerable evidence that they accumulate precisely as the body ages (Harman, 1988; Hekimi et al., 2011). Normally, body cells can independently synthesize special protective substances—antioxidants—that counteract the harmful effects of free radicals throughout life (Cutler, 1991). The body also gets antioxidants such as vitamin C, vitamin E and carotenoids from food.

The oxidative stress theory suggests that reducing free radicals in the body, switching off or correcting genes that can cause this process or increasing the levels of antioxidants can prolong a person's life. This theory certainly has considerable prospects in the fight against aging. However, everything is not so clear yet. Free radicals are obviously not only harmful, but also good for the body, and many of them are synthesized by the body itself (Halliwell, 1994; Halliwell & Gutteridge, 2015).

Another popular theory deals with telomeres (see, e.g., Aubert & Lansdorp, 2008), which we have already discussed in Chap. 8. Let us recall that according to many scientists, telomeres—the end sections of chromosomes—play the central role in the life of cells and in aging. Like caps, they 'close' each end of the chromosome, thus limiting the process of reading and copying of DNA, primarily to

[3] It is widely accepted that the regulation of these pathways is the major way how the diet influences health and longevity (Pertseva, 2016).

prevent errors resulting from excessive copying. It turns out that every time when a chromosome is copied, the number of telomeres at its ends decreases. This is one of the reasons why certain types of cells age and die when the body reaches a certain age. Perhaps, our bodies may also age for the same reason, although these are only hypotheses (Slagboom et al., 1994). However, the danger of shortening the ends of chromosomes is evidenced by the fact that the shorter the telomeres in the cells of people over the age of 60, for example, in blood lymphocytes, the higher the mortality rates from infectious and cardiovascular diseases (Moskalev, 2018). Over time, the likelihood of telomere damage increases, and this can have very serious consequences. At best, cells launch the program for self-destruction (apoptosis). At worst, they begin to replicate uncontrollably. As a result of mutations, cancer cells acquire the ability to lengthen their telomeres and replicate infinitely. Actually, the prohibition of telomere growth in most cells is associated with the body's protection against 'unauthorised' cell division.

So, the telomere theory says that uncontrolled cell division poses a serious threat to the body. And if it became possible to artificially induce cell division and renewal in the right places, this could help us in terms of resisting aging and even rejuvenation of the body.

But even with this theory, everything is not clear. Cells in the body divide at different rates, and some cells, such as skin, bone marrow, intestinal mucosa, etc., can even lengthen their telomeres and continue to divide.[4]

Of great interest is the theory of cellular debris, actively developed by the aforementioned Claudio Franceschi and the well-known biomedical gerontologist Aubrey de Grey (De Grey & Rae, 2007). Debris accumulated as a result of damage and/or death of cells and organelles (cellular debris) can lead to many problems: blockages, nerve conduction problems, including chronic inflammation. At the same time, the immune system is able to recognize debris as a dangerous element and start the process of its destruction. The imbalance between the amount of cellular debris and the body's ability to deal with it increases with age and with the weakening of the general immune system (Franceschi & Campisi, 2014; Franceschi et al., 2006, 2017). Debris, by the way, accumulates not only in cells, but also in the genetic code. Thus, a potential role of 'junk DNA' sequences in aging and cancer has been discovered (Washington State University, 2021).

It is worth mentioning the ideas of Mikhail Blagosklonny and Michael Hall (Blagosklonny & Hall, 2009). They studied the TOR protein (mTOR in mammals) (the key regulator of cell growth and metabolism) and concluded that this protein, which is necessary for development and reproduction, becomes the engine of aging upon completion of development. The TOR promotes proliferation of smooth muscle cells in arteries (a critical phase in the development of atherosclerosis), fat accumulation, tumor growth, etc. (see e.g., Wullschleger et al., 2006). In

[4] Interestingly, bacteria with circular DNA have no telomeres at all, but also no free DNA ends. Perhaps that is why they are capable of living almost for ever. On the other hand, their ability to constantly divide does not have fatal consequences.

addition, the TOR contributes to the accumulation of proteins and of dysfunctional mitochondria, which cause DNA-damaging free radicals and impair cellular energy metabolism. It also promotes the accumulation of degradation-resistant proteins in neurons, a process that plays a role in the development of Alzheimer's disease and other forms of neurodegeneration. Blagosklonny and Hall have shown that later in life, the TOR signaling can also influence the initiation/activation of the cellular aging process, which damages neighboring cells and undermines the ability of tissues to regenerate. The drug rapamycin has been shown to be able to affect the TOR protein and thus slow down some of the mechanisms of aging (Harrison et al., 2009; Sharp, 2011; Shenk, 2009). The effect of the TOR protein on aging has actually become a separate field in the study of aging. It has been suggested that this is the key to controlling aging (see, e.g., Gifford, 2015).

However, many scholars, including Blagosklonny and Hall, believe that such a position is too optimistic. The non-systemic effect of rapamycin shows that evolution has not created a mechanism for slowing down aging. Rather, the life-prolonging effects of this and other drugs, as well as certain lifestyle choices such as calorie restriction, indirectly affect the aging process, only causing it to progress more slowly than in common case (Stipp, 2012). However, to slow down aging is also extremely important.

Another important theory for understanding the aging process is the theory of mitochondrial damage (see, e.g., Bratic & Larsson, 2013). Mitochondria are responsible for cellular respiration, i.e., the conversion of oxygen and organic matter into energy. This energy powers virtually all processes in an organism. We can say that mitochondria are a kind of energy station of cells, borrowed from pre-cellular organisms. A decrease in the number of such stations, or a decrease in their activity, leads to a decrease in the total amount of energy produced. That is why, people lose strenght as they age, their immunity weakens, etc. Just as the battery in an electrical appliance runs out, so the body's energy production is depleted. Mitochondrial damage is closely linked to mutations and oxidative stress (Lenaz et al., 1998). Therefore, it is theoretically possible to slow down aging by reducing damage to mitochondria. For example, researchers are already trying to activate the mitochondria of cells with a substance called nicotinamide adenine dinucleotide (NAD). This gives hope of rejuvenating muscle tissue (Yegher, 2013).

There is an opinion that our biota, that is, all the microorganisms in our bodies, are responsible for aging. The bacterial composition of the gut allows medical researchers to determine a person's age, body type and even the region of residence (Enders, 2017). The gut microbiota undergoes significant changes throughout life as a result of age, diet, environment and disease. According to this theory, the transplantation of biota from young people to the elderly can prolong the life of the latter. We will discuss this in more detail below.

Speaking of aging theories, it is worth mentioning David Sinclair's theory (Sinclair & LaPlante, 2019). According to Sinclair, 80% of our health in old age is determined by our lifestyle and how we live, and only 20% is determined by genetics. Sinclair and his colleagues discovered a 'backup hard drive of youthfulness' in the cell. Now the researchers' goal is to access this 'hard drive' and reboot it to activate the genes for long life. Sinclair actively promotes the right way of life, which can, of course, contribute to longevity. However, in our opinion, this is still a too optimistic approach in general. In reality, genetics is extremely important, even primary, as shown by the data obtained from the study of centenarians (see below).

There is no single cause of aging, and there is no consensus on the issue of whether it is actually possible to detect the process of aging. One can make reasonable predictions that other mechanisms of aging, currently unknown or neglected, will also be discovered. It is possible that as the number of such mechanisms and factors of aging increase, it will be possible to find some common basis for all of them and to link together factors that now seem separate to us.

## 11.4 Technologies and Trends of the Fight Against Aging

**General ideas about technologies and directions**. Anti-aging, in general, will be a complex field, consisting of the search for and dissemination of medical and lifestyle ways to somehow slow down, delay and mitigate the aging process, reduce the negative effects of the age-related (dependent) conditions, etc. If we formulate this in the form of the most general directions, then we can say that these are: (1) the efforts to extend life expectancy; (2) the struggle to improve the quality of biological life in the period of old age; (3) the development of technologies to slow down aging and rejuvenate the organism. There are, of course, striking ideas that aging can be 'cancelled', but so far these are still dreams, quite probably without any basis.

In Chaps. 8 and 9, we discussed a number of modern and future biomedical technologies that contribute to increasing life expectancy and improving its biological quality. In particular, we have discussed the victory over incurable diseases, organ transplantation, and the creation of artificial organs, the use of bioprinters, nanotechnologies, disease prevention, neural interfaces, etc. In this chapter, we will focus on the fields that we have covered insufficiently or not at all.

We have not found any classification of such anti-aging fields and methods; in fact, instead of classifications, there is only a listing of these fields and technologies. Therefore, we presumed to divide all the technologies into several groups: (1) medical (geroprotectors, food supplements, "drugs for old age", etc.); (2) technologies that increase the body potential (immunity, 'hormesis', etc.); (3) cell therapy; (4) technologies of regeneration, substitution/cyborgization (organ replacement, growing organs and tissues; cyborgization); (5) genetic and molecular therapy; (6) engineering technologies; (7) a comprehensive study of longevity; (8) the organization of a life-affirming and healthy lifestyle (including, of course, the prevention of diseases and deviations). We have arranged this section following this classification. But one should clearly realize that the areas and technologies listed below are closely intertwined both with each other and with others described in Chaps. 8 and 9. And we emphasize that one can observe (and in the future it will become even more apparent) *the use of self-regulating systems (medical-biological, medical-technological, genetic, physiological, and other) in all types of technologies*[5] *as well as the combination of many technologies with self-regulating systems into techno-biological (medical) systems for monitoring and maintaining parameters, which we discussed in previous chapters.*

We believe that in the last third of the twenty-first century, a special medical-biotechnological environment will gradually develop that will reshape medicine (see Chaps. 6 and 8). It will be a system for health monitoring, control and possible assistance, in which treatment and support of the organism will be performed mainly by autonomous systems that will be able to function regularly and constantly. In particular, it will become possible to constantly scan the organism or certain organs and even transmit the relevant information to medical centers in case of potential threats or serious deterioration in health. At the same time, smart computer systems will be able to distinguish significant fluctuations in indicators and make recommendations about short- and long-term lifestyle habits. In addition, a system of almost total monitoring and control will be created not only for patients but also for different age groups (primarily the elderly, of course), and for people in different risk groups. Moreover, we argue that this special medical, biological and technological environment will become permanent. This means that life without it may already become impossible, just as we can hardly imagine our life without the Internet and gadgets. Just as many hundreds of millions of people cannot live without constant use of drugs, so in the future a large part of the population will not be able to live outside this medical-biotechnological environment for long.[6]

### 11.4.1 Geroprotectors

Here a special role is attributed to the so-called geroprotectors and to some drugs that are already in the market or are being introduced, and to bioadditives. ***Geroprotectors*** are drugs that slow down aging and increase life expectancy. However, although the main emphasis is on prolonging life, it is equally important to note that many drugs and supplements, both related and unrelated to geroprotectors, significantly improve the biological quality of life in general, and especially during aging (e.g., collagen considerably facilitates the work of the joints). It is worth

[5] Here is an interesting example of such a system. Scientists at the Rochester Clinical College, led by Jan van Dursen, have succeeded in creating a genetically modified mouse whose diet contains a special substance that kills all senescent cells. By regularly clearing the animal's body of senescent cells, it became possible to slow down the emergence of such age-related pathologies like the loss of fat and muscle tissue, the development of cataracts (Moskalev, 2018). More about senescent cells see below.

[6] However, the development of such systems will face numerous real problems related to legal, moral and political difficulties, as well as a natural reluctance to be under constant control 24 hours a day and to follow the prescriptions of soulless systems (for details see Chaps. 6, 8, 15).

mentioning metformin as a drug that is already known as a geroprotector. It is used to treat and prevent diabetes. It is remarkable that although it has been used in medicine for more than half a century, it is the only promising geroprotector so far that has not shown any particular side effects (Konstantinov, 2021; Moskalev, 2018). There are more than two hundred potential geroprotectors (Moskalev, 2018), but so far most of the geroprotectors in which researchers invest their hopes only prolong the lives of laboratory animals. Several medicines can act as geroprotectors, including some that have been known for a long time. For example, Claudio Franceschi, whom we have already mentioned, believes that even aspirin is a 'geroprotector' because it suppresses inflammation. Great hopes are placed on the immunosuppressant rapamycin.[7] Rapamycin is an extremely important geroprotector with few side effects that reduce the risk of major age-associated diseases, including cancer, Alzheimer's disease, Parkinson's disease, cardiac degeneration, type 2 diabetes, osteoporosis, etc. (Stipp, 2012). One should note, however, that these substances suppress the immune system. There is also a lot of talk about selegiline, this medicine is used to help control the symptoms of Parkinson's disease.

The famous amphibian, the naked mole rat, surprised the entire scientific world with its ability not to age (Buffenstein, 2005). Scientists started to use hyaluronic acid, which is produced by the naked mole rat's body, and it seems that its protective properties allow it to live so long. But there are serious problems with the use of this acid, which prevents some cells from dividing.[8] We need to find a way to deliver it or develop molecules that will interfere with the work of proteins that destroy hyaluronic acid in our bodies (Tian et al., 2013).

Thus, today geroprotectors seem to be the main way to fight aging. However, there are medical concerns about their use. There are some problems that need to be resolved. The point is that there are virtually no drugs without side effects, and the body can quickly adapt to the drug, reducing or eliminating its effectiveness.

### 11.4.2 Technologies that Increase the Potential of the Body

**Enhanced immunity and the concept of 'hormesis'**. The pandemic has shown once again that, despite all the advances in medicine, the immunity, both individual and collective, plays and will continue to play a dramatic role in maintaining

---

[7] Rapamycin is produced by bacteria that were found on Easter Island. They produce it to suppress the growth of fungi. As a result, the fungi stop multiplying and growing, but they start to live longer. In other words, they slow down their aging. About the possible role of rapamycin in life extension see Stipp, 2012; Gifford, 2015.

[8] The hyaluronic acid molecule has a short and a long form. The short form is used for cosmetic purposes. However, it causes inflammation. And the long form, which the naked mole rat has, on the contrary, has anti-inflammatory and, possibly, anti-cancer effect, because it reduces the ability of the cells responsible for inflammation to divide. But its molecules are too big and therefore cannot pass through cell membranes.

human health. Moreover, boosting immunity seems to be the most important task of medicine today and in the future. The strength of immunity varies greatly from person to person, so the potential for strengthening it is still very great. Therefore, the means to increase the body's resistance to the processes that cause aging seem to us very promising. However, in addition to purely medical support, early diagnosis, the use of dietary supplements and immune stimulants, this requires a combination of a number of conditions and factors: a healthy lifestyle, proper diet, the absence of bad habits, physical activity, special procedures, improved ecology, etc. This is still a long way off. Self-regulating systems that increase immunity will also be extremely important. It is possible that increasing immunity will become a state program in many countries. People will be forced to give up bad habits and take the necessary measures to live longer and have a strong immunity. At the same time, propaganda and control over the health of citizens will be strengthened through the techno-medical systems described in the previous chapters and above.

Ideas relating immunity to longevity overlap in many ways with the concept of 'hormesis'. Hormesis is the stimulating effect of low stress, whereas high stress can be destructive. It is about 'improving stamina', a training of the defense systems, only at the cellular level. Hormesis can be triggered by heavy metals, ionizing and UV radiation, hypergravity, and hyperosmotic shock (Moskalev, 2018). But for now, it is mostly just an idea, not a technology. In any case, the potential to boost immunity is still being investigated. Improving it naturally and helping the body in various biomedical ways seems to be one of the main directions for increasing human life expectancy.

### 11.4.3 Cellular Therapy

Cell therapy is primarily associated with the so-called stem cells, which have long been promoted as a powerful means of rejuvenating the body. Let us recall that stem cells are unspecialized, that is, they are a kind of blank-slate for the body, from which all other cells with specialized functions are made. The use of stem cells yields important results (see also Chap. 8). Basically, stem cells are derived from embryos aged 3 to 14 days and sometimes from adult cells (bone marrow or fat), but the latter have limited functionality. Embryos are mainly grown in laboratories up to 14 days of age, and growth beyond this is usually prohibited, although the rules have recently been relaxed somewhat. There are a number of restrictions and bans on the use of stem cells.

In addition to stem cells, cellular therapy will involve the use of a wide range of technologies in the future.

According to Moskalev (2018), the cell therapy of the future is seen as a combination of various methods:

(1) stimulation of internal cell ability to renew the body (diet, geroprotectors and nutritional supplements). Cell renewal is a natural process that in many cases

continues throughout lifetime, but can be disrupted over time. By maintaining and stimulating it, one can expect to prolong life;
(2) rejuvenating gene therapy of organs and tissues, such as neuroendocrine organs (mainly the hypothalamus);
(3) inducing the death of old cells. This is also a way of forcing cells to renew themselves;
(4) the use of special stem cells pre-treated with geroprotectors. These special cells will be used specifically to solve the problem of aging;
(5) artificial reconstruction and implantation of stem niches in the body, which themselves could produce stem cells and rejuvenate the body;
(6) replacement of defective organ parts;
(7) removal of cells that 'misbehave', like senescent cells.[9]

In the future, numerous self-cleaning systems are also possible that will monitor the maintenance of order in cells and in the whole organism, such as self-cleaning mechanisms for cellular debris or DNA repair mechanisms (Li et al., 2017).

### 11.4.4 Technologies of Regeneration and Substitution/ Cyborgization

We discussed these issues in Chap. 8. Here we would like just to add a few details.

A very promising direction is the cultivation of new organs from stem cells or from artificial tissues, including biological materials, and their subsequent implantation in place of tissues and organs that have lost their functionality. Clinics have only learned how to do this successfully with a few simple organs: the trachea, oesophagus, blood vessels; however, experiments with animals are encouraging. We have also discussed replacing organs with artificial or genetically modified ones. Cyborgization, i.e., the replacement of important elements and parts of our body with technical analogues, is also closely related to this direction; and the number of such analogues is growing (see about this in Chaps. 1 and 8). This trend will contribute to the development of all aspects of the fight against aging.

### 11.4.5 Biota

The human microbiome is the totality of the microbiota (microflora) living in human tissues and body fluids. However, it mainly refers to the biota of the gastrointestinal tract. The human microbiome consists mainly of bacteria, but also includes archaea, fungi, protozoa, and even viruses. A good, healthy microflora

[9] Senescent cells are cells that accumulate in the body with aging, send false signals that accelerate the aging process and actively destroy the surrounding tissue, cause chronic inflammation and stimulate the formation of tumors. Several laboratories around the world are currently developing methods to remove senescent cells from the human body.

reduces inflammation in several ways: it boosts immunity, in particular, by strengthening the integrity of the epithelial barrier; it displaces pathogenic bacteria, preventing them from multiplying; it promotes the development of substances involved in protection against extracellular pathogens (Pertseva, 2016). Accordingly, any positive influence on the biota balance (also with the help of drugs and bioadditives) reduces negative phenomena, eliminates or alleviates age-related conditions, and indirectly contributes to life extension. One of the most promising biota-related technologies is biota transplantation (a type of gastrointestinal transplantation mentioned above). It has been shown that transplantation of microbiota from young animals to old ones can improve and prolong the life of the latter (Bana & Cabreiro, 2019; Thevaranjan et al., 2017). It is now possible to transplant microbiota from one person to another. In this way, some inflammatory bowel diseases can be cured, and it is possible that it will also be possible to prolong life. By the way, this is a good example of a self-regulating biological system. After transplantation, the biotic system in the recipient's body acts independently and gradually restores the work of the gastrointestinal subsystem and of the whole body (Vaughn et al., 2022).

### 11.4.6 Gene Therapy

Gene therapy targets alterations in the genetic make-up of cells in order to give the body new properties or to treat diseases (this issue was also discussed in Chaps. 8 and 9). A huge amount of research has been devoted to the relationship between different genes and aging. Hundreds of genetic variants in our genome are already known to affect lifespan (see, e.g., Jazwinski, 1996), and this can have a limited beneficial result, as well as a cumulative effect. However, according to Vadim Gladyshev, not a single gene is known to make a very strong contribution to life expectancy. It is quite possible that there are no such genes, and we are talking not about a single mechanism that prolongs life, but about a whole set of genes. These technologies include studies of telomerase and telomeres, as discussed above.

In technical terms, gene therapy is the process of introducing new genes into a person or modifying existing genes (this can be performed using specially designed viral particles, lentiviral or adenoviral, such as the use of mRNA in vaccines against COVID-19). This technology has long been used to treat genetic diseases. The changes can be permanent or temporary. CRISPR/Cas9 technology—the transfer of the desired gene into the body—has great prospects and has already shown interesting results in animals, in particular suppressing aging, improving health and extending lifespan in mice (Beyret et al., 2019). CRISPR-Cas9 technology has already been used for the treatment of certain diseases. For example, a method has been developed that involves taking human stem cells (or a person's own stem cells), 'editing' the faulty gene and introducing them back (Ledford, 2020; Naddaf, 2022).

As we discussed in Chap. 9, we do not think that even in the distant future the control over the genetic program will become possible. It is not only too complicated, but also dangerous for a system as complex as the human body. Any programmer will tell you that it is impossible to seriously improve a program of many thousands of lines of code by adding or removing a tiny algorithm. The program has to be completely rewritten. But limited effects on genes and the genome are quite possible in the future. And even this will be a huge step forward (albeit full of complexities, dangers and the need to adapt to such technologies) in the entire complex of anti-aging medicine. In short, today we cannot control the processes in the brain, but we can have a limited influence on it, which allows us to treat many mental and psychological illnesses or at least to hope for the possibility of curing them.

### 11.4.7 An Engineering Approach to Aging of Cells

Moskalev (2018) has formulated the approach of the above-mentioned de Grey and Rae (De Grey & Rae, 2007), the possibilities of transition to which are distant and unclear and technologies should be developed for a long time. Such an anti-aging approach will require constant and, at some point, even total intervention in the body. Such an intervention may have unforeseen and negative effects. On the other hand, it has the advantage of being an integrative approach. In particular, they propose the following scheme:

- Cell death and cell atrophy.—To eliminate them, it is proposed to introduce stem cells and growth factors into the body.
- Mutations that lead to cancer.—To combat tumours, it is proposed to completely switch off telomerase which attaches itself to the end of the chromosome.
- Mutant mitochondria.—Mitochondria often lose important parts of their DNA, which causes their inability to make the proteins they need. Since there are very few genes in mitochondrial DNA, de Grey believes it is essential to move them to a safer place—to the cell nucleus.
- Cross-linking of extracellular proteins.—Cross-linking of the proteins in the walls of blood vessels with the connective tissue of elastin and collagen leads to atherosclerosis and reduced joint mobility. It has been proposed to develop molecules and enzymes that break those crosslinks (De Grey & Rae, 2007; see also Moskalev, 2018).

We repeat that such proposals are still quite abstract from the point of view of specific technologies, but some of them are likely to be implemented in the future.

## 11.5 Comprehensive Study of Longevity and the Creation of the Environment Necessary for Longevity

**About centenarians**. To understand the causes and mechanisms of aging, it is now believed that medical studies of centenarians, as well as the study of the conditions and histories of their lives together with psychological studies of their condition can reveal a lot. Centenarians live twenty to thirty years longer than the average life expectancy. One can get across statements that centenarians do not experience typical problems of old age until they are ninety. Of course, making everyone a centenarian is an unrealistic task, but it seems possible to significantly bring people up to the level of today's centenarians. And the threshold of longevity, which is currently set at 100+, could rise to 110+. But now only one in four or five thousand people lives to be a hundred years old. Most of them are women.[10] A comprehensive study of the phenomenon of longevity can contribute to the fight against old age.

What are the characteristics of long-livers? They can be divided into three groups, although they are closely related: genetic characteristics, physical characteristics and lifestyle (well, and, of course, there is always an element of luck, since fatal diseases and accidents take the lives of many potential centenarians long before they reach the age of 100).

**Genetic and epigenetic factors**. It is very likely that longevity is inherited, that is, it is transmitted with the genes of parents and ancestors. It may seem counterintuitive, but there are quite a few lucky people among the centenarians who are able to live to a hundred without having a healthy lifestyle, simply because their stress coping mechanisms work better—this is a phenomenon of family longevity that is passed on from generation to generation. Moreover, the longer a person lives after the age of eighty, the greater the role of genes. For example, centenarians have special protective genes that prevent the development of age-associated diseases.

So it is necessary to find these genetic features and learn how to influence certain genetic programs. However, it is important to keep in mind that we are not talking about replacing genes, but about mechanisms to activate them or to prevent their silencing, as often happens in old age (and vice versa, to block the action of genes that accelerate aging). In other words, we should follow the path of creating self-regulating genetic systems. In many ways, we are talking not only about direct genetic predispositions to longevity, but also about so-called *epi*genetic factors. The latter are associated with changes in gene activity during cell growth and division, changes in protein synthesis caused by mechanisms that do not change the nucleotide sequence in the DNA. In other words, centenarians have some genes

[10] Note that such unevenness in processes, and especially in longevity, is an important rule of natural evolution, which manifests itself almost everywhere. For example, the life spans of stars vary widely from tens of millions of years in giants to a few trillions of years in dwarfs. The lifespan definitely depends on the mass of the star, but even within similar sizes it varies greatly (Grinin, 2013).

whose functions do not switch off with age. As a result, their health and physical condition deteriorate much more slowly than that of normal people. Conversely, some of the gene functions that affect aging are blocked and switched on much later. Epigenetic features are thought to be influenced by various factors, including lifestyle, stress, diet, etc., but the significance of these factors is not entirely clear. At the same time, epigenetic changes persist through a number of mitotic divisions of somatic cells and can also be passed on to subsequent generations. Therefore, heredity can play a very important role here. However, the mechanisms of "switching on" or "switching off", of weakening or strengthening the various functions associated with genes, are still triggered by certain biochemical signals, so the study of such mechanisms may help to correct epigenetic mechanisms and thus fight against aging.

**Peculiarities of the body**. It has also become possible to find some physiological mechanisms that distinguish centenarians, in particular, in immune cells and lymphocytes (Konstantinov, 2021). And the professor at the University of Bologna, Claudio Franceschi, whom we have mentioned, in twenty years of research, studied thousands of centenarians and a certain number of super-centenarians (those who are over one hundred and ten years old) from different regions of the world.[11] It turns out that long-livers have important differences in their microbiota—the bacteria that live in their intestines.

**Lifestyle**. Each centenarian has found his or her own and somewhat unique way of long life. But nevertheless, there is a correlation between a healthy lifestyle and longevity, between the quality of life, a person's psychological state and longevity. The well-known Harvard Medical School Grant Study of a large group of people over 80 years (a 80-year longitudinal study) (Mineo, 2017; Shenk, 2009), which aimed to identify predictors of healthy aging, showed a definite correlation between individuals' happiness with good psychological conditions: a happy childhood, a large family, strong relationships, etc., on the one hand, and longevity and better health, on the other.

**Organization of a life-affirming and healthy lifestyle. Personalization and prevention**. If it appears impossible to completely defeat aging, it may be possible to adapt to it much better (see Chaps. 6 and 7). At the same time, society as a whole and its institutions will have to adapt to a new medical, and perhaps legal, institution that requires people to adhere to new medical policies directed to their lifestyle. This direction requires a separate study. First, as we said above and in Chaps. 6–8, a gradual movement toward the creation of a consistent medical and biotechnological environment for old and sick people is assumed. It will emerge through constant changes, through trial and error in order to find the best combination of safety, freedom of choice, care and independence, monitoring and privacy and so on. It is difficult to tell how successful this will be; most likely, the

[11] The difference between centenarians and super-centenarians is similar to that between normal people and centenarians. It is known that the chances of people who have reached the age of one hundred to become super-centenarians (i.e., to live at least another ten years) are only two in a thousand.

path will be rather difficult (see Chap. 15). So, although older people are increasingly adapting to new technologies, such as computers and mobile phones, and actively use them in their lives (Wilson et al., 2022), such an environment will inevitably put pressure on the psyche. Therefore, adaptive ways to mitigate the negative effects are needed.

Second, since the adaptation for people with disabilities, especially through technology (see Chap. 10; Mitzner et al., 2018), is faster than that for older people, so models of adaptation for people with disabilities will help shape models of adaptation for the elderly.

Third, the development of personalization and individualization in medicine, which today is hampered by the need to work according to medical protocols, will be extremely important. We can talk about many directions of development of individualization in relation to the elderly and the old. One of them is related to predicting and choosing a general direction for slowing down aging. As we age, we acquire very individual characteristics (or, more precisely, characteristics that are innate or acquired with age appear): some people are prone to insulin resistance, some to inflammation, and some to genetic instability that leads to cancer. In the words of Claudio Franceschi, "it is your problem areas that need to be corrected—for someone to reduce the vascular age, and for someone to lengthen the telomeres". But here too there is a danger of cliché, albeit a much more subtle one. Another possibility is the choice of an individual mode of medical care (with the help of the described medical-biotechnological systems). Of course, there is the likelihood of conflicts and reluctance (sometimes quite legitimate and fair) of patients to be under surveillance. This can reinforce non-conformist attitudes. Obviously, the transition to a real personalization is possible only with a very high level of technology and the development of self-regulating and self-managing systems, when life programs will be combined for everyone, in fact, a "nanny" will be assigned to everyone, sometimes this will be very annoying (i.e., personalization will be carried out not only with a carrot, but also with a stick, but let us hope that it will be soft enough).

One of the most important breakthroughs in increasing life expectancy will come from improvements in prevention. It is widely believed that the medicine of the future will be preventive medicine. Of course, it is unlikely that medicine will be able to prevent all problems in advance, since the human body develops in different ways. But it will certainly be able to diagnose diseases at a much earlier stage. We believe that the medicine of the future will evolve toward constant monitoring and rapid response to abnormalities. More about the future of prevention and early diagnosis is in Chap. 8.

To conclude this section, we believe that the number of anti-aging technologies will increase, as will their effectiveness (although the latter will grow more slowly than the former).

## 11.6 Why and How Anti-aging Will Become the Main Challenge for Medicine in the Future

**Aging and progress**. We would like to conclude this chapter by discussing *why we associate progress with moving along the path of development and dignifying the aging process of individuals and society as a whole.* The fact is that, in our opinion, there can be no general advance without some kind of progress in the field of demography. And there has never been in the long run. Depopulation, as a result of wars, invasions or epidemics could sometimes (but by no means always) be the starting point for progress, but not the actual progress. Depopulation is a catastrophe, after which social evolution could start searching for new paths of development, but it usually took a long time. In the late twentieth and early twenty-first centuries, the trend toward quantitative population growth weakened, as we pointed out in Chaps. 5–7. At the same time, the trend towards increasing the quality of life of the population has intensified. In modern conditions, the limits and limitations of the endless growth of consumption in the developed countries have become apparent, but the possibility of increasing life expectancy and improving its biological quality has strengthened. In this context, the effect of the aging process on society and all its structures will intensify (see Chaps. 13 and 15). Let us repeat that aging, and especially healthy aging, is a major and ongoing challenge for society. It is a kind of indicator of the well-being of society on a national and global scale. Any deviation from the normal course of life in any of its aspects immediately leads to a reduction in life expectancy and a drop in the indices of well-being and happiness. In our opinion, the aspiration to increase life expectancy and improve biological quality of life will have a particular impact on scientific and technological progress and innovative breakthroughs. We would even argue that *without efforts to increase life expectancy and improve the quality of biological life, progress, both social and technological, will be greatly slowed down.*

The increasing aging of society as a process can only be stopped when life expectancy comes to a standstill and falls for a long time, which is possible if the process of increasing the quality of life becomes negative, that is, if the economy stagnates and starts to shrink, the level of wages and incomes decreases, if medical and social services deteriorate sharply, if legalized suicides dramatically increase (see Chap. 7) and if confrontations and military conflicts become constant, life expectancy will decrease, and respectively, the number of the elderly and old people. This is, of course, possible as a result of a military, climatic, epidemiological or other global catastrophe. But this will no longer be progress, but regression, with possible depopulation. But we hope that this will not happen, or that such periods will not be long (but see Chap. 15 for negative developments). In the case of progressive development, global aging, as an associated and inevitable process, will begin to influence the development of progress, because many technological solutions will be required, without which it will be impossible to create a full-fledged environment for maintaining high life expectancy. Actually, we have already pointed out (in Chap. 7; see also Chap. 13; Grinin et al., 2022) that by the end of this century, the aging process will start to slow down technological progress. However, there is no contradiction here. Slowing down does not mean stopping. It means moving on to other tracks. Of course, on the one hand, older people tend to be more conservative (see Chaps. 7 and 13), but on the other hand, they may welcome innovations that improve their condition. Therefore, a dual process of hampering and stimulating innovation becomes possible, which may lead to the channeling in a particular direction. This will also contribute to the ability to combine stability and progress in a more organic way.

**The growing importance of anti-aging in medicine as a long-term vector of development**. As the progress of society is largely based on the process of adapting to aging, as the number of elderly increases, they and the medical establishment will inevitably put this process **at the center of a new medicine—the social medicine of the future**. Summing up this chapter, we can say that whether or not aging has a common caused, we may expect that in the future*, anti-aging, the efforts to increase life expectancy and improve its biological quality, as well as related problems, technologies and attitudes in the organization of lifestyle, will become one of the most important tasks of medicine.* Of course, this does not mean that medicine will pay no attention to other categories of people, but it does mean that many breakthrough discoveries, innovations and technologies will necessarily and actively be applied in the direction of anti-aging. They will be developed and discovered in the field of gerontology and age-related (elderly) medicine, and from there they will spread to other categories of people. This will also be facilitated by the fact that elderly people will make up an increasing proportion of the disabled people.

And this prospect of the growing importance of increasing societal aging (Kunkel & Settersen, 2021) for medicine, already outlined in the twentieth century and intensified in the early twenty-first century, will develop and be realized throughout this century.[12]

**Social and medical adaptation to aging**. Adaptation to aging, which we discussed in Chaps. 6 and 7, should lead to the fact that, on the one hand, the elderly will be better prepared to cope with the period of old age in all respects, and on the other hand, society will be better able to integrate the growing number of elderly people. Self-organization of the elderly in society will affect social mechanisms, economic, educational and political institutions, etc. Thus, the aging of society will become one of the integral elements of society, its social, political, technological, demographic, humanitarian and financial, scientific and other dimensions.

In this chapter, we are talking about *medical adaptation to aging*, specific technologies and forms, although, of course, it is intrinsically linked to other technological areas and to specific characteristics of socialization of the elderly.

One must realize that adapting medicine to aging, making significant progress in the fight against aging to increase longevity, will require not only a shift in the focus of medicine on these problems, but in general will significantly transform

---

[12] If by that time aging is recognized as a disease (see below), then medicine will all the more focus on the treatment of aging and related diseases and deformities.

the entire healthcare system, making it essentially a new system for maintaining health at a certain level. And even if, in the future, it is not the fight against old age, but other problems, such as epidemics, new diseases, environmental disasters, etc., that will temporarily dominate the agenda, the elderly population will in any case be a high-risk group and will receive special attention.

Why do we believe that anti-aging can become the most important (or at least one of the most important) goals of medicine in the future? First of all, of course, because the proportion of the elderly and old people will be so large that caring for them will inevitably change the proportions and priorities in medicine. We have talked about other reasons above and we will touch them a little below. It is important to emphasize the idea that the fight against aging will dramatically increase the already high importance of medicine as a societal institution. And if tomorrow medicine becomes an integral part of the MANBRIC convergence (see Chap. 4), then later the problems of aging and anti-aging will become an integral part of medicine.

**How will the problems of aging affect medicine? Geriatrics**. Of course, so far we can only outline some issues here. But in general, one can distinguish the following paths. First of all, there is a need for the institutionalization of anti-aging medicine within general medicine, that is, (a) the development of departments, faculties and specialities at universities and medical colleges; (b) a sharp increase in scientific research in the field of gerontology and geriatrics; (c) an increase in the number of medical specialists in the field of the old-age problems; and (d) and a corresponding sharp increase in funding and prestige. And we will also see similar processes associated with an increase in the number and quality of medical institutions and specialists in the field of aging.

Today, the field of practical medicine has already emerged—geriatrics, a separate branch of gerontology, which deals with the study, prevention and treatment of diseases of old age. It aims at supporting health by preventing and treating diseases and disabilities in older adults (FREIDA, 2020).

Geriatrics is a field of medicine that is at the intersection of medical and social problems. It studies the problems of aging and helps to solve problems related to making this difficult period of a person's life healthier and more active. Such specialists should help patients maintain their social skills for longer and ensure that they will not be deprived of the opportunity to live a full life.

It is important to emphasize that geriatrics, from the very beginning, emphasizes the specific features of health care for the elderly. This is medicine not just in a broad sense, but in a very broad sense—it covers social, medical-psychological and technological aspects. This is the path of medical adaptation to aging (part of the general process of adaptation to aging). After all, it is not only about treatment, but also about the socialization and support of the elderly. This field is at the intersection of medical and social factors affecting the well-being of senior citizens. In this sense, medicine starts to transform itself into a system of health care and support for the necessary lifestyles. In their manifestos, geriatricians (and geriatric medical teams) claim themselves to be not only doctors, but also social workers all in one, in other words, geriatrics is medicine for the elderly with a strong social bias. Notably, recent studies have shown that geriatricians are among the most satisfied among physicians when it comes to their career choice.

Successful medical adaptations require a special kind of team of geriatric specialists. The team members are often health professionals with special training in caring for older adults, including nurses, social workers, dieticians, physiotherapists, occupational therapists and consultant pharmacists. Geriatricians work closely with interdisciplinary teams (Williams et al., 2002). Such doctors consider a patient's condition in general, how it progresses or deteriorates, and monitor it over time. Such geriatricians are very much needed in old age. They should not be just narrow specialists who see only part of the problem. In this respect, geriatricians can be compared to pediatricians, who follow children from birth to adulthood. Of course, as the number of geriatricians grows, so will their specialization. But there will be a common ground that allows one to reveal common problems of old age even through the prism of your specialization.

Many universities have added geriatrics to their list of disciplines, and specialized geriatric centers are emerging in developed countries. However, in general, such doctors are still rare. As the number of such specialists (gerontologists, geriatricians, etc.) and relevant institutions increases and as society focuses on these areas, medicine will be re-configured to be more active in addressing these areas of public concern.

**Eleven reasons why the role of the anti-aging trend in medicine will increase**. Let us summarize again why this can become a reality.

1. The number of the elderly and old people will continue to grow, so the proportion of such patients will also increase.
2. In connection with the growth of this group, other branches of medicine (sports medicine, esthetic medicine, etc.) will begin to "grow old".
3. In any case, extending life expectancy requires more resources than just maintaining good health.
4. All illnesses in old age require more efforts from doctors and society, and pose a greater danger to patients.
5. The attitude of medicine toward the elderly will change not only in terms of their numbers, but also in terms of their becoming a priority group for doctors. In addition, the number of elderly doctors, who will be more loyal to their peers, will inevitably increase.
6. At the same time, the trend will be not only to treat diseases, but also to improve age-related conditions (and this, as we have seen, is largely a new phenomenon).
7. The center of medicine is generally associated with the most problematic groups, where morbidity and mortality are higher, as well as the consequences of treatment, which are higher in the elderly (the COVID-19 pandemic appeared to be a good indicator of this).
8. The center of medicine will be connected with better funding, and due to the growth in the number of this group, the funding will inevitably increase. Also,

if the process of adaption to aging and to age in general follows the scenarios described in Chap. 7, then there will be legislative or publicly approved measures to support this group, through benefits, subsidies and priorities.

9. If particular attention of medicine begins to focus on the older people, then this process will be self-reinforcing due to the activity of the elderly's self-organized groups and associated political parties (see Chap. 7), due to the allocated funds, the development of science and the growth of the elderly population, as well as due to efforts to expand the influence of those medical-pharmacological and scientific corporations that will be associated with the elderly.
10. Most of the innovations in medicine and other related fields will be somehow related to this group, that is, they will find application in this group, but at the same time, discoveries directly related to aging problems (e.g., in genetics, in geroprotectors), can give impetus to the development of medicine in general, for all population groups.
11. The fight against aging gains momentum because many wealthy people (e.g., Bezos, Musk, Gates) initiate research in this direction, considering anti-aging as their personal goal.

**Acknowledgements** This chapter has been prepared with support by the Russian Science Foundation (Project No. **23-11-00160**).

## References

Age UK. (2022). https://www.ageuk.org.uk/information-advice/health-wellbeing/conditions-illnesses/

Aubert, G., & Lansdorp, P. M. (2008). Telomeres and aging. *Physiological Reviews, 88*(2), 557–579.

Bana, B., & Cabreiro, F. (2019). The microbiome and aging. *Annual Review of Genetics, 53*(1), 239–261. https://doi.org/10.1146/annurev-genet-112618-043650

Beckman, K. B., & Ames, B. N. (1998). The free radical theory of aging matures. *Physiological Reviews, 78*, 547–581.

Beth, D. (2019). Why ageing should be classified as a disease. *The Conversation*, November 27. http://theconversation.com/why-ageing-should-be-classified-as-a-disease-126642

Beyret, E., Liao, H.-K., Yamamoto, M., Hernandez-Benitez, R., Fu, Y., Erikson, G., … & Izpisua Belmonte, J. C. (2019). Single-dose CRISPR–Cas9 therapy extends lifespan of mice with Hutchinson–Gilford progeria syndrome. *Nature Medicine, 25*(3), 419–422. http://10.1038/s41591-019-0343-4

Blagosklonny, M., & Hall, M. (2009). Growth and aging: A common molecular mechanism. *Aging, 1*(4), 357–362.

Booth, L. N., & Brunet, A. (2016). The aging epigenome. *Molecular Cell, 62*(5), 728–744.

Bratic, A., & Larsson, N.-G. (2013). The role of mitochondria in aging. *The Journal of Clinical Investigation, 123*(3), 951–957. https://doi.org/10.1172/JCI64125

Buffenstein, R. (2005). The naked mole-rat: A new long-living model for human aging research. *The Journals of Gerontology Series A: Biological Sciences and Medical Sciences, 60*(11), 1369–1377.

Chukayeva, I. I., & Larina, V. N. (2017). Vozrast-assotsiirovannyye sostoyaniya (geriatricheskiye sindromy) v praktike vracha-terapevta polikliniki. *Lechebnoye Delo, 1*, 6–15.

Chumakov, A. N., & Yurchenko, P. S. (2021). Global pandemic in terms of objective causes and subjective speculations. *Journal of Globalization Studies, 12*(2), 157–168.

Cutler, R. G. (1991). Antioxidants and aging. *The American Journal of Clinical Nutrition, 53*(1), 373–379.

Dattani, S., Spooner, F., Ritchie, H., & Roser, M. (2018). Causes of Death. URL: https://OurWorldInData.org

De Grey, A., & Rae, M. (2007). *Ending aging: The rejuvenation breakthroughs that could reverse human aging in our lifetime*. Martin's Press.

DiNicolantonio, J., & Fung, J. (2019). *The longevity solution: Rediscovering centuries-old secrets to a healthy*. Victory Belt Publishing.

Dodson, J. A., & Chaudhry, S. I. (2012). Geriatric conditions in heart failure. *Current Cardiovascular Risk Reports, 6*(5), 404–410.

Efremov, I., & Arkhangelsky, V. (2023). Global fertility changes in the context of the COVID-19 pandemic. *Journal of Globalization Studies, 14*(2), 165–170.

Enders, G. (2017). *Ocharovatel'nyy kishechnik. Kak samyy mogushchestvennyy organ upravlyayet nami*. Bombora.

Finch, C. E., Pike, M. C., & Witten, M. (1990). Slow mortality-rate accelerations during aging in some animals approximate that of humans. *Science, 249*, 902–905.

Franceschi, C., Bonafè, M., Valensin, S., Olivieri, F., De Luca, M., Ottaviani, E., & De Benedictis, G. (2006). Inflamm-aging: an evolutionary perspective on immunosenescence. *Annals of the New York Academy of Sciences, 908*(1), 244–254. https://doi.org/10.1111/j.1749-6632.2000.tb06651.x

Franceschi, C., Garagnani, P., Vitale, G., Capri, M., & Salvioli, S. (2017). Inflammaging and 'Garb-aging.' *Trends in Endocrinology & Metabolism, 28*(3), 199–212. https://doi.org/10.1016/j.tem.2016.09.005

Franceschi, C., & Campisi, J. (2014). Chronic inflammation (inflammaging) and its potential contribution to age-associated diseases. *Journals of Gerontology Series A: Biomedical Sciences and Medical Sciences, 69*(Suppl. 1), S4–S9.

FREIDA. (2020). https://freida.ama-assn.org/specialty/geriatric-medicine-fm

Gallego, J. (2016). Nanoparticles are moving us toward a cancer-free world. A new therapy gives chemo a much needed boost. *Futurism, 11*, 10–16.

Gifford, B. (2015). Does a real anti-aging pill already exist? Inside Novartis's push to produce the first legitimate anti-aging drug. *Bloomberg.com*, 12 February. https://www.bloomberg.com/news/features/2015-02-12/does-a-real-anti-aging-pill-already-exist

Grinin, A. L. (2016a). Samoregulyatsiya kak global'nyy trend megaevolyutsii. In L. E. Grinin & A. V. Korotayev (Eds.), *Evolyutsiya srezy, pravila, prognozy* (pp. 17–43). Uchitel.

Grinin, A. L. (2016b). Samoregulyatsiya kak trend Bol'shoy istorii. In L. E. Grinin & A. V. Korotayev (Eds.), *Istoriya i Matematika: Megaistoricheskiye aspekty* (pp. 128–154). Uchitel.

Grinin, L. (2020). How can COVID-19 change geopolitics and economy? *Journal of Globalization Studies, 11*(2), 121–134.

Grinin, L., Grinin, A., & Korotayev, A. (2022). COVID-19 pandemic as a trigger for the acceleration of the cybernetic revolution, transition from e-government to e-state, and change in social relations. *Technological Forecasting and Social Change, 175*, 1–15. https://doi.org/10.1016/j.techfore.2021.121348

Grinin, L. E. (2013). *Bol'shaya istoriya razvitiya mira: Kosmicheskaya evolyutsiya*. Uchitel.

Grinin, L. E. (2017). *Bol'shaya istoriya razvitiya mira: istoriya i evolyutsiya Solnechnoy sistemy*. Uchitel.

Halliwell, B. (1994). Free radicals, antioxidants, and human disease: Curiosity, cause, or consequence? *The Lancet, 344*(8924), 721–724.

Halliwell, B., & Gutteridge, J. M. C. (2015). *Free radicals in biology and medicine* (5th ed.). Oxford University Press.

Harman, D. (1988). Free radicals in aging. *Molecular and Cellular Biochemistry, 84*, 155–161. https://doi.org/10.1007/BF00421050

Harman, D. (2001). Aging: Overview. *Annals of the New York Academy of Sciences, 928*(1), 1–21.

Harraan, D. (1955). *Aging: A theory based on free radical and radiation chemistry*. University of California Radiation Laboratory.

Harrison, D. E., Strong, R., Sharp, Z. D., Nelson, J. F., Astle, C. M., Flurkey, K., Nadon, N. L., Wilkinson, J. E., Frenkel, K., Carter, C. S., Pahor, M., Javors, M. A., Fernandez, E., & Miller, R. A. (2009). Rapamycin fed late in life extends lifespan in genetically heterogeneous mice. *Nature, 460*, 392–395.

Hekimi, S., Lapointe, J., & Wen, Y. (2011). Taking a 'Good' look at free radicals in the aging process. *Trends in Cell Biology, 21*(10), 569–76. https://doi.org/10.1016/j.tcb.2011.06.008. PMID: 21824781; PMCID: PMC4074523.

Hughes, K. A., & Reynolds, R. M. (2005). Evolutionary and mechanistic theories of aging. *Annual Review of Entomology, 50*, 421–445.

Hussain, M. (2021). The changing dynamics of liberal international order: Covid-19 and its global implications. *Journal of Globalization Studies, 12*(2), 143–156.

Inouye, S. K., Studenski, S., Tinetti, M. E., & Kuchel, G. A. (2007). Geriatric syndromes: Clinical, research, and policy implications of a core geriatric concept. *Journal of the American Geriatrics Society, 55*(5), 780–91. https://doi.org/10.1111/j.1532-5415.2007.01156.x. PMID: 17493201; PMCID: PMC2409147.

Irshad, S. M. (2020). COVID-19 and global 'At-Risk Community': From benefit-sharing to risk-sharing of economic crisis. *Journal of Globalization Studies, 11*(2), 135–145.

Jasiński, A. M., & Bąkowska, A. M. (2020). Globalization and life satisfaction of the polish population. Before and during the COVID-19 pandemic. *Journal of Globalization Studies, 11*(2), 146–155.

Jazwinski, S. M. (1996). Longevity, genes, and aging. *Science, 273*(5271), 54–59.

Kakkar, S. (2021). The outbreak of sars-cov-2 (Covid-19): Impact on international tourism. *Journal of Globalization Studies, 12*(1), 145–158.

Kanasi, E., Ayilavarapu, S., & Jones, J. (2016). The aging population: Demographics and the biology of aging. *Periodontol, 72*(1), 13–18. https://doi.org/10.1111/prd.12126. PMID: 27501488

Khaltourina, D., Matveyev, Y., Alekseev, A., Cortese, F., & Ioviţă, A. (2020). Aging fits the disease criteria of the international classification of diseases. *Mechanisms of Ageing and Development, 189*, 111230.

Konstantinov, A. (2021). Kak zamedlit' stareniye. *Expert.ru*. https://expert.ru/expert/2021/37/kak-zamedlit-stareniye/?mindbox-click-id=7b3b7ede-5674-42b9-90dd-60a6af4a084f&utm_source=mindbox&utm_medium=email&utm_campaign=LuchshieStati2021-NEaktivnaya

Kunkel, S. R., & Settersten, R. A. (2021). *Aging, society, and the life course* (6th ed.). Springer. https://doi.org/10.1891/9780826180353

Larrad-Jimenez, A., Kristofic, S., Sunny, S., Martin-Duce, A., & Corpas, E. (2021). *Endocrinology of aging*. Elsevier. https://doi.org/10.1016/B978-0-12-819667-0.00013-5

Ledford, H. (2020). CRISPR gene therapy shows promise against blood diseases. researchers report early successes using genetic approaches to treat sickle-cell anaemia and β-thalassaemia. *Nature, 588*, 383. https://doi.org/10.1038/d41586-020-03476-x

Lenaz, G., Cavazzoni, M., Genova, M. L., D'Aurelio, M., Merlo Pich, M., Pallotti, F., Formiggini, G., Marchetti, M., Parenti Castelli, G., Bovina, C. (1998). Oxidative stress, antioxidant defences and aging. *BioFactors*, 8(3–4), 195–204. https://doi.org/10.1002/biof.5520080305

Li, J., Bonkowski, M. S., Moniot, S., Zhang, D., Hubbard, B. P., Ling, A. J., Rajman, L. A., Qin, B., Lou, Z., Gorbunova, V., Aravind, L., Steegborn, C., & Sinclair, D. A. (2017). A conserved NAD(+) binding pocket that regulates protein-protein interactions during aging. *Science, 355*(6331), 1312–1317. https://doi.org/10.1126/science.aad8242

López-Otín, C., Blasco, M. A., Partridge, L., Serrano, M., & Kroemer, G. (2013). The hallmarks of aging. *Cell, 153*, 1194–1217.

Mineo, L. (2017). Good genes are nice, but joy is better. https://news.harvard.edu/gazette/story/2017/04/over-nearly-80-years-harvard-study-has-been-showing-how-to-live-a-healthy-and-happy-life/

Mitzner, T. L., Sanford, J. A., & Rogers, W. A. (2018). Closing the capacity-ability gap: Using technology to support aging with disability. *Innovation in Aging, 2*(1), igy008. https://doi.org/10.1093/geroni/igy008

Moskalev, A. A. (2018). *2120 let zhizni—tol'ko nachalo. Kak pobedit' stareniye?* Eksmo.

Musa, M. (2016). Opinion: Aging, just another disease. *The Scientist*, November 17. https://www.the-scientist.com/critic-at-large/opinion-aging-just-another-disease-32602

Naddaf, M. (2022). The science events to watch for in 2023. *Nature*, December 19. https://doi.org/10.1038/d41586-022-04444-3

NCOA- The National Council on Aging. (2021). NCOA's 2021 Annual Report. URL: https://www.ncoa.org/page/ncoa-annual-report-2021

Nelson, P., & Masel, J. (2017). Intercellular competition and the inevitability of multicellular aging. *Proceedings of the National Academy of Sciences USA, 114*(49), 12982–12987. Epub 2017 Oct 30. https://doi.org/10.1073/pnas.1618854114. PMID: 29087299; PMCID: PMC5724245.

Pertseva, M. (2016). *Nutrigerontologiya: pitaniye vs. Stareniye*. https://biomolecula.ru/articles/nutrigerontologiia-pitanie-vs-starenie#source-6

Ricklefs, R. E. (2000). Intrinsic aging-related mortality in birds. *Journal of Avian Biology, 31*(2), 103–111.

Seifabadi, M. S., & Dolatabadi, A. B. (2022). Covid-19, globalization and strengthening authoritarianism in the Middle East. *Journal of Globalization Studies, 13*(1), 135–157.

Sharp, Z. D. (2011). Aging and TOR: Interwoven in the fabric of life. *Cellular and Molecular Life Sciences, 68*, 587–597.

Shenk, J. W. (2009). What makes us happy? Is there a formula—some mix of love, work, and psychological adaptation—for a good life? https://www.theatlantic.com/magazine/archive/2009/06/what-makes-us-happy/307439/

Sinclair, D. A., & LaPlante, M. D. (2019). *Lifespan: Why we age—And why we Don't have to*. Atria books.

Slagboom, P. E, Droog, S., & Boomsma, D. I. (1994). Genetic determination of telomere size in humans: A twin study of three age groups. *American Journal of Human Genetics, 55*, 876–882.

Stipp, D. (2012). A new path to longevity. *Scientific American, 306*(1), 32–39. Bibcode:2011SciAm.306f..32S. https://doi.org/10.1038/scientificamerican0112-32. PMID 22279832.

Thevaranjan, N., Puchta, A., Schulz, C., Naidoo, A., Szamosi, J. C., Verschoor, C. P., et al. (2017). Age-associated microbial dysbiosis promotes intestinal permeability, systemic inflammation, and macrophage dysfunction. *Cell Host & Microbe, 21*(4), 455–466.

Tian, X., Azpurua, J., Hine, C., Vaidya, A., Myakishev-Rempel, M., Ablaeva, J., Mao, Z., Nevo, E., Gorbunova, V., & Seluanov, A. (2013). High-molecular-mass hyaluronan mediates the cancer resistance of the naked mole rat. *Nature, 499*(7458), 346–9. https://doi.org/10.1038/nature12234. PMID: 23783513; PMCID: PMC3720720.

Tinetti, M. E., Inouye, S. K., Gill, T. M., & Doucette, J. T. (1995). Shared risk factors for falls, incontinence, and functional dependence. Unifying the approach to geriatric syndromes. *JAMA, 273*(3), 1348–1353.

Troen, B. R. (2003). The biology of aging. *Mount Sinai Journal of Medicine, 70*(1), 3–22.

UNFPA. (2012). *Ageing in the Twenty-First Century: A Celebration and A Challenge*. UNFPA and HelpAge International. https://www.unfpa.org/publications/ageing-twenty-first-century

Vaughn, B. P., Fischer, M., Kelly, C. R., Allegretti, J. R., Graiziger, C., Thomas J., McClure E., Kabage A. J., & Khoruts A. (2022). Effectiveness and safety of colonic and capsule fecal microbiota transplantation for recurrent Clostridioides difficile infection. *Clinical Gastroenterology and Hepatology, 9*. https://doi.org/10.1016/j.cgh.2022.09.008

Washington State University. (2021). Potential Role of 'Junk DNA' Sequence in Aging, Cancer. *ScienceDaily*, 23 July. www.sciencedaily.com/releases/2021/07/210723105258.htm

WHO—World Health Organization. (2020). Immunity Passports in the Context of COVID-19. https://www.who.int/news-room/commentaries/detail/immunity-passports-in-the-context-of-covid-19

WHO—World health Organization. (2022). *International Classification of Diseases (ICD)*. https://www.who.int/standards/classifications/classification-of-diseases

Widdowson, M. (2021). From Covid-19 to zero-gravity: Complex crises and production revolutions. *Journal of Globalization Studies, 12*(1), 117–144.

Williams, B. C., Remington, T., & Foulk, M. (2002). Teaching interdisciplinary geriatrics team care. *Academic Medicine: Journal of the Association of American Medical Colleges, 77*(9), 935. https://doi.org/10.1097/00001888-200209000-00040

Wilson, S. A., Byrne, P., Rodgers, S. E., & Maden, M. (2022). A systematic review of smartphone and tablet use by older adults with and without cognitive impairment. *Innov Aging, 6*(2), igac002. https://doi.org/10.1093/geroni/igac002. PMID: 35243008; PMCID: PMC8889997.

Wullschleger, S., Loewith, R., & Hall, M. N. (2006). TOR signaling in growth and metabolism. *Cell, 124*(3), 471–84. https://doi.org/10.1016/j.cell.2006.01.016. PMID: 16469695.

Yegher, K de. (2013). Omolozheniye myshey: yeshche odin shag na puti k bessmertiyu? *InoSMI*, December 29. http://inosmi.ru/world/20131229/216124322.html

Zhavoronkov, A., & Bhullar, B. (2015). Classifying aging as a disease in the context of ICD-11. *Frontiers in Genetics, 4*(6), 326. https://doi.org/10.3389/fgene.2015.00326. PMID: 26583032; PMCID: PMC4631811.

# Part IV

# Trends of Cybernetic Revolution in the Light of Technological Progress and Aging

In this part, some of the issues raised in the previous parts will be elaborated further. In particular, Grinin et al. study the evolution of the technological process from the Upper Paleolithic onwards, but now not only relying on their theories of production principles and production revolutions, but also using mathematical methods. However, the main themes of the monograph: the Cybernetic Revolution and aging remain the focus of the authors' attention.

This part

(1) will make forecasts how technological progress will develop in the twenty-first century, with a slowdown or an acceleration;

(2) will show how the change in the pace of technological progress is related to global aging.

To address the first problem, Grinin et al. study changes in the speed of the technological process since ancient times. The idea of a slowdown in technological progress is now very popular: the idea that this decelerating trend, emerging in the 1970s, will remain unchanged, and that the slowdown will continue. This approach, however, seems to be too linear, and therefore raises serious doubts: the authors' research shows that from the 2030s (or 2040s) we will see a new acceleration in technological progress. This will definitely be associated with the beginning and development of the final phase of the Cybernetic Revolution (the phase that the authors call "the epoch of self-regulating systems"). This acceleration will continue for several decades, until, toward the end of the twenty-first century, it will start to slow down again. It is worth adding that Grinin et al. originally wanted to call this part "Rise and demise of explosive technological progress", which fully reflects the essence of their conclusions. Thus, the picture of changes in the rate of technological progress is by no means linear, and it looks much more scientific.

To address the second problem, Grinin et al. investigate the effect of aging on the rate of technological progress. They come to the conclusion that this influence is non-linear. Aging will initially contribute to increase in the rate of technological progress, but then a reversal will kick in, slowing the development down. Thus global aging has a dual and contradictory effect on technological progress: positive and negative.

This part also considers (3) the changes in the consumption patterns of the future society, resulting from the aging of society; followed by (4) forecasts about how much and how radically the technological development, going hand in hand with the aging process, will change the political structure. At the same time, the experience of anti-COVID measures allows us to better see the prospects for future transformations and make important forecasts until the end of the twenty-first century.

In both cases, very radical changes are expected, both in the economic and the political model of society, the reference points for the formation of the electronic state.

# Long-Term Dynamics of Technological Growth: Would it Lead Us to the Singularity?

12

**Abstract**

In this chapter, the authors set the following tasks: (1) to make forecasts about how technological progress will develop in the twenty-first century, with a slowdown or an acceleration; (2) to show how the change in the pace of technological progress is related to global aging and the development of this process. Grinin et al. solve these problems on the basis of the theory of production revolutions and certain mathematical methods. The authors find that the general dynamics of accelerating technological growth over the past 40,000 years can be described with amazing accuracy ($R^2 = 0.99$) by a simple hyperbolic equation: $y_t = C/(t_0 - t)$. The authors' analysis shows that in the coming decades the global technological growth rate will return to a hyperbolic trajectory at the beginning of the final phase of the Cybernetic Revolution (i.e., from the 2030s–2040s). The acceleration will continue up to the late twenty-first century. Then, technological growth will gradually slow down towards the end of the twenty-first century as suggested by the singularity point in the year 2106. As the reader will see, global aging will play a leading role in the changes, first in the acceleration and then the slowdown of the rate of the technological progress.

## 12.1 What this Chapter is About

This chapter covers a very wide span from the Upper Paleolithic to the beginning of the twenty-second century, when a new era will probably start. In this chapter we consider the long-term dynamics of technological progress, presenting one of the options for measuring its speed throughout the entire historical process.

Measuring the speed of technological progress is only one and not the main goal of the chapter. The main themes of the monograph: the Cybernetic Revolution and



L. Grinin et al., *Cybernetic Revolution and Global Aging*, World-Systems Evolution and Global Futures, https://doi.org/10.1007/978-3-031-56764-3_12

global aging remain the focus of our attention. Measuring process speed allows us to make some forecasts. In this chapter, we set the following tasks:

(1) to make forecasts about how technological progress will develop in the twenty-first century, with a slowdown or an acceleration;

(2) to show how the change in the pace of technological progress is related to global aging and the development of this process.

We solve these tasks on the basis of the theory of production (or technological) revolutions and the theory of production principles, as well as involving mathematical methods. We find that the general dynamics of accelerating technological growth over the past 40,000 years can be described with amazing accuracy ($R^2 = 0.99$) by a simple hyperbolic equation: $y_t = C/t_0 - t$, where $y_t$ is the technological growth rate, measured as the number of technological transitions per unit of time, while $t_0$ and C are constants, with $t_0$ being a point of 'technological singularity'.

We show that on a macroscale there has been a tendency to an increase in technological growth, albeit with fluctuations. However, the upcoming technological growth will not be infinite. In the chapter, we describe a scenario of how, when and why this will change.

First, the slowdown in technological growth in recent decades, beginning in the 1970s, can be explained as a general fluctuation of technological growth. One views this slowdown as a necessary condition for further rapid growth.

Second, the authors' analysis shows that there are a number of reasons to expect that in the coming decades the global technological growth rate will return for some time to a hyperbolic trajectory at the beginning of the final phase of the Cybernetic revolution (i.e., from the 2030s–2040s).

Third, this acceleration will continue up to the late twenty-first century.

Fourth, technological growth will gradually slow down towards the end of the twenty-first century, as suggested by the singularity point in the year 2106. As the reader will see, global aging will play a leading role in the changes, first in the acceleration and then the slowdown of the rate of the technological progress.

Fifth, after the singularity point, the rate of technological progress will continue to slow down compared to the previous epoch, and it is difficult to predict any subsequent acceleration in the twenty-second century (at least in its first half).

Probably, the most important point is that the pattern of scientific and technological development itself will change dramatically. Thus, such a slowdown will mean that the old type of technological progress is beginning to transform into a new one, and, most likely, a transition to new forms of social relations will start.

In the modern world people deal daily with numerous achievements of scientific and technological progress. People are becoming more and more dependent on them, spending a considerable amount of time keeping up with progress. In general, the entire human history, especially in the last few centuries, may be regarded (albeit, with significant limitations) as a history of achievements in science and technology, especially information technologies (see Chap. 2; see also Bernal, 1965; Boserup, 1965; Braudel, 1973; Bunch & Hellemans, 2004; Carree, 2003; Carus-Wilson, 1987; Childe, 1952; Cipolla, 1975; Galor & Tsiddon, 1997; Grinin & Grinin, 2015c, 2016; Grinin et al., 2017a, 2020a, 2020b; Kayal, 1999;

Kremer, 1993; Kurzweil, 2001; Phillips, 2011). So, technological growth is one of the most important factors of societies' transformations and development. Thus, it is extremely important to identify some patterns in the history of technological development, and to try to anticipate forthcoming transformations in technologies and society. Regrettably, as we have pointed out in Chaps. 1–4, there are only few well-grounded studies which could describe technological development in a systematic and consistent way, providing scientific explanations of why and how the technological revolutions occur.

In this chapter we consider the long-term dynamics of technological progress, presenting one of the options for measuring its speed throughout the entire historical process. As we have already mentioned above, we solve these problems on the basis of the theory of production (or technological) revolutions and the theory of production principles, as well as applying mathematical methods. For this purpose, we use well-known mathematical equations that allow us to compare our results with the results of researchers who measure the rate of general evolution on Earth in respect of the singularity (see below). We also apply the methodology proposed by Panov (2005, 2017, 2020).

To answer the questions outlined above, we use, first, the theory of production principles described in detail above in Chap. 2 and summarized briefly below.

## 12.2 Long-Term Technological Growth: Theory

**Production Principles and Production Revolutions**. A large number of technological breakthroughs have been observed in human history. As has already been demonstrated in Chap. 2 above, among the major technological breakthroughs in history, the most important are **the three technological or production revolutions**: (1) the Agrarian Revolution (with the Neolithic revolution being the first part of it); (2) the Industrial Revolution; and (3) the Cybernetic Revolution. From our point of view, each revolution initiates a new stage in the development of the world's productive forces, as well as a transition to a new stage of technological evolution. The point is that each production revolution entails a transition to a fundamentally new production system; the beginning of each marks a border between corresponding production principles.

1. Agrarian Revolution (from 12,000–10,000 BP to 5,500–3,000 BP). This meant the transition to systematic food production using a new type of energy (the power of domestic animals), and, on this basis, the transition to a complex social division of labor. This revolution was also associated with the use of new sources of energy (animal power) and new materials.
2. Industrial Revolution (the last third of the fifteenth to the first third of the nineteenth centuries). It resulted in the concentration of production in industry, characterized by the use of machines. Not only was manual labor replaced by machines, but biological energy was replaced by water and steam energy.

3. Cybernetic Revolution (from 1950 to the 2060/2070s) has already led to the emergence of powerful information technologies, and will stimulate in the future the transition to the widespread use of self-regulating systems in different spheres of activity. These systems will be able to operate without human control and intervention. The Cybernetic Revolution is not over yet. We believe that it will provide huge steps in improving human health, quality of life, and our ability to influence and control the human body (for more details see *Part II* above). This revolution in the 1960s–1970s was sometimes called the 'scientific-technical' revolution (*e.g.*, Benson & Lloyd, 1983, Bernal, 1965). However, it would be more appropriate to call it the *Cybernetic Revolution*, since its main changes will imply increasing opportunities to control various processes by means of self-regulating/managing systems.

As we have pointed out in Chaps. 3, 4 and 10, the drivers of the final phase of the Cybernetic Revolution will be medical technologies, additive manufacturing (3D printers), nano- and biotechnologies, robotics, IT/AI and cognitive technologies—in their combination they will form a sophisticated system of self-regulating production. We denote this complex as MANBRIC convergence. Among them, medical technologies will become the integrating center (see Grinin & Grinin, 2015c, 2016, 2020; Grinin & Korotayev, 2016a, 2016b; Grinin et al., 2016; Grinin et al., 2017a, 2017b, 2020a, 2020b, 2021).

**Phases of production principles**. Every production revolution can be regarded as an integral part of the production principle. The production revolution is the first 'half' of the production principle, whereas the development of mature technologies based on the production principle occurs during the second half.[1] The cycle of a production principle can be represented by six phases. Our mathematical analysis is based on this six-phase pattern:

(1) The first phase—*starting*—is the beginning of a production revolution. A new production principle emerges in one or a few places, although in rather undeveloped, incomplete forms.
(2) The second phase is that of *primary modernization*. It is associated with a wider diffusion of new production forms, as well as with strengthening and the vigorous expansion of a new production principle.
(3) The third phase is the *completion of the production revolution*. The production principle acquires advanced characteristics.

[1] The cycle of each production revolution looks as follows: *the initial innovative phase* (emergence of a new revolutionizing production sector)—*the modernization phase* (diffusion, synthesis and improvement of new technologies)—*the final innovative phase* (when new technologies acquire their mature characteristics). See as example the scheme of the Cybernetic Revolution in Fig. 12.1. For more information about the cycle of production revolutions and their structural interconnection with production principles see Chaps. 2 and 4; also see Grinin, Grinin, and Korotayev 2017a (as well as our works mentioned above).

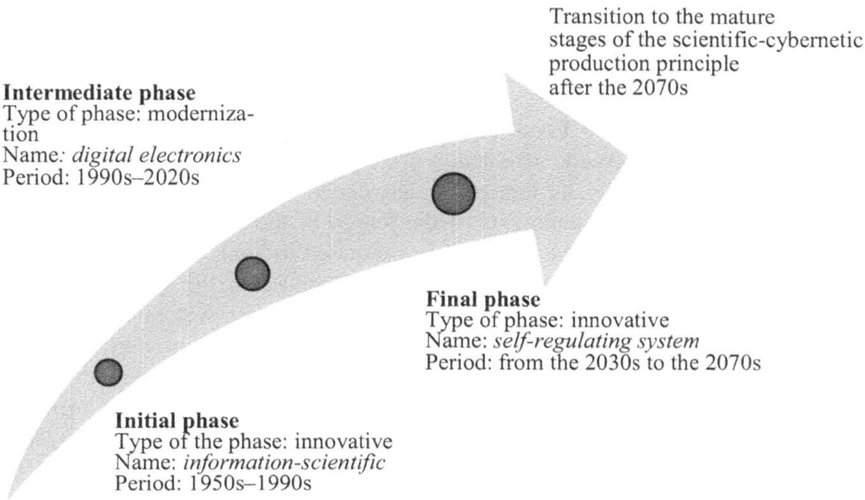


**Fig. 12.1** Phases of the cybernetic revolution

(4) The fourth phase is the phase of *maturity and expansion*. It relates to the diffusion of new technologies in most spheres and regions of the process of production. The production principle acquires its mature form, and this leads to important changes in the socio-economic sphere.
(5) The fifth phase is that of the *absolute dominance of a production principle*. It leads to the intensification of production and the full realization of the potential of the principle.
(6) The sixth phase is that of *non-system phenomena, a preparatory phase for transition to a new production principle*. New inventions and improvements in technologies lead to the emergence of non-system elements which pave the way for the formation of a new production principle.

Under favorable conditions, these elements form a new system. So, the current cycle ends, but in some societies the transition to a new production principle starts, and the cycle repeats at a new level.

Based on this six-phase cycle of the production principle, we perform our calculations of the speed of technological progress, where the transition from one stage to the next is considered as a phase transition.

For a description of the history of technological changes within the Hunter-Gatherer, Craft-Agrarian, and Trade-Industrial Production Principles see Chap. 2. A more detailed description of the Scientific-Cybernetic Production Principle and the Cybernetic Revolution can be found below in the following subsection of this chapter.

## 12.3 Scientific-Cybernetic Production Principle, Aging and its Influence on the Pace of Technological Progress

**The Scientific-Cybernetic Production Principle and the Cybernetic Revolution.** The scientific-cybernetic production principle is only in its initial stages (see Figs. 12.1 and 12.2); only its first phase has ended, and the second phase is in progress. Hence, all the calculations of the length of the forthcoming phases are highly hypothetical. These calculations are presented in Tables 12.1 and 12.2.

The *first* phase of the Scientific-Cybernetic Production Principle took place between the 1950s and the mid-1990s, when a vigorous development of information technologies and the start of real economic globalization were observed. It is also connected with the transition to scientific methods in production and distribution management. Particularly important changes took place in information technologies. In addition, this production revolution had a few other directions: in energy technologies, in synthetic materials production, automation, space exploration and agriculture. However, its main results are still forthcoming.

The *second* phase of the Scientific-Cybernetic Production Principle began in the mid-1990s in conjunction with the development and wide diffusion of user-friendly computers, AI, communication technologies such as mobile phones and so on. Medicine and biotechnologies have also made significant progress, as well as some other innovative fields (see Grinin & Grinin, 2015a, 2015b, 2015c: part 3; 2016: Chaps. 3–4; Grinin et al., 2017a). This phase continues to the present day.

The third phase of the Scientific-Cybernetic Production Principle is likely to begin around the 2030s. It will mean the beginning of the final phase of the Cybernetic Revolution, which in our view could become the epoch of 'self-regulating/self-managing systems'. The final phase of this revolution may start in the sphere of medicine and will be connected with its innovative branches; this will lead to

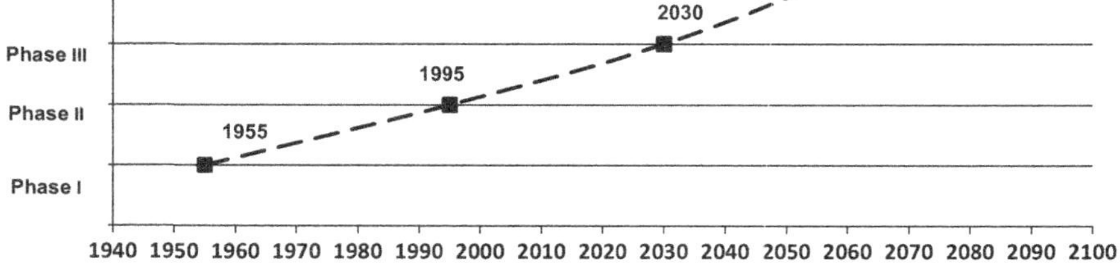


**Fig. 12.2** Development of the scientific-cybernetic production principle

**Table 12.1** Chronology of production principle phases. Figures before brackets correspond to absolute dates (BP); figures in brackets correspond to years BCE. **Bold** values indicate phase lengths (in thousands of years)

| Production principle | 1st Phase | 2nd Phase | 3rd Phase | 4th Phase | 5th Phase | 6th Phase | Overall for production principle |
|---|---|---|---|---|---|---|---|
| 1. Hunter-Gatherer | 40,000–30,000 BP | 30,000–22,000 | 22,000–17,000 | 17,000–14,000 | 14,000–11,500 | 11,500–10,000 | 40,000–10,000 |
| | (38,000–28,000 BCE) | (28,000–20,000 BCE) | (20,000–15,000 BCE) | (15,000–12,000 BCE) | (12,000–9500 BCE) | (9500–8000 BCE) | (38,000–8000 BCE) |
| | **10** | **8** | **5** | **3** | **2.5** | **1.5** | **30** |
| 2. Craft-Agrarian | 10,000–7300 | 7300–5000 | 5000–3500 | 35000–2200 | 2200–1200 | 800 CE–1430 CE | 10,000–570 |
| | (8000–5300 BCE) | (5300–3000 BCE) | (3000–1500 BCE) | (1500–200 BCE) | (200 BCE–800 CE) | **0.6** | (8000 BCE–1430 CE) |
| | **2.7** | **2.3** | **1.5** | **1.3** | **1.0** | | **9.4** |
| 3. Trade-Industrial | 1430–1600 | 1600–1730 | 1730–1830 | 1830–1890 | 1890–1929 | 1929–1955 | 1430–1955 |
| | **0.17** | **0.13** | **0.1** | **0.06** | **0.04** | **0.025** | **0.525** |
| 4. Scientific-Cybernetic | 1955–1995* | 1995–2030 | 2030–2055 | 2055–2070 | 2070–2080 | 2080–2090 | 1955–2090 |
| | **0.04** | **0.035** | **0.025** | **0.015** | **0.01** | **0.01** | **0.135–0.160** |

**Note* Starting from the second column of the row we give our estimates of the expected lengths of the phases of the Scientific-Cybernetic Production Principle.[2]

[2] For the purposes of mathematical calculations in this chapter, we take into account only one version of the chronology forecast for the Scientific-Cybernetic Production Principle (the second version can be found in Table 12.3, Chap. 2).

**Table 12.2** Production principles and their phase lengths (in thousands of years)

| Production principle | 1st Phase | 2nd Phase | 3rd Phase | 4th Phase | 5th Phase | 6th Phase | Overall |
|---|---|---|---|---|---|---|---|
| 1. Hunter-Gatherer | 10 | 8 | 5 | 3 | 2.5 | 1.5 | 30 |
| 2. Craft-Agrarian | 2.7 | 2.3 | 1.5 | 1.3 | 1.0 | 0.6 | 9.4 |
| 3. Trade-Industrial | 0.17 | 0.13 | 0.1 | 0.06 | 0.04 | 0.025 | 0.525 |
| 4. Scientific-Cybernetic | 0.04 | 0.035* | 0.025 | 0.015 | 0.01 | 0.01 | 0.135 |

*Note* * This row indicates our estimates of the expected lengths of the phases of the Scientific-Cybernetic Production Principle

serious modification of human organism and, perhaps, change its biological nature (for more detail see Chap. 8; Grinin & Grinin, 2015c, 2016; Grinin et al., 2016; Grinin et al., 2017a, 2017b, 2020a, 2020b, 2021).

As the reader may remember, the first phase of a new production principle corresponds to the initial phase of a new production revolution. So, the *first* phase of the Scientific-Cybernetic Production Principle = the *initial* phase of the Cybernetic Revolution; the *second* phase of the Scientific-Cybernetic Production Principle = the *intermediate* phase of the Cybernetic Revolution. The third phase of the Scientific-Cybernetic Production Principle = the *final* phase of the Cybernetic Revolution (see Fig. 12.1).

The fourth phase implies that, in the next two decades, the sector of self-regulating systems will rapidly expand and diffuse to various regions at enormous speed. MANBRIC technologies will reach developed forms and take a central place in the new production principle. At the same time, this will be a period of significant increase in life expectancy and, accordingly (against the background of low fertility), a period of rapid global aging that will also involve regions that are still 'young', including Sub-Saharan Africa and South Asia (Grinin & Grinin, 2015c, 2016; Grinin et al., 2016; Grinin et al., 2017a, 2017b; Grinin & Korotayev, 2015a).

The expected duration of the *fourth*, *fifth*, and *sixth* phases of the scientific-cybernetic production principle is 2055–2070, 2070–2080 and 2080–2090, respectively (see Tables 12.1 and 12.2 below, about expected innovations in these phases see Chaps. 2, 15; Chaps. 8–10; see also Grinin, 2006).

**Aging and the rate of technological progress**. There are a number of scholars who believe that the growth rate of technological progress has already passed its culmination and it is already slowing down (Maddison 2007; Teulings and Baldwin 2014; Phillips 2011; about this see also Korotayev and Bozhevolnov, 2010; LePoire and Korotayev, 2020; for recent research about decreasing number of breakthrough publications and patents see Park et al. 2023; see also Cowen and Southwood 2019); it is also possible to observe this if we compare the number of inventions per decade in 1950–1960 with that in 1970–1990 according to Bunch and Hellemans' database (2004).

However, we do not think that the speed of future technological progress will gradually slow, or that it will remain constant. For the time period our theory allows us to forecast the change in speed will be non-linear. At the first stage of the Cybernetic Revolution, the speed of technological progress accelerated, and in the second stage (which we have been in since the 1990s), it slowed down. We believe that this deceleration will not change until the early/mid-2030s or the beginning of the 2040s, after which technological growth will experience a new acceleration. There will then be a gradual slowdown up to the point of singularity, with a subsequent change in pattern (see below).

The increasing speed of future technological progress will, according to our forecasts, begin with the third phase of the Scientific-Cybernetic Production Principle (the final phase of the Cybernetic Revolution). The effect of aging on this acceleration will be very significant. As we have shown in Chap. 11, the health problems associated with aging and longer life expectancy will gradually become the leading direction in the medicine of the future. Thus, the role of medicine and the entire MANBRIC technological system, which is partly related to medicine, in the not so distant future can be associated with the problem of aging in every society and in general, with global aging. As a result, medical technologies will rapidly develop under the influence of an aging population (Phillips, 2011), and this will expand the search for opportunities to create "smart", self-regulating and self-managing systems, including robots, which can largely replace human labor, especially in the service sector (Chap. 10; Frey & Osborne, 2017), including complex services (e.g., in the fields of elder care, education, medicine, etc.) (DeCanio, 2016).

We believe that in the coming decades aging will be one of the most important drivers of scientific and technological change. Thus, until the last third of our century, the aging of the population will not impede technological and other development. On the contrary, the process of global aging will itself be the driving force for change, reform and acceleration of technological innovation.

However, by the late twenty-first century and the early next century, aging can slow down the technological growth and bring about a change in direction. What could cause such a slowdown? Among several reasons, the conservatism of the elderly, which will increase in a huge part of society, will strongly influence the development.

We seek to identify the actual mechanisms and relationships to explain the reasons for a possible acceleration and slowdown in the future speed of technological process. Thus, we associate very large changes in technological development with global aging in the future (as one of the most important results of technological progress), but, as we will see, the effect of aging on the speed of technological progress is non-linear and produces different effects at different phases.

It is important to take into consideration that (as we have already pointed out in Chaps. 6 and 7) global aging has a dual and contradictory effect on technological progress: positive and negative. On the one hand, the need to improve the biological quality of life of the elderly, the struggle to increase life expectancy, the fight against incurable diseases, the need to extend the working age of the elderly, etc. (which has been discussed in Chaps. 4, 6 and 7) spur science and technology to search for new breakthrough solutions. Moreover, the growth of financial flows and interests, which are influenced by the elderly and old people, will continue. On the other hand, old people are prone to conservatism both in terms of physiological properties, and in terms of worldview and in terms of cognitive characteristics. In addition, they already have a lot and therefore they do not need many fundamental expenses (housing, furnishings, equipment, etc.) that younger people need. All this has a negative impact on technological progress and slows it down. And these multidirectional trends will continue to influence technological progress. However, until the last third of the twenty-first century, as we believe, the first, positive trend will prevail; respectively, aging in general will affect the acceleration of technological progress. However, this trend will weaken due to the increase in the proportion of the elderly, and, accordingly, the trend toward conservatism will increase until it becomes predominant. Then, aging as a whole will affect the slowdown of technological progress. One can assume that the current consumption pattern may also change under the influence of the global aging process. And this change, in turn, will have a serious impact on the entire structure of production and on scientific-technological progress (for details on conservatism and its influence see Chap. 13).

It is worth noting that the impact of global aging on speed and direction of scientific-technological progress is understudied (Galor & Weil, 2000; Grey & Rae, 2008; Prettner, 2013; Tsirel, 2008), whereas global aging affects technological, economic, political, social and other spheres in various ways (Barsukov, 2019; Bloom & Luca, 2016; Coleman & Rowthorn, 2015; Fichtner, 2018; Fukuyama, 2002; Goldstone, 2015; Goldstone et al., 2015; Grinin & Grinin, 2015b, 2015c, 2017; Grinin & Grinin, 2016; Grinin & Korotayev, 2010, 2015b, 2015c, 2016a, 2016b; Grinin et al., 2017a; Haas, 2015; Harper, 2006; Hsu & Lo, 2019; Jackson, 2021; Jackson et al., 2008, 2013; Kapelyushnikov, 2019a, 2019b; Korotayev & Shulgin, 2018; Lee & Mason, 2011; Maestas et al., 2016; Park & Shin, 2015; Popova & Yanik, 2014; Powell & Khan, 2013; UNCTAD, 2012; WHO, 2017, 2021; Zimmer, 2016).

## 12.4 Mathematical Interpretation of Technological Progress

**On stable proportions in the cycles of the production principles**. One of the objectives of this chapter subsection is to present the following results:

(1) To show the duration of each of the four production principles and the duration of each of the six stages within every production principle. These data are presented in Tables 12.1 and 12.2, which show (a) the general time parameters of the production principles; (b) the acceleration of technological evolution both within a production principle from one stage to another, as well as in comparison with the previous and subsequent production principles; (c) data that allow

us to summarize the technological narrative and the chronological description of historical process;

(2) To show that a production principle is not just a certain stage of development of the world's productive forces, but a rather complex cycle of technical innovations and organizational-technological system rearrangements of production. It inevitably requires changes in various spheres of society and also brings new changes. Tables 12.3 and 12.4 show calculations of the relationships between the stages (and combinations of stages) within each production principle and demonstrate that each cycle of the production principle retains a surprising consistency, which cannot be accidental. For example, the duration of the first stage of each production principle ranges from 28 to 33% of the duration of the entire production principle. Recall that these are among the most important stages of production revolutions. The ratio of the duration of the stages to each other is also quite close, for example, in all four principles of production the ratio varies within a rather narrow range. There is a small scatter of proportions, oscillating around a certain value in all 19 ratios (given in Tables 12.3 and 12.4). These stable ratios reveal certain deep and fundamental patterns of technological development and technological evolution in the framework of the historical process. All of this allows us to make some predictions about the duration of the future stages of the Scientific-Cybernetic Production Principle;

(3) To calculate the acceleration of technological progress, the results of which we will present below. This subsection serves as the basis for the conclusions of the following subsections.

We present dates for all phases of all production principles in Table 12.1. However, it should be taken into account that all dates are averaged for convenience. The absolute lengths of the phases in thousands of years are given in Table 12.2 and also put in bold in Table 12.1.

Table 12.3 (Ratio of Each Phase [and Phase Combination] Length to the Total Length of Respective Production Principle) presents the results of our calculations of the ratio of the duration of each phase to the duration of the respective production principle using a rather simple methodology. The absolute length of a phase (or a sum of the lengths of two or three phases) is divided by the total length of the respective production principle. For example, if the length of the Hunter-Gatherer Production Principle is 30,000 years, the duration of its first phase is

**Table 12.3** Ratio of each phase (and phase combination) length to the total length of respective production principle (%)

| Production principle | 1 | 2 | 3 | 4 | 5 | 6 | 1–2 | 3–4 | 5–6 | 1–3 | 4–6 |
|---|---|---|---|---|---|---|---|---|---|---|---|
| 1. Hunter-Gatherer | 33.3 | 26.7 | 16.7 | 10 | 8.3 | 5 | 60 | 26.7 | 13.3 | 76.7 | 23.3 |
| 2. Craft-Agrarian | 28.7 | 24.5 | 16.0 | 13.8 | 10.6 | 6.4 | 53.2 | 29.8 | 17 | 69.1 | 30.9 |
| 3. Trade-Industrial | 32.4 | 24.8 | 19 | 11.4 | 7.6 | 4.8 | 57.1 | 30.5 | 12.4 | 76.2 | 23.8 |
| 4. Scientific-Cybernetic | 29.6 | 25.9 | 18.5 | 11.1 | 7.4 | 7.4 | 55.6 | 29.6 | 14.8 | 74.1 | 25.9 |
| Average | 31 | 25.5 | 17.6 | 11.6 | 8.5 | 5.9 | 56.5 | 29.2 | 14.4 | 74.0 | 26.5 |

**Table 12.4** Comparison of phase length ratios for each production principle (%)

| Production principle | 1:2 | 2:3 | 3:4 | 4:5 | 5:6 | (1+2):(3+4) | (3+4):(5+6) | (1+2+3):(4+5+6) |
|---|---|---|---|---|---|---|---|---|
| 1. Hunter-Gatherer | 125 | 160 | 166.7 | 120 | 166.7 | 225 | 200 | 328.6 |
| 2. Craft-Agrarian | 117.4 | 153.3 | 115.4 | 130 | 166.7 | 178.6 | 175 | 224.1 |
| 3. Trade-Industrial | 130.8 | 130 | 166.7 | 150 | 160 | 187.5 | 246.2 | 320 |
| 4. Scientific-Cybernetic | 114.3 | 140 | 166.7 | 150 | 100 | 187.5 | 200 | 285.7 |
| Average | 121.4 | 144.2 | 149.7 | 133.3 | 160.9 | 190.3 | 205.3 | 282.1 |

10,000, the second is 8,000, and the duration of the third is 5,000 years. The ratio of the duration of the first phase to the total duration of the production principle will be 33.3%; the ratio of the sum of the duration of the first and the second phases to the total duration of production principle will be 60%; and the ratio of the sum of the duration of the first, the second and the third phases to the total duration of production principle will be 76.7%.

Table 12.4 (Comparison of Phase Length Ratios for Each Production Principle) employs an analogous methodology to compare the duration of phases (and combinations of phases) within a production principle. For example, for the hunter-gatherer production principle the ratio of the first phase length (10,000 years) to the second one (8,000 years) equals 125%; whereas the ratio of the second phase to the third one (5,000 years) is 160%. In the meantime, the ratio of the sum of the duration of the first and the second phases to the sum of the duration of the third and the fourth phases (3,000 years) is 225%.

Tables 12.3 and 12.4 also present the average rates for all the production principles.

Thus, the proposed periodization demonstrates stable patterns of recurrent developmental cycles with a shortening of the period (each of which includes six phases); however, *each subsequent cycle was shorter than the previous one due to the acceleration of technological growth.* Note that these are recurrent cycles, because within each cycle the development follows the same pattern in some respects: every phase within every cycle plays a functionally similar role; what is more, the proportions of the lengths of the phases and their combinations remain stable (see Tables 12.3 and 12.4). All of this is confirmed by the calculations in Tables 12.3 and 12.4, according to which the stable proportions of phase lengths and their combinations remain intact with the change of production principles (see also Chap. 2).

Our quantitative analysis presented in the tables above indicates the following points: (a) the temporal evolution of each production principle has recurrent features; there are stable mathematical proportions between the lengths of phases and phase combinations within each production principle (Tables 12.3 and 12.4); (b) the cycle analysis clearly indicates that the technological development rate increases sharply as a result of production revolutions; (c) the analysis of the stable proportions of the production principle cycles makes it possible to propose some tentative forecasts—in particular, with respect to the lengths of the future phases of the fourth production principle, i.e., the Scientific-Cybernetic Production Principle as well as the length of the final phase of the Cybernetic Revolution.

It seems appropriate to make here a small digression. As we have discussed in Chap. 2, in the Hunter-Gatherer Production Principle, everything in society was exclusively dependent on the environment. In this respect, the question arises how to explain the quantitative proportions we have detected between the periods of the Hunter-Gatherer Production Principle which we present in Tables 12.3 and 12.4. We have empirically determined certain correlations between the duration of phases recurring within each production principle. But to what extent are these proportions relevant to the Hunter-Gatherer Production Principle, if for the identification of the beginning of its periods we involve some exogenous factors of nature and climate changes?

In fact, since the climate changes could have occurred at any other time these proportions are random to some extent. However, in general they are not random at all, and moreover, they are endogenously reasonable, since each successive cyclic change described requires more or less definite amount of time. This perfectly explains why the lengths of the given processes-stages correlate with each other in certain proportions. Second, although the climate changes can be considered as external (and therefore, random) factor for a society, the diversity of macroevolutionary lines significantly neutralizes such randomness. The idea following from the rule of the necessary diversity (see, e.g., Grinin et al., 2020) maintains that the larger is the diversity, the higher is the probability of the emergence of required randomness at the right time and in the right place. In the same way a person staking on more than one event at a time protects himself from accidents, and so, figuratively speaking, evolution with greater variability can accomplish a breakthrough if not in one place then in another. That is why, although the proportions in the correlation between the phases of the Hunter-Gatherer Production Principle can slightly shift, but they will remain almost the same since the qualitative changes, if unprepared, prevent excessive suitable cases. Meanwhile, if such a shift is delayed when a society appears ready ('overripe') for the changes necessary for a qualitative breakthrough even less suitable situations will work. In particular, along with periods of maximum cooling in some regions, i.e., from 22,000 to 17,000 BP [3] (which was, on the whole random with respect to social macroevolution at a certain point), in some places that avoided glaciation, an intensive and highly specialized gathering developed (Hall, 1986: 201; Harlan, 1986: 200; Fainberg, 1986: 185; Goring-Morris et al., 2009; Shea, 2013). That was consistent with social evolution. Consequently, the most important breakthroughs (from hunting and gathering to agriculture) could have followed the same pattern already from the period of 18,000 years ago. The start of the Agrarian Revolution might have been slightly accelerated, but, most likely, it would have delayed its transition to the second phase.

**Mathematical interpretation of the technological progress: methodology and calculations**. Each production principle is a six-phase cycle. The beginning of each phase can be considered as an important technological shift or phase transition. As a result of our periodization, 24 phases and corresponding 23 phase transitions have been identified (Table 12.5). The chronology of the identified phases is presented in the same table.

As a rule, complex and long-term processes cannot proceed evenly, an observation that also applies to technological evolution. As we note above, technological progress is a series of accelerations and decelerations in the speed of technological development. There is an idea that the mechanism of such rhythms is related to

[3] This was the period of the maximum spread of glaciers (referred to as the last glacial maximum).

**Table 12.5** Production principle phases, their dates, lengths and dynamics of technological growth rate (for empirically observed data points)

| Production principle phases | Date of the phase start | Phase length (years) | Macrotechnological growth rate (frequency of phase transitions per year) |
|---|---|---|---|
| Hunter-gatherer 1 | 40000 BP | 10,000 | $1.0 \times 10^{-4}$ |
| Hunter-gatherer 2 | 30000 BP | 8,000 | $1.3 \times 10^{-4}$ |
| Hunter-gatherer 3 | 22000 BP | 5,000 | $2.0 \times 10^{-4}$ |
| Hunter-gatherer 4 | 17000 BP | 3,000 | $3.3 \times 10^{-4}$ |
| Hunter-gatherer 5 | 14000 BP | 2,500 | $4.0 \times 10^{-4}$ |
| Hunter-gatherer 6 | 11500 BP | 1,500 | $6.7 \times 10^{-4}$ |
| Craft-agrarian 1 | 10,000 BP | 2,700 | $3.7 \times 10^{-4}$ |
| Craft-agrarian 2 | 5300 BCE | 2,300 | $4.3 \times 10^{-4}$ |
| Craft-agrarian 3 | 3000 BCE | 1,500 | $6.7 \times 10^{-4}$ |
| Craft-agrarian 4 | 1500 BCE | 1,300 | $7.7 \times 10^{-4}$ |
| Craft-agrarian 5 | 200 BCE | 1,000 | $1.0 \times 10^{-3}$ |
| Craft-agrarian 6 | 800 CE | 630 | $1.6 \times 10^{-3}$ |
| Trade-industrial 1 | 1430 | 170 | $5.9 \times 10^{-3}$ |
| Trade-industrial 2 | 1600 | 130 | $7.7 \times 10^{-3}$ |
| Trade-industrial 3 | 1730 | 100 | $1.0 \times 10^{-2}$ |
| Trade-industrial 4 | 1830 | 60 | $1.7 \times 10^{-2}$ |
| Trade-industrial 5 | 1890 | 39 | $2.6 \times 10^{-2}$ |
| Trade-industrial 6 | 1929 | 26 | $3.8 \times 10^{-2}$ |
| Scientific-cybernetic 1 | 1955 | 40 | $2.5 \times 10^{-2}$ |
| Scientific-cybernetic 2 | 1995 | | |

the slowdown of progress due to constant obstacles, such as lack of knowledge (Kayal, 1999). Of course, this is true in general. However, according to our theory, the acceleration and deceleration of technological progress depends on the functional features of each time phase within a super-long cycle of technological change (or a production principle). At some stages there is a kind of 'explosion' of innovations, where one can observe an acceleration of technological progress (e.g., the first and the third stages of production principle), while at others, these innovations are improved and diffused, thus slowing down (e.g., at the second stage of the production principle). At some stages (e.g., the fifth) a powerful expansion of the production principle occurs, while at others, a slowdown is observed under the influence of crises (e.g., the last stage).

To calculate the technological growth rate, we apply the methodology proposed by Alexander Panov (2005, 2017, 2020), according to which the temporal distance between phase transitions (temporal length of the phases) is recalculated as the frequency of phase transitions = number of phase transitions per year = macroevolutionary growth rate. In Panov's case, this was the speed of planetary macroevolutionary development; in our case, this variable can be interpreted as the technological growth rate within historical process (equally appropriately called the macrotechnological growth rate). It is noticeable that, as in Panov's time series—like in the similar time series of Modis (2002, 2003), Kurzweil (2001, 2005) and LePoire (2009, 2013) (see Korotayev, 2018, 2020a, 2020b for an analysis of these time series)—the temporal length of the phases in the time series systematically decreases, whereas the macrotechnological growth rate increases in a similarly systematic way, following a rather remarkable pattern (see Table 12.5).

**Calculation of the singularity given the continous nature of the Scientific-Cybernetic Production Principle**. It is important to note that the singularity does not indicate the point at which the value of a given variable actually becomes infinite; rather, it indicates the point before which the hyperbolic shape of the curve should change to a different *trajectory implying a certain slowdown* of the processes that have been observed in recent decades (Huebner, 2005; LePoire, 2005; Phillips, 2011; Korotayev, 2018, 2020a, 2020b). Below, we will discuss the possibility of a new acceleration of technological growth.

We believe that the calculation of the singularity can be processed either by referring only to the empirically observed data, or by using some data points predicted on the basis of a theoretical analysis of estimated technological development. That is why we use a dual approach to determining the singularity. In the first case, we show that if we use today/the present as the last point for calculations, the result will be close to the predictions by Kurzweil, Modis and Panov, which shows that our mathematical apparatus is quite adequate. However, only a mathematical apparatus without an essential theoretical part is obviously not enough. And since we have shown—hopefully convincingly—that the slowdown and acceleration of the technological process are cyclical, we give below the calculation of the singularity in accordance with the forecast of the expected acceleration of the technological process after the 2030s and the 2040s. In fact, this calculation of the singularity constitutes an important part of this chapter.

The graphic presentation of the macrotechnological growth rate detected in our time series is shown in Fig. 12.3.

It is not difficult to see that the general shape of the resultant curve is unmistakably hyperbolic, and it is well-known that the hyperbolic function has an explicit mathematical singularity.

Let the X-axis represent the time before the singularity (whereas the Y-axis will represent the technological growth rate)—and calculate the singularity date by obtaining a hyperbolic curve that describes our time series in the most accurate way. The results of this analysis are presented in Fig. 12.4 (note that our mathematical analysis has identified the singularity date for this time series as 2018 CE).

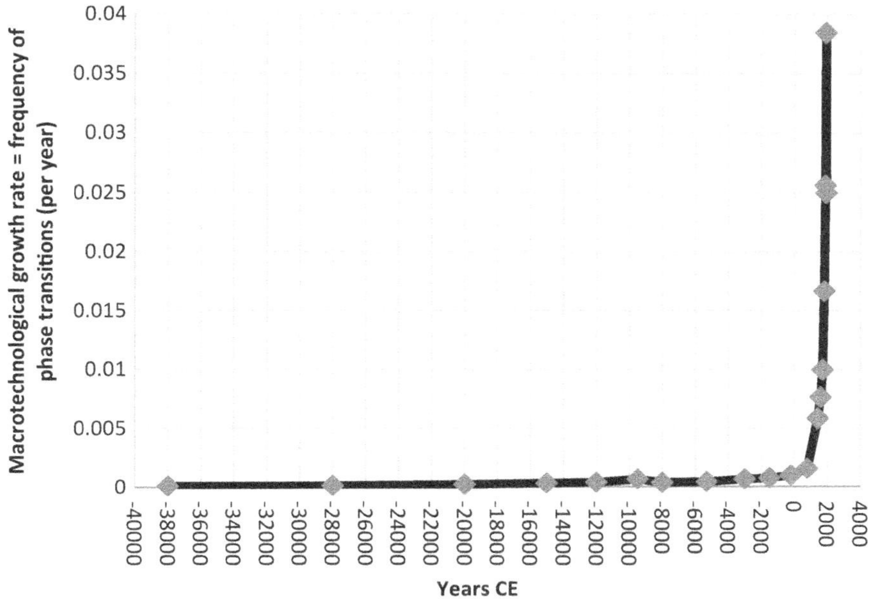


**Fig. 12.3** Dynamics of the global macrotechnological growth rate (frequency of phase transitions per year), 40000 BP to the late twentieth century

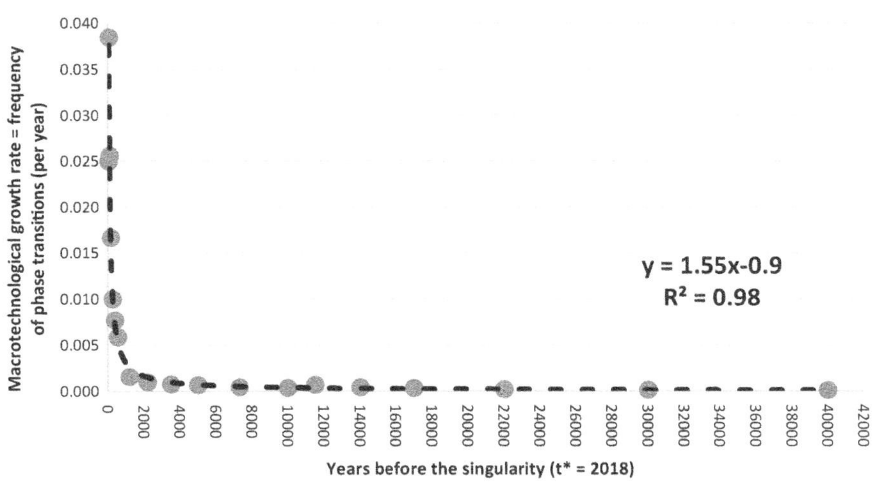


**Fig. 12.4** Scatterplot of the phase transition points described in Table 12.5 with the power-law regression line (fitted with the least-squares method) that identifies the date of singularity as 2018 CE (natural scales)

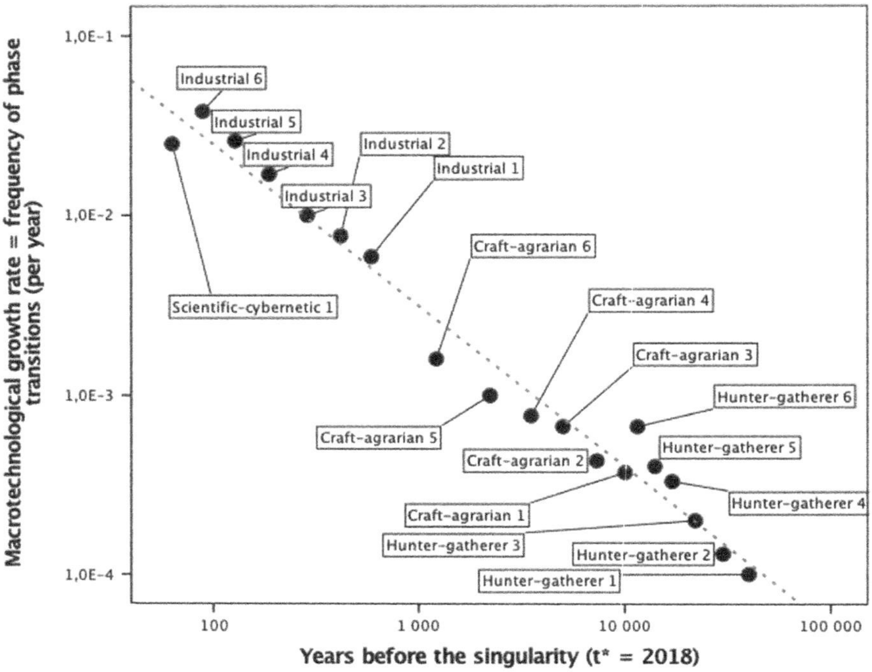


**Fig. 12.5** Scatterplot of the phase transition points described in Table 12.5 with the power-law regression line (fitted with the least-squares method) identifying the date of singularity as 2018 CE (*double logarithmic scale*)

The same figure is presented on a double logarithmic scale in Fig. 12.5.

Let us now analyze the results. As we see, our power-law regression of the technological growth phase transitions provides a best-fit equation (Eq. 12.1) describing this time series in a rather accurate ($R^2 = 0.98$) way:

$$V_t = \frac{1.55}{x^{0.9}}, \tag{12.1}$$

where $V_t$ is the global macrotechnological development rate, $x$ is the time remaining to the singularity and 1.55 and 0.9 are constants.

Note that the denominator's exponent (0.9) turns out to be rather close to 1; hence, there are some grounds to use this equation in the following simplified form:

$$V_t = \frac{1.55}{x}, \tag{12.2}$$

where $V_t$ is the global macrotechnological development rate, $x$ is the time remaining to the singularity and 1.55 is a constant.

Of course, $x$ (the time remaining until the singularity) at time $t$ is equal to $t^* - t$, where $t^*$ is the time of the singularity. Thus,

$$x = t * -t.$$

Hence, Eq. 12.2 can be re-written as Eq. 12.3:

$$V_t = \frac{1.55}{t^* - t}, \tag{12.3}$$

where $V_t$ is the global macrotechnological development rate at time $t$, $t^*$ is the time of the singularity, and 1.55 is a constant,

Finally, let us recollect that our least-squares analysis of the phase transition points described in Table 12.5 has identified the singularity date as 2018 CE. Thus, Eq. 12.3 can be further re-written as:

$$V_t = \frac{1.55}{2018 - t}. \tag{12.4}$$

Of course, in a more general form it should be written as

$$V_t = \frac{C}{t^* - t}, \tag{12.5}$$

where $C$ and $t^*$ are constants.

Note that an algebraic equation of the type

$$y_t = \frac{C}{t^* - t} \tag{12.6}$$

can be regarded as the solution of the following differential equation:

$$\frac{dy}{dt} = \frac{y^2}{C} \tag{12.7}$$

(see, *e.g.*, Korotayev et al., 2006a: 118–120).

Thus, the acceleration pattern implied by Eq. (12.4) can be spelled out as follows:

$$\frac{dV}{dt} = \frac{V^2}{1.55} \approx 0.65V^2. \tag{12.8}$$

Thus, the overall pattern of acceleration of the global technological growth rate that described rather accurately the data points of the technological growth phase transitions presented above in Table 12.5 with model (4)/(5) can be spelled out as follows: *throughout most of human history (at least since the Upper Paleolithic Revolution, i.e., c. 40,000 ya), the increase in macro-technological growth rate a* *times tended to be accompanied by an $a^2$ increase in its rate of acceleration; thus, a twofold increase in macrotechnological development rate tended to be accompanied by a fourfold increase in the rate of acceleration of this development rate; a tenfold increase in macrotechnological development rate tended to be accompanied by a 100-fold increase in the acceleration speed of this development rate; and so on...* The past tense is used in the statement above because the global technological growth does not seem to have followed this pattern in recent decades due to the slowdown mentioned above (otherwise, incidentally, it would have become infinite in 2018). Below, we will discuss the possibility and implications of a new acceleration in global technological growth.

**Notes about acceleration patterns**. Note that a rather similar acceleration pattern has been earlier detected for the Modis–Kurzweil series of 'canonical milestones/complexity jumps' (Modis, 2002, 2003; Kurzweil, 2005) as well as for the Panov series of 'global phase transitions/biospheric revolutions' (Panov, 2005, 2017, 2020; Korotayev, 2018, 2020a, 2020b). Incidentally, the Modis–Kurzweil series starts with the origin of the Milky Way 10 billion years ago and ends with the emergence of the Internet and human genome sequencing around 1995, whereas the Panov series begins with the origin of life on the Earth 4 billion years ago and ends with information globalization, which Panov dates to 1991 CE.

Indeed, the acceleration pattern detected in the Modis–Kurzweil series is described with 99.89 % accuracy by Equation 12.9:

$$y = \frac{2.054}{(2029 - t)^{1.003}}, \tag{12.9}$$

where $y$ is the global macrodevelopment rate (number of phase transitions per a unit of time), and 2029 CE is the best-fit singularity point estimate,

The simplified version of this model is

$$y_t = \frac{2.054}{2029 - t}, \tag{12.10}$$

this algebraic expression can be regarded as a solution for the following differential equation:

$$\frac{dy}{dt} = \frac{y^2}{2.054} \approx 0.5y^2. \tag{12.11}$$

On the other hand, the acceleration pattern detected in the Panov series is described with 99.91% accuracy by Eq. 12.12 (Korotayev, 2018, 2020a, 2020b):

$$y = \frac{1.886}{(2027 - t)^{1.01}}. \tag{12.12}$$

The simplified version of this model is given by Eq. 12.12,

$$y = \frac{1.9}{2027 - t}. \tag{12.13}$$

An algebraic equation of this type can be regarded as a solution of the following differential equation, which is very similar to the one we obtained above for the Modis–Kurzweil series, as well as for our series of technological phase transitions:

$$\frac{dy}{dt} = \frac{y^2}{1.9} \approx 0.5y^2. \tag{12.14}$$

As we can see, all three series are described accurately by similar mathematical models with similar parameters, including $t^*$ (the singularity time point); for the explanation of this phenomenon see Korotayev 2020a, 2020b.

## 12.5 Technological Singularity

**The pace of technological progress and the singularity**. As we have mentioned in Chap. 1, many researchers suppose that we have been moving toward considerable transformations and that in the next decades, human civilization will undergo dramatic changes. It is even possible that the very nature of development of humankind as a biological species will change qualitatively (Moiseev, 2000). They often use the term singularity to point out a certain unprecedented level of development after which we face a new developmental phase of humans and nature.[4] Indeed, the mathematicians denote the state which we are approaching as a result of accelerating changes as singularity: the rate of technological progress will increase so much that the denominator in the formula will contain zero, when divided by the rate of progress which pushes—theoretically—to infinity. Nobody knows how this will work in practice. This gives rise to many speculative theories about possible supertransformations, as not least the transition of humankind to immortality. On the one hand, it is worth separating mysticism from reality, in the sense that one should not expect anything supernatural after reaching the conditional singularity point. On the other hand, one should understand that the rate of change cannot increase endlessly, and therefore, the trend curve will change its direction after a while. This means that the processes will proceed in a significantly

[4] The issue of a global history singularity has been discussed quite actively for more than two decades (see, e.g., Diakonoff, 1999: 348; Panov, 2005, 2017, 2020; Kurzweil, 2005; Ayres, 2006; Modis, 2006, 2020; Muehlhauser & Salamon, 2012; Magee & Devezas, 2011; Eden et al., 2012; Grinin et al., 2013; Tsirel, 2014; Shanahan, 2015; Grinin & Grinin, 2015; Callaghan et al., 2017; Khaltourina et al., 2006; Korotayev, 2005, 2007, 2012, 2013, 2018, 2020a, 2020b; Korotayev et al., 2006; Korotayev & Malkov, 2015; LePoire & Korotayev, 2020; Panov et al., 2020; Nazaretyan, 2015, 2017, 2018, 2020; LePoire, 2020; Grinin et al., 2020a, 2020b).

different way, that is, already along some new (mostly unknown) trajectories. However, we think that the "Singularity Hypothesis" may be quite useful for the theory of the historical process and technological progress. Let us reiterate. The singularity does not indicate the point at which the value of a given variable actually becomes infinite; rather, it indicates the point before which the hyperbolic shape of the curve should change to a different trajectory implying a certain slowdown of the processes that have been observed in recent decades.

Concerning technological singularity, it is worth mentioning that already in the 1950s, the most farsighted academics spoke about a certain point of inevitable trend change—an assessment based on the speed of technological breakthroughs within the started technological revolution which we call the Cybernetic Revolution. The rapid change leading to "some essential singularity" was articulated earlier by John von Neumann in the 1950s. He maintained that the ever-accelerating progress of technology and changes in the mode of human life conveys the impression of approaching some essential singularity in the history of the race, beyond which human affairs, as we know them, could not continue (Ulam, 1958: 5). However, later (after the 1980s) the pace of technological change slightly slowed down. As we said above, today many scientists believe that the growth rate of technological progress has already passed its peak (for recent research on the decreasing number of breakthrough publications and patents see Park et al. 2023; see also Cowen and Southwood 2019). However, in our interpretation the current slowdown is just 'a calm before the storm' and we should expect a new powerful technological wave and an acceleration in the pace of technological progress. This wave is already starting. That is why we have a strong feeling that the world is in a state of a dramatic change, which we discussed above. Above we explain the role that aging will play in this acceleration and then slow down. However, even with this slow down the technological progress continue for two or three decades. It is only by the late twenty-first–the early twenty-second century, that we expect the technological development to pass its singularity point and experience an essential slowing down of the growth rate. Moreover, since technological development is in many respects a key point for the social advancement, we believe that it will be followed by a slowdown in other processes as well.[5]

**Calculation of the singularity, taking into account the predicted phases of the Scientific-Cybernetic Production Principle**. As we have mentioned above, there are different ways to estimate the singularity point in respect with the theoretical approaches to forecasting the future development of technological progress. Note that Eq. 12.1 has been calculated solely on the basis of empirically observed data points. However, the theory of production principles allows us to forecast a few more data points.

[5] The above said, however, does not mean the impossibility of a new acceleration in the twenty-second or twenty-third centuries. But in any case, those will be processes of a different type in many respects.

Indeed, we have shown above that there are grounds to expect that the *second* phase of the Scientific-Cybernetic Production Principle or the intermediate phase of the Cybernetic Revolution (that began in the mid-1990s) will continue until around 2030, when there are grounds to expect the third phase—the beginning of its final phase, which we expect may become the epoch of 'self-regulating/ managing systems', the vast expansion of opportunities to purposefully influence and direct various natural and production processes. This phase is expected to continue until around 2055, when there are grounds to forecast the start of the fourth phase.[6] This phase implies that the sector of self-regulating systems will rapidly improve over the next two decades, and will diffuse to various regions at enormous speed. At the same time, this should be a period of significant growth in life expectancy. The duration of the last two phases has been estimated above to be around 20 years. This allows us to forecast a few data points to add to the list of empirically estimated data points, resulting in Table 12.6.

We also pay some attention to the question of a singularity in acceleration patterns, since there are reasons to expect that its detection could help identify important inflection points in the processes under study.

A mathematical analysis of the resultant time series yields the results in Fig. 12.6.

The same figure is presented on a double logarithmic scale below (Fig. 12.7).

Let us analyze these new results. As we see, our power-law regression on the technological growth phase transitions including the data points presented in Table 12.6 (which includes four data points projected on the basis of the assumed new wave of acceleration of global macrotechnological growth rate forecast by the production principle theory) has identified the following best-fit equation describing this time series in a rather accurate ($R^2 = 0.98$) way:

$$V_t = \frac{3.32}{x^{0.98}}, \tag{12.15}$$

where $V_t$ is the global macrotechnological development rate, $x$ is the time remaining until the singularity and 3.32 and 0.98 are constants.

Note that the denominator's exponent (0.98) is much closer to 1 than in Eq. 12.1; hence, there are even stronger grounds to use this equation in a simplified form,

$$V_t = \frac{3.32}{x}, \tag{12.16}$$

where $V_t$ is the global macrotechnological development rate, $x$ is the time remaining until the singularity and 3.32 is a constant.

---

[6] However the third phase can begin later according to the second version of unfolding of the Cybernetic Revolution, i.e., in the 2060s and consequently the fourth phase can last until the 2090s (for details see Table 12.3 in Chap. 2).

**Table 12.6** Production principle phases, their dates, lengths and dynamics of technological growth rate (for the empirically observed and forecast data points)

| Production principle phases | Date of the phase's start | Phase length (years) | Macrotechnological growth rate (frequency of phase transitions per year) |
|---|---|---|---|
| Hunter-gatherer 1 | 40000 BP | 10,000 | $1.0 \times 10^{-4}$ |
| Hunter-gatherer 2 | 30000 BP | 8,000 | $1.3 \times 10^{-4}$ |
| Hunter-gatherer 3 | 22000 BP | 5,000 | $2.0 \times 10^{-4}$ |
| Hunter-gatherer 4 | 17000 BP | 3,000 | $3.3 \times 10^{-4}$ |
| Hunter-gatherer 5 | 14000 BP | 2,500 | $4.0 \times 10^{-4}$ |
| Hunter-gatherer 6 | 11500 BP | 1,500 | $6.7 \times 10^{-4}$ |
| Craft-agrarian 1 | 10000 BP | 2,700 | $3.7 \times 10^{-4}$ |
| Craft-agrarian 2 | 5300 BCE | 2,300 | $4.3 \times 10^{-4}$ |
| Craft-agrarian 3 | 3000 BCE | 1,500 | $6.7 \times 10^{-4}$ |
| Craft-agrarian 4 | 1500 BCE | 1,300 | $7.7 \times 10^{-4}$ |
| Craft-agrarian 5 | 200 BCE | 1,000 | $1.0 \times 10^{-3}$ |
| Craft-agrarian 6 | 800 CE | 630 | $1.6 \times 10^{-3}$ |
| Trade-industrial 1 | 1430 | 170 | $5.9 \times 10^{-3}$ |
| Trade-industrial 2 | 1600 | 130 | $7.7 \times 10^{-3}$ |
| Trade-industrial 3 | 1730 | 100 | $1.0 \times 10^{-2}$ |
| Trade-industrial 4 | 1830 | 60 | $1.7 \times 10^{-2}$ |
| Trade-industrial 5 | 1890 | 39 | $2.6 \times 10^{-2}$ |
| Trade-industrial 6 | 1929 | 26 | $3.8 \times 10^{-2}$ |
| Scientific-cybernetic 1[7] | 1955 | 40 | $2.5 \times 10^{-2}$ |
| Scientific-cybernetic 2 | 1995 | 35 | $2.9 \times 10^{-2}$ |
| Scientific-cybernetic 3 | 2030 | 25 | $4.0 \times 10^{-2}$ |
| Scientific-cybernetic 4 | 2055 | 15 | $6.7 \times 10^{-2}$ |
| Scientific-cybernetic 5 | 2070 | 10 | $1.0 \times 10^{-1}$ |
| Scientific-cybernetic 6 | 2080 | | |

[7] Let us recall again that in this chapter we take into account only one version of the Scientific-Cybernetic Production Principal chronology forecast (for the second version, see Table 12.3 in Chap. 2). Accordingly, the calculation of the singularity for the second version of the singularity would be different.

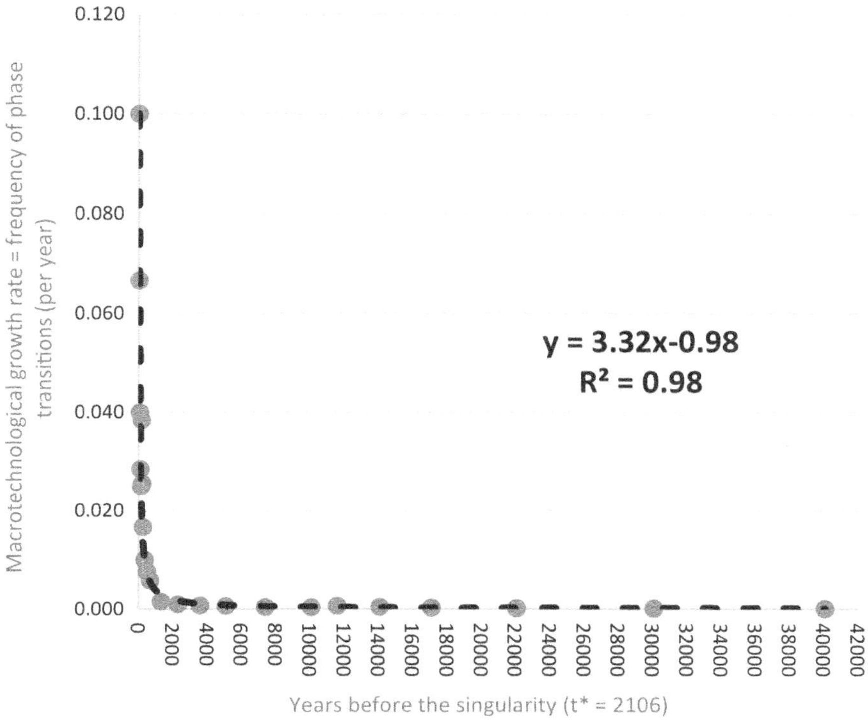


**Fig. 12.6** Scatterplot of the phase transition points (both empirically estimated and forecast) described in Table 12.6 with the power-law regression line (fitted with the least-squares method) identifying the singularity date as 2106 CE (natural scales)

Finally, as our least-squares analysis of the phase transition points described in Table 12.6 has identified the singularity date as 2106 CE, Eq. 12.15 can be further re-written:

$$V_t = \frac{3.32}{2106 - t}. \tag{12.17}$$

Thus, if our forecast based on the production principle theory is justified, there are grounds to expect that the global macrotechnological growth rate will return for some time in the forthcoming decades to a hyperbolic trajectory—this time with the singularity parameter equal to 2106. This implies that in the late twenty-first and the early twenty-second centuries the rate of global macrotechnological growth will experience one more decline, and there are some grounds (to be discussed in the next chapter) to expect that this decline will be much more pronounced than that of recent decades.

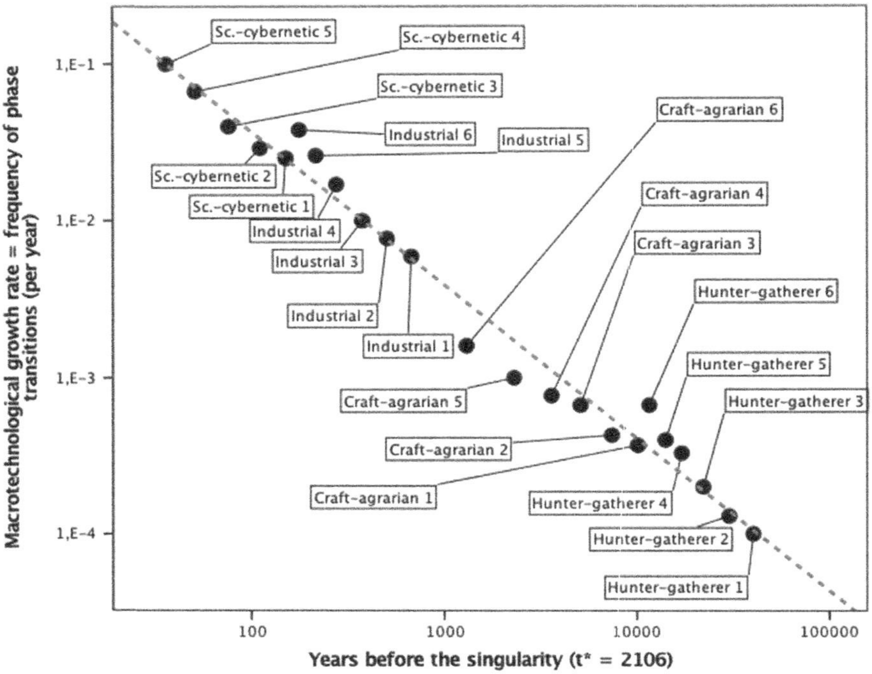


**Fig. 12.7** Scatterplot of the phase transition points (both empirically estimated and forecast) described in Table 12.6 with the power-law regression line (fitted with the least-squares method) identifying the singularity date as 2106 CE (double logarithmic scale)

## 12.6 Conclusion

In this chapter we consider the long-term dynamics of technological progress, presenting one option for measuring its speed throughout the entire historical process based on the theory of production revolutions and the theory of production principles. Thus, our research covers a very large span of 40,000 years between the Upper Paleolithic and the beginning of the twenty-second century.

We find that the general dynamics of accelerating technological growth over the last 40,000 years can be described with amazing accuracy ($R^2 = 0.99$) by a simple hyperbolic equation: $y_t = C/(t_0—t)$, where $y_t$ is the technological growth rate, measured as a number of technological phase transitions per unit of time, while $t_0$ and $C$ are constants; $t_0$ can be interpreted as a 'technological singularity' point. This approach allows us to measure the speed of technological progress as well as to make some predictions.

In this chapter, we have set and, we hope, have satisfactorily solved the following tasks:

(1) to offer forecasts about how technological progress will develop in the twenty-first century, with a slowdown or acceleration;

(2) to show how the change in the pace of technological progress is associated with global aging and the development of this process.

Concerning the first objective we would like to mention again that so far there has been a tendency to an increase in technological growth on the macroscale, albeit with fluctuations. However, as we have seen, future technological growth will not be infinite. In this chapter, we describe a scenario of how, when and why this rate will change. We show that the rate of technological progress is non-linear and will change dramatically. First, the slowdown in technological growth in recent decades, beginning in the 1970s, can be explained as a general fluctuation in technological growth. We view this slowdown as a necessary precondition for further rapid growth. Second, our analysis shows that there are a number of reasons to expect that the global technological growth rate will return to a hyperbolic trajectory for some time in the forthcoming decades, corresponding to the beginning of the final phase of the Cybernetic Revolution. Perhaps, the great acceleration in the development of new Large Linguistic Models (or new forms of AI) which many analysts of AI (e.g., IBM, 2023) have been talking about after the emergence of ChatGTP would be a signal of such a turn from a slowdown in the technological progress to its acceleration. Third, this acceleration will continue up to the late twenty-first century. Fourth, technological growth at the end of the twenty-first century will gradually slow down to the singularity point, in the year 2106.

Concerning the second objective, it is important to summarize that just the factor of aging will play a leading role. The novelty of our research lies in our attempt to present a short framework of qualitative model in which the rate of change in the population age structure correlates with the development of future technologies. We obtain a nontrivial result, according to which the process of global aging can accelerate and change the direction of technological progress in the coming decades, and then the aging of society can slow down the scientific-technological progress in the late twenty-first and early twenty-second centuries. We will continue these studies in the next chapter.

To conclude this chapter, it is worth pointing out that, after the singularity point, the rate of technological progress will slow down compared to that of the previous epoch. Definitely, the rate of technological growth will experience a slowdown. However, it is difficult to predict if there will be any subsequent acceleration in the twenty-second century (at least, in its first half). Probably the most important point is that the pattern of scientific and technological development itself will change dramatically. Thus, such a slowdown will mean that the old type of technological progress will begin to transform into a new one, and, most likely, a transition to new forms of social relations will begin.

**Acknowledgment** This chapter has been prepared with support by the Russian Science Foundation (Project No. **23-11-00160**).

## References


Ayres, R. U. (2006). The singularity is near: When humans transcend biology by Ray Kurzweil (book review). *Technological Forecasting and Social Change, 73*(2), 95–127.

Barsukov, V. N. (2019). From the demographic dividend to population ageing: World trends in the system-wide transition. *Economic and Social Changes: Facts, Trends, Forecast, 12*(4), 167–182. https://doi.org/10.15838/esc.2019.4.64.11

Benson, I., & Lloyd, J. (1983). *New technology and industrial change: The impact of the scientific-technical revolution on labour and industry*. Kogan Pages.

Bernal, J. D. (1965). *Science in history*. (3rd ed.). Hawthorn Books.

Bloom, D. E., & Luca, D. L. (2016). *The global demography of ageing: Facts, explanation, future*. Institute for the Study of Labor.

Boserup, E. (1965). *The conditions of agricultural growth: The economics of agrarian change under population pressure*. Aldine.

Braudel, F. (1973). *Capitalism and material life, 1400–1800*. Harper and Row.

Bunch, B. H., & Hellemans, A. (2004). *The history of science and technology: A browser's guide to the great discoveries, inventions, and the people who made them, from the dawn of time to today*. Houghton Mifflin Harcourt.

Callaghan, V., Miller, J., Yampolskiy, R., & Armstrong, S. (2017). *Technological singularity*. Springer. https://doi.org/10.1007/978-3-662-54033-6

Carree, M. A. (2003). Technological progress, structural change and productivity growth: A comment. *Structural Change and Economic Dynamics, 14*(1), 109–115. https://doi.org/10.1016/S0954-349X(02)00035-8

Carus-Wilson, E. (1987). The Woolen Industry. In: Postan, M. M., Miller E., (eds.), The Cambridge Econimic History of Europe. Vol. 2. Trade and Industry ibn the Middle Ages (pp. 614–692). Cambridge University Press.

Childe, G. (1952). *New light on the most ancient East*. (4th ed.). Routledge & Paul.

Cipolla, C. M. (Ed.). (1975). *The Fontana economic history of Europe*. Collins/ Fontana Books.

Coleman, D., & Rowthorn, R. (2015). Population decline—Making the best of inevitable destiny? In L. E. Grinin & A. V. Korotaaayev (Eds.), *History & mathematics: Political demography and global aging* (pp. 26–41). Uchitel.

Cowen, T., & Southwood, B. (2019). *Is the rate of scientific progress slowing down?* George Mason University.

DeCanio, S. J. (2016). Robots and Humans -complements or substitutes? *Journal of Macroeconomics, 49*, 280–291.

de Grey, A. B, & Rae, M. (2008). *Ending aging: The rejuvenation breakthroughs that could reverse human aging in our lifetime*. St. Martin's Press.

Diakonoff, I. M. (1999). *The paths of history*. Cambridge University Press.

Eden, A. H., Moor, J. H., Soraker, J. H., & Steinhart, E. (Eds.). (2012). *Singularity hypotheses: A scientific and philosophical assessment*. Springer-Verlag. https://doi.org/10.1007/978-3-642-32560-1

Fainberg, L. A. (1986). Rannepervobytnaya obshchina okhotnikov, sobirateley, rybolovov. In Y. V. Bromley (Ed.), *Istoriya pervobytnogo obshchestva. Epokha pervobytnoy rodovoy obshchiny* (pp. 130–235). Nauka.

Fichtner, J. J. (2018). Global ageing and public finance. *Business Economics, 53*(2), 72–78.

Frey, C. B., & Osborne, M. A. (2017). The future of employment: How susceptible are jobs to computerisation? *Technological Forecasting and Social Change, 114*, 254–280.

Fukuyama, F. (2002). Our post-human future: Consequences of the bio-technology revolution. Farrar, Straus, and Giroux, New York.

Galor, O., & Tsiddon, D. (1997). Technological progress, mobility and economic growth. *The American Economic Review, 87*(3), 363–382.

Galor, O., & Weil, D. N. (2000). Population, technology, and growth: From Malthusian stagnation to the demographic transition and beyond. *American Economic Review, 90*(4), 806–828.

Goldstone, J. A. (2015). Population aging and global economic growth. In J. Goldstone, L. E. Grinin, & A. Korotayev (Eds.), *History and mathematics: Political demography and global aging* (pp. 147–155). Uchitel.

Goldstone, J. A., Grinin, L. E., & Korotayev, A. V. (2015). Introduction, research into global aging and its consequences. In J. A. Goldstone, L. E. Grinin, & A. V. Korotayev (Eds.), *History & mathematics: Political demography and global aging* (pp. 5–9). Uchitel Publishing House.

Goring-Morris, A. N., Hovers, E., & Belfer-Cohen, A. (2009). The dynamics of Pleistocene and early Holocene settlement patterns and human adaptations in the Levant—An overview. In J. J. Shea & D. E. Lieberman (Eds.), *Transitions in prehistory: Essays in honor of Ofer Bar-Yosef* (pp. 185–252). Oxbow Books for the American School of Prehistoric Research.

Grinin, L. (2006). *Proizvoditelnye sily i istoricheskiy protsess.* 3rd edition. KomKniga.

Grinin, L. E., & Korotayev, A. V. (2010). Will the global crisis lead to global transformations? 1, The global financial system: Pros and Cons. *Journal of Globalization Studies, 1*(1), 70–89.

Grinin, L. E., & Korotayev, A. V. (2015). Global population aging, the sixth Kondratieff wave, and the global financial system. In J. A. Goldstone, L. E. Grinin, & A. V. Korotayev (Eds.), *History and mathematics: Political demography and global aging* (pp. 81–106). Uchitel Publishing House.

Grinin, L. E., & Korotayev, A. V. (2015). Global'noe starenie naseleniya, shestoj tekhnologicheskij uklad i mirovaya finansovaya Sistema. In L. E. Grinin, A. V. Korotayev, & V. M. Bondarenko (Eds.), *Kondrat'evskie volny: nasledie i sovremennost'* (pp. 107–132). Uchitel Publishing House.

Grinin, L., & Korotayev, A. (2016). Global population aging, the sixth Kondratieff wave, and the global financial system. *Journal of Globalization Studies, 7*(2), 11–31.

Grinin, L. E., Grinin, A. L., & Korotayev, A. (2017). Forthcoming Kondratieff wave, cybernetic revolution, and global aging. *Technological Forecasting & Social Change, 115*, 52–68.

Grinin, L., Grinin, A., & Korotayev, A. (2020a). A quantitative analysis of worldwide long-term technology growth: From 40,000 BCE to the early 22nd century. *Technological Forecasting and Social Change, 155*(June), 119955. https://doi.org/10.1016/j.techfore.2020.119955

Grinin, L., Grinin, A., & Korotayev, A. (2021). Global trends and forecasts of the 21st century. *World Futures, 77*(5), 335–370. https://doi.org/10.1080/02604027.2021.1949939

Grinin, L. E., & Grinin, A. L. (2015a). The cybernetic revolution and historical process. *Social Evolution and History, 14*(1), 125–184.

Grinin, L. E., & Grinin, A. L. (2015b). Global technological perspectives in the light of cybernetic revolution and theory of long cycles. *Journal of Globalization Studies, 6*(2), 119–142.

Grinin, L. E., & Grinin, A. L. (2015c). *Ot rubil do nanorobotov, Mir na puti k epokhe samoupravlyaemyh sistem (istoriya tekhnologij i opisanie ih budushchego.* Moscow Branch of the Uchitel Publishing House, Moscow.

Grinin, L. E., & Grinin, A. L. (2016). *The cybernetic revolution and the forthcoming epoch of self-regulating systems.* Moscow Branch of the Uchitel Publishing House.

Grinin, L., & Grinin, A. (2020). The cybernetic revolution and the future of technologies. In A. V. Korotayev, & D. LePoire (Eds.), *The 21st century Singularity and global futures. A big history perspective* (pp. 377–396). Springer, Cham. https://doi.org/10.1007/978-3-030-33730-8_17

Grinin, L., & Korotayev, A. (2015a). *Great divergence and great convergence. A global perspective.* Springer.

Grinin, L. E., & Korotayev, A. V. (2016b). Introduction, global evolution and global aging. In L. E. Grinin, & A. V. Korotayev (Eds.), *Evolution and big history: Dimensions, trends, and forecasts* (pp. 5–17). 'Uchitel' Publishing House.

Grinin, L., Markov, A., & Korotayev, A. (2013). On similarities between biological and social evolutionary mechanisms: mathematical modeling. *Cliodynamics, 4*(2), 185–228. https://doi.org/10.21237/C7clio4221334

Grinin, L. E., Korotayev, A. V., & Tausch, A. (2016). *Economic cycles, crises, and the global periphery.* Springer, Cham. https://doi.org/10.1007/978-3-319-41262-7

Grinin, L. E., Korotayev, A., & Markov, A. (2020b). Big history and macro-evolution. Evolutionary principles and mechanisms at biological and social phases of the big history. In C. Benjamin,

E. Quaedackers, & D. Baker (Eds.), *The Routledge companion to big history* (pp. 72–105). Routledge.

Haas, M. L. (2015). Population aging and the future of the great powers. In J. A. Goldstone, L. E. Grinin, & A. V. Korotayev (Eds.), *History & mathematics: Political demography & global aging* (pp. 133–146) 'Uchitel' Publishing House.

Hall, F. (1986). Proiskhozhdeniye i razvitiye zemledeliya. In V. M. Masson (Ed.), *Drevniye tsivilizatsii Vostoka* (pp. 201–204). Tashkent FAN.

Harlan, D. R. (1986). Resursnaya baza osnovnykh rastitel′nykh kul′tur Iranskogo plato i sosednikh regionov. In V. M. Masson (Ed.), *Drevniye tsivilizatsii Vostoka* (pp. 199–201). Taskent FAN.

Harper, S. (2006). Addressing the implications of global aging. *Journal of Population Research, 23*(2), 205–223.

Hsu, Y. H., & Lo, H. C. (2019). The impacts of population ageing on saving, capital formation, and economic growth. *American Journal of Industrial and Business Management, 9*(12), 2231–2246. https://doi.org/10.4236/ajibm.2019.912148

Huebner, J. (2005). A possible declining trend for worldwide innovation. *Technological Forecasting and Social Change, 72*, 980–986.

IBM. (2023). *Augmented work for an automated, AI-driven world. AI won't replace people—but people who use AI will replace people who don't.* https://www.ibm.com/thought-leadership/institute-business-value/en-us/report/augmented-workforce

Jackson, R., Howe, N., Strauss, R., & Nakasima, K. (2008). *The greying of the great powers. Demography and geopolitics in the 21st century*. Center for Strategic & International Studies.

Jackson, R., Macaranas, R., & Peter, T. (2013). *U.S. Development Policy in an Ageing World. New Challeges and New Priorities for a New Demographic Era. A Research Report from Project on U.S. Leadership*. Center for Strategic and International Studies.

Jackson, R. (2021). *The shape of things to come. The macro challenges of population ageing*. Concord Coalition & the Global Ageing Institute.

Kapelyushnikov, R. (2019a). Fenomen stareniya naseleniya: ekonomicheskiye effekty. *Ekonomicheskaya politika, 14*(2), 8–63.

Kapelyushnikov, R. (2019b). *Fenomen stareniya naseleniya: ekonomicheskiye effekty*. Institut ekonomiki RAN.

Kayal, A. (1999). Measuring the pace of technological progress: Implications for technological forecasting. *Technological Forecasting and Social Change, 60*(3), 237–245.

Khaltourina, D., Korotayev, A., & Malkov, A. (2006). A compact macromodel of the world system demographic and economic growth, 1–1973 CE. In R. Trappl (Ed.), *Cybernetics and systems* (Vol. 1, pp. 330–335). Austrian Society for Cybernetic Research.

Korotayev, A. (2005). A compact macromodel of world system evolution. *Journal of World-Systems Research, 11*(1), 79–93. https://doi.org/10.5195/jwsr.2005.401

Korotayev, A. (2007). Compact mathematical models of world system development, and how they can help us to clarify our understanding of globalization processes. In G. Modelski, T. Devezas, & W. R. Thompson (Eds.), *Globalization as evolutionary process: Modeling global change* (pp. 133–160). Routledge.

Korotayev, A. (2013). Globalization and mathematical modeling of global evolution. In L. E. Grinin & A. V. Korotayev (Eds.), *Evolution: Development within big history, evolutionary and world-system paradigms* (pp. 69–83). Uchitel.

Korotayev, A. V., & Bozhevolnov, J. V. (2010). Nekotorye obschie tendentsii ekonomicheskogo razvitiya Mir-Sistemy. In A. A. Akaev, A. V. Korotayev, & G. G. Malinetsky (Eds.), *Prognoz i modelirovanie krizisov i mirovoy dinamiki* (pp. 161–172). LKI/URSS.

Korotayev, A., Malkov, A., & Khaltourina, D. (2006). *Introduction to social macrodynamics: Compact macromodels of the world system growth.* KomKniga/URSS.

Korotayev, A. V., & Markov, A. V. (2015). Mathematical modeling of biological and social phases of big history. In L. E. Grinin, I. V. Ilyin, P. Herrmann, & A. V. Korotayev (Eds.), *Globalistics and globalization studies: Big history & global history* (pp. 319–343). 'Uchitel' Publishing House.

Korotayev, A. V., & Shulgin, S. G. (2018). *Vliyaniye stareniya naseleniya na global'nuyu sistemu tsennostey i politicheskuyu dinamiku*. RANEPA. https://doi.org/10.2139/ssrn.3139641

Korotayev, A. (2012). Globalization and mathematical modeling of global development. In L. E. Grinin, I. V. Ilyin, & A. V. Korotayev (Eds.), *Globalistics and globalization studies* (pp. 148–158). 'Uchitel' Publishing House.

Korotayev, A. (2018). The 21st century singularity and its big history implications: A re-analysis. *Journal of Big History, 2*(3), 73–119. https://doi.org/10.22339/jbh.v2i3.2320

Korotayev, A. V. (2020a). How singular is the 21st century singularity? In A. V. Korotayev, & D. LePoire (Eds.), *The 21st century singularity and global futures. A big history perspective* (pp. 571–595). Springer, Cham. https://doi.org/10.1007/978-3-030-33730-8_26

Korotayev, A. V. (2020b). The 21st century singularity in the big history perspective. A re-analysis. In A. V. Korotayev, & D. LePoire (Eds.) *The 21st century singularity and global futures. A big history perspective* (pp. 19–75). Springer, Cham. https://doi.org/10.1007/978-3-030-33730-8_2

Kremer, M. (1993). Population growth and technological change: One million B.C. to 1990. *The Quarterly Journal of Economics, 108*(3), 681–716. https://doi.org/10.2307/2118405

Kurzweil, R. (2005). *The singularity is near: When humans transcend biology*. Viking Penguin.

Kurzweil, R. (2001). The law of accelerating returns. *KurzweilAI.net* 3-7-2001. http://www.kurzweilai.net/articles/art0134.html?printable=1

Lee, R., & Mason, A. (2011). *Population aging and the generational economy: A global perspective*. Edward Elgar.

LePoire, D., & Korotayev, A. V. (2020). Conclusion. In A. V. Korotayev, & D. LePoire (Eds.), *The 21st century singularity and global futures. A big history perspective* (pp. 599–620). Springer, Cham. https://doi.org/10.1007/978-3-030-33730-8_27

LePoire, D. J. (2005). Application of logistic analysis to the history of physics. *Technological Forecasting and Social Change, 72*(4), 471–479. https://doi.org/10.1016/S0040-1625(03)00044-1

LePoire, D. J. (2009). *Exploration of connections between energy use and leadership transitions, systemic transitions*. Palgrave Macmillan. https://doi.org/10.1057/9780230618381_10

LePoire, D. (2013). Potential economic and energy indicators of inflection in complexity. In L. E. Grinin & A. V. Korotayev (Eds.), *Evolution: Development within big history, evolutionary and world-system paradigms* (pp. 108–118). 'Uchitel' Publishing House.

LePoire, D. J. (2020). Exploring the singularity concept within big history. In A. V. Korotayev, & D. LePoire (Eds.), *The 21st century singularity and global futures. A big history perspective* (pp. 77–97). Springer, Cham. https://doi.org/10.1007/978-3-030-33730-8_3

Maddison, A. (2007). *Contours of the world economy, 1–2030*. Oxford University Press.

Maestas, N., Mullen, K. J., & Powell, D. (2016). *The effect of population ageing on economic growth, the labor force and productivity*. NBER.

Magee, C. L., & Devezas, T. C. (2011). How many singularities are near and how will they disrupt human history? *Technological Forecasting and Social Change, 78*(8), 1365–1378. Retrieved January 15, 2019, from https://www.sciencedirect.com/science/article/pii/S0040162511001661

Modis, T. (2002). Forecasting the growth of complexity and change. *Technological Forecasting and Social Change, 69*(4), 377–404. https://doi.org/10.1016/S0040-1625(01)00172-X

Modis, T. (2003). The limits of complexity and change. *The Futurist, 37*(3), 26–32.

Modis, T. (2006). The singularity myth. *Technological Forecasting & Social Change, 73*(2).

Modis, T. (2020). Forecasting the growth of complexity and change—an update. In A. V. Korotayev, & D. LePoire (Eds.), *The 21st century singularity and global futures. A big history perspective* (pp 101–104). Springer, Cham. https://doi.org/10.1007/978-3-030-33730-8_4

Moiseev, N. N. (2000). *Sudba tsivilizatsii. Put' razuma*. Yazyki russkoi literatury.

Muehlhauser, L., & Salamon, A. (2012). Intelligence explosion: Evidence and import. *Singularity hypotheses: A scientific and philosophical assessment*. Springer. https://link.springer.com/chapter/10.1007/978-3-642-32560-1_2

Nazaretyan, A. P. (2015). Megahistory and its mysterious singularity. *Herald of the Russian Academy of Sciences, 85*(4), 352–361. https://doi.org/10.1134/S1019331615040061

Nazaretyan, A. P. (2017). Mega-history and the twenty-first century singularity puzzle. *Social Evolution & History, 16*(1), 31–52.

Nazaretyan, A. P. (2018). The polyfurcation century: Does the evolution on Earth have a cosmological relevance? *Journal of Big History, 2*(1), 27–41. https://doi.org/10.22339/jbh.v2i1.2253

Nazaretyan, A. (2020). The 21st century's "mysterious singularity" in the light of the big history. In A. V. Korotayev, D. LePoire (Eds.) *The 21st century singularity and global futures. A big history perspective* (pp. 345–362). Springer, Cham. https://doi.org/10.1007/978-3-030-33730-8_15

Panov, A., LePoire, D. J., & Korotayev, A. (2020). The 21st century singularity in the big history perspective. An overview. In Korotayev A, & LePoire D (Eds.), *The 21st century Singularity and global futures. A Big History perspective* (pp. 1–18). Springer, https://doi.org/10.1007/978-3-030-33730-8_1

Panov, A. D. (2005). Scaling law of the biological evolution and the hypothesis of the self-consistent galaxy origin of life. *Advances in Space Research, 36*(2), 220–225. https://doi.org/10.1016/j.asr.200503.0.01

Panov, A. D. (2017). Singularity of evolution and post-singular development. In B. Rodrigue, L. Grinin, A. Korotayev (Eds.), *From big bang to galactic civilizations. A big history anthology. Volume III. The ways that big history works: Cosmos, life, society and our future* (pp. 370–402). Primus Books.

Panov, A. D. (2020). Singularity of evolution and post-singular development in the big history perspective. In A. V. Korotayev, D. LePoire (Eds.), *The 21st century singularity and global futures. A big history perspective* (pp. 439–465). Springer. https://doi.org/10.1007/978-3-030-33730-8_20

Park, D., & Shin, K. (2015). Impact of population aging on Asia's future growth. In J. A. S. Goldstone, L. E. Grinin, & A. V. Korotayev (Eds.), *History & mathematics: Political demography and global aging* (pp. 107–132). Uchitel.

Park, M., Leahey, E., & Funk, R. J. (2023). Papers and patents are becoming less disruptive over time. *Nature, 613*(7942), 138–144.

Phillips, F. (2011). The state of technological and social change: impressions. *Technological Forecasting and Social Change, 78*(6), 1072–1078. https://doi.org/10.1016/j.techfore.2011.03.020

Popova, S., & Yanik, A. (2014). Problemy global'nogo stareniya naseleniya: Analiz dokumentov i strategii OON. *Mezhdunarodnoye pravo i mezhdunarodnyye organizatsii, 3*, 429–443.

Powell, J. L., & Khan, H. T. (2013). Aging and globalization: A global analysis. *Journal of Globalization Studies, 4*(1), 137–146.

Prettner, K. (2013). Population aging and endogenous economic growth. *Journal of Population Economics, 26*(2), 811–834.

Shanahan, M. (2015). *The technological singularity*. MIT Press.

Shea, J. (2013). *Stone tools in the Paleolithic and Neolithic Near East: A guide*. Stony Brook; NY: Stony Brook University.

Teulings, C., & Baldwin, R. (Eds.). (2014). *Secular stagnation: Facts, causes, and cures*. CEPR.

Tsirel, S. V. (2008). Zametki ob istoricheskom vremeni i putyah istoricheskoj evolyucii. In L. E Grinin, A. V. Korotayev, & S. Y. Malkov (Eds.), *Istoriya i matematika: Modeli i teorii* (pp. 246–278). LKI/URSS.

Tsirel, S. V. (2014). O tom, chto budet posle tochki singulyarnosti. In L. E. Grinin, A. V. Korotayev, & A. V. Markov (Eds.), *Evolyutsiya: ot protozvezd k singulyarnosti*? (pp. 340–381). Volgograd: Uchitel.

Ulam, S. (1958). Tribute to John von Neumann. *Bulletin of the American mathematical society, 64*(3), 1–49.

UNCTAD. (2012). *Doklad o mirovykh investitsiyakh, 2012 god. Konferentsiya Organizatsii ob"yedinennykh natsiy po torgovle i razvitiyu*. https://unctad.org/system/files/official-document/wir2012overview_ru.pdf

WHO—World Health Organization. (2017). *Global strategy and action plan on ageing and health*. World Health Organization.

WHO—World Health Organization. (2021). *Ageing and health*. World Health Organization. https://www.who.int/news-room/fact-sheets/detail/ageing-and-health

Zimmer, Z. (2016). *Global aging in the twenty-first century: Challenges, opportunities and implications*. Routledge.

# Will Global Aging Change the Pace of Technological Progress and Create a New Consumption Model?

13

**Abstract**

This chapter discusses the mode and role of global aging in the future technological growth and model of consumption. Grinin et al. develop the idea that global aging has a dual and contradictory effect on technological progress: positive and negative. On the one hand, the need to improve the biological quality of life of the elderly, the struggle to increase life expectancy or against incurable diseases etc. spur science and technology to search for new breakthrough solutions. On the other hand, older people tend to be more conservative than the younger and middle-aged groups, leading to a slowdown in the rate of technological progress. In this chapter, Grinin et al. analyze these two trends and show how they and their relationship will affect the rate of technological progress in the twenty-first century. The other goal of this chapter is to show that one can assume that the current consumption pattern may also change under the influence of the global aging process. Thus, Grinin et al. suppose that global aging can change people's needs that will lead to the transition to a new economic model, one that is not based on an endless increase of consumption.

## 13.1 What This Chapter is About

In this chapter, we discuss how the pace of technological progress and the economic model of society will change under the influence of the development of the global aging process. Thus, the chapter continues the discussion of the issues raised in the previous chapter.

We believe that global aging is one of the most important factors affecting society in the coming decades. This chapter continues the discussion on the mode and role of global aging and will be connected with the research questions which we try to answer in the present book, that is:



L. Grinin et al., *Cybernetic Revolution and Global Aging*, World-Systems Evolution and Global Futures, https://doi.org/10.1007/978-3-031-56764-3_13

- What could be the driver of an acceleration of technological growth in the forthcoming decades?
- Is a slowdown of technological growth likely to occur?
- What could cause such a slowdown?
- How could such a slowdown affect socio-economic and socio-political relations?
- What is the relationship between global aging and technological progress?
- Why is global aging likely to be one of the most important factors affecting technological growth in the near future?

We continue to develop the idea raised in the previous chapters that global aging has a dual and contradictory effect on technological progress: positive and negative. On the one hand, the need to improve the biological quality of life of the elderly, the struggle to increase life expectancy, the fight against incurable diseases, the need to extend the working age of the elderly, etc. spur science and technology to search for new breakthrough solutions.

On the other hand, older people tend to be more conservative than the younger and middle-aged groups, leading to a slowdown in the rate of technological progress. In this chapter, we analyze these two trends and show how they and their relationship will affect the rate of technological progress in the twenty-first century. So, we obtain a non-trivial result, that the process of global aging can accelerate and change the direction of technological progress in the coming decades, and then the aging of society can slow down scientific-technological progress in the late twenty-first and early twenty-second centuries.

However, slowing down does not mean stopping. We argue that technological and economic development will not come to an end. It will be even unreasonable to assume that technological progress will slow down in a major way. Otherwise, economic growth will critically slow down or stop, and then society may plunge into a crisis and slide into a rough distribution system.

The other goal of this chapter is to show that one can assume that the current consumption pattern may also change under the influence of the global aging process. This change, in turn, will have a serious impact on the entire production structure and on scientific-technological progress (as we will discuss later). The conservatism of the elderly may lead not only to a slowdown in the pace of development but also a transition to a different economic system.

The current model is associated with an increasing consumption. Consuming more today than yesterday and more tomorrow than today is a model that leads nowhere and is actually absurd. Also, sometimes the desire for sustainable GDP growth seems absurd. However, it works and will continue to work for decades, which is especially important for poor countries where large parts of the population are dissatisfied with their level of consumption, and many even lack access to the supply of basics.

So, it is a hard task to optimize consumption. However, we suppose that global aging can change people's needs, especially under the condition of the stabilization or reduction of the population. As a result, under the influence of the various future changes, the focus on measuring economic growth by reference to the GDP growth alone must be replaced.

Overall, the Cybernetic Revolution and global aging should ultimately lead to the transition to a new economic model, one that is not based on an endless increase of consumption. If the growth model of the economy differs from today's paradigm, parameters of quality and longevity of life may play an impotnant role. Accordingly, business models may change, although it is not very clear how this will happen.

## 13.2 Global Aging and Technological Progress: Introduction

In this monograph, the authors discuss the role of global aging as well as of societal aging in particular societies as one of the main factors of both the technological acceleration and the subsequent deceleration. The coming decades will be a period of significant increase in life expectancy and, as a consequence (against the background of low fertility), a period of rapid global aging that will also involve regions that are still 'young', including sub-Saharan Africa and South Asia (see Chaps. 6 and 7; see also Grinin & Grinin, 2015c, 2016; Grinin & Korotayev, 2015a; Grinin, Korotayev & Tausch, 2016; Grinin et al., 2017a, 2017b). Aging and life expectancy are tightly connected with medical innovations. As we have already mentioned in the previous chapters (Chaps. 3, 4 and 8), the final phase of the Cybernetic Revolution is likely to begin around the 2030s. The final phase will become the epoch of 'self-regulating and self-managing systems'; it may start in medicine and will be connected with its innovative branches. The drivers of the final phase of the Cybernetic Revolution will be medical technologies, additive manufacturing (3D printing), nano- and biotechnologies, robotics, IT, as well as cognitive technologies (connected with AI), which will combine to form a sophisticated system of self-regulating production. Among these, medical technologies will become the main integrating center (Chaps. 3, 4 and 8; see also Grinin & Grinin, 2015c, 2016, 2020; Grinin & Korotayev, 2016a, 2016b; Grinin, Korotayev & Tausch, 2016; Grinin, Grinin & Korotayev, 2017a). As a result, we may witness serious modifications of human organisms and, perhaps, changes in their biological nature (for more details, see Chaps. 3, 4, 8 and 11; see also Grinin & Grinin, 2015c, 2016; Grinin, Korotayev & Tausch, 2016; Grinin et al., 2017a).

Thus, generally speaking, we have to recognize that the development of aging, including 'healthy aging', growth of biological life, quality and life expectancy are interconnected with technological progress and its pace. This chapter is dedicated to this subject. It continues the discussion on the mode and role of global aging and will be connected with the research questions that we try to answer in our book, that is: What could be the driver of an acceleration of technological growth in the coming decades? Is a slowdown in technological growth likely to occur? What could cause such a slowdown? How could such slowdown affect socio-economic and socio-political relations? What is the relationship between global aging and technological progress? Why is global aging likely to be one of the most important factors affecting technological growth in the near future?

These are important questions but their importance becomes even higher because the impact of global aging on the зace and direction of scientific-technological progress is understudied (Bös & Weizsäcker, 1989; Galor & Weil, 2000; Grey & Rae, 2008; Peterson, 1999; Prettner, 2013; Tsirel, 2008; WHO 2017),[1] whereas global aging affects technological, economic, political, social and other spheres in various ways and its influence has serious macroeconomic consequences (Coleman & Rowthorn, 2015; Fukuyama, 2002; Goldstone et al., 2015; Goldstone, 2015; Haas, 2015; Harper, 2006; Jackson et al., 2008, 2013; Kapelyushnikov, 2019a, 2019b; Lee & Mason, 2011; Morgan & Kunkel, 2016; Park & Shin, 2015; Powell & Khan, 2013; Zimmer, 2016; Zimmer, 2016).

We have shown different consequences of global aging in different spheres (see Chaps. 6 and 7). The problem of the present chapter, however, is connected not only with the consequences of global aging but essentially with the identification of the mechanisms of complex interaction between societal aging and technological progress. The very consequences may appear in a new light if we understand the above-mentioned connections with their positive and negative feedbacks. So, the problem is very important.

One of the goals of the present chapter is to establish a correlation between the progress of societal aging, i.e., the rate of change in the population age structure, and the development of future technologies. We obtain a nontrivial result, according to which the process of global aging can accelerate and change the direction of technological progress in the coming decades, and then the aging of society can slow down scientific-technological progress in the late twenty-first and early twenty-second centuries.

The other goal is to show that one can assume that the current consumption pattern may also change under the influence of the global aging process. This change, in turn, will have a serious impact on the entire structure of production and on scientific-technological progress (as we will discuss below; about the rate of technological progress, see Chap. 12).

## 13.3 On Conservatism of the Elderly

**Is Conservatism a Negative or a Positive Quality?** Conservatism is often seen as immanent in aging societies (see in particular Jackson, et al., 2008: 188; Lambert-Pandraud & Laurent, 2010; Tsirel, 2008; Phua et al., 2020; see pros and cons Glenn, 1974; Peterson et al., 2020).[2] It is a sort of conventional or folk wisdom that certain dimensions of biological, social and psychological aging contribute to some

[1] See also Grinin and Grinin 2015c, 2017; Grinin and Grinin 2016; Grinin and Korotayev 2010, 2015b, c, 2016a, b; Grinin, Grinin and Korotayev 2017a, 2020.

[2] See also Grinin and Grinin 2017; Korotayev et al., 2017; Korotayev and Shulgin 2018; Shulgin et al., 2019; Korotayev et al., 2019, 2023a, 2023b.

kind of conservatism (Glenn, 1974; Peterson et al., 2020). In many respects, we agree with this statement, albeit some qualifications must be made (see below); in any case, there are many contradictions and empirical evidence is not as definitive as it was fifty years ago (Glenn, 1974). It is important that if conservatism is a peculiar quality of age, this quality will be of great influence. We note, by the way, that both the media and their ideological orientation in terms of condemning or approving conservatism can change significantly.

One of the reasons why empirical evidence is not definitive, in our opinion, lies in the fact that the concept of conservatism explicitly, and more often implicitly, has a negative connotation. That is, conservatism is considered as a non-positive quality. Therefore, in many studies, the authors use euphemisms to refer to conservatism, but call it differently in particular if attitudes are seen as age-related - thus changing the terminology has the aim to avoid offending this age group. For example, it is said that elderly consumers tend to be more cautious in making decisions (Phillip & Sternthal, 1977); mature consumers are more aware of the risk associated with purchases (Dychtwald & Gable, 1990); note such features of the elderly consumers as iresistance to innovation/technology (Bae et al., 2021; Gilly & Zeithaml, 1985; Hough & Kobylanski, 2009). Below we give more examples. In the present context, we do not attach any evaluative meaning to the concept of conservatism; we would like just to emphasize that the elderly are more committed to the prevailing stereotypes in their lifestyles and habits; respectively, they are less prone to change than the young and middle-aged groups; furthermore, they are also all the least prone to make drastic changes in their lives and to engage in any kind of risky actions.[3] Aging people may become more conservative in the sense that their attitudes and values become more resistant to change (Glenn, 1974: 176; Korotayev et al., 2023a, 2023b).

It should be borne in mind that, firstly, we are talking about a greater or lesser prevalence of relevant qualities in large population groups, which cannot be automatically transferred to specific people (their conservatism or desire for change at any age may depend on personal qualities and circumstances). Secondly, such adherence to habits and stereotypes in lifestyle can be considered differently in different situations, that is, as a positive, neutral or negative quality, and therefore there is not and cannot be a single assessment. Thirdly, what has been said about conservatism can be connected to a lesser extent with political preferences, and even more related to possible assessments of the situation.[4] Fourthly, it is important to keep in mind that increasing life expectancy changes the perception of the

[3] However, old people can and often do such actions when they feel that life is over and they can become a burden for relatives. Is not this the reason for such difficult decisions as transferring one's property to one's relatives during one's lifetime, or, even more so, deciding on euthanasia? (about the latter see Chap. 7 and below).

[4] However, as Peterson et al. (2020: 600) pointed out: we also find support for the folk wisdom: on those occasions when political attitudes do shift over the lifespan, liberals are more likely to become conservatives than conservatives are to become liberals, suggesting that folk wisdom has some empirical basis, even if it overstates the degree of change.

qualities of people of a certain age. There is apparently less conservatism among sixty-year-olds today than there was thirty years ago. And this is some guarantee that the conservatism of the elderly will not become an absolute momentum. Fifthly, we do not claim that older people are absolutely conservative, there may be situations and areas where this is not the case. For instance, in areas such as medicine and pharmaceuticals, older people may tend to innovate more often than younger people. This is an important point, we will elaborate on it in the next section.[5] There are studies whose authors claim that the change will be greater for cohorts in later stages of life (aged 60 and older) than for cohorts in earlier stages (Danigelis et al., 2007). However, this may be true only for some narrow areas.

**Differences Between the Young and the Old and Economic Growth**. However, on the whole, it would be wrong to deny that the physiology, the freshness of perception and energy, the capacity for imitation, the flexibility of memory and body are very different between younger and older people; and also, the psychology of older people is different from the psychology of young people. If they are healthy, the mental abilities of the elderly, as well as their ability to learn and adapt to change, may in some respects be as good as those of the young. However, objectively, the memory of young people is more active, stronger and more adaptive to information. All this affects the different effectiveness of teaching the young and the elderly.[6] In general, acquiring new skills is more difficult for older people than for young people (e.g., Zemlyakova and Pomuran 2014; for an example, about the difficulties experienced by elderly people in adapting to the Internet, see Neskromnykh and Mamadaliev 2017). However, it is often said that older people have advantages when it comes to grasping complex issues; also, if it is about the need for experience-based performance, the elderly are mostly in an advantageous position.

In sum, it is justified to say that the aging of the labor force objectively leads to a slowdown in economic growth. In addition, older people are generally less productive than young people—e.g., people aged 40–65 compared to workers aged 20–40 (Goldstone, 2015), whose productivity tends to increase rapidly with increasing experience and education (Lee & Mason, 2011),[7] not to mention people over 65. As we show in Chap. 7, in general, the aging of the labor force impedes both the introduction of innovations and the retraining of labor force, as well as opportunities for increasing labor productivity. It also reduces the mobility of labor force (Kapelyushnikov, 2019b: 23).[8] There is a negative impact of aging

[5] In Chap. 7, we have also found that the global aging will have both positive and negative influence on the development of human capital.

[6] A good example is foreign languages leraning. It is well-known that children and teenagers learn foreign languages more easily than elderly people.

[7] According to other research, labor productivity peaks between the ages of 35 and 54 (Park & Shin, 2015: 109).

[8] Although the impact of the increasing aging of the workforce can be positive in certain respects (due to their greater experience and matching with jobs [Kapelyushnikov, 2019b: 23, etc.]).

on economic growth rates caused by its multiple impact on society, in particular, due to a sharp reduction in the number of working hours (Goldstone, 2015; Kapelyushnikov, 2019b: 42; see also Grinin, Grinin & Korotayev, 2020).

As for consumerism, older people who have already acquired, experienced and seen a lot have largely lost their desire to pursue new things and become less active than the young. The current situation in the Japanese economy, where the proportion of the elderly is growing, and the proportion of the young is decreasing, confirms this fact. This demographic structure of the population cannot contribute to noticeable economic growth. The Japanese economy has been suffering from a weak period of GDP growth, with two and a half decades of deflation due to an aging population that does not want to spend much money and prefers to save instead. No wonder that mood in Japan is rather depressed (Coleman & Rowthorn, 2015: 31; Coulmas, 2007; Grinin & Korotayev, 2014, 2018; Ogawa et al., 2005).[9]

**Conservatism and Its Impact on Innovation and Consumption.** It is important to emphasize that in this chapter, as well as in the monograph as a whole, we do not set ourselves the task of analyzing the manifestations or non-manifestations of conservatism in the elderly in all aspects. Our study focuses on narrower aspects: the desire for technological innovation and consumption of new goods and the adaptability to them.

In this regard, it can be argued that

- an increase in the proportion of older people in the number of employees objectively leads to a slowdown in economic growth and its innovative development. As a result, the activity of a very large part of the employees, their creative potential, is likely to be reduced compared to today and, even more so compared to a period of 20–30 years earlier;
- from the point of view of adaptation to technological progress, in the pursuit of changes and the rate of acquisition of new skills, older people are much less adaptive than their younger counterparts;
- young people are more eager for change than older people because the latter generally have fewer needs and less desire for innovations than young people. If it were the other way around, society would have ceased its development long ago.
- The growth of conservatism in aging societies is also connected to the fact that young people are not dominated by their life experience, they see many things differently and are generally more open to change. Of course, a strong need also forces older people to change, but in one way or another, an increase in the cohorts of older people leads to a slowdown of developing and related phenomena.
- There is a slowdown consumption in an older society, a diminishing need for career advancement, well-being and success. All these factors that have been

[9] In 2022, deflationary processes in this country were replaced by inflationary ones under the influence of global processes.

and still are the most important engine of development, fade away and will continue to diminish. With a decrease in the number of children, investments in the younger generation and the need to provide for, another important factor in development, will begin to weaken.

We will continue with this analysis in the following sections, since, paradoxically, aging can have not only a negative but also a very positive effect on technological progress.

## 13.4 Global Aging as a Factor Influencing the Technological Growth Rate

### 13.4.1 Aging and Technological Progress: A Positive Feedback

In the previous chapters (Chaps. 4, 7, 8, 10 and 11), we have already shown why the global aging will be one of the most important factors in the coming decades (see Grinin et al., 2017a; see also Grinin & Grinin, 2015a, 2015b; Grinin & Korotayev, 2015b, 2015c, 2015d, 2015e, 2016a, 2016b; Grinin & Grinin, 2015c, 2015d, 2017).

In this subsection, we will look at how and why global aging in the coming decades could become one of the most important drivers of technological breakthroughs through the 2070 and 2080s. In the next subsection, we will discuss why global aging will later become an obstacle to technological progress.

As we expect (see Chap. 4), a new technological breakthrough will begin around the 2030s, starting in new branches of medicine and related fields: bio- and nano-technologies, additive and cognitive technologies and some others. It will also mark the beginning of the final phase of the Cybernetic Revolution. As we have pointed out, for such breakthroughs in the sphere of medicine to begin in the 2030s, the world will need the following prerequisites: an explosive growth in the proportion of the elderly people in the population; a growing need for labor force in the economy and an interest on the part of the state in increasing the working capacity of older people, as well as a growing number of well-to-do and educated people concerned about their health. Huge financial resources will also be accumulated for technological breakthroughs: pension funds, which will increase; government funding for health and social needs; increased spending on health by an aging population and a growing world middle class. *All these resources can attract large investment in various venture capital projects, and, in the long term, a very large demand for innovative medical technologies.*

We have also noted above that the MANBRIC complex will be formed in the process of the Cybernetic Revolution, and that new medical technologies will play an integrating role (see Chaps. 3 and 4). This will have a twofold effect: on the one hand, it will affect the increasing life expectancy, its physiological quality and increase the age limit for physical activity. On the other hand, the problem of the explosive growth of the number of older people and especially the problem of rising pension costs and labor shortages. As already mentioned, supporting an increasing number of elderly people will require an increase of the upper working-age limit due to a 10–20 year increase in working capacity and a greater involvement of people with disabilities into the labor process due to new technical means and medical advances.

In fact, in the coming decades, aging societies in developed and less developed countries (like Japan, Germany, the USA, China, Russia, Poland, and so on) will face an existential problem: either they raise the retirement age for aging cohorts, or they condemn pensioners to an existence that is hardly compatible with the possibilities of increasing life expectancy.[10] A pension crisis is almost inevitable in the face of a shrinking working population and slowing economic growth. The normal way out of this situation may be the development of medicine to a degree that will increase the working age, while developing labor-saving and labor-replacing technologies that can, firstly, reduce the shortage of labor, and secondly, reduce labor costs, spending on caring for the elderly and the sick. But this is possible only with the acceleration of technological progress for a sufficiently long time. Another option seems inhumane—to deprive the aging population of sufficient means for a normal and healthy existence, to cut pensions and to develop the institution of euthanasia instead of treating the elderly patients (which is already actively developing in some countries, like Canada), in a word, to 'fight' for a reduction in life expectancy in every possible way in order to remove financial burden from the state.[11] Here, by the way, new pandemics could 'help'. But the realization of such a negative scenario would testify to a deep crisis in the society that will allow it. We hope that it will not materialize. Moreover, the elderly are a very large group of voters, in particular, it is very influential in Japan. And although the possibility of such a negative scenario should by no means be ignored, we still do not consider it here.

As a result, with the development of a positive scenario, medical technologies will rapidly advance under the influence of an aging population (Phillips, 2011), and this will expand the search for opportunities to create "smart", self-regulating/managing systems, including robots, which can largely replace human labor, especially in the service sector, including complex services, e.g., in the fields of elderly care, education, medicine, etc. (Chap. 10; DeCanio, 2016; Frey & Osborne, 2017).[12]

---

[10] Of course, raising the retirement age is a very painful social issue and one that will be hotly contested. The events of 2023 in France illustrate this well. However, in a society where women do not want to have children, there is no other option at the moment (migration in any case is limited and connected with growing problems). The problem of pension provision will affect more and more new societies.

[11] This danger is not imaginary. Otherwise, there would be no discussion of Yusuke Narita, a Yale professor, who suggested mass suicide for old people in Japan to ease the burdens of Japan's rapidly aging society (Rich & Hida, 2023).

[12] It is not possible to consider within this chapter the effects of global aging and increasing labor shortages (see Chaps. 4, 6 and 10; see also Grinin & Korotayev, 2016a; Grinin, Grinin & Korotayev,

*Thus, until the last third of our century, the aging of the population will not impede technological and other development. On the contrary, the process of global aging itself will be the driving force for change, reform and acceleration of technological innovation.*

### 13.4.2 Global Aging and Technological Progress in the Last Third of the Twenty-First Century and the Early Twenty-Second Century: A Possible Negative Feedback

This period corresponds to the fifth and sixth phases of the Scientific-Cybernetic Principle (the 2070s–2090s), characterized by numerous changes. In particular, a growing number of social self-regulating systems will mostly operate autonomously, regulating the behavior of large numbers of people in certain situations. They can be used to create positive or negative behavioral stimuli (carrot and stick method) to regulate human behavior (see Chap. 14). However, an especially significant impulses will come form global aging, which by this time will affect all countries of the world. We believe that under the influence of aging and the increasing level of complexity of self-regulating systems, we will face the transition to a new economic system (see below). Concerning the relationship between global aging and technological progress, it is worth repeating that it is non-linear. At some point, the positive feedback we have mentioned above is likely to be replaced by negative feedback. Why? As we have discussed above, older people are more conservative. Above we set out the reasons why this is not just a popular belief, but a fact confirmed by rigorous scientific research.

It is important to take into consideration (as we have pointed out in Chaps. 6 and 7) that both biological and societal aging have a dual and contradictory effect on technological progress: positive and negative. On the one hand, the need to improve the biological quality of life of the elderly, the struggle to increase life expectancy, the fight against incurable diseases, the need to extend the working age of the elderly, etc. (which was discussed above and in Chaps. 4, 6 and 7) spur science and technology to search for new breakthrough solutions. Also, the growth of financial flows and interests, which are influenced by the elderly and old people, is and will continue. On the other hand, old people are prone to conservatism

2017a, 2017b). Of course, in the process of development of the Scientific-Cybernetic Production Principle, there will be significant changes in the number of people employed in various professions, as well as major changes in the professional nomenclature; some of these professions will begin to disappear under the influence of new technologies (including new chats and robotization). In our opinion, unqualified services will be particularly vulnerable. At the same time, the domain of qualified and highly qualified services will undergo considerable changes (see Grinin & Grinin, 2015c for more details). However, as a result of the rapid aging of the population, we expect that the total amount of labor resources will decrease faster (at least initially) than the technological progress of MANBRIC technologies will reduce jobs (see Grinin & Grinin, 2015c; Grinin & Korotayev, 2016a; Grinin, Grinin & Korotayev, 2017a, 2017b).

in different respects, in particular, in terms of physiological characreistics, world-view and cognitive features (see above). In addition, they already do not need many basic expenses (housing, furniture, equipment, etc.) that are necessary for younger people and that the elderly already have. All this has a negative impact on technological progress, slowing it down.

Thus, both tendencies operate simultaneously. This is quite understandable because within an elderly society progress and conservatism are combined in different ways. In Chaps. 6 and 8, we explained how the conservatism of the elderly can be combined with the continuation of the trend toward technological progress, using the example of the future self-managing supersystem of the medical–biotechnological environment. We assumed that maintaining healthy aging and longevity would require special environment, in which an increasing part of the population would spend more and more time. This supersystem will constantly monitor the state of health, control all the most important physiological and psychological processes and provide prescriptions.[13] The creation of such a medical–biotechnological system should come to life precisely through technological development, which is in itself a powerful source of development. At the same time, this system will contribute to maintaining the stability of a habitual way of life.

So, these multidirectional trends do not disappear, they continue and will continue to influence technological progress. Here we can speak about the unity and struggle of opposites. However, until the last third of the twenty-first century, we believe that the first, positive trend will prevail; respectively, aging in general will have a positive affect on the acceleration of technological progress. However, this trend will be weakened by the increase in the proportion of the elderly, and, accordingly, the trend toward conservatism will increase until it becomes predominant. That is why in the mature phases of the Scientific-Cybernetic Production Principle, aging will become a factor that will start to slow down scientific and technological progress. However, the slowdown doesn't mean the stop (see below).

What factors will strengthen the second trend—conservatism?

(1) This is a change in the age structure of the population. For example, according to some studies, a 1 percentage point increase in the share of the elderly (60+) in the total population by 1 percentage point tends to be accompanied by decreases in the growth rate of GDP per capita, the share of investment in GDP, the rate of personal savings, and the number of hours worked, respectively, by

[13] As we have pointed out in Chap. 8, just as many hundreds of millions of people cannot live without lifelong medication, so in the future a huge part of the population will be unable to stay outside this techno-medical environment for a long time. This system implies not only a breakthrough in medical technologies but also a powerful breakthrough in care technologies, continuous monitoring of the condition of people (not only of the elderly and disabled), control, remote medicine, psychological support, prediction, prescriptions and urgent care. But along with increasing length and comfort of biological life, this, of course, will bring a lot of dangers and problems, which we discuss in detail in the present chapter.

−0.14; −0.20; −0.59; −0.89% points. And the inflation rate increases by 0.17% points (Aksoy et al., 2015).

(2) Stabilization of the population in the World System as a whole by the late twenty-first century (either a slight increase or some reduction in the population is possible).[14] This will greatly reduce the opportunities for economic growth and hence, innovation. It is very important as population growth has been the most important driver of technological development throughout human history (Kuznets, 1960; Aghion & Howitt, 1998; Boserup, 1965; Dolgonosov, 2016; Fomin, 2020; Grinin, 2006; Grinin & Grinin, 2015c, 2016, 2017; Grinin et al., 2015; Grossman & Helpman, 1991; Jones, 2005; Kapitza, 1996; Korotayev, Malkov & Khaltourina, 2006; Korotayev & Markov, 2015; Korotayev, 2005, 2007, 2012, 2013, 2018, 2020; LePoire & Korotayev, 2020; Simon, 2000; Turchin et al., 2021).

(3) The situation when technology will be able to solve a number of important problems that have driven technology in the past decades, together with the movement toward greater orderliness of development and regulation of life, will reduce the sources of further growth. As a result of the above-described processes, society will become much more orderly, stochastic moments will play a smaller role and a different development algorithm will begin to form, which will be associated with greater predictability and smaller scales of fluctuations, but also with a slower movement.

*Thus, it is possible that the aging of society, with the end of global population growth and improved social planning capabilities, will contribute to the transition of individual societies and global society as a whole to a more comfortable and slower development (the so-called sustainable development, which is much discussed) by the end of this century or in the early twenty-second century.* We call this type of society "cybernetic society" (see Chap. 15 for details). Of course, other scenarios are also possible—for example, if the climate continues to deteriorate, some societal degradation may also occur.

However, we argue that technological or economic development will hardly stop. We speak about a slowdown in the innovation process in terms of the emergence of radical and rapid new technological breakthroughs. On the other hand, the innovations established in the previous decades during a powerful stream will be very widely implemented with their noticeable improvements and modifications. There will also emerge a lot of upgrading inventions. Still, at the same time, many problems will emerge, searching for solutions in terms of adapting these inventions to the societal life, to rights and freedoms, to people's free will, especially

[14] In the meantime, many countries, which now have low natural growth, will experience population decline (see UN Population Division, 2022a, 2022b) while in societies where high fertility rates are observed at present, i.e., in Tropical African countries and some countries of Asia (see Chapter 6; see also Korotayev & Zinkina, 2014, 2015; Zinkina & Korotayev, 2014a, 2014b; Korotayev et al., 2016; 2022a, 2022b, 2023a, 2023b; Grinin & Grinin, 2015c, 2017; Grinin & Korotayev, 2023) the population growth will slow down and society will become increasingly older.

free will of the elderly. It goes without saying that each of these issues will require major discourse, not least ethical one.

## 13.5 Transition to a New Economic Model

Thus, the above-mentioned conservatism of the elderly may lead to a slowdown in technological and economic development. It entails a transition to a different economic system. This will be discussed later.

The current model depends on increasing consumption. 'Consume more today than yesterday, and more tomorrow than today', is a model, which leads largely to the dead end and becomes basically absurd. The desire for exponential GDP growth seems absurd. However, it works and will work for decades, which is particularly important for poor countries whose populations are not satisfied with their level of consumption (for the current consumption model see Grinin & Grinin, 2021).

In any case, the modern economy can only grow if the necessity and importance of ever-increasing consumption is strengthened in people's minds. This creates a dead end: you cannot increase consumption all the time, but without economic growth, society stagnates. However, the directions of development can and should be gradually changed, maintaining the pace of economic growth while eliminating the most odious things and injustices.

Much has been said about endless and senseless consumption, about consumerism, about the lack of spirituality associated with this, about the decline in culture, etc. And it has been rightly said, but it has not been possible to resolve this contradiction. To solve social problems, economic growth is needed, otherwise, the volume of resources is compressed and it has to be rationed. And economic growth requires demand is needed, which is supported by an increase in consumption. By the way, it is supported by advertising, which also causes harm, but without which it is difficult, if not impossible, to promote goods in a consumer society. It is possible and even necessary to act in terms of minimizing the harm from such institutions of a market society, but only by carefully channeling the directions of development and gradually changing priorities. Thus, unfortunately, so far and for the next, at least two decades, it will not be possible to radically change the model of consumption in the modern economy and society, if only due to direct restrictions, bans and regulations (such ideas are popular enough). It is especially difficult that most of humankind does not consume enough and it will take decades to achieve a satisfactory living standard for these billions of people. However, it is possible to prepare society and business for such a change, but only by avoiding drastic measures that, instead of good, will bring more harm and reduce real standard of living—especially taking into account that most countries of the world are still very far from normal first-world level of consumption.

As for the modern economy is concerned, our position is that, to a certain level, the correction of senseless growth, the race for prestigious consumer goods and the morally and legally unlimited greed for profit are very useful. A moderate movement toward a better system of distribution in a society in which inequality has become excessive is also very desirable. This will only improve the society and give it an impetus for further development.

But, we would stress this again, such limitation and self-restraint, canalization and redistribution are useful only up to a certain level, beyond which they cause more harm than good. Where such a level lies is a very important question, but it is rather not ideological, but empirical and must be established in the course of cautious socio-political experiments.

For the elaboration of such a systematic approach to the theoretical optimization of consumption, several lines of approach are relevant. Among them are:

First, where should consumption be placed in relation to the needs and wants? A tentative view of Abraham Maslow's (1943) hierarchy of needs can serve as a starting point (see in this context also Doyal & Gough, 1991; Gough & McGregor, 2007).

Second, what is the relationship between use value and exchange value and to what extent is growth guided by a strict orientation towards use value? Two subsets are relevant here, namely (i) societal and (ii) personal use value.

Third, and closely linked to the second point—it makes a difference in which "area" growth takes place. Reference can be made to Marx's definition of two departments, later extended by Luxemburg and Herrmann (see, e.g., Herrmann, 2012, in particular: 49 f.); relevant is in this context is also the three-sector approach to the economy, now extensively elaborated, which refers, for instance, to seven sectors (see Sanguin, 1976).

Fourth, of course, such an optimization will require a scrupulous understanding of differentiation between age groups as well as many other things.

However, the main idea of this chapter is that, in the long run, debates on the problem of reducing consumerism do not take into account the role of the aging of society although this process can greatly transform the consumption model.

It makes sense to point out that since the second half of the twentieth century, and especially since the 1970s, older consumers and buyers began to be actively studied; the elderly as consumers became an object of special interest (Abdel-Ghany & Sharpe, 1997; Berg & Liljedal, 2022; Boyd Thomas & Lee Okleshen Peters, 2009; Drolet et al., 2007; Ensley & Pride, 1991; Fregolente et al., 2019; Goldberg, 2009; Goldstein, 1968; Halwani, 2021; Ilicic et al., 2016; John & Cole, 1986; Kang & Ridgway, 1996; Linden, 1976; Meneely et al., 2009; Nunan & Di Domenico, 2019; Oates et al., 1996; Phillips & Sternthal, 1977; Reisenwitz & Wimbish, 1997; Schewe, 1988; Towle & Martin, 1976; Williams & Drolet, 2005; Yoon et al., 2009;). For example, in 1990, Dychtwald and Gable (1990) maintained: "The clear and unmistakable implication is that business is about to face a larger, more powerful mass of mature consumers as we enter the 1990s and beyond" (p. 62). In 2013, the situation already became quite obvious. In Western societies, aging consumers will be the dominant consumer segment and a new and challenging business opportunity for retailers in the near future (Kohijoki & Marjanen, 2013). Elderly buyers have their own characteristics, they are more cautious and conservative. Sheau-Fen et al. (2012, p. 52) maintain: "In line with

Phillip and Sternthal's (1977) argument that elderly consumers tend to be more cautious in making decisions, Srinivasan and Ratchford (1991) found age to be positively related to perceived risk in the purchase of new automobiles. Dychtwald and Gable (1990) note that mature consumers are more conscious about the risk associated with the purchase decisions they make due to the psychological difference between the younger and older age groups. Simcock et al. (2006) and Laforet (2008) took this view further by examining the impact of age on several risk perceptions". Similar conclusions have been arrived at by many other researchers who note such characteristics of the elderly consumers as resistance to innovation/technology (Bae et al., 2021; Gilly & Zeithaml, 1985; Hough & Kobylanski, 2009), conservatism (Lambert-Pandraud & Laurent, 2010; Phua et al., 2020), risk aversion (Reisenwitz et al., 2007) and technology anxiety (Berg & Liljedal, 2022; Chaouali & Souiden, 2019; Meuter et al., 2003).

Indeed, aging can change people's needs, especially in conditions of the stabilization of the population or its decline. As a result, under the influence of all the above-mentioned future changes, the model measuring economic growth in terms of GDP should be replaced (Coleman & Rowthorn, 2015: 37). The modern consumption model will also change.

The transformation of the modern consumption model will be a complex process, possibly changing a variety of important aspects of our lives. Above, we mentioned Japan as an example of an aging society. It also provides an example of development without GDP growth alongside scientific and technological development. In the 2010s, the 'Japanese disease' spread to European countries (before 2020), partly due to the aging of their populations (there are other reasons as well, although we do not address them in this chapter; for more detail see Grinin & Korotayev, 2014, 2017b, 2018).[15]

Some contribution to reducing consumerism old be made by the proliferation of the Sun City-type settlements for the elderly (see Chap. 7 above for details). We do not rule out the prospects of such an institution. Such settlements contribute to the formation of the necessary institutions and the involvement of older people in active life, although age segregation also has negative effects, such as separating the elderly population from the mainstream of life and strengthening their conservatism (see, e.g., Asadzadeh et al., 2022; Kuznetsov, 2012; Lyubimov, 2013;).

In Chaps. 6 and 8, we have shown that, as a result of society's adaptation to societal aging, society should significantly change the distribution precisely in the direction of increasing the share of benefits and funds allocated to the elderly and the disabled. Adaptation to aging and disability, creation of an appropriate techno-biological environment, health monitoring, treatment, care, etc.—are all very expensive. Thus, the transition to this trend has actually started, but still it proceeds with difficulty. Any change in distribution is a conflictual and

[15] Only Covid-19, lockdowns and huge emission of money stop disinflation and has started high inflation.

complicated matter fraught with strong fluctuations. It is worth taking into consideration that the matter is not only in the pension system but in the general revision of the ideology of public spending, which should eventually be expressed in institutionalized agism (see Chap. 7).

On the accumulation side, society needs to increase savings for the future, in order to support an ever-growing number of elderly people, on the one hand. It is precisely the scarcity of savings that poses one of the main threats to the pension system of an aging society (in addition to the problems which we discuss in Chap. 6: the reduction of labor force and the lack of mechanisms to protect pension savings from inflation and other losses). On the other hand, the system is developing in such a way that relations will gradually require people to save less at the expense of non-market and social support systems (this is driven by aging). Already today, the ideology of rejecting ownership in favor of rented things (so-called 'sharing') has started to spread rapidly. But currently, it has rather ideological and political roots, reflecting the interests of certain power strata. Hence, the desire for a rapid implementation of this trend seems extremely dangerous. Yet, in the future, these processes will be intensively supported by global aging, since from a certain age people may be less inclined to saving.

It is also important to note that the development of societies within the framework of the unfolding of the Cybernetic Production Principle toward a higher degree of planning will also contribute to the reduction of consumerism (due to the spread of self-regulating systems [Chaps. 4 and 15; Grinin, Grinin & Korotayev, 2021, 2022; Grinin & Grinin, 2019]). An increasing amount of planning also brings more restrictions for people, and here compromises are needed since conflicts are inevitable. The market mechanisms will shrink due to increasing planning and the growing share of public goods in the overall system of distribution and consumption, and other factors, in particular, due to the possible guaranteed income in society. But it is extremely dangerous to make haste of burying capitalism, the market and so on; their important role will be preserved for a long time and is unlikely to ever disappear (Grinin et al., 2024).

So, we suggest that the late twenty-first and early twenty-second centuries will be decisive for human civilization. Fundamentally new socio-economic relationships will begin to emerge, the outlines of which are not yet clear. As a result, the modern consumption model will start to change dramatically. We have offered some ideas about possible reasons and trends for such transformations. However, this subject requires much more indepth research.

**Acknowledgements** This chapter is the output of a research project implemented as part of the Basic Research Program at HSE University in 2024 with support from the Russian Science Foundation (Project No. 23-18-00535).

## References


Abdel-Ghany, M., & Sharpe, D. L. (1997). Consumption patterns among the young-old and old-old. *Journal of Consumer Affairs, 31*(1), 90–112. https://doi.org/10.1111/j.1745-6606.1997.tb00828.x

Aghion, P., & Howitt, P. (1998). *Endogenous growth theory*. MIT Press.

Aksoy, Y., Basso, H. S., & Smith, P. R. (2015). *Demographic Structure and Macroeconomic Trends*. Banco de Espana.

Asadzadeh, M., Maher, A., Jafari, M., Mohammadzadeh, K. A., & Hosseini, S. M. (2022). A review study of the providing elderly care services in different countries. *Journal of Family Medicine and Primary Care, 11*(2), 458–465.

Bae, H., Jo, S. H., & Lee, E. (2021). Why do older consumers avoid innovative products and services? *Journal of Services Marketing, 35*(1), 41–53. https://doi.org/10.1108/JSM-10-2019-0408

Berg, H., & Liljedal, K. T. (2022). Elderly consumers in marketing research: A systematic literature review and directions for future research. *International Journal of Consumer Studies, 46*(5), 1640–1664.

Bös, D., & von Weizsäcker, R. K. (1989). Economic consequences of an ageing population. *European Economic Review, 33*(2–3), 345–354.

Boserup, E. (1965). *The conditions of agricultural growth: the economics of agrarian change under population pressure*. Aldine, Chicago.

Boyd Thomas, J., & Lee Okleshen Peters, C. (2009). Silver seniors: Exploring the self-concept, lifestyles, and apparel consumption of women over age 65. *International Journal of Retail & Distribution Management, 37*(12), 1018–1040. https://doi.org/10.1108/09590550911005001

Chaouali, W., & Souiden, N. (2019). The role of cognitive age in explaining mobile banking resistance among elderly people. *Journal of Retailing and Consumer Services, 50*, 342–350. https://doi.org/10.1016/j.jretconser.2018.07.009

Coleman, D., & Rowthorn, R. (2015). Population decline—making the best of inevitable destiny? In J. A. Goldstone, L. E. Grinin, & A. V. Korotayev (Eds.), *History & mathematics: political demography and global aging* (pp. 26–41). Uchitel.

Coulmas, F. (2007). *Population decline and aging in Japan—The social consequences*. Routledge.

Danigelis, N. L., Hardy, M., & Cutler, S. J. (2007). Population aging, intracohort aging and sociopolitical attitudes. *American Sociological Review, 72*(5), 812–830.

De Grey, A., & Rae, M. (2008). *Ending aging: The rejuvenation breakthroughs that could reverse human aging in our lifetime*. St. Martin's Press.

DeCanio, S. J. (2016). Robots and humans—complements or substitutes? *Journal of Macroeconomics, 49*, 280–291.

Dolgonosov, B. M. (2016). Knowledge production and world population dynamics. *Technological Forecasting and Social Change, 103*, 127–141.

Doyal, L., & Gough, I. (1991) *A theory of human need*. Macmillan Press.

Drolet, A., Williams, P., & Lau-Gesk, L. (2007). Age-related differences in responses to affective vs. rational ads for hedonic vs. utilitarian products. *Marketing Letters*, 18(4), 211–221. https://doi.org/10.1007/s11002-007-9016-z

Dychtwald, K., & Gable, G. (1990). Portrait of a changing consumer. *Business Horizons*, 33(1), January–February, 62–73. https://doi.org/10.1016/0007-6813(90)90073-K

Ensley, E. E., & Pride, W. M. (1991). Advertisement pacing and the learning of marketing information by the elderly. *Psychology & Marketing, 8*(1), 1–20. https://doi.org/10.1002/mar.4220080102

Fomin, A. (2020). Hyperbolic evolution from biosphere to technosphere. In A. V. Korotayev, & D. LePoire (Eds.) *The 21st century singularity and global futures. A Big History perspective* (pp. 105–118). Springer, Cham. https://doi.org/10.1007/978-3-030-33730-8_5

Fregolente, A., Junqueira, I., & Medeiros, P. (2019). Active and wealthy Brazilian older adults: Identity and consumption motivations. *Journal of Consumer Marketing, 36*(5), 633–642. https://doi.org/10.1108/JCM-12-2017-2496

Frey, C. B., & Osborne, M. A. (2017). The future of employment: How susceptible are jobs to computerisation? *Technological Forecasting and Social Change, 114*, 254–280.

Fukuyama, F. (2002). *Our post-human future: Consequences of the bio-technology revolution*. Farrar, Straus, and Giroux, New York.

Galor, O., & Weil, D. N. (2000). Population, technology, and growth: From Malthusian stagnation to the demographic transition and beyond. *American Economic Review, 90*(4), 806–828.

Gilly, M. C., & Zeithaml, V. A. (1985). The elderly consumer and adoption of technologies. *Journal of Consumer Research, 12*(3), 353–357. https://doi.org/10.1086/208521

Glenn, N. D. (1974). Aging and Conservatism. *The Annals of the American Academy of Political and Social Science, 415*, 176–186.

Goldberg, M. E. (2009). Consumer decision making and aging: A commentary from a public policy/marketing perspective. *Journal of Consumer Psychology, 19*(1), 28–34. https://doi.org/10.1016/j.jcps.2008.12.005

Goldstein, S. (1968). The aged segment of the market—1950 and 1960. *Journal of Marketing, 32*, 62–68.

Goldstone, J. A. (2015). Population aging and global economic growth. In J. Goldstone, L. E. Grinin, & A. Korotayev (Eds.), *History and mathematics: Political demography and global aging* (pp. 147–155). Uchitel.

Goldstone, J. A., Grinin, L. E., & Korotayev, A. V. (2015). Introduction, research into global aging and its consequences. In J. A. Goldstone, L. E. Grinin, & A. V. Korotayev (Eds.), *History & mathematics: Political demography and global aging* (pp. 5–9). Uchitel Publishing House.

Gough, I., & McGregor, J. A. (2007). *Wellbeing in developing countries: From theory to research*. Cambridge University Press.

Grinin, L. E., Grinin, A. L. (2015a). The cybernetic revolution and historical process. *Social Evolution and History, 14*(1), 125–184.

Grinin, L. E., & Grinin, A. L. (2015b). Kiberneticheskaya revolyutsiya, global'noe starenie i shestoj tekhnologicheskij uklad. In V. M. Bondarenko (Ed.), *XXIII kondrat'evskie chteniya: Tupiki global'noj ekonomiki, poisk novoj nauchnoj paradigmy* (pp. 89–105). Kondratieff Foundation.

Grinin, L. E., & Grinin, A. L. (2015c). *Ot rubil do nanorobotov, Mir na puti k epokhe samoupravlyaemyh sistem (istoriya tekhnologij i opisanie ih budushchego*. Moscow Branch of the Uchitel Publishing House.

Grinin, L. E., & Grinin, A. L. (2016). *The cybernetic revolution and the forthcoming epoch of self-regulating systems*. Moscow Branch of the Uchitel Publishing House.

Grinin L. E., & Grinin, A. L. (2017). Vliyanie protsessa global'nogo stareniya na tempy nauchno-tekhnicheskogo progressa i izmenenie modeli potrebleniya. In: Bondarenko VM (ed) *Nauchnoe nasledie N.D. Kondrat'eva i sovremennost'. Mezhregional'naya obshchestvennaya organizatsiya sodejstviya izucheniyu, propagande nauchnogo naslediya N.D. Kondrat'eva* (pp. 98–111). Moscow.

Grinin, L., & Grinin, A. (2020). The cybernetic revolution and the future of technologies. In A. V. Korotayev, & D. LePoire (Eds.), *The 21st century singularity and global futures. A Big History perspective* (pp. 377–396). Springer, Cham. https://doi.org/10.1007/978-3-030-33730-8_17

Grinin, L., & Grinin, A. (2021). Razmyshleniya ob ekonomicheskom roste i budushchem. Chast' 2. Global'nyye problemy i ekonomicheskiy rost. *Filosofiya i Obshchestvo, 4*, 5–38. https://doi.org/10.30884/jfio/2021.04.01

Grinin, L. E., Grinin, A. L., & Korotayev, A. (2017a). Forthcoming Kondratieff Wave, cybernetic revolution, and global aging. *Technological Forecasting & Social Change, 115*, 52–68.

Grinin, L., Grinin, A., & Korotayev, A. (2017b). The MANBRIC-technologies in the forthcoming technological revolution, In: T. Devezas et al. (Eds.), *Industry 4.0—Entrepreneurship and structural change in the new digital landscape: what is coming on along with the fourth industrial revolution* (pp. 243–261). Springer. https://doi.org/10.1007/978-3-319-49604-7_13

Grinin, L. E., & Korotayev, A. V. (2010). Will the global crisis lead to global transformations? 1, The global financial system: Pros and Cons. *Journal of Globalization Studies, 1*(1), 70–89.

Grinin, L., Grinin, A., & Korotayev, A. (2021). Global trends and forecasts of the 21st century. *World Futures, 77*(5), 335–370. https://doi.org/10.1080/02604027.2021.1949939

Grinin, L., Grinin, A., & Korotayev, A. (2022). COVID-19 pandemic as a trigger for the acceleration of the Cybernetic Revolution, transition from E-government to E-state, and change in social relations. *Technological Forecasting & Social Change, 175*, 1–10, https://doi.org/10.1016/j.techfore.2021.121348

Grinin, L. E., & Grinin, A. L. (2019). Technological Dimension of Big History and the Cybernetic Revolution. *Evolution, 7*, 250–277.

Grinin, L., Grinin, A., Malkov, S., & Korotayev, A. (2024). Capitalism's Unclear Futures. *World Futures*, https://doi.org/10.1080/02604027.2024.2335359.

Grinin, L. E., Korotayev, A. V., & Tausch, A. (2016). *Economic cycles, crises, and the global periphery*. Springer. https://doi.org/10.1007/978-3-319-41262-7

Grinin, L., & Korotayev, A. (2014). The inflation and deflationary trends in the global economy, or 'the Japanese disease' is spreading. *Journal of Globalization Studies, 5*(2), 154–173.

Grinin, L., & Korotayev, A. (2015a). *Great divergence and great convergence, a global perspective*. Springer. https://doi.org/10.1007/978-3-319-17780-9.

Grinin, L. E., & Korotayev, A. V. (2015b). Global population aging, the sixth Kondratieff Wave, and the global financial system. In J. A. Goldstone, L. E. Grinin, & A. V. Korotayev (Eds.), *History and mathematics: Political demography and global aging* (pp. 81–106). Uchitel Publishing House.

Grinin, L. E., & Korotayev, A. V. (2015c). Global'noe starenie naseleniya, shestoj tekhnologicheskij uklad i mirovaya finansovaya Sistema. In L. E. Grinin, A. V. Korotayev, & V. M. Bondarenko (Eds.), *Kondrat'evskie volny: Nasledie i sovremennost'* (pp. 107–132). Uchitel Publishing House.

Grinin, L. E., & Korotayev, A. V. (2015d). Global'noye stareniye naseleniya, shestoy tekhnologicheskiy uklad i mirovaya finansovaya Sistema. In L. E. Grinin & A. V. Korotaev (Eds.), *Istoriya i matematika: Futurologicheskiye i metodologicheskiye aspekty* (pp. 31–56). Uchitel.

Grinin, L. E., & Korotayev, A. V. (2015e). *Kiberneticheskaya revolyutsiya, global'noye stareniye i shestoy tekhnologicheskiy uklad, XXIII kondrat'yevskiye chteniya: tupiki global'noy ekonomiki, poisk novoy nauchnoy paradigmy, Sbornik statey uchastnikov konferentsii* (pp. 89–105). Moskva: Izdatel'stvo: Mezhregional'naya obshchestvennaya organizatsiya sodeystviya izucheniyu, propagande nauchnogo naslediya N.D. Kondrat'yeva.

Grinin, L., & Korotayev, A. (2016a). Global population aging, the sixth Kondratieff wave, and the global financial system. *Journal of Globalization Studies, 7*(2), 11–31.

Grinin, L., Tsirel, S., & Korotayev, A. (2015). Will the explosive growth of China continue?. *Technological Forecasting and Social Change, 95*, 294–308. https://doi.org/10.1016/j.techfore.2014.06.023

Grinin, L. E. (2006a) *Proizvoditel'nyye sily i istoricheskiy protsess*. 3-ye izd. Uchitel'.

Grinin, L. E., & Korotayev, A. V. (2016b). Introduction, global evolution and global aging. In L. E. Grinin & A. V. Korotayev (Eds.), *Evolution and big history: dimensions, trends, and forecasts* (pp. 5–17). 'Uchitel' Publishing House.

Grinin, L. E., & Korotayev, A. V. (2018). The future of the global economy in the light of inflationary and deflationary trends and long cycles theory. *World Futures, 74*(2), 84–103. https://doi.org/10.1080/02604027.2017.1357934

Grinin, L., & Korotayev, A. (2023). Africa—The continent of the future. Challenges and opportunities. In V. Sadovnichy et al. (Eds.), *Reconsidering the limits to growth. A report to the Russian Association of the Club of Rome* (pp. 225–238). Springer, 2023. https://doi.org/10.1007/978-3-031-34999-7_13

Grossman, G., & Helpman, E. (1991). *Innovation and growth in the global economy*. MIT Press.

Halwani, L. (2021). Heritage luxury brands: Insight into consumer motivations across different age groups. *Qualitative Market Research: An International Journal, 24*(2), 161–179. https://doi.org/10.1108/QMR-07-2019-0092

Haas, M. L. (2015). Population aging and the future of the great powers. In: J. A. Goldstone, L. E. Grinin, & A. V. Korotayev (Eds.), *History & mathematics: Political demography & global aging* (pp. 133–146). 'Uchitel' Publishing House.

Harper, S. (2006). Addressing the implications of global aging. *Journal of Population Research, 23*(2), 205–223.

Herrmann, P. (2012). Rights—Developing ownership by linking control over space and time; writings on philosophy and economy of power Vol. 3; Bremen: Europaeischer Hochschulverlag.

Hough, M., & Kobylanski, A. (2009). Increasing elder consumer interactions with information technology. *Journal of Consumer Marketing, 26*(1), 39–48. https://doi.org/10.1108/07363760910927037

Ilicic, J., Baxter, S. M., & Kulczynski, A. (2016). The impact of age on consumer attachment to celebrities and endorsed brand attachment. *Journal of Brand Management, 23*(3), 273–288. https://doi.org/10.1057/bm.2016.5

Jackson, R., Howe, N., Strauss, R., & Nakasima, K. (2008). *The greying of the great powers. Demography and geopolitics in the 21st century*. Center for Strategic & International Studies.

Jackson, R., Macaranas, R., & Peter, T. (2013). *U.S. development policy in an ageing world. New challenges and new priorities for a new demographic era. A Research Report from Project on U.S. Leadership*. CSIS (Center for Strategic and International Studies).

John, D. R., & Cole, C. A. (1986). Age differences in information processing: Understanding deficits in young and elderly consumers. *Journal of Consumer Research, 13*(3), 297–315. https://doi.org/10.1086/209070

Jones, C. I. (2005). The shape of production functions and the direction of technical change. *The Quarterly Journal of Economics, 120*, 517–549.

Kang, Y. S., & Ridgway, N. M. (1996). The importance of consumer market interactions as a form of social support for elderly consumers. *Journal of Public Policy & Marketing, 15*(1), 108–117. https://doi.org/10.1177/074391569601500110

Kapelyushnikov, R. (2019a). Fenomen stareniya naseleniya: Ekonomicheskiye effekty. *Ekonomicheskaya Politika*, 14(2), 8–63.

Kapelyushnikov, R. (2019b). *Fenomen stareniya naseleniya: ekonomicheskiye effekty*. Institut ekonomiki RAN.

Kapitza, S. P. (1996). The phenomenological theory of world population growth. *Physics-Uspekhi, 39*(1), 57–71.

Kohijoki, A.-M., & Marjanen, H. (2013). The effect of age on shopping orientation-choice orientation types of the ageing shoppers. *Journal of Retailing and Consumer Services, 20*(2), 165–172. https://doi.org/10.1016/j.jretconser.2012.11.004

Korotayev, A. (2005). A compact macromodel of world system evolution. *Journal of World-Systems Research, 11*(1), 79–93. https://doi.org/10.5195/jwsr.2005.401

Korotayev, A. (2007). Compact mathematical models of world system development, and how they can help us to clarify our understanding of globalization processes. In G. Modelski, T. Devezas, & W. R. Thompson (Eds.), *Globalization as evolutionary process: Modeling global change* (pp. 133–160). Routledge.

Korotayev, A. (2012). Globalization and mathematical modeling of global development. In L. E. Grinin, I. V. Ilyin, & A. V. Korotayev (Eds.), *Globalistics and globalization studies* (pp. 148–158). 'Uchitel' Publishing House.

Korotayev, A. (2013). Globalization and mathematical modeling of global evolution. In L. E. Grinin & A. V. Korotayev (Eds.), *Evolution: Development within big history, evolutionary and world-system Paradigms* (pp. 69–83). Uchitel.

Korotayev, A. (2018). The 21st century singularity and its big history implications: A re-analysis. *Journal of Big History, 2*(3), 73–119. https://doi.org/10.22339/jbh.v2i3.2329

Korotayev, A. V. (2020). The 21st century singularity in the big history perspective. A re-analysis. In A. V. Korotayev, & D. LePoire (Eds.), *The 21st century singularity and global futures. A Big History perspective* (pp. 19–75). Springer, Cham. https://doi.org/10.1007/978-3-030-33730-8_2

Korotayev, A., Butovskaya, M., Shulgin, S., & Zinkina, J. (2023). How can the global ageing affect the global value system? An evolutionary perspective. *Social Evolution & History, 22*(1), 57–76. https://doi.org/10.30884/seh/2023.01.03

Korotayev, A., Malkov, A., & Khaltourina, D. (2006a). *Introduction to social macrodynamics: Compact macromodels of the world system growth.* KomKniga/URSS.

Korotayev, A., Malkov, A., & Khaltourina, D. (2006b). *Introduction to social macrodynamics: Compact macromodels of the world system growth.* KomKniga/URSS.

Korotayev, A. V., & Markov, A. V. (2015). Mathematical modeling of biological and social phases of big history. In L. E. Grinin, I. V. Ilyin, P. Herrmann, & A. V. Korotayev (Eds.), *Globalistics and globalization studies: Big history & global history* (pp. 319–343). 'Uchitel' Publishing House.

Korotayev, A., Novikov. K., & Shulgin, S. (2019). Values of the elderly people in an aging world. *Sociology of Power, 31*(1), 114–142. https://doi.org/10.22394/2074-0492-2019-1-114-142

Korotayev, A. V., & Shulgin, S. G. (2018). *Vliyaniye stareniya naseleniya na global'nuyu sistemu tsennostey i politicheskuyu dinamiku.* RANEPA. https://doi.org/10.2139/ssrn.3139641

Korotayev, A., Zinkina, J., Goldstone, J., & Shulgin, S. (2016). Explaining current fertility dynamics in tropical Africa From an anthropological perspective: A cross-cultural investigation. *Cross-Cultural Research, 50*(3), 251–280. https://doi.org/10.1177/1069397116644158

Korotayev, A., Shulgin, S., Zinkina, J., & Slav, M. (2022a). Estimates of the possible economic effect of the demographic dividend for sub-saharan Africa for the period up to 2036. *Vostok (Oriens), 2*, 108–123. https://doi.org/10.31857/S086919080019128-7

Korotayev A. V., Ustyuzhanin V. V., Zinkina J. V., Shulgin S. G., & Grinin L. E. (2022b). Towards mathematical modeling of the political and demographic future of Africa. *Sistemnyi Monitoring Globalnyh i Regionalnyh Riskov = Systemic Monitoring of Global and Regional Risks, 13*, 271–321. https://doi.org/10.30884/978-5-7057-6184-5_09

Korotayev, A., Shulgin, S., Ustyuzhanin, V., Zinkina, J., & Grinin, L. (2023). Modeling social self-organization and historical dynamics. Africa's futures. In V. Sadovnichy et al. (Eds.), *Reconsidering the limits to growth. A report to the Russian Association of the Club of Rome* (pp. 461–490). Springer. https://doi.org/10.1007/978-3-031-34999-7_20

Korotayev, A., & Zinkina, J. (2014). How to optimize fertility and prevent humanitarian catastrophes in tropical Africa. *African Studies in Russia, 6*, 94–107.

Korotayev, A., & Zinkina, J. (2015). East Africa in the Malthusian Trap? *Journal of Developing Societies, 31*(3), 1–36. https://doi.org/10.1177/0169796X15590322

Korotayev, A., Zinkina, J., Shulgin, S., & Bykanova, D. (2017). Pochemu pozhilye lyudi bolee religiozny, chem molodye? Kogortnye i vozrastnye faktory, ili budush-chee religioznyh cennostej v ekonomicheski razvityh stranah. *Religiovedenie, 3*(3), 134–144.

Kuznets, S. (1960). Population change and aggregate output. In *Demographic and economic change in developed countries* (pp. 324–340). Princeton University Press.

Kuznetsov, A. (2012). *Poselki dlya pensionerov.* https://xn----dtbhaacat8bfloi8h.xn--p1ai/social-settlements

Laforet, S. (2008). Retail brand extension—perceived fit, risks and trust. *Journal of Consumer Behaviour: An International Research Review, 7*(3), 189-209.

Lambert-Pandraud, R., & Laurent, G. (2010). Why do older consumers buy older brands? The role of attachment and declining innovativeness. *Journal of Marketing, 74*(5), 104–121. https://doi.org/10.1509/jmkg.74.5.104

LePoire, D., & Korotayev, A.V. (2020). Conclusion. In A. V. Korotayev, D. LePoire (Eds.), *The 21st century singularity and global futures. A big history perspective* (pp. 599–620). Springer, Cham. https://doi.org/10.1007/978-3-030-33730-8_27

Lee, R., & Mason, A. (2011). *Population aging and the generational economy: A global perspective.* Edward Elgar.

Linden, F. (1976). Consumer markets: Age and income—1985. *The Conference Board Record, 13*(June), 8–13.

Lyubimov, M. (2013). *Pensionnyye poseleniya v SSHA: gde dostoyno vstretit' starost'.* https://tranio.ru/articles/pensionnye_poseleniya_v_ssha_gde_dostoyno_vstretit_starost/

Maslow, A. H. (1943). A theory of human motivation. *Psychological Review, 50*(4), 370–396.
Meneely, L., Burns, A., & Strugnell, C. (2009). Age associated changes in older consumers retail behaviour. *International Journal of Retail & Distribution Management, 37*(12), 1041–1056. https://doi.org/10.1108/09590550911005010
Meuter, M., Ostrom, A., Bitner, M., & Roundtree, R. (2003). The influence of technology anxiety on consumer use and experiences with selfservice technologies. *Journal of Business Research, 56*(11), 899–906. https://doi.org/10.1016/S0148-2963(01)00276-4
Morgan, L. A., & Kunkel, S.R. (2016). *Aging, society, and the life course* (5th ed.). Springer. https://connect.springerpub.com/content/book/978-0-8261-2173-8
Nunan, D., & Di Domenico, M. (2019). Older consumers, digital marketing, and public policy: A review and research agenda. *Journal of Public Policy & Marketing, 38*(4), 469–483. https://doi.org/10.1177/0743915619858939
Oates, B., Shufeldt, L., & Vaught, B. (1996). A psychographic study of the elderly and retail store attributes. *Journal of Consumer Marketing, 13*(6), 14–27. https://doi.org/10.1108/07363769610152572
Ogawa, N., Kondo, M., & Matsukura, R. (2005). Japan's transition from the demographic bonus to the demographic onus. *Asian Population Studies, 1*(2), 207–226.
Park, D., & Shin, K. (2015). Impact of population aging on Asia's future growth. In J. A. S. Goldstone, L. E. Grinin, & A. V. Korotayev (Eds.), *History & mathematics: Political demography and global aging* (pp. 107–132). Uchitel.
Peterson, P. G. (1999). Gray down: The global ageing crisis. *Foreign Affairs, 78*(1), 42–55.
Peterson, J. C., Smith K. B, & Hibbing J. R. (2020). Do people really become more conservative as they age? *The Journal of Politics, 82*(2), 600–611.
Phillips, F. (2011). The state of technological and social change: Impressions. *Technological Forecasting and Social Change, 78*(6), 1072–1078. https://doi.org/10.1016/j.techfore.2011.03.020
Phillips, L., & Sternthal, B. (1977). Age differences in information processing: A perspective on the aged consumer. *Journal of Marketing Research, 14*(4). https://doi.org/10.1177/002224377701400402
Phua, P., Kennedy, R., Trinh, G., Page, B., & Hartnett, N. (2020). Examining older consumers' loyalty towards older brands in grocery retailing. *Journal of Retailing and Consumer Services, 52*, 101893. https://doi.org/10.1016/j.jretconser.2019.101893
Powell, J. L., & Khan, H. T. (2013). Aging and globalization: A global analysis. *Journal of Globalization Studies, 4*(1), 137–146.
Prettner, K. (2013). Population aging and endogenous economic growth. *Journal of Population Economics, 26*(2), 811–834.
Reisenwitz, T., Iyer, R., Kuhlmeier, D. B., & Eastman, J. K. (2007). The elderly's internet usage: An updated look. *Journal of Consumer Marketing, 24*(7), 406–418. https://doi.org/10.1108/07363760710834825
Reisenwitz, T. H., & Wimbish, J. G., Jr. (1997). The purchase decision process and involvement of the elderly regarding nonprescipton products. *Health Marketing Quarterly, 15*(1), 49–68. https://doi.org/10.1300/J026v15n01_04
Rich, M., & Hida, H. (2023). A Yale professor suggested mass suicide for old people in Japan. What did he mean? *The New York Times*. Retrieved February 12, 2023, from https://www.nytimes.com/2023/02/12/world/asia/japan-elderly-mass-suicide.html
Sanguin, A. L. (1976). Une nouvelle distinction des secteurs d'activité en géographie économique. *Bulletin De L'association De Géographes Français, 53*(437), 289–294.
Schewe, C. D. (1988). Marketing to our aging population: Responding to physiological changes. *Journal of Consumer Marketing, 5*(3), 61–73. https://doi.org/10.1108/eb008232
Shulgin, S., Zinkina, J., & Korotayev, A. (2019). Religiosity and aging: Age and cohort effects and their implications for the future of religious values in high-income OECD countries. *Journal for the Scientific Study of Religion, 58*(3), 591–603. https://doi.org/10.1111/jssr.12613
Simon, J. (2000). *The great breakthrough and its cause*. Ann Arbor: University of Michigan Press.
Simcock, P., Sudbury, L., & Wright, G. (2006). Age, perceived risk and satisfaction in consumer decision making: a review and extension. *Journal of Marketing Management, 22*(3-4), 355-377.

Towle, J., & Martin, C. (1976). The elderly consumer: one segment or many. In *Proceedings, Spring Business Conference* (pp. 463–468). American Marketing Association.

Tsirel, S. V. (2008). Zametki ob istoricheskom vremeni i putyah istoricheskoj evolyucii. In L. E Grinin, A. V. Korotayev, & S. Y. Malkov (Eds.), *Istoriya i matematika: Modeli i teorii* (pp. 246–278). LKI/URSS.

Turchin, P., Hoyer, D., Korotayev, A., Kradin, N., Nefedov, S., et al. (2021). Rise of the war machines: Charting the evolution of military technologies from the Neolithic to the Industrial Revolution. *PLoS ONE, 16*(10), e0258161. https://doi.org/10.1371/journal.pone.0258161

UN Population Division. (2022a). *United Nations Population Division Database*. United Nations, Department of Economics and Social Affairs, Population Division. http://www.un.org/esa/population

UN Population Division. (2022b). *World population prospects 2022*. United Nations, Department of Economics and Social Affairs, Population Division.

WHO—World Health Organization. (2017). *Global strategy and action plan on ageing and health*. World Health Organization.

Williams, P., & Drolet, A. (2005). Age-related differences in responses to emotional advertisements. *Journal of Consumer Research, 32*(3), 343–354. https://doi.org/10.1509/jmkg.69.2.97.60757

Yoon, C., Cole, C. A., & Lee, M. P. (2009). Consumer decision making and aging: Current knowledge and future directions. *Journal of Consumer Psychology, 19*(1), 2–16. https://doi.org/10.1016/j.jcps.2008.12.002

Zimmer, Z. (2016). *Global aging in the twenty-first century: challenges, opportunities and implications*. Routledge.

Zinkina, J., & Korotayev, A. (2014a). Explosive population growth in tropical Africa: Crucial omission in development forecasts (emerging risks and way out). *World Futures, 70*(4), 271–305. https://doi.org/10.1080/02604027.2014.894868

Zinkina, J., & Korotayev, A. (2014b). Projecting Mozambique's demographic futures. *Journal of Futures Studies, 19*(2), 21–40.

# 14 The Cybernetic Revolution, COVID-19 and the E-state

**Abstract**

Grinin et al. show the impact of the pandemic on the course and the pace of the Cybernetic Revolution and make forecasts about very large changes not only in technology, but also in the socio-political sphere. Anti-COVID measures have had a number of side negative effects, which Grinin et al. discuss in detail. However, the growth of the role of medicine due to COVID strongly confirms Grinin et al.'s ideas of the integrating role of medical technologies in the MANBRIC-convergence and has made significantly clearer the mechanisms that can provide such a breakthrough in medicine and related technologies. COVID-19 has become a powerful impetus in further changing socio-administrative relations. The authors pay great attention to the role of a special type of self-managing systems, by means of which it is possible to regulate our behavior without human participation just by means of AI. The authors are talking about socio-technical self-managing systems (SSSs). SSSs with the help of AI will regulate various social and administrative relations (i.e., control, verification, security, checking, identification, issue of documents, etc.). Grinin et al. forecast that the development of SSSs pushes society towards the formation of an electronic state (e-state) with its positive and strongly negative sides.

## 14.1 What This Chapter is About

We show the impact of the pandemic on the course and the pace of the Cybernetic Revolution and make forecasts about very large changes not only in technology but also in the socio-political sphere. The 2020–2022 pandemic has become one of the very large and unexpected events of the twenty-first century. The COVID-19 pandemic has significantly changed the public's focus; these changes are likely



L. Grinin et al., *Cybernetic Revolution and Global Aging*, World-Systems Evolution and Global Futures, https://doi.org/10.1007/978-3-031-56764-3_14

to persist for a long time. Ultimately, this may lead to significant technological transformations, which, in our opinion, are rather underestimated.

The COVID-19 pandemic has become a phenomenon that, for the first time in many decades or even centuries, placed healthcare problems at the center of both intrasocial and global relations. The socio-political and economic role of medicine has sharply increased due to the large number of COVID infections.

One of the main ideas of this chapter is that the COVID-19 pandemic has become a trigger for the acceleration of the Cybernetic Revolution. In this respect, the pandemic has confirmed our forecasts, also allowing us to make these forecasts more accurate and more detailed. The COVID-19 pandemic has confirmed that the final phase of the Cybernetic Revolution will very likely begin in the 2030s at the intersection of a number of medical, nano, bio, artificial intelligence and other technologies, which we denote as "MANBRIC convergence".

The growing role of medicine due to COVID strongly confirms our ideas about the integrating role of medical technologies in the MANBRIC convergence and implies practical recommendations regarding the most promising directions of the technological development in the nearest future. At this point in time, the fight with the pandemic has significantly clarified the mechanisms that can provide such a breakthrough in medicine and related technologies.

It is very important that COVID-19 has become a powerful impetus also for changing socio-political (and socio-administrative) relations in the forthcoming decades. Therefore, combining technological and social forecasts in a single study appears all the more important, since a systematic research into the relationships and interactions remains insufficient.

The COVID-19 pandemic has revealed some mechanisms of such system connections, thus helping to improve forecasting. What do we mean? As already mentioned in many chapters of this monograph, the most important outcome of the Cybernetic Revolution will be the development of self-regulating/self-managing systems in the production, economy and life, which will flourish during the Cybernetic Revolution.

In this chapter, we pay great attention to the role of a special type of self-managing systems whose the importance and influence on society has grown dramatically in recent years and will continue to grow rapidly. We speak about socio-technical self-regulating/self-managing systems (SSSs). With the help of artificial intelligence, SSSs will regulate various social and administrative relations.

The SSSs perform social and administrative functions (i.e., control, verification, distribution, security, control, identification, issuance of documents, rating and other functions) using a set of technologies with little or no involvement of officials and specialists. Thus, they can be used by authorities at different levels and by the state as a whole (as well as by the management of service centers where the authorities consider such regulation necessary: airports, crowded places, etc.).

That is why the development of SSSs, in one way or another, pushes society toward the formation of an electronic state. What exactly is an electronic state? And how radical will the changes associated with it be? We understand the concept of an e-state as a state with a significantly reduced number of state control bodies, mainly based on SSS technology.

In politics and administration, the development of the e-state will entail very significant, one might even say revolutionary, changes in the field of governance, as many social and administrative relations will be regulated by these systems. This will take place both at the level of individual administrative units and cities (so-called smart cities) and at the level of the state as a whole.

However, the development of SSSs and the formation of the e-state may lead to the total unification of people's behavior which will suppress the individuality and special abilities of people, as well as serious changes in the political and social spheres. This can accelerate the change in social relations, as well as increase confrontation in society because the use of self-regulating systems in governance will lead to increased technological and political control. This is all the more dangerous because such systems, although not yet widespread in private life, are very likely to become so. There are also many other new ideas and forecasts.

## 14.2 Introductory Remarks: Multidimensional Impact of COVID-19 and Its Influence on Technological and Social Development

By now, there have been quite a number of studies about the coronavirus and its effect on society.[1] In addition to research into possible strategies for dealing with the pandemic (including its consequences) and various medical aspects, many other issues are being studied, from the economic consequences to the impact on the climate (Aslam et al., 2020; Belhadi et al., 2020; Brammer et al., 2020; Chakraborty & Maity, 2020; Forster et al., 2020; Yoo & Managi, 2020; Bhat et al., 2021; Ehnts & Paetz, 2021; Jones et al., 2021; Usman et al., 2021; Choi et al., 2022; Kumar et al., 2022; Ray et al., 2022; Saengtabtim et al., 2022, see also below).

However, so far, there have been few works with a systematic analysis of the impact of COVID on our future.[2] In fact, the study of this new phenomenon still seems to be largely unsystematic. However, one may argue that the COVID-19 pandemic has become an important driver of even larger changes (e.g., Brammer et al., 2020; Brem et al., 2021; Hofbauer and Komlosy 2020; Amankwah-Amoah

[1] E.g., Atkeson, 2020; Baker et al., 2020; Bortoló et al., 2023; Cruz-Cárdenas et al., 2021; Chumakov & Yurchenko, 2021; Damaševičius & Zailskaitė-Jakštė 2023; Efremov & Arkhangelsky, 2023; Fabris 2022; Grinin, 2020; Grinin et al., 2022; Grix et al., 2021; Hantrais et al., 2021; Hussain, 2021; Irshad 2020; Jackson et al., 2022; Jasiński and Bąkowska 2020; Kakkar, 2021; Kumar et al., 2021a, 2021b; Mahagamage & Marasinghe 2023; McKeown et al. 2022; Sarfraz et al., 2021; Seifabadi & Dolatabadi, 2022; Walby 2022; Widdowson, 2021;Wood 2022; Xie et al., 2022; Zimmerling & Chen, 2021; see also further references below.

[2] One of the exception is Schwab and Malleret 2020, though, of course, their point of view is a very controversial.

et al., 2021; Court et al., 2021), whose nature and consequences are still mostly unclear. We believe that the COVID-19 pandemic has significantly accelerated the processes that we have already discussed in this monograph and in our other works (Grinin and Grinin, 2015; 2016; Grinin et al., 2017a; 2017b; 2020; Grinin and Korotayev 2015, 2020; Grinin et al., 2016). It has also speeded up some other processes that deserve close attention.

The COVID-19 pandemic has significantly changed the public's focus; and these changes are likely to persist for a long time. Ultimately, this may lead to significant technological transformations, which, in our opinion, are rather underestimated. Among other valuable papers on this topic, the article by Brem et al. (2021) deserves special mention, as it provides an overview of technologies related to the coronavirus treatment, and adaptation in times of crisis. We also rely on a number of other works describing the use of different technologies during the pandemic (Abi Younes et al., 2020; Amankwah-Amoah et al., 2021; Basu et al., 2020; Bortoló et al., 2023; Choong et al., 2020; Court et al., 2021; Estrada & Arturo, 2020; Fabris, 2022; Hantrais et al., 2021; Jat & Singh, 2020; Jung & Lim, 2020; Musyuni et al., 2020; Sarfraz et al., 2021; Shah et al., 2020; Skorup & Haaland, 2020; Xie et al., 2022; Zimmerling & Chen, 2021). Of course, we were also interested in works analyzing vaccination and other important medical and biotechnological issues, as well as the integration of these technologies (e.g., Brem et al., 2021; Chung et al., 2020; Musyuni et al., 2020; Kumar et al., 2021a, 2021b; Bortoló et al., 2023; Brüssow, 2022), as well as general issues of technological forecasting and social change (Brem et al., 2021; Phillips, 2007, 2011; Zimmerling et al., 2021).

In general, and specifically for our monograph, it is important to understand that the COVID-19 pandemic has become a phenomenon that for the first time in many decades or even centuries placed healthcare issues at the center of both intrasocial and global relations. The pandemic has affected almost every area of life and human activities (e.g., Atkeson, 2020; Baker et al., 2020; Grinin, 2020; Irshad, 2020; Jasiński & Bąkowska, 2020; Cruz-Cárdenas et al., 2021; Grix et al., 2021; Hantrais et al., 2021; Kumar et al., 2021a, 2021b; Sarfraz et al., 2021; Zimmerling & Chen, 2021; Fabris, 2022; Jackson et al., 2022; McKeown et al. 2022; Walby, 2022; Wood, 2022; Xie et al., 2022; Bortoló et al., 2023; Damaševičius & Zailskaitė-Jakštė, 2023; Mahagamage & Marasinghe, 2023). This has led to noticeable changes in technologies and their diffusion, as well as changes in the economy, politics and the life of society. It has also spawned a number of projects that can significantly accelerate the development of certain technologies, as well as change socio-political relations. In general, the analysis of data in different areas over recent years allowed us to identify the following important changes related to the pandemic.

**The Focus of this chapter**. In the present chapter, we would like to focus on several issues. In particular: How has the coronavirus pandemic affected the speed and canalization of the approaching final phase of the Cybernetic Revolution, as well as the acceleration of the MANBRIC convergence? How does COVID-19 influence the formation of socio-technical self-regulating systems and how can their development affect people's social behavior and also the formation of an e-state in the future? Despite the magnitude of these problems, they are closely interconnected, so we believe that they should be studied as a single complex.

We want to show that the COVID-19 pandemic has become a powerful trigger which will both accelerate technological development in medicine and other areas—especially those that form the MANBRIC-system[3]—and, at the same time, catalyze the convergence of these areas. This is greatly supported by the sharply increased demand for the development of a number of fields of medicine, bio- and nanotechnology, artificial intelligence and others.

On the whole, the pandemic has not only confirmed our forecasts, but has also allowed us to significantly refine them, to make them more precise, to enrich them, and to see them in greater detail. In particular, it has considerably confirmed the ideas of the integrating role of medical technologies in the MANBRIC convergence and implies practical recommendations regarding the most promising directions of technological development in the nearest future. The experience of coping with the COVID-19 pandemic has confirmed that the final phase of the Cybernetic Revolution will begin at the intersection of a number of medical, bio, digital and several other technologies. As a result, the fight against the pandemic has significantly clarified the mechanisms which can provide such a breakthrough in medicine and related technologies. Among the many impacts that the pandemic of coronavirus has had and will continue to have on society and the World System as a whole, one of the most important will be the acceleration of the start of a new technological wave—the final phase of the Cybernetic Revolution (in particular see Fig. 14.2).

The chapter also contributes to the emerging theory of e-government by introducing the notion of the e-state, its future influence and some practical recommendations for the regulation of this particular sphere of development. As we have already shown, the unfolding Cybernetic Revolution will be associated with the widespread introduction of self-regulating and self-managing systems in production and a wide variety of services and also in life in general. Self-regulating/ self-managing systems are systems that can function with minimal or no human involvement (see Chaps. 3 and 4; see also Grinin and Grinin, 2015; 2016; Grinin et al. 2017a; 2017b; 2020, 2021a, 2021b; Grinin and Korotayev, 2015; Grinin et al., 2016). In this chapter, we argue that among numerous self-regulating systems in production, the economy and everyday life that will flourish during the final phase of the Cybernetic Revolution, the socio-technical self-regulating/ managing systems (SSSs) will play a special role. These are technologies designed to regulate administrative and social relations in society by combining AI with other technologies; in some respects, these combinations of technologies are able to perform the functions of administrative bodies and law enforcement.

[3] Let us recollect that MANBRIC is an acronym for a complex of technologies: Medical–Additive–Nano–Bio–Robotics–Information and Cognitive technologies.

It is worth clarifying the difference between socio-technical *self-regulating* and socio-technical *self-managing* systems. As we explain in Chap. 3, self-regulating systems can be divided into two sub-types: properly self-regulating and self-managing ones. In this respect, biological, physiological, genetic, many biotechnological (such as self-purification of waste water), nanotechnological and other natural systems that are being introduced or will be introduced during the Cybernetic Revolution can hardly be called "self-managing systems", since they are generally designed to regulate a certain level of health, norms, etc. That is, they remain self-regulating systems. But AI-based technical systems can rather be called self-managing systems. So, future universal robots that are able to move, speak, make decisions, react to dangers, protect people and so on, as well as fully self-driving cars are self-managing systems. Concerning SSSs, they can be both self-regulating and self-managing systems depending on their functions. So a complex autonomous system that will regulate traffic or so-called smart houses are self-regulating systems because their main functions are limited regulation. However, a team of robot-policemen patrolling streets is rather a self-managing system because their actions are unpredictable, so they have to constantly make decisions.

During the pandemic, there appeared an urgent need for the regulation of public life, which served as an impetus for the development of such systems. Meanwhile, the use of socio-technical self-regulating/self-managing systems (SSSs) will almost inevitably lead to a noticeable change in social and even political relations in society, shifting them toward e-government and an e-state. The present chapter also devotes considerable attention to the analysis of the latter, while addressing such an issue as the rapidly growing threat to privacy and personal data.

Combining technological and societal forecasting in a single study seems all the more important, because systematic research into the relationships and interactions between the rapid development of various technologies, especially AI technologies, medicine and biotechnology, on the one hand, and social and political relations on the other, is regrettably still insufficient. The COVID-19 pandemic has revealed some of the mechanisms of such systemic connections, thus helping to improve forecasting.

**COVID-19, Technologies and Aging**. In the analysis of global aging, we rely on the studies mentioned in Chaps. 6 and 7, including demographic forecasts, all of which predict an increase in this process within the World System in this century, especially in high and middle-income countries (Kaneda et al., 2016; Lutz et al., 2018; UN Population Division, 2019, 2022a, 2022b; Vollset et al., 2020; Wittgenstein Center 2024). In this context, it is worth noting that the coronavirus pandemic has exposed the problem of protecting the elderly as the most vulnerable part of the population (see Brem et al., 2021; Fraser et al., 2020; Agrawal et al., 2021; see also Chaps. 6 and 7). In Chaps. 5–7, 11 and 13, as well as in our own

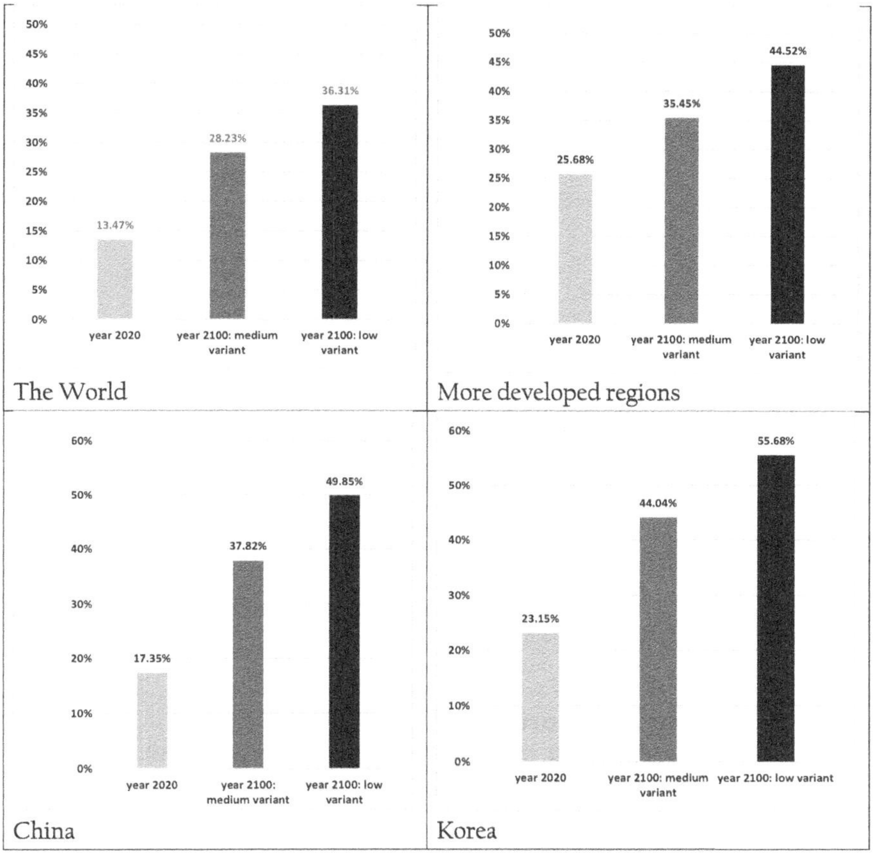


**Fig. 14.1** Dynamics of the share of older people (60+) in the total population (%) in the twenty-first century according to the medium and low scenarios of the UN population division forecast. *Source* The authors' calculation on the basis of the forecast estimates provided by the UN population division (2022a). Low variant produces a lower trajectory for the world population as it assumes a more pronounced fertility decline, and, hence, corresponds to a higher share of older people (60+) in the total population. The lower variant seems more realistic due to considerations spelled out by Vollset et al. (2020)

works,[4] we have shown in detail the further progression of the global aging process (Fig. 14.1 is presented for an additional illustration).

[4] Grinin et al. 2017a, 2017b, 2020, 2021a, 2021b; see also Grinin and Grinin, 2015; 2016; Grinin and Korotayev 2015; Grinin et al. 2016; Goldstone 2015; Goldstone et al., 2015.

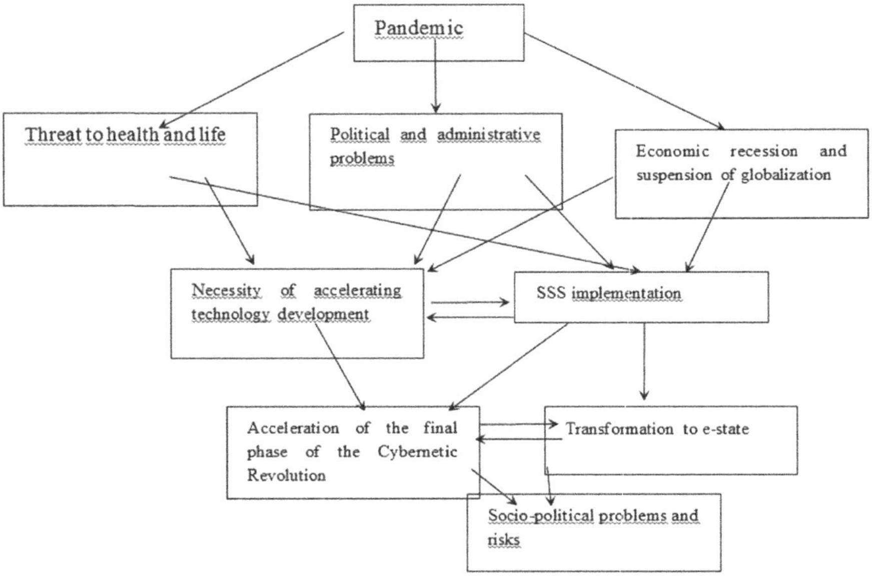


**Fig. 14.2** Impact of the pandemic on the acceleration of the cybernetic revolution and the transformation of e-government into e-state

## 14.3 How Can the Coronavirus Affect the Speed and Channeling of Advancement of the New Technological Wave?

Above we have argued that tackling the COVID-19 pandemic has acted as a powerful trigger accelerating the Cybernetic Revolution (see also Fig. 14.2). According to our forecast, and also due to the COVID-19 pandemic, the final phase of the Cybernetic Revolution will begin in the 2030s, that is, in the next 10 to 15 years.

But why do we consider the pandemic as a phenomenon that will inevitably require certain changes? There are two main reasons. First, the pandemic has generated a deep anxiety about the possibility of future pandemics. Second, the pandemic has also affected the development of MANBRIC technologies.

### 14.3.1 The Coronavirus Affects the Speed and Channeling of the Final Phase of the Cybernetic Revolution

The fact is that the COVID-19 pandemic has both undermined the idea of public safety and revealed the vulnerability of modern society to further pandemics. Also, it has not only posed additional threats to people's health and lives, especially to the elderly, but it has strongly begun to depress the economy more than other economic crises and recessions (see Abi Younes et al., 2020; Grinin, 2020;

Martin et al., 2020; OECD, 2020; Khan et al., 2021; Delardas et al., 2022). In fact, the fight against the pandemic has set back the economy and social well-being, while at the same time retarded globalization by suspending the travel and tourism industry: flights, trade, etc. Moreover, it has revealed the need for changes at the world-system level. Suddenly it has become clear that the pandemic threatens the development of humankind on many fronts; breaking this deadlock requires additional development of various technologies.

As we have seen, until now the above-mentioned problems are not easily resolved and will become more acute in the coming years. In addition, the outbreak of another pandemic caused by new pathogens is highly probable, and it is necessary to be ready for it. In fact, there is a dilemma: deal with the destructive impact of the pandemic, and generate herd immunity, or develop various technological, social and socio-technical innovations that will help to cope not only with the current pandemic but also with new ones. It is quite obvious that countries and the World System in general will choose the second option. Accordingly, as we have seen above and will see further, such technological needs for innovations are relevant to those directions which form the MANBRIC complex (Chauhan et al., 2020; Choong et al., 2020; Kimmig et al., 2020; Palestino et al., 2020; Patel & Gohil, 2021; Tang et al., 2021; Weiss et al., 2020; Yang, 2021). Both the pandemic and the need to overcome its negative impact, as well as the possibility of a new pandemic, will strongly stimulate a technological breakthrough and bring closer the beginning of the final phase of the Cybernetic Revolution. In particular, Abi Abi Younes et al. (2020) write that the pandemic has shown the need for coordination and openness at all stages of the research and product development process. We agree that scientific openness can accelerate the development breakthroughs in science and technology.

Let us explain why this can and should happen in the set of directions we have described and presented in Fig. 14.2. First, the fight against the pandemic and the need for testing and vaccination concentrate enormous financial and other resources, comparable to what we observe in the area of care for the elderly. And where large financial expenditures and interests are usually concentrated, there is a focus of political and economic life. Second, the pandemic demanded immediate solutions that became urgent priorities for governments (e.g., the supply of medicines or an increase in hospital bed capacity) and increased funding for the medical sphere. Third, this means involving the whole society, not just part of it. And this was almost immediately accepted by politicians and put on the agenda of their programs, which, in one way or another, began to be implemented. All these accelerated technological development and the final phase of the Cybernetic Revolution. The process of development and implementation of socio-technical self-regulating systems (SSSs), the way in which these systems will support the formation of the e-state, as well as the risks of these transformations will be discussed below.

### 14.3.2 The Growing Impact of MANBRIC Technologies on Development

**Medicine and Public Health Policy**. Let us recollect what we have maintained in Chaps. 4 and 8 about the integrating role of medical technologies in the MANBRIC convergence. There we connected the integrative importance of medicine in the MANBRIC complex with a number of factors. Namely, with its growing significance for aging society, with an increasing labor shortages, with the need to extend the working-age, and rehabilitate disabled people; with rapidly growing pension obligations (medicine can help here by extending the working age and improving the quality of biological life); with the growth of the middle class in developing countries and, in general, with an increase in living standards and education, which in turn leads to an increase in the medical expenditure. Even before the pandemic, medical research had a very prominent place in the public funding of science in many countries, for example, in the USA (Research America, 2017), Sweden (Statistics Sweden, 2020) and others. However, the COVID-19 pandemic has greatly increased the focus on health care. We will show below that the pandemic has acted as a trigger increasing the importance of medical and related technologies (see also Fig. 14.2).

The fight against the coronavirus has raised questions about a shortage of medical staff and the prompt treatment of patients. Robots in medicine, in particular robotic surgery (e.g., Kimmig et al., 2020), can by useful in this case (see Chap. 10). The safety of doctors can also be achieved through the development of robotics (Kimmig et al., 2020; Palestino et al., 2020; Tang et al., 2021) and telemedical technology (Houlding et al., 2021).

Urgent solutions to healthcare problems are becoming a factor of global, political and economic scale, without which the development of international contacts will stall. Countries need to unite in the fight against pandemics (Fisher & Wilder-Smith, 2020). The measures taken in this direction actually bring us closer the beginning of a new technological wave.

The COVID-19 pandemic, by placing medical technology, the organization of healthcare and medical control in society at the center of public attention, has stimulated change and acceleration not only in technological innovation but also in social, administrative and even political relations. All this represents in many respects a single complex of transformations that can be productively studied as a system (see Fig. 14.2). The socio-political and economic role of medicine has sharply increased due to the large number of COVID infections. This significantly changed the government's attitude to healthcare services and its funding, including the redistribution of the budget in favor of medical expenses (Abi Younes et al., 2020; Anderson et al., 2020; Hartman et al., 2022), which in turn led to an increase in the technological component of this set of transformations (Abi Younes et al., 2020; Brem et al., 2021; Research 2021). In some cases, the redistribution of funds for R&D has favored a number of medical and biotechnology industries at the expense of other S&T sectors (e.g., Abi Younes et al., 2020; Agarwal & Gaule, 2022; Basu et al., 2020); which is also in line with the WHO guidelines (see, e.g., WHO, 2020a, 2020b). Thus, the needs created by the healthcare crisis have dramatically accelerated the adoption of a wide range of technologies, and many companies have moved quickly in this direction (Schwab and Malleret 2020; Amankwah-Amoah et al., 2021).

Above (see also Chap. 4), we have listed the main factors that would allow medicine, as an integrating part of the MANBRIC convergence, to take center stage in both the technological wave and the solution of social problems (including acute labor shortages, pressure on pension and social obligations etc.). All these remain and will have their effect. However, because of the COVID-19 pandemic, the problems of health care unexpectedly came to the center of public attention earlier and appeared to be more acute than they seemed before 2020. The pandemic has intensified the role of medicine as an integrating part of MANBRIC convergence.

**Bio-, Nano- and Additive Technologies**. The role of red biotechnology has also increased dramatically. We have discussed in Chaps. 8 and 9 the role of new vaccines as an impetus for innovative breakthroughs in red biotechnology and genetic engineering (see also, e.g., Abi Younes et al., 2020; Brüssow, 2022). Funding and application of these biotechnologies, especially vaccines, have dramatically increased (Chung et al., 2020; Zimmer et al., 2020; M. Kumar et al., 2021a, 2021b). It is of the utmost importance that by the end of 2020 several vaccines from different countries and producers had become available.[5] Competition among countries to develop vaccines has intensified, which is positive and will contribute to the development of medicine and biotechnology in general. The urgency of safety issues and the occurrence of emergency situations have helped to remove some of the obstacles to the development of science and technology, especially to the approval of drugs and vaccines nationwide and around the world. The approval system for new drugs and vaccines is changing dramatically as a result of the COVID-19 crisis, and it became much simpler and faster. However, as we pointed out in Chaps. 8 and 9, COVID-19 vaccines have had an ambiguous effect. One should note that as to these vaccines, we observe significant progress, since vaccines have been developed and introduced based on new principles, and this, undoubtedly, contributes to the development of red biotechnology (Brem et al. 2021; Chung et al., 2020; Musyuni et al., 2020; Abi Younes et al., 2020; Zimmer et al., 2020). But at the same time, this progress also has come at a very high price because of excess mortality and a huge number of side effects due to a great speed with which this progress has been achieved. However, we suppose that, one way or another, this largest experiment in history will sooner or later result in the development of biotechnological science. But there is no doubt that the price paid for this new science could have been much lower. Thus, although they have not yet shown their effectiveness and have had very large and ever-increasing side effects,

[5] It is worth bearing in mind that before the COVID-19 more than a hundred different vaccines were already being developed by dozens of companies in the world (Corey et al., 2020; Jeyanathan et al., 2020; Remmel 2021).

we believe that they open a path forward to new technologies that will hardly be smooth (as we have already observed), but which will finally have a considerable importance in the future.

The pandemic has given impetus to the development of a number of other technologies—in particular, additive technologies (Choong et al., 2020; Patel & Gohil, 2021), which also have important medical applications (Choong et al., 2020). The development of nanotechnology development has also been stimulated (Tang et al., 2021; Yang, 2021), since the need for new materials (including for the protection of doctors) impacts the development of these materials. Nanotechnology is especially urgent in the search for new principles for the development of vaccines and drugs. Since modern biotechnology operates at the nanoscale, the development of effective nanocarriers is extremely important for overcoming the limitations of traditional antiviral therapy (Chauhan et al., 2020). In the future, cognitive technologies may also be widely used in medicine and other areas.

**Information Technologies and AI**. The need for isolation and social distancing has led to a dramatic rise in online technologies (Renu, 2021). This has led to an increase in the use of information technologies and their qualitative growth, including in areas like remote/telemedicine healthcare services (e.g., Basu et al., 2020; Shah et al., 2020), or remote psychological support (Liu et al., 2020a, 2020b), the use and development of biometrics, including a contactless form of biometrics (Kumari & Seeja, 2021). Let us also mention distance learning (Rapanta et al., 2020), which has already been actively developing (Ashman et al., 2014), but has now become a necessity (Dhawan, 2020). One can agree with the forecast of the significant impact of the COVID-19 crisis on the speed, implementation and direction of ICT (Abi Younes et al., 2020). Internet commerce has also demonstrated explosive growth which has increased the demand for innovations in robotics, including drones (Estrada & Arturo, 2020; Jat & Singh, 2020; Skorup & Haaland, 2020).

In addition, it is important to note that the transition to anti-crisis management of society and monitoring of the implementation of anti-epidemic regulations, including the tracking of movements and contacts, has caused a surge in demand for artificial intelligence technologies, especially for medical administrative measures. This has led to the introduction of techno-social innovations into society's life. We believe that these measures can accelerate the development of social-technical self-regulating/self-managing systems, change social relationships, as well as increase confrontation in society (see below).

We would like to conclude this section with the idea that the above-described phenomena have increased the interdependence between the branches of MANBRIC, in particular medicine, nanotechnology biotechnology, AI and additive technologies (Chauhan et al., 2020; Choong et al., 2020; Palestino et al., 2020; Tang et al., 2021; Weiss et al., 2020).

Further, we want to consider (1) the development of socio-technical self-regulating systems (SSS) and the way these systems can change social and administrative relations and support the development of the e-state; (2) some prospects and risks of these transformations, accelerated by the COVID-19.

## 14.4 Socio-technical Self-regulating/Self-managing Systems (SSSs) and the E-state

### 14.4.1 The Impact of COVID-19 on the Development of Socio-technical Self-regulating Systems (SSS) and How This May Affect Social Behavior and the Emergence of the E-state

In recent decades, the development of information technologies and artificial intelligence has produced a noticeable impact on social relations. This impact can be traced in many areas, in particular, in the structure of employment, the disappearance of some occupations and the emergence of others (see also Chap. 10). Opportunities for distance learning, jobs and services, including health care, are also expanding rapidly. In addition, much has changed in terms of the creation, distribution, transformation and analysis of information and control over information flows. The development of technologies for tracking human activity leads to the commercialization and actual alienation of human behavior and privacy that were previously considered inalienable (desires, inclinations, habits, etc.).[6] It has long been observed that a new structure of power over people is emerging on the Internet. This state of affairs has serious implications for democracy (Schwartz, 1999). Today's AI capabilities already pose systemic dangers to society and human rights.

As mentioned in the introduction to this chapter, there is a growing problem of intrusion into people's private lives, this intrusion occurs with the help of artificial intelligence technologies, which will jeopardize people's privacy through constant access to personal (including intimate) information. Moreover, personalization as an important characteristic of the Cybernetic Revolution is rapidly developing (see Chap. 3; see also Ashman et al., 2014; Grinin and Grinin, 2016; Grinin et al., 2021a, 2021b), which exacerbates the problem of privacy breaches. Therefore, it is extremely important to analyze the risks of this process. One should realize that information privacy is a value that helps shape the society in which we live and our individual identity (see Chaps. 1 and 10; Schwartz, 1999; Ashman et al., 2014; Cecere et al., 2015; Moustaka et al., 2019; Schwartz, 1999; Solove, 2008; Schwab, Malleret 2020; see also Brammer et al., 2020; Alharbi, 2020). The urgent need to monitor compliance with security measures during the pandemic exacerbated this problem and at the same time showed that further development in this direction will lead to qualitative changes. The fact is that new combined technological systems are being created based on AI technologies that collect, store and analyze information about billions of people, as well as networked ICT. Such systems are aimed at administrative, legal, social and even political regulation and control over the behavior of individuals, social groups and even society as a whole, up to and including the regulation of the World System. SSS will be used to regulate many

---

[6] The term alienation is borrowed from terminology of Karl Marx (see about types of alienation in Chap. 7).

legal aspects of human existence, sometimes also hierarchical relations in society, and may further cause the formation and widespread use of social ratings (similar to the Social Credit System in China). In short, there is a tendency to delegate more and more tasks from the authorities to socio-technical systems (Plebe & Perconti, 2020). At present, such systems are actively implemented, for example, in face recognition, location tracking systems, traffic control (e.g., Transparency Market Research, 2020), imposition of fines, electronic registration, issuance of documents, and many others that were previously the prerogative of the authorities.

*Socio-technical self-regulating/self-managing systems (SSS)* perform social and administrative functions (that is, control, verification, distribution, security, rating and other functions) using a set of technologies in the absence or little participation of officials and specialists. The development of SSS is progressing rapidly. In some cases, they are beginning to be used in order to impose certain patterns of behavior on people. The most notable of these is China, where officials have developed a Social Credit System for individuals, businesses and the government to control and assess their reliability, especially to control non-urban residents (Kuznetsova & Mashkina, 2020). Many projects of the Social Credit System have been implemented only as regional pilot programs, although there are plans to distribute them countrywide (Chin & Wong, 2016; Creemers, 2018), and which also serve as a model for other countries (Síthigh & Siems, 2019). The year 2020 was in some ways a turning point in this regard. For example, the use of facial recognition systems (including masked faces [Sulochanan Karthick Ramanathan et al., 2021]) increased by around 7% in 2021 and by further 12% in 2022. Famous VPN company reported the presence of location tracking systems already in 42 messenger apps with at least 187 million downloads (Sean, 2021) and it has rapidly increased during the pandemic (Stanley & Granick, 2020).

### 14.4.2 SSS and the Growth of E-regulations

The pandemic and the emerging need to control anti-epidemic measures have dramatically accelerated the development of such SSS. In particular, electronic passport systems, movement control, penal systems, accounting systems, systems of businesses control and many other systems emerged as a result of the lockdowns and they can be considered as prototypes of SSS (Lyon, 2001; Ram & Gray, 2020; Richards, 2012; Wang et al., 2020; Zhuang et al., 2021). Various projects began to appear: electronic passports (COVID immunity passports) (Hasan et al. 2020; WHO, 2020a, 2020b), QR codes which are supposed to record all vaccinations (Barnes, 2021; Mbunge et al., 2022), etc. Thus, the coronavirus pandemic has shown that the introduction of SSS can become a fast-developing process which will take 10–20 years. At the same time, the idea of creating immunity passports goes beyond the national level in cooperation with the largest IT companies (Leswing, 2021). In the future, one may expect that the development of the socio-technical self-regulating systems will proceed in different directions while the development in the field of medicine will continue. As we discuss in Chaps. 6,

7 and 8, super-complex self-regulating systems may emerge, which we call a medico-biotechnological environment, for continuous monitoring of health and critical parameters of not only of patients in hospitals but also of people at home and in various places, including the chronically ill and elderly, as well as many other people.

The need to monitor compliance with COVID-19 restrictions led to the active use of SSS technologies by the authorities. However, their introduction has not only improved the control but has also created some problems, e.g., related to the techno-social ranking of people (this issue has been discussed for a long time [but with reference to commercial companies] [see, e.g., Lyon, 2001]). Anyway, the positive aspects of SSS are quite obvious, which is why they are spreading. SSS can significantly improve the social environment in many different ways. Some evidence of this has emerged already during the COVID-19 pandemic (Mhlanga, 2022). For example, in China several AI programs were used to treat mental health crises during the pandemic (Zhu et al., 2022). As a result, people at risk of suicide can be identified using special AI software that tracks and analyzes posts on Weibo and alerts specific volunteers to act accordingly (Liu et al., 2020a, 2020b).

Since such SSSs can regulate an entire society, it is already extremely important today to clearly regulate the possibilities for manipulating personal data and using these data for social regulation. It is necessary to take into account the existence of different types of behavior that have the right to exist in society. Meanwhile, the SSSs can lead to total unification, thereby suppressing the individuality and special abilities of people. The emergence and development of SSSs will lead to serious changes in the political and social spheres. In particular, we also believe that the widespread introduction of SSSs can accelerate the change in social relations, as well as increase confrontation in society, because the use of self-regulating systems in governance will lead to increased technological and political control. This is all the more dangerous because, although such systems are not yet widespread in private life, they will be. Sooner or later, therefore, both the introduction of socio-technical self-regulating systems and the development of the global aging process will have to be adapted and institutionalized and then anchored in laws or stable relationships in such a way as to optimize interests and minimize dangerous consequences (for details see Chap. 15). If such an institutionalization does not take place, then the following will intensify: (a) social tensions and (b) the negative impact of the growing proportion of the elderly on democratic institutions, which, as mentioned above, will be more noticeable in the direct functioning of democracy.

The idea of e-government (see below) and the application of AI and ICT are closely related to the impact that science and technology can have on privacy standards taking into account the opportunities to access and collect personal information and use it against the interests of people. However, this is not a new issue, as it appeared even before the emergence of Internet. Alan Westin, in his book *Science, Privacy, and Freedom* (Westin, 1966), analyzed the conflict between privacy and the use what was then the latest surveillance technologies. Since that time, communication technologies, data analysis and surveillance technology have advanced considerably. As a result, the problem has become even much more acute. A number of works have already been devoted to the analysis of its various aspects in the present and future (e.g., Ashman et al., 2014; Cecere et al., 2015; Moustaka et al., 2019; Schwartz, 1999; Solove, 2008; see also Brammer et al., 2020; Alharbi, 2020).

### 14.4.3 SSSs: Risks and Protests

However, such a rapid and uncontrolled implementation also carries serious risks. Obviously, there is a danger of increasing standardization of people and a loss of their individuality. One of the main problems is that over many decades people have become accustomed to freedom of behavior, and this kind of regulations has been causing protests since 2020 (Brennan, 2020; Kowalewski 2021). As a result, SSS can lead to the emergence of a considerable behavioral stratification in society, which will negatively affect the consolidation of society. This is especially true with respect to the elderly due to the fact that they are particularly ulnerable to COVID-19 (Dadras et al., 2022; Fraser et al., 2020). But let us point to a serious contradiction here. On the one hand, it seems logical that tighter restrictions are required for the elderly, but on the other hand, given the fact that, as we saw in Chap. 13, the older population is generally more conservative in their habits, all kinds of SSS implementations that reduce individualization can be perceived by them as particularly painful. Accordingly, the level of frustration among this population will be noticeably higher.

In general, it is not surprising that we observe a sharp increase in public and governmental anxiety and negative reactions on part of large sectors of society (Brammer et al., 2020) even resulting in massive protests against the COVID-19 restrictions. At the same time, we observe the division of society into groups caused by medical measures (Brennan, 2020; Carothers and Press 2020; Kowalewski 2021), and also cases of discrimination in relation to medical care, vaccines, etc. (Parolin & Lee, 2022; Sohil & Sohail, 2022).

By some estimates, since March 2020, there have been more than 100 mass protests in 56 countries against the coronavirus restrictions. Such demonstrations occur not only in developed countries such as Australia, Germany, Canada and the USA but also in poorer countries such as Brazil, Mexico and Nigeria, and in many other regions of the world. Moreover, opposition to isolation occurs in states with very different political regimes—both in consolidated autocracies and in liberal democracies (Carothers and Press 2020; Wood et al., 2022). In January 2021 quite violent protests against a new wave of restrictions took place in a number of countries, for example, in the Netherlands and Georgia. Especially large protests of this type were observed in China between November 2 and December 5, 2022 (Burki, 2023).

And since very powerful forces and huge financial and political interests will be concentrate on innovations related to medical control of (and influence over) particular populations, these innovations can cause serious conflicts and may affect all societies.

Nevertheless, most people are likely to agree with this form of regulation, although there is no doubt that such an attack on human rights (Gerstenfeld, 2020; see also Kowalewski, 2021) can not only trigger a wave of protests but also split society, cause strong behavioral stratification, social tensions and turbulence (Wood et al., 2022). However, there may appear a large number of outsiders, that is, a mass of people who fail to fit into the rather strict requirements of the SSS and the electronic state. There will also be an increase in deviant behaviour, such as rejection of such requirements (e.g., Kowalewski, 2021).

The processes of active implementation of SSS and other information and digital technologies, as described above combined with the COVID-19 restrictions, have intensified the role of the state, which to some extent is beginning to transform into an electronic/digital state (or e-state). However, the adaptation to such an e-state will be difficult and conflict-prone.

### 14.4.4 Toward Electronic State?

The formation of SSSs will lead to serious changes in the political spheres: (a) as a result of the spread of SSSs, the number of officials will decrease, interaction between citizens and state bodies will become possible bypassing local and intermediary bodies; in general, this will greatly affect the number of people employed in public administration; (b) the state will become cheaper, but at the same time digital control will increase, which can cause strong behavioral stratification and protests. As a result, such developments and changes will lead to the formation of an e-state.

What is e-state? Traditionally (according to the United Nations e-government development database), e-government has been defined as the use of ICTs to improve the efficiency of government institutions and deliver government services online (UNeGovDD, 2013). In recent decades, many aspects of e-government and its prospects have been studied (e.g., Malodia et al., 2021; Pérez-Morote et al., 2020; Silcock, 2001), as well as problems in its development, citizens' attitude toward it, and the political and legal framework that determine the level of e-government (Glyptis et al., 2020). In many respects, we already are living in a society with an e-government and feeling the impact of e-government on our privacy standards. Thus, when we talk about the future, it is important that the concept of *e-state* is much more advanced in this respect than e-government. We understand e-state as a *state with a significantly reduced number of state supervisory bodies, mainly based on SSS technology*. What will be the changes associated with the establishment of the e-state?

- One of these major changes will be the transformation of public administration toward the increasing use of electronic, automated forms of interaction and control. In the language of cybernetics, SSS will create a new communicative circuit in the management of society, or more precisely, it will change the contours of the relationship between the center and the periphery of society. In particular, it can increase the opportunities for rapid feedback from the public. The influence of the SSS will be perceived and will significantly change or even revolutionize the nature of governance.

- Major changes can be expected in the structure of the administrative apparatus. Some administrative units (e.g., local governments, officials, special control bodies, etc.), which are important for the interaction between the central government and the population, will disappear or will be transformed. As a result, the facilitated interaction between a citizen and the state will affect many officials and may deprive them of their jobs, that is, in many respects, the establishment of the e-state will be a revolutionary transformation.
- This will significantly reduce the cost of the management process, so the state will need far less money for its maintenance. This will be, of course, a positive result.
- However, at the same time, the e-state will significanly increase in size and extent causing alienation and frustration in society. The fact is that the power of the state will increase, and management will become less flexible since it will largely depend on technology and the human factor of programmers. A historical analogy suggests itself. In the first half of the nineteenth century, workers had to adapt their physical, psychological and mental abilities to the needs of machine production. This led not only to the alienation of labor (according to Marx) but also to a large extent to the alienation of the personality of the worker (see Grinin and Grinin, 2015). Like this described process, the technological features of social control in the future may force people to adapt to them, causing alienation and frustration. This will also be enhanced by the fact that electronically organized authorities will look like an impersonal, unusual and, therefore, alien government. It should be added that, in addition to AI, robots will be actively involved in the work of the SSS and certain parts of the e-state work. Robots can partially or completely replace certain groups of civil servants (policemen, firefighters, etc. [see Chap. 10)]), and at the same time, their interaction not only with other robots, but also with humans and other SSSs will be vital. But in this case, the difficulties of adapting people to SSSs and the e-state will naturally increase significantly, and the feeling of alienation and loss of personalization will intensify.
- The introduction of the e-state can affect democratic procedures. In particular, it is quite possible that a transition from representative to direct democracy will take place. Direct democracy is more efficient in getting feedback from the population quickly, but it can undermine the stability of governance (especially in large or highly divided societies) because of the rapid and frequent changes in the public mood. Moreover, it will become difficult to make long-term decisions. It should be noted that direct democracy can increase the influence of the growing older part of society.

## 14.5 Concluding Remarks

The COVID-19 pandemic has become a phenomenon that, for the first time in a long time, has drawn public attention to health care problems. It has become a powerful trigger accelerating technological development in medicine and related areas (what we call MANBRIC—convergence: medicine—additive—bio-nano-info technologies—robotics—cognitive technologies) and, at the same time, catalyzing the convergence of these directions and changes in socio-political (and socio-administrative) relations. Our analysis of the COVID-19 pandemic as a trigger for the acceleration of the start of the final phase of the Cybernetic Revolution confirms the idea of the integrating role of medical technologies in the MANBRIC convergence.

The measures taken against the coronavirus pandemic, together with new technologies applied and the SSSs, will undoubtedly provide many tools to tackle new pandemics with unknown pathogens. If such negative events occur, it can be expected that they will also increase the role of medicine as a technological and social factor and channel this process in terms of increasing attention to technological development and, sooner or later, minimizing the risks associated with this development. This has a number of practical implications.

However, a technological revolution is always associated with radical socio-political changes. In the present chapter, we have discussed two major changes of this kind: (1) and development of the SSSs. These are a specific type of self-regulating systems, namely social-technical self-regulating/self-managing systems, which help to regulate social, administrative or even political relations. They can be used by the authorities at different levels and by the state as a whole (as well as by the management of service centers where the authorities consider such regulation necessary: airports, crowded places, etc.). (2) Formation of the electronic state (e-state). The development of the SSSs, in one way or another, pushes society toward the formation of an e-state, that is, a state with a greatly reduced number of administrative bodies, which relies mainly on technologies in the form of SSS and AI for its functioning. Since the measures taken against the pandemic have accelerated the development of SSS and increased the demand for such a technology, it has been shown that the COVID-19 pandemic was able to accelerate important innovations, not only related to medicine, biotechnology and some other technologies but also related to the development of social relation technologies and the e-state. Consequently, societies and the World System have received a strong impetus to move in the direction we have defined.

However, let us reiterate that such a transformation of society can be quite painful, it inevitably causes and will cause strong social and political tensions, protests and conflicts, while trust in governments may decline (Strandberg, 2020; Vieten, 2020). We witnessed this in 2020–2022 (Brennan, 2020; Kowalewski, 2021; van der Zwet et al., 2022), while protests often took the form of violent clashes, occasionally escalating to full-scale battles (Carothers and Press 2020; ACLED 2020). We observed a return to the pre-pandemic logic of protest that under lines the depth of the crisis of power revealed by the COVID-19, during which inequality increased (Gerbaudo, 2020). Therefore, it is important to study these types of social and political risks, which in some cases can lead a society into a deep crisis. Due to the negative impact of the SSSs on the violation of citizens' rights, at first, the expansion of the employment of SSSs may cause an additional growth of frustration, alienation and conflicts in society. We believe that due to the mutual adjustment of the SSSs and the mentality of the society, the former will gradually improve and adapt more and more to the positions of different strata within society until the use of SSSs reaches the optimum. However, the process of adjustment of new socio-political institutions and technologies, as well as of the e-state, on the one hand, and the interests of different social strata, constitutional rights, freedoms and mentality, on the other, will not be easy (for the arguments in favor of increasing unrest, see Reicher & Stott, 2020). It is necessary to develop completely new laws that would optimize socio-technological interventions in people's lives, especially intrusion in their privacy. It will be a difficult process of trial and error to move forward on the path towards harmonizing technological processes with social life.

**Acknowledgements** This chapter is an output of a research project implemented as part of the Basic Research Program at HSE University in 2024 with support from the Russian Science Foundation (Project No. 23-18-00535).

## References

Abi Younes, G., Ayoubi, C., Ballester, O., Cristelli, G., de Rassenfosse, G., Foray, D., et al. (2020). COVID-19: Insights from innovation economists. *Science and Public Policy, 47*(5), 733–745. https://doi.org/10.1093/scipol/scaa028

ACLED [= Armed Conflict Location & Event Data Project]. (2020). *COVID-19 and Political Unrest in Northeast India*. ACLED. https://www.jstor.org/stable/resrep26621?seq=1#metadata_info_tab_contents

Agarwal, R., & Gaule, P. (2022). What drives innovation? Lessons from COVID-19 R&D. *Journal of Health Economics, 82*, 102591.

Agrawal, S., Makuch, S., Dróżdż, M., Strzelec, B., Sobieszczańska, M., & Mazur, G. (2021). The impact of the COVID-19 emergency on life activities and delivery of healthcare services in the elderly population. *Journal of Clinical Medicine, 10*(18), 4089.

Alharbi, F. S. (2020). Dealing with data breaches amidst changes in technology. *International Journal of Computer Science and Security (IJCSS), 14*(3), 108–114.

Amankwah-Amoah, J., Khan, Z., Wood, G., & Knight, G. (2021). COVID-19 and digitalization: The great acceleration. *Journal of Business Research, 136*, 602–611.

Anderson, J., Bergamini, E., Brekelmans, S., Cameron, A., Darvas, Z., Jíménez, M. D., & Midões, C. (2020). The fiscal response to the economic fallout from the coronavirus. *Bruegel datasets*.

Ashman, H., Brailsford, T., Cristea, A. I., Sheng, Q. Z., Stewart, C., Toms, E. G., & Wade, V. (2014). The ethical and social implications of personalization technologies for e-learning. *Information & Management, 51*(6), 819–832. https://doi.org/10.1016/j.im.2014.04.003

Aslam, F., Aziz, S., Nguyen, D. K., Mughal, K. S., & Khan, M. (2020). On the efficiency of foreign exchange markets in times of the COVID-19 pandemic. *Technological Forecasting and Social Change, 161*, 120261. https://doi.org/10.1016/j.techfore.2020.120261

Atkeson, A. (2020). *What will be the economic impact of COVID-19 in the US? Rough estimates of disease scenarios* (No. w26867) (p. w26867). National Bureau of Economic Research. https://doi.org/10.3386/w26867

Baker, S. R., Bloom, N., Davis, S. J., & Terry, S. J. (2020). *Covid-induced economic uncertainty*. National Bureau of Economic Research. https://www.nber.org/papers/w26983

Barnes, R. (2021). Microsoft, oracle and other tech giants team up on Covid-19 vaccine "passports"—Tech Compsmag. *News—Tech News, Latest News, Business News and world News*. Retrieved February 3, 2021, from https://www.compsmag.com/news/technology/microsoft-oracle-and-other-tech-giants-team-up-on-covid-19-vaccine-passports/

Basu, S., Phillips, R. S., Phillips, R., Peterson, L. E., & Landon, B. E. (2020). Primary care practice finances in the United States Amid the COVID-19 pandemic: Study estimates the potential impact of COVID-19 on operating expenses and revenues of primary care practices. *Health Affairs, 39*(9), 1605–1614. https://doi.org/10.1377/hlthaff.2020.00794

Belhadi, A., Kamble, S., Jabbour, C. J. C., Gunasekaran, A., Ndubisi, N. O., & Venkatesh, M. (2020). Manufacturing and service supply chain resilience to the COVID-19 outbreak: Lessons learned from the automobile and airline industries. *Technological Forecasting and Social Change, 163*, 2–19. https://doi.org/10.1016/j.techfore.2020.120447

Bhat, S. A., Bashir, O., Bilal, M., Ishaq, A., Dar, M. U. D., Kumar, R., & Sher, F. (2021). Impact of COVID-related lockdowns on environmental and climate change scenarios. *Environmental Research, 195*, 110839.

Bortoló, G. M., Valdés, J. Á., & Nicolas-Sans, R. (2023). Sustainable, technological, and innovative challenges post Covid-19 in health, economy, and education sectors. *Technological Forecasting and Social Change*, 122424.

Brammer, S., Branicki, L., & Linnenluecke, M. (2020). COVID-19, societalization and the future of business in society. *Science of the Total Environment, 34*(4), 2–7. https://doi.org/10.5465/amp.2019.0053

Brem, A., Viardot, E., & Nylund, P. A. (2021). Implications of the coronavirus (COVID-19) outbreak for innovation: Which technologies will improve our lives? *Technological Forecasting and Social Change, 163*, 120451. https://doi.org/10.1016/j.techfore.2020.120451

Brennan, E. (2020). *Coronavirus and protest: How Covid-19 has changed the face of American activism*. United States Studies Centre. https://www.ussc.edu.au/analysis/coronavirus-protest-how-covid-19-has-changed-the-face-of-american-activism.

Brüssow, H. (2022). MRNA vaccines against COVID-19: A showcase for the importance of microbial biotechnology. *Microbial Biotechnology, 15*(1), 135–148.

Burki, T. (2023). Moving away from zero COVID in China. *The Lancet Respiratory Medicine, 11*(2), 132.

Carothers, T., & Press, B. (2020). *The global rise of anti-lockdown protests—and what to do about it*. World Politics Review, Benjamin Press. *Retrieved* November 6, 2020, from https://www.worldpoliticsreview.com/articles/29137/amid-the-covid-19-pandemic-protest-movements-challenge-lockdowns-worldwide.

Cecere, G., Le Guel, F., & Soulié, N. (2015). Perceived Internet privacy concerns on social networks in Europe. *Technological Forecasting and Social Change, 96*, 277–287. https://doi.org/10.1016/j.techfore.2015.01.021

Chakraborty, I., & Maity, P. (2020). COVID-19 outbreak: Migration, effects on society, global environment and prevention. *Science of the Total Environment, 728*, 2–7. https://doi.org/10.1016/j.scitotenv.2020.138882

Chauhan, G., Madou, M. J., Kalra, S., Chopra, V., Ghosh, D., & Martinez-Chapa, S. O. (2020). Nanotechnology for COVID-19: Therapeutics and vaccine research. *ACS Nano, 14*(7), 7760–7782. https://doi.org/10.1021/acsnano.0c04006

Chin, J., & Wong, G. (2016). China's new tool for social control: A credit rating for everything. *The Wall Street Journal, 28*. https://www.wsj.com/articles/chinas-new-tool-for-social-control-a-credit-rating-for-everything-1480351590.

Choi, Y., Kim, H. J., & Lee, Y. (2022). Economic consequences of the COVID-19 pandemic: Will it be a barrier to achieving sustainability? *Sustainability, 14*(3), 1629.

Choong, Y. Y. C., Tan, H. W., Patel, D. C., Choong, W. T. N., Chen, C.-H., Low, H. Y., et al. (2020). The global rise of 3D printing during the COVID-19 pandemic. *Nature Reviews Materials, 5*(9), 637–639. https://doi.org/10.1038/s41578-020-00234-3

Chumakov, A. N., & Yurchenko, P. S. (2021). Global pandemic in terms of objective causes and subjective speculations. *Journal of Globalization Studies, 12*(2), 157–168.

Chung, Y. H., Beiss, V., Fiering, S. N., & Steinmetz, N. F. (2020). COVID-19 vaccine frontrunners and their nanotechnology design. *ACS Nano, 14*(10), 12522–12537. https://doi.org/10.1021/acsnano.0c07197

Corey, L., Mascola, J. R., Fauci, A. S., & Collins, F. S. (2020). A strategic approach to COVID-19 vaccine R&D. *Science, 368*(6494), 948–950. https://doi.org/10.1126/science.abc5312

Court, E. L., Lennon, N. A., Zarka, A., Pryor, M., Lawrence, S., & Alexander, R. J. (2021). COVID as a driver of innovation: getting the right patient to the right place at the right time. A novel registrar-led approach to minimising unnecessary footfall in the surgical assessment unit of a district general hospital. *British Journal of Surgery, 108*(Supplement_7), znab361–177.

Creemers, R. (2018). *China's social credit system: An evolving practice of control (SSRN Scholarly Paper No. ID 3175792)*. Social Science Research Network. https://doi.org/10.2139/ssrn.3175792.

Cruz-Cárdenas, J., Zabelina, E., Guadalupe-Lanas, J., Palacio-Fierro, A., & Ramos-Galarza, C. (2021). COVID-19, consumer behavior, technology, and society: A literature review and bibliometric analysis. *Technological Forecasting and Social Change, 173*, 121179.

Dadras, O., SeyedAlinaghi, S., Karimi, A., Shamsabadi, A., Qaderi, K., Ramezani, M., & Voltarelli, F. A. (2022). COVID-19 mortality and its predictors in the elderly: A systematic review. *Health Science Reports, 5*(3), e657.

Damaševičius, R., & Zailskaitė-Jakštė, L. (2023). Impact of COVID-19 pandemic on researcher collaboration in business and economics areas on national level: A scientometric analysis. *Journal of Documentation, 79*(1), 183–202.

Delardas, O., Kechagias, K. S., Pontikos, P. N., & Giannos, P. (2022). Socio-economic impacts and challenges of the coronavirus pandemic (COVID-19): An updated review. *Sustainability, 14*(15), 9699.

Dhawan, S. (2020). Online learning: A panacea in the time of COVID-19 crisis. *Journal of Educational Technology Systems, 49*(1), 5–22. https://doi.org/10.1177/0047239520934018

Efremov, I., & Arkhangelsky, V. (2023). Global fertility changes in the context of the COVID-19 pandemic. *Journal of Globalization Studies, 14*(2), 165–170.

Ehnts, D., & Paetz, M. (2021). COVID-19 and its economic consequences for the Euro Area. *Eurasian Economic Review, 11*, 227–249.

Estrada, R., & Arturo, M. (2020). *The uses of drones in case of massive epidemics contagious diseases relief humanitarian aid: Wuhan-COVID-19 crisis* (SSRN Scholarly Paper No. ID 3546547). Social Science Research Network. https://doi.org/10.2139/ssrn.3546547.

Fabris, N. (2022). Impact of COVID-19 pandemic on financial innovation, cashless society, and cyber risk. *Economics, 10*(1), 73–86.

Fisher, D., & Wilder-Smith, A. (2020). The global community needs to swiftly ramp up the response to contain COVID-19. *The Lancet, 395*(10230), 1109–1110. https://doi.org/10.1016/S0140-6736(20)30679-6

Forster, P. M., Forster, H. I., Evans, M. J., Gidden, M. J., Jones, C. D., Keller, C. A., et al. (2020). Current and future global climate impacts resulting from COVID-19. *Nature Climate Change, 10*(10), 913–919. https://doi.org/10.1038/s41558-020-0883-0

Fraser, S., Lagacé, M., Bongué, B., Ndeye, N., Guyot, J., Bechard, L., et al. (2020). Ageism and COVID-19: What does our society's response say about us? *Age and Aging, 49*(5), 692–695. https://doi.org/10.1093/aging/afaa097

Gerbaudo, P. (2020). The pandemic crowd. *Journal of International Affairs, 73*(2), 61–76.

Gerstenfeld, M. (2020). *How coronavirus emergency measures threaten civil rights* (pp. 68–71). Begin-Sadat Center for Strategic Studies. Retrieved January 20, 2021, from https://www.jstor.org/stable/resrep26356.18

Glyptis, L., Christofi, M., Vrontis, D., Giudice, M. D., Dimitriou, S., & Michael, P. (2020). E-Government implementation challenges in small countries: The project manager's perspective. *Technological Forecasting and Social Change, 152*, 2–11. https://doi.org/10.1016/j.techfore.2019.119880

Goldstone, J. (2015). Population aging and global economic growth. In J. Goldstone, L. Grinin, & A. Korotayev (Eds.), *History & mathematics: demography & aging* (pp. 147–155). Uchitel.

Goldstone, J., Grinin, L., & Korotayev, A. (2015). Research into global aging and its consequences. In J. Goldstone, L. Grinin, & A. Korotayev (Eds.), *History & mathematics: Political demography & global aging* (pp. 5–9). Uchitel.

Grinin, L. (2020). How can COVID-19 change geopolitics and economy? *Journal of Globalization Studies, 11*(2), 119–132.

Grinin, A., & Grinin, L. (2015). The cybernetic revolution and historical process. *Social Evolution and History, 14*(1), 125–184.

Grinin, L., & Korotayev, A. (2015). *Great divergence and great convergence a global perspective.* Springer.

Grinin, L., Grinin, A., & Korotayev, A. (2017a). Forthcoming Kondratieff wave, cybernetic revolution, and global aging. *Technological Forecasting and Social Change, 115*, 52–68. https://doi.org/10.1016/j.techfore.2016.09.017

Grinin, L., Grinin, A., & Korotayev, A. (2020). A quantitative analysis of worldwide long-term technology growth: From 40, 000 BCE to the early 22nd century. *Technological Forecasting and Social Change, 155*, 119955. https://doi.org/10.1016/j.techfore.2020.119955

Grinin, L., Grinin, A., & Korotayev, A. (2021a). COVID-19 Pandemic as a trigger for the acceleration of the Cybernetic Revolution, transition from E-Government to E-State, and change in social relations. *Technological Forecasting & Social Change, 175*, 1–16. https://doi.org/10.1016/j.techfore.2021.121348.

Grinin, L., Grinin, A., & Korotayev, A. (2021b). Global Trends and forecasts of the 21st Century. *World Futures, 77*(5), 335–370. https://doi.org/10.1080/02604027.2021.1949939.

Grinin, L., Grinin, A., & Korotayev, A. (2022). COVID-19 pandemic as a trigger for the acceleration of the cybernetic revolution, transition from E-government to E-state, and change in social relations. *Technological Forecasting and Social Change, 175*, 1–15. https://doi.org/10.1016/j.techfore.2021.121348

Grinin, L., & Grinin, A. (2016). *Ot Rubil Do Nanorobotov, Mir Na Puti k Epokhe Samoupravlyaemyh Sistem (Istoriya Tekhnologij i Opisanie Ih Budushchego).* Moscow Branch of the Uchitel Publishing House.

Grinin, L., & Korotayev, A. (2020). *COVID-19 pandemic, geopolitics, and recession.* International Center for Education and Social and Humanitarian Studies. https://doi.org/10.13140/RG.2.2.13998.54089.

Grinin, L., Korotayev, A., & Tausch, A. (2016). *Economic cycles, crises, and the global periphery.* Springer, Cham. https://doi.org/10.1007/978-3-319-41262-7

Grinin, L., Grinin, A., & Korotayev, A. (2017). The MANBRIC-technologies in the forthcoming technological revolution. In *Industry 4.0: Entrepreneurship and Structural Change in the New Digital Landscape* (pp. 243–261). Springer, Cham. https://doi.org/10.1007/978-3-319-49604-7_13

Grix, J., Brannagan, P. M., Grimes, H., & Neville, R. (2021). The impact of Covid-19 on sport. *International Journal of Sport Policy and Politics, 13*(1), 1–12.

Hantrais, L., Allin, P., Kritikos, M., Sogomonjan, M., Anand, P. B., Livingstone, S., & Innes, M. (2021). Covid-19 and the digital revolution. *Contemporary Social Science, 16*(2), 256–270.

Hartman, M., Martin, A. B., Washington, B., Catlin, A., & National Health Expenditure Accounts Team. (2022). National health care spending in 2020: Growth driven by federal spending in response to the COVID-19 pandemic: National health expenditures study examines US health care spending in 2020. *Health Affairs, 41*(1), 13–25

Hasan, H. R., Salah, K., Jayaraman, R., Arshad, J., Yaqoob, I., Omar, M., & Ellahham, S. (2020). Blockchain-based solution for COVID-19 digital medical passports and immunity certificates.

*IEEE Access, 8*, 222093–222108. Presented at the IEEE Access. https://doi.org/10.1109/ACCESS.2020.3043350

Hofbauer, H., & Komlosy, A. (2020). Corona-Krise: Anschub für eine kybernetische Wende. *Telepolis*. https://www.heise.de/tp/features/Corona-Krise-Anschub-fuer-eine-kybernetische-Wende-4723168.html

Houlding, E., Mate, K. K., Engler, K., Ortiz-Paredes, D., Pomey, M. P., Cox, J., & Lebouché, B. (2021). Barriers to use of remote monitoring technologies used to support patients with COVID-19: Rapid review. *JMIR mHealth and uHealth, 9*(4), e24743.

Hussain, M. (2021). The changing dynamics of liberal international order: Covid-19 and its global implications. *Journal of Globalization Studies, 12*(2), 143–156.

Irshad, S. M. (2020). Covid-19 and global 'at-risk community': From benefit-sharing to risk-sharing of economic crisis. *Journal of Globalization Studies, 11*(2), 132–142.

Jackson, T., Steed, L., Pedruzzi, R., Beyene, K., & Chan, A. H. Y. (2022). COVID-19 and behavioral sciences. *Frontiers in Public Health, 9*, 2335.

Jasiński, A. M., & Bąkowska, A. M. (2020). Globalization and life satisfaction of the polish population. Before and during the Covid-19 pandemic. *Journal of Globalization Studies, 11*(2), 146–155.

Jat, D. S., & Singh, C. (2020). Artificial intelligence-enabled robotic drones for COVID-19 outbreak. In A. Joshi, N. Dey, & K. C. Santosh (Eds.), *Intelligent systems and methods to combat Covid-19* (pp. 37–46). Springer. https://doi.org/10.1007/978-981-15-6572-4_5

Jeyanathan, M., Afkhami, S., Smaill, F., Miller, M. S., Lichty, B. D., & Xing, Z. (2020). Immunological considerations for COVID-19 vaccine strategies. *Nature Reviews Immunology, 20*(10), 615–632. https://doi.org/10.1038/s41577-020-00434-6

Jones, C. D., Hickman, J. E., Rumbold, S. T., Walton, J., Lamboll, R. D., Skeie, R. B., & Ziehn, T. (2021). The climate response to emissions reductions due to COVID-19: Initial results from CovidMIP. *Geophysical Research Letters, 48*(8), e2020GL091883.

Jung, J. H., & Lim, D.-G. (2020). Industrial robots, employment growth, and labor cost: A simultaneous equation analysis. *Technological Forecasting and Social Change, 159*, 120202. https://doi.org/10.1016/j.techfore.2020.120202

Kakkar, S. (2021). The outbreak of sars-cov-2 (Covid-19): Impact on international tourism. *Journal of Globalization Studies, 12*(1), 145–158.

Kaneda, T., Greenbaum, C., & Patierno, K. (2016). *2018 world population data sheet*. Population Reference Bureau. https://www.prb.org/2016-world-population-data-sheet/

Khan, A., Khan, N., & Shafiq, M. (2021). The economic impact of COVID-19 from a global perspective. *Contemporary Economics, 15*(1), 64–76.

Kimmig, R., Verheijen, R. H. M., & Rudnicki, M. (2020). Robot assisted surgery during the COVID-19 pandemic, especially for gynecological cancer: a statement of the society of European robotic gynaecological surgery (SERGS). *Journal of Gynecologic Oncology, 31*(3). https://doi.org/10.3802/jgo.2020.31.e59

Kowalewski, M. (2021). Street protests in times of COVID-19: Adjusting tactics and marching 'as usual.' *Social Movement Studies, 20*(6), 758–765. https://doi.org/10.1080/14742837.2020.1843014

Kumar, V., Alshazly, H., Idris, S. A., & Bourouis, S. (2021a). Evaluating the impact of covid-19 on society, environment, economy, and education. *Sustainability, 13*(24), 13642.

Kumar, M., Kumari, N., Thakur, N., Bhatia, S. K., Saratale, G. D., Ghodake, G., & Chung, S. M. (2021b). A comprehensive overview on the production of vaccines in plant-based expression systems and the scope of plant biotechnology to combat against SARS-CoV-2 virus pandemics. *Plants, 10*(6), 1213.

Kumar, A., Singh, P., Raizada, P., & Hussain, C. M. (2022). Impact of COVID-19 on greenhouse gases emissions: A critical review. *Science of the Total Environment, 806*, 150349.

Kumari, P., & Seeja, K. R. (2021). A novel periocular biometrics solution for authentication during Covid-19 pandemic situation. *Journal of Ambient Intelligence and Humanized Computing*. https://doi.org/10.1007/s12652-020-02814-1

Kuznetsova, V., & Mashkina, O. (2020). Trudovaya migratsiya i problemy obucheniya detey rabochikh-migrantov v Kitaye. *Istoriya i Sovremennost', 2*(36), 46–69.

Leswing, K. (2021). Companies are racing to build digital passports for people to prove they've had the Covid vaccine. *CNBC*. Retrieved February 4, 2021, from https://www.cnbc.com/2021/01/24/vaccine-passes-could-get-us-back-in-bars-.html

Liu, J., Chang, H., Forrest, J.Y.-L., & Yang, B. (2020a). Influence of artificial intelligence on technological innovation: Evidence from the panel data of China's manufacturing sectors. *Technological Forecasting and Social Change, 158*, 120142. https://doi.org/10.1016/j.techfore.2020.120142

Liu, S., Yang, L., Zhang, C., Xiang, Y.-T., Liu, Z., Hu, S., & Zhang, B. (2020b). Online mental health services in China during the COVID-19 outbreak. *The Lancet Psychiatry, 7*(4), e17–e18. https://doi.org/10.1016/S2215-0366(20)30077-8

Lutz, W., Goujon, A., Kc, S., Stonawski, M., & Stilianakis, N. (2018). *Demographic and human capital scenarios for the 21st century: 2018 assessment for 201 countries*. Publications Office of the European Union.

Lyon, D. (2001). *Surveillance society: Monitoring everyday Lifl*. McGraw-Hill Education.

Mahagamage, Y., & Marasinghe, K. (2023). The socio-economic effects of covid-19. *Saúde e Sociedade*, 32.

Malodia, S., Dhir, A., Mishra, M., & Bhatti, Z. A. (2021). Future of e-Government: An integrated conceptual framework. *Technological Forecasting and Social Change, 173*, 121102.

Martin, A., Markhvida, M., Hallegatte, S., & Walsh, B. (2020). Socio-economic impacts of COVID-19 on household consumption and poverty. *Economics of Disasters and Climate Change, 4*(3), 453–479.

Mbunge, E., Ndumiso, N., Kavu, T. D., Dandajena, K., Batani, J., & Fashoto, S. G. (2022). Towards QR code health systems amid COVID-19: Lessons learnt from other QR code digital technologies. *The science behind the COVID pandemic and healthcare technology dolutions* (pp. 129–143). Springer International Publishing.

McKeown, J. S., Bista, K., & Chan, R. Y. (2022). *Global higher education during COVID-19: Policy, society, and technology*. STAR Scholars.

Mhlanga, D. (2022). The role of artificial intelligence and machine learning amid the COVID-19 pandemic: What lessons are we learning on 4IR and the sustainable development goals. *International Journal of Environmental Research and Public Health, 19*(3), 1879.

Moustaka, V., Theodosiou, Z., Vakali, A., Kounoudes, A., & Anthopoulos, L. G. (2019). Enhancing social networking in smart cities: Privacy and security borderlines. *Technological Forecasting and Social Change, 142*, 285–300. https://doi.org/10.1016/j.techfore.2018.10.026

Musyuni, P., Aggarwal, G., Nagpal, M., & Goyal, R. K. (2020). A case study: Analysis of patents on coronaviruses and Covid-19 for technological assessment and future research. *Current Pharmaceutical Design*. https://doi.org/10.2174/1381612826666200720233947

OECD. (2020). *COVID-19 and fiscal relations across levels of government* (p. 45). OECD. Retrieved January 12, 2021, from https://read.oecd-ilibrary.org/view/?ref=129_129940-barx72laqm&title=COVID-19-and-Fiscal-Relations-across-Levels-of-Government

Palestino, G., García-Silva, I., González-Ortega, O., & Rosales-Mendoza, S. (2020). Can nanotechnology help in the fight against COVID-19? *Expert Review of Anti-Infective Therapy, 18*(9), 849–864. https://doi.org/10.1080/14787210.2020.1776115

Parolin, Z., & Lee, E. K. (2022). The role of poverty and racial discrimination in exacerbating the health consequences of COVID-19. *The Lancet Regional Health-Americas, 7*, 100178.

Patel, P., & Gohil, P. (2021). Role of additive manufacturing in medical application COVID-19 scenario: India case study. *Journal of Manufacturing Systems, 60*, 811–822.

Pérez-Morote, R., Pontones-Rosa, C., & Núñez-Chicharro, M. (2020). The effects of e-government evaluation, trust and the digital divide in the levels of e-government use in European countries. *Technological Forecasting and Social Change, 154*, 119973. https://doi.org/10.1016/j.techfore.2020.119973

Phillips, F. (2007). On S-curves and tipping points. *Technological Forecasting and Social Change, 74*(6), 715–730.

Phillips, F. (2011). The state of technological and social change: Impressions. *Technological Forecasting and Social Change, 78*(6), 1072–1078. https://doi.org/10.1016/j.techfore.2011.03.020

Plebe, A., & Perconti, P. (2020). Plurality: The end of singularity? In A. V. Korotayev & D. J. LePoire (Eds.), *The 21st century singularity and global futures* (pp. 163–184). Springer International Publishing. https://doi.org/10.1007/978-3-030-33730-8_8

Ram, N., & Gray, D. (2020). Mass surveillance in the age of COVID-19. *Journal of Law and the Biosciences, 7*(lsaa023). https://doi.org/10.1093/jlb/lsaa023

Rapanta, C., Botturi, L., Goodyear, P., Guàrdia, L., & Koole, M. (2020). Online university teaching during and after the Covid-19 crisis: Refocusing teacher presence and learning activity. *Postdigital Science and Education, 2*(3), 923–945. https://doi.org/10.1007/s42438-020-00155-y

Ray, R. L., Singh, V. P., Singh, S. K., Acharya, B. S., & He, Y. (2022). What is the impact of COVID-19 pandemic on global carbon emissions? *Science of the Total Environment, 816*, 151503.

Reicher, S., & Stott, C. (2020). On order and disorder during the COVID-19 pandemic. *British Journal of Social Psychology, 59*(3), 694–702. https://doi.org/10.1111/bjso.12398

Remmel, A. (2021). COVID vaccines and safety: What the research says. *Nature, 590*(7847), 538–540.

Renu, N. (2021). Technological advancement in the era of COVID-19. *SAGE Open Medicine, 9*, 20503121211000910.

Research America. (2017). *U.S. Investments in medical and health research and development* (p. 16). Research America. https://www.researchamerica.org/sites/default/files/Publications/InvestmentReport2019_Fnl.pdf

Richards, N. M. (2012). The dangers of surveillance. *Harvard Law Review, 126*, 1934–1963.

Saengtabtim, K., Leelawat, N., Tang, J., Suppasri, A., & Imamura, F. (2022). Consequences of COVID-19 on health, economy, and tourism in Asia: A systematic review. *Sustainability, 14*(8), 4624.

Sarfraz, Z., Sarfraz, A., Iftikar, H. M., & Akhund, R. (2021). Is COVID-19 pushing us to the fifth industrial revolution (society 5.0)?. *Pakistan journal of medical sciences, 37*(2), 591.

Schwab, K., & Malleret, T. (2020). *Covid-19: The great reset.* WEF: Edition.

Schwartz, P. M. (1999). Internet privacy and the state. *Connecticut Law Review, 32*, 815.

Sean. (2021). We analyzed 450 apps and found location trackers in every one.| ExpressVPN. *Home of internet privacy.* Retrieved February 3, 2021, from https://www.expressvpn.com/blog/digital-security-lab-location-trackers-smartphone-apps-research/

Seifabadi, M. S., & Dolatabadi, A. B. (2022). Covid-19, globalization and strengthening authoritarianism in the middle east. *Journal of Globalization Studies, 13*(1), 135–157.

Shah, S., Diwan, S., Kohan, L., Rosenblum, D., Gharibo, C., Soin, A., et al. (2020). The technological impact of COVID-19 on the future of education and health care delivery. *Pain Physician, 23*, 14.

Silcock, R. (2001). What is E-government. *Parliamentary Affairs, 54*(1), 88–101. https://doi.org/10.1093/pa/54.1.88

Síthigh, D. M., & Siems, M. (2019). The Chinese social credit system: A model for other countries? *The Modern Law Review, 82*(6), 1034–1071. https://doi.org/10.1111/1468-2230.12462

Skorup, B., & Haaland, C. (2020). How drones can help fight the coronavirus. *SSRN Electronic Journal.* https://doi.org/10.2139/ssrn.3564671

Sohil, F., & Sohail, M. U. (2022). Global discrimination of COVID-19 vaccine. *Human Vaccines & Immunotherapeutics, 18*(1), 2028515.

Solove, D. J. (2008). *Understanding privacy.* Harvard University Press.

Stanley, J., & Granick, J. S. (2020). The limits of location tracking in an epidemic. 9.

Statistics Sweden. (2020). Government budget allocations for R&D (GBARD) 2020. *Statistiska Centralbyrån.* Retrieved January 13, 2021, from http://www.scb.se/en/finding-statistics/statistics-by-subject-area/education-and-research/research/government-budget-allocations-for-rd/pong/statistical-news/government-budget-allocations-for-rd-gbard-2020/.

Strandberg, T. (2020). Coronavirus: US and UK governments losing public trust. *The Conversation.* http://theconversation.com/coronavirus-us-and-uk-governments-losing-public-trust-137713.

Sulochanan Karthick Ramanathan, S., Fathima Kamal Basha, R., & Banu, A. (2021). A novel face recognition technology to enhance health and safety measures in hospitals using SBC in pandemic prone areas. *Materials Today: Proceedings*. https://doi.org/10.1016/j.matpr.2020.11.336

Tang, Z., Zhang, X., Shu, Y., Guo, M., Zhang, H., & Tao, W. (2021). Insights from nanotechnology in COVID-19 treatment. *Nano Today, 36*, 101019. https://doi.org/10.1016/j.nantod.2020.101019

Technavio. (2020). *Global 3D facial recognition systems market 2018–2022*. Retrieved February 3, 2021, from https://www.technavio.com/report/global-3d-facial-recognition-systems-market-analysis-share-2018

Transparency Market Research. (2020). *Impact of COVID-19 on adaptive traffic control system market growth, segments, size, industry analysis and opportunities 2030*. Retrieved February 3, 2021, from https://ksusentinel.com/2021/01/11/impact-of-covid-19-on-adaptive-traffic-control-system-market-growth-segments-size-industry-analysis-and-opportunities-2030/

UN Population Division. (2019). *World population prospects 2019: Volume I: Comprehensive tables*. UN Population Division.

UN Population Division (2022a). *United Nations population division database*. United Nations, Department of Economics and Social Affairs, Population Division. http://www.un.org/esa/population

UN Population Division (2022b). *World population prospects 2022*. United Nations, Department of Economics and Social Affairs, Population Division.

UNeGovDD. (2013). E-Government. The United Nations E-Government Development Database. Retrieved February 4, 2021, from https://publicadministration.un.org/egovkb/en-us/About/UNeGovDD-Framework

Usman, M., Husnain, M., Riaz, A., Riaz, A., & Ali, Y. (2021). Climate change during the COVID-19 outbreak: Scoping future perspectives. *Environmental Science and Pollution Research, 28*(35), 49302–49313.

van der Zwet, K., Barros, A. I., van Engers, T. M., & Sloot, P. (2022). Emergence of protests during the COVID-19 pandemic: Quantitative models to explore the contributions of societal conditions. *Humanities and Social Sciences Communications, 9*(1), 1–11.

Vieten, U. M. (2020). The "new normal" and "pandemic populism": The COVID-19 crisis and anti-hygienic mobilisation of the far-right. *Social Sciences, 9*(9), 165. https://doi.org/10.3390/socsci9090165

Vollset, S. E., Goren, E., Yuan, C.-W., Cao, J., Smith, A. E., Hsiao, T., Murray, C. J. L., et al. (2020). Fertility, mortality, migration, and population scenarios for 195 countries and territories from 2017 to 2100: A forecasting analysis for the global burden of disease study. *The Lancet*. https://doi.org/10.1016/S0140-6736(20)30677-2

Walby, S. (2022). Crisis and society: Developing the theory of crisis in the context of COVID-19. *Global Discourse, 12*(3–4), 498–516.

Wang, S., Ding, S., & Xiong, L. (2020). A new system for surveillance and digital contact tracing for COVID-19: Spatiotemporal reporting over network and GPS. *JMIR mHealth and uHealth, 8*(6), e19457.

Weiss, C., Carriere, M., Fusco, L., Capua, I., Regla-Nava, J. A., Pasquali, M., et al. (2020). Toward nanotechnology-enabled approaches against the COVID-19 pandemic. *ACS Nano, 14*(6), 6383–6406. https://doi.org/10.1021/acsnano.0c03697

Westin, A. F. (1966). Science, privacy, and freedom: Issues and proposals for the 1970's. Part I–The current impact of surveillance on privacy. *Columbia Law Review, 66*(6), 1003–1050.

WHO. (2020a). *A coordinated global research roadmap* (p. 68). WHO. Retrieved February 2, 2021, from https://www.who.int/publications/m/item/a-coordinated-global-research-roadmap

WHO. (2020b). *Immunity passports in the context of COVID-19*. Retrieved February 3, 2021, from https://www.who.int/news-room/commentaries/detail/immunity-passports-in-the-context-of-covid-19

Widdowson, M. (2021). From Covid-19 to zero-gravity: Complex crises and production revolutions. *Journal of Globalization Studies, 12*(1), 117–144.

Wittgenstein Centre. (2024). *Wittgenstein centre human capital data explorer*. Retrieved January 2, 2024, from http://dataexplorer.wittgensteincentre.org/wcde-v2/
Wood, S. (2022). How consumers behave in a crisis: International lessons (and Innovations) from COVID-19. *Journal of International Marketing, 30*(2), 5–12.
Wood, R., Reinhardt, G. Y., RezaeeDaryakenari, B., & Windsor, L. C. (2022). Resisting lockdown: The influence of COVID-19 restrictions on social unrest. *International Studies Quarterly, 66*(2), sqac015.
Xie, X., Ramakrishna, S., & Manganelli, M. (2022). Smart building technologies in response to COVID-19. *Energies, 15*(15), 5488.
Yang, D. (2021). Application of nanotechnology in the COVID-19 pandemic. *International Journal of Nanomedicine, 16*, 623.
Yoo, S., & Managi, S. (2020). Global mortality benefits of COVID-19 action. *Technological Forecasting and Social Change, 160*, 120231. https://doi.org/10.1016/j.techfore.2020.120231
Zhu, Y., Janssen, M., Wang, R., & Liu, Y. (2022). It is me, chatbot: Working to address the COVID-19 outbreak-related mental health issues in China. User experience, satisfaction, and influencing factors. *International Journal of Human–Computer Interaction, 38*(12), 1182–1194.
Zhuang, Q., Sun, G., Zhang, F., Zhang, J., You, L., Chen, S., & Pan, H. (2021). How Internet technologies can help hospitals to curb COVID-19: PUMCH experience from China. *Health Information MaNagement Journal, 50*(1–2), 95–98.
Zimmer, C., Corum, J., & Wee, S.-L. (2020). Coronavirus vaccine tracker. *The New York Times*. https://www.nytimes.com/interactive/2020/science/coronavirus-vaccine-tracker.html. Accessed 2 February 2021.
Zimmerling, A., & Chen, X. (2021). Innovation and possible long-term impact driven by COVID-19: Manufacturing, personal protective equipment and digital technologies. *Technology in Society, 65*, 101541.

# Part V
# Conclusion

# Conclusion. Toward Cybernetic Society

15

**Abstract**

In this Conclusion, Grinin et al. undertake two tasks simultaneously: (1) to summarize a large part of the ideas and forecasts which have been presented in this monograph and (2) to develop a portrait of the new cybernetic society, in which most of the processes analyzed in this book will take on new qualities. The authors call this new unprecedented type of society "cybernetic", because its formation will be inextricably linked to the completion of the Cybernetic Revolution. The authors forecast that it will be a society, which will be very different from all previous societies. Why? On the one hand, it will be closely connected to very profound technological changes in the management and regulation of our life at all levels. On the other hand, it is inextricably linked to the process of global aging; moreover, aging becomes an integral part of society. Cybernetic society must go through a difficult path of social frictions, conflicts, and optimizations before many things become institutionalized. The horizon of forecasts is sometimes very long—up to the first third of the twenty-second century. The authors describe changes awaiting humanity that are very serious and uncertain, but at the same time hopeful.

## 15.1 What This Chapter Is About

In this concluding chapter, we summarize the main ideas of this monograph and the numerous forecasts made in it.

In the chapters of this volume, we discuss the possible social changes in a future society. The horizon of predictions is sometimes very long, even considering some possible changes that could be fully realised in the first third of the twenty-second century. We describe changes awaiting humanity that are very serious and uncertain, but at the same time hopeful.

Two main themes have been explored in this monograph:



L. Grinin et al., *Cybernetic Revolution and Global Aging*, World-Systems Evolution and Global Futures, https://doi.org/10.1007/978-3-031-56764-3_15

(1) The Cybernetic Revolution, which has been developing new technologies and innovative fields for decades, and its impact on the future society. The Cybernetic Revolution is a fundamental transition from the Industrial Production Principle to the production of services and goods based on the widespread introduction of self-regulating/self-managing systems, that is, systems that can not only function in the absence (or with minimal participation) of people but can also make complex decisions independently;
(2) The process of societal aging and, in general, global aging in the World System and the impact of these processes on the future society. Both of these topics, as the reader have seen, turned out to be closely intertwined, because these two powerful processes are closely and sometimes unexpectedly interconnected and integrated.

In this concluding chapter, we especially focus on trying to foresee how these two very important processes of development will form a new type of society in the late twenty-first century.

Thus, in this concluding chapter we finish our respective forecasts of what the future society could be like, a society which will emerge as a result of numerous transformations. We call this new unprecedented type of society "cybernetic", because its formation will be inextricably linked with the completion of the Cybernetic Revolution.

We forecast that it will be a society, which will have fundamentally new characteristics and consequently will be very different from all previous societies. Why? On the one hand, it will be closely connected to very deep technological changes in the management and regulation of our life at all levels. On the other hand, it is inextricably linked to the process of global aging; moreover, aging has become an integral part of society. As it develops and society's institutions adapt to it, no sphere of society will remain unchanged, including technology, economy, consumption, social sphere, ideology, ecology, international relations etc.

In short, this cybernetic society is an elderly society with the institutionalization of age, relying almost entirely on smart (cybernetic) technologies and self-regulating/self-managing systems. Accordingly, this society will have its strengths and weaknesses, successes and even triumphs, but it will not be free from some serious problems and tensions. Cybernetic society must go through a difficult path of social frictions, conflicts and optimizations before many things become institutionalized. Techno-social regulation has important advantages, but it also includes coercive violence against people's will and freedom. Therefore, a long period of adjustment will be required to optimize regulations in order to take into account a variety of needs, desires and problems.

Thus, in this Conclusion, we undertake two tasks simultaneously: (1) to summarize a large part of the ideas and forecasts which have been presented in this monograph and (2) to develop a portrait of the new cybernetic society in which most of the processes analyzed in this book will take on a new quality.

## 15.2 Introductory Notes

In this Conclusion, we try to present our vision of the future society, as well as the prerequisites for its formation and the main processes that prepare its coming. Of course, this is only one of the options for the possible evolution of society in the current century. Also, negative scenarios are not excluded (see at the end of the chapter). Nevertheless, the characteristics of the future society presented in this chapter seem to us to be quite reasonable, based on powerful and well-defined objective processes, which we spoke about in detail in this monograph.

We call the future society 'cybernetic society', because of its tight connection with the Cybernetic Revolution. Cybernetics is the science of managing complex systems. We call this revolution 'cybernetic' because it will bring fundamental changes not just in technology, but also in management processes and procedures in all spheres of life due to the wide distribution of self-regulating and self-managing systems. In other words, we will see tremendous changes in the nature and methods of managing and controlling processes in the broadest sense of the word—from production to everyday life, from medicine to social behavior, from planning to consumption, from transport to cooking, from caring for the elderly to communications, from resting to trading.[1] Moreover, revolutionary changes will affect social, economic, administrative and even political regulation. Accordingly, cybernetic society is also connected to the notion 'cybernetic' because it will be more stable, regulated and manageable than the present one. As we see further, this cybernetic society is a society that will take shape (and is already taking shape) as a result of the completion of the Cybernetic Revolution and the further advancement of technology after it, as well as the development of all other spheres of society: those spheres are demographic, economic, political and administrative, international relations, social, environmental, technological, economic, consumption, ideological spheres, etc. Each of them will be analyzed further.

This society is the result of several processes and factors (see below and also Fig. 15.1). However, this cybernetic society is based on two main foundations: (1) the high level of technology that will be created by the Cybernetic Revolution and its further impact on the future society and (2) the aging process that has reached maturity, based on the increase in life expectancy. The synthesis of the two will result in a new type of society with its advantages and probable shortcomings and problems that we will also try to anticipate. Thus, it will be an elderly society, with the institutionalization of age differences, relying almost entirely on smart (cybernetic) technologies and AI-based self-regulating/self-managing systems.

Both processes have been analyzed in the previous chapters and we will summarize the main theses about them below. However, with regard to the Cybernetic Revolution (since many chapters have been devoted to it), we will limit ourselves to briefly reminding the reader of the main ideas about the Cybernetic Revolution

[1] Today, no one is surprised that the main stock exchange trading is carried out by automated programs. And this has greatly changed the work of exchanges (see Chap. 3).

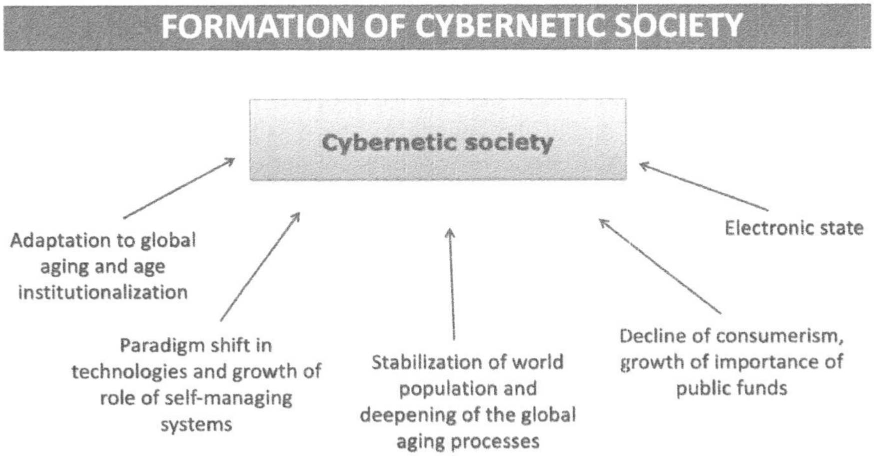


**Fig. 15.1** Factors influencing the formation of cybernetic society

and the period after this revolution (i.e., the mature phases of the Scientific-Cybernetic Production Principle). On the other hand, with respect to the aging process and our predictions about its development and its comprehensive impact on society, we will have to repeat quite a lot of what we have already said in previous chapters, especially in Chap. 7. However, this is necessary in order to understand our ideas about cybernetic society.

In this way, we will solve two problems at once: why will explain the ideas necessary for understanding our vision of the future cybernetic society, and then we will then summarize most of the ideas that have been presented in this monograph.

This way we hope to help the reader to see a new society.

## 15.3 Cybernetic Revolution and Scientific-Cybernetic Production Principle: How Technology Will Develop in the Twenty-First Century

**Production Principles and Production Revolutions**. For the purpose of analyzing the course of technological transformations in history and explaning the current and forthcoming technological changes we have developed the theory of Production Principles and Production Revolution. We single out four production principles: 1. Hunter-Gatherer; 2. Craft-Agrarian; 3. Trade-Industrial; 4. Scientific-Cybernetic. The change in production principles is connected with production revolutions. Among all major technological breakthroughs in history, the most important are the three production revolutions: (1) the Agrarian Revolution (10–3 millennia BCE); (2) the Industrial Revolution (fifteenth–nineteenth centuries); and (3) the Cybernetic Revolution (1950– up to 2070). For details see Chap. 2.

Every major technological change, not to mention the production revolution, inevitably (albeit with a noticeable time lag) changes society; its socio-professional characteristics, and somewhat later its cultural, ideological and even political life.

**The Cybernetic Revolution and Its Phases.** Thus according to our concept, *The Cybernetic Revolution* is the last of the largest production (technological) revolutions in all history following the Agrarian and Industrial ones. It is a great transition from the industrial production to the production and service sector based on the implementation of self-regulating/self-managing systems. The first phase of this revolution started in the 1950s and brought the development of powerful information technologies, new materials, and sources of energy, as well as a distribution of automation. We consider the period from the 1990s to the 2030s as the middle or modernization phase of the Cybernetic Revolution. The initial and middle phases of the Cybernetic Revolution have significantly changed society, which today is markedly different from what it was in the mid-twentieth century. Society has changed not only in technological but also in demographic, social, mental and cultural aspects, as well as in terms of family and sexual relations—not to mention financial, economic and consumer relations, communications, the way we express our thoughts and emotions and the way we receive and comprehend information.

Between the 2030s and 2070s, the final phase of this revolution will lead to a new level of self-operated control in technical systems, namely, to the level of self-regulating and self-managing systems. For details of this change see Chap. 3. Accordingly, the final phase of the Cybernetic Revolution as well as the mature phases of the Scientific-Cybernetic Production Principle, should also greatly change future society.

**Self-Regulating and Self-Managing Systems** are systems that by means of the embedded programs and a set of smart components can regulate or even manage themselves, responding in an intelligent way to feedback from the environment as well as operate and make decisions or suggest the best solutions independently in a wide range of variations or cases of changing of context with minimal or no human intervention. Of utmost importance is the fact that the self-regulating systems will become a major part of technological process. Self-regulating systems can have different levels of complexity of organization—from smart systems with elements or limited programs of self-managing to systems with artificial intelligence that allow making decisions within a fairly wide framework (self-driving cars of the future; housekeeping robots; a system that independently selects plant varieties suitable for particular regions and climates etc.). High-level self-regulating systems can be called self-managing systems. By self-regulating systems we mean not only machines and other technical systems but a wider range of systems and control processes, including those of biological, physiological, techno-biological, social and other nature, which are and will be implemented in different areas (including medicine, genetic engineering, robotics, social relations). For details see Chap. 3.

It is worth making an additional clarification about the difference between *self-regulating* and *self-managing* systems. As we explain in Chap. 3, self-regulating systems are divided into two sub-types. Properly self-regulating and self-managing ones. In this respect, biological, physiological, genetic, many biotechnological (such as self-purification of wastewater), nanotechnological and other natural systems that are being introduced or will be introduced during the Cybernetic Revolution could hardly be called 'self-managing systems' since they are generally designed to regulate a certain level of health, norms, etc. That is, they remain self-regulating systems. But AI-based technical systems (such as self-driving cars, drones, robots etc.) can rather be called self-managing systems.

**Self-Regulating Systems and AI**. There are differences between self-regulating/self-managing systems and artificial intelligence. AI in a strict sense is a special technology and not an autonomous technological (or production) system. That is why self-regulating/self-managing systems and AI are different classes of objects. In other words, the definitions of self-regulating systems and AI are not synonymous, they lie in somewhat different dimensions, although being closely related and interconnected. In other words, *AI and self-regulating systems are related but not synonymous concepts, and are actually concepts defining different classes of objects.* So, with respect to self-regulating systems, we may speak about the role of information as the most important substance and resource, while AI is a technological component at all stages of the process. AI is a special universal technology, which is (in a great number of cases) an indispensable part of self-managing systems, just as is an electric motor in electric machines; or as is an internal combustion engine (ICE) in cars, tractors and other machines. Hundreds and thousands of different machines are based on AI. However, there are classes of self-regulating systems (biological, physiological etc.) in which AI generally doesn't play a central role. Just as an internal combustion engine or an electric motor does not determine the operating principle of all machines, but only of their large class. So, AI does not determine the functions and principle of operation of all self-regulating systems but does characterize their large class. For details see Chaps. 3 and 10.

**MANBRIC convergence**. Among numerous technological directions of the future, we identify seven leading ones. These directions are emerging and in the future they will eventually combine into a single complex, which we call the MANBRIC convergence. The abbreviation MANBRIC is an acronym for the names of the seven most active new areas of technological development, which will eventually constitute a powerful system, moreover convergence. They are respectively: Medicine, Additive technologies, Nanotechnologies, Biotechnologies, Robotics, Information technologies (of course, including AI) and Cognitive technologies. It is this complex that will become the core of future technological development. (For details see Chaps. 4 and 10; later we analyze every branch of the MANBRIC convergence with some forecasts of their development in the 2030s to the 2070s.)

**Medicine** will be the first sphere to start the final phase of the Cybernetic Revolution, because among other areas, medicine, will be an integrating part of MANBRIC for several important reasons. In particular, the medical sphere has unique opportunities to combine the above-mentioned technologies into a single complex. Even today, all these areas of MANBRIC are very actively used in modern medicine, but in the future, their relationship with medicine will increase noticeably, and at some point fundamentally. In addition, the process of society's aging, which is currently developing rapidly, creates an acute demand for labor-saving technologies, as well as giving a powerful impulse to medicine. Progress in the latter will help to extend the working age and improve the health-related (biological) quality of life. All this has made health issues the subject of interest for the whole society, placing medicine and biotechnology at the forefront of technological development. The leading role of medicine in the final phase of the Cybernetic Revolution (at least in the 2030–2040/50s; when other areas may come to the fore) is also explained by the resulting economic and social consequences connected with a growing shortage of labor. For details see Chaps. 4 and 8.

**Scientific-Cybernetic Production Principle in its maturity**. After the completion of the Cybernetic Revolution, its innovative technologies based on self-regulating systems/self-managing systems will be both widespread and improved. The period of the mature phases of the Scientific-Cybernetic Production Principle will begin. Its fourth phase[2] will begin in the 2060s–2080s (depending on different scenarios of the completion of the Cybernetic Revolution [see Chap. 2]). The fourth phase is the phase of maturity and expansion. It refers to the diffusion of new technologies to most regions and spheres of production. The production principle will then acquire its mature form, and this leads to important changes in the socio-economic sphere. So this phase implies that in its time the sector of self-regulating/self-managing systems will rapidly improve and diffuse to various regions at an enormous speed. MANBRIC technologies will finally emerged and take a central place in the new production principle. We believe that during this period, some technologies, in particular biotechnologies, will achieve very great success, thanks to which many environmental problems can be solved or it will become clearer how they can be solved (see Chap. 9). The process of reducing manual and unskilled labor in the service sector (and, accordingly, reducy the number of jobs) will continue due to the development of robotics and other MANBRIC technologies. However, we do not expect excessive unemployment, as by this time, due to the reduction in the share of the working population (see Chap. 6), the shortage of labor compensates for the loss of jobs. Such processes will be even more active in the next phase, in which, in particular, we expect a very active introduction of robots and other self-managed systems in the field of medicine, care for the sick and the elderly.[3] We also expect (under the influence of various factors, including the new system of technologies) that during this phase the completion of the formation of a system of a new world order based on greater stability and a broader consensus of the leading world forces will occur (see Grinin et al., 2021b, 2023b). Cybernetic society will actively form during this period, although many of its features will not yet be sufficiently expressed, especially since such regions as Africa will still remain relatively young.

[2] The first–third phases are the period of the Cybernetic Revolution.

[3] Of course, these processes, as well as those described below, will not start suddenly, they will increase starting from 2040 and 2050. But they will be especially powerful at the time described.

The fifth phase is that of the absolute dominance of a production principle. It leads to an intensification of production and the full realization of the potential of the principle. It will presumably take place in the 2070s and the 2080s. Accordingly, it can be expected that during this period those areas that could lag behind during the Cybernetic Revolution period, in particular nanotechnology, possibly nanomechanics and nanomachine building (which have been discussed in Chap. 10), as well as robotics, will achieve significant development. Self-driving cars will become mainstream. The fifth phase is that of non-system phenomena. Here, the most important subsystems and institutions of cybernetic society will begin to actively improve. In particular, medicine in symbiosis with other technologies will increasingly focus on anti-aging. During this period, the influence of conservatism, which will increase in connection with the growth of the proportion of the elderly population, will become noticeably stronger than before. As a result, consumerism will begin to decline, and a new consumption model will be actively formed (with less emphasis on market relations and a larger share of public funds, including medical ones).[4] Correspondingly, the pace of technological progress will begin to slow down (see Chap. 13). As a result of the development of socio-technical self-managing systems, the electronic state will finally emerge (see Chap. 14; see also below).

The sixth phase is even more characterized by the emergence of non-systemic elements; accordingly, during this period, cybernetic society will acquire its most mature features. There will be new inventions and improvements in technologies leading to the emergence of non-systemic elements (which pave the way for the formation of a new production and social system). Perhaps at that time, there will be major breakthroughs in the energy sector, and it will finally be possible to master new types of energy. This is expected to take place in the 2080s and the 2090s.[5] Together with the continued improvement of self-regulating systems, which will eventually become self-managing, all this will greatly increase the level of complexity and autonomy in terms of technology, amount of energy and the level of resource-saving, which—together with the rise in the level of development in currently underdeveloped societies—will significantly reduce demands on the pace of economic growth; accordingly, a further slowdown in the pace of technological progress can be compensated for by the rate of savings, wastefulness and depth of processing, as well as cheaper (but not free) energy. The latter will make it possible to implement a number of grandiose projects, it will also expand the possibilities of space exploration.

Their fifth and sixth phases imply the beginning of the transition to a new economic system (see Chap. 13 and below), due to the increasing level of complexity of self-managing systems and serious advances in medicine. By this time, the process of global aging will have reached all countries of the world. There will be

[4] See Chap. 13 for details; see also below.

[5] The dates of the fourth–sixth phases will depends on when the Cybernetic Revolution will begin and finish, and consequently the dates the previous phases.

different scenarios of how society will change under the influence of aging. However, in any case, it will be a period of strong adaptation to societal aging, which will greatly contribute to the creation of cybernetic society. Also, there will be a growing number of socio-technical self-regulating/ managing systems that will mostly operate autonomously and regulate the behavior of large numbers of people in certain situations.

During the fourth, and especially the fifth and sixth phases of the Scientific-Cybernetic Production Principle, we expect significant growth in life expectancy and rapid global aging (see Chaps. 6 and 7). As a result of the decline in the birth rate, the trend toward stabilization of the world population (and in many countries toward depopulation), as well as the deepening of the process of global aging, a new TPR (type of population reproduction) will finally develop (see Chaps. 5 and 7; see also below). This TRP will inolve total technological support for health and quality of biological life with the stabilization (or reduction) of the population, with a very large proportion of the elderly population.

**Scientific-Cybernetic Production Principle, rate of technological progress and aging**. Technological progress in the twenty-first century will develop at an uneven rate. From the 2030s, we will see a new acceleration in technological progress which has begun to slow down since the 1970s. And it will definitely be associated with the beginning and development of the final phase of the Cybernetic Revolution. This acceleration will continue for several decades, until, toward the late twenty-first century, it will start to slow down again. It is very interesting and important that these changes in the speed of technological progress will be tightly connected with the maturity of the aging process. This influence of aging will be non-linear. Aging will initially contribute to an increase in the rate of technological progress, but then it will begin to have the opposite effect, slowing it down (see Chaps. 12 and 13 for details). However, this slowdown doesn't mean a complete stop; we argue that technological or economic development will hardly stop. In fact, it will be unreasonable not to allow the technological progress to slow down too much. In any case, the deeper the process of global aging goes, the more it depends on the development of medicine, biotechnology and other technologies;

So, societal aging has a dual and contradictory effect on technological progress: positive and negative. On the one hand, the need to improve the biological quality of life of the elderly, the struggle to increase life expectancy, the fight against incurable diseases and the need to extend the working age of the elderly, spurs science and technology to search for new breakthrough solutions. However, on the other hand, older people tend to be more conservative than the younger and middle-aged categories, leading to a slowdown in the rate of technological progress. So, we obtain a non-trivial result, according to which the process of global aging can accelerate and change the direction of technological progress in the coming decades, and then the aging of society can and will slow down scientific-technological progress in the late twenty-first and early twenty-second centuries (see Chap. 13).[6]

Thus, we can observe the mutual influence between global aging and technological change within the Cybernetic Production Principle. The aging process itself is due to the increase in life expectancy and can only develop if there is a corresponding or even accelerated development of technologies. We show the direct impact of technology on changes in the organization of the state, political and social life. Accordingly, such a complex synthesis cannot but affect deeply and systematically the shape of the future society. But on the other hand, this enlarged life duration will create new serious problems, tensions and contradictions in a more graying and technological society, about which see below.

Thus, the combined effects of technological innovation and global aging will have a transformative impact on society in the twenty-first century.

## 15.4 Aging and Social Transformations

**Global Aging: Society's Challenges and Transformations**. The process of global aging over several of the coming decades will increasingly influence both individual societies and the World System in general as well as the further change of the type of population reproduction.[7] In Chaps. 6 and 7, we present different statistical data related to global aging. We just remind some figures. Already today, the share of the population aged 65 + amounts to 20–28% in 22 countries and regions. By 2050, the number of people aged 65 and over is projected to make 16 percent worldwide, i.e., more than twice the number of children under 5 and about the same as the number of children under 12. For the first time in human history, people aged 65 and over will outnumber children (see Shcherbakova, 2019; Kamiya & Hertog, 2020; Medici, 2021; UN Population Division, 2022a, 2022b).

[6] In a narrow, mathematical sense, this also means the onset of a singularity at some point. According to our calculations, this conditional singularity point for the rate of technological progress will come in 2106. In Chaps. 1 and 12, we have explained in sufficient detail that the movement towards the singularity point is characteristic of many processes. One should understand that the change rate cannot increase endlessly, and therefore, the trend curve will change its direction after a while. This means that the processes will proceed in a significantly different way, that is, already along some new (mostly unknown and even opposite) trajectories (Grinin et al., 2013; Grinin, Grinin & Korotayev 2020a, 2020b; Korotayev & Malkov 2016; Korotayev 2005, 2007, 2018, 2020a, 2020b; Korotayev et al., 2006a, 2006b; LePoire & Korotayev 2020; Panov et al., 2020).

[7] See, e.g., Clark and Spengler 1980; Pifer, Bronte, 1986a, 1986b; Lee et al., 1988; Bös and Weizsäcker 1989; Peterson 1999; Fukuyama 2002; Jackson et al., 2008, 2013; Lee, Mason, and Cotlear 2010; UN, 2013, 2020; Popova, Yanik, 2014; Goldstone, 2015; Goldstone et al., 2015; Orlická 2015; Shcherbakova 2019; Bloom, Luca, 2016; Lee 2016; Maestas, Mullen, and Powell 2016; Morgan and Kunkel 2016; van Dullemen, 2017; Fichtner, 2018; Randers et al., 2018; Yaşar and Yaşar, 2018; Barsukov, 2019; Hsu and Lo, 2019; Kapelyushnikov, 2019a; Mitchell & Walker, 2020; Jackson, 2021; Medici, 2021.

Such a development, without continuous and systematic adaption of society to the aging process increases, the probability of various socio-political risks and problems associated with population aging. Among them, one finds economic slowdowns, pension crises, problems of caring for the growing numbers of disabled and frail people, maintaining the financial savings of the elderly, the issue of agism and the vulnerability of the elderly in times of crisis and pandemics and consequently the maintainance of life expectancy, as well as a stable democracy and even a stable world order. These risks and problems are of particular concern given that by the mid-twenty-first century, two-thirds of people over 65 will live in middle and low-income countries (WHO, ). However, although the influence of the aging on various areas of social activity will be dramatic in most countries, the ability of modern society to cope with increasing risks associated with this process is problematic.

So even now global aging is an extremely important problem and, it will become the most crucial issue in the future while insufficient attention to it threatens with societal tensions, crisis and intergenerational conflicts as well as political and social instability.

Among the challenges that society already faces and will face in the future are: a sharp increase in the number of people suffering from dementia and other incurable age-related diseases in less developed countries, the struggle for resources due to the increasing number of the elderly and sick people, agism, increasing conservatism and a possible slowdown in technological progress, an aggravation of intergenerational contradictions and conflicts, a possible decline in democracy due to population aging and many others (for details see Chap. 7).

Obviously, the process of global aging (a) perades the entire society, affects all its processes and its influence will be constantly increasing; (b) will not be able to develop in the future without profound changes in society. In other words, the processes of global aging and increasing life expectancy will entail very serious changes in socio-political life; (c) the development of these trends can cause serious tensions and internal conflicts. In general, one can expect a rather difficult and conflicting path of adaptation to aging (see below).

We believe that whether earlier or later, easier or harder and with less or more expense society can find an appropriate way to both manage and adapt to aging through the transformation of a number of its institutions. In Chap. 7, we also analyze different ways and forms of adaptation to aging that must be elaborated in the future in order to mitigate frictions and tensions that could impede societal development. Below in a special section, we will discuss in detail the process of institutionalizing old age.

**Aging and a New Type of Population Reproduction**. As we have said above, as a consequence of the increase in life expectancy the main features of a new TPR (type of population reproduction) will occur. This TPR is named the social-qualitative type of population reproduction, which, in our opinion, should take shape by the end of the present century (see Chaps. 5 and 7). This will provide a transition to the model of total technological support for health and quality of biological life and will occur with the stabilization (or reduction) of the population numbers, with a very large proportion of an elderly population. To support a very high life expectancy in this society, very complex self-managing systems will be required. We denote them as a continuous medical–biotechnological environment (MBTE; see below for details). In this way, a new demographic and technological system will promote the formation of a new societal type, which can be denoted as *cybernetic society*.

## 15.5 Agism and Institutionalization of Aging

**What is Agism?** In any society, there is a problem of agism, since people of different ages coexist in it. Agism is understood as discrimination or other infringement on the rights and opportunities of people based on the stereotypes about their age in economic, legal, domestic, or psychological forms as well as in medical, social and other services and also in their consumption (for details see Chap. 7). It manifests itself in various forms and situations in everyday life.

In the context of the process of global aging, agism is becoming a serious problem today. And it is usually associated with discriminatory attitudes toward the elderly.[8] However, agism can occur in relation to different age groups, including children and young people and on the part of different age groups. For example, young people can experience problems trying to get jobs, having lower wages, limited opportunities for social mobility and careers due to the dominance of the одвук people in high positions, etc.

As for agism towards the elderly in modern society, it can intensify (including self-agism, that is, elderly people underestimation of their own abilities) as a result of the fact that life expectancy has increased very significantly, while the ideas about the appropriate rights, duties and opportunities of each age have largely remained at the same level. Meanwhile, people who were considered old eighty years ago (e.g., at the age of 65) today can be considered to be in a late stage of maturity. It is worth noting that there will also be big changes in the perception of age in the future.

**Agism as a Complex and Ambiguous Social Phenomenon**. Agism is usually interpreted as a negative factor, as a stable stereotype or as an established form of age discrimination. In Chap. 7, we show in detail that along with problematic agism, good, correct agism is also very widespread, associated with respect for old age, with an understanding of the complexities of adolescence or other transitional periods, the difficulties of the young people's lives, etc. It is estimated that every second person in the world holds ageist views (WHO, 2021). Yet, here one should have in mind that if such views become so widespread, then they do make sense. In any case, agism is not a superficial phenomenon. We believe it has

---

[8] "There are three discrete ageism predictors: age discrimination, defined as detrimental treatment of older persons; negative age stereotypes, defined as the negative beliefs of older persons about older people in general; and negative self-perceptions of aging, defined as the negative beliefs of older persons about their own aging" (Levy et al. 2020: 175).

strong foundations in society, since these ideas make people a) match their own desires and interests with their age position; b) match other people's oppotunities with their age status and c) facilitate communication, subordination in society. It is impossible to change half of humankind.

Moreover, it is fundamentally important to realize that the problem of agism will persist in society for a long time due to global aging, and we must learn how to live with it. Of course, we must try to mitigate the effects of agism, e.g., by encouraging older people to increase their self-esteem and also by trying to awaken respect for older age in younger age groups. But it is impossible to solve this problem so completely that it disappears. Therefore, it is perhaps much more reasonable to channel agism than to try to eliminate it, and tackle only the extreme and most negative manifestations of agism. Agism will be institutionalized in cybernetic society in the frame of institutionalization of age (see below).

**The Role of Age Differences Will Increase**. For many reasons, we believe this, as well as that in the long run social division in cybernetic society will be increasingly linked to age differences. It will be the completion of a long process that has been ongoing already for seventy years. Since the second half of the twentieth century, the role of the differences related to gender, age and other anthropological characteristics has become noticeably greater, whereas class (economic) differences have been decreasing in the social struggle. It is worth mentioning that from the very beginning of the Cybernetic Revolution, that is, in the 1950s–1970s, women's and youth movements, as well as movements related to gender identity (LGBTQ + ), have become more active. The struggle for the rights of racial and ethnic minorities has also become more active. Thus, while at the beginning of the Cybernetic Revolution, the problem of the relationship between the generations of parents and children was the most important, and it can be assumed that in the final phase of the Cybernetic Revolution (in the 2030s–2070s), the elderly will come to the fore.

*In connection with the growing importance of the issue of age, it will manifest itself in new aspects and directions, greatly changing society.*

**Institutionalization of Age**. So, as the importance of age categories in society will increase, in our opinion, it is more correct to institutionalize some aspects of these categories than to try to provide everyone with equal rights, because the latter may be unrealistic. How might that be? First of all, we assume that agism will go through several stages in its development, including quite conflicting ones. But the laws, rules and institutions will be gradually worked out by experience and will lead to a situation where a system will arise for taking into account the interests of different age groups, fixing their rights, duties and restrictions, up to quotas in political representation and in holding posts (as is observed today in relation to quotas for women and people of color).

Every society, at each of its developmental stages must start to actively integrate this problem into everyday life in order to accelerate the processes of adaptation to aging and the institutionalization of the changing age structure of society.

Thus, agism *can be gradually institutionalized into a certain and systematic consideration of the interests of different age groups, establishing their rights, duties and restrictions, perhaps, even up to quotas for political representation, holding positions,* etc.

In addition, institutionalization of age can develop in different spheres, forms and patterns, in particular through:

(a) De jure or de facto different age-specific rights. Of course, such a difference in rights exists today, in relation to minors, as well as the rights of people from a certain age to receive pensions and some other benefits. But otherwise, as a whole, legally adult people are formally equal in rights. However, we believe that in the future the number of rights or restrictions associated with age will increase greatly. This will not necessarily be expressed in legal forms but may take place in the form of stable public perceptions, judicial precedents and the like.[9] Accordingly, there can be a lot of combinations of rights, restrictions and obligations, depending on the society and its cultural and historical characteristics, the social position, the economic context and much more. In addition, there will be an expansion of judicial and legal activity to protect the age rights of all people;
(b) The structuring of the electorate by age categories and, accordingly, the formation of political and legal programs. Also the self-organization of age groups of citizens in various forms ranging from clubs and mutual aid societies to world organizations; from providing mutual assistance to the lobbying of interests;
(c) The strengthening of the ideological orientation when discussing the rights and obligations of age groups (including all kinds of campaigns—as we see in the feminist movement);
(d) The inclusion of the age aspect in all types of planning and activities (e.g., in the planning of buildings, in legal acts and laws and in requirements for employers and landlords). Here, to a certain extent, the institutionalization of disabled people today can serve as a guideline for the development of the institutionalization of old age tomorrow.[10]

## 15.6 The Frame of Cybernetic Society and Its Subsystems

Thus, we come to the conclusion that cybernetic society will emerge as a result of the completion of the Cybernetic Revolution and the maturity of aging. This society will be (a) super-technological; (b) socio-technologically regulated at all levels and (c) a society in which the division into age categories will become a much more important social marker than it is today. The latter, in turn, will lead to a noticeable increase in the social and political significance of these age cohorts. Now it is time to look at how such society might be organized. In Fig. 15.1, one can see what factors will influence the formation of this society.

---

[9] See in Chap. 7 comparison of men and women rights as a clarification of this idea.

[10] Here, to a certain extent, the institutionalization of disability (i.e. special right of disable people) can serve as a guideline.

In what follows, we will show how the factors indicated in the diagram, as well as others (in particular, the slowing down of economic growth, the movement toward optimization of consumption, the production based on self-managing systems, as well as increasing conservatism due to the aging process), will affect the formation of cybernetic society. This society is likely to be more stable, with more guarantees and a longer planning horizon, but will be more regulated, with less innovation and more restrictions on freedom than in modern society. See also Fig. 15.3.

### 15.6.1 The Frame of Cybernetic Society

First, let us briefly consider the expected shape of cybernetic society in its subsystems (see Fig. 15.2).

In the **technological sphere**, the increased rate of self-regulating and self-managing systems will make it possible: a) to ensure a high life expectancy and high quality of biological life, which will support the process of global aging; b) to cover the technological regulation of many political, administrative, managerial, social and economic relations (see Chaps. 11, 12 and 13).

In the **demographic sphere**, one may expect that cybernetic society will be characterized by: (a) a stabilization of the population (or its decline) on a global scale; (b) a steady increase in the proportion of the elderly population in societies (see Chaps. 6 and 7); (c) as a result, the formation of a new—social-qualitative—type of population reproduction.

In the **economic sphere**, we expect that the aging of the population will (a) transform an economic model of supply, demand and consumption toward a less

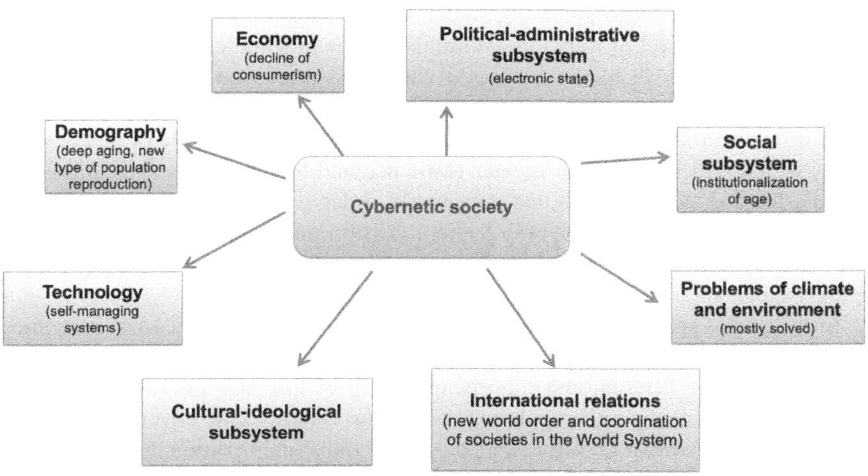


**Fig. 15.2** Subsystems of cybernetic society

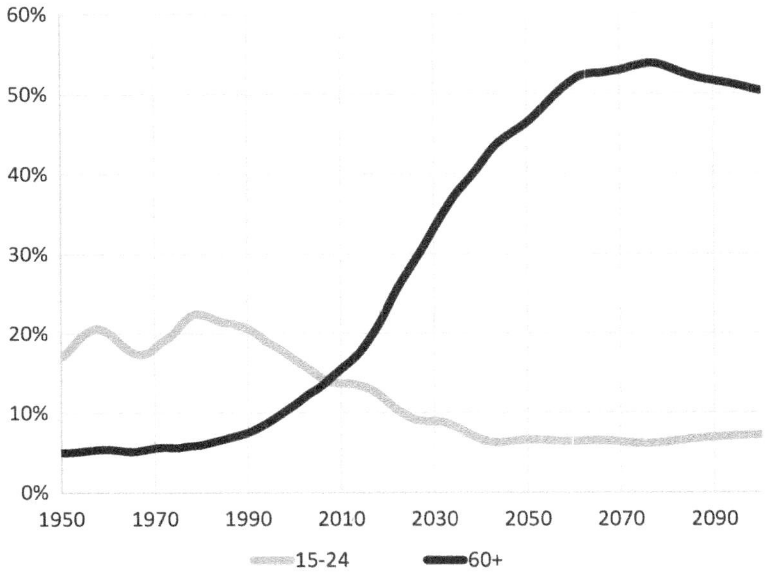


**Fig. 15.3** "Youth bulge" and "elderly bulge" in South Korea: the dynamics of the share of population aged 15–24 and 60 + in total population, 1950–2021 with the UN medium projection till 2100—"South Korean Cross". *Data source:* authors' own calculations on the basis of materials provided by the UN Population Division database

consumerist and more optimal model; (b) reduce the size of the market area and increase public funding (see Chap. 13).

In **politics and administration**, many social and administrative relations will be regulated by specific social-technical self-managing systems. This will take place both at the level of individual administrative units and cities (so-called smart cities) and at the level of the state as a whole. As a result, the "electronic state" will emerge (see Chap. 14).

In **international relations,** we expect the formation in full scale of a new world order, which will be based on greater resilience and broader consensus among the world's leading powers; also this new world order will involve a greater coordination of actors within the World System (Grinin and Korotayev, 2023a). There may also be significant changes in international relations in connection with global aging (see Chap. 7).

In the **social sphere**, it can be expected that: (a) age categories will become a very important social marker and will be institutionalized; (b) in this respect, social tensions between different age cohorts and aging will be also institutionalized; (c) regulation through the self-regulating/managing systems will cover society from top to bottom, which in turn can cause not only stabilization but also social tensions.

**The cultural and ideological subsystem** will be characterized by major changes in the system, content and technologies of education, in which robots and AI will begin to actively participate. Culture will also change significantly as machines will finally become, if not smarter, then 'more educated' than we are. One can assume this social culture to be a culture of greater interaction against the background of the growth of the ideology of stability.

In the **field of ecology**, we assume that a stable global ecology should already be a prerequisite for cybernetic society to reach its maturity.

### 15.6.2 Technological and Demographic Subsystems of Cybernetic Society

As we have already mentioned above, the development of technological innovation will, in the long run, create a super-technological society in which self-regulating and self-managing systems will support high life expectancy (perhaps, even its gradual increase) and improve the quality of biological life; they also will encompass technological regulation of many political, administrative, managerial, social and economic relations. There will be many different technological innovations in MANBRIC technologies and other fields. For details see Chaps. 8–10. In this section, we will discuss only some of the directions of change in medicine that will merge a number of different innovative branches (AI, Robotics, Bio and Nano technologies, Additive and Cognitive technologies as well as transport, perhaps new energy sources etc.). Due to the process of aging a new type—a social-qualitative type—of population reproduction will be established (see above and below). About the influence on political, administrative, managerial, social and economic relations, see in the following sections as well as in Chaps. 12–14.

**Anti-Aging as the Leading Trend of Medicine of the Future**. This phrase is the title of Chap. 11 which well expresses one of the main ideas of our book. It is important to point out that in cybernetic society anti-aging research will reach its full potential. We believe that anti-aging technologies, that is, technologies for slowing down and ennobling aging, as well as for stopping and alleviating the problems associated with it, will become not only an integral part of the future of medicine but will become its system-forming direction. More and more attention will be paid not only to the treatment of diseases of old age but also to the elimination or alleviation of the so-called geriatric syndromes (or age-associated conditions). A very noticeable part of medicine will revolve around rejuvenation, plastic surgery and other anti-aging corrective procedures (which, perhaps, will become as commonplace and regular as a vacation in a sanatorium). The perception of aging in society will also change (see Chap. 7); that will be an important part of the adaptation to aging and the institutionalization of age (see above). Thus, the refocusing of medicine on the problems of aging in line with innovations in anti-aging technologies will imply a process of medical adaptation to aging.

**Medical and Biotechnological Environment as One of the Important Institutions of Cybernetic Society**. In the context of such a transformation of medicine alongside the increasing life expectancy and aging of society by the late twenty-first century will lead to the creation of a super-complex system, which we may define as a special medical–biotechnological environment (see Chaps. 6, 7 and 8). So in the future, a huge part of the population will be unable to stay outside this medical–biotechnological environment. This environment will gradually combine various self-regulating/self-managing and smart systems to create a super-complex system and will become permanent. In our opinion, only in such an environment will it be possible to move toward increasing life expectancy.

What will the MBTE (medical–biotechnological environment) be like? This super-complex system will be a system for health monitoring, control and possible help; where the treatment and support of the organism will be performed mostly by autonomous systems which will be able to function regularly and constantly. In particular, a constant scanning of the organism, including certain organs, tissues, vessels, functions, biometric parameters, etc. will be this autonomous check. Built-in sensors will provide control and regulation of all vital processes (down to cellular level). One can imagine that such systems will be able to detect potentially dangerous situations and quickly respond to critical situations in a different way. This will be a good example of a self-managing system that makes complex decisions autonomously. At the same time, these systems will be able to monitor significant fluctuations of indicators and make recommendations about short- and long-term living habits as well as trying to prevent deviations from the norm and deviations in behavior. Perhaps, it will be important to indicate the exact time of taking drug and the dosage, or the time of physical activities and required exercises with the account of different circumstances and then recommend the most appropriate diet, etc. Of course, these functions may result in rather contradictory and conflicting options.

Such a system, having perceived disturbing signals in the mood of the observed individuals, can transfer the communication with them directly to a specialist (of course, by recording the conversation and analyzing it via neuro-computer methods or other analysis). Forecasting will allow propaedeutic testing of people in order to identify certain deviations, including the mental ones that threaten others with outbreaks of violence (shooting, etc.). An urgent help may be also provided in some critical situations, when a delay can endanger the life of the observed individual. This means that the technologies will allow the system to carry out some actions under certain circumstances, for example, to give an injection, artificial respiration, provide first aid or something like that. Or to give instructions to other people (systems) to perform such actions, as well as to urgently contact medical-technological centers.

As we have said above, this *special medical, biological and technological environment* will become permanent. This means that life may already become impossible without it, similar as how today we can hardly imagine ourselves without the Internet, Wi-Fi and gadgets. Just as many hundreds of millions of people cannot live without constant medication, so in the future, a huge part of the population will be unable to live outside this medical–biotechnological environment (MBTE) for a long time.

This system of almost total monitoring and control will serve not only patients but also different age groups (primarily the elderly and the disabled, of course) as well as people from different risk groups. Millions and perhaps even billions of people will live within this environment. Gradually, it will expand and, apparently, merge with the system of the so-called smart homes.

This system implies not only a breakthrough in medical technologies but also a powerful breakthrough in care technologies, in the continuous monitoring of people's condition (not only of the elderly and the disabled), control, telemedicine, psychological assistance, prediction, prescriptions and emergency care. The psychological counseling (both therapeutic and urgent), also probably carried out in most cases with the help of robotic systems, will become essential (e.g., in the case of a suicide threat).

However, along with the increased duration and comfort of biological life, there are likely to emerge many dangers and challenges (including cohabitation, coexistence with robots and with self-managing systems in general, most of which will have annoying AI). In addition, humanity may face complicated ethical and legal problems with regard to such monitoring since there is always the danger that watching 'Big Brother' will take full advantage of the situation. Of course, people's free agency will be restricted, because sometimes it is more difficult to resist machines than human wishes. At the same time, certain health imperatives will emerge. In fact, everyone will have their own electronic nurse. By the way, this could be especially important for monitoring children and caring for the sick people who stay at home. If there emerge some relatively cheap multifunctional robots able to flexibly react to changes, then people's lives will become much more comfortable (but in that case, their independence will decrease).

The development of such systems will face numerous real problems related to legal, moral and political difficulties, as well as with natural unwillingness to be under constant 24-hour control and to follow the prescriptions of soulless systems. These problems will hopefully be mitigated by a wider and more complex social-technological environment (see below) and due to institutionalization of aging.

**Demographic Subsystem of Cybernetic Society**. As has been already mentioned, in the demographic sphere we expect:

(a) Graying society, in which the proportion of the elderly population will be unusually high and may even constantly increase (see, e.g., Fig. 15.3):

It will lead to a noticeable increase in the social and political significance of these age which we discuss below;

(b) Stabilization of the population (or movement toward it) on a global scale due to decline in birth rate in Africa. Perhaps we will witness even depopulation not only in the developed countries but on the world-system scale. However, one should note that as society ages and adapts to aging, different scenarios may become possible in the future: from minor and controlled growth and complete stabilization of the population to a more or less slow but stable depopulation.[11] It is also not impossible to move toward technologies of (semi)artificial reproduction, at least one can expect a very wide distribution of improved IVF technologies, that could, to some extent, change the demographic picture.[12] Thus, in cybernetic society, we will witness the emergence of the new social-qualitative type of population reproduction.[13] As shown above, this will be a transition to the model of total technological support for health and quality of biological life with a stabilization (or reduction) of the population, with a very large proportion of the elderly population.

### 15.6.3 Economic Subsystem

**New Economic Model of Consumption**. The growth of the share of elderly in the population in cybernetic society will lead to certain conservatism in needs and demand (Chap. 13; Grinin & Korotayev, 2020a, 2020b). This will noticeably reduce consumerism and reorganize the economic model of supply, demand and consumption. In other words, the current wasteful consumption pattern may also change under the influence of the global aging process. And this change, in turn, will have a serious impact on the entire production structure. The current model is associated with an increasing consumption. Consuming more today than yesterday, and more tomorrow than today is largely an absurd model. Also, sometimes the desire for endless growth seems unrealistic. Although it works and will continue to work for decades, especially for poor countries whose populations are not satisfied with their level of consumption, early or later aging can change people's needs, especially in conditions of the stabilization of the population or its decline. It should ultimately lead to the transition to a new economic model without an endless increase in consumption. Accordingly, business models may change, although it is not very clear how this will happen. So we expect that the modern model of constant growth of GDP will be strongly transformed in cybernetic society as well as the indicator of growth itself, which will be more connected with qualitative parameters. In such a case, the model of constant growth of the economy should differ from today, it is likely to include some parameters of quality and longevity of life.[14]

---

[11] For demographic forecasts and other details see in Chaps. 6 and 7.

[12] Today it is difficult to expect radical innovations in reproductive technologies (see Chap. 7). However, the situation may be different within 6–7 decades.

[13] As we said in Chaps. 5 and 7, this TPR will form after the completion of the Cybernetic Revolution at the mature stages of the Cybernetic Production Principle, i.e., in the late twenty-first century.

[14] However, we suppose that GDP will be the most relevant indicator long enough. The GDP indicator is often subject to generally fair criticism (see Costanza et al., 2014; Easterlin 2010; Goldsmith 2019; Kapoor and Debroy 2019; McElwee and Daly 2014). The matter is that, although this criticism is generally true, it is intended to hide the fact that the potential of developed countries has weakened, as well as to cast doubt on the need for economic growth in general; however, the

What the optimization of consunption can mean is, of course, not completely clear today, just as to how much this optimization will be "optimal". Probably, we will move in this direction towards a model that will combine market and state regulation in various proportions (not in the same way as it was in socialist countries, but, perhaps, in a similar way). Thus, capitalism as an economic structure will persist, while market relations will change toward greater stability and predictability (that is, market volatility will decrease) for several reasons, including the fact that self-managing systems will be the leading technologies, and by their very nature require a certain degree of stability (see, e.g., Grinin Malkov et al., 2022).

The optimization of consumption will be facilitated by (a) ideological attitudes (the prototypes of which we see today in the attempts to reduce consumption in order to reduce the carbon footprint) and norms associated with the moral condemnation of excessive consumption; (b) the probable promotion of an economy (i.e., in some form, the revival of the ideology of economy and thrift, which was very common during the period of young capitalism, especially in Protestant countries); (c) the development of waste-free production, in the promotion of which biotechnology can play a special role (see Chap. 10). Significantly, as we have emphasized more than once, one of the key characteristics of the Cybernetic Revolution is the constant saving of resources and energy (see Chap. 3).

The likely decline in market-acquired goods may be accompanied by an increase in public funds, including those used to improve biological quality of life and healthy lifestyles. Below we will add with another important aspect to the picture related to the weakening of market relations.

**Aging and the Transformation of Labour Relations**. In modern society, there are many social guarantees and types of insurance against various problems and misfortunes that can happen to a person (unemployment, accidents, disability, depreciation of professional skills, etc.), as well as developed pension and social security systems. Thanks to an enormous amount of resources, society turned out to be able to support the weak. This gives *an opportunity to better demonstrate one's skills, qualifications, talents and to receive many times more than the social minimum.* Consequently, the coercive factors of labour become weaker, while the incentives to activate people's talents and energy increase (for more details see Grinin, 2006, 2015b). Such a socio-economic model has evolved throughout the Cybernetic Revolution.[15] The scarcity of labor resources significantly contributes to the development of this economic model, because the pressure of unemployment is greatly weakened, so the hard economic reasons for coercion to labor are weakening. In the future, it is quite likely the introduction of some kind of guaranteed income cannot be ruled out, which has been spoken about in recent years. As the

rejection of growth as a goal objectively leads to the stagnation of society (for details see Grinin 2021; Grinin, Malkov et al., 2022).

[15] In Chap. 7, we have called it as the economic type of alienation when we discussed types of alienation and their correlation with Production Principles. For our concept, we use the Hegel-Marx term "alienation".

number of elderly people will dramatically increase, the future *Cybernetic Society will fully correspond to this economic model.* The reasons are the following. Older people have a guaranteed income in the form of pensions and savings. In addition, unlike young people, they have already acquired a lot of basic necessities, so they will need to borrow less, etc., that is, they will be less dependent on employers. Moreover, the development of self-managing systems will partly solve the problems of labor shortages, both in simple jobs (especially in services) and in increasingly complex jobs. And it will probably also lead to the possibility of a shorter working week (at least for the working elderly people). Thus, *the greater the proportion of elderly and old people in the population who are employed in the economy, the more noticeable will be the features of the economic model that requires less dependence on the labor market than is common today.* In other words, people in the cybernetic society will be much less dependent on the economic needs to earn money for living due to pensions and other guarantees. However, we predict that there will be other economic restrictions and regulations that will be difficult enough to adapt to, for example, limitations related to age quotas for jobs, service and positions.

### 15.6.4 Political Subsystem in the Light of Self-Regulating/Self-Managing Systems

**E-state and the Change of the Political Regime**. In politics and administration, the development of social-technical self-regulating systems (*SSSs*), as we argued in Chap. 14, will entail very significant, one might even say revolutionary, changes in the field of governance because many social and administrative relations will be regulated by these systems. This will take place both at the level of individual administrative units and cities (so-called smart cities) and at the level of the state as a whole. As a result, what we have called the "electronic state" (e-state) will emerge. This will lead to a sharp reduction in the number of bureaucrats, as well as to the "cheapening" of the state, and to a reduction in the vices associated with management (corruption, red tape, nepotism, etc. (see Chap. 14; also Grinin, Korotayev, 2021a)).[16] But this will also create a number of problems. On the one hand, with the development of social-technical self-managing systems (SSSs) a transition to direct democracy through permanent electronic voting is very likely. But on the other hand, the use of such self-managing systems in governance will lead to increased technological and political control. We anticipate that this will result in a particular type of political regime, something like *democratic authoritarianism.* Finally, the role of the elderly in politics will increase dramatically and,

[16] But, nevertheless, it is impossible to assert one hundred percent that the number of civil servants in general will sharply decrease. It is likely that, according to the laws of Cyril Parkinson, even in such a technological society as the cybernetic one, officials will find a job for themselves.

accordingly, age will become an important line of political agitation (for more details see Grinin, 2023a; see also Grinin et al., 2021a).

**Technology, Management and Stability**. We assume that, firstly, the use of SSSs in management will give a great impetus to the development of so-called smart cities that are so much discussed. But it is not so often understood that regulation in such cities should be precisely techno-social. In addition, the development of self-driving cars, drones, robots, etc. (see Chap. 10 for more on these) should also fit into the governance of these smart cities and beyond.

Secondly, the development and implementation of the SSSs at different levels of government will contribute to a certain degree of stability ingovernance, as changing management algorithms will become more difficult than in the current political and bureaucratic process. This is especially noticeable after the change of ruling parties in elections. Similarly, for example, with regard to the idea of protected budget items that are not reviewed annually, there may be algorithms, the revision of which will require special cases and procedures.

Thirdly, the role of the elderly people will sharply increase in politics (this is already clearly visible today, in particular in the USA), and, accordingly, age will become an important political dividing line.

It can be assumed that, in general, society will become more solidary, but the contradictions between age groups, between conservatism and the need for technological progress and between the rights to personal freedom and socio-technological restrictions will be felt quite strongly.

### 15.6.5 International Relations

**The Growth of Societies' Coordination**. In international relations, we may see the maturation of a new world order, which will increase the coordination of societies. This would be the optimal scenario for maintaining a high life expectancy and a stable society. It is also very likely that countries with older populations will become more peaceful, with less ambition to wage war. However, of course, we cannot deny the likelihood of less positive or even negative scenarios, which we mentioned in Chap. 7. In international relations, the optimal scenario for maintaining a high life expectancy and stability of society could be realized as a result of changes in individual societies and the maturity of a new world order, due to which the coordination of societies will increase.

This may also be due to the further development of supranational and even global institutions (AI and SSSs may well be a very important element here, providing the necessary level of information for management). By the way, agreements on the development/non-development of a general type of AI may well become one of the most important moments in international relations.

**A Crisis of Democracy?** There will also be signs of a crisis of democracy in its modern form (on the problems of democracy see, e.g., Tsirel, 2020; Grinin et al., 2021b; Grinin et al., 2023b), which will require its modernization, especially at the supranational level. Democracy, having certain advantages at the national level (but not everywhere and not always), may lose these advantages at the supranational level. For example, the European Union, which once seemed to be a model of supranational entities, no longer looks so due to a number of serious problems. In particular, super-democratic procedures and the requirement of consensus create a controversial de jure and de facto governance (in the latter case, stronger states 'twist the arms' of weaker ones, but at the same time, external forces, by manipulating these weak forces, can affect a whole union of nation states). Therefore, the principle of democracy is likely to be only a part (and not always the leading one) of such unions (see: Grinin, 2016; Grinin, Korotayev, Tausch 2019 for more details).

**Expertocracy?** We do not rule out that along with the change in democratic structures, new authorities may emerge in the form of expert councils (armed with the most powerful AI).

The rejection of democracy as a system of supranational governance needs some new forms. One of our assumptions is the growing role of experts and international expert bodies. Even now, their de facto role is sometimes very significant, as, for example, the role of rating agencies. That is why the development of a kind of an expertocracy cannot be excluded as a possibility. Accordingly, there may be expert authorities who can play a political role,—for example, if the nomination of political experts based on their professional qualities becomes the prerogative of national academies (scientific or other institutions) or the result of public elections of selected experts by professional communities. The purpose of such expert meetings is to conduct (directly or with the help of appropriate expert organizations) an expertise and a scientific analysis of various acts and decisions in terms of their compliance with all scientific achievements. Today many decisions are made on an emotional basis, without calculations and understanding of the consequences (e.g., climate legislation, medical regulation, family laws, parenting, technological regulations etc.). It is important to take into account the constant changes in the development of science and technology in order to be able to regulate these processes. Even given the inevitable limitations (and possible biases) of experts' decisions, they will still be more balanced and better adapted to the needs of modern society. Thus, experts will become something like another branch of government, their influence will increase and the positions of expert organizations will remotely resemble those of the workshops and guilds of the Middle Ages.

Experts, unlike politicians, have a different kind of authority, so they can act without regard to voters and society and, moreover, they are often transnational. Experts from different nations (but united by a single professional ethos) are often closer to each other than to representatives of their countries with a low level of education (Grinin, 2016; Grinin et al., 2021).

### 15.6.6 Social Subsystem and Ecological Environment

**Stratification and Relations**. In the social sphere, it is expected that age will become a very important social marker, while regulation by self-regulating systems will cover society from top to bottom. Our idea is that the role of generations in the social division of cybernetic society will grow, and, accordingly, the positions of different generations should be more clearly taken into account in various relations, institutions and, possibly, laws. All this will create new relationships, social-age strata and forms of interaction. However, age categories will create new social minorities and other problems. In particular, social tensions between different age groups are very likely, which may be weakened by the institutionalization of age. It can be assumed that society as a whole will become more solidary because age markers are not constant ones and in any case are less antagonistic than property or even educational status; however, agism (even institutionalized one) will be felt. Also, as we have mentioned above the contradictions between age strata, between the rights to personal freedom and socio-technological restrictions will definitely be percieved strongly enough.

Thus, in cybernetic society, due to the increasing social significance of age differences, self-organization around age lines may increase. For example, self-help groups are already emerging, in which the "younger elderly" or stronger people help the "older elderly" or people whose abilities are limited. There are examples of this in countries with different levels of development (UN Population Division, 2015). Accordingly, different variants of such social self-organization or institutionalization around the age-defined boundaries are possible. Other variants of this type of social self-organization are also possible (see Chap. 7).

Of course, in connection with the development of technology, there will be big changes in the socio-professional composition, as well as in the field of education. For example, previously we talked about reduction in the number of civil servants due to the development of the SSSs. Also, many professions will become rarer. This can affect both creative professions, for example, designers and journalists and service sector professions (cashiers, waiters, cooks, etc.), as well as drivers, teachers, doctors, etc.

**Socio-Technological Symbiosis and Its Consequences**. As argued above, technology will regulate many more relationships in the described society than it does today; people's dependence on technology will increase, including in the area of health care. As a result, together with the medical–biotechnological environment (MBTE), a kind of socio-technological sphere will emerge, in which people will be forced to stay almost continuously. Thus, on the basis of MBTE, smart sites, SSSs and similar systems of the e-state, other self-regulating systems *arise*.

It can also be assumed that the development of technology, as well as the involvement of more and more elderly age groups of the population in labour force, will reduce the working week, for example, to 30 h, that is, to a 6-h working day.

**Ecological Environment**. In the field of ecology and the impact on the natural environment, an important role will be played by the understanding that an increase in life expectancy and its biological quality is possible only under certain environmental standards; otherwise, the harm to health from adverse ecology will undermine all the successes of medicine. Thus, we assume that a good ecological context should already be a prerequisite for cybernetic society to reach its maturity. The possibilities of future biotechnologies, in our opinion, can largely solve these problems (see Chap. 9). Of course, one cannot rule out a scenario of worsening climate and environmental conditions. No doubt, this will definitely delay the transition to cybernetic society.

**Cultural and Ideological Subsystem.** Making predictions about culture and ideology is especially difficult. Based on technological trends, one can assume great changes in the system and technologies of education, which will become more mobile and not as rigidly professional as it is today will occur. We have already talked about the trend toward universalization, which leads to the fact that with the help of AI, self-managed systems and other technologies, a person will be able to master many skills that were previously (and still are) the domain of professionals. So while self-managed systems will do a lot, other skills, abilities and competencies will be required. And this trend will permeate primary and secondary schools. At the same time, there are great fears that many of the mental skills that are generally necessary for a person in any society may be poorly developed in children and people of the future (due to the competencies of self-managing systems).

As for culture, there will also be many changes here; in fact, the notion of a "cultured person" is likely to change, especially against the backdrop of omniscient robots, AI and other systems. With regard to public culture, there may be different options, one of which is a culture of greater interaction and greater solidarity in society, although selfishness and individualism will not disappear. Due to the greater manageability and stability of society, one can assume (see above) the growth of the ideology of stability.

We hope that positive changes will take place in public morality, which is weakening for a number of reasons, among which an important place is occupied by the speed of technological and other changes. The faster technology develops, the less morality is recognized because it does not find a new balance. Also, it is rather possible that beyond some limit of speed of scientific and technical development, a noticeable destruction of morals or their disintegration into numerous group versions will begin. As Stanisław Lem (1968) rightly observes, the intensity with which the facilitation by technical means of performing tasks undermines values is positively correlated with the efficiency with which tasks are performed. This means that the more effectively technologies solve certain problems, the more they change society and its moral and legal pattern, the consequences of which we will begin to realize much later. However, because cybernetic society will be more stable, we believe that there will also be some stabilization of public consciousness, social psychology and moral norms, since the far too rapid pace of technological change is currently literally undermining the stability of an increasing number and variety of moral standards.

### 15.6.7 Summary

**A More Stable and More Regulated Society**. In general, we expect a movement toward a certain equilibrium associated with the reduction in the rate of technological progress analyzed in Chap. 12 (see also above), the widespread introduction of self-managing systems (including the socio-technical ones mentioned above) and, accordingly, greater predictability of development. The slowdown in economic growth rates and the stabilization of the population, as well as the movement toward optimization of consumption (the new consumption model described in Chap. 13 and above) will inevitably create a new development model, with relatively low growth rates, but with a smaller range of fluctuations. In particular, there is some reason to believe that the regulation of technological innovations, which have become too rapid, uncontrollable and even without clear explanations of the algorithms for their work, will also increase. This is especially true for AI, cognitive technologies and a number of other technologies, in particular those areas related to biotechnology. Moreover, today many technological innovations that have a major impact on society are either not properly regulated or directly contradict fundamental legal norms. We believe that this imbalance will be reduced in the cybernetic society.

Perhaps, all this will allow the implementation of the idea of sustainable development, which today seems more like a wish. Thus, we assume that the cybernetic society will be a society with more guarantees and more stability, but it will also be noticeably more regulated than today. Production based on self-managing systems and the growth of conservatism due to the aging process will make life more stable and predictable. However, this can weaken the society's capacity to innovativeness and lead to restrictions on freedom. Moreover, stability within societies is unlikely to exist without a more or less stable foreign policy. It is likely that the ideology of stability (as a very important value) will also be strengthened, as opposed to the modern ideology, which claims that changes must go on continuously and in an ever-increasing volume.

Thus, self-managing systems in the economy and administrative management will lead to greater stability and greater order; stabilization of the population, the reduction of economic growth rates, conservatism of the majority of the population, a new model of "optimization" of consumption, as well as the need to maintain high life expectancy—all these and a number of other factors require greater regulation, predictability and stability. Below, in Fig. 15.4, we graphically show what factors in cybernetic society will influence the strengthening of its stability and the growth of regulation.

Of course, a society cannot exist in a relatively stable state for too long. One way or another it will be taken out of balance, but we believe that such a period can be stable for several decades (at the end of the twenty-first and the first decades of the twenty-second century).

We do not seek to draw an ideal society, but rather to show the logic of development that is potentially foreseeable today. In the description of this society,

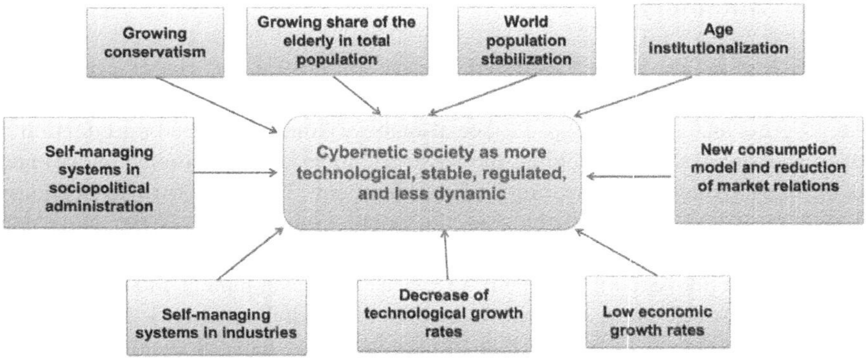


**Fig. 15.4** Factors defining the cybernetic society's characteristics

as well as throughout the entire monograph, we try to show that every improvement or achievement in technology and other areas, especially a radical innovation, always has its negative sides. The question is how to balance them. Achieving a better balance largely depends on the position of our and future generations and their willingness to work for a better future. It can be assumed that different societies will differ quite a lot from each other in terms of the balance of rights and restrictions on age categories, just as they differ today in the balance of social guarantees.

**Post-Human Revolution?** In the Introduction, we have devoted much attention to the idea of a post-human revolution and the unfolding course toward the increasing use of non-biological or artificial biological tissues and organs for the successful development of medical care, the extension of life expectancy and the improvement of the quality of human biological life. We called this process cyborgization. And it will continue for quite a long time. Of course, in any case, the trends of global aging, the introduction of new technologies and the growing ability to change the biological nature of human beings (see Chaps. 3 and 8; also Grinin, 2015b; Grinin et al., 2017a) will create certain ethical and legal difficulties, which will be even more serious than today's gender problem. However, we believe that although cybernetic society will undoubtedly move much further along this path than the modern one, even so, this society will not be post-human, but will be quite human, only older and with much greater capabilities of medicine and other technologies. In other words, if the post-human revolution ever happens, cybernetic society will still be on the verge of becoming post-human. Despite the exceptionally rapid development of AI, we also believe that cybernetic society, which will be much more regulated than the current one, will not turn into something like Kurzweil's idea of 'matrix' (not to mention the society of the science fiction film "The Matrix"), in which the thinking of the inhabitants will be directly controlled by a global AI, although some essential elements of this may take shape.

### 15.6.8 Postscript

We have considered a vision of the future of society, which seems to us quite possible. However, new types of societies do not emerge by themselves. The future is being created every day, as there is a constant confrontation between various forces seeking to secure their positions, interests and ideologies. Many things change in the process of such a confrontation. Therefore, the future is not predetermined, even if some trends seem to be almost one hundred percent predictable. Take such a large-scale process as, for example, global democratization. In the 1990s, when Huntington wrote about the third wave of democratization (Huntington, 1993), it seemed that this wave would "flood" the World System, making authoritarian systems completely outcast. But since then, there has been a wave of increasing authoritarianism (see Selbin, 2022). And now it is not at all obvious that stable democracy will be established everywhere. Not at all, rather the new world order will be a recognition of the legitimacy of the existence of states with different regimes (see Grinin, et al., 2023b). We pointed out the same thing in relation to globalization, which after achieving very great success, suddenly began to recede, a process due to the resistance of those who stood at its origins and violently pushed it. The trend toward the creation of more and more integrated supranational associations, of which the EU was a model, is also stalling today, being replaced by situational unions (see Grinin, 2009; Grinin, Korotayev 2010, 2011; Grinin et al., 2023b).

The same applies to more objective processes. Thus, the trend toward economic growth (and the associated rise in the standard of living) seemed unshakable, but now this possibility is increasngly being debated (primarily because growth has stalled). Is growth necessary? Is it worth striving to improve the standard of living?[17]

Thus, every process has its own limitation (as every action has its own reaction). These limits can be in the form of time, which is invisible or insufficiently aware of counteractions that begin to slow down the process; either in the form of social forces that try to reverse the process or as a result of discoveries and innovations (as happened with demographic growth as a result of the invention of contraception; see Chap. 5). The reader remembers that we have paid enough attention to the so-called, singularity, as a result of which (i.e., when the limits of development are reached) processes change their direction or even become their antipodes.

All of the above fully applies to the process of global aging. Demographic processes lend themselves very well to forecasting, but these are, as a rule, inertial forecasts that do not take qualitative changes into account. Therefore, we do not exclude at all the possibility that the aging process may slow down or even reverse at some point (as we saw during the COVID pandemic). Well, if the situation changes as a result of some breakthrough reproductive innovations (see

[17] For details see Grinin et al., 2023a, 2023b, 2023c, 2023d, 2023e; Grinin 2021a, 2021b.

above), then the demographic structure of society will become more proportional: the elderly will no longer be such a large part of the population, there will be many children and young people. If a "cross" similar to the South Korean one we have cited (see Fig. 15.3) does not disappear, then at least the demographics dominated by a prominent older age structure will not be so frightening. And this will reduce the conservatism of society. Although, of course, artificial reproduction will create completely new and very complex problems.[18] But it is much worse if the future society turns into an anti-gerontocratic one and practices, on an increasing scale, the forcing of old people to legalized suicide (euthanasia). We wrote about the growing scale of this latter possibility in Chap. 7.

Or there will break out a new natural or man-made pandemic, more terrible than COVID, which will wipe away the elderly (we emphasize that epidemiologists and the WHO constantly talk about the threat of a new pandemic). It is even more likely that the deterioration of international relations, weak economic dynamics, complicated by the desire for an early green transition, military tensions, etc. will sharply reduce the standard of living. This deterioration, especially in connection with marked inflation, rising public debt and attempts to control pension savings, will hit pensioners hardest in the long run. Accordingly, the volume and quality of medical care will deteriorate. The decline in the standard of living can in no way ensure that life expectancy will increase but even remain at the current level. This means that, along with the continuing decline in the birth rate, many societies risk experiencing pronounced depopulation and the consequent degradation of nations. And, in this case, it is very likely that the threat of "The Decline of the West", expressed by Oswald Spengler more than a hundred years ago (1991) can become a reality.

There may be other negative trends, some of which we cannot even imagine yet. It would be a pity if humanity lost a unique chance to create a society of centenarians, people who would have much more time to fulfil their potential and their dreams.

However, we believe that although no one can guarantee that the cybernetic society we described will actually emerge, the chances of its creation are high and it is likely to appear.

**Acknowledgements** This chapter is an output of a research project implemented as part of the Basic Research Program at HSE University in 2024 with support from the Russian Science Foundation (Project No. 23-18-00535).

[18] For example, it can be used for geopolitical purposes. Indeed, why do not some political elites use new reproductive technologies? Therefore, it is quite possible that in the future political elite will be able to use 'industrial' reproductive technologies for their geopolitical purposes. For example, they can launch a population growth race. But if some countries try to solve the problem of shortage of children by incubating them in artificial placentas, the race for 'child reproduction' will inevitably start, and nobody knows what it will bring (see also Grinin & Grinin 2016: 142).

## References

Akaev, A., & Davydova, O. (2023). Climate and Energy. Scenarios of Energy Transition and Global Temperature Changes Based on Current Technologies and Trends. In: V. Sadovnichy et al. (Eds.), *Overcoming the Limits to Growth. A Report to the Russian Association of the Club of Rome*. Springer. https://doi.org/10.1007/978-3-031-34999-7_4.

Akaev, A., Malkov, S., Davydova, O., Kovaleva, N., Malkov, A., Grinin, L., & Korotayev, A. (2023). Modeling Social Self-Organization and Historical Dynamics. Industrial Society. In: V. Sadovnichy et al. (Eds.), *Overcoming the Limits to Growth. A Report to the Russian Association of the Club of Rome* (pp. 387–417). Springer. https://doi.org/10.1007/978-3-031-34999-7_18.

Akaev, A., Malkov, S., Grinin, L., Bilyuga, S., Davydova, O., Grinin, A., Kovaleva, N., Malkov, A. (2023). Modeling Social Self-Organization and Historical Dynamics. Modern Society and Problems of Global Transition. In: V. Sadovnichy et al. (Eds.), *Overcoming the Limits to Growth. A Report to the Russian Association of the Club of Rome* (pp. 461–490). Springer. https://doi.org/10.1007/978-3-031-34999-7_20.

Alkema, L., Raftery, A., Gerland, P., Clark, S., Pelletier, F., Buettner, T., & Heilig, G. (2011). Probabilistic projections of the total fertility rate for all countries. *Demography, 48*(3), 815–839. https://doi.org/10.1007/s13524-011-0040-5

Allen, R. C. (2009). *The British Industrial Revolution in Global Perspective*. Cambridge University Press.

Allen R. C. (2011). *Global Economic History: A Very Short Introduction*. Oxford University Press.

Andersen, L.R. (2020). *Bildung—Keep Growing. A report to the Club of Rome. A report to the Club of Rome*. Nordic Bildung.

Bardi, U. (2017). *The Seneca Effect. Why Growth is Slow but Collapse is Rapid. A report to The Club of Rome*. Springer. https://doi.org/10.1007/978-3-319-57207-9.

Barnes, R. (2021). Microsoft, Oracle and other tech giants team up on Covid-19 vaccine 'passports'. *Tech Compsmag. News—Tech News, Latest News, Business News and world News*. https://www.compsmag.com/news/technology/microsoft-oracle-and-other-tech-giants-team-up-on-covid-19-vaccine-passports/.

Barsukov, V. N. (2019). From the Demographic Dividend to Population Aging: World Trends in the System-Wide Transition. *Economic and Social Changes: Facts, Trends, Forecast, 12*(4), 167–182. https://doi.org/10.15838/esc.2019.4.64.11.

Bellwood P. (2004). *First Farmers: The Origins of Agricultural Societies*. Wiley-Blackwell.

Benson, I., & Lloyd, J. (1983). *New Technology and Industrial Change: The Impact of the Scientific-Technical Revolution on Labour and Industry*. Kogan Page.

Berg, C. (2019). *Sustainable Action: Overcoming the Barriers*. Routledge.

Bernal, J. D. (1965). *Science in History*. 3rd ed. Hawthorn Books.

Bloom, D. E., & Luca, D. L. (2016). *The Global Demography of Aging: Facts, Explanation*. Elsevier.

Bös, D., & Von Weizsäcker, R. K. (1989). Economic consequences of an aging population. *European Economic Review, 33*(2–3), 345–354.

Bozesan, M. (2020). *Intergal Investing. From Profit to Prosperity. A report to the Club of Rome*.Springer. https://doi.org/10.1007/978-3-030-54016-6.

Cipolla, C. M. (Ed.). (1976). *The Industrial Revolution, 1700–1914*. Harvester Press.

Chase-Dunn, C., Kim, J., & Alvarez, A. (2020). *Comparing Waves of Structural Deglobalization: A World-Systems Perspective*. University of California-Riverside.

Childe, V. G. (1948). *What Happened in History*. Penguin Books.

Chin, J., & Wong, G. (2016). China's new tool for social control: A credit rating for everything. *The Wall Street Journal, 28*. https://www.wsj.com/articles/chinas-new-tool-for-social-control-a-credit-rating-for-everything-1480351590.

Clark, R. J., & Spengler, J. J. (1980). *The Economics of Individual Population Aging*. Cambridge University Press.

Costanza, R., Kubiszewski, I., Giovannini, E., Lovins, H., McGlade, J., Pickett, K. E., Ragnarsdóttir, K. V., Roberts, D., De Vogli, R., Wilkinson, R. (2014). Development: Time to leave GDP behind. *Nature News*, 505(7483), 283–285. URL: https://doi.org/10.1038/505283a.

Creemers, R. (2018). China's social credit system: an evolving practice of control. *Social Science Research Network*. https://doi.org/10.2139/ssrn.3175792

Easterlin, R. A. (2010). Well-Being, Front and Center: A Note on the Sarkozy Report. *Population and Development Review*, 36(1), 119–124. URL: https://doi.org/10.1111/j.1728-4457.2010.00320.x.

Fichtner, J. J. (2018). Global aging and public finance. *Business Economics, 53*(2), 72–78.

Foerster, H. von, Mora, P. M., & Amiot, L. W. (1960). Doomsday: Friday, 13 November, A.D. 2026: At this date human population will approach infinity if it grows as it has grown in the last two millennia. *Science, 132*(3436), 1291–1295. https://doi.org/10.1126/science.132.3436.1291.

Fücks, R. (2013). *Intelligent Wachsen*. Carl Hanser Verlag.

Fukuyama F. (2002). *Our Post-Human Future: Consequences of the Bio-Technology Revolution*. Farrar, Straus, and Giroux.

Gallagher J. (2020). Fertility rate: 'Jaw-dropping' global crash in children being born. *BBC*. https://www.bbc.com/news/health-53409521.

Goldsmith, C. (2019). Why GDP is no longer the most effective measure of economic success. *World Finance*. URL: https://www.worldfinance.com/strategy/why-gdp-is-no-longer-the-most-effective-measure-of-economic-success.

Goldstone, J. A. (2015). Population Aging and Global Economic Growth. In J. A. Goldstone, L. Grinin, & A. Korotayev (Eds.), *History & Mathematics: Demography & Aging* (pp. 147–155). Uchitel.

Goldstone, J. A., Grinin, L., & Korotayev, A. (2015). Research into Global Aging and Its Consequences. In J. A. Goldstone, L. Grinin, & A. Korotayev (Eds.), *History & Mathematics: Political Demography & Global Aging* (pp. 5–9). Uchitel.

Grinin, L. (2006). *Proizvoditel'nyye sily i istoricheskiy protsess*. URSS.

Grinin, L. E. (2009). The State in the Past and in the Future. *Herald of the Russian Academy of Sciences*, 79 (5): 480–486. http://www.maik.ru/contents/herald/herald5_9v79cont.pdf.

Grinin, L. (2011). *Gosudarstvo i istoricheskiy protsess. Epokha formirovaniya rannego gosudarstva: Obshchiy kontekst sotsial'noy evolyutsii pri obrazovanii gosudarstva*. LKI.

Grinin, L. (2021). How globalization shifts the global economic and political balance. *Journal of Chinese Philosophy, 48*(2), 222–234.

Grinin, L., & Grinin, A. (2015a). Kiberneticheskaya revolyutsiya i shestoy tekh-nologicheskiy uklad. *Istoricheskaya Psikhologiya i Sotsiologiya Istorii, 8*(1), 172–197.

Grinin, L., & Grinin, A. (2015b). *Ot rubil do nanorobotov. Mir na puti k epokhe samoupravlyayemykh sistem (istoriya tekhnologiy i opisaniye ikh budushchego)*. Uchitel.

Grinin, A., & Grinin, L. (2015c). The cybernetic revolution and historical process. *Social Evolution and History, 14*(1), 125–184.

Grinin, L., & Grinin, A. (2016). *The Cybernetic Revolution and the Forthcoming Epoch of Self-Regulating Systems*. Uchitel.

Grinin, A., & Grinin, L. (2021a). Razmyshleniya ob ekonomicheskom roste i budushchem. Stat'ya pervaya. Globalizm vs rost VVP i zakat Zapada. *Filosofiya i Obshchestvo, 3*, 5–34. https://doi.org/10.30884/jfio/2021.03.01

Grinin, A., & Grinin, L. (2021b). Razmyshleniya ob ekonomicheskom roste i budushchem. Stat'ya vtoraya. Globalizm, neosotsializm, spaseniye klimata i ekonomicheskiy rost. *Filosofiya i Obshchestvo, 4*, 5–38. https://doi.org/10.30884/jfio/2021.04.01

Grinin, L., & Grinin, A. (2023a). Analyzing Social Self-Organization and Historical Dynamics. Future Cybernetic W-Society: Socio-Political Aspects. In: V. Sadovnichy et al. (Eds.), *Overcoming the Limits to Growth. A Report to the Russian Association of the Club of Rome*. Springer. https://doi.org/10.1007/978-3-031-34999-7_22.

Grinin, L., & Grinin, A. (2023b). Technology. Limitless Possibilities, Effective Control. In: V. Sadovnichy et al. (Eds.), *Overcoming the Limits to Growth. A Report to the Russian Association of the Club of Rome*. Springer. https://doi.org/10.1007/978-3-031-34999-7_8.

Grinin, L., Grinin, A., & Korotayev, A. (2017a). Forthcoming Kondratieff wave, Cybernetic Revolution, and global aging. *Technological Forecasting and Social Change, 115*, 52–68. https://doi.org/10.1016/j.techfore.2016.09.017

Grinin, L., Grinin, A., & Korotayev, A. (2017b). The MANBRIC-Technologies in the Forthcoming Technological Revolution. In Devezas, T., Leitão, J., Sarygulov, A. (eds.), *Industry 4.0—Entrepreneurship and Structural Change in the New Digital Landscape: What is Coming on Along with the Fourth Industrial Revolution* (pp. 243–261). Springer. https://doi.org/10.1007/978-3-319-49604-7_13.

Grinin, L., Grinin, A., & Korotayev, A. (2020a). A Quantitative analysis of worldwide Long-Term technology growth: From 40,000 BCE to the Early 22nd Century. *Technological Forecasting and Social Change, 155*, 1–16. https://doi.org/10.1016/j.techfore.2020.119955

Grinin L, Grinin A, Korotayev AV (2020b) Dynamics of technological growth rate and the forthcoming singularity. In: Korotayev AV, LePoire D (eds) *The 21st century Singularity and global futures. A Big History perspective* (pp 287–344). Springer, Cham. https://doi.org/10.1007/978-3-030-33730-8_14.

Grinin, L., Grinin, A., & Korotayev, A. (2021a). COVID-19 Pandemic as a trigger for the acceleration of the Cybernetic Revolution, transition from E-Government to E-State, and change in social relations. *Technological Forecasting & Social Change, 175*, 1–16. https://doi.org/10.1016/j.techfore.2021.121348.

Grinin, L., Grinin, A., & Korotayev, A. (2021b). Global Trends and forecasts of the 21st Century. *World Futures, 77*(5), 335–370. https://doi.org/10.1080/02604027.2021.1949939

Grinin, L., Grinin, A., & Korotayev, A. (2023a). Future Political Change. Toward a More Efficient World Order. In: V. Sadovnichy et al. (eds.), *Overcoming the Limits to Growth. A Report to the Russian Association of the Club of Rome* (pp. 191–206). Springer. https://doi.org/10.1007/978-3-031-34999-7_11.

Grinin, L., Grinin, A., & Korotayev, A. (2023b). Global Aging—an Integral Problem of the Future. How to Turn a Problem into a Development Driver? In: V. Sadovnichy et al. (Eds.), *Overcoming the Limits to Growth. A Report to the Russian Association of the Club of Rome* (pp. 117–135). Springer. https://doi.org/10.1007/978-3-031-34999-7_7.

Grinin, L., Grinin A., & Malkov, S. (2023a). Economics. Optimizing Growth. In: V. Sadovnichy et al. (Eds.), *Overcoming the Limits to Growth. A Report to the Russian Association of the Club of Rome*. Springer. https://doi.org/10.1007/978-3-031-34999-7_9.

Grinin, L., Grinin, A., & Malkov, S. (2023b). Socio-political transformations. A difficult path to cybernetic society. In V. Sadovnichy et al. (Eds.), *Reconsidering the limits to growth. A report to the Russian Association of the Club of Rome*. Springer. https://doi.org/10.1007/978-3-031-34999-7_10.

Grinin, L., & Korotayev, A. (2010). Will the global crisis lead to global transformations? The Coming Epoch of New Coalitions. *Journal of Globalization Studies, 1*(2), 166–183.

Grinin, L. E., & Korotayev, A. V. (2011). The coming epoch of new coalitions: possible scenarios of the near future. *World Futures, 67*(8), 531–563. https://doi.org/10.1080/02604027.2011.625749

Grinin, L., & Korotayev, A. (2014). Globalization shuffles cards of the world pack: In which direction is the global economic-political balance shifting? *World Futures, 70*(8), 515–545. https://doi.org/10.1080/02604027.2014.982474

Grinin, L., & Korotayev, A. (2015). *Great Divergence and Great Convergence*. Springer. https://doi.org/10.1007/978-3-319-17780-9

Grinin, L., & Korotayev, A. (2016). Global population aging, the sixth kondratieff wave, and the global financial system. *Journal of Globalization Studies, 7*(2), 11–31.

Grinin, L., & Korotayev, A. (2021). Seven Weaknesses of the U.S., Donald Trump, and the Future of American Hegemony. *World Futures, 77*(1), 23–54. https://doi.org/10.1080/02604027.2020.1801309.

Grinin, L., Korotayev, A., & Tausch, A. (2016). Economic cycles, crises, and the global periphery. *Springer*. https://doi.org/10.1007/978-3-319-41262-7
Grinin, L., Grinin, A., & Korotayev, A. (2021). Global Trends and Forecasts of the 21st Century. *World Futures, 77*(5), 335–370. https://doi.org/10.1080/02604027.2021.1949939
Grinin, L., Malkov, S., Grinin, A., & Korotayev, A. (2022). Will capitalism die? Reflections on the Capitalism of the Past, Present and Future. *Sotsiologicheskiy Zhurnal = Sociological Journal, 28*(2), 100–130. https://doi.org/10.19181/socjour.2022.28.2.8988.
Grinin, L., Malkov, S., & Korotayev, A. (2023). Developed and Developing Countries. Toward a Common Goal at Different Speeds. In: V. Sadovnichy et al. (Eds.), *Overcoming the Limits to Growth. A Report to the Russian Association of the Club of Rome* (pp. 207–224). Springer. https://doi.org/10.1007/978-3-031-34999-7_12.
Grinin, L., Markov, A., & Korotayev, A. (2013). On similarities between biological and social evolutionary mechanisms: mathematical modeling. *Cliodynamics* 4 (2): 185–228. https://doi.org/10.21237/C7clio4221334.
Harris, D., & Hillman, G. (Eds.). (1989). *An Evolutionary Continuum of People—Plant Interaction*. Unwin Hyman.
Hasan, H. R., Salah, K., Jayaraman, R., Arshad, J., Yaqoob, I., Omar, M., & Ellahham, S. (2020). Blockchain-Based solution for COVID-19 digital medical passports and immunity certificates. *IEEE Access, 8*, 222093–222108. https://doi.org/10.1109/ACCESS.2020.3043350
Hsu, Y. H., & Lo, H. C. (2019). The Impacts of Population aging on saving, capital formation, and economic growth. *American Journal of Industrial and Business Management, 9*(12), 2231–2246. https://doi.org/10.4236/ajibm.2019.912148
Huntington SP (1993). *The Third Wave: Democratization in the Late Twentieth Century*. University of Oklahoma Press.
Jackson, R. (2021). *The Shape of Things to Come. The Macro Challenges of Population Aging*. Concord Coalition & the Global Aging Institute.
Jackson, R., Howe, N., Strauss, R., & Nakasima, K. (2008). *The Graying of the Great Powers. Demogarphy and Geopolotitics in the 21st Century*. Center for Strategic & International Studies.
Jackson R., Macaranas R., & Peter T. (2013). *U.S. Development Policy in an Aging World. New Challeges and New Priorities for a New Demographic Era. A Research Report from Project on U.S. Leadership*. Center for Strategic & International Studies.
Jaspers, K. (1953). *The Origin and Goal of History*. Yale University Press.
Kalinovskaya, K. (1976). *Vozrastnyye gruppy narodov Vostochnoy Afriki*. Nauka.
Kalinovskaya, K., & Markov, G. (1987). Obshchestvennoye razdeleniye truda u sko-tovodcheskikh narodov. Azii i Afriki. *Vestnik MGU. Seriya Istoriya, 6*, 56–69.
Kamiya, Y., & Hertog, S. (2020). *Measuring household and living arrangements of older persons around the world*. United Nations, Department of Economics and Social Affairs, Population Division.
Kaneda, T., Greenbaum, C., & Patierno, K. (2016) *World Population Data Sheet*. Population Reference Bureau. https://www.prb.org/2016-world-population-data-sheet/.
Kapelyushnikov, R.I. (2019). *Fenomen stareniya naseleniya: ekonomicheskiye effekty*. Institut ekonomiki RAN.
Kapitza, S. (2006). *Global population blow-up and after. Report to the Club of Rome*. Global Marshall Plan Initiative.
Kapitza, S., (2009). *Obshchaya teoriya rosta chelovechestva. Kak ros i kuda idet mir cheloveka*. Nikitsky Klub.
Kapoor, A., Debroy, B. (2019). GDP Is Not a Measure of Human Well-Being. *Harvard Business Review*. URL: https://hbr.org/2019/10/gdp-is-not-a-measure-of-human-well-being.
Kirdina, S. (2001). *Institutsional'nyye matritsy i razvitiye Rossii*. IEiOPP SO RAN.
Kirdina, S. (2004). *X- i Y-ekonomiki: Institutsional'nyy analiz*. Nauka.
Kirdina, S. (2014). Institutional matrices theory, or X&Y theory: The main provisions and applications. *Journal of Institutional Studies, 6*(3), 13–33.
Korotayev, A. (2003). Unilineal descent groups and deep christianization: A Cross-Cultural comparison. *Cross-Cultural Research, 37*(1), 132–156. https://doi.org/10.1177/1069397102238925

Korotayev, A. (2004). *World religions and social evolution of the old world Oikumene civilizations: A cross-cultural perspective*. Edwin Mellen Press.

Korotayev, A. (2005). A compact macromodel of world system evolution. *Journal of World-Systems Research, 11*(1), 79–93. https://doi.org/10.5195/jwsr.2005.401

Korotayev, A. (2006). The World System Urbanization Dynamics: A quantitative analysis. In P. Turchin, L. Grinin, V. C. de Munck, & A. Korotayev (Eds.), *History & Mathematics: Historical Dynamics and Development of Complex Societies* (pp. 44–62). KomKniga/URSS.

Korotayev, A. (2007). Compact Mathematical Models of World System Development, and How They Can Help Us to Clarify Our Understanding of Globalization Processes. In G. Modelski, T. Devezas, & W. R. Thompson (Eds.), *Globalization as Evolutionary Process: Modeling Global Change* (pp. 133–160). Routledge.

Korotayev A (2018). The 21st Century Singularity and Its Big History Implications: A Re-analysis. *Journal of Big History,* 2(3), 73–119. https://doi.org/10.22339/jbh.v2i3.2320.

*Korotayev AV (2020a). How singular is the 21st century Singularity?. In: Korotayev AV, LePoire D (eds) The 21st century Singularity and global futures. A Big History perspective (pp. 571–595). Springer, Cham.* https://doi.org/10.1007/978-3-030-33730-8_26.

Korotayev AV (2020b). The 21st century Singularity in the Big History perspective. A re-analysis. In: Korotayev AV, LePoire D (eds) *The 21st century Singularity and global futures. A Big History perspective* (pp. 19–75). Springer, Cham. https://doi.org/10.1007/978-3-030-33730-8_2.

Korotayev, A., Malkov, A., & Khaltourina, D. (2006a). *Introduction to Social Macrodynamics: Compact Macromodels of the World System Growth*. KomKniga/URSS.

Korotayev, A., Butovskaya, M., Shulgin, S., & Zinkina, J. (2021). Impact of Population Aging on the Global Value System. *Vek Globalizatsii, 4*, 69–80. https://doi.org/10.30884/vglob/2021.04.05.

Korotayev, A., & Malkov, A. (2016). A compact mathematical model of the World System economic and demographic growth, 1 CE—1973 CE. *International Journal of Mathematical Models and Methods in Applied Sciences, 10*, 200–209.

Korotayev, A., Malkov, A., & Khaltourina, D. (2006). *Introduction to Social Macrodynamics: Secular Cycles and Millennial Trends*. KomKniga/URSS.

Korotayev, A., Malkov, S., & Musieva, J. (2023). Demography. Toward Optimization of Demographic Processes. In: V. Sadovnichy et al. (Eds.), *Overcoming the Limits to Growth. A Report to the Russian Association of the Club of Rome* (pp. 97–116). Springer. https://doi.org/10.1007/978-3-031-34999-7_6.

Kovaleva, N. (2023). Ecology. Life in the "Unstable Biosphere". In: V. Sadovnichy et al., (Eds.) *Overcoming the Limits to Growth. A Report to the Russian Association of the Club of Rome*. Springer. https://doi.org/10.1007/978-3-031-34999-7_5.

Kuenkel, P. (2019). *Stewarding Sustainability Transformations*. Springer. https://doi.org/10.1007/978-3-030-03691-1

Kuznetsova, V., & Mashkina, O. (2020). Trudovaya migratsiya i problemy obucheniya detey rabochikh-migrantov v Kitaye. *Istoriya i Sovremennost'*, 2(36), 46–69.

Lee R. D. (2016). *Macroeconomics, Ageing, and Growth*. Cambridge (Mass.): NBER. NBER Working Paper № 22310.

Lee, R. D., Arthur, W. B., & Rodgers, G. (1988). *Economics of Changing Age Distributions in Developed Countries*. Oxford University Press.

Lee, R., & Mason, A. (2011). *Population Aging and the Generational Economy: A Global Perspective*. Edward Elgar.

Lee R. D., Mason A., & Cotlear D. (2010). *Some Economic Consequences of global ageing. A discussion note for the World Bank*. World Bank.

Lem, S. (1968). *Summa Technologiae*. Mir.

LePoire D, Korotayev AV (2020). Conclusion. In: Korotayev AV, LePoire D (eds), *The 21st century Singularity and global futures. A Big History perspective* (pp. 599–620). Springer. https://doi.org/10.1007/978-3-030-33730-8_27.

Levy, B. R., Slade, M. D., Chang, E.-S., Kannoth, S., Shi-Yi, W. (2020). Ageism amplifies cost and prevalence of health conditions. *The Gerontologist*, 60(1), 174–181. https://doi.org/10.1093/geront/gny131.

Lilley, S. (1976). Technological Progress and the Industrial Revolution, 1700–1914. In: C. M. Cipolla (Ed.), *The Industrial Revolution, 1700–1914. The Fontana Economic History of Europe* (pp. 187–254). The Harvest Press.

Lovins, L. H., Wallis, S., Wijkman, A., & Fullerton, J. (2018). *A Finer Future. Creating an Economy in Service to Life.* New Society Publishers.

Lutz, W., Goujon, A., Kc, S., Stonawski, M., & Stilianakis, N. (2018). *Demographic and human capital scenarios for the 21st century: 2018 assessment for 201 countries.* Publications Office of the European Union.

Maddison Project (2022). *Maddison Project database.* http://www.ggdc.net/maddison.

Maestas N., Mullen K.J., & Powell D. (2016). *The Effect of Population Aging jn Economic Growth, The laboe Force and Prodactivity.* NBER.

Malkov, S. (2009). *Sotsial'naya samoorganizatsiya i istoricheskiy protsess: voz-mozhnosti matematicheskogo modelirovaniya.* Librokom.

Malkov, S. (2021). Razmyshleniya na temu: Yest' li budushcheye u kapitalizma? *Informatsionnyye Voyny, 4*(60), 41–49.

Malkov, S., Grinin, L., Grinin, A., Musieva, J., & Korotayev, A. (2023). Modeling Social Self-Organization and Historical Dynamics. Global Phase Transitions. In: V. Sadovnichy et al. (Eds.), *Overcoming the Limits to Growth. A Report to the Russian Association of the Club of Rome* (pp. 387–417). Springer. https://doi.org/10.1007/978-3-031-34999-7_19.

Malkov, S., Kovaleva, N., Grinin, L., & Korotayev, A. (2023). Modeling Social Self-Organization and Historical Dynamics. Agrarian Society. In: V. Sadovnichy et al. (Eds.), *Overcoming the Limits to Growth. A Report to the Russian Association of the Club of Rome* (pp. 309–335). Springer. https://doi.org/10.1007/978-3-031-34999-7_17.

Malkov, S., & Maksimov, A. A. (2018). *Kontury Gryadushchego Miroporyadka. Istoriya i Sovremennost, 4*(30), 3–29.

McDaniel, S. A., & Zimmer, Z. (2016). *Global aging in the twenty-first century: Challenges, opportunities and implications.* Routledge.

McElwee, S., Daly, L. (2014). Why we should abolish the GDP. *Washington Post.* https://www.washingtonpost.com/posteverything/wp/2014/06/05/why-we-should-abolish-the-gdp/.

Medici, A. C. (2021). *Health sector challenges and policies in the context of ageing populations.* United Nations, Department of Economics and Social Affairs, Population Division.

Mitchell, E., & Walker, R. (2020). Global ageing: successes, challenges and opportunities. *British Journal of Hospital Medicine, 81*(2), 1–9. https://doi.org/10.12968/hmed.2019.0377.

Mokyr J. (1999). The British Industrial Revolution: An Economic Perspective. Westview Press.

Mokyr, J., & Voth, H.-J. (2010). Understanding Growth in Europe, 1700–1870: Theory and Evidence. In: S. Broadberry & K. H. O'Rourke (Eds.), *The Cambridge Economic History of Modern Europe 1700–1870* (vol. 1, pp. 7–42). Cambridge University Press.

Morgan, L. A., Kunkel, S.R. (2016). *Aging, Society, and the Life Course.* 5th Edition. New York, Springer. https://connect.springerpub.com/content/book/978-0-8261-2173-8.

Orlická, E. (2015). Impact of population aging and elderly poverty on macroeconomic aggregates. *Procedia Economics and Finance, 30*, 598–605.

Panov A, LePoire DJ, Korotayev A (2020). The 21st century Singularity in the Big History perspective. An overview. In: Korotayev A, LePoire D (eds) *The 21st century Singularity and global futures. A Big History perspective* (pp 1–18). Springer. https://doi.org/10.1007/978-3-030-33730-8_1.

Pereira, C. A., Bardi, U., Bateson, N., Bologna, G., Cheng, Y.H., van Dieren, W., Dixson-Declève, S., et al. (2022). *Limits and Beyond: 50 years on from The Limits to Growth what did we learn and what's next?* Exapt Press.

Perissi, I., & Bardi, U. (2021). *The Empty Sea.* Springer. https://doi.org/10.1007/978-3-030-51898-1

Peterson, P. G. (1999). Gray down: the global aging crisis. *Foreign Affairs, 78*(1), 42–55.

Philipson, M. (Ed.) (1962). *Automation: Implications for the Future*. Vintage.

Pifer A., Bronte L. (1986a). Introduction: Squqring the Pyramid. In Pifer A., Bronte L. (eds) *Our Ageing Society: Paradox and Promise*. New York: W.W.Norton.

Pifer, A., & Bronte, L. (Eds.). (1986b). *Our Ageing Society: Paradox and Promise*. W.W.Norton.

Popova, S., & Yanik, A. (2014). Problemy global'nogo stareniya naseleniya: Analiz dokumentov i strategii OON. *Mezhdunarodnoye Pravo i Mezhdunarodnyye Organizatsii, 3*, 429–443.

Raftery, A. E., Li, N., Ševčíková, H., Gerland, P., & Heilig, G. K. (2012). Bayesian probabilistic population projections for all counties. *Proceedings of the National Academy of Sciences, 109*(35), 13915–13921. https://doi.org/10.1073/pnas.1211452109

Ram, N., & Gray, D. (2020). Mass surveillance in the age of COVID-19. *Journal of Law and the Biosciences, 7*(1), 1–17. https://doi.org/10.1093/jlb/lsaa023

Randers, J. (2012) *2052: A Global Forecast for the Next Forty Years. Report to the Club of Rome*. Chelsea Green Publishing.

Randers, J., Rockström, J., Stoknes, P. E., Golüke, U., Collste, D., & Cornell, S. (2018). *Transformation is feasible. A report to the Club of Rome*. Stockholm Resilience Centre.

Reed, C. A. (Ed.). (1977). *The Origins of Agriculture*. Mouton.

Rifkin, J. (2011). *The Third Industrial Revolution. How Lateral Power Is Transforming Energy, the Economy, and the World*. Palgrave Macmillan.

Schwab, K. (2016). *The Fourth Industrial Revolution*. World Economic Forum

Schwab, K., & Malleret, T. (2020). *COVID-19: The Great Reset*. Agentur Schweiz.

Sean (2021). We Analyzed 450 Apps and Found Location Trackers in Every One. *ExpressVPN* 28.01.2021. https://www.expressvpn.com/blog/digital-security-lab-location-trackers-smartphone-apps-research/.

Selbin E (2022). All Around the World: Revolutionary Potential in the Age of Authoritarian Revanchism. In: Goldstone JA, Grinin L, Korotayev A (eds) *New Waves of Revolutions in the 21st Century—Understanding the Causes and Effects of Disruptive Political Changes*. Springer.

Shcherbakova, Y. (2019). Stareniye naseleniya mira po otsenkam OON 2019 goda. *Demoskop Weekly*, 837–838. http://demoscope.ru/weekly/2019/0837/barom01.php.

Sheffield, J., Grinin, L., & Korotayev, A. (2013). *Globalization: Yesterday, today, and tomorrow*. Emergent Publications.

Spengler, O. (1991). *[1918])*. Oxford University Press.

Sylvester E., and Klotz L. C. (1983). *The Gene Age: Genetic Engineering and the Next Industrial Revolution*. Scribner.

Síthigh, D. M., & Siems, M. (2019). The chinese social credit system: a model for other countries? *The Modern Law Review, 82*(6), 1034–1071. https://doi.org/10.1111/1468-2230.12462

Stanley, J., & Granick, J. S. (2020). *The Limits of Location Tracking in an Epidemic*. ACLU.

Technavio. (2020). *Global 3D Facial Recognition Systems Market 2018–2022*. Technavio.

Tsirel, S. (2020). Future technological achievements as a challenge for post-singularity human society. In A. V. Korotayev, & D. LePoire (Eds.), *The 21st century Singularity and global futures. A Big History perspective* (pp. 419–437). Springer.

UN Population Division. (2013). *World Population Aging 2013*. United Nations, Department of Economics and Social Affairs, Population Division.

UN Population Division. (2015). *World Population Ageing 2015*. United Nations, Department of Economics and Social Affairs, Population Division.

UN Population Division (2019). *World population prospects 2019: volume I: Comprehensive tables*. UN Population Division.

UN Population Division (2020). *World Population Aging 2020*. United Nations, Department of Economics and Social Affairs, Population Division.

UN Population Division (2022a). *United Nations Population Division Database*. United Nations, Department of Economics and Social Affairs, Population Division. http://www.un.org/esa/population.

UN Population Division (2022b). *World population prospects 2022*. United Nations, Department of Economics and Social Affairs, Population Division.

UNFPA (2012). *Aging in the Twenty-First Century: A Celebration and A Challenge.* UNFPA and HelpAge International. https://www.unfpa.org/publications/aging-twenty-first-century.
UNeGovDD (2013). *E-Government. The United Nations E-Government Development Database.* https://publicadministration.un.org/egovkb/en-us/About/UNeGovDD-Framework.
van Dullemen C.E. (2017). The Politics of Aging, Global Challenges and National Solutions. *Journal of Aging and Neuropsychology. 01*(2017), 1–4. https://doi.org/10.29011/JANP-104.100004.
Vishnevskiy, A. (2005). *Izbrannyye demograficheskiye trudy.* Nauka.
Vollset, S. E., Goren, E., Yuan, C.-W., Cao, J., Smith, A. E., Hsiao, T., Murray, C. J. L., et al. (2020). Fertility, mortality, migration, and population scenarios for 195 countries and territories from 2017 to 2100: A forecasting analysis for the Global Burden of Disease Study. *The Lancet, 396*(10258), 17–23. https://doi.org/10.1016/S0140-6736(20)30677-2
von Weizsäcker, E.U., & Wijkman, A. (2018) *Come on! Capitalism, Short-termism, Population and the Destruction of the Planet. A report to the Club of Rome.* Springer. https://doi.org/10.1007/978-1-4939-7419-1.
Wang, S., Ding, S., & Xiong, L. (2020). A new system for surveillance and digital contact tracing for COVID-19: Spatiotemporal reporting over network and GPS. *JMIR Health and Health, 8*(6), e19457.
Weber, M. (2002). *[1905]).* Penguin Books.
WHO. (2020). Immunity passports" in the context of COVID-19. https://www.who.int/news-room/commentaries/detail/immunity-passports-in-the-context-of-covid-19.
WHO—World Health Organization (2021a). *Ageing and health.* https://www.who.int/news-room/fact-sheets/detail/ageing-and-health.
WHO—World Health Organization (2021b). *Ageism is a global challenge.* https://www.who.int/news/item/18-03-2021-ageism-is-a-global-challenge-un.
Wijkman, A., & Skånberg K., (2017). *The Circular Economy and Benefits for Society. Jobs and Climate Clear Winners in an Economy Based on Renewable Energy and Resource Efficiency. A report to The Club of Rome.* Club of Rome.
Wittgenstein Center (2020). *Wittgenstein Center Human Capital Data Explorer.* http://dataexplorer.wittgensteincentre.org/wcde-v2/.
Yakushev, Y., Sinyavskaya, O., & Voron, O. (2020). *COVID-19 i smertnost' v domakh prestarelykh v mire i v Rossii.* Institut sotsial'noy politiki NIU VSHE.
Yaşar E., & Yaşar M. (2018). Global Wealth and Aging Population. In: R. Yilmaz & G. Löschnigg (Eds.), *Studies on Balkan and Near Eastern Social Sciences* (vol. 2, pp. 47–59). Peter Lang.
Zinkina, J., Christian, D., Grinin, L., Ilyin, I., Andreev, A., Aleshkovski, I., Shulgin, S., & Korotayev, A. (2019). A big history of globalization: the emergence of a global world system. *Springer.* https://doi.org/10.1007/978-3-030-05707-7

**SPRINGER NATURE**

# GPSR Compliance

*The European Union's (EU) General Product Safety Regulation (GPSR) is a set of rules that requires consumer products to be safe and our obligations to ensure this.*

*If you have any concerns about our products, you can contact us on ProductSafety@springernature.com*

In case Publisher is established outside the EU, the EU authorized representative is:

Springer Nature Customer Service Center GmbH
Europaplatz 3
69115 Heidelberg, Germany